Biodiversity and Business

Laladhas Krishna Panicker · Prakash Nelliyat ·
Oommen V. Oommen
Editors

Biodiversity and Business

Bio Prospecting and Benefit Sharing

Springer

Editors
Laladhas Krishna Panicker
Computational Biology & Bioinformatics
University of Kerala
Thiruvananthapuram, Kerala, India

Prakash Nelliyat
Centre for Biodiversity Policy & Law
National Biodiversity Authority
Chennai, Tamil Nadu, India

Oommen V. Oommen
Centre for Venom Informatics
University of Kerala
Thiruvananthapuram, Kerala, India

ISBN 978-3-031-71676-8 ISBN 978-3-031-71674-4 (eBook)
https://doi.org/10.1007/978-3-031-71674-4

This Springer imprint is published by the registered company Springer Nature Switzerland AG
The registered company address is: Gewerbestrasse 11, 6330 Cham, Switzerland

This book is dedicated to Prof. M.S. Swaminathan for his contributions to the Green Revolution in India aimed towards irradicating poverty eventually and Prof. M.S. Valiathan for integrating modern medicine with Ayurveda, the traditional form of healthcare system. Their contributions enhanced the scope of biodiversity in business severalfold in the past few decades.

Foreword

One of the biggest paradoxes of our global order lies in the proliferation of issues that require multilateral cooperation in an era where the appetite for that kind of cooperation is decreasing. There is no bigger example of that than the discourse on climate and biodiversity, where every reputable expert is worried about our achievement of climate goals and the systematic deterioration of our environment, besides the mechanisms to negotiate those obstacles. Having been in the multilateral system through a nearly three-decade-long stint in the United Nations, I am one of the many who are worried about the long-term impacts of the deadlocks in these negotiations on the quality of life our future generations will live, and the sustainability of our resource systems.

As the world began to wax eloquent about 'Sustainable Development' in the late 20th century, an urgent need was realized to fill the academic void that existed on knowledge about the environment in which we live, and how it might be hurtling down a dangerous path. In this backdrop, Dr. Oommen's anthology, which situates itself at the interface of biodiversity and business, is a much-needed and rigorously researched intervention. With fascinating insights and contributions from both established experts and up-and-coming researchers, this anthology spans a huge corpus and would be of great interest to those who are interested in the existential question we must address in our times.

I anticipate that the detailed academic overview provided in this book will go a long way in educating the stakeholders of both the present and the future about the challenges, priorities, and opportunities that confront experts as they pave the complex path ahead.

July 2024

Dr. Shashi Tharoor
Member of Parliament, Former
Minister of State for Human Resource
Development and Minister of State for
External Affairs to the Government
of India and Under-Secretary-General
of the United Nations
New Delhi, India

Preface

One of the most discussed topics among research and industrial communities and policymakers in the twenty-first century is biological diversity and sharing of benefits derived from it. Biodiversity provides several goods and innumerable services to the people. Bioprospecting 'the search for valuable genetic and biochemical information found in wild animals, plants or microbial organisms' for product development as a purely scientific and commercial endeavor is attracting worldwide attention. Developing nations rich in biological resources and traditional knowledge have become the source for access to genetic resources with potential applications in diverse sectors. Bio-Prospecting raises complex legal and ethical questions, particularly concerning the protection of traditional knowledge and the equitable sharing of benefits. The Convention on Biological Diversity (CBD), together with its Nagoya Protocol, creates an internationally accepted regulatory framework for access to genetic resources and fair and equitable sharing of benefits arising out of their utilization.

The book **Biodiversity and Business** has its genesis in this background and is organized into 5 sections, namely Biodiversity and Business Reciprocity, Accessing Biodiversity Stock for Sustainable Use, Corporate/Business Responsibilities in the Application of Life Sciences, Fostering Partnership in Bioprospecting and IPR in the Context of CBD. The 27 chapters contributed by eminent scientists and policy makers in this field cover all sectors as Cosmaceuticals, Pharmaceuticals and Food and Nutrition. The exploration of biological resources ranging from plants to marine organisms for potential commercial applications is detailed. New concepts such as Managing invasive alien species through ABS, Digital sequence information and ABS, ABS in protected areas are introduced. The importance of addressing issues of benefit sharing and the protection of traditional knowledge when commercializing products derived from indigenous practices is highlighted in the section "IPR in the context of CBD". Issues such as sharing the benefits equitably with local communities and case studies of successful models of ABS worldwide are also included.

Personal experiences as well as scientific knowledge of bioprospecting and ABS are given significance. The book concludes with highlighting that ensuring the availability of biodiversity's resources in regenerative and customized medical treatments is a scientific pursuit and an ethical obligation. Policymakers in all nations should ensure that access to genetic resources and traditional knowledge should protect the interests of the local community and also safeguard welfare of all human beings.

The contributing authors present views from the academic, and the policymaker perspectives, for promoting biodiversity conservation, sustainable utilization of bioresources and ABS. This book is useful to researchers, policymakers and students alike and the readers will be benefitted from the wide range of subjects covered.

Thiruvananthapuram, India — Laladhas Krishna Panicker
Chennai, India — Prakash Nelliyat
Thiruvananthapuram, India — Oommen V. Oommen

Acknowledgements

The editors acknowledge the great contributions of all the authors and the editing support rendered by Dr. N. Preetha and Prof. Dr. M.A. Akbarsha. The editorial and technical supports rendered by Mr. Vinod M.P., Ms. Neenu Mohan, Dr. Sherin K. and Mrs. Swathy K., Researchers in our Department and for proofreading by Avantika George, Jacksonville, USA, during the publication process are acknowledged. The facilities of the Department of Computational Biology and Bioinformatics, University of Kerala, accredited with A++ grade by NAAC, Government of India, are also appreciated with gratitude. We are immensely honoured to acknowledge the academic and editorial guidance received from team Springer-Nature, particularly those of Dr. Margaret Deignan and Ms. Aarthi Padmanabhan.

Contents

Biodiversity and Business Reciprocity

Benefit Sharing by Bio-Prospectors: Can It Resolve the Global Biodiversity Crisis?

Prakash Nelliyat, Oommen V. Oommen, and Laladhas Krishna Panicker

Introduction

The underlining philosophy of 'Benefit Sharing' has a strong advocacy with bio-trade and bio-prospecting. Biological resources and their associated Traditional Knowledge (TK) are the major sources or inputs for developing a wide range of consumer products such as: modern drugs, botanical medicines, new seed varieties, and ornamental horticulture, crop protection, healthcare, agriculture, personal care, and cosmetics etc., Historically, these products and their manufacture play a significant role in enhancing the economy through business and human welfare. However, in recent decades, the emergence of biotechnology in developed nations accelerated bio-prospecting at different levels, which leads to unsustainable harvest or extraction of genetic or biological resources from the developing countries and leads to bio-piracy. The extraction of renewable biological resources more than their regeneration, leads to species' extinction and imbalance of the ecosystem. When the services provided by ecosystems/biodiversity are impaired, there is a reduction in productivity which ultimately impacts the economy and the wellbeing of the people who depend on the ecosystems for their livelihood.

At present, biodiversity, a critical global asset, is under real threat. This resulted the formation of the Convention on Biological Diversity (CBD) and the origin of the Access and Benefit Sharing (ABS) principle. The United Nations' conference on Trade and Development (UNCTD) has also come up with seven bio-trade principles and criteria—including 'fair and equitable sharing of benefits'—for terrestrial,

P. Nelliyat (✉)
Centre for Biodiversity Policy and Law, National Biodiversity Authority, Taramani, Chennai 600113, India
e-mail: prakashnelliyat@gmail.com

O. V. Oommen · L. Krishna Panicker
Department of Computational Biology & Bioinformatics, University of Kerala, Kariyavattom, Thiruvananthapuram, Kerala 695581, India

L. Krishna Panicker et al. (eds.), *Biodiversity and Business*,
https://doi.org/10.1007/978-3-031-71674-4_1

marine and other aquatic biodiversity-based products. No doubt the principle of benefit sharing has a huge scope for stabilizing the global ecosystem or biodiversity goods and services in such a manner as to enhance human welfare for ever.

Benefit sharing occurs when the beneficiaries or users of biological resources (bio-prospectors) make payments to the providers of those services. It can be considered as a predominant approach that aims to manage natural resources and ecosystem stewardship, wherein the users compensate for the loss of the biodiversity. This payment (benefit sharing amount), acts as an incentive for the community to conserve and sustainably use the biological resources in a cyclical manner. In brief, this introductory chapter examines the linkages of biodiversity business or bio-prospecting with the benefit sharing mechanics of the CBD.

The Nagoya Protocol on ABS: A Pathway for Sustainable Bio-Prospecting and Bio-Business

The Nagoya Protocol on 'Access to Genetic Resources and Fair and Equitable Sharing of Benefits Arising from the Utilization', to the CBD was adopted at the tenth meeting of the Conference of Parties (CoP) on 29th October, 2010 in Nagoya, Japan. This Protocol represents the concrete picture on one of the three objectives of the Convention, namely "the fair and equitable sharing of the benefits arising from the utilization of genetic resources". The Protocol has come up with a series of binding obligations which will result in smooth, more transparent and more equitable partnerships between the users (bio-prospectors) and providers (local communities) of genetic resources and associated TK.

The underlying principle is the Nagoya Protocol, by which the business engagement has been encouraged, where the business community should understand its dependencies and impacts on biodiversity/ecosystem goods and services, and to mainstream the efforts to mitigate the negative effects. According to Braulio (2015), "the Nagoya Protocol opens an important opportunity for the business sector to show progressive leadership in the area of bio-entrepreneurship and the CBD will provide a supportive role in these endeavours". In this context, programmes such as the 'Global Partnership for Business and Biodiversity' can ensure the necessary information and tools to maximize the effectiveness of businesses.

Presently, 196 countries are Parties in the CBD and 142 countries have ratified the Nagoya Protocol on ABS (CBD, 2024). It implies that there is a universal acceptability on ABS as emerging options for biodiversity conservation. However, many Governments have a limited understanding of the scientific, technological and commercial realities of bio-discovery or bio-prospecting and the factors that influence corporate behaviour. This is the fact that industry engagement with ABS and the CBD process as a whole still varies both across and within sectors, and this means that many companies lack a basic understanding of the issue (Normand, 2015).

Broadly, the Nagoya Protocol provides clear and transparent procedural guidelines to the Government and bio-prospectors about the operation of the ABS. It advocates that the Governments should establish clear rules for users, when accessing genetic/biological resources and associated TK. Under this new regime, countries that require prior informed consent (PIC) in order to allow access to their genetic resources must also provide for the legal certainty, clarity and transparency of their domestic ABS requirements. In addition, these countries must provide information on how to apply for PIC, provide a clear and transparent written decision in a reasonable period of time, and for the issuance of a permit as evidence that PIC was granted and the Mutually Agreed Terms (MAT) established. They must also enact clear rules and procedures for the establishment of ABS agreements.

Governments also have the obligation to establish national focal points in order to provide information on the national procedures for ABS and they are obligated to establish competent national authorities with the responsibility for granting access to genetic resources (CBD, 2011). The Protocol also establishes a number of requirements for Governments to establish clear national processes for persons or entities interested in accessing the TK associated with genetic resources. In this way, Parties are to encourage the development of community protocols, the minimum requirements for mutually agreed terms and model contractual clauses for benefit-sharing.

The key innovative provisions of the Nagoya Protocol are those related to compliance. These provisions require Governments to adopt measures that ensure that users within their jurisdiction respect the ABS requirements of provider countries. It also provides for the monitoring of genetic resources once they have left the provider country through the establishment of checkpoints and of an internationally recognized certificate of compliance. Through the implementation of this monitoring system, the concerns of provider countries, especially misappropriation or misuse of their genetic resources, can be addressed effectively and thus reduce the need for the development of more stringent or restrictive ABS measures (Normand, 2015).

The Nagoya Protocol recognizes that different bio-prospecting sectors may use genetic resources for different purposes and thus apply different ABS practices, which is very well captured in this book. Further, the development of model contractual clauses, codes of conduct, guidelines, best practices and standards is encouraged by the Protocol. The development of these tools can be an effective way for business communities or bio-entrepreneurs to inform Governments about their practices, to demonstrate their willingness to respect the principles of ABS and the Nagoya Protocol, and to provide practical approaches to ABS in line with the business needs and the particular requirements of their work.

The ABS Clearing-House is another landmark of the Nagoya Protocol which acts as a common platform for the parties to share information related to the ABS. This information hub is a major source for bio-prospectors to explore ABS and its progress at the global level. As per the Protocol, all Parties have the obligation to use the ABS Clearing-House information by their National Focal Points, Competent National Authorities and relevant national ABS measures (CBD, 2011). In addition, the Parties must inform the Clearing-House about the permits that they have issued.

Besides, the Non-Parties to the Protocol are also encouraged to share all relevant ABS information through the Clearing-House. Therefore, those seeking access to genetic resources and associated TK, will find reliable information, in a standardized and organized manner, in the ABS Clearing-House.

The ABS Clearing-House also supports legal certainty for users accessing genetic resources. When a provider country grants access to a genetic resource and publishes the information on the permit or its equivalent in the ABS Clearing-House, the Clearing-House automatically generates an Internationally Recognized Certificate of Compliance (IRCC) that is then sent to thc user. The certificate provides evidence that the genetic resource has been accessed in accordance with prior informed consent and that mutually agreed terms have been established, as required by the ABS measures of the provider country (Normand, 2015).

In addition, the ABS Clearing-House hosts a variety of resources and information relevant to ABS, including capacity-building tools, videos, reports and articles, and is a repository for easy access to many ABS resource materials. This is a ready reference to the bio-entrepreneurs and other ABS stakeholders to explore more about the ABS. Further, businesses and research institutions can contribute to the ABS Clearing-House by submitting these types of tools as well as other relevant resources and materials (Normand, 2015).

Parties of the CBD, including India, submitted their "Interim National Report on the Implementation of the Nagoya Protocol" to the CBD on 2017 (CBD, 2017). The interim national report is a useful tool for both the Parties and non-Parties to the Nagoya Protocol to assess the level of implementation of the Protocol, as well as gaps, and needs in terms of capacity. India has made considerable progress in implementing the Nagoya Protocol (Nelliyat, 2022); however, there is a long way to go.

In brief, the Nagoya Protocol creates a new enabling environment and concrete platform, which can contribute to building trust between the users and the providers of genetic /biological resources and associated TK, and therefore, build a sound foundation for future ABS partnerships and new business opportunities. Currently, a number of countries are revising or developing their national ABS legal frameworks in order to meet their obligations under the Protocol. What is needed is to establish better linkages and understanding between the private sector and the Governments for the establishment of regulatory frameworks conducive to the creation of new ABS based (win–win) partnerships, which is the emphasis of this book.

Bio-Prospectors' Myth About the ABS

The above discussion clearly reveals that the Nagoya Protocol provides more transparency of the ABS. Further, it comes up with a number of tools for the smooth operation of the ABS as well as to handle the practical challenges in its implementation. However, after 3 decades of the emergence of CBD and 13 years of the Nagoya Protocol, the track record of ABS is not as it is supposed to be due to multiple reasons. It includes: lack of clarity about the ABS provisions among the stakeholders, lack of awareness about the ABS related commitments among the genetic/biological resource providers (local and indigenous communities) and users (bio-prospectors), weak enforcement on the ABS provisions or legal norms, and non-cooperative approach from bio-entrepreneurs in ABS follow-ups and compliance.

Even though, the Nagoya Protocol and other CBD documents strongly advocate the role of business in ABS, it is doubtful to what extent the business sector has realised this fact. From the literature one can infer that 'the bio-entrepreneurship or bio-business should follow ABS and ABS facilitates bio-entrepreneurship or bio-business'. We carried out a series of interviews with the bio-entrepreneurs from different sectors (pharmaceuticals/modern drugs, botanical medicines, agricultural seeds and crop protection products, ornamental horticultural products, health or personal care/cosmetic products, food and beverages etc.) in India and the inference reveals that bio-prospectors understood ABS from the wrong perspectives. Further, they substantiate their views against ABS or benefit sharing in a strong manner, which is highly unacceptable.

Bio-entrepreneurs generally perceived the ABS amount as a tax imposed by the Government or the enforcement agencies (in the case of India, the National Biodiversity Authority and the State Biodiversity Boards) on the users of biological or genetic resources. Normally, a taxpayer cannot anticipate any benefit from the Government on account of the tax he paid. This wrong notion (that ABS is a tax) should be rectified or re-conveyed. Most of the bio-entrepreneurs do not have a complete clarity on the entire ABS process and its objectives as well as its necessity in their business.

In the case of ABS, the royalty collected from the bio-entrepreneurs will be transferred to the local community (in India's case through the Biodiversity Management Committees) and it acts as an incentive for them to conserve and sustainably use the biological resources for ever. In this process, bio-entrepreneurs can anticipate the biological resources' continuous availability for business. In other words, when the community conserves and sustainably uses the biological resources, the concept of "raw material security" may be achieved. Unfortunately, this fact is not conveyed to the bio-entrepreneurs in a convincing manner.

Need for a Synergy Between Business and Biodiversity or Mainstreaming the Biodiversity in Business

All business progress in this world depends on biodiversity and ecosystem services directly or indirectly. However, businesses also impact nature or biodiversity negatively. The goods and services provided by the biodiversity sector are the inputs (raw-materials) for the industries. But, the over-extraction of the natural resources including the biological resources (beyond their re-generative capacity) and the indiscriminate discharge of waste (beyond the assimilative capacity of nature) hampers the ecosystem, which the world is experiencing today.

With this realization, the Parties to the CBD have been exploring ways to enhance private-sector collaboration for achieving the CBD goals. The eighth meeting of the Conference of the Parties (COP 8) in Brazil (2006) addressed this need seriously and COP 9 in Germany (2008) invited Parties to promote public–private partnerships. In 2009, the Jakarta Charter on Business and Biodiversity was released focusing on the sustainable use of biodiversity. At the COP 10 in Japan (2010) a decision for the direct engagement with the business sector was adopted with the intension to meet the 'Strategic Plan for Biodiversity 2011–2012'. The COP 11 in Hyderabad (2012) explored the ways of strengthening this engagement further (Hem Pande, 2012). Subsequent COPs also emphasised on the practicability of industries' cooperation in biodiversity management.

Over the years, even if the instantaneous pressure from the CBD and the Government exists, there has been a limited interest from the private sector in undertaking eco-friendly activities. However, a comprehensive awareness regarding the economic benefits obtained through biodiversity conservation activities (or on the other hand, how biodiversity degradation will affect them) among the mass business community has to be generated.

Genetic or biological resources are renewable natural resources and the raw material for industries. In the early days of human history, industries' extraction of biological resources (may be in a limited manner) may not have hampered their regeneration or the ecosystem/biodiversity. But, the rapid population growth, development in biotechnology and business expansion changed the scenario. Resources are experiencing over-extraction, which affects their stock and leads to ecological instability. Besides, industrialization/urbanization related mass land use change and pollution also affects the biodiversity.

Therefore, in recent years, especially after the boom in biotechnology in developed nations, the bio-wealth from the biodiversity rich parts of the world (developing countries) is being drained substantially. The current bio-businesses are at the cost of damage to biodiversity, which should be addressed. Ideally, through the operationalization of ABS, we are mainstreaming the biodiversity in business or creating a synergy/harmony between business and biodiversity. This harmony is urgently needed to safeguard biodiversity and enhance human welfare in the world. In brief, biological resources' extraction related costs or damages are to be internalized through the ABS money collected from the bio-prospectors.

In brief, the book is an authenticated and a legitimate attempt to rectify the bio-prospectors' myth about the ABS as well as the need for mainstreaming biodiversity in business. Various chapters of the book have directly and indirectly addressed the different dimensions of this aspect in the conceptual framework of 'ethical' and 'equity' concerns of bio-prospecting.

Thematic Focus of the Book

The chapters of the book are broadly categorised under different thematic sub-headings and their essence is discussed below:

Biological Resources and Associated Traditional Knowledge for Bio-Prospecting: Inequality Concerns and Scope of ABS

Bioprospecting is the process of scouring biodiversity for novel organic compounds with potential socio-economic value. Even though ancient societies extensively studied and utilized the world's biological resources for human welfare, our understanding of biodiversity is still quite limited. Biological resources, particularly the plant variety has enormous potential for the generation of more beneficial natural products but is currently little understood and insufficiently recorded. Hence, the scope for further bio-prospecting is enormous. In this situation, ABS and its strict enforcement can help the preservation of biodiversity. ABS promotes responsible innovation and economic progress via empowering communities, encouraging cooperation, and offering legal clarity.

The use of animals in experimentation in the contexts of toxicology and biomedical sciences is primarily for (a) finding the toxic inflictions of chemicals on humans, and (b) drug discovery and vaccine testing. The variety of animals used by the researchers and bio-prospectors in huge numbers in dissections, cytology and physiology, in the manner practiced in the developing countries, has been connected to biodiversity loss as revealed in the case of frogs, lizards, insects, etc. and is a concern warranting legal protection of several species. Thus, animal use in dissection, physiology and pharmacology in India is banned and replaced by digital learning, an aspect of New Approach Methodologies (NAMs), saving millions of animals in the wild and has contributed to the conservation of biodiversity.

NAMs encompass in vitro (cell-based assays: 2D culture, 3D culture, organoids, organs-on-chip, human-on-chip, microphysiological systems, etc.), computational/in silico tools, *in chemico* techniques, model organisms of lesser sentience (zebra fish, *Caenorhabditis elegans*, Drosophila, etc.) and the high-throughput testing or omics technologies, which not only save biodiversity but improve precision, might be appropriate. The new technologies further facilitate commercial ventures of large-scale production and supply of advanced in vitro, in silico, and *in chemico* tools and the alternative model organisms.

In bio-prospecting, the biological resources associated TK plays a significant role. TK is a knowledge created, maintained, and passed down from generation to generation within a community, becoming a part of their cultural or spiritual identity. TK provides a definite clarity to the bio-entrepreneurs through the information on how to utilize specific biological resources for product manufacturing. However, there are major shortcomings in the assessment of TK related to ABS instances from the standpoint of evolving institutions. The bio-entrepreneurs from developed countries are engaged in 'free riding the genetic resources and TK and technologies of the developing countries' with the realization of the enormous economic potential of these resources and information. TK based bio-prospecing, bio-survey and product development and the ethical perspectives of benefit sharing in different countries are examined in detail in the book.

The biological resources associated TK has been lost over the past few centuries. Corporations that engage in bio-survey and bio-prospecting may concentrate on finding unique substances, genes, or living things with advantageous traits. Bioprospecting and product creation based on biological resources and associated TK require careful consideration of ethical issues, such as equitable benefit-sharing with local communities, Intellectual Property Rights (IPR) protection, and the promotion of sustainable practices. TK may aid in bio-prospecting by offering insights into the possible applications of certain plants or natural resources, which are beneficial for sustainable resource management. Through the integration of scientific precision and ethical accountability, bio-surveys and bio-prospecting provide the possibility of unleashing the latent potential of nature and guaranteeing its enduring sustainability for future generations.

No doubt that, biological resources are "Green Gold" and the associated TK with local and indigenous communities has a significant role in profit making to the bio-prospectors. Therefore, exploring and further mining of the vastness of its potential has been of great interest to biotechnologists, especially, food processers, health care specialists and others. However, in reality, the owners and providers of genetic or biological resources and associated TK have always had the worst of the bargain and their share in the benefits has remained unjust and iniquitous and in this way their right or entitlement over the biological resources has been denied. Since traditional practices and knowledge systems are in the public domain, the Intellectual Property Law has no place or space for them. For concretising the communities' rights, the emergence of CBD was the major legal development and this path-breaking initiative's conscious and legally crystallised effort over the last three decades is transparent. However, past experience reveals that these efforts are almost inconsequential in effect and have hardly arrested the trend of perpetuation of the inequity and injustice to the biological resource, the community and its customary practices. This realization is the key for filling the lacuna in the Nagoya Protocol Implementation by the Parties of the CBD.

Sectoral/Industry and Species Specific Bio-Prospecting and ABS

Biological resources based industrial sectors are divergent and use various biological resources from different ecosystems. There are fundamental disparities in different industries' approaches and dependencies on biodiversity in terms of: types of biological resources accessed, methods and sources of collection, quantities used, Research and Development (R&D), biotechnology applications, nature of production, and dependencies on biological resources associated Traditional Knowledge (TK), cost benefit ratio and others. Sometimes, the same biological resources may generate different values in different sectors/industries. Hence, the sector wise assessment of bio-prospecting is a prerequisite for framing effective ABS policies, which is discussed extensively by various chapters in the book.

In the case of the cosmetics industry they investigate natural sources including plants, marine life, and microbes in search of novel bioactive chemicals for use in cosmetic formulas. For bio-prospecting, the biological resources associated TK existing with the local and indigenous communities are widely used by the cosmetic industries in different parts of the world and their achievements with the existing ABS arrangements and their proposed plan is examined as case studies in the book. Through innovation and the creation of products with distinct skincare advantages, cosmetic enterprises may leverage the abundance of resources present in plants, marine organisms, and microbes. However, Bioprospecting in other sectors also, must be carried out in an ethical and responsible manner, taking into account the indigenous groups' rights and biodiversity conservation activities. ABS in different sectors' bio-prospecting has the potential to significantly boost innovation, assist local people' livelihoods, and promote biodiversity conservation as the demand for natural and eco-friendly cosmetic goods rises.

Research done over the past few decades has established valuable insights into the biological properties of various natural sources, including fruits, vegetables, spices, and millets, as well as their bioactive constituents, with the aim of improving human health. Numerous bioactive compounds (nutraceuticals) have been isolated from these natural sources. Compounds such as curcumin, piperine, resveratrol, isothiocyanates, EGCG (Epigallocatechin gallate), quercetin, and sulphoraphane, among others, have demonstrated significant disease-preventive and therapeutic properties against a wide spectrum of human ailments. Nutraceutical agents hold immense potential for the prevention and treatment of diseases. Considering the significance of nutraceuticals, especially in promoting human well-being, their sustainability with respect to the raw-materials security has to be ensured; this is where ABS plays a significant role.

Medicinal plants are valued sources of herbal medications worldwide, with significant untapped commercial pharmaceutical potential. Globally, between 50,000 and 80,000 plant species are utilized for therapeutic purposes. Generally, developing nations which are rich in genetic/biological resources and associated TK have become the epicentre of research by the developed nations. Various factors are affecting bio-prospecting and benefit sharing in developing nations, such as: the sustainability of medicinal plant resources, intricate trade networks, problems with traceability, focus on academic research and publications, lack of Harmonized System Codes (HSC) for the export of medicinal plants, etc. Hence, developing countries like India should develop strategies for the medicinal plants' conservation. In this regard they should build their technical capability, add value to the resource, and create opportunities for benefit sharing with the local communities, who can conserve these resources.

Research on deriving bio-active compounds from marine biological resources are progressing significantly in recent years. Bacteria and bacterial associates, fungi and fungal associates, microalgae, seaweeds, corals, sponges, polychaetes, crustaceans, molluscs, echinoderms, tunicates, ascidians, fishes, mammals etc. are promising groups for yielding compounds. The compounds derived from the marine organisms are effective as: anticancerous, anti-inflammatory, neurological, anti-oxidant, anti-malarial, and anti-bacterial etc., including Alzheimer's disease. Presently, bio-prospecting on a large scale from the wild has depleted the resources and at the same time deteriorated the marine environment. Hence, precautionary measures are necessary for the sustainability of these precious resources. Judicious exploitation of biological resources by keeping the ocean ecosystem intact is needed, where ABS should be an effective tool.

Globally, Invasive Alien Species (IAS) have been identified as the second biggest threat to biodiversity. However, its management is a cumbersome task, as existing management options are ineffective in many cases. Nowadays, the benefits offered by the IAS, through livelihood options for the local communities and the raw-material source for many industries, are under due consideration, which is also reducing the spirit of the policymakers to take stringent action against the IAS invasion. However, in a biodiversity rich country like India, IAS management is extremely needed. The proposed strategy is that the fund mobilized through ABS (when industries use IAS for commercial purpose) can be used by the enforcement agencies for the conservation of the native flora and fauna as well as to halt deeper invasion of IAS through its continuous extraction for industrial purposes. In this way, a win–win situation for ecosystem/biodiversity (irradication of IAS and promotion of native species) and economy/business and bio-prospectors (newly promoted native species' business scope or commercial utilization and sustainability) can be achieved.

Traditional Knowledge: A Hidden Wealth in Bio-Prospecting

TK reflects the cumulative wisdom passed down through generations within indigenous societies. This knowledge is deeply intertwined with the local biodiversity, encompassing the properties and uses of plants, fungi, animals, and other biological resources unique to specific ecosystems. TK and the rich genetic/biological resources available in the rich bio-diverse countries in the world, including India, are significantly exploited for bio-prospecting, especially after the emergence of modern biotechnology. However, there is no doubt that the intersection of TK and bio-prospecting holds immense potential for shaping the bio-economy of the future. It is imperative to recognize and respect the contributions of indigenous and local communities to scientific advancements. Upholding the principles of equity, fair benefit sharing, and sustainable resource management will be crucial in harnessing the full potential of bio-prospecting while safeguarding the rights and interests of TK holders. By adopting collaborative partnerships and embracing innovative models of benefit sharing, a bio-economy can be created that not only drives economic growth but also promotes cultural preservation and environmental sustainability for generations to come.

In TK, plant associated TK is predominant. From time immemorial, plants are used by mankind to fulfil their primary health-care needs. Historically, such knowledge is being transferred through generations orally and documented on various platforms, which facilitates further investigation. Still humanity is using many therapeutically active phytochemical constituents of traditional origin. The curative ability of plants and use of traditional medicinal knowledge is exposed to the public on many critical occasions, including the COVID 19 pandemic. During the post-COVID times also, the scientific world gained interest to explore more from TK to treat such life-threatening diseases and to boost immunity in the long run. Screening, extraction, identification, isolation and purification of bioactive phyto-constituents and structural elucidation are performed to validate TK associated with plant species, which eventually results in commercial products.

No doubt that the ethno-botanical bio-prospecting has contributed largely to healthcare around the globe. Compared with the total plants that exist in the world, the number of plant species examined for their pharmaceutical products are a minimum. Hence, a series of bio-prospecting studies could be performed on codified and non-codified TK which potentially yields novel drugs. Over-exploitation of biodiversity and subsequent species loss may hamper the natural product based drug discovery. However, modern drug discovery approaches such as, synthetic biology; combinatorial chemistry and rational drug design are emerging, though plant species harvested from natural habitats are safe, affordable and valuable. Conservation efforts and responsible access of potential plant species and/or associated TK has to go hand in hand.

In the globalized world, multinational corporations from industrialized nations exploit the biological resources and TK in developing countries and are engaged in bio-prospecting for valuable bio-products. This is an important issue as it is a threat to biodiversity and TK. Among the 17 globally significant mega-diverse nations; India, China, Philippines, Indonesia, and Malaysia exemplify unique examples in Asia. They represent dynamic challenges and opportunities in ABS. Bio-prospecting of the developed nations is targeting these countries' biological resources for valuable products manufacturing. The genetic/biological resources from the mega-diverse countries experience bio-piracy especially due to the western countries' unethical bio-prospecting. Bio-piracy involves the unauthorized patenting of genetic resources or TK from developing countries, lacking prior consent, recognition, or fair compensation. To proactively address potential problems related to bio-piracy, developing countries are advised to establish a searchable database of TK and genetic resources like TKDL employed by India. This measure facilitates global access for patent examiners, minimizing the risk of granting patents for undeserving claims. Ultimately, this initiative aims to safeguard the heritage of the developing countries from western companies' exploitation.

Intellectual Property Right (IPR),and Digital Sequence Information (DSI) in Bio-Prospecting and Its ABS

Bio-prospecting, (the exploration of biodiversity for potential commercial applications) raises complex legal and ethical questions, with respect to the protection of TK and the equitable sharing of benefits arising out of it. In a country like India, where biodiversity and TK are rich and diverse, numerous legal disputes and controversies are arising. There are notable cases that have shaped the landscape of IPR and bio-prospecting in India. It investigates the cases involving patents, trademarks, and copyrights related to products derived from Indian biological resources, as well as instances of bio-piracy where foreign entities have been accused of exploiting indigenous knowledge and genetic resources without proper authorisation. In this regard, the analysis carried out in one of the chapters in the book underscores the significance of India's legal and policy framework, including the Biological Diversity Act of 2002 and 2023 as well as enumerating and examining these IPR-related cases. It provides a deeper understanding of the evolving legal dynamics surrounding bio-prospecting in India, as well as informs the policymakers, researchers, and stakeholders about the complexities and ongoing developments in this critical intersection of law, biological resources, and intellectual property.

The CBD and the Nagoya Protocol succeeded in the creation of an internationally accepted regulatory framework for access to genetic resources and fair and equitable sharing of benefits arising out of their utilization, particularly for the physical genetic resources and the resultant commercial profit. Here, a bilateral provider-user approach that emphasized on the sovereign rights of nations to regulate access to their genetic resources was the criterion. However, during the last seven years, the focus of CBD negotiations on ABS focus towards a multilateral approach also with its regulatory framework with the generation, access and use of nucleotide sequence data, termed as digital sequence information.

The COP-15 showcases the profound impact of synthetic biology on the international legal and policy landscape of ABS for the utilization of genetic resources. The multilateral benefit sharing mechanism would be operational under the CBD for dematerialized genetic resources or the DSI. This multilateral approach to benefit sharing does not seem to be based on the reciprocity of access and the points that would trigger benefit sharing obligations are yet to be finalized. The decoupling of access from benefit sharing has catered to a decade long demand of the scientists to maintain open access to data. Despite this global recognition of a benefit-sharing obligation, the success of this mechanism hinges on several critical factors such as the access and use patterns over DSI by researchers, their responsiveness to the benefit-sharing obligations, and the overall perception of how such commitments might impact scientific research.

The recent advancements in the scientific field and biotechnology sector have a significant impact on the regulatory framework of biodiversity. Researchers, scientists, and multinational corporations favour open access to sequencing data and push for information sharing through public databases. However, the provider group, indigenous people and the local community, support the benefit-sharing commitment with regard to DSI. In order to share both monetary and non-monetary benefits with those who provide genetic resources, a systematic method of ABS needs to be implemented. The non-traceability of the country of origin for DSI and the challenges in regulating its access and use are significant issues that need immediate attention. The increased use of DSI for patenting is going to cause concerns in the IPR system also. The countries that have an effective disclosure requirement will have to see whether the same needs to be extended to DSI as well. The tracking and monitoring of DSI will require innovative solutions and improvements in the patent system. There are still unsettled matters like triggering points of obligations, management of the funds, rights of the aboriginal and local people, relations with the Nagoya protocol, etc., that need to be negotiated further and resolved through discussions in CBD forums.

Bio-Prospecting and ABS Challenges for Areas Beyond National Jurisdiction (ABNJ) and Protected Areas

The conservation of the marine biodiversity in Areas Beyond National Jurisdiction (ABNJ) took a new turn as the global leaders felt their aspiration to develop a specific legal framework. To initiate the process an ad hoc open-ended informal working group was established by the United Nations General Assembly in 2004. Subsequently, in June 2023, a treaty for the conservation of marine biological diversity in ABNJ exists. The Treaty has blended both binding and voluntary measures, specifically upgrading the legal framework of the United Nations Convention on Law of the Sea (UNCLOS). The ABNJ Treaty is expected to accomplish multiple aspirations of various stakeholders by establishing a new International Authority, backed by scientific and technical committees that will watch over the diminishing and vulnerable biodiversity in ABNJ.

However, the ABNJ Treaty in its current format is yet to advance novel solutions for the conservation of marine biodiversity in Areas Beyond National Jurisdiction. The problems highlighted are; (a) the Treaty has failed to alter the rigid premise of state sovereignty and the sovereign rights of powerful nations and (b) complex issues like conservation and management of common resources leave a narrow space of elbowing for all nations irrespective of their position in international politics.

Globally, protected areas are broadly considered as one of the best options for managing important ecosystems with rich biodiversity through strict legal and institutional measures by the Government with limited or prohibited public accessibility. Hence, accessing genetic and biological resources is practically impossible from the protected areas. However, they are buffer zones of genetic materials for bio-prospectors and informally drain them to industrialists in many countries or are experiencing bio-piracy. This scenario encourages the arguments for the ABS concept for the genetic/biological resources drain from the protected areas, in a restrictive manner. Further, as protected areas' management required huge finance and its source is limited, particularly in developing countries, the money derived through ABS becomes an effective financial source for the protected areas' management. Hence, a win–win situation for both the parties (protected area managers and the bio-prospectors) will emerge through the implementation of ABS in protected areas.

In India, 27% of its geographical area's biodiversity is conserved as protected areas through different legal norms. The biological resources in this vast area and its ABS potential are real concerns and may be the situation in other countries too. Hence, appropriated policy decision towards developing ABS as an inevitable part of protected areas' management is urgently needed.

Bio-Business Management Strategies and ABS Potential Estimation for Biodiversity Conservation

Since industrialisation, biological resources and ecosystems based businesses ended up in promoting unsustainable consumption of the supply source services and the collapse of many ecosystems. Commercial production of plantation crops like coffee, tea, cocoa, rubber, teak, and oil palms by clearing natural forests and adopting unscientific land uses stand out as the primary drivers for the extinction and endangerment of many tropical forest-dwelling species and traditionally conserved plant genetic resources. Likewise, the seafood industry and aquaculture market are growing much faster, creating innumerable impacts on the ecosystem.

Here, the development of a strategy, which reduces the pressure on biodiversity and developing a sustainable commodity supply chain, is required. In this context, the do-how strategy for reciprocating business and biodiversity through a Business & Biodiversity 5-P Platform with five interrelated strategic objectives and pathways: Production sector, Partnership platforms, Promotion and public outreach mechanisms, Policymaking system and Political offices is proposed. The primary objective of this platform is to strengthen the joint management of the source of biological resources in trade and help to achieve the 2050 global vision of people living in harmony with Nature. One of the chapters in the book describes the need and the necessary steps required to organize and manage such a Platform of the 5 P actors to ensure strong programmatic support by developing and implementing effective policies, regulations, and frameworks for achieving the triple objectives of the CBD.

Currently, implementing the Nagoya Protocol on ABS is a serious attempt focussed on by the Parties of the Convention. Some of the countries signed the ABS agreement and are using the amount for biodiversity conservation and others are in the facilitation process. However, many countries don't have any idea about their ABS potential and no one has attempted its estimation, as it is a cumbersome and complex task. At present, a majority of the bio-prospecting industries do not comply with the ABS provisions in many countries or share the benefits to the community.

Hence, a significant portion of finance (ABS amount) which is to be assigned for biodiversity conservation is being lost for the country. In this context, one of the chapters in the book attempted the estimation of the ABS potential. This attempt may reveal the spectrum of ABS and enhance its scope, as an effective financial source for biodiversity conservation, and help the Governments to take appropriate policy measures for implementing the ABS. However, this exercise is a preliminary attempt in ABS estimation based on the available data and the methodology has been designed by the author in consultation with experts. The authenticity of this exercise and assessment methodology should be re-examined further.

Need of the Hour

Even if the CBD emerged before 3 decades and with universal memberships, and 142 countries ratified the Nagoya Protocol on ABS (CBD, 2024), the implementation of the ABS principle is a slow process which is a serious concern. Still, a majority of the Parties is only with the preparatory work for the implementation of the Protocol. But on the ground, biodiversity degradation is multiplying, which has rightly been pointed out in the recent report of the Intergovernmental Science-Policy Platform on Biodiversity and Ecosystem Services (IPBES).This global assessment revealed a warning picture on the deterioration of the health of ecosystems (United Nation, 2019). We need to realize the fact that, if biodiversity and species are once lost, they are an irreversible loss forever (Nelliyat, 2017).

Hence, any action towards safeguarding biodiversity should not be delayed. This dichotomy must seriously think about the Parties of the CBD and appropriate action should be taken without further delay. However, the major challenge here is the mobilization of the required money for the conservation of biodiversity. A recent study by the Biodiversity Finance Initiative (BIOFIN), with the support of the UNDP, pinpointed that around US$ 52 billion is being spent on biodiversity per year globally. This is against an estimated annual required fund of between US$ 150–440 billion. According to the report, this funding gap is a challenge in achieving the CBD's Strategic Plan and the Sustainable Development Goals—SDGs (UNDP, 2018). However, the report claimed that: 'it can be overcome as it is only between 0.2 and 0.6% of global GDP'.

The Nagoya Protocol is an opportunity for the business sector to be creative and forward looking, and to propose approaches to the implementation of ABS that balances their requirements with the principles of equity and fairness established by the CBD. We should realise the fact that for the successful operationalisation of the ABS (implementation of the Nagoya Protocol) for biodiversity conservation and its sustainable use, cooperation from the bio-entrepreneurs (business sectors) is extremely significant and the government should work for that. By engaging in dialogue with policy makers and making constructive proposals on the best route to implementation, the business community can help to ensure that the international system being set up is functional, minimizes transaction costs and contributes to building the trust for the establishment of ABS partnerships that can bear fruit for both users and providers in the years to come (Normand, 2015).

In brief, most of the Parties of the CBD are in the initial stage of the implementation of the Nagoya Protocol on ABS. However, they should accelerate the process without any delay. In a broader sense, for implementing the Nagoya Protocol, the unanimous cooperation of all the parties is needed. If the user countries do not fulfill their obligation, the provider countries may face its adverse impacts. Similarly, if the provider countries do not enforce their genetic or biological resources' access regulations, the user countries may exploit the benefit. Hence, along with the strict enforcement of the legal norms by the Government, the business communities' rework is urgently required, as discussed in the forthcoming chapters of this book.

Conflict of Interest The authors have no conflict of interest to declare.

References

Braulio, D. (2015) . The Nagoya Protocol and business. In S. David, H. Johan & N. Kristina (Eds.), *Business 2020—A magazine on business and biodiversity*. Convention on Biological Diversity.

CBD. (2011). *Nagoya Protocol on access to genetic resources and the fair and equitable sharing of benefits arising from their utilization to the convention on biological diversity*. Convention on Biological Diversity.

CBD. (2017). *Interim national report on the implementation of the Nagoya Protocol*. https://absch.cbd.int/en/search?schema=absNationalReport&currentPage=3&rowsPerPage=25

CBD. (2024). *The clearing-house mechanism (CHM) of the convention on biological diversity*. https://chm.cbd.int/database?_gl=1*gzvwer*_ga*ODg1MTY1NDU5LjE0ODAxMzkxMDg.*_ga_7S1TPRE7F5*MTcxNDA0MjE5NS42LjEuMTcxNDA0MjcwMi40NS4wLjA

Hem Pande. (2012). Business and biodiversity in India. In S. David, J. Jameson, H. Johan & B. Michael (Eds.), *Business 2020 (special focus on India) A magazine on business and biodiversity*. Convention on Biological Diversity.

Nelliyat, P. (2017). Bio-resources valuation for ensuring equity in access and benefit Sharing: Issues and challenges. In K. P. Laladhas, P. Nilayangode, & O. V. Oomen (Eds.), *Biodiversity for sustainable Development*. Springer International Publishing.

Nelliyat, P. (2022). Implementation of the Nagoya Protocol in India: Experience, challenges and learning. In O. V. Oomen, K. P. Laladhas, P. Nelliyat, & B. Pisupati (Eds.), *Biodiversity conservation through access and benefit sharing*. Springer International Publishing.

Normand, V. (2015). The Nagoya Protocol and the business community. In S. David, H. Johan, & N. Kristina (Eds.), *Business 2020—A magazine on business and biodiversity*. Convention on Biological Diversity.

UNDP. (2018). *The biodiversity finance initiative workbook 2018: Finance for nature*. BIOFIN—United Nations Development Programme, New York.

United Nation. (2019). *UN report: Nature's Dangerous decline 'unprecedented': Species extinction rates 'accelerating'*. United Nations Sustainable Development. https://www.un.org/sustainabledevelopment/blog/2019/05/nature-decline-unprecedented-report/

Fostering Bio-Partnerships: A 5-P Alliance for Environmentally and Socially Responsible Bio-Businesses

N. Anil Kumar and Gouri Anil

Introduction

Target 13 of the Kunming-Montreal Global Biodiversity Framework (KM-GBF) seeks to facilitate a significant increase in the benefits shared from the use of biodiversity by 2030 through legally complied mechanisms and strategically organised partnership pathways for access to genetic resources and associated traditional knowledge (CBD/COP/Dec/15/4). As a follow-up to this and by recognising the need for responsible biodiversity-business partnerships, Target 15 of the KM-GBF encourages the business sector and biodiversity management actors to perform their roles effectively by calling for a cross-sectoral approach, regular monitoring, assessment, and disclosure of sector-specific business risks and impacts on biodiversity.

In the current and predicted scenario of a widening economic divide and increasing ecological destruction, socially and environmentally responsible businesses become the front-line sector in operationalising the sustainable use and benefit-sharing frameworks of the Convention on Biodiversity and creating green transition plans and bio-economy (UNEP, 2020). The World Economic Forum (WEF) estimates that nature-dependent businesses, such as those in construction, agriculture, and food and beverages, generate approximately $44 trillion in economic value annually (WEF, 2020). As part of the KM-GBF implementation commitments, the governments of developed countries and growing economies like Brazil, China, and India earmarked

N. A. Kumar (✉)
M. S. Swaminathan Research Foundation, Chennai 600113, India
e-mail: anilmaruthur@gmail.com

G. Anil
Department of Geography and Environment, London School of Economics and Political Science (LSE), London WC2A 2AE, UK

N. A. Kumar
Kerala State Biodiversity Board, Thiruvananthapuram 695003, India

L. Krishna Panicker et al. (eds.), *Biodiversity and Business*,
https://doi.org/10.1007/978-3-031-71674-4_2

new funding and incentives to mobilise the business sector towards green-transition plans, bio-economy innovations, and strategies.

Since industrialisation, businesses that depend on nature have promoted unsustainable consumption of supply source services, leading to the loss of biodiversity and ecosystem degradation (Bishop et al., 2010). Implementers of economic development interventions often fail to foresee environmental consequences and compromise legal, environmental, and social responsibilities. As an example, India, one of the 17 mega biodiversity countries in the world, has seen 45 million hectares of forests being denuded to various extents, and even some of the National Parks and Protected Areas of the country are considered to be endangered (IUCN, 1990; Swaminathan, 2010). The scenario continues in many biodiversity-rich but economically disadvantaged tropical countries, most evidently in the food and agriculture sectors.

The commercial production of plantation crops like coffee, tea, cocoa, rubber, teak, and oil palm by clearing natural forests and adopting unscientific land uses stands out as the primary driver for the extinction and endangerment of several tropical forest-dwelling species as well as quite a lot of traditionally conserved plant genetic resources (Gibbs et al., 2010; Kissinger et al., 2014). Nearly 90% of the world's demand for palm oil is met by two biodiversity-rich tropical countries, Indonesia and Malaysia. This industry is rapidly expanding to other countries, including India (where palm oil's current annual consumption is about 10 million metric tonnes), West and Central Africa, Latin America, and Papua New Guinea. Similarly, the seafood industry, the world's last major hunter-gatherer food system and aquaculture market, is growing unsustainably much faster than the agriculture sector (Branch et al., 2011; Bourillon & Allison, 2014).

Following the food and agriculture sector, the pharmaceutical industry, including nutraceuticals, cosmetics, biotech, and traditional medicine (TRM), emerged as a major exploitative biodiversity-dependent sector (Young, 1999). The pharmaceutical industry significantly depends on and influences numerous forms of biodiversity, ecosystems, and genetic resources, ranging from forests, desserts, wetlands, and seas to vascular plants, ferns, seaweeds, algae, fungi, microbial life, and many forms of ecosystem services, including water, soil, and mineral products (Afzal et al., 2023; Neergheen-Bhujun et al., 2017). Over the past 10–15 years, a wider supply of medicinally or nutritionally important biodiversity by the corporate pharma sector worldwide has been reported. For example, out of the 1,562 drugs approved by the USFDA between 1981 and 2014, 320 were derived from plants and natural products, 141 were herbal mixtures, 64 were pure natural products, and 61 were synthetic drugs that have origins in the pharmacophores of natural products (Newman & Cragg, 2016).

The growing demand for herbal health products fuels the continued exploration and exploitation of natural compounds for drug discovery, aided by advancements in molecular biology and digital technologies that provide unprecedented opportunities to uncover novel active molecules from the biological world. Chaachouay and Zidane (2024), in their extensive review of the global literature on plant-derived natural products for drug development, concluded that developing drugs using plant-based pharmaceutical technology offers an efficient, cost-effective, and safe alternative to conventional procedures that use animal cell cultures or microbial fermentation. Notwithstanding, the critical importance of the role of the business sector in promoting the sustainable use of biodiversity and benefit-sharing, only about 5% of businesses understand how they impact biodiversity and nature or recognise and influence the rights and roles of local and Indigenous communities who safeguard biodiversity and add value to genetic resources (WBA Nature Benchmark Report, 2022). The Customer Data Platform Data of 2022 reveals that 70% of the member companies failed to assess the impact of their business on biodiversity (TNFD, 2023).

The World Economic Forum outlines five action steps for the business sector to address this shortcoming and engage in biodiversity conservation and sustainable use as a lead partner in the implementation of the KM-GBF. Among these five-pronged action steps, the advocacy for "don't do it alone" stands out for its strong conviction that isolated efforts are ineffective. While there are global joint initiatives for biodiversity and business collaboration in general at the global level, there are few notable industry-specific "do-it-together" coalitions known at either the global, national, or sub-national level. For instance, initiatives like the Taskforce on Nature-related Financial Disclosures (TNFD) and the India Business and Biodiversity Initiative (IBBI) aim to help companies and financial institutions incorporate nature into their decisions, but their reach extends only to a few transnational corporations or big companies.

Recognising the TNFD's and IBBI's limited reach and drawing on our experience and partnerships with diverse communities of a socio-economic system in India's poverty-climate-biodiversity hotspots, we propose creating and nurturing a 5-P Biodiversity and Business Partnership ecosystem to achieve the KM-GBF 2030 Goals and the provisions of the Nagoya Protocol on Access and Benefit Sharing (ABS) (IBBI Working Groups - CII-ITC-CESD, 2023). This proactive alliance is suggested to integrate and harness the power of five critically important communities in the socio-economic and political system in the manner of natural ecosystem communities. Such a systemic approach will foster the CBD strategy of whole-of-government and whole-of-society involvement in implementing the NBSAPs, the ABS commitments, and supporting partner countries to eliminate the fear of bio-piracy.

We elaborate on the description, rationale, structure, and member entity's role in the partnership and discuss the operational plan for this ecosystem of communities by taking India's Pharma business as an example in the following sections.

Understanding the 5-P Biodiversity-Business Partnerships

Five communities that have a high and critical stake in defining, managing, and shaping a biodiversity-business management paradigm are: '*Producers*' comprising farmers, fishermen, individual manufacturers, workers, healers, artisans, miners,

Indigenous People, and others at one end, and companies and enterprises on the other end that engage in making consumable products out of cultivated or wild biodiversity by utilising their traditional knowledge and innovations. '*Partners*' dominated at different levels, ranging from local community institutions, NGOs, bio-trading partners, conservation organisations, and research collaborators who are the primary associates of producers in the value chain. '*Promoters*' comprise individuals, institutions, citizen groups, journalists, religious leaders, and celebrities who advocate and push for policies, principles, and practices for bio-economy and sustainable development agenda. '*Policymakers*' are mainly those government department officials who are responsible for protecting and enhancing nature-human-social systems, the economy, biodiversity, and the environment. Finally, and importantly, 'Politicians,' including the ministers and other leaders of ruling and opposition parties in a democracy, or those heads in autocracies who provide guidance and vision, ensure policies and programmes needed for sustainable and equitable use of biodiversity for resilient development.

The 5-P bio-business partnership can be nurtured through a concerted long-term effort benefiting every partner in a mutually reinforcing manner, akin to a *food web* in a natural ecosystem. In both the natural and social ecosystems, the base 'Producer' community creates consumable products for themselves from biodiversity, ecosystem services, and others by harnessing resources and skills. The 'Partner' groups that are analogous to organisms engage in symbiotic relationships with the producers in different ways, such as mutualism, where partners benefit mutually from the collaboration, or commensalism, where one partner benefits while the other is neither helped nor harmed, or parasitism, where one benefits at the expense of the other. The third group, 'Promoters,' are similar to facilitator species like the Pollinators and Seed distributors in the ecosystem that create beneficial conditions for the overall food web and sometimes contribute to creating enabling conditions for symbiotic collaborations between species indirectly. The next two groups at the apex, the 'Policymakers' and 'Politicians,' are similar to keystone species and ecosystem engineers in Nature and have the power to provide leadership, control, connect, change the system, and bring progress.

These five interrelated groups significantly impact the operational ecosystem's structure, function, and health and collectively lead to the system's sustainability and regenerative capability. In a natural ecosystem, nature performs this job in a balanced and collective manner, benefiting all the actors and ensuring regenerative growth for the system, whereas in the human social system, it is imbalanced, benefiting only a few who exploit the individual benefits and ultimately degenerative growth of the system.

A vast network of multi-stakeholders operates at global to national levels for conservation and sustainable use of biodiversity, to name a few, CBD, IUCN, ACTO, CMS, National Park Service, The Nature Conservancy, The Tropical Forest Alliance (TFA), ASEAN Centre for Biodiversity (ACB), and South Asia Cooperative Environment Programme (SACEP), with the principles and promises to tackle the issues related to free, informed, fair and equitable benefit sharing. The World Economic Forum, global initiatives like the Taskforce on Nature-related Financial Disclosures

(TNFD), and the national initiative of the India Business and Biodiversity Initiative (IBBI) are two major examples that specifically target business and biodiversity reciprocity. However, most of these institutional networks operate at the international, regional, and national levels, often disconnected from ground-level realities. This disconnect frequently results in failures in biodiversity management and sustainable business value-chain development operations. This issue is particularly evident in large federal countries like India, despite national-level legally binding commitments to the sustainable use of biodiversity and the fair and equitable sharing of benefits derived from it. (MEA, 2005). To address this problem, coordinated actions at the sub-national level involving state and local governments or ecological-zone-specific efforts are needed. Such an effort would also broaden the customer base and enhance biodiversity-business reciprocal actions.

The perspectives of and solutions from key actors in this partnership can be streamlined into five major action pathways. *Pathway 1 is* **Sustainable Production and Consumption** of bioresource supplies, including knowledge, skills, and innovations; *Pathway 2* is **Empowered Partners** for learning networks and collaborative action in Sustainable Biodiversity Management; *Pathway* 3 is **Active Promotion** of Sustainable Biodiversity Supplies for building up a bio-business economy and green development; *Pathway 4* **Enabling Policies** for biodiversity conservation, sustainable use, and fair and equitable benefit-sharing operations; and *Pathway* 5 **Political Will** for mobilizing resources (financial, capacities, and science, and technologies) for regenerative and sustainable biodiversity conservation and bio-business development. These pathways can be integrated as a major strategy to achieve the desired outcomes that deal with business and biodiversity reciprocity concerns and contributions. See Fig. 1 for an illustration of how these relationship pathways connect and function within a bio-business partnership and produce results.

Operating the 5-P Platform for Responsible Pharma Business

The pharmaceutical and traditional medicine sector businesses in India are intensively gearing up to tap the emerging opportunity to produce novel plant-made pharmaceuticals (PMPs), phytopharmaceutical drugs (PPDs), and herbal therapeutics. According to the Ministry of Ayush, India's export of medicinal plant-derived extracts rose from \$364 million in 2015–2016 to \$401.68 million in 2016–2017, \$456.12 million in 2017–2018, and \$628.25 million during 2022–2023. The primary objective of the 5-P platform concerned with the Pharma sector is to strengthen the joint management of the source of phytochemical raw drugs in trade and help develop a sustainable value chain for genetic resource use. With an active 5-P approach, the sustainability of the supply source of bioresources can be ensured, and it alone can only help companies protect the long-term viability of their businesses and secure a social license to operate in the market.

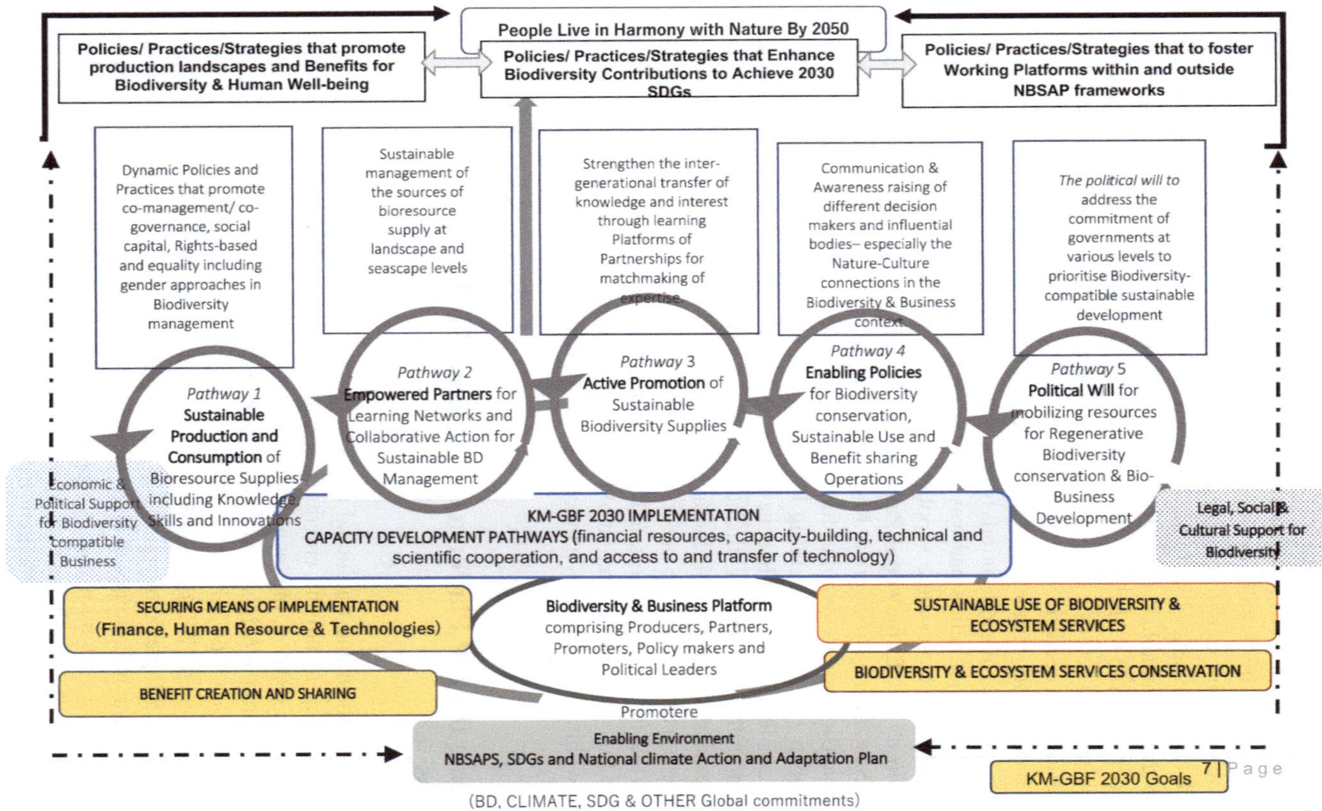

Fig. 1 The 5-P biodiversity-business partnership framework for implementing the sustainable Bio-trade policies and practices

The Production Sector Commitments

A significant obstacle to establishing an environmentally and socially responsive pharma sector with equitable benefits to nature and people lies in the current patterns of thinking, behaviour, and action, especially on the part of the major players within the system. The Indian pharma sector, with around 10,500 companies, dominates the global generic medicine and vaccine markets, meeting over 60% of the demand for DPT, BCC, and Measles vaccines (Ganguly, 2024). As the patents on many drugs on the market expire by 2030, Indian companies are poised for further growth by tapping the emerging opportunity for new drug production. To meet the growing export demand and enhance the quality branding of this commodity, the Government of India has implemented several measures. As per the Ministries of Commerce and Ayush, these include establishing the Pharmaceutical Export Promotion Council, offering financial incentives for exporters to participate in trade fairs, organising international business meets and conferences, and taking specialised actions for the medicinal plant sector. Further, the "Voluntary Certification Scheme for Medicinal Plants Produce" (VCSMPP), along with quality labels such as the Ayush Mark and Premium Mark, has been introduced to help the industry establish quality standards through Good Agricultural Practices (GAPs) and Good Field Collection Practices (GFCPs). In India, the Central Drugs Standard Control Organization (CDSCO) has enhanced monitoring through amended regulations in 2023 to assess the authenticity of extracts processed from natural products and herbals.

The Pharma industry needs to balance both the profit and people dimensions, nurturing human and herbal health and supporting environmental sustainability. Achieving long-term sustainability for the pharmaceutical industry as well as for nature and people requires the systemic perspective of One Health. The One Health approach recognises the interconnectedness of human, animal, and environmental health, specifically biodiversity, climate, and land–water systems, emphasising the need for a holistic perspective (Campbell, 2024). Though the One Health approach has to be the norm within the Pharma production and consumption sector, their understanding of the intricate nexus between the components of this system is limited or not reflected in their action. See Fig. 2, which depicts the linkages within the health system pentagon (human health, animal health, land and water system health, climate system health, and biodiversity) and the priority sciences that need to be harnessed for promoting sustainable production and consumption behaviours and practices.

The Partner Sector Commitments

Among the partner organisations, science and technology institutions stand out considering their competitive advantage in business development. India is emerging as a global scientific powerhouse, as evidenced by its significant research output, ranking third globally in quantity and 13th in quality of science publications as of

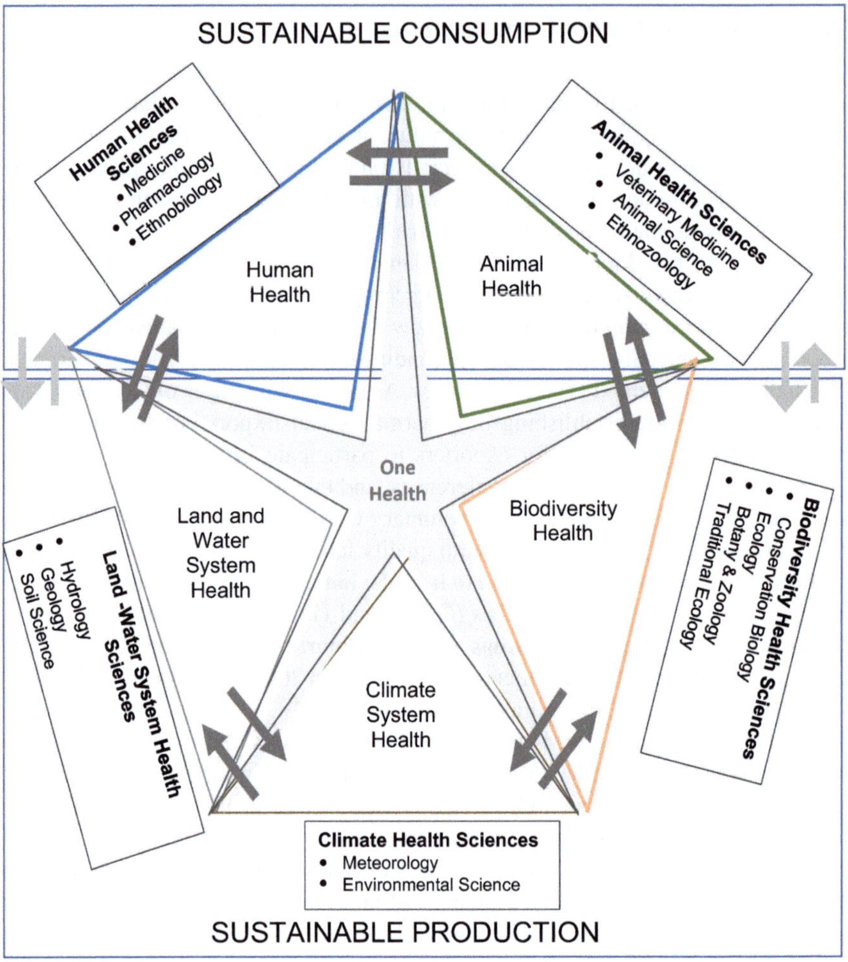

Fig. 2 One health: The biodiversity & business pentagon

2014, despite India's spending on research and development be just 0.64% of the GDP and private sector involvement not balanced (*Nature* **628**, 473 (2024)). India is now the world's third-largest pharmaceutical supplier, renowned for its affordable medicines and generic drugs (*India More Than Ready to Take Steps to Become Science Powerhouse: Nature*, n.d.). According to the Nature Index 2014, India is among the top ten leading countries in chemistry as measured by weighted fractional count (WFC). There is also a steady increase in the establishment of universities and the number of prestigious institutions such as the Indian Institutes of Technology (IITs) and Indian Institutes of Science Education and Research (IISERs). However, India's engagement in life sciences is comparatively less than that of other Asian nations such as China, Singapore, Taiwan, and South Korea.

On one side, over 80% of the known global biodiversity remains under-explored or unexplored for its bioactive molecules, and on the other side, addressing concerns about overexploitation of wild sources of medicinal plants and drug quality improvement, the need for public–private-producer knowledge partnerships in pharma research and development remains a significant challenge. The demand for value-added extracts of medicinal plants is also increasing in the global market, particularly in Europe, alongside collaborative efforts. Chaachouay and Zidane (2024), emphasise the importance of collaboration among researchers from fields such as ethnobotany, botany, taxonomy, pharmacognosy, pharmaceutics, medicinal chemistry, organic chemistry, molecular biology, microbiology, pharmacology, and plant ecology to make the discovery of new chemical components in medicinal plants more efficient and cost-effective.

However, a multidisciplinary collaboration among academics alone cannot achieve total system sustainability and social equity improvement in this field. To ensure a sustainable supply of medicinal plants, drug quality, and efficacy improvement of phytopharmaceutical drugs (PPDs), and herbal therapeutics, cross-sectoral collaboration between academics, the private and public pharmaceutical and health industries, traditional herbal healers, and knowledge holders from the local communities becomes crucial.

Promoters and Public Outreach Commitments

For the Pharma Business, the major promoters are the communicators, influencers, or media that help spread ideas, innovations, and best practices in green health promotion or access and benefit sharing related to the conservation and sustainable use of medicinal plant genetic resources. They interact closely with producers, partner groups, politicians, and policymakers to ensure the widespread adoption of policies, regulations, and practices supporting innovation and collaboration. They enhance the spread and success of certain products or processes, like birds, winds, or water dispersing seeds or helping with timely pollination or breeding.

Knowledge, including traditional knowledge and innovations from indigenous and local communities, as well as the scientific and product/process understanding of researchers and managers, is a critical component of the pharmaceutical industry value chain (Ramy et al., 2020). The value of a genetic resource to an industry is derived from the knowledge possessed by individuals or societies. The positive correlation between traditional knowledge of the medicinal properties of genetic resources and the efficiency of drug discoveries in modern laboratories is well-established (Ahtiainen & Pouta, 2011; Taubman, 2008).

The legally binding Convention on Biological Diversity (CBD) enabled the Nagoya Protocol on Access to Genetic Resources, and the Fair and Equitable Sharing of Benefits (ABS) came into force ten years ago. This protocol addresses crucial issues related to managing knowledge associated with the conservation and sustainable use of biodiversity. To operationalize the Nagoya Protocol, the contracting

Parties need to develop a professional Knowledge Management System. Such a system can facilitate organizational learning and archive the wealth of knowledge embedded across multiple stakeholders. Currently, in India, there are attempts to pool knowledge resources across states under the People's Biodiversity Register tool provision in Biodiversity (Amendment) Act 2023 (Mohan, 2023). However, efforts to use this knowledge to create new insights, drive innovations, and influence research, development, and policy interventions are limited. A systemic approach is required to define, capture, analyse, and communicate knowledge resources across geographies for diverse purposes using various media.

Formalising both popular communication systems and scientific peer-reviewed publications is necessary to ensure transparency in plans, decisions, instructions, and the use and benefits of bioresources. This should, however, be supported by a bottom-up feedback mechanism. Concerted action from every stakeholder is required to develop and adopt a communication plan that effectively communicates India's core competencies, achievements, and innovations in R&D methods, and builds replicable models in integrated bioresource use for different stakeholder groups at national and international levels.

India has the potential to transform into a one-stop Knowledge Centre for Pharma, Traditional Medicine, and integrated health development. A clearinghouse mechanism specifically for the medicinal plant use and development sector must be developed, utilising the services of universities and national research organisations. This clearinghouse mechanism should have a formal, professional knowledge communication system as its foundation to build and retain a strong customer base. Customers in the pharma, cosmetics, and traditional medicine sectors include key stakeholders in sustainable bioresource utilisation and health and wellness. To expand further, it is crucial to (i) retain critical stakeholder groups such as taxpayers, donors, scientists, practitioners, policymakers, students, teachers, farmers, NGOs, and individuals interested in green growth, and (ii) attract new customers, especially from the private sector globally. An appropriate body with strong leadership and effective communication capabilities is required to manage the KMS related to the Pharma industry on a pan-India level.

Policymaking System and Good Governance Commitments

Within the legally binding Corporate Social Responsibility (CSR) regime, some Indian companies find enhanced opportunities for cooperation and partnerships across diverse stakeholders to implement the 2030 sustainable development agenda, green development protocols, and biodiversity framework (Singh & Rahman, 2021). Poverty researchers and poverty reduction experts are just beginning to understand the dynamics between ecosystems and human behaviour, and there are few efforts to achieve "win–win" solutions, such as bioprospecting, ecotourism, integrated conservation and development projects, and payments for ecosystem services (Barrett et al., 2011). To sustain and up-scale such a win–win scenario, the need for functional cross-sectoral stakeholder groups becomes very critical as does the need to realise the KM-GBF targets.

In the global pharma industry, India, one of the 17 mega biodiversity countries, is an active player with a current market value of $65 billion, projected to reach $130 billion by 2030, according to records from the Ministry of Commerce, India, thanks to the policies and programmes of the governments of India. With this growth trajectory, the Commerce Ministry wants to place India as the global medical and wellness tourism hub by 2030. According to Ayush records, India's herbal pharmaceutical sector has expanded by 18.5% growth, with over 8100 plant-based drug manufacturing units in 2022. The National Medicinal Plant Board's data shows, that 78% of the 7500–8000 plants of recorded medicinal uses in Traditional Medicine (TRM), including Ayurveda, Siddha, Unani, and Homeopathy, are met from the wild, of which 242 species were traded over 100 metric tonne per year in 2014. Production and consumption persist, yet efforts to sustain the sourcing and supply of raw materials from the wild remain minimal on the part of the industry.

Where does the pharma get the medicinal plant raw materials? Historically, all habitats housed plants of medicinal importance to humans, and many such species were conserved in human settlements, particularly in rural landscapes, encompassing diverse ecosystems such as culturally protected forest patches, mountains, wetlands, and legally protected areas. However, conservation planning and governance often exclude these community-protected areas. (Srivathsa et al., 2023), in their article Prioritizing India's Landscapes for Biodiversity, Ecosystem Services and Human Well-being | Nature Sustainability, highlight the importance of considering both protected and non-protected areas in conservation efforts within India. Their study found that, despite protected areas covering about 5% of the land, 85% of top-priority sites for conservation were outside these zones. This underscores the need for a landscape-level approach that integrates human-use landscapes and ecosystems critical for ecosystem services.

The WBA Nature Benchmark Report of 2022 reveals that 97% of the 389 companies analysed globally are yet to commit to a nature-positive trajectory by 2030, and less than 13% of them have a clear commitment to respect Indigenous People's roles and rights over protecting and enhancing biodiversity. In India, some estimates show that in 2016, over 1000 medicinal plant species were threatened across different ecosystems, mainly due to the industry's over-exploitation of wild sources (Gowthami et al. 2021).

Exceptions are reported in India, where a few pharmaceutical companies conduct their businesses with a heightened sense of social responsibility. As per the India corporate social responsibility (CSR) reports, through such initiatives, companies like Aurobindo, Cipla, Divi's Laboratories Ltd, Lupin Ltd, Mankind Pharma, and Dr Reddy's Laboratories actively support local communities in securing health, education, disaster responsiveness, sustained water availability, and more (https://thecsrjournal.in/pharmaceutical-pharma-companies-india-csr-corporate-social-responsibility/). Despite such efforts towards social upliftment and attempts to transition the source of raw material supply from wild to cultivated spaces by the pharma and herbal industries in India, the literature available is insufficient to demonstrate how many of these industries have analysed their impact on the availability and ensured sustainable supply of medicinal plant diversity or such raw materials sourced from the wild.

Political Offices and Political Will

The role and responsibilities of political offices and their leadership are pivotal in the conservation and sustainable use of biodiversity, with political will being crucial in driving forward initiatives that support biodiversity conservation. Within ten years of the Convention on Biological Diversity (CBD) coming into force, thanks to the political leadership, the Government of India enacted the necessary legal framework in 2002 and subsequently established a three-tier implementation structure to achieve the objectives of biodiversity conservation, sustainable use, and sharing of benefits arising from such use with the providers of genetic resources and associated traditional knowledge. This three-tiered implementation body comprises the national-level National Biodiversity Authority, state-level biodiversity boards, and local-level biodiversity management committees. According to the National Biodiversity Authority's website, as of January 2024, there are 277,688 Biodiversity Management Committees working in coordination with 28 State Biodiversity Boards and 8 Union Territories in India.

The National Biodiversity Strategy and Action Plan and targets, along with the indicators developed every five years in compliance with CBD provisions, including the Nagoya Protocol (2010–2014), are major tools to set the programmes of action and monitoring mechanisms. At the national level, India has initiated several programmes to enhance biodiversity conservation, ecosystem services, and sustainable livelihoods. Key initiatives include:

- **National Mission for Biodiversity and Human Well-Being (NMBHWB):** Aimed at enhancing biodiversity conservation and ecosystem services.
- **Green India Mission (GIM):** Part of the National Action Plan on Climate Change (NAPCC), focusing on increasing forest and tree cover, restoring ecosystems, and enhancing biodiversity.
- **Biodiversity Finance Initiative (BIOFIN):** In partnership with the United Nations Development Programme (UNDP), aiming to identify and mobilize financial resources for biodiversity conservation.
- **National Adaptation Fund for Climate Change (NAFCC):** Supports adaptation projects that integrate biodiversity conservation to enhance ecosystem resilience.

Despite these efforts, progress has been sluggish in enacting programmes that promote the other two goals: sustainable use and benefit sharing with the providers of knowledge, generally local community men and women. A major reason is the slow progress in the operationalization of the Nagoya Protocol (2014), which sets guidelines for utilizing genetic resources for a broad spectrum of users, ranging from academics and other non-commercial researchers to commercial companies. There is inadequate awareness among users of the procedures for compliance with access and benefit sharing.

Political leaders must effectively engage all critical stakeholders to overcome these challenges. This requires political offices to:

- **Promote Capacity Development:** Implement appropriate training programs for local communities, researchers, and businesses on the requirements and benefits of the Nagoya Protocol.
- **Enhance Awareness Campaigns:** Educate the public and stakeholders about the importance of genetic resource conservation and equitable benefit-sharing.
- **Facilitate Partnerships:** Create avenues for collaboration between local communities, knowledge providers, government agencies, the private sector, and non-governmental organizations to promote efforts in biodiversity conservation and benefit-sharing.
- **Ensure Transparency:** Monitor compliance with the Nagoya Protocol through reporting requirements and audits, and provide publicly accessible information on the process.

The significance of the People's Biodiversity Register, an instrument advocated in the Biodiversity Act, cannot be overstated. With the National Green Tribunal orders in August 2019 and March 2020, the pace of forming Biodiversity Management Committees (BMCs) and preparing the Biodiversity Registers has improved. However, this progress will yield results at the outcome level only when BMCs become active in their localities and engage effectively in the conservation and sustainable management of biodiversity. Currently, only less than ten to twenty per cent of these BMCs are active in the country. Respective local governments and offices must take proactive steps to ensure the activation and engagement of the BMCs.

A Case Study on Bio-Partnership: MSSRF's Community Agrobiodiversity Centre

MSSRF's **Community Agro-biodiversity Centre** (CAbC) was started in 1997 in the Wayanad district of Kerala. The Centre's activities centre on the conservation and sustainable management of AGRO-BIODIVERSITY in strong partnership with local communities, mainly tribe communities like *Kurichiya, Kuruma, Paniya* and *Kattunaikka.* The activities fall into three Operational Areas - **Conservation** (Agrobiodiversity Conservation and Enhancement); **Education** (Agrobiodiversity Education, Communication & Training) and **Livelihoods** (Agrobiodiversity based sustainable Livelihoods and Food Security) with notable results.

The CAbC now emerged as a model Centre for mainstreaming biodiversity conservation in tribal/rural agricultural development. The Centre is working with over

20,000 households across Wayanad, Idukki and Ernakulam districts covering about 25,000 ha of land, with field-level interventions in Livelihood Development, Food and Nutrition and Conservation programmes focusing on education and awareness programmes.

The Centre has formed and nurtured two Community Institutions, viz., Wayanad Tribal Development Committee; and SEED Care- a Traditional Rice Farmers Forum. The team works with 89 women Self Help Group, and 7 Farmers Clubs, apart from several local Self Government Institutions.

The Centre has perfected four streams of sustainable agricultural and rural development methods: *conservation, cultivation, consumption,* and *commerce,* and is developing the additional resources to serve as a Resource Centre.

CAbC's over 25 years of efforts in the Wayanad District of Kerala helped to promote sustainable management of many "Farmers' varieties" and traditional germplasm in PGRs like Rice, Roots and Tubers and Beans. The research and extension programmes on Farmers' varieties have resulted in:

- Legal recognition of Farmers' Varieties (For the first time in India, community-conserved farmers' varieties of rice got legal recognition through the Protection of Plant Varieties and Farmers Rights Act 2001(PPV&FRA 2001).
- On-farm conservation of Farmers' Varieties (Wayanad became the first district in the state to have the largest number of traditional rice varieties in cultivation).
- A Lead Botanical Garden for Rare Endemic Threatened (RET) in Kerala. The Centre became a conservatory for RET plant species found in the Western Ghats (Since 2010, nearly 100,000 seedlings of 160 RET species have been multiplied and distributed to those who are interested in growing such species).
- *Plant Genome Saviour Award* to a grass-root institution promoted by MSSRF and represented by the tribal communities of Wayanad that are being nurtured by the Centre.
- GIAHS status for Kuttanad Below -Sea level Farming. (The Centre in association with the state agricultural department of Kerala helped the Kuttanad region of the state get the global recognition of FAO promoted GIAHS status for the Below -Sea level Farming).
- The Centre's work in the promotion of family farming, integrated tribal development, RET plant conservation, and nurturing 'child scientists' getting widely recognized in the state.

In conclusion, the local community and other stakeholder partnerships under the CAbC umbrella helped the Wayanad district emerge as a model for community agrobiodiversity management. Key impacts of this effort include the recovery of 300+ RET plants and eco-restoration strategies by involving integrated conservation methods, which have included addressing the ecosystem connectivity challenges faced by many local governments within a fragmented production landscape of mountains in the Malabar region. The 4 C Approach (Conservation-Cultivation-Consumption-Commerce) for the sustainable value chain development of millet, and Neglected and Underutilised Species, and wild food becomes pivotal. Efforts in capacity-building for PRIs, SHGs, NGOs, and the Biodiversity Management Committees became a model for mainstreaming biodiversity.

Conclusion and Recommendations

Unlocking the potential of responsible biodiversity-based businesses becomes crucial for achieving sustainable development goals, especially for a growing economy like India. It, however, demands creating active, inclusive, and gender-responsive partnership avenues at subnational, national, and international stages. Such avenues act as the engines that drive the adoption of biodiversity conservation and sustainable use practices and ensure equitable benefit-sharing through Access and Benefit-Sharing (ABS) mechanisms. Promoting collaboration within specific sectors allows for targeted solutions that address the unique challenges and opportunities confronted by each industry. This is where the 5-P approach becomes relevant to reach out to every key actor in the conservation and utilisation of biodiversity, including the smallholder farming and indigenous people communities and the medium, small, and micro-enterprises.

In India, the appropriate body to formulate and nurture the 5-P partnership discussed is the National Biodiversity Authority (NBA) through the State Biodiversity Boards and the Biodiversity Management Committees and in close collaboration with business firms by utilising their provisions such as Corporate Social Responsibility (CSR) and Corporate Environmental Responsibility (CER). Transparent information sharing, capacity-building initiatives, robust conflict resolution mechanisms, and rigorous monitoring and evaluation processes are the cornerstones of such partnerships.

By emphasizing the Indian Pharma business and biodiversity management context, we recommend the following five-pronged action plan to propel responsible biodiversity-based businesses and ensure the flourishing of Access and Benefit-Sharing (ABS) mechanisms.

1. **Foster Multidisciplinary and Cross-Sectoral Collaboration:** Collaborative efforts between academics, private and public pharmaceutical and health industries, traditional herbal healers, farmers, local communities, and local leaders are crucial for ensuring a sustainable supply of medicinal plants and improving drug quality. Public-private-producer knowledge partnerships should be strengthened to address overexploitation and enhance research and development.
2. **Embracing the One Health Paradigm:** The pharmaceutical industry stands to gain significant advantages by wholeheartedly adopting the One Health approach. This philosophy emphasizes cross-sectoral collaboration, recognizing the intricate web that connects human, animal, and environmental health. By embracing this approach, the industry can cultivate a holistic perspective, fostering an integrated scientific approach that draws upon the strengths of various academic

disciplines. This collaborative spirit can pave the way for the development of more comprehensive and effective medications, diagnostics, and preventive measures.

3. **Building Bridges for Sustainable Solutions:** Ensure a sustainable supply of high-quality medicinal plants by creating a symphony of collaboration across diverse stakeholders. Academics, private and public pharmaceutical and health industries, traditional herbal healers, farmers, local communities, and local leaders must all come together in a collaborative effort. Strengthening public-private-producer knowledge partnerships will be key to addressing overexploitation and propelling research and development efforts. This collaborative spirit can not only secure a sustainable future for medicinal plants but also unlock breakthroughs in drug quality and efficacy.
4. **Develop a Professional Knowledge Management System (KMS):** To truly unlock the potential of biodiversity conservation and sustainable use, a robust professional Knowledge Management System is essential. This system should act as a central hub, facilitating the pooling of diverse knowledge resources. By fostering this collaborative knowledge exchange, we can generate new insights, propel innovations, and ultimately influence research, development, and policy interventions. Formalizing communication systems within the KMS is crucial, ensuring transparency and fostering trust. Peer-reviewed publications and robust feedback mechanisms will be the cornerstones of this system, guaranteeing the quality and reliability of the knowledge shared.
5. **Responsible Management Through Stakeholder Engagement:** Effective stakeholder engagement lies at the heart of responsible biodiversity management. Governments must prioritize capacity building and awareness campaigns to empower stakeholders. Political offices must promote capacity development, enhance awareness campaigns, facilitate partnerships, and ensure transparency in compliance with the Nagoya Protocol. The People's Biodiversity Register should be actively used to engage local communities in biodiversity conservation and sustainable management. Efforts should be made to increase awareness among users of the procedures for compliance with access and benefit sharing.

Conflict of Interest The authors have no conflict of interest to declare.

References

Afzal, I., Habiba, U., & Yasmeen, H. (2023). Review on therapeutic potential of phytochemicals from medicinal plants. *Journal of Bioresource Management, 10*(4). ISSN 2309-3854.

Ahtiainen, H., & Pouta, E. (2011). The value of genetic resources in agriculture: A meta-analysis assessing existing knowledge and future research needs. *International Journal of Biodiversity Science, Ecosystem Services & Management, 7*(1), 27–38. https://doi.org/10.1080/21513732.2011.593557

Barrett, C. B., Travis, A. J., & Dasgupta, P. (2011). Biodiversity conservation and poverty traps. *Proceedings of the National Academy of Sciences, 108*(34), 13907–13912. https://doi.org/10.1073/pnas.1011521108

Bourillon, E., & Allison, E. H. (2014). Fulfilling future food security—The contribution of marine resources. *Food Policy, 44*, 39–54.

Branch, T. A., Jensen, O. P., Ricard, D., Ae, Y., & Hilborn, R. (2011). Contrasting global trends in marine fishery status obtained from catches and from stock assessments. *Conservation Biology, 32*(4), 780–787.

Campbell, T. (2024). *The interconnectedness of human, animal and environmental health.* Pursuit by the University of Melbourne. https://pursuit.unimelb.edu.au/articles/the-interconnectedness-of-human-animal-and-environmental-health

Chaachouay, N., Zidane, L. (2024). Plant-Derived natural products: A source for drug discovery and development. *Drugs Drug Candidates, 3*, 184–207. https://doi.org/10.3390/ddc3010011

Ganguly, N. (2024). *The Indian pharma industry: Shaping a healthier world.* ETHealthworld.com. https://health.economictimes.indiatimes.com/news/industry/the-indian-pharma-industry-shaping-a-healthier-world/109950548

Gibbs, H. K., Rulli, M. C., Morton, D. C., Noojiem, P., Walker, N. F., Busch, J., Lyson, P. A., et al. (2010). Tropical forests and readily available agricultural land. *Current Opinion in Environmental Sustainability, 2*(1), 49–53.

Gowthami, R., Sharma, N., Pandey, R., & Agrawal, A. (2021). Status and consolidated list of threatened medicinal plants of India. *Genetic Resources and Crop Evolution, 68*(6), 2235–2263.

IBBI Working Groups - CII-ITC-CESD. (2023). *CII-ITC-CESD.* https://sustainabledevelopment.in/ibbi-working-groups/

IUCN (1990). 1990 IUCN red list of threatened animals. IUCN Library System. Retrieved from https://portals.iucn.org/library/node/590

Kissinger, G., Herold, M., & vanden Berg, E. (2014). Drivers of deforestation and land use change in tropical forests. *Proceedings of the National Academy of Sciences, 111*(19), 6964–6969.

Mohan, V. (2023). India aims to have a People's Biodiversity Register in every village, a verification campaign launched in Goa. *The Times of India.* https://timesofindia.indiatimes.com/india/india-aims-to-have-peoples-biodiversity-register-in-every-village-verification-campaign-launched-in-goa/articleshow/100453996.cms

Nagoya Protocol. (2014). Nagoya protocol on access to genetic resources and the fair and equitable sharing of benefits arising from their utilization to the convention on biological viversity. Retrieved from https://www.cbd.int/abs/doc/protocol/nagoya-protocol-en.pdf

NDTV (n.d.). *India more than ready to take steps to become science powerhouse: Nature.* NDTV.com. https://www.ndtv.com/india-news/india-more-than-ready-to-take-steps-to-become-science-powerhouse-nature-5477288

Neergheen-Bhujun, V., Awan, A.T., Baran, Y., Bunnefeld, N., Chan, K., Dela Cruz, T.E., Egamberdieva, D., Elsässer, S., Johnson, M.V., Komai, S., Konevega, A.L., Malone, J.H., Mason, P., Nguon, R., Piper, R., Shrestha, U.B., Pešić, M., & Kagansky, A. (2017). Biodiversity, drug discovery, and the future of global health: Introducing the biodiversity to biomedicine consortium, a call to action. *Journal of Global Health, 7*(2), Article 020304. https://doi.org/10.7189/jogh.07.020304

Newman, D.J., Cragg G.M. (2020). Natural Products as Sources of New Drugs over the Nearly Four Decades from 01/1981 to 09/2019. *Journal of Natural Products. 83*(3), 770-803. https://doi.org/10.1021/acs.jnatprod.9b01285. PMID:32162523.

Ramy, A., Af Ragab, M., & Arisha, A. (2020). Knowledge management in the pharmaceutical industry between academic research and industry regulations. *Knowledge Management Research & Practice, 20*(2), 202–218. https://doi.org/10.1080/14778238.2020.1767517

Singh, A. P., & Rahman, Z. (2021). Integrating corporate sustainability and sustainable development goals: towards a multi-stakeholder framework. *Cogent Business & Management, 8*(1). https://doi.org/10.1080/23311975.2021.1985686

Srivathsa, A., Vasudev, D., Nair, T., Chakrabarti, S., Chanchani, P., DeFries, R., Deomurari, A., Dutta, S., Ghose, D., Goswami, V. R., Nayak, R., Neelakantan, A., Thatte, P., Vaidyanathan, S., Verma, M., Krishnaswamy, J., Sankaran, M., & Ramakrishnan, U. (2023). Prioritizing India's landscapes for biodiversity, ecosystem services and human well-being. *Nature Sustainability, 6*(5), 568–577. https://doi.org/10.1038/s41893-023-01063-2

Swaminathan, M. S. (2010). *From green to evergreen revolution, Indian Agriculture: Performance and challenges* (pp.70–71). New Delhi: Academic Foundation.

Taskforce on Nature-Related Financial Disclosures (TNFD) (2023). *Executive summary of the recommendations of the TNFD.* tnfd.global. https://tnfd.global/wp-content/uploads/2023/09/Executive_summary_of_the_TNFD_recommendations.pdf?v=1695117009

Taubman, A. (2008). Traditional knowledge and genetic resources. ResearchGatc. https://www.researchgate.net/publication/283482712_Traditional_Knowledge_and_Genetic_Resources

TEEB for Business [Bishop, J., Bertrand, N., Evison, W., Gilbert, S., Grigg, A., Hwang, L., Kallesoe, M., Vakrou, A., van der Lugt, C., & Vorhies, F.]. (2010). *TEEB—The economics of ecosystems and biodiversity report for business—Executive summary* (p. 27). Bonn, Germany: TEEB.

The World Benchmarking Alliance (WBA). (2022). *Nature benchmark report 2022.*

UN Environment Programme World Conservation Monitoring Centre (UNEP-WCMC). (2020). Biodiversity measures for business: Corporate biodiversity measurement and disclosure within the current and future global policy context. UNEP-WCMC.

World Economic Forum & AlphaBeta. (2020). The future of nature and business. World Economic Forum.

Young, R. N. (1999). Importance of biodiversity to the modern pharmaceutical industry. *Pure and Applied Chemistry, 71*(9), 1655–1661.

Analysing the Possibilities of Managing the Invasive Alien Plants of India Through Access and Benefit Sharing Mechanism: A Review

S. Sandilyan, Prakash Nelliyat, and G. Nagendra Prabhu

Introduction

Biodiversity is the foundation of all life on our planet. It not only sustains the air we breathe, the food we eat, and the water we drink but also provides us with invaluable medicines and other vital resources essential for all living organisms. In essence, biodiversity plays a central role in ensuring the successful and harmonious functioning of every ecosystem upon which mankind depends. In addition to its ecological significance, it is crucial to acknowledge the immense commercial and economic value offered by existing biodiversity. Countless national and international research organizations have emphasized this fact time and again. The diverse array of species that exist in our ecosystems presents us with unlimited potential for scientific discoveries, technological advancements, and sustainable business opportunities. However, despite its undeniable importance, our planet's biodiversity has been facing numerous threats for the past few centuries. Habitat destruction, pollution, overexploitation, and climate change are just some challenges. Further, in the wake of the Industrial, Green, and Blue revolutions, as well as the forces of globalization, many ecosystems are facing unprecedented threats. These threats have resulted in a significant loss of biodiversity worldwide.

On the other hand, it is heartening to note that numerous international treaties and conservation organizations have been tirelessly working towards halting this loss.

S. Sandilyan (✉)
Former Fellow on Invasive Alien Species (IAS), Chennai, India
e-mail: ssandilyan@gmail.com

S. Sandilyan · P. Nelliyat
Former Fellow on Access and Benefit Sharing (ABS), Centre for Biodiversity Policy and Law (CEBPOL), National Biodiversity Authority (NBA), Taramani, Chennai 600113, India

G. N. Prabhu
Centre for Research On Aquatic Resources, Sanatana Dharma College, University of Kerala, Alleppey, Kerala 688003, India

L. Krishna Panicker et al. (eds.), *Biodiversity and Business*,
https://doi.org/10.1007/978-3-031-71674-4_3

Despite these efforts, recent reports have highlighted the relentless damage caused by ever-increasing anthropogenic pressures on existing species diversity. Fresh studies have pointed out that, among the list of biodiversity threats, invasive alien species (IAS) have been identified as the second biggest threat to global biodiversity. The recent global assessment report by IPBES (Intergovernmental Science-Policy Platform on Biodiversity and Ecosystem Services) highlighted that nearly 37,000 alien species are reported from various ecosystems across the countries. Among them, 3500 species have played a major role in the extinction of 60% of the global species. Besides, IAS also affects global and regional food security, the economy, and human well-being (Sandilyan & van't Klooster, 2016; Sandilyan et al., 2018d; Surendra et al., 2013; www1).

Considering the deleterious impacts of IAS, several biodiversity-rich countries have tried different methods to prevent entry and to halt the post-entry damages caused by IAS, but their success is still limited. Every year the problem is spiralling up across the continents (Sandilyan & van't Klooster, 2016; Sandilyan et al., 2018c, 2018d). Further, recent ecological modelling studies on biological invasion highlighted that, in the coming days, developing countries will face more disasters than developed countries (Early et al., 2016). India is one of the developing countries that is highly prone to multispecies invasion (Adhikari et al., 2015).

Countries across the continents are coming up with different management strategies (chemical, biological, and physical removal) to restrict further the deep invasion of IAS in different ecosystems. However, effective control/eradication/management is lacking due to several reasons, including insufficient manpower and money power and poor scientific approaches/policies. Some of the recent studies shed light on the socioeconomic benefits rendered by some IAS, especially how it helps the livelihood of local communities (Sandilyan & van't Klooster, 2016).

In this context, this chapter examines the possibilities of IAS management through the Convention on Biological Diversity (CBD)'s Access and Benefit Sharing (ABS) principle. Here the concerned industrial sector that uses IAS for their production should share a portion of their benefits (as prescribed in the legal norms of the country that ratified the Nagoya Protocol on ABS) with the enforcement agencies and the local community (CBD, 2011a, 2011b). Here, the enforcement agencies or the local community cannot directly use the money obtained through ABS for the conservation of IAS, but it can be used for eradicating/managing the IAS spread as well as promoting the native species or enhancing the local biodiversity. A horizon of new ABS mechanisms can emerge for eradication or managing the spread of IAS by simultaneously identifying native species with a good commercial scope or industrial use in the invaded area and constantly promoting cultivation and sustainable harvesting.

IAS: Indian Scenario

In the past, several thousands of alien species have been deliberately introduced into India by colonial rulers. The British especially introduced several thousand species of flora and fauna in different landscapes for sports, agriculture, ornamental, and aquaculture purposes (Joshi et al., 2018; Narayanan et al., 2007; Roxburgh, 1814; Royle, 1840). The majority of the introduced species easily acclimatized to Indian habitats/ecosystems and subsequently, a commendable number of species have shown exponential growth in diverse landscapes. Researchers mentioned that prevailing species-friendly climatic conditions, unique landscapes, and fertile soil varieties always support all kinds of life forms in this peninsula (Prathapan & Dharmarajan, 2009). These earlier introductions drastically changed the natural diversity, economy, food habits, and aesthetic and cultural values of this nation (Antani & Mahapatra, 2022; Joshi et al., 2018; Narayanan et al., 2007). In addition, a recent survey revealed that more than 50% of Indian hotspots are highly vulnerable to new multispecies invasions (Adhikari et al., 2015). Research related to eradication, control, and management of exotic invasive species in India is extremely poor when compared to several countries with less biodiversity and infrastructure (Adhikari et al., 2015).

A recent report from the Centre for Biodiversity Policy and Law (CEBPOL) housed at the National Biodiversity Authority (NBA) mentioned that there is no exclusive national policy, strategies, or legal framework in India to control and manage the IAS (Sandilyan et al., 2018d). CEBPOL also highlighted that biological invasion is still in its infancy in India and has not received adequate attention from stakeholders, managers, policymakers, researchers, and the common public so far (Sandilyan et al., 2018a, 2018d). There were hardly 150 scientific publications/reports till 2000 relating to IAS (Hiremath & Sundaram, 2013). Interestingly in the past two decades, several research organizations and universities have started to address the issue on different platforms; however, most of the information is about the different invasive species in various habitats and very few have discussed the management strategies. Based on 30 years of IAS publication from India, CEBPOL prepared a comprehensive list of Indian IAS with the help of an expert team and released it in 2018. However, the apex bodies (NBA and Ministry of Environment, Forest and Climate Change) failed to update the list periodically. Though there are thousands of alien species reported from India, the CEBPOL and the national-level expert committee have declared only 155 species as invasive after a careful examination based on the robust criteria developed by CEBPOL (Sandilyan et al., 2018f).

Further, from the list of 100 of the world's worst IAS found in the global invasive species database (www2), 19 species (6 animals and 13 plants) were reported in India which includes giant African snail (*Achatina fulica*), banana bunchy top virus (BBTV), sweet potato whitefly (*Bemisia tabaci*), wild carp, grass carp (*Cyprinus carpio*), mosquito fish (*Gambusia affinis*), common tilapia (*Oreochromis mossambicus*), black wattle (*Acacia mearnsii*), giant cane (*Arundo donax*), siam weed (*Chromolaena odorata*), koster's curse (*Clidemia hirta*), water hyacinth (*Eichhornia *crassipes*), cogon grass (*Imperata cylindrica*), lantana (*Lantana camara*), leucaena (*Leucaena leucocephala*), bitter vine (*Mikania micrantha*), touch me not (*Mimosa pigra*), prickly pear (*Opuntia stricta*), giant salvinia (*Salvinia molesta*), gorse (*Ulex europaeus*). Unfortunately, there is no awareness about the aforesaid species among policymakers and researchers in India.

IAS Management

Globally there are very few examples of successful eradications of IAS and that too in an area of a few square kilometers such as small islands or landlocked areas (Mack & Lonsdale, 2002; Moore et al., 2011). Most of the countries including India have failed to identify the gaps in the strategic approaches and control measures, resulting in great disappointments. The lack of coordination and inadequate awareness among different stakeholders, managers, and the common public are the other important reasons for the failures in invasive species management (Sandilyan et al., 2018e). Control of IAS is not a short-term or isolated effort; managers, policymakers, and researchers need to have a profound knowledge of the issue before developing a management/control/eradication program (Bibhuti et al., 2011).

India has made several chemical, mechanical, and biological attempts to eradicate the most virulent plant species such as *L. camara*, *Prosopis juliflora*, *Parthenium hysterophorus*, *Mimosa diplotricha*, and *E. crassipes*. Ironically, the majority of the attempts across the habitats have been futile (see Sandilyan & van't Klooster, 2016; Sandilyanet al., 2018d). India needs to find effective alternative ways for better management/eradication of the aforementioned invasive species.

Socioeconomic Benefits of IAS

Naturally, all the plants including invasive species have commendable commercial values (e.g. food, medicine, fodder, firewood, timber, etc.). In this context, it is wise to identify the commercial values offered by the most problematic IAS and try to examine the possibility of the ABS mechanism with special reference to IAS management which is highly encouraged.

In India, several IAS are reported to be used to prepare food, medicine, furniture, clothes, biofuel, fodder, and biofertilizer. At the same time, economic valuations of those IAS are due in India (Raghubanshi et al., 2005; Sandilyan & van't Klooster, 2016). The commercial utilization of all IAS needs to be encouraged region-wise large-scale screening is needed. Subsequent advertising of the benefits offered by an IAS will lead to an increasing demand on local markets and industries and in due course of time the deep invasion will be gradually halted due to overharvesting by the common public and stakeholders. The intervention of the ABS mechanism at this juncture can be encouraged.

It is evident that developing and implementing a robust ABS framework can have numerous benefits. Not only does it help to conserve the diverse native flora and fauna, but it also aids in effectively managing invasive species. Establishing clear guidelines for accessing IAS resources and sharing of benefits ensures that the native biodiversity is protected in the invaded area.

Our current review aims to analyze the possible ways in which we can enhance the ABS framework by examining existing practices and identifying areas for improvement. After a series of discussions with subject experts, we have selected water hyacinth (*E. crassipes*) for the current review. In this article, we screened the ABS potential of this plant which has invaded and infested several important wetlands across India.

Water Hyacinth: A Case Study

Ecology

E. crassipes, commonly called the water hyacinth, is a native species of Amazonia (Brazil and Argentina). It is a perennial, herbaceous aquatic plant belonging to the family Pontederiaceae. In an ideal condition, the plant can grow up to 65 cm in height with dark green smooth, glossy, circular waxy leaves, having a diameter of 5–10 cm. The kidney-shaped lamina of the leaves is spongy, bulbous, and highly helpful as a buoyant. The matured plants have elongated leaves (see Sandilyan et al., 2018d).

E.crassipes can reproduce both sexually (seed propagation) and asexually (stolon vegetative reproduction), which helps in successful and aggressive invasion (Harun et al., 2021). During the breeding season, a matured plant develops an erect stalk that supports a single spike of 8 to 15 violet and yellow flowers; in some of the habitats, the flowers are light to darker blue-purple. Naturally, the flowers can be self-fertilized and the inflorescence produces 3000–4500 egg-shaped seeds, which can reach a length of up to 1.5 mm and remain viable for 5–20 years. A single rosette can produce several inflorescences and several thousands of seeds in a season (Barrett, 1980; Gopal, 1987; Hill, 1999). At the time of flower withering, naturally, the stalk of the flower slowly bends near to the water surface and releases the seed, this will go up to 18 days. The water current takes the seeds to a new destination, and sometimes the seeds also attach to the boats and travel a long distance and reach new places (Sandilyan et al., 2018d).

On the other hand, vegetative reproduction is highly reported from this plant. Vegetative reproduction is largely responsible for the rapid increase of the biomass of the species (Barrett, 1980). Daughter plants are produced by vegetative methods; from the horizontal stolons they develop their roots and later on separate from the mother plant. In warm climates, vegetative reproduction is rapid and enables the formation of large, dense mats within a short time. Besides, the young plants are detached from the mother plant during heavy wind, flood, fishing, and other associated aquatic activities (Sandilyan et al., 2018d).

Impacts of the Invasion

Globally, several studies and research reports have highlighted the negative impacts of water hyacinth on water property, native biodiversity, livelihood, and regional economy. Especially large-scale invasion affects the inland aquatic transport and reduce the fishing ability and fish catches (Sandilyan et al., 2018d).

The invasion of water hyacinth largely affects the growth, distribution, and abundance of phytoplanktons and zoo planktons, which ultimately affects the food resources of fishes and also alters the energy flow of the particular wetland. For instance, fish catch rates in the Kenyan section of Lake Victoria decreased by 45% due to this weed invasion. On the other hand large-scale invasion drastically reduced the dissolved oxygen level (Maceina et al., 1992; Mironga et al., 2011; Sandilyan et al., 2018d; Villamagna & Murphy, 2010). Further, the invasion frequently causes disturbances to the hydroelectric operations. In addition, the small irrigational canals are frequently choked by this weed, causing artificial flooding and it also affects agriculture productivity. Moreover, it constantly increases the evapotranspiration rate. The aforesaid disturbances also affect millions of people directly and indirectly (Sandilyan et al., 2018d).

Apart from the above, the invasion encourages vector multiplication in the wetlands. Especially it turned into an ideal breeding ground for the organisms causing malaria, schistosomiasis, encephalitis, filariasis, and cholera (FAO, 2002).

Economic Importance of *E. Crassipes*

Besides the detrimental effects of *E. crassipes* infestation, the plant has several commercial uses that indirectly help us to control the infestation. The rapid growth and biomass are helpful to produce:

1. Biofuel/Biogas (methane): The high cellulose content and fast growth rate of weeds make them suitable feedstock for biofuel production, which can provide an economic incentive for their cultivation and harvesting.
2. Animal feed: It can be used as nutritious animal feed especially for pigs and ducks, reducing the reliance on traditional feed crops and providing farmers with an additional income stream.
3. Phytoremediation: This species is effective at absorbing and removing heavy metals and other pollutants from contaminated aquatic medium and making them useful for environmental clean-up efforts.

4. Medicinal compounds: This species contains valuable medicinal compounds that can be extracted and used in pharmaceutical applications, creating a market demand for their cultivation (Sandilyan et al., 2018d).

However, special technical skill is required to obtain biogas and briquettes from water hyacinth (Rakotoarisoa, 2017). In the same way, manufacturing furniture and handicrafts from this weed is a viable option but it requires adequate market outlets (Patel, 2012). Further, recently researchers have derived sugar molecules like lignin, cellulose, and hemicellulose from the weed and have converted it to polyhydroxybutyrate, which is a raw material to prepare biodegradable plastic (Cherly, 2013).

In India, several products have been derived from water hyacinth. In Andhra Pradesh, Allika, a social enterprise, is effectively manufacturing several products with the help of rural women and successfully marketing them in several parts of the country. Their products range from handbags, sling bags, and laptop bags, to bowls, hats, trays, cups, lampshades, and clutches and wallets (Sandilyan et al., 2018d).

By leveraging the economic potential of weeds, we can encourage their controlled cultivation and harvesting, which can ultimately help to manage and mitigate the detrimental effects of weed infestation.

Kerala as a Model

In Kerala, commendable efforts have been made in the past for using these weeds for beneficial purposes. Especially the researchers at the Centre for Research on Aquatic Resources (CRAR), Sanatana Dharma College, Alleppey, Kerala, reported that all parts of the plant can be used in one way or similar to the coconut palm.

A Community Training Centre has been established at the Centre to transfer the technologies to local entrepreneurs, especially women and others from all walks of life. Two student start-ups, namely Ecoloop 360 and EichhoTech, were also established to produce and market the various value-added products from water hyacinth using the technologies developed at CRAR.

Products Manufactured

Research at CRAR has led to the development of several innovative processes and products from water hyacinth. The most important ones are briefly described below:

Mushroom cultivation—The stem (petiole) and leaves of freshly harvested water hyacinth were used as bedding material for the cultivation of different species of mushrooms. The common mushrooms grown over a bed of pretreated water hyacinth included *Pleurotus florida* and *Hypsizygus ulmarius* (Anoop Kumar et al., 2014). Control bed was prepared using paddy straw and the results indicated that water hyacinth was a suitable substrate for the cultivation of common mushrooms.

Modified biomass briquettes—It is common practice in many parts of the world to dump the harvested water hyacinth on the banks of lakes and rivers near the infested sites. These plants will dry and pose multiple disposal problems to the people. Such dried water hyacinth biomass was powdered and used for the preparation of briquettes. Briquettes are compressed organic matter that is used as fuels and can bc prepared by hand or using simple to complex machines. The approach at CRAR was to use the dried water hyacinth powder as briquettes along with cow dung using the traditional method or using simple hand extruders used in kitchens. We were successful in developing briquettes from dried water hyacinth biomass for use as cheap fuel (unpublished data).

Pulp-based products—Freshly harvested stems and leaves of water hyacinth were pulped using a kitchen grinder and mixed with the pulp from used paper and appropriate binders. The resultant pulp was used to prepare multi-purpose products ranging from special canvas for paintings, biodegradable nursery pots, handicraft items, egg and fruit tray models, non-load-bearing bricks, partition boards, etc. (Nagendra Prabhu, 2016).

Hand-made paper—The pulp from freshwater hyacinth is also an excellent raw material for the preparation of hand-made paper. The research at CRAR has resulted in the development of hand-made paper with different ratios of water hyacinth pulp along with used newspapers, discarded office papers, or pulp from various vegetable and agricultural wastes. The papers thus produced were used to prepare visiting cards, wedding invites, art, and craft works, and as a background for unique works of art such as dry stem art and single rope art using dried water hyacinth stem. Such innovative products were designed by the student start-up called EichhoTech which was founded during 2019–2020 in the college.

Success Stories

Since the majority of the products developed from water hyacinth were innovative and new, there was no established "market rate" as the case with other products. The strategy followed by the group was to concentrate on "high-value low-volume" products that were unique, eco-friendly, and customized to suit the specific needs of the identified or potential customers. Such products could earn a premium price from niche markets such as high-end lakefront resorts, and enlightened customers preferring innovative and green products. The student start-up secured funding from the Kerala Development and Innovation Strategic Council (KDISC) under its inaugural Young Innovators Programme (https://yip.kerala.gov.in/yip-2019-2022) as well as the Kerala University Business Innovation and Incubation Centre (https://kubiic.in/portfolio/eichhotech/).

In 2021, the idea of biodegradable nursery pots proposed by EichhoTech was declared a winner under the Student Category of the India Plastic Challenge Hackathon organized by the Ministry of Environment, Forests and Climate Change (https://www.linkedin.com/posts/india-plastic-challenge-35498a213_winner-announcement-activity-6883800770500808704-I1Bn/ and https://pib.gov.in/PressReleasePage.aspx?PRID=1813806). Also, in 2023, they were selected as one of the three winners of the nature-based solutions entrepreneurship challenge 2023 (https://www.linkedin.com/posts/kerala-institute-of-local-administration-kila-_nbsentrepreneurchallenge-naturebasedsolutions-activity-7034044573567897600-mSLF/).

They could also secure business orders from the Kerala Start-up Mission for the Huddle Global 2022 worth more than INR 50,000. (https://www.thehindu.com/news/national/kerala/student-startup-converts-water-hyacinth-into-eco-friendly-gift-hamper/article66288748.ece). They were also selected to supply mementos, invitation letters, writing pads, and seed pens for the Nehru Trophy Boat Race during 2022 and 2023, for which they received a total of INR 75,000. (https://www.thehindu.com/news/national/kerala/alappuzha-gearing-up-for-69th-edition-of-nehru-trophy-boat-race/article67150580.ece). The student start-up supplied vising cards to many high-ranking officials of the Government of India using the hand-made paper from water hyacinth (https://www.onmanorama.com/lifestyle/news/2023/12/11/transforming-nature-overgrowth-into-eco-friendly-visiting-cards.html). Further, products like paintings on eco-friendly canvas from water hyacinth pulp and customized bookmarks were displayed and sold to individual customers at several exhibitions and awareness sessions during the past few years. Recently, wedding invitation cards printed on water hyacinth hand-made paper created ripples in social media and also in the print media. (https://www.onmanorama.com/lifestyle/women/2024/01/04/kuttanad-girl-s-green-wedding-initiative-sheds-light-on-the-mena.html).

Current Status of Ecoloop 360 and EichhoTech

Ecoloop360 worked briefly on the production of briquettes from water hyacinth and secured seed fund from KDISC (https://englisharchives.mathrubhumi.com/agriculture/agri-news/value-added-products-from-water-hyacinth-gives-hope-to-lake-revival-1.3437603; https://www.thehindu.com/news/cities/Thiruvananthapuram/turning-a-menace-into-a-boon/article26208094.ece). However, later their focus shifted to other areas and they are currently not working on water hyacinth.

EichhoTech, however, has made tremendous progress since its establishment in 2019, as mentioned above. Currently, it is being incubated at the Technology Business Incubator in the College.

IAS Management Through ABS: What Is Needed?

IAS management through ABS is a new thought we propose in this chapter. Its operationalization must be a challenge and sometimes difficult to accept. However, the current financial need for biodiversity management is huge. A recent study by the UNDP-sponsored Biodiversity Finance Initiative (BIOFIN) claimed that US$ 52 billion is spent on biodiversity per year globally. This is against an estimated annual financing need of between US$ 150 and 440 billion. This funding gap is a challenge in achieving the CBD's Strategic Plan and the SDGs (UNDP, 2018).

Here, we are assuming that ABS might be one of the effective innovative financial options for mobilizing the cost of biodiversity conservation. After the commencement of the CBD, the transformation of biodiversity from a public good to the sovereign rights of nations occurred, where the genetic and biological resources' emergence/existence is important. In this context, particularly from a management perspective, the IAS's origin and sovereign rights need to be considered as a topic of debate even in the sections of the Conference of the Parties (COP-15)of the CBD.

ABS might be a workable conservation and sustainable use strategy for the case of native genetic or biological resources having commercial potential. In this case, the ABS money can be channeled for the conservation and sustainable use of those biological resources (species). But this philosophy needs some twist when it comes to IAS. Here, rather than using the ABS money for conserving the ABS-induced biological resources (here IAS), we suggest using it for restricting that particular IAS' further spread and gradual eradication (we cannot propose IAS conservation through the ABS money, instead we propose restoration of the degraded ecosystem and proliferation of native species in that habit). Further, instead of IAS, we strongly propose the conservation and sustainable use of native species having high commercial and market potential. In other words, the strategy is 'replacing the IAS with native species'. Afterward, the ABS can operate in its normal form.

In this process, the enforcement agencies, who are responsible for implementing the Nagoya Protocol on ABS in their respective countries (the concerned Ministry or the Competent National Authority of the country to the CBD), should come up with guidelines on IAS ABS with consideration of the following provisions:

1. Documentation and mapping of ecosystem-wise IAS in the country (state, districts, and blocks/panchayat level)
2. Identifying the utility of the species with a focus on the current commercial use as well as the proposed future marketing scope
3. Examining the supply chain and value addition of commercially significant IAS
4. List out the industries/traders who use/handle the IAS (national and international)
5. Estimate the ABS potential of the IAS
6. Come up with innovative and scientific methods for IAS harvest and managing its eradication. Identify suitable native species for replacing the IAS and its conservation and sustainable use
7. Specify the local institution's (such as Biodiversity Management Committees) role in IAS management and promotion of native species.

Conclusion

Most of the Indian inland wetlands have been drastically devastated by the invasion of water hyacinth; however, to date, there is no comprehensive data available about the biodiversity, economic, and livelihood loss caused by the species across India (Sandilyan et al., 2018d). On the other hand, global studies on invasive species eradication highlighted the practical bottlenecks of the process, and several studies have discussed the field challenges in aquatic invasive management (e.g. Harun et al., 2021). The eradication of water hyacinth species from Indian wetlands needs regional cooperation, huge money, and manpower with a continuous monitoring system. Practically it is a high burden for a large country like India which supports 15.98 million hectares of wetlands across different landscapes. Moreover, the manmade wetlands of India are 1.8 times bigger than the natural wetlands, which interlace with several million peoples' religious and cultural faiths (Sandilyan, 2022; MoEF, 1995).

Studies from India related to IAS management shed some light on the benefits offered by some invasive plants (Sandilyan & van't Klooster, 2016). Interestingly some of the studies highlighted the commercial benefits of water hyacinth. Especially studies from Kerala give strong hope for the chance of the ABS mechanism. It is the need of the hour to identify different stakeholders, policymakers, and local people in and around Indian wetlands and educate them about the need for the removal of water hyacinth biomass to process different products and also for native biodiversity management.

Regular harvesting will lead to checking the proliferation and halting further invasion of water hyacinth. The involvement and participation of residents also need to be encouraged by establishing village-level committees with the help of the State Biodiversity Boards (SBBs) and Biodiversity Management Committees (BMCs).

On the other hand, the ABS enforcement agencies need to develop a special cell for assessing the ABS potential of water hyacinth in India, for which a pilot survey/study can be carried out in association with Sanatan Dharma College of Alleppey, Kerala. Based on the result it can be extended to other invasive plants of India.

Conflict of Interest The authors have no conflict of interest to declare.

References

Adhikari, D., Tiwary, R., & Barik, S. K. (2015). Modelling hotspots for invasive alien plants in India. *PLoS ONE, 10*(7), e0134665. https://doi.org/10.1371/journal

Anoop Kumar, V., Sreelakshmi, T. P., Azmi, T., Bindu, P., Unnikrishna Pillai, P. R., & Nagendra Prabhu, G. (2014). *Mushroom cultivation using aquatic weeds of Kerala*. Proceedings of national symposium on emerging trends in biotechnology (pp. 166–176). ISBN 978-93-80095-47-9.

Antani, V., & Mahapatra, S. (2022). Evolution of Indian cuisine: A socio-historical review. *Journal Ethnic Foods, 9*, 15. https://doi.org/10.1186/s42779-022-00129-4

Barrett, S. C. H. (1980). Sexual reproduction in *Eichhornia crassipes* (water hyacinth). II. Seed production in natural populations. *Journal of Applied Ecology, 17*, 113–124.

Bibhuti, P. L., Talukdar, B. K., & Sarma, P. (2011). Invasive species in grassland habitat: An ecological threat to the greater one-horned rhino (*Rhinoceros unicornis*). *Pachyderm, 49*, 33–39.

CBD. (2011a). Convention on biological diversity text and annexes. Secretariat of the Convention on Biological Diversity, Montreal.

CBD. (2011b). Nagoya Protocol on access to genetic resources and the fair and equitable sharing of benefits arising from their utilization to the convention on biological diversity. Convention on Biological Diversity, United Nations.

Cherly, N. F. A. (2013). Assessment and utilization of water hyacinth in the water bodies of tamil Nadu. *International Journal of Scientific Research and Reviews., 2*(1), 58–77.

Early, R., et al. (2016). Global threats from invasive alien species in the twenty first century and national response capacities. *Nature Communication.* https://doi.org/10.1038/ncomms12485

FAO. (2002). *Management of problematic aquatic weeds in Africa FAO efforts and achievements during the period 1991–2001.* Rome, Italy.

Gopal, B. (1987). *Aquatic plant studies 1. Water Hyacinth.* Elsevier Publishing (Original not referred).

Harun, I., Pushiri, H., Amirul-Aiman, A. J., & Zulkeflee, Z. (2021). Invasive Water Hyacinth: Ecology, impacts and prospects for the rural economy. *Plants, 10*, 1613. https://doi.org/10.3390/plants10081613

Hill, M. (1999). The world's worst aquatic weed. *Pesticide Outlook*, 58–61(Original not referred).

Hiremath, A. J., & Sundaram, B. (2013). Invasive plant species in Indian protected areas: Conserving biodiversity in cultural landscapes. In L. C. Foxcroft et al. (Eds.), Plant invasions in protected areas: Patterns, problems and challenges. *InvadingNature—Springer Series in Invasion Ecology, 7.* https://doi.org/10.1007/978-94-007-7750-7_12. © Springer Science+Business MediaDordrecht 2013.

Joshi, A. A., Sankaran, M., & Ratnam, J. (2018). Foresting the grassland: Historical management legacies in forest grassland mosaics in southern India and lessons for the conservation of tropical grassy biomes. *Biological Conservation, 224*, 144–152.

Maceina, M. J., Cichra, M., Betsill, R., & Bettoli, P. (1992). Limnological changes in a large reservoir following vegetation removal by grass carp. *Journal of Freshwater Ecology, 7*, 81–93.

Mack, R. N., & Lonsdale, W. M. (2002). Eradicating invasive plants: Hard-won lessons for islands. In C. R. Veitch, & M. N. Clout (Eds.), Turning the tide: The eradication of invasive species (pp. 164–172). IUCN, Gland: IUCN SSC Invasive Species Specialist Group.

Mironga, J. M., Mathooko, J. M., & Onywere, S. M. (2011). The effect of Water Hyacinth (*Eichhornia crassipes*) Infestation on Phytoplankton productivity in Lake Naivasha and the status of control. *Journal of Environmental Science and Engineering, 5*(10), 1252–1261.

MOEF. (1995). *Wetlands of India.* Ministry of Environment and Forest, New Delhi India.

Moore, J. L., Runge, M. C., Webber, B. L., & Wislon, J. R. (2011). Contain or eradicate? Optimising the management goal for Australian acacia invasions in the face ofuncertainty. *Diversity and Distributions, 17*, 1047–1059.

Nagendra Prabhu, G. (2016). *Journal of Aquatic Biology & Fisheries, 4*, 8–14.

Narayanan, S. P., Anitha, K., Prasad, & Narendra, S. N. S. (2007). Water hyacinth (*Eichhornia crassipes*) management of an Invasive weed. The Indian scenario. *Sarovar Saurabh, 3*(4), 4–7.

Patel, S. (2012). Threats, management and envisaged utilizations of aquatic weed Eichhornia crassipes: An overview. *Reviews in Environmental Science and Bio/Technology, 11*, 249–259. https://doi.org/10.1007/s11157-012-9289-4

Prathapan, K., & Dharmarajan, P. (2009). Biological Diversity act, 2002: Threat to agricultural production and food security. *Current Science, 97*(5), 626–629.

Raghubanshi, A. S., Rai, L. C., Gaur, J., & Pand Singh, J. S. (2005). Invasive alien speciesand biodiversity in India. *Current Science, 88*(4), 25.

Rakotoarisoa, T. (2017). Use of water hyacinth (*Eichhornia crassipes*) in poor and remote regions—A case study from LakeAlaotra, Madagascar (M.Sc., thesis). Universität Hildesheim, Madagascar.

Roxburgh, W. (1814). Hortus bengalensis*: A catalogue of the plants growing in the honourable East India company botaniuc garden at Calcutta* (p. 40). Mission Press.

Royle, J. F. (1840). *Essay on productive resources of India.* Botanic garden established at Calcutta. WH Allen.

Sandilyan, S., & Sand Charlotte I. E. A. van't Klooster. (2016). The other sides of invasive alien plants of India—With special reference to medicinal values. *Journal for Nature Conservation, 31*, 16–21.

Sandilyan, S., et al. (2018a). *A review on impacts of invasive alien species on Indian inland aquatic ecosystem.* Published by National Biodiversity Authority, Chennai, India. ISBN: 978-81-932753-3-7.

Sandilyan, S., et al. (2018b). *A review on impacts of invasive alien species on Indian coastal ecosystem.* Published by National Biodiversity Authority, Chennai, India. ISBN: 978-81-932753-1-3.

Sandilyan, S., et al. (2018c). *Impacts of invasive alien species on Island ecosystems of India with special reference to Andaman group of Islands.* Published by National Biodiversity Authority, Chennai, India. ISBN: 978-81-932753-5-1.

Sandilyan, S., et al. (2018d). *Strategies for Control and management of some selective Invasive alien plant species endangering Indian biodiversity.* Published by National Biodiversity Authority, Chennai, India. ISBN : 978-81-940589-8-4.

Sandilyan, S., et al. (2018e). *Guidelines for prioritization of invasive alien plants of India for management.* Published by National Biodiversity Authority, Chennai, India. ISBN: 978-81-932753-4-4.

Sandilyan, S., et al. (2018f). *Invasive Alien Species of India.* Published by National Biodiversity Authority, Chennai, India

Sandilyan. S. (2022). Alien fish species in Indian inland wetlands: current status and future challenges. *Wetlands Ecol Manage* 30, 423–437. https://doi.org/10.1007/s11273-022-09870-8.

Surendra, B., Muhammed, A. A., & Raju, S. A. J. (2013). Invasive Alien species assessment in urban ecosystem: A case study from Andhra University, Visakhapatnam, India. *International Research Journal of Environment Sciences, 2*(5), 79–86.

Villamagna, A., & Murphy, B. (2010). Ecological and socio-economic impacts of invasive water hyacinth (*Eichhornia crassipes*): A review. *Freshwater Biology, 55*, 282–298.

UNDP. (2018). 'The Biodiversity Finance Initiative Workbook 2018: Finance for Nature", BIOFIN, United Nations Development Programme, New York.

Net Based References

www1. https://indianexpress.com/article/india/biodiversity-loss-37000-alien-species-introduced-by-human-activities-says-report-8924756/#:~:text=%E2%80%9CClimate%20change%20is%20also%20predicted,the%20impacts%20of%20climate%20change. Accessed 15 January 2024.

www2. https://www.iucngisd.org/gisd/100_worst.php. Accessed 20 January 2024.

Biosurvey & Bioprospecting

J. R. Rani, Laladhas Krishna Panicker, V. S. Soumya, Aswini Jayachandran, M. P. Vinod, P. Sunitha, and Oommen V. Oommen

Introduction

In biology and environmental science, the terms "bioprospecting" and "biosurvey" are synonymous, yet they refer to distinct facets of investigation and discovery. The term "bioprospecting" describes the process of looking for, removing, and using biological resources such as plants, animals, microbes, and their genetic material for a range of uses in industry, agriculture, and medicines. Finding novel biological substances, genes, or organisms with possible commercial or therapeutic value is the main objective of bioprospecting. To find novel species or bioactive chemicals with special qualities, researchers engaged in bioprospecting may comb through a variety of habitats, such as seas, rainforests, and other places rich in biodiversity.

In contrast, a biosurvey is a methodical investigation of a particular region to evaluate the biological variety and richness of that area. Numerous taxonomic groupings, including fungi, animals, plants, and microbes, may be covered by the survey. Documenting and comprehending the distribution and health of various species within an ecosystem is the primary goal of a survey. Not only is this data useful for conservation initiatives, but it may also be used to track changes in biodiversity over time. Field sampling, data gathering, and analysis are frequently used in biosurveys to give a

J. R. Rani
School of Biotechnology & Pharmaceutical Sciences, Vignan's Foundation for Science, Technology & Research, Guntur, Andra Pradesh, India

L. Krishna Panicker (✉) · A. Jayachandran · M. P. Vinod · P. Sunitha · O. V. Oommen
Department of Computational Biology & Bioinformatics, University of Kerala, Thiruvananthapuram, Kerala, India
e-mail: laladhas@gmail.com

L. Krishna Panicker · O. V. Oommen
Center for Sustainable Environment-India Biodiversity Action, Thiruvananthapuram, Kerala, India

V. S. Soumya
St. Stephens College, Kollam, Kerala, India

L. Krishna Panicker et al. (eds.), *Biodiversity and Business*,
https://doi.org/10.1007/978-3-031-71674-4_4

thorough picture of the biological populations in a given area. Although the investigation and study of biodiversity is a common component of both bioprospecting and biosurvey, their goals and approaches are different. While biosurvey is mainly concerned with recording and comprehending the natural elements of a specific place, bioprospecting focuses on finding materials with commercial or medical value. Nonetheless, both endeavors advance our knowledge of the natural world and its potential advantages for a range of human needs. To maintain biodiversity protection, these activities must be carried out responsibly and sustainably.

Corporations that engage in biosurvey and bioprospecting may concentrate on finding unique substances, genes, or living things with advantageous traits. To guarantee moral and sustainable operations, these companies frequently work with scientists, environmentalists, and local communities. Collaboration between scientific researchers, local populations with traditional knowledge, and industry participants is frequently necessary for the integration of these concepts. Bioprospecting and product creation based on biological resources and traditional knowledge requires careful consideration of ethical issues, such as equitable benefit-sharing with local populations, protection of intellectual property rights, and the promotion of sustainable practices. Traditional knowledge may aid in bioprospecting by offering insights into the possible applications of certain plants or natural resources, which is beneficial for sustainable resource management. Achievements in India's' ABS mechanism has been reported elsewhere (Laladhas et al., 2013, Laladhas et al., 2023a; Laladhas et al., 2023b, Nelliyat et al., 2023, Nilayangode et al., 2023) This chapter focuses on a few initiatives that were created using traditional knowledge about natural substances and how benefit sharing was carried out in consideration.

ABS Formulation Following a Biosurvey of Kenya's Fungal Diversity to Find Novel Antibiotics

The goal of the study, "ASAFEM - The Search for New Antibiotics from Tropical Fungi," was to find anti-quorum sensing substances and antibiotics. The research specifically examined secondary metabolites from Kenyan medicinal plants- fungi and mushrooms in the Kakamega and Mount Elgon areas to explore their potential for the synthesis of novel antibiotics for use in human medicine. Secondary metabolites are organic substances that are not necessary for an organism's growth, development, or reproduction but they may be able to defend against other organisms (i.e., have antibacterial action). Endophytic fungi are isolated from plant material, and macrofungi (macromycetes) are collected for studies on fungal variety, antimicrobial secondary metabolites generated by fungi, and anti-quorum sensing inhibitory compounds that may be useful for the synthesis of antibiotics are noteworthy (https://bccm.belspo.be/newsletter/03/search-new-antibiotics-tropical-fungi).

The initiative supported the sufficient and efficient transfer of relevant information and technology in line with ABS objectives and also developed the skills of young

Kenyan scientists in the fields of drug research and biodiversity. To promote innovation and sustainable growth in science and technology, the initiative encouraged greater cooperation and funding of scientific research among research institutes from Africa and the European Union.

The resource supplier and management of the Kakamega Nature Reserve is the Kenya Wildlife Service (KWS), situated atop Mount Elgon. Biological Scientists and Chemists worked together to extract active substances and characterize molecules. Accessing agency for bioresources is the Helmholtz Centre for Infection Research (HZI), located in Braunschweig, Germany. HZI's research efforts are primarily focused on investigating biodiversity globally, with a particular focus on bacteria and fungi, as well as purifying new chemicals possessing biological activity and prospective application in the development of antitubercular medications. (https://www.helmholtz-hzi.de/en/research/research-topics/antiinfectives/microbial-drugs/our-research).

The initiative started primarily with fundamental research, but if active molecules with therapeutic properties are found, research and development might lead to future commercial exploitation of the study outcomes. The benefit-sharing arrangement is in line with Kenya's ABS framework that was signed by the accessor and the supplier (authorities). Egerton University asked for prior informed consent (PIC) in 2013 before beginning to negotiate mutually agreed terms (MAT) and a material transfer agreement (MTA) (http://www.egerton.ac.ke/).

A community representative assisted in the PIC's preparation. All project participants then signed MAT and MTA, which governed the distribution of monetary and non-monetary rewards. Negotiations were held with Kenya Wildlife Service (KWS) on PIC, MTA, and MAT papers (http://meas.nema.go.ke/abs/).

The partner institution in Kenya oversaw all other necessary licenses for export, research, and access to protected regions. The parties also signed a grant agreement. Fieldwork licenses for Mount Elgon and Kakamega were available. Community members served as the forest guides throughout fieldwork. Since the initiative was initially backed by the local university and Kenyan government, European partners had no issues with the ABS procedures.

Benefits from monitoring include cash for research and infrastructure that Kenya may use to expand its capabilities and research facilities. Only over two-thirds of the € 100,000 that the Kenyan government had authorized were spent. Through HZI, the BMBF (ERAfrica) provided € 30,000 to buy research supplies in Kenya. (http://www.abs-initiative.info/partner-countries/kenya/).

Benefits that are not monetary include collaboration on research, including assistance for the development of infrastructure and the sharing of knowledge and resources about biodiversity; and many collaborative publications, was also the result from student research funded by DAAD funding. One way to increase capacity is through three-month internships. Postgraduate students from Kenya were offered travel to Germany to gain knowledge of research techniques (http://www.biochemie.tu-berlin.de). Later, the Masters students involved were able to get financial supporting for their PhD thesis from the DAAD. After graduation, a solid academic partnership has been maintained with the German universities.

ABS in the Assessment of Biodiversity and Traditional Knowledge in the Okavango Region Africa

The Okavango is a globally significant system of wetlands, floodplains, and forest savannas that supports biological variety. The "The Future Okavango" (TFO) project investigates the Okavango region's biodiversity to compile data that will assist local populations in managing their resources and using their land sustainably, biosurvey using molecular methods of higher plants, birds, and soil microbes. Plant ethnobotanical research and traditional knowledge evaluation, which involves speaking with local populations and indigenous people, are conducted. For example, Angolan partners were mostly responsible for gathering traditional knowledge related to biological and genetic resources. The first phase of the project was funded by the German Ministry of Education and Research (BMBF) and ran from 2010 to 2016 with funding from SASSCAL ("Southern African Science Service Centre for Climate Change and Adaptive Land Management"), sectoral follow-up studies are still underway. Angola, Botswana, Namibia, South Africa, Germany, and Zambia are all partners in this program.

The project promoted sustainable land use and resource management by local communities and contributed to nature conservation in the Okavango Basin. Biosurvey and ethnobotanical assessment are expected to pave way for advanced research in bioprospecting. The sharing of benefits arising from such studies were channelized to biodiversity conservation and livelihood improvement of local people. Project activities also provided economic benefits by way of job creation, money from local resources, etc., to the local communities. Economic benefits were obtained by local residents through work as Para-ecologists and helpers at study locations. There is sufficient research funding available for scientists in all of the donor nations. Non-financial benefits include student scholarships, capacity building in academic and non-academic areas, collaborative research programs, more than 200 joint publications, sharing of information and data acquired, and costs associated with local specimen collections. (http://www.future-okavango.org/).

Long-Term Research Partnership for Bio-Survey from the Tropical Rainforest of Sumatra, Indonesia

Changes in land use patterns i.e., monoculture of cash crops such as rubber and oil palm transformed the lowlands of Jambi province of Indonesia in the last few decades. The ecological and socioeconomic implications of land alterations are examined by "EFForTS," a conservation project for the Ecological and Socioeconomic Functions of Tropical Lowland Rainforest Transformation System.

The basic objective of the project is to give science-based information on how to conserve the degrading ecosystem and methods for improving ecological functions of the tropical forest. This project is funded by the German Research Foundation

(DFG) in a long-term phased manner now extended up to 2023. A detailed biosurvey constitutes systematic research on biodiversity thorough evaluation of the diversity of animals, plants, and soil microbes; the gathering of genetic material for barcoding; and the evaluation of the economic and environmental implications. A cooperative venture of participation occurred involving holders of small farms, local scientists, other stakeholders, and competent authorities involved in the ABS process.

A good amount of money— € 42.7 million—was transferred to Indonesian partners through more than 160 research grants, 16 capacity-building grants, and new infrastructure, including research stations, a herbarium, labs for soil ecology, hydrology, and palynology, among other facilities. Non-financial gains from a long-term collaboration between 200 project and technical specialist members from Germany and Indonesia resulted in 230 publications, of which 34 had an Indonesian first author and 72% were co-authored with peers. Fellowships to research students, training, seminars, and exchange of researchers were a part of the project. The collected samples were deposited in national collection facilities in Indonesia. DFG developed a transparent guideline for scientists on how to use ABS-related funds allotted to the program (Claudino-Sales et al., 2019).

Latin America's Access to Genetic Resources in a Panamanian Nature Reserve for Economic Purposes

A UK-based firm called Indigena Biodiversity Ltd. carried out a floral biosurvey on the Punta Patiño Reserve in Panama of the available species. A native of Panama and Latin America, the *Simabacedron* is a bush or shrub-like tree belonging to the Simaroubaceae family that has been recognized as a useful bioresource for more study. Indigena Biodiversity recognized the importance of bioresources and suggested a methodical study that would result in the creation of biomolecules with sufficient potential for industrial use. *Simabacedron's* components are amenable to alteration to increase its capacity as an antimalarial agent. A Panamanian environmental non-governmental organization namely ANCON- (National Association for the Conservation of Nature), owns and manages the conservation of the Natural Reserve Punta.

Financial gains go to all partners, including NDICASAT; ANCON, the National Research Institute of Panama; and the Ministry of Environment of Panama will split the portion of worldwide sales if a commercial product is established (ANCON, 2017). The conditions of each party's share of the advantages from any potential commercial malaria product created as a consequence of this joint initiative are outlined in the agreement. A community fund will be developed from the benefits acquired for the livelihood improvement of the local people. Also, there was an agreement for collaborative research programs in Panama-based bioresources. Non-monetary benefits include research collaboration, and exchange of the developed information to the counter-institutions (Kursar et al., 2007).

Bio-Survey of TK-Associated Medicinal Plants Conducted by the Locals in Moquegua, Peru's Highlands

Royal Botanic Gardens, Kew, United Kingdom along with the National collaborators which include Universidad Nacional Mayor de San Marcos and Instituto Nacional de Salud (INS) are the technology-enhanced research groups involved in the Bio-survey of medicinal plants from the highlands of Moquegua, Peru. Traditional knowledge associated with the medicinal plants was received from the Communities in the District of Yunga, Moquegua, Peru.

By gathering seeds and preserving them in officially recognized seed banks in Peru, this project, named "Enhancing conservation and access to medicinal plants in Peru," seeks to preserve the endemic and natural genetic diversity of medicinal plants. The bio-survey created a seed bank for wild medicinal plants in Peru and served as a focal point for researchers at the botanic garden and herbarium of the National Centre for Intercultural Health (CENSI). The goal of this research is to improve the conservation of priceless medicinal plants and investigate the feasibility of their possible influence on the availability of therapeutic characteristics. By the ABS framework of Peru, INS gathers related traditional knowledge for documentation, and duplicate plant specimens are stored at the Kew Herbarium for taxonomic confirmation. Kew conducted preliminary studies on seed germination, lifespan, and other related topics with the assistance of Peruvian experts and students. The initiative ran from February 2018 until December 2020.

Monetary benefits include only basic funds for research and are aimed at the preliminary results for subsequent monetary support. Non-monetary support includes training facilities for the students and researchers for world-renowned conservationists in the Kew Gardens. Training, workshops, conservation know-how for seed banks, seed preservation, germination, and viability assessment for the communities were received. A long-term scientific collaboration with technological and infrastructural facilities gives effective scientific results and better ABS outcomes.[1]

Preserving Wild Edible Plants to Restore the Traditional Mediterranean Diet

The project "Restoring the traditional Mediterranean diet through the conservation of wild edible plants" aims to support the sustainable use of the traditional Eastern Mediterranean diet, which was recognized by UNESCO in 2010 as an intangible cultural heritage of humanity. It does this by preserving traditional knowledge and using science-based conservation techniques to protect wild edible plants in Jordan and Lebanon. It seeks to preserve and study wild plants that are consumed by the

[1] ABS in Peru. Overview of Legal Requirements Procedures and Relevant Authorities. static1.squarespace.com/static/58bfcaf22994ca36885f063e/t/63ff510c3a855752cf25f624/1677676818564/ABS+in+Peru.pdf

surrounding populations and to find ways that these plants may be grown instead of being picked from the wild. Information gathered is disseminated to communities and local partners. It encourages locals to use these plants sustainably and commercialize them, as well as to preserve traditional knowledge and ensure food security in this region.

Wild edible species are still collected and used in regional cuisine in Jordan and Lebanon. In these nations, they improve nutrition and dietary variety. However, over-collection and habitat loss have put some of these tasty species at risk of extinction. There is little information available on their cultivation, and it is typically unknown how important these plants are for nutrition and overall health. Under the Jordanian and Lebanese-led Millennium Seed Bank Partnership (MSBP or MSB), this initiative has been created collaboratively. Its primary components include gathering ethnobotanical information about wild edible plant species found in the Eastern Mediterranean and prioritizing those that are most significant to the local community.

Within the context of the Useful Plants Experiment, a one-year pilot experiment has been sponsored. Initial research on *Gundelia. tournefortii*, sometimes referred to as "Akkoub" locally, was conducted in 2019 and 2020. As per the nations' ABS regulations, the project's execution has required gathering and gaining access to plant material of wild edible species. The nutritional value and seed biology of these plants are being studied using a variety of research methods. A database including details on the region's wild edible plants, including those used medicinally, has also been created. This assists in guiding the most interesting species to collect and where to put them in relation to other Mediterranean-related plant collection, and diet as well as identifying gaps in information.

The project will soon entail seed collecting, conservation, and plant propagation operations (pending further funding). It will also help the creation of marketable food and goods as a result of their promotion. This initiative connects to ongoing efforts in Lebanon by the Shouf Biosphere Reserve concerning the restoration of historic agricultural terraces and the sustainable collection of edible wild plants, along with their synthesis using other indigenous crop kinds.To ensure that all national legal and ethical obligations are met, activities are carried out under access and benefit-sharing agreements between RBG Kew and partners in addition to a grant agreement with each provider. The ABS agreements were in place when the project started since NARC in Jordan and the LARI in Lebanon have been cooperating with RBG Kew under the MSBP since the early 2000s. Hundreds of seed collections have been made possible by the cooperation thus far, and 636 native plant species from Jordan and 924 from Lebanon have been protected and saved.

The partnership's current initiative is mostly focused on the preservation of wild edible species. Grant agreements were formed expressly for this pilot project with NARC (formerly known as the National Centre for Agricultural Research and Extension (NCARE)) in Jordan and SBR in Lebanon, and the ABS agreements were later renewed. While LARI continues to work with SBR to provide technical assistance for seed collection and conservation, RSCN has been linked to the project in Jordan to support the work in the communities. Fieldwork and the gathering of plant material are subject to permits that are overseen by the appropriate authorities and are necessary in protected areas (Ulian et al., 2019).

Representing the Concepts of ABS in a Caucasus Biodiversity Catalogue

Scientists from the Zoological Research Museum Alexander Koenig (ZFMK) in Germany and Ilia State University in Georgia began working together in 2017 to develop the Georgian-German Biodiversity Centre (GGBC). Examining the species and genetic variety of plants and animals in the Caucasus region, a worldwide biodiversity hotspot is the aim of the program. The project was supported by the German Federal Ministry of Education and Research (BMBF) and ran until the autumn of 2020.

Due to the initial pilot project's success, BMBF funded a larger, follow-up initiative called the Caucasus Barcode of Life project (CaBOL). CaBOL involves other partner institutions from Armenia (Yerevan State University) and Georgia (Agricultural University, Tbilisi). The goal of this initiative is to use molecular and classical methodologies to perform extensive and applicable biodiversity studies. Species' identified DNA barcodes are kept in a reference database and made accessible to the general audience.

The research is based on the knowledge German scientists from ZFMK gained while working on the German Barcode of Life (GBOL1), a prior effort. The programs promote international collaboration, enhance research infrastructures, include education, and investigate biodiversity in the Caucasus region by ABS principles. Both initiatives help Georgia and Armenia develop their capacities (e.g., by educating researchers in molecular methods and developing databases). There are also citizen science components in the GGBC and CaBOL initiatives.

No special licenses were needed to access genetic resources in Georgia or Armenia in the absence of national ABS laws. In Georgia, permissions for collection were approved in 2018 and 2019 by the Agency for Protected Areas and the Ministry of Agriculture and Environmental Protection. Furthermore, Ilia State University supplied clearance letters in 2018 and 2019 which functioned as export licenses. The Ministry of Environment in Armenia issued collection permits for Caucasus Barcode of Life (CaBOL) (https://ggbc.eu/).

An Agreement in ABS in Bhutan Asia for the Usage of Orchids

Bhutan's great degree of diversity and comparatively good condition of preservation make its biological diversity of worldwide relevance and enormous scientific worth. The Himalayan state is home to 469 species of orchids, many of which are native to Bhutan. In Bhutan, orchids are utilized for many reasons. While certain species are used as food items, many others are used in traditional medicine, which is a vital component of healthcare in rural communities. Protected by the Forest Act, wild orchid species are another draw for eco-tourists.

The Bhutanese Constitution and national policy framework have a strong emphasis on protecting the natural environment. Sustainable usage and biodiversity protection are overseen by the National Biodiversity Centre (NBC). It encourages teamwork for cooperative research, technical capacity building, and technology transfer. NBC's endeavors have led to the economic value acquisition of species with unknown medicinal or other qualities, including the edible orchid *Cymbidium erythraeum*. The Ministry of Agriculture and Forests received an application from Quantum Pharmaceuticals Limited (QPL) in 2009 for authorization to conduct research on and utilize the *Cymbidium erythraeum* species for commercial purposes.

The goal of the study was to determine whether the cellulose found in orchid cells might be utilized to make a cosmetic cream. The demand for orchids expanded as a result of the commercial possibilities. Previously, the *Cymbidium erythraeum* species was mostly taken from the wild rather than being grown. A Memorandum of Agreement (MoA) between NBC and QPL was signed in compliance with the Biodiversity Act of Bhutan. The MoA sought to protect orchids while permitting business ventures that shared advantages with nearby populations. Eighteen families in Samtengang, chosen by NBC, began cultivating *Cymbidium erythraeum* as a component for cosmetics under the terms of the MoA.

From the time of the first access request until the ABS agreement's final talks, NBC assisted the ABS process. Following the identification of a place for the orchids' sourcing, a participatory rural evaluation and engagement process was conducted to develop the "Samtengang Orchid Management Group," a community organization dedicated to the propagation and maintenance of orchids. The local community worked with the required officials to create a community by-law, commonly referred to as community procedure. It described how to get mutually agreed terms (MAT), prior informed consent (PIC), and benefit-sharing while using local resources.

On behalf of the community, NBC mediated the final ABS agreement and assisted in the communication between the corporation and the community organization. The project advanced only due to the local community's support. The Samtengang Orchid Management Group's eighteen families joined together to supply QPL with *Cymbidium erythreaum* blossoms.

At the time of the lawsuit, in 2009, there were no established processes for ABS agreements. QPL as the user and NBC signed a memorandum of understanding. The NBC approved a commercial use permit at QPL's request because an agreement had already been inked. The terms of the agreement between the corporation and NBC included information about the advantages to be shared with the Samtengang Orchid Management Group, as well as specifications and quantities of genetic resources to be obtained.

According to the agreement, royalties from the sale of the cosmetic product and a premium price when genetic resources are tapped would be given to each home. Therefore, the money received from the propagation house's orchid blooms is handled through a community fund that is overseen by the recently created bylaw (a community protocol). QPL further consented to contribute to the Bhutan Access and Benefit Sharing (BABS) Fund. The fund was established to improve rural livelihoods and to support the protection and sustainable use of Bhutan's biodiversity. Additionally, the fund balances and supplements Bhutan's expenses associated with conservation (Noltie, 1994).

Conclusions

Finally, in our effort to comprehend and utilise the diversity of life on earth for the benefit of both people and the environment, biosurveys, and bioprospecting are indispensable instruments. We may develop important insights on the distribution, abundance, and variety of living things through systematic biosurveys, which are necessary for conservation and management plans that are successful. Meanwhile, a route to finding new substances, genes, and organisms with potential uses in business, agriculture, and medicine is provided by bioprospecting. However, these actions must be carried out in an ethical manner, considering the rights of indigenous peoples and the preservation of ecosystems. Through the integration of scientific precision and ethical accountability, biosurveys and bioprospecting provide the possibility of unleashing the latent potential of nature and guaranteeing its enduring sustainability for future generations.

Acknowledgements The authors express their gratitude to the Department of Computational Biology & Bioinformatics at the University of Kerala and Soumya (SRF) thank the Council of Scientific and Industrial Research (CSIR), New Delhi, India, for Senior Research Fellowship (08/666(0003)/2019-EMR-I).

Conflict of Interest The authors have no conflict of interest to declare.

References

Access and Benefit-Sharing—Information portal for Kenya: http://meas.nema.go.ke/abs/

ANCON. (2017). Catalogue of native plants with potential for biotrade and bioprospecting of the private nature reserve of Punta Patiño, Darién—Panama. Ministry of Environment, GIZ, UNDP, GEF, IDIAP, Panama.

Caucasus Barcode of Life (CaBOL) https://ggbc.eu/
Claudino-Sales, V., & Claudino-Sales, V. (2019). Tropical rainforest heritage of Sumatra, Indonesia. Coastal world heritage sites, pp.563–569.
Kursar, T. A., Caballero-George, C. C., Capson, T. L., et al. (2007). Linking bioprospecting with sustainable development and conservation: The Panama case. *Biodiversity and Conservation, 16*, 2789–2800. https://doi.org/10.1007/s10531-007-9214-2
Laladhas, K. P., Preetha, N., BaijuLal, B., & Oommen, O. V. (2013). Region-specific biodiversity education-the role of People's Biodiversity Register. *Current Science (00113891), 105*(12).
Laladhas, K. P., Rani, J. R., Dionysius, J. M., Nair, A. S., Sudhakaran, P. R., & Oommen, O. V. (2022). Achievements in India's ABS mechanism. In O. V. Oommen, K. P. Laladhas, P. Nelliyat, & B. Pisupati (Eds.), *Biodiversity conservation through access and benefit sharing (ABS)* (pp. 37–53). Springer. https://doi.org/10.1007/978-3-031-16186-5_3
Laladhas, K. P., Fila, M., Nair, A., & Oommen, O. (2023a). Joint biodiversity management committees: A mechanism to manage and conserve eco-regions. *Advances in Environmental and Engineering Research, 4*, 1–25. https://doi.org/10.21926/aeer.2303042
Laladhas, K., Rani, J. R., Sree, A., Nair, A., & Oommen, O. (2023b). *Statutory and obligatory responsibilities of state biodiversity boards for the conservation of indigenous biodiversity and ABS, biodiversity conservation through access and benefit sharing (ABS) Himalayas and Indian sub-continent* (pp. 55–77). Springer International Publishing. https://doi.org/10.1007/978-3-031-16186-5_4
Nelliyat, P., Pisupati, B., Laladhas, K. P., & Oommen, O. V. (2023). Convention on biological diversity, access and benefit sharing: A glance. In O. V. Oommen, K. P. Laladhas, P. Nelliyat, & B. Pisupati (Eds.), *Biodiversity conservation through access and benefit sharing (ABS)*. Springer. https://doi.org/10.1007/978-3-031-16186-5_1
Nilayangode, P., Laladhas, K. P., & Oommen, O. V. (2023). Strengthening ABS implementation in India. In O. V. Oommen, K. P. Laladhas, P. Nelliyat, & B. Pisupati (Eds.), *Biodiversity conservation through access and benefit sharing (ABS)*. Springer. https://doi.org/10.1007/978-3-031-16186-5_12
Noltie, H. J. (1994). *Flora of Bhutan: Including a record of plants from Sikkim and Darjeeling*. Royal Botanic Garden Edinburgh.
The ABS capacity development initiative. http://www.abs-initiative.info/partner-countries/kenya/
The Future Okavango. Findings, scenarios, and recommendations for action. http://www.future-okavango.org/
Ulian, Tiziana Hani, Nizar Abulaila, Khaled Mattana, Efisio Howes, Melanie-Jayne Gomez Barreiro, Pablo van Slageren, Michael Breman, Elinor Liu, Udayangani Cossu, Tiziana Breidy, Joelle Vilanova, Santiago Pironon, Samuel Ondo, Ian. (2019). Restoring the traditional Mediterranean diet through the conservation of wild edible plants. https://doi.org/10.13140/RG.2.2.18913.04965.
Website of BCCM. https://bccm.belspo.be/newsletter/03/search-new-antibiotics-tropical-fungi
Website of HZI. https://www.helmholtz-hzi.de/en/research/research-program
Website of the Egerton University. http://www.egerton.ac.ke/
Website of the Institute of Chemistry, TU Berlin, Germany. http://www.biochemie.tu-berlin.de

Bioprospecting, Synthetic Biology, Digital Sequence Information and ABS: A Review on Global Outlook in the Context of the CBD

Prabha S. Nair and Nishmma James

Introduction

The intricate web called biodiversity encompasses a variety of species, ecosystems and genetic diversity that knits together the life on earth. A significant part of our environment consists of animals, plants and microorganisms. The geographical pattern of the Earth varies from continent to continent, and the biodiversity differs from region to region. Human survival relies on biodiversity for food, water, medicines, and raw materials. Our dependence on biodiversity is growing with a shift from general resources to harnessing biodiversity for precision products. The focus of this transition is bent towards development of highly targeted and personalized solutions in medicine, nutrition, and various other fields (Co-operation on Health and Biodiversity, October 2010). The Millennium Ecosystem Assessment and the Intergovernmental Panel on Climate Change have highlighted biodiversity's underappreciated and unexplored potential for human health and well-being (Co-operation on Health and Biodiversity, October 2010). These biological resources have the potential to provide raw materials for the creation of medicines and for use in agriculture and industrial applications. The foundation of our food security lies in the genetic diversity within crops, fostering resilience against pests and environmental stresses (Esquinas-Alcázar, 2005). Traditional medicine, especially in developing nations, relies heavily on the vast pharmacological potential hidden within wild species (Esquinas-Alcázar, 2005). Examples like Ziconotide, a painkiller derived from cone snail venom, and Lanoxin, a cardiac drug from the Digitalis plant, showcase how nature's bounty has already yielded lifesaving medicines (Esquinas-Alcázar, 2005).

P. S. Nair (✉) · N. James
School of Law, Christ (Deemed to be) University, Delhi - NCR, Ghaziabad, India
e-mail: prabha.s.nair@gmail.com

N. James
Junior Intellectual Property Law Attorney, Puthran and Associates, Thiruvananthapuram, Kerala, India

L. Krishna Panicker et al. (eds.), *Biodiversity and Business*,
https://doi.org/10.1007/978-3-031-71674-4_5

Biotechnology has played a key role in rendering biological diversity useful in these ways for a better living (Ashish Swarup Verma et al., 2011).

Biotechnology has existed since the beginning of civilisation, with early applications such as fermentation, selective breeding and vaccine development.[1] The development of biotechnology in terms of technological advancements has been divided into ancient, classical and modern (Ashish Swarup Verma et al., 2011). The ancient period of biotechnology predates the 1800s, which are basically the discoveries or developments such as food preservation and storage methods and fermented products such as cheese and curd and cross breeding processes for human benefits, all rooted in common observations about nature (Ashish Swarup Verma et al., 2011). The classic biotechnology period spanned the period from 1800 to the mid-twentieth century. During this period, crucial observations backed by scientific evidence laid the foundation for subsequent discoveries. Gregor Mendel's groundbreaking work with the pea plant uncovered the laws of inheritance, identifying invisible units later known as genes. Concurrent with the scientific validation of Mendel's work by the later scientists, early discoveries in molecular biology, including the identification of the nucleus by Robert Brown and the isolation of nucleic acid by Friedrich Miescher (Ralf Dahm, 2005), set the stage for understanding DNA as genetic material. Noteworthy advancements included Robert Koch's description of bacterial colonies on a solid medium and the coining the term "chromosome" by Heinrich Von Waldeyer-Hartz (Akhtar et al., 2011). Vaccines against smallpox and rabies, developed by Edward Jenner and Louis Pasteur, respectively, marked significant milestones. T.H. Morgan's work with fruit flies in 1926 further elucidated the role of chromosomes in inheritance, coinciding with the coinage of the term "gene" by Wilhelm Johannsen (Akhtar et al., 2011). Simultaneously, Alexander Fleming's discovery of antibiotics, particularly penicillin, revolutionised medicine, proving to be a significant breakthrough for humanity. Vaccines and antibiotics emerged as pivotal contributors to human well-being during this transformative period.

Succeeding classic biotechnology has seen modern biotechnology's growth since the aftermath of World War II. During this stage, the biotechnology field has experienced phenomenal growth and development from its early roots. The key path-breaking change includes the 1953 discovery of the double helix structure of DNA, the creation of monoclonal antibodies in 1975, and the sequencing of the human genome in 2001, which led to the path for life-changing applications, such as vaccines, diagnostics, and gene therapy. The 1970s saw the emergence of recombinant DNA technology, leading to the creation of the first monoclonal antibodies and genetically engineered organisms. The 1980s witnessed the Human Genome Project's launch and the approval of the first genetically engineered vaccine. The 1990s brought about gene therapy advancements, the sequencing of the first complete organism's genome, and the development of DNA chips. The 2000s saw the completion of the human genome sequence, the creation of embryonic stem cells, and the first synthetically replicated organism. The mid-twentieth century has seen the shift from

[1] https://www.bio.org/what-biotechnology.

fundamental discoveries of DNA structure and function to groundbreaking achievements in gene editing and synthetic biology; the past few decades have witnessed a continuous surge in knowledge and application (Bhatia, 2018). Key milestones like the human genome sequencing, the creation of Dolly, the cloned sheep, and the development of personalised medicine therapies stand as testaments to this remarkable progress. These milestones, along with countless others, have revolutionized our understanding of life and paved the way for new diagnostics, treatments, and technologies that continue to shape the future of biotechnology. Other examples of biotechnology applications in industry include the cultivation of plant tissues, culture of mammalian cells, biocatalysts, treatment and reuse of waste products via biotechnological methods (bioremediation), fermentations, obtaining biotechnology of fuel and organic raw material as an alternative to petroleum, and biotechnological process engineering. Advancements in biotechnology have further accelerated this journey of discovery.

Bioprospecting and Biotechnology

Biotechnology has also led to the development of precision products in medicine, such as vaccines, by utilizing genetic resources from diverse ecosystems, to the development of biomaterials, such as polymers, proteins, and nucleic acids.[2] In 1989, members of the International Society for Chemical Ecology (ISCE) adopted a resolution in Göteborg, Sweden to accelerate the search for new compounds from nature in collaboration with the developing countries in the tropical regions rich in biodiversity (Osslon, 2017) and this was based on the belief that the chemical or genetic structure of natural products could yield more leads in product discovery with the help of combinatorial chemistry.[3] The process of exploring the genetic and biochemical traits of biological materials for commercially valuable product development is known as bioprospecting (Bavikatte & Tvedt, 2015). Bioprospecting activities primarily target a wide range of genetic resources and are used by different sectors, including public and private R&D agencies in agriculture, healthcare, and manufacturing. The process involves collection of biological resources, scientific investigation over them, screening, research and analysis followed by product development and commercialization (Beattie et al., 2011). A specific example of bioprospecting is the use of neem tree *(Azadirachta Indica)* for preparation of malaria medicines and insecticides and the use of the active compound Azadirachtin in developing cancer medicines. A novel compound, 2′-3′-dehydrosalannol (DHS) isolated from *A. indica* uncrushed leaves, inhibited growth and induced apoptosis in Triple-negative breast cancer (TNBC) cell lines (Sofowora et al., 2013). The scientists have found 60 different biochemicals purified from this neem overall. Another example to cite is the rose periwinkle (*Catharanthus roseus*), which is the source of anticancer drugs

[2] https://www.bio.org/what-biotechnology.

[3] https://www.millenniumassessment.org/documents/document.279.aspx.pdf.

like vincristine and vinblastine. The active compounds of rose periwinkle have been patented and utilized in developing drugs that treat various types of cancer (Elujoba et al., 2005).

The advent of molecular biotechnology and advanced techniques such as recombinant DNA technology has made bioprospecting much faster and easier (Khan et al., 2016). Cellular and molecular biology has offered new avenues for biotechnological research and gene editing tools such as CRISPR has made manipulation of genetic make-up speedy and cost-effective. Biotechnology is unlocking a new era of precision in utilizing biodiversity. Fermentation, a cornerstone of many cultures, is now employed for advanced bioproducts like biofuels and biodegradable plastics (Khan et al., 2016). Vaccines, once developed through serendipitous discoveries, are now being engineered with pinpoint accuracy, targeting specific diseases and saving countless lives. Nature-based products find their way into nutrition with personalized food formulations enriched with specific nutrients or probiotics. Even the cosmetics industry is embracing the power of nature, utilizing bio-active compounds from plants and marine life to create sustainable and effective products (Fonseca et al., 2023).

These scientific and technological developments in biotechnology and the commercial returns garnered by its products continue to raise many parallel concerns among the international community. Biodiversity was believed to be a domain of open access or 'heritage of mankind' devoid of regulatory requirements, thus offering unhindered access for research and development (Pauchard, 2017). The early days of drug discovery using bioprospecting were largely dependent over ethnobotanical knowledge used in indigenous medicines (Cordell, 2024). The open access notion totally disregarded the stewards of biodiversity including the biodiversity-rich countries from where such materials were accessed and the communities and the local and indigenous people who have been maintaining and preserving biological resources and the knowledge associated with them for generations (Shengji, 2001). While acknowledging the benefits offered by the technology, the Southern part of the Globe still remains worried about the ever-widening gaps with the Northern countries in terms of access to technology and its benefits and to financial resources for conservation and technological applications (Cordell, 2024; Domingo-Fernandez et al., 2023). In the aftermath of rDNA technology, the South was particularly skeptical about the increased enclosure of biological materials through Intellectual Property Rights guarding the products of biotechnology with high price and restricted access for further research and development (Cordell, 2024). The consistent demand of the European Countries for stringent plant variety protection on plant genetic resources for food and agriculture with scant attention to the rights of farming communities raised further concerns of food security (Tilford, 1998). The polarization in terms of technology in the North and biodiversity in the Southern part of the globe demanded equitable solutions that could create economic incentives for biodiversity conservation, serve as the foundation for the modern biotechnology industry in developing countries, and provide traditional people with ways to benefit from their unique understanding of the environment in which they live (Wiess & Eisner, 1998). The international political diplomacy included the concerns over biodiversity and its relationship with IPR and biotechnology as part of negotiations for species conservation and thus the umbrella Convention on Biological Diversity (CBD) signed in the year 1992 became the cornerstone of biodiversity regulation premised over sovereign rights of nations, prior informed consent and mutually agreed terms. These principles laid the foundation for access and benefit sharing mechanism under the CBD that had set international norms in ensuring regulated access to genetic resources and associated traditional knowledge.

Regulation of Bioprospecting Under the CBD Framework

A Bilateral Framework

The language of the CBD may prompt one to consider it as a Convention drafted for species conservation. A few countable provisions in the CBD lay the foundation for recognition of stewardship over biological diversity and associated traditional knowledge. The CBD stands for three major objectives, viz., conservation of biological diversity, sustainable utilization of its components and fair and equitable sharing of benefits arising from the utilization of such components (Article 1, CBD). The CBD asserts that meeting of these triple objectives would also include appropriate access to genetic resources and appropriate transfer of relevant technologies giving due consideration to all rights over those resources and to technologies, and appropriate funding requirements (Article 1, CBD). In its Preamble, the Convention reaffirmed the sovereign rights of countries over their biological resources and provided that due to the increased dependence of traditional communities on biological diversity, it is desirable that they also have a recognized right to share the benefits whenever their knowledge, innovations and practices are used by others (Preamble, CBD). The concern on traditional knowledge-holders reflected in the preambular paragraph reappears in Article 8(j) in the form of a flexibility offered to countries for incorporating traditional knowledge within their domestic legislations for regulated access subject to approval of the communities and fair and equitable benefit sharing (Article 8(j) CBD). It further provides that, founded upon the sovereign rights granted over natural resources, the Member Countries shall have the authority to determine access to their genetic resources pursuant to a domestic law created for the purpose (Article 15 CBD). It further requires that the conditions created for access to genetic resources shall be for environmentally sound uses and for promoting the objectives of the CBD (Article 15 CBD). Article 15 further asserts that the right to regulate access to genetic resources is available only to countries of origin where the genetic resources can be found in *in-situ* conditions (Article 2 CBD definition of country of origin). Further, Article 15 conditions the grant of access to genetic resources by way of prior informed consent of and mutually agreed terms with the country of origin.

Depending on the national legislation, prior informed consent and mutually agreed terms from the local people and indigenous communities can be a legitimate requirement under the CBD system when read with Article 8(j). Article 16 of the CBD emphasizes the need to ensure access to and transfer of technology, including those protected with IPR, on favourable terms to developing countries and Article 19 reasserts the need to transfer the benefits of biotechnology to developing countries on equitable basis. The Convention opens the door for regulating the interaction between genetic resource producers and users by creating restrictions on the free access to genetic resources and associated traditional knowledge. The two operational principles, Prior Informed Consent and Mutually agreed terms lay the foundation of the Access and Benefit Sharing regime under the CBD. The CBD employs a bilateral, contract negotiation approach for the fair and equitable sharing of benefits with providers of genetic resources when access to genetic resources is sought for by different users. PIC and MAT are made the prerogatives not only of the country of origin, but also extended to the indigenous and local communities who are directly involved in the maintenance of biodiversity related traditional knowledge and conservation of the genetic resources of the region.

Recognising that there are multiple players and stakeholders involved in the process of ABS, the voluntary Bonn Guidelines was brought in 2000 in order to assist the States to develop and implement their own national framework. Despite its voluntary nature, this is considered to be the first step for implementing the ABS provisions of the CBD.[4] The realization of the international community that there is a long way to achieve the third objective of the CBD, the Nagoya Protocol came into force in 2014, as a supplementary Protocol to the CBD. Currently, the Protocol forms the central framework for ABS with dedicated provisions detailing obligations of user and provider countries in terms of ABS. It elaborates the procedures for obtaining Prior Informed Consent (PIC) and Mutually Agreed Terms from the country of origin and the local and indigenous communities before accessing their biological resources (Article 6, Nagoya protocol) and traditional knowledge (Article 7, Nagoya protocol). Article 5 of the Protocol reiterates that the benefits arising from the commercial utilization of genetic resources, their subsequent applications and commercialization shall be subject to fair and equitable benefit sharing with the country of origin or the local and indigenous communities who have established rights over such genetic resources based on mutually agreed terms. The major contribution of Nagoya Protocol to the CBD framework is the obligation imposed on the countries (user countries) where the genetic resources would be utilized to ensure that such genetic resources utilized in their respective jurisdictions have been accessed in compliance with the domestic ABS legislations of the provider countries (Article 15 Nagoya protocol).

The bilateral contractual relationship established under the Convention on Biological Diversity (CBD) is, thus, supported by the Bonn Guidelines and the Nagoya Protocol. The CBD, along with its supplementary Nagoya Protocol, is based on a bilateral player-to-player exchange of resources, emphasizing the importance of

[4] https://www.cbd.int/abs/infokit/revised/web/factsheet-bonn-en.pdf.

exchange protocols and access (Coolsaet et al., 2013). The Bonn Guidelines provide guidance on the implementation of ABS provisions, assisting countries in the development of national ABS frameworks and the establishment of mutually agreed terms.[5] The Nagoya Protocol further supports the bilateral contractual relationship by providing a framework for the implementation of ABS provisions, aiming to ensure the fair and equitable sharing of benefits arising from the utilization of genetic resources (Morgera et al., 2015). It sets out obligations for both providers and users of genetic resources and includes provisions related to prior informed consent and mutually agreed terms. It operates on the PIC and MAT principles laid down by Article 15 of the CBD and allows traditional knowledge associated with genetic resources as a subject matter to be regulated under the domestic ABS framework (Article 7, Nagoya protocol). The CBD and the Nagoya Protocol obligations are left to the discretion of providing countries to a greater extent (Medaglia et al., 2013). Fair and equitable benefits sharing obligations are created only at the point of access to genetic resources and traditional knowledge and they are expected to follow the value chain until commercialization. Thus, the edifice of the entire CBD system is physical access to genetic resources by the user. Even though, the CBD and Nagoya offer two different regulated subject matter, viz., the genetic resources and traditional knowledge and both come with their own baggage of issues to be resolved in a regulatory context, this chapter is limited to genetic resources in the crossroads of scientific and technological advancements posing challenges to the ABS regulatory framework.

Genetic Resources: An Unresolved Puzzle

The CBD defines genetic resources as "genetic material of actual or potential value" (Article 2, CBD) and further defines genetic material as "any material of plant, animal, microbial or other origin containing functional units of heredity" (Article 2, CBD). Genetic resources form part of biological resources (Article 2 CBD) which can include any biotic component of the ecosystem with actual or potential value for humanity. Schei and Tvedt (2010) extensively discuss the term 'genetic resources' in the CBD as evolving and responding to the advancements in biotechnology and agree that the emphasis on physicality can be challenging to an international framework on access and benefit sharing. According to them, the discussions in the CBD on the benefit sharing aspects of the term genetic resources was mostly in the context of the developments in rDNA technology (Schei & Tvedt, 2010). These authors believe that the drafters of the Convention avoided the commodity aspect of the term genetic material, and relied on the 'functional units of heredity', trying to capture even the uses of materials such as mitochondrial DNA and chloroplast DNA (Schei & Tvedt, 2010). The CBD's intention of including the term "genetic resources" seems to create benefit sharing obligations on the physically accessed genetic resources from a broad set of modern biotechnological uses (Schei & Tvedt, 2010). It remains silent

[5] https://www.cbd.int/decision/np-mop/?id=13408.

on access and benefit sharing obligations where the genetic resources are indirectly used without directly accessing them. However, the Nagoya protocol has made an unsuccessful effort to capture such indirect uses in its international regime on access and benefit sharing.

The Nagoya protocol deals with this challenge by including two terms, viz., 'derivative' and "utilization of genetic resources" in its definition clauses. The Protocol defines derivative as "a naturally occurring biochemical compound resulting from the genetic expression or metabolism of biological or genetic resources, even if it does not contain functional units of heredity" (Article 2(e) Nagoya Protocol). Further, it defines "Utilization of genetic resources" as conducting 'research and development on the genetic and/or biochemical composition of genetic resources, including through the application of biotechnology' (Article 2(c) Nagoya protocol). The Protocol brought within its scope genetic resources, associated traditional knowledge and the benefits arising from their utilization (Article 3 Nagoya Protocol). In its benefit sharing clause (Article 5, Nagoya protocol), the Protocol extends the fair and equitable benefit sharing obligation to benefits arising from the utilization of genetic resources as well as their subsequent applications and commercialization. However, while addressing the process of access to genetic resources (Article 6, Nagoya Protocol) and traditional knowledge (Article 7, Nagoya Protocol) in their respective provisions, the Protocol mandates prior informed consent from the country of origin and prior approval and involvement of the traditional knowledge holders at the time of physical access to the genetic resources and traditional knowledge as the case may be. Thus, the effort to cover indirect uses of genetic resources remains unsuccessful since all benefit sharing obligations have to be created at the time of physical access to genetic resources.

The terms of the bilateral contract between the proposed user and access provider could impose benefit sharing obligations on the direct and indirect uses of the genetic resources based on the proposed use. Indirect use as to include the use of a "derivative" calls for a physical, naturally occurring part of genetic material retained by the user(s) and thus totally dependent on the physical or material aspect of genetic resources. The term utilization of genetic resources can still be extended to indirect uses of genetic resources as to mean any kind of biotechnological R&D applications on the genetic or biochemical compositions of genetic resources. However, this kind of utilizations to fall within the scope of benefit sharing obligations have to be included at the time of physical access enabled through PIC and MAT by the provider country. Thus, the starting point of benefit sharing obligations under the Nagoya protocol is the physical access to genetic resources and any claim for benefit sharing on subsequent utilizations and commercialization by different players should be traceable through the first access. Once the chain is broken, the benefit sharing obligations under the Nagoya would be nearly impossible to follow. This challenge has been further widened by the technological advancements such as DNA sequencing and the storage facilities of gene sequence data banks conferring genetic information a separate existence from the physical genetic material.

Technological Challenges to the CBD Nagoya Framework

We have seen that the CBD and the Nagoya protocol operate on the definition of genetic resources in the sense that they are genetic material containing functional units of heredity. When we delve on the functional units in genetic resources, it carries a meaning synonymous to DNA since the DNA sequences contain or encode the instructions for cell development, reproduction and survival (Elena Bencurova et al., 2023). Even though the three-dimensional structure of the DNA was resolved by Watson and Crick in 1953, the scientific community acquired the capacity to read the DNA almost after a decade only in 1965 when Robert Holley and his colleagues developed the first whole nucleic acid from the yeast *Saccharomyces cerevisiae* followed by the parallel efforts of Fred Sanger and colleagues on sequencing (James M. Heather & Benjamin Chain, 2016). Improvements made to the Sanger sequencing resulted in the first-generation automated DNA sequencing machines which could read only less than one kilobase of DNA fragments (James M. Heather & Benjamin Chain, 2016). This required scientists to use techniques such as shotgun sequencing to clone and sequence DNA fragments separately and later assemble them to long contiguous sequences (James M. Heather & Benjamin Chain, 2016). Sequencing was further augmented by technologies such as polymerase chain reaction and rDNA technology by providing high concentrations of pure DNA for sequencing (James M. Heather & Benjamin Chain, 2016). The combination of rDNA technology with first generation sequencing technology resulted in a genomic revolution, the improvements wherein helped analysing hundreds of samples simultaneously for the Human Genome Sequencing Project (James M. Heather & Benjamin Chain, 2016). Subsequent developments in the DNA sequencing technologies came to be generally referred to as second and next generation sequencing allowing for high throughput sequencing. The term throughput refers to the amount of DNA that can be processed through the automated sequencers at a given point of time. Sequencing by ligation, sequencing by synthesis, etc., are the major mechanisms currently available for high throughput sequencing on commercial scale (Ong'era et al., 2023).

The sequence of each DNA fragment is known as a read and in high throughput sequencing, millions of reads get sequenced in one shot (Alexis N. Burian et al., 2021). This is made possible with the help of advanced automated sequencers using computational tools. The user gets the reads in a plain text file format. Submission of DNA sequence data to genome banks is a practice started the second half of 1980s and is currently a mandate for publication in this domain. The researchers publishing results of determined sequences in different journals would deposit the sequences in DNA sequence data banks for presenting the sequence with greater data integrity (Christian Burks & Laurie Tomlinson, 1989). The extensive amount of data thus generated through sequencing projects resulted in the formation of genome databases operating on efficient algorithms and software facilitating quick data assimilation, better usable formats and effective biological data management (Bayat, 2002). Such databases were developed by public and private enterprises and currently, the International Nucleotide Sequence Collaboration created through the partnerships of the

European Molecular Biology Laboratory, the DNA Data Bank of Japan (DDBJ) and the GenBank, USA, allow free access to genomic data through internet. Such data is not limited to DNA sequences alone, but extends to macro-molecular data such as proteomes, transcriptomes etc. Thus, over the past three decades, biotechnology has moved from the material aspects of biology to informational aspects, heavily relying upon advanced applications of mathematics, engineering and computer software. This dependence gave a totally new dimension to the discipline of bio-informatics which was initially used for the structural analysis of proteins in biochemistry (Jeff Gauthier et al., 2019). The manual extraction of information from DNA through comparisons, calculations, pattern matching and sequence analysis was taken over by computers from humans with the help of bioinformatics (Jeff Gauthier et al., 2019). The earlier efforts to chemically synthesise DNA was further augmented by new genome editing tools like CRISPR facilitating the writing of new sequences (Hoose et al., 2023). These capabilities have been contributing to the development of cutting-edge technologies, widely known as synthetic biology, for manipulating DNA, designing and assembling of new sequences, creation of new biological materials, genetic circuits and metabolic pathways, thus allowing humans to engineer biological systems and entire organisms (Randall A. Hughes & Andrew D. Ellington, 2017).

Synthetic biology focuses on the application of engineering principles to biology, allowing scientists to construct biological components and systems that can be used for various purposes, such as producing sustainable materials, developing new medicines, and improving agricultural processes. In simpler terms, synthetic biology can be thought of as "rewriting the code of life" by designing and building biological systems with specific functions, similar to how a computer programmer writes code to create software.[6] Synthetic biology emerges with huge potential to provide a wide variety of academic and industrial applications such as making of novel therapeutics, materials, biosensing, biofuels, manufacturing-capabilities and improving agricultural processes. According to Hughes and Ellington (2017), the two foundational technologies in synthetic biology are the DNA reading by next generation sequencing and modern DNA synthesis technologies bringing in greater predictability and reliability to engineered biological systems. However, the ability of genomic engineering or synthetic biology to build upon large DNA sequence data sets is primarily dependent on the unhindered access to biological sequence data accumulated as a result of high throughput sequencing in the genomic repositories.

Synthetic biology tests the foundations of the CBD – ABS framework operating on PIC and MAT aimed at regulating the physical access to genetic resources. The existence of large sets of biological data in the genome banks reduces the dependence of scientists over physical genetic materials for R&D in synthetic biology (The International Civil Society Working Group on Synthetic Biology, 2016). With the help of sequencing technologies, genetic information contained in the genetic resources are dematerialized and externally stored in genomic databases enabling open access to

[6] https://www.genome.gov/about-genomics/policy-issues/Synthetic-Biology. Accessed 14 February 2024.

researchers (Akpoviri et al., 2023). Due to the later advent of ABS laws, a large chunk of the data available in the genomic databanks may not contain accurate information that could trace the country of origin and verify the claims for benefit sharing over commercial utilization of biotechnology applications using them (Akpoviri et al., 2023). In synthetic biology, multiple sequences from different organisms are used for R&D which makes tracing of the country of origin an almost impossible task (Rourke, 2020). At the same time, the technology has wider applications to promote the first two objectives of the CBD, i.e., conservation and sustainable use of biological diversity (Prip et al., 2015). The CBD addresses the diverse challenges posed by synthetic biology to its third objective, fair and equitable benefit sharing, using the terminology 'digital sequence information'.

Digital Sequence Information in the CBD/ Nagoya Framework—shift to Multilateralism

The implementation of the CBD is supported by an open-ended intergovernmental scientific advisory body known as the Subsidiary Body on Scientific, Technical and Technological Advice (SBSTTA), established under Article 25 of the CBD. Identification of new and emerging issues relevant to the conservation and sustainable use of biological diversity is one among the different tasks of the SBSTTA. In the 9th Conference of Parties (COP), the Secretariat of the CBD requested SBSTTA to identify new and emerging issues relevant for the CBD.[7] Synthetic biology was one among the four different issues identified by the SBSTTA in this regard. The 16th SBSTTA demanded a detailed consideration of the social, economic, cultural and biosafety dimensions of synthetic biology vis-à-vis the three objectives of the CBD followed by an analysis of the possible gaps and overlaps of the provisions of the CBD and its Protocols in relation to this technology.[8] Pursuant to this, the 11th COP of CBD directed to gather relevant information in this regard and make it available to the SBSTTA for further consideration.

The 18th SBSTTA reported that there could be conflicting interpretations whether the digital information of DNA sequences used in synthetic biology can be treated as a genetic resource attracting fair and equitable benefit sharing obligations under Article 15 of the CBD[9] The report further stated that the possibility of considering the components, products and organisms as genetic resources could form another mooting point which would further raise challenges to the premise of sovereign rights. In relation to the Nagoya Protocol, the report provided that the definition of utilization when read with the definition of biotechnology might tend to assume that synthetic biology is within the scope of the Protocol. The report further added that

[7] UNEP/CBD/COP/9/29 dated 9 October 2008 at COP/9/29 Para 13, p. 198.

[8] UNEP/CBD/SBSTTA/REC/XVI/12 dated 21 June 2012 Option 2, Para 3 and 3 bis.

[9] https://www.cbd.int/doc/meetings/sbstta/sbstta-18/information/sbstta-18-inf-04-en.pdf at para 170–180.

the term 'derivative' on the protocol might also raise multiple concerns in deciding the ambit of the protocol over derivatives produced by synthetic organisms modelled after natural organisms. The report noted that the benefit sharing obligations under Article 5 of the protocol extends to the utilization of genetic resources and subsequent applications and commercialization raising questions on the extent of modification that would make benefit sharing applicable to natural genetic resources to bind digital information based products also. Thus, the report concluded that the application of Nagoya protocol to synthetic biology remains ambiguous and that the technology might negatively affect the benefit sharing objective of the CBD.

Based on the report of the 18th SBSTTA, the 12th COP of the CBD in 2014 constituted an Ad-Hoc Technical Expert Group on Synthetic Biology (AHTEG) to report to SBSTTA considering the potential benefits and risks of synthetic biology in relation to the objectives of the CBD and its protocols[10] In 2015, the AHTEG on Synthetic Biology reported that the rise of synthetic biology and its use of digitized sequence data could potentially disrupt the fair and equitable sharing of benefits derived from physical access and utilization of genetic resources, ultimately leading to inequity as use of the digital sequence data would be without a proper benefit-sharing mechanism.[11] Based on this finding, the 20th SBSTTA recommended to the 13th COP that ABS related aspects of the use of digital sequence information (DSI) need to be addressed by the Conference of Parties Serving as Meeting of Parties (COP-MOP) of the Nagoya Protocol.[12] Therefrom, the international community started using the term 'Digital Sequence Information', particularly in relation to the ABS aspects of synthetic biology.

COP-13 decided to bring in a coordinated and non-duplicated approach between COP and COP-MOP and COP decisions on DSI were decided to be binding for COP-MOP too.[13] It further recognized DSI as a crosscutting issue pertinent to all three objectives of the CBD. It deliberated on the equivalence between sequence information and genetic resources per se and the suitability of the terminology of "DSI" as recommended by the 20th SBSTTA meeting. It further made the potential implications of DSI for all the three objectives of the CBD as an agenda for its 14th meeting and to facilitate the same, to constitute an AHTEG on DSI which would report to SBSTTA which in turn would make recommendations to the COP and the COP-MOP. COP-14 emphasized on the need to address divergence among the members on benefit sharing concerns related to the use of DSI on genetic resources.[14] It established a science and policy forum comprised of the Member States and other stakeholders

[10] https://www.cbd.int/doc/decisions/cop-12/cop-12-dec-24-en.pdf.

[11] (UNEP/CBD/SYNBIO/AHTEG/2015/1/2 dated 4 September 2015 Para 15 (updated report and synthesis of views in response to paragraph 7(b) OF DECISION XII/24 on new and emerging issues: synthetic biology Ad hoc Technical Expert Group on synthetic biology Montreal, Canada, 21–25 September 2015).

[12] https://www.cbd.int/doc/recommendations/sbstta-20/sbstta-20-rec-08-en.pdf.

[13] https://www.cbd.int/doc/decisions/cop-13/cop-13-dec-16-en.pdf.

[14] https://www.cbd.int/doc/decisions/cop-14/cop-14-dec-20-en.pdf.

including NGOs, intergovernmental organisations and local and indigenous communities. An Extended AHTEG on DSI was also established in this meeting to explore various dimensions of DSI such as its concept, scope, terminology, benefit-sharing, etc.

Subsequently, based on the later SBSTTA recommendations, the COP at its 15th meeting in 2022, decided to use DSI as placeholder term for the current discussions within the CBD given the lack of consensus on an appropriate terminology.[15] It could also generate consensus on bringing the use of DSI on genetic resources under the fair and equitable benefit sharing obligation. The agreement was to devise a benefit sharing solution that could be effective, efficient, feasible, and practical, featuring legal certainty and clarity for the providers and users of DSI. The COP emphasized on the need to maintain *status quo* on open access to DSI. Towards realizing the new benefits sharing commitment, the COP-15 decided to deviate from the bilateral approach of the CBD to a multilateral approach while recognizing the possibilities of alternate options and exceptions from multilateralism. Thus, it was agreed upon that a multilateral benefit sharing mechanism for the use of DSI would be established as part of the Kunming Global Biodiversity Framework adopted by the CBD. An Ad-hoc Open-ended Working Group on Benefit Sharing was also established to develop this mechanism further. The major elements of the multilateral mechanism as identified in the annex of decision 15/9 to be deliberated by the AHTEG-Benefit Sharing include the benefit sharing triggers, criteria for disbursement of benefit sharing which might embrace the geographical origin of genetic resource, alternate policy options for benefit sharing from the use of DSI, interface between the multilateral benefit sharing mechanism and existing national legislations that include benefit sharing obligations for the use of DSI, relationship between the Nagoya Protocol and the proposed multilateral mechanism, traditional knowledge and the roles, rights and interests of local and indigenous communities vis-à-vis DSI, and the possibilities for voluntarily extending the multilateral mechanism to genetic resources and biological diversity.

In November 2023, in Geneva, the first meeting of this AHTEG was held with the agenda to discuss the major elements of the multilateral system as identified above. At the meeting a Committee of the Whole was created to deliberate on the elements of the multilateral mechanism.[16] The Co-chairs of the Committee presented a draft that listed the possible elements of the multilateral mechanism for benefit sharing from the use of DSI on genetic resources, and those elements were grouped into five clusters during the meeting for ease of deliberations. The draft also contained the possibility of establishing a fund for multilateral benefit sharing. The clusters are contributions to the fund, disbursement of the fund, non-monetary benefit sharing, governance and relations to other approaches and systems for further intersessional work.

[15] https://www.cbd.int/doc/decisions/cop-15/cop-15-dec-09-en.pdf.

[16] https://www.cbd.int/doc/c/8531/d3e3/739fe0f4c5931b46dab505f3/wgdsi-01-03-en.pdf.

Conclusion

The developments at the COP-15 showcase the profound impact of synthetic biology on the international legal and policy landscape on access and benefit sharing for the utilization of genetic resources. It resulted in two divergent approaches in relation to fair and equitable benefit sharing under the CBD, a pre-existing bilateral approach and a new multilateral approach. To be more precise, for physical genetic resources where physical access is required, the pre-existing bilateral framework elaborated in the Nagoya Protocol would apply. PIC and MAT would remain the essential elements of this approach. Parallel to this, the multilateral benefit sharing mechanism would be operational under the CBD in less than a decade for dematerialized genetic resources or the DSI. This multilateral approach to benefit sharing does not seem to be based on the reciprocity of access and the points that would trigger benefit sharing obligation are yet to be finalized. The decoupling of access from benefit sharing has catered to a decade long demand of the scientists to maintain open access to data. Despite this global recognition of a benefit-sharing obligation, the success of this mechanism hinges on several critical factors, such as the access and use patterns over DSI by researchers, their responsiveness to the benefit-sharing obligations, and the overall perception of how such commitments might impact scientific research.

Conflict of Interest The authors state no conflict of interest.

References

Akhtar, A., Fuchs, E., Mitchison, T. J. (2011). A decade of molecular cell biology: Achievements and challenges, 12. https://doi.org/10.1038/NRM3187

Akpoviri, F. I., Baharum, S. N., & Zainol, Z. A. (2023). Digital Sequence Information and the Access and Benefit-Sharing Obligation of the Convention on Biological Diversity. *Nanoethics, 17*(1). https://doi.org/10.1007/s11569-023-004363

Bavikatte, K. S., (2015). Tvedt MW Beyond the thumb rule approach: Regulatory Innovations for Bioprospecting in India. *Law Environment and Development Journal, 11* (1). http://www.lead-journal.org/15001.pdf

Bayat, A. (2002, Apr 27). Science, medicine, and the future: Bioinformatics. *BMJ, 324*(7344), 1018–1022. https://doi.org/10.1136/bmj.324.7344.1018. PMID: 11976246; PMCID: PMC1122955.

Beattie, A. J., Hay, M. E., Magnusson, B. (2011). Ecology and bioprospecting, 36. https://doi.org/10.1111/J.1442-9993.2010.02170.X

Cordell, G.A. (2024). The contemporary nexus of medicines security and bioprospecting: A future perspective for prioritizing the patient. *Natural Products and Bioprospecting, 14*(11). https://doi.org/10.1007/s13659-024-00431-5

Elena Bencurova et al. (2023). DNA storage—From natural biology to synthetic biology.

Synthetic Biology. In: National Human Genome Research Institute. (2024) https://www.genome.gov/about-genomics/policy-issues/Synthetic-Biology Accessed 14 Feb 2024

Ashish Swarup Verma et al. (2011, July–September). Biotechnology in the Realm of History. *Journal of Pharmacy and Bioallied Sciences, 3*(3), 321–323

CBD/COP/DEC/14/20 Para 6, dated 30 November 2018, Decision adopted by COP to CBD, Digital sequence information, Fourteenth meeting, Sharm El-Sheikh, Egypt, 17–29 November 2018, https://www.cbd.int/doc/decisions/cop-14/cop-14-dec-20-en.pdf

CBD/COP/DEC/15/9, dated 19 December 2022, Decision adopted by COP to the CBD, Digital sequence information on genetic resources, fourteenth meeting- Part II, Montreal, Canada, 7–19 December 2022, https://www.cbd.int/doc/decisions/cop-15/cop-15-dec-09-en.pdf

CBD/WGDSI/1/3, dated 18 November 2023, Report of the Open-ended Working Group on Benefits sharing from the use of Digital Sequence Information on genetic resources on its first meeting, Geneva, 14–18 November 2023, https://www.cbd.int/doc/c/8531/d3e3/739fe0f4c593 1b46dab505f3/wgdsi-01-03-en.pdf

Beattie, A. J. (2024). Chapter 10: New products and industries from biodiversity. Millennium assessment org. https://www.millenniumassessment.org/documents/document.279.aspx.pdf

The Convention on Biological Diversity, 1992.

Coolsaet, Brendan, Tom Dedeurwaerdere, & John Pitseys. (2013). The Challenges for Implementing the Nagoya Protocol in a Multi-Level Governance Context: Lessons from the Belgian Case. *Resources, 2*(4), 555–580. https://doi.org/10.3390/resources2040555

The International Civil Society Working Group on Synthetic Biology, Synthetic Biology and the CBD Five key decisions for COP 13 & COP-MOP 8 (2016), 1–8

Cordell, G. A. (2024). The contemporary nexus of medicines security and bioprospecting: A future perspective for prioritizing the patient. *Natural Products and Bioprospecting, 14*(11). https://doi.org/10.1007/s13659-024-00431-5

Ralf Dahm. (2005). Friedrich Miescher and the discovery of DNA. *Developmental Biology, 278* (2), 274–288. https://doi.org/10.1016/j.ydbio.2004.11.028

CBD/COP/DEC/XIII/16, dated 16 December 2016 Decision adopted by COP to CBD on Digital sequence information on genetic resources, thirteenth meeting Cancun, Mexico, 4–17 December 2016.

Domingo-Fernandez, D., Gadiya, Y. et al. (2023). Modern Drug Discovery using ethnobotany: A large-scale cross-cultural analysis of traditional medicine reveals common therapeutic uses, 26. https://doi.org/10.1016/j.isci.2023.107729

Elujoba, A. A., Odeleye, O. M., Ogunyemi, C. M. (2005). traditional medical development for medical and dental primary healthcare delivery system in Africa. *African Journal of Traditional, Complementary and Alternate Medicine, 2*(1), 46–61.

Esquinas-Alcázar, J. T. (2005). Protecting crop genetic diversity for food security: Political, ethical and technical challenges. *Nature Reviews Genetics, 6*(12), 946–953. https://doi.org/10.1038/nrg 1729

Shengji, P. (2001). Ethnobotanical Approaches of Traditional Medicine Studies: Some Experiences from Asia, 39. https://doi.org/10.1076/PHBI.39.S1.74.0005

Fonseca, S., Amaral, M. N., Reis, C. P. (2023). Custódio L Marine Natural Products as Innovative Cosmetic Ingredients, 21. https://doi.org/10.3390/md21030170

Jeff Gauthier, Antony T Vincent, Steve J Charette, Nicolas Derome. (2019, November). A brief history of bioinformatics. *Briefings in Bioinformatics, 20*(6), 1981–1996. https://doi.org/10.1093/bib/bby063

Alexis N. Burian, Wufan Zhao, Te-Wen Lo, & Deborah, M. Thurtle-Schmidt. (2021, September–October). Genome sequencing guide: An introductory toolbox to whole-genome analysis methods. *Biochemistry and Molecular Biology Education, 49*(5), 815–825. Published online 2021 August 11. https://doi.org/10.1002/bmb.21561

Bhatia, S. (2018). History, scope and development of biotechnology. *IOP* Publishing Ltd.

Hoose, A., Vellacott, R., Storch, M., Freemont, P. S., Ryadnov, M. G. (2023). DNA synthesis technologies to close the gene writing gap. *Nature Reviews Chemistry, 7*(3), 144–161. https://doi.org/10.1038/s41570-022-00456-9. Epub 2023 Jan 23. Erratum in: Nat Rev Chem. 2023 Aug 7(8), 590. PMID: 36714378; PMCID: PMC9869848.

https://www.bio.org/what-biotechnology

https://www.cbd.int/abs/infokit/revised/web/factsheet-bonn-en.pdf

https://www.cbd.int/decision/np-mop/?id=13408
https://www.cbd.int/doc/decisions/cop-12/cop-12-dec-24-en.pdf
https://www.cbd.int/doc/decisions/cop-13/cop-13-dec-16-en.pdf
https://www.cbd.int/doc/meetings/sbstta/sbstta-16/official/sbstta-16-13-en.pdf
https://www.cbd.int/doc/meetings/sbstta/sbstta-18/information/sbstta-18-inf-04-en.pdf
https://www.millenniumassessment.org/documents/document.279.aspx.pdf

Hughe, R. A., & Ellington, A. D. (2017). Synthetic DNA synthesis and Assembly: Putting the Synthetic in Synthetic Biology, PUBMED Central.

James M. Heather & Benjamin Chain. (2016). The sequence of sequencers: The history of sequencing DNA.

Khan, S., Ullah, M. W. (2016). Siddique R Role of Recombinant DNA Technology to Improve Life. https://doi.org/10.1155/2016/2405954

Medaglia, J.C., Tvedt, M.W., Perron-Welch, F., Jørem, A., Phillips, F. K. (2013). The Interface between the Nagoya Protocol on ABS and the ITPGRFA at the International Level. Potential Issues for Consideration in Supporting Mutually Supportive Implementation at the National Level, Fridtjof Nansen Institute: Lysaker, Norway.

Morgera, E., Tsioumani, E., & Buck, M. (2015). Unraveling the Nagoya Protocol: A Commentary on the Nagoya Protocol on Access and Benefit-sharing to the Convention on Biological Diversity. Brill. http://www.jstor.org/stable/10.1163/j.ctt1w76vvq

Olsson, S.B. (2017). The Bengaluru Consortium. New Frontiers for Chemical Ecology: Reaffirming a Commitment to the Göteborg Resolution. *Journal of Chemical Ecology, 43* (2–3). https://doi.org/10.1007/s10886-016-0813-9

Ong'era, E. M. et al. (2023). High-throughput sequencing approaches applied to SARS-CoV-2.

Pauchard, N. (2017) Access and Benefit Sharing under the Convention on Biological Diversity and Its Protocol: What Can Some Numbers Tell Us about the Effectiveness of the Regulatory Regime? 6. https://doi.org/10.3390/RESOURCES601001

Prip, C., Rosendal, G. K., Tvedt M. W. The state of technology transfer obligations in global environmental governance and law: biodiversity conservation and sustainable use. Centre for Biodiversity Policy and Law, NBA, Chennai. FNI Report 4/2015. https://www.sprep.org/attachments/VirLib/Global/state-technology-transfer.pdf

UNEP/CBD/SBSTTA/REC/XX/8 dated 2 May 2016 recommendations adopted by the Subsidiary Body on Scientific, Technical and Technological Advice xx/8. synthetic biology subsidiary body on scientific, technical and technological advice Twentieth meeting Montreal, Canada, 25–30 April 2016, https://www.cbd.int/doc/recommendations/sbstta-20/sbstta-20-rec-08-en.pdf

Rourke, M. (2020). Access and benefit-sharing DNA Componentry for plant synthetic biology: Bioparts expressed in plant chassis. https://nph.onlinelibrary.wiley.com/doi/pdf/10.1002/ppp3.10184

Schei & Tvedt. (2010). 'Genetic Resources' in the CBD: The Wording, the Past, the Present and the Future.

Sofowora, A., Ogunbodede, E., Onayade, A. (2013). The role and place of medicinal plants in the strategies for disease prevention. *African Journal of Traditional, Complementary and Alternative Medicines*. https://doi.org/10.4314/ajtcam.v10i5.2

The Importance of Biodiversity to Human Health. (2010). In cohabinitiative Co-operation on Health and Biodiversity. https://www.cbd.int/doc/health/cohab-policy-brief1-en.pdf. Accessed February 5, 2024.

David S. Tilford. (1998). Saving the Blueprints: The international legal regime for plant genetic resources. *Case Western Reserve Journal of International Law, 30*, 373. https://scholarlycommons.law.case.edu/jil/vol30/iss2/2

Christian Burks & Laurie Tomlinson (1989). Submission of data to GenBank. https://www.researchgate.net/publication/20243846_Submission_of_data_to_GenBank?enrichId=rgreq-f987cd2f7b12cd0bf6b89a14d5531f16-XXX&enrichSource=Y292ZXJQYWdlOzIwMjQzODQ2O0FTOjEwMjc2Mjc3NTg0MjgyNkAxNDAxNTExOTYzOTI4&el=1_x_2&_esc=publicationCoverPdf

UNEP/CBD/COP/9/29 dated 9 October 2008 at COP/9/29 Para 13, p.198

UNEP/CBD/SBSTTA/REC/XVI/12 dated 21 June 2012 Option 2, Para 3 and 3bis.

UNEP/CBD/SYNBIO/AHTEG/2015/1/2 dated 4 September 2015 Para 15 (updated report and synthesis of views in response to paragraph 7(b) OF DECISION XII/24 on new and emerging issues: synthetic biology Ad hoc Technical Expert Group on synthetic biology Montreal, Canada, 21–25 September 2015

Verma, A. S., Agrahari, S., Rastogi, S., Singh, A. (2011, July). Biotechnology in the realm of history. *Journal of Pharmacy and Bioallied Sciences, 3*(3), 321–323. https://doi.org/10.4103/0975-7406.84430. PMID: 21966150; PMCID: PMC3178936.

Weiss, C. (1998). Eisner T Partnerships for value-added through bioprospecting. *Technology in Society, 20*, 481–498. https://doi.org/10.1016/s0160-791x(98)00029-3

Accessing Biodiversity Stock for Sustainable Use

Marine Bioprospecting a Promising Future Venture and Precautionary Measures to Maintain Its Sustainability

K. V. Jayachandran, L. Divya, and Oommen V. Oommen

Introduction

Marine genetic diversity is huge and as per the latest data on World Register of Marine Species (2024, http://www.marinespecies.org/) around 2,45,000 (2,45,644 in 2014) valid species has been on record (Kingdoms Animalia—2,06,378; Chiromista—20,381; Plantae—10,418; Protozoa—638; Archaea—116; Fungi—1,378; Viruses—115) (Bouchet et al., 2023). Organisms related to 34 phyla (77 classes) are represented in this ecosystem (Briggs, 1995). Utilization of the marine bio-resources for food and other general purposes has been as old as mankind. However, more recently industry-based compounds have been derived, namely, pharmaceuticals, food and beverages, cosmetics, agriculture-based products, bio-fertilizers, biotechnology products, unique gene isolation, biomolecules, biochemical compounds, biofuels and so on. The term Nutraceuticals was derived from words, 'Nutrition' and 'Pharmaceuticals' and was first defined by Dr. Stephen De Felice in 1989 (Torabally & Rahmanpoor, 2019) and defined as "food, or parts of a food, that provide medical or health benefits including the prevention and treatment of diseases". It is classified into three: Dietary supplements, functional

K. V. Jayachandran (✉)
Department of Aquatic Biology & Fisheries, University of Kerala, Thiruvananthapuram, Kerala, India
e-mail: jayachandrankvdr@gmail.com

O. V. Oommen
Department of Computational Biology & Bioinformatics, University of Kerala, Thiruvananthapuram, Kerala, India

L. Divya
Department of Zoology, University of Calicut, Malappuram, Kerala, India

O. V. Oommen
Center for Sustainable Environment-India Biodiversity Action, Thiruvananthapuram, Kerala, India

L. Krishna Panicker et al. (eds.), *Biodiversity and Business*,
https://doi.org/10.1007/978-3-031-71674-4_6

food and medicinal food. Bioactive components like vitamins, minerals, amino acids, fatty acids, and antioxidants make up the **dietary supplements**; Polyphenols,Anthocyanidins, phytoestrogens, terpenoids, carotenoids, limonoids, phytosterols, glucosinolates, and isoflavonoids are included in the class of phytochemicals; Probiotics of live microbial feed supplement for improving its intestinal microbial balance; Fermented dietary ingredients known as prebiotics enable certain alterations in the gastrointestinal microbiota's composition and/or activity, which improves the host's overall health and well-being; Dietary fibers containing non-starchy polysaccharides, including resistant dextrins, cellulose, hemicelluloses, gum, and pectin; Nutraceutical enzyme foods that are naturally sourced, nutrient-enriched, and fortified with vital nutrients (fortified nutritional supplements) are referred to as **functional foods**. The unique nutritional requirements for each medicinal food, are medically evaluated and are created for the dietary management of a particular disease or conditions and are intended to be ingested under a physician's supervision (Torabally & Rahmanpoor, 2019). Marine ecosystem is a treasure for these compounds and hence research activities are steadily increasing (Avhad & Bhangale, 2023; ba Zodape, 2018; Bhakuni & Ranat, 2005; Bhatnagar & Kim, 2010; Chi et al., 2014; Carroll et al., 2020; Ghallab et al., 2024; Karthikeyan, 2022; Lindequist, 2016; MAREX, 2014; Oli et al., 2023; Ramesh et al., 2021; Rigogliuso et al., 2023; Wildman et al., 2006).

In terms of history, Tyrian purple was a fabric dye produced by the Phoenicians as early as 1200 BC from the mucus of many species of Murex sp. snails. The Greeks and Romans continued to produce it until 1453 AD, with the fall of Constantinople. Fish and marine algal metabolites made up the marine natural products for a very long period. The vitamins A and D from fish liver oil, marine biopolymers like agar and carrageenan, and polyunsaturated fatty acids like eicosapentaenoic acid and docosahexaenoic acid are notable examples. The real marine drug development started with the discovery of spongothymidine and spongouridine in 1950s from the Caribbean sponge, Tectitethya crypta (de Laubenfels) (Newman & Cragg, 2018). The creation of the antibiotic class known as cephalosporins began with cephalosporin C, which was produced by the fungus Acremonium chrysogenum (Thirumalachar & Sukapure, 1966) that was isolated from Mediterranean Sea water samples close to Sardinia (König, 1992; Lin & Kück, 2022). A high prostaglandin concentration was noted by Weinheimer and Spraggins (1969) for the gorgonian Plexaura homomalla (Esper).

Cytarabine, a nucleoside derived from sponge, approved by FDA in 1969 was the first marine drug; followed by Vadarabine, a nucleoside from sponge, approved by FDA in 1976; Fludarabine, a nucleoside from sponge, approved by FDA 1992 (EMA, 1994); Ziconotide (Prialt), a peptide derived from a tropical cone snail, Conus magus Linnaeus, 1758, approved by FDA in 2004 (EMA 2005); Omega-3 acid ethyl esters, PUFA from fish and approved by FDA 2204 (EMA 2005); Nelarabine, a nucleoside from sponge, approved by FDA 2015 (EMA 2007); Yondelis (trabectedin) an alkaloid from tunicate, approved by FDA 2015 (EMA 2007); Erubulin, a macrolide derived from sponge, approved by FDA 2010 (EMA 2011); Brentuximab Vedotin, ADC derived from mollusc and cyanobacterium, approved by FDA 2011

(EMA 2012); Lurbinectedin, an alkaloid from tunicates, approved by FDA 2019; Polatuzumab vedotin, ADC from mollusc and cyanobacterium, approved by FDA 2019 (EMA 2020); Enfortumab vedotin, ADC derived from mollusc and cyanobacterium, approved by FDA 2019 (EMA 2021) and Belantamab mafodotin, ADC derived from mollusc and cyanobacterium, approved by FDA 2020 (EMA 2020) are the drugs derived from marine bioresources in chronological order (Cappello & Nieri, 2021). Since 1990, almost half of all new natural products derived from animals have originated from sponges (Porifera). Corals and jellies (Cnidaria) have contributed 30%, while sea stars and other Echinodermata, mollusks, and vertebrates (Chordata) have contributed 5% each to the natural products. Additionally, chemicals have been extracted from the following sources: lamp shells (Brachiopoda), crustaceans (Arthropoda), moss animals (Bryozoa), flatworms (Platyhelminthes), acorn worms and their cousins (Hemichordata), and polychaetes (Annelida) (Snelgrove, 2016). Several studies have demonstrated that natural compounds, usually derived from microorganisms, most commonly bacteria or cyanobacteria, are taken from higher marine species, including marine invertebrates. But because these microorganisms live in symbiotic relationships with larger species, they are usually not able to be produced in a pure culture on their own.

A look at patent applications on marine natural products reveals a strong increase since middle of 1980s (Bongiorni & Pietra, 1996). Roughly 23,570 marine natural products have been reported from marine organisms (microorganisms, red algae, brown algae, green algae, sponges, cnidarians, bryozoans, molluscs, echinoderms, and tunicates) during 1950 has reported to be growing at a rate of 4 per cent per year to the current report of 40,000 (Arrieta et al., 2010; Carroll et al., 2020; Leal & Calado, 2012; Liu et al., 2019).

Recent research activities revealed that deep oceans, more specifically, hydrothermal vents and cold seeps harbour immense biodiversity. The organisms in these areas have developed some unique and highly valuable biochemical and genetic properties because of the peculiar climatic conditions prevailing in these areas. Predominant stress factors at deep depth of oceans are extreme conditions, very high pressure and temperature, perpetual darkness, absence of oxygen and so on. These stressful conditions have paved way for the evolution of many species and are rare in these regions (Bekiari, 2023). These species do possess wonderful bioactive compounds which are often characterized by structural novelty, complexity, and diversity (Matz, 2002). Even though very few of these products have made it to market, marine bioprospecting offers substantial financial potential.

The global biotechnology industry was estimated to be worth USD 1.55 trillion in 2023, according to the Marine industry Analysis Report (MAR) 2023 and is projected to expand at a compound annual growth rate (CAGR) of 13.96% between 2024 and 2030. Credence Research, Inc. reports that the Global Marine Biotechnology Market (MBM) on aspects, growth, future prospects, and competitive Analysis, 2022—2030 the global demand for marine biotechnology was valued at USD 6215.8 Million in 2022 and is expected to reach USD 12,204.7 Million in 2030, 8.8% growth CAGR 2023 and 2030. The ever-growing demand for marine bioactive compounds for human welfare compels for the consolidation of data on marine bio-resources, their products and possible action may help for further research in the specific area. Detailed reviews on the subject are available (Chari & Berde, 2024; Hamed et al., 2015; Perera et al., 2023) In this chapter the bio-resources dealt with are marine bacteria, sponge associated bacteria and bio-active compounds, secondary metabolites from marine sources, fungi, microalgae, sea weeds, cnidarians, sponges, annelids, crustaceans, Mollusca, echinoderms, ascidians and tunicates, fishes, and drugs of marine origin (already developed and approved, under phase III, II, I, Preclinical trials).

Marine Bacteria, Bio-Active Compounds, and Potential Activity

Several bio-active compounds (pigments) have been isolated from marine bacteria. There are bacteria associated with many other organisms including sponges. The following table provides a detailed account on these bio-resources (Table 1).

Secondary Metabolites from Marine Bio-Resource

The secondary metabolites directed connected with anticancer, anti-bacterial/ anti-fungal, anti-inflammatory, neurogenic activity, anti-tuberculosis activities are dealt in this section (Debbab et al., 2010) (Table 2).

Marine Bacteria, Bio-Active Compounds, and Potential Activity

Marine fungi are very important bio-resource from where several drugs have been derived. Even antifungal derivatives have been developed from this bio-resource and details are included in Table 3.

Marine Microalgae, Bio-Active Compounds, and Potential Activity

Microalgae proliferate and produce large biomass and support the productivity. It is "single cell protein". The derivatives of microalgae are used for: dyes for food (ice cream, jelly, candies and juice), food industry (functional food, food additives, beverages etc.), Pharmaceutical industry (antibiotics and anti-microbial drugs) and

Table 1 Biologically active compounds isolated from marine bacteria and sponge associated bacteria

Pigment	Bacterial strain	Potential activity	Reference
Tauramamide (Lipopeptide)	Brevibacillus laterosporus PNG276	Gram-positive, human pathogen Enterococcus sp.	Debbab et al. (2010)
2-alkylidene-5-alkyl-4-oxazolidinones, lipoxazolidinone A	Marinispora (strain NPS008920)	Antimicrobial	Debbab et al. (2010)
Undecylprodigiosin	Streptomyces rubber	Anticancer	https://www.ncbi.nlm.nih.gov/pmc/articles/PMC3794959/
Cycloprodigiosin	Pseudoalteromonas denitrificans	Immuno-suppresant; anticancer; antimalarial	https://pubs.acs.org/doi/https://doi.org/10.1021/acs.orglett.5b01527
Heptyl prodigiosin	α-Proteobacteria	antiplasmodial	https://pubmed.ncbi.nlm.nih.gov/12503881/
Prodigiosin	Pseudoalteromonas rubra Hahella chejuensis	Antibacterial; anticancer; algicidal	https://www.ncbi.nlm.nih.gov/pmc/articles/PMC9342244/
Astaxanthin (carotene)	Agrobacterium aurantiacum; Pseudoalteromonas luteoviolacea	Antioxidation	https://doi.org/10.3390/md12010128

(continued)

Table 1 (continued)

Pigment	Bacterial strain	Potential activity	Reference
Violacein	Pseudoalteromonas tunicate; Pseudoalteromonas sp. 520PI Collimonas CT	Antibiotic; Antiprotozoan; Anticancer	https://jbioleng.biomedcentral.com/articles/, https://doi.org/10.1186/s13036-021-00262-9
Methyl suphenate (phenazine derivative)	Pseudonocardia sp.	Antibiotic	https://www.mdpi.com/2073-4395/10/5/655
Phenazine derivates	Bacillus sp.	Cytotixic	https://pubmed.ncbi.nlm.nih.gov/17615672/
Pyocyanin and Pyorubrin	Pseudomonas aeruginosa	Antibacterial	https://pubmed.ncbi.nlm.nih.gov/6432831/
Phenazine-1-carboxylic acid	Pseudomonas aeruginosa	Antiobiotic	https://pubchem.ncbi.nlm.nih.gov/compound/Phenazine-1-carboxylate
5,10-dyhydrophencomycin methyl ester	Streptomycete sp.	Antibiotic	https://www.pharmaffiliates.com/en/52448-10-9-cytidine-5-monophosphate-methyl-ester-pa280201011.html

(continued)

Table 1 (continued)

Pigment	Bacterial strain	Potential activity	Reference
Fridamycin D, Himalomycin A, B	Streptomycete sp. B6921	Antibacterial	https://pubmed.ncbi.nlm.nih.gov/14763560/
Chinikomycin A and Chinikomycin B, Manumycin A	Streptomycete sp. B045	Anticancer	https://doi.org/10.1021/np030518r
Tambjamines (BE-18591, Pyrrole and synthetic analog of the same	Pseudoalteromonas tunicata	Antibiotic, anticancer	https://www.ncbi.nlm.nih.gov/pmc/articles/PMC6457199/
Melanins	Alteromonas nigrifaciens; Shewanella colwelliana; Cellulophaga tyrosinoxydans	Defense against UV radiation	https://en.wikipedia.org/wiki/Melanin
Scytonemin	Cyanobacteria	Protection from UV irradiation; anti-inflammatory; antiproliferative	https://www.sciencedirect.com/science/article/abs/pii/S2211926422000492
Tryptanthrin	Cytophaga/ Flexibacteria AM13, I strain	Antibiotic	https://www.ncbi.nlm.nih.gov/pmc/articles/PMC9861433/

Sponge associated bacterial bioactive compounds

Compounds	**Bacterial source and association**	**Activity**	**Reference**
2-undecyl-4-quinolinone	Pseudomonas sp. 1537-E7 on Homophymia sp.	Antimalarial and anti-HIV	https://pubchem.ncbi.nlm.nih.gov/compound/2-Undecyl-4_1H_-quinolinone

(continued)

Table 1 (continued)

Pigment	Bacterial strain	Potential activity	Reference
3,6-diisopropypiperazine-2,5-dione	Pseudomonas sp. NJ6-3-1 on Hymeniacidon perlevis	Anti-microbial	https://pubchem.ncbi.nlm.nih.gov/compound/3_6-Dimethylpiperazine-2_5-dione
Andrimid	Vibrio sp. M22-1 on Hyatella sp.	Anti-bacterial	https://www.ncbi.nlm.nih.gov/pmc/articles/PMC2533188/
2,4,4’-trichloro-2’-hydroxy-diphenyl ether	Micrococcus luteus R-1588-10 on Xestospongia sp.	Antibacterial	https://pubmed.ncbi.nlm.nih.gov/4920280/
Cyclo-(l-pro-l-phe)	Alcaligenes faecalis A72 on Stellate tenuis	Anti-bacterial	https://pubmed.ncbi.nlm.nih.gov/34436256/
Iturin	Bacillus subtilis A202 on Aplysia aerophoba	Antifungal	https://pubmed.ncbi.nlm.nih.gov/33983074/
Manzamine A	Micromonospora sp. on Acanthostrongylophora sp.	Antimalarial	https://www.sciencedirect.com/topics/chemistry/manzamine-a
Surfactin, Iturin and Fengycin	Bacillus subtilis A184 on Aplysia aerophoba	Antifungal	https://pubmed.ncbi.nlm.nih.gov/27021396/
Trisindoline	Vibrio sp. on Hyrtios altum	Antibacterial	https://pubmed.ncbi.nlm.nih.gov/7697760/

Table 2 Additional information on the Bioactive Secondary metabolites from marine sources (Based on Karthikeyan et al., 2022)

Secondary metabolites	Marine source	Applications
Anticancer		
Aureoverticillactam	*Streptomyces aureoverticillatus*	Cytotoxicity, various cell line tumors
Caprolactones	*Streptomyces* sp.	Anti-cancer cell lines
Chinikomycins	Streptomyces sp.	Cytotoxicity, various cell line tumors cancer cell lines in humans
IB-00208	Actinomadura sp.	Cytotoxicity and against Gram-positive bacteria
Salinosporamide A (NPI-0052)	Salinispora tropica	Cytotoxicity,proteosome inhibition and NF-κB activation
Urdamycin	Streptomyces fradiae	strong antibacterial and anti-cancer activity
Himastatin	Streptomyces hygroscopicus	antibiotic with antitumor and antibacterial activity
Daryamide D	Streptomyces strain CNQ-085	Cytotoxicity, antifungal against Candida albicans and human colon carcinoma
Marinomycin	Marinispora sp. strain CNQ-140	Against cancer cell proliferation
Manumycin	Streptomyces sp. M045	Antitumor activity of human cancer cell lines
Marmycin	Streptomyces sp.	Cytotoxicity of tumor cells and cell arrest during the G1 cell cycle process
Nonactin	Streptomyces griseus, Streptomyces werraensis, Streptomyces chrysomallus and Streptomyces tsukubaensis	Strong anti-neoplastic and anti-bacterial activity
Chartreusin	Streptomyces chartreusis	Active against certain gram-positive bacteria
Altemicidin	Streptomyces sioyaensis SA-1758	Antitumor and acaricidal activity
Streptochlorin	Streptomyces sp.	Effective for the treatment of cholangiocarcinoma
Marineosins	Streptomyces sp.	Effective to control human colon carcinoma
Ammosamides	Steptomyces variabilis	Against MIA PaCa-2 pancreatic cancer cell line

(continued)

Table 2 (continued)

Secondary metabolites	Marine source	Applications
Caboxamycin	Streptomyces sp. NTK 937	Against Gram positive bacteria, selected human tumor cell lines and enzyme phosphodiesterase
Hoiamide D	Symploca sp	Inhibitor to 53 Mdm2 interaction
Niphateolide	Niphates olemda	Inhibitor of p53-Hdm2/MDM2 interaction
Hexylitaconic acid	Arthrinium sp	Blocks p53/MDM2 binding
Lissoclinidine B	Lissoclinum cf. badium	Inhibition of GST fused human Hdm2
Himeic acid A	Aspergillus sp.	Inhibition of Ubiquitin-Activating Enzyme (E1)
Girolline	Cymbastela cantharell and Axinella brevistyla	Arresting G2/M cell cycle in cancer cells
Leucettamol A	Leucetta aff. microrhaphis	Against the ubiquitin E2 enzymes Ubc13 and Uev1A by 50%
Dysidiolide	Dysidea etheria	Inhibitory action of Cdc25 protein phosphatase
Sulfircin	Ircinia sp	inhibit Cdc25 phosphatase
Coscinosulfate	Coscinoderma mathewsi	inhibitory action towards Cdc25A
Halenaquinone	Xestospongia carbonaria	irreversible inhibitor of recombinant human Cdc25B phosphates
Secalonic acid D	Penicillium oxalicum	inhibits the course of the cell cycle in human embryonic palatal mesenchymal cells
Stellettin B	Jaspis stellifera	inhibit Cdk action and enhance in p27 expression
Antibacterial		
Abyssomicins	Verrucosispora sp.	Inhibits the pathway between chorismate and para-aminobenzoic acid
Frigocyclinone	Streptomyces griseus	Lead molecule against Kaposi's Sarcoma Associated Herpes Virus
Gutingimycin	Streptomyces sp.	Antibacterial antifungal and antimicroalgal activities
Helquinoline	Janibacter limosus	Antibacterial antifungal and antimicroalgal activities
Himalomycins	Streptomyces sp.	Antimicrobial activity against Gram-positive bacteria

(continued)

Table 2 (continued)

Secondary metabolites	Marine source	Applications
Lajollamycin	Streptomyces nodosus	Antimicrobial activity against drug-sensitive and -resistant Gram-positive bacteria and inhibited the growth of B16-F10 tumor cells in vitro
Tylosin	Streptomyces fradiae	Potential for the treatment of respiratory and other infections caused by Mycoplasma species
Maklamicin	Micromonospora sp. GMKU326	Antimicrobial activity against Gram-positive bacteria
Lobophorin K	Streptomyces sp. M-207	Antibiotic activity against pathogenic Gram-positive bacteria
Asenjonamide C	Streptomyces asenjonii strain KNN 42	Antimicrobial activity against Gram-positive bacteria
Gilvocarcin HE	Streptomyces sp.	More significant cytotoxicity and antimicrobial activity due to the vinyl side chain
Zunyimycins	Streptomyces sp. FJS31-2	Inhibits the proliferation of lung cancer cells by the activation of apoptosis by an AKT pathway
Formicamycins A	Streptomyces formicae	Potent antibacterial activity against clinical MRSA and VRE isolates
Allocyclinones	Actinoallomurus	Possess activity against various Gram-positive bacteria, including antibiotic-resistant strains, with increasing antibacterial potency with the number of chlorine substitutes
Ageloline A	Streptomyces sp. SBT345	Able to reduce oxidative stress and genomic damage induced by the oxidative stress inducer 4-nitroquinoline-1-oxide (NQO)
Buanmycin	Streptomyces sp	Inhibition of sortase A, which is a promising target for antibiotic Discovery
Citreamicin	Streptomyces caelestis	Citreamicins exhibit cytotoxic activity against HeLa and Hep62 cells in addition to their potent antibiotic activity
Kocurin	Kocuria sp. strain MI-67-EC3-038	Kocurin is active against methicillin-resistant Staphylococcus aureus (MRSA)

(continued)

Table 2 (continued)

Secondary metabolites	Marine source	Applications
Fijimycins	Streptomyces sp.	Possess significant in vitro antibacterial activity against methicillin-resistant Staphylococcus aureus (MRSA) strains
Arylomycin	Streptomyces sp. HCCB10043	Inhibits a promising antimicrobial target, type I signal peptidase (SPase)
Antibacterial/anticancer		
1-Hydroxy-1-Norresistomycin	Streptomyces variabilis	Potent cytotoxic activity against cell lines viz. HMO2 (gastric adenocarcinoma) and HePG2 (hepatic carcinoma) in vitro
Antibacterial / antifungal		
Bonactin	Streptomyces sp.	Bonactin displayed antimicrobial activity against both Gram-positive and Gram-negative bacteria as well as antifungal activity
Anti-inflammatory		
Diazepinomicin (ECO-4601)	Micromonosproa sp	Preclinical broad-spectrum antitumor potential, antioxidant and anti-protease activities
Cyclomarins	Streptomyces sp.	Interesting lead structures for the development of drugs against tuberculosis and malaria
Neurogenic activity		
Komodoquinone A	Streptomyces sp. KS3	Induces differentiation of neuronal cells in the neuroblastoma cell line, Neuro 2A and arrests the cell cycle in step G1
Granaticins	Streptomyces lateritius	Granaticinhas significant antitumor activity against P-388 lymphocytic leukaemia in mice and cytotoxicity against KB cells
Hymenialdisine	Agelasidae, Axinellidae, and Halichondriidae	suppresses many pro inflammatory cytokines (IL-1, IL-2, IL-6, and NO) by inhibition of NF-kB signalling pathway
Antifungal		
Amphotericin B	*Streptomyces nodosus*	Treatment of most systemic fungal infections
Anti-tuberculosis		

(continued)

Table 2 (continued)

Secondary metabolites	Marine source	Applications
Rifampicin	-	Rifampicin had immunomodulatory effects through its ability to modify human monocyte production of measured cytokines
Streptomycin	-	Antibiotic activity against gram-positive and gram- negative bacteria
Amikacin	-	Potent activity against antibiotic-resistant clinical Isolates
Viomycin	-	Tuberculostatic agent active against both streptomycin—sensitive and streptomycin—resistant strains
Capreomycin	-	Bactericidal in vitro among the anti-TB drugs against non-replicating tubercle bacilli
Kanamycin	-	Kanamycin intramuscular administration had a satisfactory effect in mice with staphylococci, pneumococci, and Ulebsiella pneumonia infections
Cycloserine	-	Antituberculosis activity invitro and invivo

animal husbandry (fodder, food supplement and nutrition). These are a rich source of bioactive chemicals, including astaxanthin, sterols, polyunsaturated fatty acids, β-carotene, polysaccharides, eicosapentaenoic acid (EPA), sterols, and docosahexaenoic acid (DHA), and biologically active molecules and hence their market value is unimaginably high. These biomolecules and metabolites have antibiotic, antiviral, anti-inflammatory, anticancer, and neuroprotective activities (Zhao et al., 2017). Blue-green algae contain abundant PUFAs and pigments such as carotenoids and chlorophylls -chlorophyll a, phycocyanins, and phycoerythrin (Chen et al., 2018a). Table 4 contains selected microalgae bio-resources and their bioactive compounds.

Algal oils: Algal oil known as DHA Gold is produced by Crypthecodinium cohnii and Schizochytrium sp. It is added to food and beverages as an adult dietary supplement. It is also given as food for pregnant and nursing women and is advantageous for cardio-vascular health (Cohen & Ratledge, 2010).

Spirulina sp. is a filamentous alga having a large number of advantageous benefits: rich in nutrients (high protein, thiamine, riboflavin, magnesium, Potassium, Niacin, Copper, Iron, Omega-3 fatty acids and Omega-6, essential amino acids); with antioxidant and anti-inflammatory compounds; protects LDL cholesterol from oxidation; lowers cholesterol and triglycerides; could control blood sugar;may have anticancer properties; could be effective against anaemia; may improve muscle strength

Table 3 Information on selected marine fungi bio-resource and bioactive compounds derived from it

Name of Fungi	Compound	Bioactivity	Source of information
Myrothecium roridum	Verrucarin A	Anti-tumor	Hasan et al. (2015)
Penicillium sp.	Chromone derivative	Anti-tumor	
Penicillium sp. (Strain DG M3) 6'C (from ascidian Didemnum granulatum)	1 3-desoxy-phomenone	Anti-tumor	Hasan et al. (2015)
Penicillium paxilli (Strain Ma G)K (from sponge Mycale angulosa)	Pyrenocine A	Mammalian cancer cell; HCT8 human colon, CNS295 CNS cells	Hasan et al. (2015)
Trichoderma virens (from ascidian Didemnum molle)	Trichodermamide A, B	Human colon cancer	Hasan et al. (2015)
Leptosphaeria sp. (marine alga Sargassum tortile)	Epipolysulfany (dioxopiperazine cytotoxins 5)	Anti-tumor	Hasan et al. (2015)
Zygosprorium masonii	Zygosporamide	Agent 60 Cancer cell lines in NCL	Hasan et al. (2015)
Aspergillus fumigatus OPUST 106B-5 (from marine fish Mugil cephalus)	Cepalimysin A	Human HL-60 Leukemia	Hasan et al. (2015)
Penicillium chrysogenum QEN-245 (from red alga, Laurencia)	Diterpene, conidiogenone C	HL-60	Hasan et al. (2015)
Fusarium spp. (from sea fan Annella sp.)	Fusarathraquinone, Fusarathraquinone A-C Fusarone	Anti-tumor	Hasan et al. (2015)
Cryptosphaeria sp. (from unidentified ascidian)	Cryptosphaerolide	HCT-116 cells	Hasan et al. (2015)
Paecilomyces sp. (from mangrove)	Poeciloxocins A, B	Hep G2 cell lines	Hasan et al. (2015)

(continued)

Table 3 (continued)

Name of Fungi	Compound	Bioactivity	Source of information
Penicillium paxilli (strain MaG) K	4-methoxy-5-(3-methoxybut-1-enyl)-6-methyl-2H-pyran-2-one	Anti-tumor	Hasan et al. (2015)
Penicillium raistrikii (on spone Axinella corrugate)	Norliquexanthone	Anti-tumor	Hasan et al. (2015)
Aspergillus fumigatus	(2–3,3-dimethylprop-1-ene)-costaclavine (2–3,2-dimethylprop-1-ene-epicosta-clavine	Anti-tumor	Hasan et al. (2015)
Aspergillus sp.	Chondrimanin B	Antibacterial	Hasan et al. (2015)
Myceliopphthora lutea Costantin	Isolates	Antibacterial, cytotoxic	Hasan et al. (2015)
Aspergillus protuberus sp.1	Extract	Anticancer	Hasan et al., 2015
Aspergillus fumigatus	Protease Alkaline Protease 1 (Alp 1)	Fungal asthma	https://www.ncbi.nlm.nih.gov/pmc/articles/PMC7345148/
Penicillatum citrinum	Alkaloids	Anticancer	Hasan et al. (2015)
Fusarium sp.	Alkaloids	Anticancer	Hasan et al. (2015)
Apiospora montagnei	Alkaloids	Anticancer	Hasan et al. (2015)
Ascothyta salicorniae	Polyketide Pyrrolidinone-A40	Antimalarial	Hasan et al. (2015)
Dichotomyces cejpii	Secondary metabolite	Treat Alzheimer dementia	Hasan et al. (2015)
Nigrospora sp. (deep sea)	Secondary metabolite	Anti-microbial and application in cosmetics	Hasan et al. (2015)
Aspergillus unguis (from sponge)	Secondary metabolite	Antioxidant	Hasan et al. (2015)
Aspergillus sp. (from sponge Tethya aurantium	Secondary metabolite	Cytotoxicity	Hasan et al. (2015)

(continued)

Table 3 (continued)

Name of Fungi	Compound	Bioactivity	Source of information
Chaetomium sp.	Chaetominedione	Enzyme inhibition	https://pubchem.ncbi.nlm.nih.gov/compound/Chaetominedione
Ascomycetous 222	Balticolid	Antiviral	https://pubmed.ncbi.nlm.nih.gov/21673893/
Fusarium sp.	3-O-methylfusarubin; 4-Hydroxyphenethyl methyl succinate	Cytotoxicity	https://doi.org/10.1016/j.tet.2011.07.078
Penicillium griseofulvum Y19-07	4-Hydroxy-phenethyl 2–4-hydroxyphenyl acetate	Cytotoxicity	https://pubmed.ncbi.nlm.nih.gov/20183253/
Penicillium viridicatum	Fumigaclavine	Antifungal; Antibacterial	https://pubchem.ncbi.nlm.nih.gov/compound/Fumigaclavine-C
Penicillium expansum	Expansol A, B,	Cytotoxicity	https://pubchem.ncbi.nlm.nih.gov/compound/Expansol-A; https://pubchem.ncbi.nlm.nih.gov/compound/Expansol-B
Penicillium citrinum	(3s)-3–5-dimethyl-8-methoxy-3–4-dihydro-1Hisochromen-6-ol	Cytotoxicity; anticoccidial activity	https://pubchem.ncbi.nlm.nih.gov/compound/14582205
Trichoderma koningii	Trichodermaketone A	Antifungal	https://pubchem.ncbi.nlm.nih.gov/compound/46833401

(continued)

Table 3 (continued)

Name of Fungi	Compound	Bioactivity	Source of information
Hypocrea vinosa	Sterigmatocystin, O-methylsterigmatocystin, chaetochromin, chaetocin, chetomin, cochliodinols, and mollicellin G	Cytotoxicity	https://pubmed.ncbi.nlm.nih.gov/7296410/
Aspergillus versicolor	Averatin, Methyl-averatin, Nidurufin, anthraquinones	Antibacterial, antiviral, algicide, antifungal, immunostimulant, anticancer, cytotoxic	https://www.mdpi.com/2072-6651/12/11/714
Aspergillus fumigatus	Bismethylthio) gliotoxin 6-methoxyspirotryprostatin B	Antiparasitic	https://academicjournals.org/journal/AJBR/article-full-text-pdf/1B084C241856
Aspergillus ostianus	Circundatin A* Circundatin B* Circundatin E* Circundatin H*	Antibacterial	https://academicjournals.org/journal/AJBR/article-full-text-pdf/1B084C241856
Aspergillus flavus	Desmethylnomifensine	Antibacterial	https://academicjournals.org/journal/AJBR/article-full-text-pdf/1B084C241856
Cladosporium sp. F14	Cinnamic acid, 2-phenylethano cyclo-phe-Pro), Cyclo-Val-Pro)	Antibacterial, antioxidant	https://academicjournals.org/journal/AJBR/article-full-text-pdf/1B084C241856
Fusarium sp	Fusaquinon A Fusaquinon B Fusaquinon C	Cytotoxicity	https://academicjournals.org/journal/AJBR/article-full-text-pdf/1B084C241856

(continued)

Table 3 (continued)

Name of Fungi	Compound	Bioactivity	Source of information
Nodulisporium sp. CRIF1	Nodulisporacid A Vermelhotin*	Antiplasmodial	https://academicjournals.org/journal/AJBR/article-full-text-pdf/1B084C241856
Penicillium sp.	Chromanone A	Cytotoxicity	https://academicjournals.org/journal/AJBR/article-full-text-pdf/1B084C241856
Chrysosporium	Chrysosporium synchronum 1-O-a-syl) chlorogentisyl alcohol	Scavenging	https://academicjournals.org/journal/AJBR/article-full-text-pdf/1B084C241856
Chondrostereum sp.	Hirsutanol A	Cytotoxicity	https://academicjournals.org/journal/AJBR/article-full-text-pdf/1B084C241856
Aspergillus sp.	Asperxanthone Asperbiphenyl	Antiviral	https://academicjournals.org/journal/AJBR/article-full-text-pdf/1B084C241856
Aspergillus sp. SF-5044	Protuboxepin A	Cytotoxicity	https://academicjournals.org/journal/AJBR/article-full-text-pdf/1B084C241856

(continued)

Table 3 (continued)

Name of Fungi	Compound	Bioactivity	Source of information
Rhizopus sp.	2-PDA-61 Aspericin A Aspericin B Aspericin C	Cytotoxicity	https://academicjournals.org/journal/AJBR/article-full-text-pdf/1B084C241856
Aspergillus sp.	SF-5044 Protuboxepin A	Cytotoxicity	https://academicjournals.org/journal/AJBR/article-full-text-pdf/1B084C241856
Aspergillus sp.	Aspergilone A	Cytotoxicity antifouling	https://academicjournals.org/journal/AJBR/article-full-text-pdf/1B084C241856
Cladosporium sp.	F14 Cladospolide E	Antibacterial	https://academicjournals.org/journal/AJBR/article-full-text-pdf/1B084C241856
Coniothyrium cereale	Z)-coniosclerodinol coniosclerodione coniolactone	Antibacterial	https://academicjournals.org/journal/AJBR/article-full-text-pdf/1B084C241856
Phomopsis sp.	ZH-111 3R,4S)-3,4-Dihydro-4,5,8-trihydroxy-3-methylisocoumarin	Cytotoxicity	https://academicjournals.org/journal/AJBR/article-full-text-pdf/1B084C241856

(continued)

Table 3 (continued)

Name of Fungi	Compound	Bioactivity	Source of information
Penicilium citreonigrum	–	Neuroactivity	https://academicjournals.org/journal/AJBR/article-full-text-pdf/1B084C241856
Phoma sp., Phomaexigua var exigua, Scopulariopsis sp., Chrysosporium queenslandicum, Pycnidiophora dispersa	Secondary metabolites	Neuroactivity	https://academicjournals.org/journal/AJBR/article-full-text-pdf/1B084C241856
Penicillium chrysogenum QEN-24S	Penicisteroid A	Cytotoxicity Antifungal	https://academicjournals.org/journal/AJBR/article-full-text-pdf/1B084C241856
Antifungal compounds			
Penicillium cf. montanense & Neopetrosia chaliniformis	Xestodecalactone B	Anti-fungal	https://pubchem.ncbi.nlm.nih.gov/compound/Xestodecalactone-B
Paraphaeosphaeria sp.	Ascosalipyrrolidinone A	Anti-fungal	https://doi.org/10.1021/jo000307g
Ascochyta salicorniae	2,3-Dihydro-2-hydroxy-2,4-dimethyl-5-transpropenyfuran-3-one; 3,6,8-trihydroxy-3-(3 5-dimethyl-2-oxo-3	Anti-fungal	https://doi.org/10.1021/jo000307g
Keissleriella sp. YS4 108	E)-heptenyl)-2 3-dihydronaphthelen-14H)-one	Anti-fungal	https://www.ncbi.nlm.nih.gov/pmc/articles/PMC10607872/

(continued)

Table 3 (continued)

Name of Fungi	Compound	Bioactivity	Source of information
Zopfiella latipes	Zoptiella latipes A, B (Pyrrolidinone derivative)	Anti-fungal	https://www.lunduniversity.lu.se/lup/publication/db99e06a-a754-432f-baff-b4403ef1e4e7
Microsphaeropsis sp. (from sponge *Myxilla incrustans*)	Microsphaeropsin 41 (Eremophilane derivative)	Anti-fungal	https://www.sciencedirect.com/topics/chemistry/eremophilane
Bacillus licheniformis	Ieodoglucomide C ieodoglycolipid	Anti-microbial	https://doi.org/10.3390/toxins15020093
Bacillus laterosporus	Basiliskamides A Basiliskamides B	Anti-microbial	https://www.ncbi.nlm.nih.gov/pmc/articles/PMC9966175/
Haliangium luteum	Haliangicin	Anti-microbial	https://www.ncbi.nlm.nih.gov/pmc/articles/PMC9966175/
Bacillus subtilis 109GGC020	Gageostatin A	Anti-microbial	https://www.ncbi.nlm.nih.gov/pmc/articles/PMC9966175/
Epilithic cyanobacteria	Hassallidin A	Anti-microbial	https://www.ncbi.nlm.nih.gov/pmc/articles/PMC9966175/

Table 4 Details of compounds and their benefits from selected microalgae (adopted from Mobin & Alam, 2017)

Name of speicies	Compounds	Benefit
Arthrospira (Spirulina) platensis	Phycocyanin, biomass	Health food, cosmetics
Arthrospira (Spirulina)	Protein, vitamin B12	Antioxidant capsule, immune system
Aphanizomenon flos-aquae	Protein, essential fatty acids, β-carotene	Health food, food supplement
Chlorella spp.	Biomass, carbohydrate extract	Animal nutrition, health drinks, food supplement
Dunaliella salina	Carotenoids, β-carotene	Health food, food supplement
Haematococcus pluvialis	Carotenoids, astaxanthin	Health food, food supplement
Odontella aurita	Fatty acid, EPA	Pharmaceuticals, cosmetics, anti-inflammatory
Porphyridium cruentum	Polysaccharides	Pharmaceuticals, cosmetics
Isochrysis galbana	Fatty acids	Animal nutrition
Phaeodactylum triconutum	Lipids, fatty acids	Nutrition, fuel production
Lyngbya majuscule	Immune modulators	Pharmaceuticals, nutrition
Schizochytrium sp.	DHA and EPA	Food, beverage, and food supplement
Crypthecodinium cohnii	DHA	Brain development, infant health, and nutrition
Nannochloropsis oculata	Biomass	Food for larval and juvenile marine
Spirulina sp.	Bioactive compounds	Food and many other benefits
Crypthecodinium cohnii	DHA	Infant formula
Schizochytrium	DHA	Cardio-vascular health

and endurance. Ahmed (2024) has given a comprehensive account on bio-active compounds of microalgae.

Marine Seaweeds, Bio-Active Compounds, and Potential Activity

Seaweeds contain proteins, vitamins, minerals and trace elements and rich source of iodine. From this bio-resource many bioactive compounds important as antioxidant, anti-inflammatory and anti-cancer and anti-diabetic properties have been derived (Park et al., 2023). Table 5 provides general characteristic of various mineral content in seaweeds and Table 6 elaborates on bio-active compounds from seaweeds.

Table 5 Generalized picture of mineral content in some sea lettuce (mg/100g Edible portion) (Mac Artain et al., 2007; McDermid & Stuercke, 2003; Ratana-Arporn & Chirapart, 2006)

Minerals	*Ulva lactuca*	*Ulva reticulata*	*Ulva fasciata*
Ca	32.5	140	0.47
K	24.5	1540	2.87
Mg	46.5	140	2.19
Na	34.0	-	-
Cu	0.03	0.6	5
Fe	1.53	174.8	86
I	0.16	1.124	–
Zn	0.09	3.3	9
M		–	3.62
P		180	0.22
S		–	5.24
B		–	77
Mn		48.1	12

Table 6 Details of bio-active compounds from selected sea weeds and their activity

Name of species	Compounds	Activity	Reference
Cymopolia barbata	Cymopol	Antioxidant	https://www.ncbi.nlm.nih.gov/pmc/articles/PMC9004519/
Avraiinvillea longicaulis	Avrainvilleol		
Halimeda opuntia	Polyophenols		
Undaria pinnatifida	Fucoxanthin		
Laureafincia claviformis	Pacifenol	Anti-inflammatory	https://www.ncbi.nlm.nih.gov/pmc/articles/PMC3985253/
Stypopodium flabelliforme	Styfotriol		
Undaria pinnatifida	Fucoidon		
Gracilaria corticata	Polysaccharides	Anti-cancer	https://www.ncbi.nlm.nih.gov/pmc/articles/PMC4178489/
Gracilaria tenuistipitata			
Plocamium telfairiae			
Capsosiphon fulvescens			
Sargassum oligocystum			
Enteromorpha flexuosa	Bioactive compound	Anti-viral	Velmurugan et al., 2015

Marine Cnidaria, Bio-Active Compounds and Their Potential Activity

Sea anemones, corals, sea pens, jellyfishes etc. belong to Cnidarians and among which corals are more important in biomaterial production. Of the marine natural products derived from this group, terpenoids constitute to 67%, followed by steroids (21%), and, to a much lower extent, alkaloids (5%) and other compounds (6%) (Romano et al., 2022). Currently, 5761 compounds have been described from cnidarians of which 42% showed cytotoxicity properties, 29% anti-inflammatory, 8% antimicrobial, 7% anti-fouling and 4% anti-viral (Collins et al., 2013). Green et al. (2013) pointed out how well coral skeleton elements work to repair broken bones. The table below gives details of selected species and their bioactive compounds (Table 7).

Marine Sponges, Bio-Active Compounds, and Their Potential Activity

Sponges are rich in bio-active compounds (McConnell et al., 1994) and the research interest started with the first introduction of first drug (Cytosar-U®) on market (Cytarabine) derived from it (Bergmann & Feeney, 1951; Romano et al., 2022). It was derived from the Caribbean sponge *Tectitethya crypta.* Since then, roughly 2700 or more natural products have been isolated from marine sponges. Chitin, a polysaccharide is produced by *Lanthella labyrinthus, L. flabelliformis, L. basta, Pseudoceratina purpurea, P. arabica, Mycale euplectelloides, Acarnus wolffgangi, Echinoclathria gibbosa, Aplysina aerophoba, A. cavernicola, A. archeri, A. cauliformis, A. fulva, Aiolochroia crassa, Plysina aerophoba.* Hydroxyapatite, a calcium phosphate mineral, produced by *Spongia agaricina, S. officianalis, S. zimocca, Petrosia ficidormis, Chondrosia reniformis, Agelas oroides. Spongia lamella, S. officianalis, Hippospongia communis, Sarcotragus spinosulus, Chondrosia remiformis, Biemna* fortis, Ircinia fusca and members of family Callyspongiidae produce collagen (see Romano et al., 2022). Over sixty-five percent of the biomaterials were initially classified as bioactive. Terpenes (35%) alkaloids (29%) and lipids (22%) were the three most common structural groups, with peptides, polyketides, and miscellaneous molecules making up the remaining 14% of the compounds (Kuznetsov et al., 2009; Costantini et al., 2017). It is highly interesting to point out that about 50% of the bioactive compounds showed cytotoxicity towards tumor cell lines (Romano et al., 2022). Varijakzhan et al. (2021) have reported that they also produce antimicrobial compounds. Additional information is listed in the Table 8.

Table 7 Consolidation of Bioactive compounds with medicinal qualities established in selected coral species (various sources)

Species	Molecule	Beneficial use and reference (mainly with experimental animals and man)
Sinularia flexibilis	Flexibiliquinone Flexibilin D	Anti-inflammation (Hu et al., 2013; Lin et al., 2013)
	11-dehydrosinulariolide	Promising molecule for the treatment of Parkinson's disease through its anti-apoptotic and anti-inflammatory action via PI3K signaling (Chen et al., 2011)
S. querciformis	11-epi-Sinulariolide Acetate	Anti-inflammation and anti-bone-loss (Lin et al., 2013)
	Silurin	Antinociception and Antineuro inflammation (Huang et al., 2012)
S. gaweli	5α,8α-Epidioxysterol	Cytotoxicity (Yen et al., 2013)
S. crassa	Crassalone A	Cytotoxicity (Cheng et al., 2012)
S. granosa	9,11-Secosterol	Cytotoxicity and anti-inflammation (Huang et al., 2012)
S. crassa	Crassalone A	Cytotoxicity (Cheng et al.,2012)
S. granosa	9,11-Secosterol	Cytotoxicity and anti-inflammation (Huang et al., 2012)
Sinularia sp.	5-Epsinuleptolide acetate (5EPA)	Cytotoxicity (Huang et al., 2013)
Scleronephthya gracillimum	Sclerosteroids	Cytotoxicity and anti-inflammation (Fang et al., 2013)
Cladiella krempfi	Kremptielins	Cytotoxicity and anti-inflammation (Tai et al., 2013)
Cladiella australis	Austrasulfone	Neuroprotective qualities via antiinflammatory pathways; atherosclerosis and multiple sclerosis (Cooper et al., 2014)
Paraminabea acronocephala	Paraminabic acid	Cytotoxicity and anti-inflammation (Chao et al., 2013)
Capnella imbricata	Capnellenes	Antineuroinflammation and antinociception (Jean et al., 2009)
Lemnalia cervicomi tenuis	Lemnalol	Anti-inflammation and antinociception (Lee et al., 2013)
Echinomuricea sp.	Echinoclerodane A	Cytotoxicity and anti-inflammation (Cheng et al., 2012)
Nephthea chabrolii	Nebrosteroids Q, R, and S	Cytotoxicity (Wang et al., 2013)
Sarcophyton crassocaule	13-acetoxysarcocrassolide	Cytotoxicity (Su et al., 2011a, 2011b)

(continued)

Table 7 (continued)

Species	Molecule	Beneficial use and reference (mainly with experimental animals and man)
Lobophytum michaelae	Michaolides and lobomichaolide	Antitumor capabilities, activity against cytomegalovirus (HCMV), related to herpesviruses (Wang & Duh, 2012)
Pleasistrea versipora	Clotrimazole	Fungicide https://www.webmd.com/drugs/2/drug-4316/clotrimazole-topical/details
Palythoa tuberulosa	Pyrazine moiety, tricyclic tryptamine derivative	Anti-cancer and anti-lymphangiogenesis (Chen et al., 2018a, 2018b)
Klyxum simplex	Simplexin E	Anti-inflammatory, antitumor (https://pubs.acs.org/doi/https://doi.org/10.1021/np900064a)
Lophytum sp.	Lobophytene, Lobohedleolide, (7Z) dimethylamino lobohedleolide, 17-dimethyllamino lobohedleolide	Antitumor; Anti-HIV https://pubchem.ncbi.nlm.nih.gov/compound/6475568
Sarcophyton crassocaule	Crassocolides H-M	Antitumor https://www.ncbi.nlm.nih.gov/pmc/articles/PMC6723858/

Marine Annelida, Bio-Active Compounds, and Their Potential Activity

Under Annelida the most studied group for nutraceuticals is Polychaeta and families. Chemical aspects of products and their applications has been studied in the families: Amphinomidae (4 species), Capitellidae (7), Chaetopteridae (2), Cirratulidae (3), Eunicidae (2), Glyceridae (4), Lumbrineridae (2), Nephytidae (1), Nereididae (5), Omphidae (3) Orbiniidae (1), Pectinaridae (1), scalibregmatidae (1), Spionidae (5), syllidae (1) and Terebellidae (6). From these species the chemical compounds derived are: Alkaloids, Halogenated aromatics Peptide, Enzyme, Pyrrole ring substances, Glycoprotein, Marine amines, Derived from lumazine, Enzyme. These bioactive compounds are effective for antipredation, antiepibiotic, antisettlement, alelochemical, antimicrobial, antirecruitment, catalyst, antipredation, antibacterial, neurotoxic, and enzymatic, bioluminescence, allelochemical, antifungal (Coutinho, 2017).

Table 8 Details of bioactive compounds derived from selected sponges and their bio-activity

Species	Compound	Bioactivity and reference
Luffarella variabilis Cocospongia scalaris	Sesterpenoid antibiotic	Anti-microbial (see Datta et al., 2015)
Tectitethya crypta	C-nucleosides, spongouridine, spongothymidine	Antiviral; synthetic analog for anti-cancer (McConnell et al., 1994)
Agelas nakamurai	Monocyclic and bicyclic diterpenoids	Inhibitory effect on Na, K-ATPase (see Datta et al., 2015)
Discodermia dissolute	Discodermlide	Anti-cancer drug (see Datta et al., 2015)
Hemiasterella minor	HTI-286 (synthetic analog)	Anti-cancer drug (see Datta et al., 2015)
Cymbastela sp.	Halichondrin B	Anti-cancer property (see Datta et al., 2015
Toxadocia zumi	Sterol	Anti-bacterial (*Staphylococcus aureus, Bacillus subtitis*) (see Datta et al., 2015)
Plakina sp.	Plakinamine A, B	Anti-microbial (see Datta et al., 2015)
Disidea avara, D. pallescens	Sesquiterpenoid and avarol	Anti-microbial and against AIDS virus (see Datta et al., 2015)
Halichondria panacea	Triprenyl phenols	Anti-microbials (see Datta et al., 2015)
Phyllospongia foliascens	Sesterterpene, Phyllofenone A	Antifungal (see Datta et al., 2015)
Aplysinopsis reticulata	Methyl-aplysinopsin	Inhibitor of monoamine oxidase (see Datta et al., 2015)
Theonella swinhoei	Bioactive compounds	Antibacterial, antiviral, and antifungal (Varijakzhan et al., 2021)
Dactylospongia elegans	Sesquiterpenoid aminoquinone nakijiquinone V, illimaquinone, smenospongine, dyctioceratine C	Anti-microbial and cytotoxicity (Balansa et al., 2019)
Xestospongia sp.	5-O-Acetyl-Renieramycin T	Induces Lung Cancer Stem Cell Apoptosis (Chantarawong et al., 2019)
Raspailia bouryesnaultae	Clerodane diterpenes	Anti-proliferative effects; anti-herpes activity (Lhullier et al., 2019)

Marine Crustacea, Bio-Active Compounds, and Their Potential Activity

Chitosan and chitin: Chitosan is the term used when the Nitrogen content in chitin is more than 7% by weight or the degree of deacetylation is more the 60% (Gagne & Simpson, 1993). Chitin is abundant in all crustaceans and cuttlefish. It is known for its biodegradability, non-toxicity, physiological inertness, anti-bacterial properties, hydrophilicity, gel-forming properties (Se-Kwon, 2010). Chitosan has applications in areas, such as, food and nutrition, biotechnology, material science, drugs and pharmaceuticals, dental and surgical appliances, agriculture, environmental protection such as removal of toxic heavy metals, industrial effluent treatment, wine clarification, and many other areas (Se-Kwon, 2010). A natural biopolymer made of N-acetyl-d-flucosamine and β (1–4) glyosidic links, chitin is a cationic amino polysaccharide.

Glucosamine Hydrochloride: It is obtained from the exoskeleton of crustaceans. It achieved importance on realizing that this biomolecule has promising properties for treatment of osteoarthritis, knee pain, and back pain in addition to dietary supplementation (Houpt et al., 1999). Nagaoka et al. (2011) reported that exhibits properties like anti-cancer and anti-inflammatory effects.

Marine Mollusca, Bio-Active Compounds, and Their Potential Activity

Among molluscs, cephalopods, bivalves, and gastropods are reported to contain bioactive compounds. In 1970 research started on the gastropod, *Conus,* to derive bioactive compounds and the first venom component derived characterized as alpha-conotoxin GI which proved to be competitive nicotinic antagonist. Olivera (2000) estimated more than 80,000 natural conotoxins that might exist in various cone shells around the world and each *Conus* species may possess on an average of 100–200 of the same. Chakraborty and Joy (2020) consolidated around 1334 bioactive metabolites in molluscs based on the literature analysis during the period, 1984–2019. Terpene comprised 55% (from gastropods) and sterols, 41% (from bivalves) and compounds with anti-inflammatory, cytotoxic, and anti-proliferative properties against mammalian cell lines. Ziconotide was used as analgesic whereas elisidepsin for antineoplastic treatment. The period 2016–2019 showed increasing importance of bioactive compounds from molluscs.

Cephalopods were accounted for many minerals, vitamins along with essential amino acids including anti-oxidative selenium. It is rich in protein, EPA, DHA (Chakraborty et al., 2016; Zlatanos et al., 2006). The findings indicated that oysters, mussels, and squid had greater levels of PUFA and EPA. It has been established that EPA and DHA are essential for many physiological and metabolic processes in humans. Long chain n-3 PUFAs were found to be inevitable to reduce the health problems, such as hypertension, asthma, cardiovascular illnesses, inflammation, type-2 diabetes, and cancer (Simopoulos, 2009).

Marine Echinoderms, Bio-Active Compounds, and Their Potential Activity

A wide range of bioactive substances, including glycosides, acid polysaccharides, sphingoid, glycolipids, sulphate, polysaccharides, and phospholipids, are found in this phylum. Various biological properties, including antibacterial, anti-inflammatory, and anticancer effects, have been demonstrated by the extracts. The most studied bioactive compounds were the triterpene glycosides and steroids extracted from the starfishes and sea cucumbers (Gomes et al., 2016).

Marine Ascidians and Tunicates, Bio-Active Compounds, and Their Potential Activity

Ascidians are chemically prolific marine animals because more than 1200 interesting natural products, such as alkaloids (Didemnidines, Meridianins, Herdmanines), polypeptides (Vitilevuamide, Diazonamides, Chondromodulin-1 (ChM-1), Didemins (Didemins A-C, Patellamides) and polyketides (Palmerolide A, Mandelides, Phosphoeleganin). Some have been produced as lead compounds or very effective medications. It is of interest to note that the symbionts (Proteobacteria, Cyanobacteria, Prochloron sp., Actinomycetes, fungi) of ascidians do produce very promising bio-active compounds, such as, alkaloids (Ecteinascidins, Eusynstyelanides, Sesbanimides, LL-141352 α and β), Polyketides (Patellazoles, Arenimycin). Compounds with cytotoxic and antineoplastic properties isolated from ascidians belong to disparate chemical classes, and three of them have been entered into clinical trials (Duo & Dong, 2019).

Tunicates are prolific producers of marine natural products. The species are known to produce bioactive compounds, such as, Polycarpine, Tuberatolides, Sodium 1-(12-hydroxy) octadecanyl sulfate, Patellazole B, C, Ascididemin, Trabectedin, Cystodytin D, E, F, G, H, I, Diplosoma zylidene 1, 2, Lejimalide A, B, Lissoclibadin 1, 2, 3, Lamellarin alpha 20-sulfate, Plitidepsin, GnRHI, II, Retinoic acid and are effective to cytotoxicity, antiviral and antifungal, Furnesoid X receptor antagonists, matrix metalloproteinase 2 inhibitor, Regeneration of gut activities (Chen et al., 2018b; Ramesh et al., 2021).

Marine Fishes, Bio-Active Compounds, and Their Potential Activity

Marine fish are regarded as being extremely significant for human health as they are a nutritious food option and supply our diet with molecules that play a critical therapeutic role in the prevention of human diseases. Even the waste from sea food can include valuable ingredients like collagen, peptides, and enzymes that could improve human health or be utilized to create new drugs (Šimat, 2021). It is necessary for the world economy. The nutritional advantages of eating fish are attributed to the availability of minerals (such as calcium, iron, zinc, and selenium), proteins, vitamins (such as B3, B6, and B12), and unsaturated essential fatty acids (Yamada et al., 2014). It has also been shown that peptides produced from fermented fish following enzymatic treatment are useful treatments for conditions like hypertension, cancer, viral infections and Alzheimer's disease (Suleria et al., 2015).

Fish derivatives have the potential to target cancer cells without harming normal tissues, prefer it for cancer prevention and therapy (Asih et al., 2018; Mirunalini & Maruthanila, 2022). More details about fishes are exemplified in Tables 9, 10.

Drugs Derived from Marine Origin

The following Table summarizes the drug derived from marine sources. Status of approval also given in the table (Table 11).

Marine Drugs for Cancer Therapy

Zuo and Kwok (2021) gave an excellent review on marine drugs for cancer therapy. It is reproduced here as it is important in this chapter (Table 12).

Consolidation of Selected Bioactive Compounds from Marine Bioresources and Their Effects

The data provided above shows the importance of marine bio-resources for derivation of various bioactive compounds effective in controlling diseases and hence a few important ones are consolidated below.

Anticancer: Aureoverticillactam, Caprolactones, Chinikomycins, IB-00208, Salinosporamide A (NPI-0052), Urdamycin, Himastatin, Daryamide D, Marinomycin, Manumycin, Marmycin, Nonactin, Chartreusin, Altemicidin, Streptochlorin,

Table 9 Proximate, mineral and fish oil components as a representation of content of a typical fish (from Awuchi et al., 2022)

FISH	Proximate composition	Moisture (59.1–87.8%)	
		Protein (9–24%)	
		Lipid (1.6–6.2%)	
		Ash	
	Mineral Composition/100g	Sodium (190 mg)	
		Magnesium (38 mg)	
		Zinc (72 mg)	
		Pottassium (384 mg)	
		Phosphorous (210 mg)	
		Iodine (30 mcg)	
		Calcium (15 mg)	
		Irom (0.3 mg)	
		Selenium (40–92 mcg)	
Composition of fish oil	Fatty acid composition	PUFA	Eicosapentaeonic acid
			Docosahexaenoic acid
			Arachidonic acid
		MUFA	Gondoic acid
			Palmitoleic acid
			Oleic acid
		SFA	Steric acid
			Palmictic acid
	Other composition	Sterols (Cholesterol, Brassicasterol, desmosterol)	
		Vitamins A, E	
		Minerals (Calcium, Potassium, Magnesium, Selenium etc.)	
		Polyphenols (flavonoids)	
		Pigments (Carotenoids)	

Marineosins, Ammosamides, Caboxamycin, Hoiamide D, Niphateolide, Hexylitaconic acid, Lissoclinidine B, Himeic acid A, Girolline, Leucettamol A, Dysidiolide, Sulfircin, Coscinosulfate, Halenaquinone, Secalonic acid D, Stellettin B, Ulapualide-A, Dercitin, Halichondrin-B, Onnamide-A, Cribostatin, Undecylprodigiosin, Prodigiosin, 1-Hydroxy-1-Norresistomycin, Violacein, Phenazine derivates, Chinikomycin A and Chinikomycin B, Manumycin A.

Anati-inflammatory: Diazepinomicin (ECO-4601), Cyclomarins, 3-O-b-glucopyranosylstigmasta-5,25-diene, Polysaccharides, lysolecithin, phenolic components.

Neurogenic activity: Komodoquinone A, Granaticins, Hymenialdisin.

Table 10 Details of nutraceuticals from selected marine fishes and their importance for health (based on Awuchi et al., 2022)

Nutraceutical compounds	Name of fish species	Health importance	References
Omega 3	Spiny dogfish, mackerel, salmon, sardines	Anti-inflammatory and cardio-protective effects, and effective for various cancers, improves insulin sensitivity	Johnson and Chastity (2014)
Omega-6	Sardines, anchovies	Effective to arthritis, hypertension, platelet aggregation, and cardiovascular problems	Johnson and Chastity (2014) Ashraf et al. (2020)
Iron and zinc	Sperata seenghala, Rita rita	Improvement of immune system, wound healing, breakdown of carbohydrate	Mohanty et al. (2016)
Vitamin A Vitamin D	Amblypharyngodon mola	Helps in bone growth and improves eyesight; avoid rickets	Tilami and Sampels (2018)
Arginine, isoleucine, tyrosine, proline	Oncorhynchus mykiss	Muscle development, detoxification of ammonia, precursor to enzymatic reactions	Mohanty et al. (2014)
Aspartic and glutamic acid	Labeo niloticus	Improves immunity and redeem chronic fatigue and flavour enhancer	Saad et al. (2015)
Methionine	Stolephorus waitei	Cellular metabolism	Hesse et al. (2004)
Serine	Mugil cephalus	Homeostasis	Sinha et al. (2020)
Alanine	Polypedates maculates	Biosynthesis of proteins	Mohanty et al. (2014)
Leucine	Lethrinus harak	Energy for protein synthesis	Duan et al. (2016)

Anti-tuberculosis: Rifampicin, Streptomycin, Amikacin, Viomycin, Capreomycin, Kanamycin, Cycloserine.

Anti-oxidant: Astaxanthin (carotene).

Antibiotic: Violacein, Methyl suphenate (phenazine derivative), Phenazine-1-carboxylic acid, 5, 10-dyhydrophencomycin methyl ester, Tambjamines (BE-18591, Tryptanthrin.

Antifungal: Amphotericin B, Bonactin, Iturin, Surfactin, Iturin and Fengycin.

Table 11 Approved Drugs from marine source

Compound name	Source	Chremical class	Disease area	Status
Compounds targeting ion channels				
Zinconotide (PrialtTM)	Cone snail	Peptide	Chronic pain	FDA approved 2004; EMA 2005
GTS-21	Nemertine worm	Anabaseine-derivative	Schizophrenia	Phase II (Academic) completed
GTS-21	Nemertine worm	Anabaseine-derivative	Alzheimer/ ADHD	Phase II
Compounds targeting enzymes				
Protein Kinase inhibitors				
Bryostatin-1	Byozoans	Polyketide	Cancer	No current clinical trials and Potential interest in Alzheimers
Proteasome inhibitor				
NPI-0052	Bacteria (Actinomycete)	Beta-lactone-gammalactum	Cancer	Phase 1
Microtubule interacting agents				
Dolastatin-10	Sea slug	Paptide	Cancer	Yet to be finalized
ILX-651	Sea slug	Peptide	Cancer	Preclinical as 4/2008, oral formulation being developed
E7389 (Eribulin)	Sponge	Halichondrin B analog	Cancer	Phase III for breast cancer USA & EU Phase II (USA) NSCLC; Phase II (US/EU) prostate, Phase II (EU) sarcoma
NPI-2358	Fungi	Diketopiperazine	Cancer	Phase II
TZT-1027 (Soblidotin) aka YHI-501	Sea slug	Peptide	Cancer	Phase III (possibly)—linked to monoclonals (Auristatin PE)
E-7974	Sponge	Tripeptide	Cancer	Phase I

(continued)

Table 11 (continued)

Compound name	Source	Chremical class	Disease area	Status
DNA and transcription interactive agents				
Ecteinascidin-743-Trabectedin (Yondelis ®)	Mollusc/sponge	Tetrahydroisoquinolone alkaloid	Cancer	Approved by EMA, Sept. 2007 Phase III ovarian concer completed file FDA/EMA Fall 08; Phase II prostate cancer ongoing; Phase II breast/lung cancer ongoing
PM00 104 (Zalypsis)	Mollusc/sponge	Synthetic alkaloid based on Jorumycin	Cancer	Late Phase I development
VEGF interacting agent				
Plitidepsin (Aplidin)	Sea squirt	Cyclic depsipeptide	Cancer	Phase II multiple myeloma completed; Phase II in T-NHL ongoing; combination studies going on
Lysosomotropic and ErbB interacting agent				
Kahalalide F	Sea slug/Algae	Cyclic depsipeptide	Cancer	Phase II completed insolid tumours; evaluated in severe psoriasis
PM02734 (Irvalec®)	Sea slug	Depsipeptide	Cancer	Phase I ongoing Phase I/II in combination with Erlontinib in lung under implementation
Compounds with unknown mechanism of action				
IPL-512602	Sponge	Steroid	Inflammation/ asthma	No further information
Compounds targeting GPCRs				
Pseudopterosin A-methyl ether (TMO)	Soft coral	Diterpene glycoside	Wound healing	Phase II

Table 12 Approved and/or Phase III, II, I and preclinical trial drugs from marine sources

Name of compound	Marine source	Chemical class	Area of activity	Reference
Clinical status: Approved				
Cytarabine (Ara-C, cytosine arabinoside)	Sponge	Nucleoside	Cancer, Leukemia	Wu et al. (2017)
Brentuximab vedotin (SGN-35)	Mollusc/ cyanobacterium	ADC (MMAE)	Cancer, Lymphoma	Katz et al. (2011)
Vidarabine, ara-A	Sponge	Nucleoside	Anti-viral	Shen (2009)
Omega-3 acid ethyl esters	Fishes	Omega-3 fattyacid	Hypertri-glyceridemia	Myo Clinic (2024)
Ziconotide	Cone snail	Peptide	Pain	McGivern (2007)
Eribulin mesylate (E7389)	Sponge	Macrolide	Breast cancer	Twelves (2010)
Trabectedin (ET-743)	Tunicate	Alkaloid	Cancer	D'Incalci and Galmarini (2010)
Polatuzumab vedotin	Mollusk/ cyanobacterium	ADC (MMAF)	Non-Hodgkin lymphoma, chronic lymphocytic leukemia, lymphoma,B-cell lymphoma, folicular	Zuo and Kwok (2021)
Enfortumab vedotin	Mollusk/ cyanobacterium	ADC (MMAF)	Metastatic urothelial cancer	Zuo and Kwok (2021)
Belantamab mafodotin	Mollusk/ cyanobacterium	ADC (MMAF)	Relapsed/refractory multiple myeloma	Zuo and Kwok (2021)
Lurbinectedin	Tunicate	Alkaloid	Metastatic small-cell lung cancer	Zuo and Kwok (2021)
Fludarabine	Sponge	Nucleoside	Leukemia	Cappello and Nieri (2021)

(continued)

Table 12 (continued)

Name of compound	Marine source	Chemical class	Area of activity	Reference
Nelarabine	Sponge	Nucleoside	Leukemia	Cappello and Nieri (2021)
Phase III clinical trials				
Plitidepsin	Tunicate	Depsipeptide	Cancer, antiviral against COVID	Papapanou (2021)
Tetrodotoxin	Puffer fish	Guanidnium alkaloid	Chronic pain As an analgesic to cancer patient	Lago et al. (2015)
Soblidotin (TZT 1027)	Bacterium	Peptide	Cancer	Watanabe (2006)
Marizomib	Bacterium	β-lactone-γ lactam	Non-small-cell lung cancer, pancreatic cancer, melanoma, lymphoma, multiple myeloma	Zuo and Kwok (2021)
Plinabulin	Fungus	Diketopiperazine	Non-small-cell lung cancer, brain tumor	Zuo and Kwok (2021)
Phase II Clinical trials				
DMXBA (GTS-21)	Nemertid Worm	Alkaloid	Cognition, Alzheimer's disease, Schizophrenia	Kem (2006)
Plinabulin (NPI-2358)	Fungus	Diketopiperazine	Cancer	Al-Shakliah et al. (2022)
Glembatumumab vedotin	Mollusc/cyanobacterium	ADC (MMAE)	Breast cancer, melanoma	Vahdat, et al. (2021)
Elisidepsin	Mollusc	Depsipeptide	Cancer	Serova, et al. (2013)
PM1004	Nudibranch, *Jorunna funebris*	Alkaloid	Cancer	Fontana et al. (2000)

(continued)

Table 12 (continued)

Name of compound	Marine source	Chemical class	Area of activity	Reference
Tasidotin, synthadotin (ILX-651)	Bacterium	Peptide	Cancer	Ebbinghaus et al. (2004)
Pseudopterosins	Soft coral	Diterpene glycoside	Would healing	Caplan (2016)
W0101	Mollusk/ cyanobacterium	ADC (MMAE)	Advanced or metastatic solid tumors	Zuo and Kwok (2021)
CX-2029	Mollusk/ cyanobacterium	ADC (MMAE)	Solid tumor, head and neck cancer, non-small-cell lung cancer, pancreatic cancer, diffuse large B-cell lymphoma	Zuo and Kwok (2021)
CAB-ROR2	Mollusk/ cyanobacterium	ADC (MMAE)	Solid tumor, non-small-cell lung cancer, triple-negative breast cancer, soft tissue sarcoma	Zuo and Kwok (2021)
RC48	Mollusk/ cyanobacterium	ADC (MMAE)	Urothelial carcinoma, advanced cancer, gastric cancer, HER2-over expressing gastric carcinoma, advanced breast cancer, solid tumors	Zuo and Kwok (2021)
Enapotamab vedotin	Mollusk/ cyanobacterium	ADC (MMAE)	Ovarian cancer, cervical cancer, endometrial cancer	Zuo and Kwok (2021)
Telisotuzumab vedotin	Mollusk/ cyanobacterium	ADC (MMAE)	Solid tumors	Zuo and Kwok (2021)
Ladiratuzumab vedotin	Mollusk/ cyanobacterium	ADC (MMAE)	Breast cancer	Zuo and Kwok (2021)
Tisotumab vedotin	Mollusk/ cyanobacterium	ADC (MMAE)	Prostate cancer (CRPC), cancer of head and neck (SCCHN), Ovary cancer, cervix cancer, endometrium cancer, bladder cancer, esophagus cancer, lung cancer (NSCLC)	Zuo and Kwok (2021)
AGS-16C3F	Mollusk/ cyanobacterium	ADC (MMAE)	Renal cell carcinoma	Zuo and Kwok (2021)
Plocabulin	Sponge	Polyketide	Solid tumors	Zuo and Kwok (2021)

(continued)

Table 12 (continued)

Name of compound	Marine source	Chemical class	Area of activity	Reference
Phase I Clinical trials				
Bryostatin 1	Bryozoa	Polyketide	Multiple sclerosis,stroke, traumatic brain damage, Alzheimer's disease, fragile X syndrome,and depression	Tian et al. (2023)
Pinatuzumab vedotin (DCD-2980S) and (DCD-4501A)	Mollusc/ cynanobacterium	ADC (MME)	Non-Hodgkin Lymphoma, Chronic lymphocytic leukemia	Advani et al. (2017)
Hemiasterlin (E7974)	Sponge	Tripeptide	Cancer	Galina et al. (2009)
HuMax ®—TF-ADC	Mollusc/ cyanobacterium	ADC (MMAE)	Ovarian cancer, endometrium, cervix, prostate	Zuo and Kwok (2021)
MORAb-202	Sponge	ADC (macrolide)	Solid tumors	Zuo et al., 2021
XMT-1536	Mollusc/ cyanobacterium	ADC (dolaflexin)	Solid tumors	Zuo et al., 2021
RF06804103	Mollusc/ cyanobacterium	ADC (auristatin variant)	Carcinoma, non-small-cell lung cancer, stomach neoplasms, esophagogastric junction neoplasms, and breast neoplasms	Zuo et al., 2021
ARX-788	Mollusc/ cyanobacterium	ADC (MMAE)	Breast cancer, gastric cancer	Zuo and Kwok (2021)
ALT-P7	Mollusc/ cyanobacterium	ADC (MMAE)	Breast cancer, gastric cancer	Zuo and Kwok (2021)
ZW49	Mollusc/ cyanobacterium	ADC (auristatin variant)	HER2-expressing cancers	Zuo and Kwok (2021)

(continued)

Table 12 (continued)

Name of compound	Marine source	Chemical class	Area of activity	Reference
Marizomib (salinosporamide A)	Bacterium	Beta-lactone-gamma lactam	Potential anticancer agent	https://en.wikipedia.org/wiki/Salinosporamide_A
Preclinical				
Chrysophaentin A	Alga (*Halobacillus salinus*)	Shikimate		Plaza et al. (2010)
Phenethylamine	Bacterium Lyngbyoic acid	Shikimate		https://en.wikipedia.org/wiki/Phenethylamine
Geodisterol sulfates	Sponge	Peptide		Costantino et al. (2017)
Pseudoalteromonas sp. metabolites	Bacterium	Polyketide		Offret et al. (2016)
Peziza vesiculosa β-carboline	Bryozoa	Alkaloid		Malve (2015)
Bromophycolides	Alga	Terpene		Cabrita et al. (2010)
Plakortin	Sponge	Polyketide		Sala et al. (2013)
Homogentisic acid	Sponge	Shikimate		Lebouvier et al. (2009)
Cladonia cervicornis diterpene	Alga	Terpene		Chen et al., (2018a, 2018b)
Hymenidin	Sponge	Alkaloid		Kobayashi (1986)
Gyrosanols	Soft coral	Terpene		https://pubmed.ncbi.nlm.nih.gov/20499851/
Dysidine	Sponge	Terpene		https://pubchem.ncbi.nlm.nih.gov/compound/Dysidine
Arenamides A and B	Bacteria	Peptide		www.sciencedirect.com/science/article/abs/pii/S0957416614000317
Capnellene	Soft coral	Terpene		https://en.wikipedia.org/wiki/Capnellene

(continued)

Table 12 (continued)

Name of compound	Marine source	Chemical class	Area of activity	Reference
Floridosides	Alga	Glycolipid		https://amb-express.springeropen.com/articles/https://doi.org/10.1186/s13568-016-0244-6
Grassystatins A-C	Bacteria	Peptide		Kwan et al. (2009)
Callyspongidiol	Sponge	Polyketide		https://www.ncbi.nlm.nih.gov/pmc/articles/PMC9964934/
Calyculin A	Sponge	PKS/NRPS		https://en.wikipedia.org/wiki/Calyculin
Pulicatin A	Bacteria	Alkaloid		https://pubchem.ncbi.nlm.nih.gov/compound/Pulicatin-A
Dysideamine	Sponge	Terpene		Suna et al. (2009)

Antimicrobial: Abyssomicins, Frigocyclinone, Gutingimycin, Helquinoline, Himalomycins, Lajollamycin, Tylosin, Maklamicin, Lobophorin K, Asenjonamide C, Gilvocarcin HE, Zunyimycins, Formicamycins A, Allocyclinones, Ageloline A, Buanmycin, Citreamicin, Kocurin, Fijimycins, Arylomycin, 1-Hydroxy-1-Norresistomycin, Prodigiosin, Bonactin, 2-alkylidene-5-alkyl-4-oxazolidinones, lipoxazolidinone A, Pyocyanin and Pyorubrin, Fridamycin D, Himalomycin A., B., 3,6-diisopropypiperazine-2,5-dione, Trisindoline.

Antimalarial: Cycloprodigiosin, 2-undecyl-4-quinolinone, Manzamine A.

Measures for the Sustainability

Growing biotechnological and scientific advancements are the driving force behind the establishment of the new and developing field of marine bioprospecting. This attempt holds great promise and presents a formidable task since it has the potential to enhance the standard of living by creating new goods and applications, creating jobs, and stabilizing the economy. However, it can also harm marine habitats in ways that cannot be repaired. Bioprospecting on a large scale from wild can lead to depletion of resources and at the same time can deteriorate the environment.

While collecting from deep sea, the periodicity of collection, duration of collection activity, movement of water in relation to collection activity, noise, and light, shifting of benthic and suspended matters can become major factors attributing to the slowly reparable precious environment. Resources will be depleted and the environment will worsen as a result. Environmental damage can be easily caused by submersible movement, drone use, unintentional oil spills, hazardous waste discharge, and ballast water discharge. For the sole purpose of manufacturing products, large-scale, aggressive, intensive, and frequently uncontrolled harvesting may wipe out organisms and their communities.

UNESCO's Ocean Decade slogan is 'The Science we need for the Ocean we want'. To protect the Oceans 10 thematic areas has been proposed and thoroughly discussed at various levels. All these efforts are to be made fruitful for conserving the values of oceans and the services we gain. One of the major precautionary measures is judicious exploitation of bio-resources by keeping our ocean ecosystem ecology intact. Access and benefit sharing should be logical, legally framed and by ensuring that its share is reached up to the lowest level (Jayachandran, 2022). In the marketing system of products, the lower strata are never considered for benefit. Jayachandran and Saiju (2022) proposed a new terminology, 'Inclusive Agile Marketing' as a new method of marketing of product. It is essential. By ensuring this our bio-resources can be protected, and Bioprospecting can be made a sustainable venture. Human being will ever be benefitted from this new technology.

Conflict of Interest The authors have no conflict of interest to declare.

References

Advani, R. H., Lebovic, D., Chen, A., Brunvand, M., Goy, A., Chang, J. E., Hochberg, E., Yalamanchili, S., Kahn, R., Lu, D., Agarwal, P., Dere, R. C., Hsin-Ju, H., Jones, S., Yu-Waye, C., & Cheson, B. D. (2017). Phase I Study of the Anti-CD22 Antibody-Drug Conjugate Pinatuzumab Vedotin with/without Rituximab in Patients with Relapsed/Refractory B-cell Non-Hodgkin Lymphoma. *Clinical Cancer Research, 23*(5), 1167–1176. https://doi.org/10.1158/1078-0432.CCR-16-0772

Ahmed, N., Sheikh, M. A., Ubaid, M., Chauhan, P., Kumar, K., & Choudhary, S. (2024). Comprehensive exploration of marine algae diversity, bioactive compounds, health benefits, regulatory issues, and food and drug applications. Measurement: Food, p. 100163.

Al-Shakliah, N. S., Kadi, A. A., Al-Salahi, R., & Rahman, A. F. M. M. (2022). In Vitro Identification of Potential Metabolites of Plinabulin (NPI 2358) in Hepatic Preparations Using Liquid Chromatography-Ion Trap Mass Spectrometry. *ACS Omega, 7*(25), 21465–21472. https://doi.org/10.1021/acsomega.2c00929

Arrieta, J., Arnaud-Haond, S., & Duarte, C. (2010). What lies underneath: Conserving the oceans genetic resources. *Proceedings of the National Academy of Sciences of the United States of America, 107*, 1838–18324.

Ashraf, S. A., Adnan, M., Patel, M., et al. (2020). Fish-based bioactives as potent nutraceuticals: Exploring the therapeutic perspective of sustainable food from the sea. *Marine Drugs, 18*(5), 265. https://doi.org/10.3390/md18050265

Asih, E. N. N., Kawaroe, M., & Bengen, D. H. (2018). Biomaterial compounds and bioactivity of horseshoe crabs, Carcinoscorpius rotundicauda biomass harvested from the Madura Strait. *Conference Series: Earth and Environmental Science, 141*, 012004. https://doi.org/10.1088/1755-1315/141/1/012004

Avhad, A. B., & Bhangale, C. J. (2023). Marine natural products and derivatives. *RPS Pharmacy and Pharmacology Reports, 2*, (2), rqad008. https://doi.org/10.1093/rpsppr/rqad008

Awuchi, C. G., Chukwu, C. N., Iyiola, A. O., Noreen, S., Morya, S., Adeleye, A. O., Twinomuhwezi, H., Leicht, K., Mitaki, N. B., & Okpala, C. O. R. (2022). Bioactive compounds and therapeutics from fish: revisiting their suitability in functional foods to enhance human wellbeing. *BioMed Research International, 2022*, 3661866. https://doi.org/10.1155/2022/3661866

Balansa, W., Mettal, U., Wuisan, Z. G., Plubrukarn, A., Ljong, F. S., Liu, Y., & Schäberle, T. F. (2019). A new sesquiterpenoid aminoquinone from an Indonesian Marine Sponge. *Marine Drugs, 17*(3), 158. https://doi.org/10.3390/md17030158

Bekiari, M. (2023). Marine Bioprospecting: Understanding the Activity and Some Challenges Related to Environmental Protection, Scientific Reseach, Ethics and the Law. In M. d. G. Garcia, A. Cortês (Eds.), *Blue Planet Law, Sustainable Development Goals Series*, 237–252, https://doi.org/10.1007/978-3-031-24888-7_19.

Bergmann, W., & Feeney, R. J. (1951). Contributions to the Study of Marine Products. 32. *The Nucleosides of Sponges*. 1. *The Journal of Organic Chemistry, 16*(6), 981–987. https://doi.org/10.1021/jo01146a023.

Bhakuni, D. S. & Ranat, D. S. (2005). *Bioactive Mrine Natural Products. Springer & Anamya publishers*, p. 382. https://rushim.ru/books/biochemie/bioactive-marine-natural-products.pdf.

Bhatnagar, I., & Kim, S.-K. (2010). Immense essence of excellence: marine microbial bioactive compounds. *Marine Drugs, 8*(10), 2673–2710. https://doi.org/10.3390/md8102673

Bongiorni, L., & Pietra, F. (1996). Marine natural products for industrial appli-cations. *Chemistry & Industry, 74*, 54–58.

Bouchet, P., Decock, W., Lonneville, B., Vanhoorne, B., & Vendepitte, L. (2023). Marine biodiversity discovery: The metrics of new species description. *Frontiers in Marine Science, 10*. https://doi.org/10.3389/fmars.2023.929989

Briggs, J. C. (1995). Species diversity: land and sea. In J. C. Briggs (Ed.), *Developments in Paleontology and Stratigraphy* (Chap. 13, Vol. 14, pp. 371–389). Elsevier. https://doi.org/10.1016/S0920-5446(06)80063-4

Cabrita, M. T., Vale, C., & Rauter, A. P. (2010). Halogenated compounds from Marine Algae. *Marine Drugs, 8*(8), 2301–2317. https://doi.org/10.3390/md8082301

Caplan, S. L., Zheng, B., Dawson-Scully, K., White, C. A., & West, L. M. (2016). Pseudopterosin A: protection of synaptic function and potential as a neuromodulatory agent. *Marine Drugs, 14*(3), 55.

Cappello, E., & Nieri, P. (2021). From life in the sea to the clinic: The Marine drugs approved and under clinical trial. *Life (Basel), 11*(12), 1390. https://doi.org/10.3390/life11121390

Carroll, A. R., Copp, B. R., Davis, R. A., Keyzers, R. A., & Prinsep, M. R. (2020). Marine Natural Products. Natural Product Reports, *37*, 175–223. https://doi.org/10.1039/C9NP00069K

Chakraborty, K., & Joy, M. (2020). High-value compounds from the molluscs of marine and estuarine ecosystems as prospective functional food ingredients: An overview. *Food Research International, 137*, 109637. https://doi.org/10.1016/j.foodres.2020.109637

Chakraborty, K., Joy, M., & Vijayagopal, P. (2016). Nutritional qualities of common edible cephalopods at the Arabian Sea. *International Food Research Journal, 23*(5), 1926–1938.

Chantarawong, W., Chamni, S., Suwanborirux, K., Saito, N., & Chanvorachote, P. (2019). 5-O-acetyl-renieramycin T from blue sponge Xestospongia sp. induces lung cancer stem cell apoptosis. *Marine Drugs, 17*(2), 109.

Chao, C. H., Wu, Y. C., Wen, Z. H., & Sheu, J. H. (2013). Steroidal carboxylic acids from soft coral Paraminabea acronocephala. *Marine Drugs, 11*(1), 136–145.

Chari, B. & Berde, C.V., 2024. New Vistas and Frontiers of Marine Bioactive Molecules in Biomedical and Pharmacotherapeutic Applications In: Marine Bioactive Molecules for Biomedical and Pharmacotherapeutic Applications. 0.1007/978-981-99-6770-4_1.

Chen, J., Hong Li., Zishuo Zhao, Xue Xia, Bo Li, Jinrong Zhang, & Xiaojun Yan. (2018a). Diterpenes from the Marine Algae of the Genus Dictyota. *Marine Drugs, 16*(5), 159. https://doi.org/10.3390/md16050159.

Chen, S. R., Wang, S. W., Su, C. J., Hu, H. C., Yang, Y. L., Hsieh, C. T., Peng, C. C., Chang, F. R. & Cheng, Y. B. (2018b, February). Anti-Lymphangiogenesis Components from Zoanthid Palythoa tuberculosa. *Marine Drugs, 16*(2), 47. https://doi.org/10.3390/md16020047

Chen, W. -F., Chakraborty, C., Sung, C. -S., Feng, C. -W., Jean, Y. -H., Lin, Y. -Y., Hung, H. -C., Huang, T. -Y., Su, T. -M., Sung, P. -J., Sheu, J. -H., Wen, Z. -H. (2011). Neuroprotection by marine-derived compound, 11-dehydrosinulariolide, in an in vitro Parkinson's model: a promising candidate for the treatment of Parkinson's disease. *Naunyn-Schmiedeberg's Archives of Pharmacology, 385*(3), 265–275.

Cheng, C.-H., Lin, Y.-S., Wen, Z.-H., & Su, J.-H. (2012). A new cubitane diterpenoid from the soft coral sinularia crassa. *Molecules, 17*(9), 10072–10078. https://doi.org/10.3390/molecules170910072

Chi, C. F., Cao, Z. H., Wang, B., Hu, F. Y., Li, Z. R., & Zhang, B. (2014). Antioxidants and functional properties of collagen hydrolysates from Spanish mackerel skin and influenced by average molecular weight. *Molecules, 19*(8), 11211–11230.

Cohen, Z., & Ratledge, C. (2010). *Single Cell Oils: Microbial and Algal Oils: Second Edition Inc.* https://doi.org/10.1016/C2015-0-02455-8

Collins, M. N., & Birkinshaw, C. (2013). Hyaluronic acid based scaffolds for tissue engineering—A review. *Carbohydrate Polymers, 92*, 1262–1279. https://doi.org/10.1016/j.carbpol.2012.10.028

Cooper, L. E., Hirabayashi, K., Kevin B., Strychar, B. K., & Sammarco, W. P., (2014). Corals and their potential applications to integrative medicine. *Hindawi Publishing Corporation Evidence-Based Complementary and Alternative Medicine*, 184959, 9. https://doi.org/10.1155/2014/184959.

Costantini, S., Guerriero, E., Teta, R., Capone, F., Caso, A., Sorice, A., Romano, G., Ianora, A., Ruocco, N., Budillon, A., Costantino, V., & Constantini, M. (2017). Evaluating the effects of an organic extract from the mediterranean sponge geodia cydonium on human breast cancer cell lines. *International Journal of Molecular Sciences, 18*(10), 2112. https://doi.org/10.3390/ijms18102112

Coutinho, M. C. L., Teixeira, V. L., & Santos, C. S. G. (2017). A review of "polychaeta" chemicals and their possible ecological role. *Journal of Chemical Ecology*. https://doi.org/10.1007/s10886-017-0915-z

Datta, D., Talapatra, S. N., & Swarnakar, S. (2015). Bioactive compounds from marine invertebrates for potential medicines—An overview. *International Letters of Natural Sciences, 34*, 42–61. https://doi.org/10.56431/p-122ej9

Debbab, A., Aly, A. H., Lin, W. H., & Proksch, P. (2010). Bioactive compounds from Marine bacteria and fungi. *Microbial Biotechnology*. https://doi.org/10.1111/j.1751-7915.2010.00179.x

D'Incalci, M., & Galmarini, C. M. (2010). A review of trabectedin (ET-743): A unique mechanism of action. *Molecular Cancer Therapeutics, 9*(8), 2157–2163. https://doi.org/10.1158/1535-7163

Duan, Y., Li, F., Li, Y., et al. (2016). The role of leucine and its metabolites in protein and energy metabolism. *Amino Acids, 48*(1), 41–51. https://doi.org/10.1007/s00726-015-2067-1

Duo, X., & Dong, B. (2019). origins and bioactivities of natural compounds derived from marine ascidians and their symbionts. *Marine Drugs, 17*(12), 670. https://doi.org/10.3390/md17120670

Ebbinghaus, S., Hersh, E., & Cunningham, C. (2004). Phase II study of synthadotin (SYN-D; ILX651) administered daily for 5 consecutive days once every 3 weeks (qd × 5q3w) in patients (Pts) with inoperable locally advanced or metastatic melanoma. *Journal of Clinical Oncology, 23*(14_suppl.). https://doi.org/10.1200/jco.2004.22.90140.7530.

Fang, Y., Ahmed, S., Liu, S., Wang, S., Lu, M., & Jiao, Y. (2013). Optimization of antioxidant exopolysaccharidess production by Bacillus licheniformis in solid state fermentation. *Carbohydrate Polymers, 98*, 1377–1382. https://doi.org/10.1016/j.carbpol.2013.07.076

Fontana, A., Cavaliere, P., Wahidulla, S., Naik, C. G., & Cimino, G. (2000). A new antitumor isoquinoline alkaloid from the marine nudibranch jorunna funehris. *Tetrahedron, 56*(37). https://doi.org/10.1016/S0040-4020(00)00629-3

Gagne, N., & Simpson, B. K. (1993). Use of proteolytic enzymes to facilitate the recovery of chitin form shrimp wastes. *Food Biotechnology, 7*(3), 253–263.

Kuznetsov, G., TenDyke, K., Towle, M. J., Cheng, H., Liu, J., Marsh, J. P., Schiller, S. E. R., Spyvee, M. R., Yang, Hu., Seletsky, B. M., Shaffer, C. J., Marceau, V., Yao, Ye., Suh, E. M., Campagna, S., Fang, F. G., Kowalczyk, J. J., & Littlefield, B. A. (2009). Tubulin-based antimitotic mechanism of E7974, a novel analogue of the marine sponge natural product hemiasterlin. *Molecular Cancer Therapeutics, 8*(10), 2852–2860. https://doi.org/10.1158/1535-7163.MCT-09-0301

Ghallab, D. S., Ibrahim, R. S., Mohyeldin, M. M., & Shawky, E. (2024). Marine algae: A treasure trove of bioactive anti-inflammatory compounds. *Marine Pollution Bulletin, 199*, 116023. https://doi.org/10.1016/j.marpolbul.2023.116023

Gomes, A. R., Freitas, A. C., Duarte, C. A., & Rocha-Santos, T. A. P. (2016). Chapter 1—Echinoderms: A Review of Bioactive Compounds with Potential Health Effects. *Studies in Natural Products Chemistry, 49*, 1–54. https://doi.org/10.1016/B978-0-444-63601-0.00001-6.

Green, D., Padula, M., Santos, J., Chou, J., Milthorpe, B., & Ben-Nissan, B. (2013). A therapeutic potential for marine skeletal proteins in bone regeneration. *Marine Drugs, 11*(4), 1203–1220.

Hamed, I., Özogul, F., Özogul, Y., & Regenstein, J. M. (2015). Marine Bioactive Compounds and Their HealthBenefits: A Review. *Comprehensive Reviews in Food Science and Food Safety, 14*(4). https://doi.org/10.1111/1541-4337.12136

Hasan, S., Ansari, M. I., Ahmad, A., & Mishra, M. (2015). Major bioactive metabolites from marine fungi. *Bioinformation, 11*(4), 176–181. https://doi.org/10.6026/97320630011176

Hesse, H., Kerft, O., Maimann, O., Zeh, M., & Hoefgen, R. (2004). Current Understanding of the Regulation of Methionine Biosynthesis in Plants. *Journal of Experimental Botany, 55*, 1799–1808.

Houpt, J. B., McMillan, R., Wein, C., & Paget-Dellio, S. D. (1999). Effect of glucosamine hydrochloride in the treatment of pain osteoarthritis of the knee. *The Journal of Rheumatology, 26*(11), 2423–2430.

Hu, L. C., Yen, W. H., Su, J. H., Chiang, M.Y.-N., Wen, Z.-H., Chen, W.-F., Lu, T.-J., Chang, Y.-W., Chen, Y.-H., Wang, W.-H., Wu, Y.-C., & Sung, P.-J. (2013). Cembrane derivatives from the soft corals, sinularia gaweli and sinularia flexibilis. *Marine Drugs, 11*(6), 2154–2167. https://doi.org/10.3390/md11062154

Huang, K. J., Chen, Y. C., El-Shazly, M., Du, Y. C., Su, J. H., & Tsao, C. W. (2013). 5-episinuleptolide acetate, a norcembranoidal diterpene from the formosan soft coral Sinularia sp., induces leukemia cell apoptosis through hsp90 inhibition. *Molecules, 18*(3), 2924–2933.

Huang, S.-Y., Chen, N.-F., Chen, W.-F., Hung, H.-C., Lee, H.-P., Lin, Y.-Y., Wang, H.-M., Sung, P.-J., Sheu, J.-H., & Wen, Z.-H. (2012). Sinularin from Indigenous Soft Coral Attenuates Nociceptive Responses and Spinal Neuroinflammation in Carrageenan-Induced Inflammatory Rat Model. *Marine Drugs, 10*(12), 1899–1919.

Jayachandran, K. V. (2022). A revisit to access and benefit sharing (ABS) with reference to aquatic systems and its bio-resources. In O.V. Oommen, K. P. Laladhas, P. Nelliyat, & B. Pisupati (Eds.), *Biodiversity Conservation Through Access and Benefit Sharing* (ABS) (pp. 229–245). Springer, Cham. https://doi.org/10.1007/978-3-031-16186-5_11, First Online: 19 January 2023.

Jayachandran, K. V., & Saiju, S. (2022). Ch.1. Agile Marketing Two Pragmatic Working Models from Kerala with Thrust on Human Resource Development and Coining a New Terminology. In S. Jayadev, et al. (Eds.), *Agile Marketing Practices, Innovations and Emerging Applications* (pp. 7–14). Infinity Indica Publishers. ISBN: 978-93-91700-45-4.

Jean, Y.-H., Chen, W.-F., Sung, C.-S., Duh, C. Y., Huang, S. Y., Lin, C. S., Tai, M. H., Tzeng, S. F., & Wen, Z. H. (2009). Capnellene, a natural marine compound derived from soft coral, attenuates chronic constriction injury-induced neuropathic pain in rats. *British Journal of Pharmacology, 158*(3), 713–725.

Johnson, M., & Chastity, B. (2014). Omega-3, omega-6 and omega-9 fatty acids: Implications for cardiovascular and other diseases. *Journal of Glycomics & Lipidomics, 4*(4), 1000123. https://doi.org/10.4172/2153-0637.1000123

Karthikeyan, A., Joseph, A., & Nair, B. G. (2022). Promising bioactive compounds from the marine environment and their potential effects on various diseases. *Journal, Genetic Engineering & Biotechnology, 20*, 14. https://doi.org/10.1186/s43141-021-00290-4

Katz, J., Janik, J. E., & Younes, A. (2011). Brentuximab Vedotin (SGN-35). *Clinical Cancer Research, 17*(20), 6428–6436. https://doi.org/10.1158/1078-0432.CCR-11-0488

Kem, W., Soti, F., Wildeboer, K., LeFrancois, S., Dong-Qing, W., Kuo-Chen, C., & Arias, H. R. (2006, April). The nemertine Toxin anabaseine and its derivative DMXBA (GTS-21): Chemical and pharmacological properties. *Marine Drugs, 4*(3), 255–273.

Kobayashi, J., Ohizumi, Y., Nakamura, H., & Hirata, Y. (1986). A novel antagonist of serotonergic receptors, hymenidin, isolated from the Okinawan marine sponge Hymeniacidon sp. *Experientia, 42*(10), 1176–1177. https://doi.org/10.1007/BF01941300

König, G. M. (1992). Meeresorganismen als Quelle pharmazeutisch bedeutsamer Naturstoffe. *Deutsche Apotheker-Zeitung, 132*, 673–683.

Kwan, J. C., Eksioglu, E. A., Liu, C., Paul, V. J., & Luesch, H. (2009). Grassystatins A-C from Marine Cyanobacteria, Potent Cathepsin E Inhibitors that Reduce Antigen Presentation. *Journal of Medicinal Chemistry, 52*(18), 5732–5747. https://doi.org/10.1021/jm9009394

Lago, J., Rodríguez, L. P., Blanco, L., Vieites, J. M., & Cabado, A. G. (2015). Tetrodotoxin, an extremely potent marine neurotoxin: Distribution, toxicity Origin and Therapeutical Uses. *Marine Drugs, 13*(10), 6384–6406. https://doi.org/10.3390/md13106384

Leal, M. C., Madeira, C., Brandão, C. A., Puga. J., & Calado, R. (2012). Bioprospecting of marine invertebrates for new natural products—A chemical and zoogeographical perspective. *Molecules, 17*, 9842–9854. https://doi.org/10.3390/molecules17089842

Lebouvier, N., Jullian, V., Desvignes, I., Maurel, S., Parenty, A., Dorin-Semblat, D., Doerig, C., Sauvain, M., & Laurent, D. (2009). Antiplasmodial activities of homogentisic acid derivative protein kinase inhibitors isolated from a Vanuatu marine sponge Pseudoceratina sp. *Marine Drugs, 7*(4), 640–653. https://doi.org/10.3390/md7040640

Lee, H. P., Huang, S. Y., Lin, Y. Y., Wang, H. M., Jean, Y. H., Wu, S. F., Duh, C.-Y., & Wen, Z.-H. (2013). Soft coral-derived lemnalol alleviates monosodium urate-induced gouty arthritis in rats by inhibiting leukocyte infiltration and iNOS, COX-2 and c-Fos protein expression. *Marine Drugs, 11*, 99–113. https://doi.org/10.3390/md11010099

Lhullier, C., Tabalipa, E. de Oliveira, Sardá, F. N., Sandjo, L. P., Schneider, N. F. Z., Carraro, J. L., Simões, C. M. O., & Schenkel, E. P. (2019). Clerodane Diterpenes from the Marine Sponge Raspailia bouryesnaultae Collected in South Brazil. *Marine Drugs, 17*(1), 7. https://doi.org/10.3390/md17010057

Lin, X., & Kück, U. (2022). Cephalosporins as key lead generation beta-lactam antibiotics. *Applied Microbiology and Biotechnology, 106*(24), 8007–8020. https://doi.org/10.1007/s00253-022-12272-8

Lin, Y. Y., Jean, Y. H., Lee, H. P., Chen, W. F., Sun, J. H., Su, J. H., Lu, Y., Huang, H. C., Sung, P. J., Sheu, J. H., & Wen, Z. H. (2013). A soft coral-derived compound, 11-epi-sinulariolide acetate suppresses inflammatory response and bone destruction in adjuvant-induced arthritis. *PloS ONE, 8*(5), Article ID e62926.

Lindequist, U. (2016). Marine-derived pharmaceuticals—challenges and opportunities. *Biomolecules and Therapeutics (Seoul), 24*(6), 561–571. https://doi.org/10.4062/biomolther.2016.181

Liu, L., Yao-Yao, Z., Chang-Lun, S., & Chang-Yun, W. (2019). Metabolites from marine invertebrates and their symbiotic microorganisms: Molecular diversity discovery, mining, and application. *Marine Life Science & Technology, 1*, 60–94. https://doi.org/10.1007/s42995-019-00021-2

MacArtain, R., Gill, C. I. R., Brooks, M., Campbell, R., & Rowland, I. R. (2007). Nutritional value of edible seaweeds. *Nutrition Reviews, 65*, 535–543.

Malve, H. (2015). Exploring the ocean for new drug developments: Marine pharmacology. *Journal of Pharmacy and Bioallied Sciences, 8*(2). https://doi.org/10.4103/0975-7406.171700.

MAR (2023) Biotechnology Market Size, Share & Trend Analysis By Technology (Nanobiotechnology, DNA Sequencing, Cell-based Assays), By Application (Health, Bioinformatics), By Region, And Segment Forecasts, 2024—2030. Report ID: 978-1-68038-134-4.

MAREX. (2014). Exploring Marine Resources for Bioactive Compounds: From Discovery to Sustainable Prouction and Industrial Application. Final Report (Executive summary) No.KBBE-2009-3-2-01, p. 9. https://cordis.europa.eu/project/id/245137/reporting/es

Matz, N. (2002). Marine biological resources: Some reflections on concepts for the protection and sustainable use of biological resources in the deep sea. *Non-State Actors and International Law, 2*, 279–300.

MBM. (2023). Marine Biotechnology Market by Product Type (Marine-Derived Pharmaceuticals, Marine-Derived Nutraceuticals, Marine Biomaterials, Marine Bioproducts) By Application (Pharmaceuticals and Healthcare, Nutraceuticals and Food, Bioremediation and Environmental Monitoring, Aquaculture and Fisheries, Bioenergy and Industrial Processes, Biotechnology Research)—Growth, Future Prospects & Competitive Analysis, 2016—2030 https://www.credenceresearch.com/report/marine-biotechnolog-market

McConnell O., Longley R. E., Koehn F. E. (1994). The discovery of Natural products with therapeutic potential. In V.P Gullo (Eds.) (pp. 109–174). Butterworth-Heinemann.

McDermid, K. J., & Stuercke, B. (2003). Nutritional composition of edible Hawaiian seaweed. *Journal of Applied Phycology, 15*, 513–524.

McGivern, J. G. (2007). Ziconotide: A review of its pharmacology and use in the treatment of pain. *Neuropsychiatric Disease and Treatment, 3*(1), 69–85. https://doi.org/10.2147/nedt.2007.3.1.69

Mirunalini, S., & Maruthanila, V. L. (2022). The impact of bioactive compounds derived from Marine Fish on Cancer. *Bentham Science, 22*(15), 2757–2765. https://doi.org/10.2174/1871520622666220330142442

Mobin, S., & Alam, F. (2017). Some promising microalgal species for commercial applications: A review. *Energy Procedia, 110*, 510–517.

Mohanty, B. P., Ganguly, S., Mahanty, A., et al. (2016). DHA and EPA content and fatty acid profile of 39 food fishes from India. *BioMed Research International, 2016*, 14. https://doi.org/10.1155/2016/4027437.4027437

Mohanty, B., Mahanty, A., Ganguly, S., et al. (2014). Amino acid compositions of 27 food fishes and their importance in clinical nutrition. *Journal of Amino Acids, 2014*, 7. https://doi.org/10.1155/2014/269797.269797

Myo Clinic. (2024). https://www.mayoclinic.org/drugs-supplements/omega-3-acid-ethyl-esters-oral-route/description/drg-20066973

Nagaoka, I., Igarashi, M., Hua, J., Ju, Y., Yomogida, S., & Sakamoto, K. (2011). Recent aspect of the anti-inflammatory actions of glucosamine. *Carbohydrate Polymers, 36*(4), 471–477.

Newman, D. J., & Cragg, G. M. (2018). Marine Natural Products with Pharmacological Properties. In *Chemical Ecology* (pp. 1–25). CRC Press.

Offret, C., Florie Desriac,† Patrick Le Chevalier, Jérôme Mounier, Camille Jégou, & Yannick Fleury (2016). Spotlight on antimicrobial metabolites from the marine bacteria pseudoalteromonas: chemodiversity and ecological significance. *Marine Drugs, 14*(7), 129. https://doi.org/10.3390/md14070129

Oli, A. K., Shivshetty, N., Javaregowda, P. K., & Revanna, K. C. (2023). Extraction and Purification of Bioactive Peptide with Antimicrobial Properties from Horseshoe Crab. *Bentham Science, 19*(3). https://doi.org/10.2174/1573407218666220606142023.

Olivera, B. M. (2000). ω-Conotoxin MVIIA: From Marine Snail Venom to Analgesic Drug. In N. Fusetani (Ed.), *Drugs From The Sea* (pp. 74–85). Karger.

Papapanou, M., Papoutsi, E., Giannakas, T., & Katsaounou, P. (2021). Plitidepsin: Mechanisms and clinical profile of a promising antiviral agent against COVID-19. *J Pers Med., 11*(7), 668. https://doi.org/10.3390/jpm11070668

Park, J.-S., Han, J.-M., Shin, Y.-N., Park, Y.-S., Shin, Y.-R., Park, S.-W., Park, S.-W., Roy, V. C., Lee, H.-J., Kumagai, Y., Kishimura, H., & Chun, B.-S. (2023). Exploring bioactive compounds in brown seaweeds using subcritical water: A comprehensive analysis. *Marine Drugs, 21*(6), 328. https://doi.org/10.3390/md21060328

Perera, R. M. T. D., Herath, K. H. I. N. M., Sanjeewa, K. K. A. & Jayawardena, T. U. (2023). Recent reports on bioactive compounds from Marine Cyanobacteria in relation to human health applications. *Life (Basel), 13*(6), 1411. https://doi.org/10.3390/life13061411

Plaza, A., Keffer, J. L., Bifulco, G., Lloyd, J. R., & Bewley, C. A. (2010). Chrysophaentins A-H, antibacterial bisdiarylbutene macrocycles that inhibit the bacterial cell division protein FtsZ. *Journal of the American Chemical Society, 132*(26), 9069–9077. https://doi.org/10.1021/ja102100h

Ramesh, C., Tulasi, B. R., Raju, M., Thakur, N., & Dufoss, L. (2021). Marine natural products from tunicates and their associated microbes. *Marine Drugs, 19*, 308. https://doi.org/10.3390/md19060308

Ratana-Arporn, P., Chirapart A. (2006). Nutritional evaluation of tropical green seaweeds. *Caulerpa lentillifera* and *Ulva reticulata. The Kasetsart Journal, 40*(SI), 75–83.

Rigogliuso, S., Campora, S., Notarbartolo, M., & Gheersi, G. (2023). Recovery of bioactive compounds from marine organisms: Focus on the future perspectives for pharmacological, biomedical and regenerative medicine applications of marine collagen. *Molecules*. https://doi.org/10.3390/molecules28031152

Romano, G., Almeida, M., Coelho, A. V., Cutignano, A., Gonçalves, L. G., Hansen, E., Khnykin, D., Mass, T., Ramšak, A., Miguel, S. R., Silva, T. H., Sugni, M., Ballarin, L., & Anne-Marie, G. (2022). Biomaterials and bioactive natural products from marine invertebrates: From basic research to innovative applications. *Marine Drugs, 20*(4), 219. https://doi.org/10.3390/md20040219

Saad, V., Wicklow, B., Wittmeier, K., et al. (2015). A clinically relevant method to screen for hepatic steatosis in overweight adolescents: A cross sectional study. *BMC Pediatrics, 15*(1), 151. https://doi.org/10.1186/s12887-015-0465-x

Sala, G. D., Hochmuth, T., Costantino, V., Teta, R., Gerwick, W., Gerwick, L., Piel, J., & Alfonso Mangoni, A. (2013). Polyketide genes in the marine sponge Plakortis simplex: A new group of mono-modular type I polyketide synthases from sponge symbionts. *Environmental Microbiology Reports, 5*(6), 809–818. https://doi.org/10.1111/1758-2229.12081

Se-Kwon, K. (2010). *Chitin, chitosan, oligosaccharides and their derivatives: Biological activities and application*. (1st ed., p. 666). CRC Press-Taylor & Francis Group. https://doi.org/10.1201/EBK1439816035

Serova, M., De Gramont, A., Bieche, I., Riveiro, M. E., Galmarini, C. M., Aracil, M., Jimeno, J., Faivre, S., & Raymond, E. (2013). Predictive factors of sensitivity to elisidepsin, a novel Kahalalide F-derived marine compound. *Marine Drugs, 11*(3), 944–959.

Shen, W., Jae-Seung, K., Mitchell, S., Kish, P., Kijek, P., & Hilfinger, J. (2009). 5'-O-D-valyl ara A, a potential prodrug for improving oral bioavailability of the antiviral agent vidarabine. *Nucleosides, Nucleotides & Nucleic Acids, 28*(1), 43–55. https://doi.org/10.1080/15257770802581757

Šimat, V., Čagalj, M., Skroza, D., Gardini, F., Tabanelli, G., Montanari, C., Hassoun, A., & Ozogul, F. (2021). Sustainable sources for antioxidant and antimicrobial compounds used in meat and seafood products. *Advances in Food and Nutrition Research, 97*, 55–118. https://doi.org/10.1016/bs.afnr.2021.03.001

Simopoulos, A. P. (2009). Omega-6/omega-3 essential fatty acids: Biological effects. *World Review of Nutrition and Dietetics, 99*, 1–16. https://doi.org/10.1159/000192755

Sinha, T., Ikelle, L., Naash, M. I., & Al-Ubaidi, M. R. (2020). The intersection of serine metabolism and cellular dysfunction in retinal degeneration. *Cell, 9*(3), 674. https://doi.org/10.3390/cells9030674

Snelgrove, P. V. (2016). An ocean of discovery: Biodiversity beyond the Census of Marine Life. *Planta Medica, 82*, 790–799. https://doi.org/10.1055/s-0042-103934

Su, C. C., Su, J. H., Lin, J. J., Chen, C. C., Hwang, W. I., Huang, H. H., & Wu, Y. J. (2011a). An investigation into the cytotoxic effects of 13-acetoxysarcocrassolide from the soft coral Sarcophyton crassocaule on bladder cancer cells. *Marine Drugs, 9*(12), 2622–2642.

Su, C. C., Su, J.-H., Lin, J. J., et al. (2011b). An investigation into the cytotoxic effects of 13-acetoxysarcocrassolide from the soft coral Sarcophyton crassocaule on bladder cancer cells. *Marine Drugs, 9*(12), 2622–2642.

Sukapure, R. S., & Thirumalachar, M. J. (1966). Conspectus of species of Cephalosporium with particular reference to Indian species. *Mycologia, 58*(3), 351–361.

Suleria, H. A. R., Osborne, S., Masci, P., & Gobe, G. (2015). Marine-based nutraceuticals: An innovative trend in the food and supplement industries. *Marine Drugs, 13*, 6336–6351. https://doi.org/10.3390/md13106336

Suna, H., Arai, M., Tsubotani, Y., Hayashi, A., Setiawan, A., & Kobayashi, M. (2009). Dysideamine, a new sesquiterpene aminoquinone, protects hippocampal neuronal cells against iodoacetic acid-induced cell death. *Bioorganic & Medicinal Chemistry, 17*(11), 3968–3972. https://doi.org/10.1016/j.bmc.2009.04.025

Tai, C. J., Su, J.-H., Huang, C.-Y., Huang, M.-S., Wen, Z.-H., Dai, C.-F., & Sheu, J.-H. (2013). Cytotoxic and Anti-Inflammatory Eunicellin-Based Diterpenoids from the Soft Coral *Cladiella krempfi*. *Marine Drugs, 11*(3), 788–799. https://doi.org/10.3390/md11030788

Tian, Z., Xin-Tong, L., Jiang, X., & Tian, J. (2023). Bryostatin-1: a promising compound for neurological disorders. *Frontiers in Pharmacology, 14*. https://doi.org/10.3389/fphar.2023.1187411.

Tilami, K. S., & Sampels, S. (2018). Nutritional value of fish: Lipids, proteins, vitamins, and minerals. *Reviews in Fisheries Science & Aquaculture, 26*(2), 243–253. https://doi.org/10.1080/23308249.2017.1399104

Torabally, N. B., & Rahmanpoor, H. A. (2019). Nutraceuticals: Nutritionally functional foods—An overview. *Biomedical Journal of Scientific Research, 15*(4), 1–3. https://doi.org/10.26717/BJSTR.2019.15.002728

Twelves, C., Cortes, J., Vahdat, L. T., Wanders, J., Akerele, C., Kaufman, P. A., & P.A. (2010). Phase III trials of eribulin mesylate (E7389) in extensively pretreated patients with locally recurrent or metastatic breast cancer. *Clinical Breast Cancer, 10*(2), 160–163. https://doi.org/10.3816/CBC.2010.n.023

Vahdat, L. T., Schmid, P., Forero-Torres, A. et al. (2021). Glembatumumab vedotin for patients with metastatic, gpNMB overexpressing, triple-negative breast cancer ("METRIC"): a randomized multicenter study. *npj Breast Cancer, 7*(57). https://doi.org/10.1038/s41523-021-00244-6

Varijakzhan, D., Jiun-yan, L., Wai-Sum, Y., Yusoff, K., Seboussi, R., Swee-Hua, E. L., Kok-Song, L., & Chou-Min, C. (2021). Bioactive Compounds from Marine Sponges: Fundamentals and Applications. *Marine Drugs, 27, 19*(5),246. https://doi.org/10.3390/md19050246

Velmurugan, S., Jerin, N., Michaelbabu, M., Bindhu, F., Dhas, S. A., & Citarasu, T. (2015). Screening and characterization of antiviral compounds from Enteromorpha flexuosa against white spot syndrome virus (WSSV) and its in vivo influence on Indian white shrimp Fenneropeneaeus indicus. *Aquaculture International, 23*, 65–80. https://doi.org/10.1007/s10499-014-9798-y

Wang, S.-K., & Duh, C.-Y. (2012). New cytotoxic cembranolides from the soft coral Lobophytum michaelae. *Marine Drugs, 10*(2), 306–318.

Wang, S. K., Puu, S. Y., & Duh, C.-Y. (2013). New steroids from the soft coral Nephthea chabrolii. *Marine Drugs, 11*(2), 571–580.

Watanabe, J., Minami, M., & Kobayashi, M. (2006). Antitumor activity of TZT-1027 (Soblidotin). *Anticancer Research, 26*(3A), 1973–1981.

Weinheimer, A. J., & Spraggins, R. L. (1969). The occurrence of two new prostaglandin derivatives (15-epi-PGA2 and its acetate, methyl ester) in the gorgonian Plexaura homomalla chemistry of coelenterates. *XV. Tetrahedron Lett., 10*, 5185–5188. https://doi.org/10.1016/S0040-4039(01)88918-8

Wildman, R. E. C., Wildman, R., & Wallace, T. C. (2006). *Handbook of Nutraceuticals and Functional Foods* (2nd ed., p. 560). CRC Press. https://doi.org/10.1201/9781420006186

Wu, D., Duan, C., Chen, L., et al. (2017). Efficacy and safety of different doses of cytarabine in consolidation therapy for adult acute myeloid leukemia patients: A network meta-analysis. *Science and Reports, 7*, 9509. https://doi.org/10.1038/s41598-017-10368-0

Yamada, S., Yamamoto, K., Ikeda, T., Yanagiguchi, K., & Hayashi, Y. (2014). Potency of fish collagen as a scaffold for regenerative medicine. *BioMed Research International, 2014*, 302932. https://doi.org/10.1155/2014/302932

Yen, W. -H., Chen, W. -F., Cheng, Ç. -H., Dai, C. -F., Lu, M. -C., Su, J. -H., Su, Y. -D., Chen, Y. -H., Chang, Y. -C., Chen, Y. -H., Sheu, J. -H., Lin, C. -H., Wen, Z -H., & Sung, P. –J. (2013). A new 5α, 8α-epidioxysterol from the soft coral Sinularia gaweli. *Molecules, 18*(3), 2895–2903. https://doi.org/10.3390/molecules18032895

Zhao, C., Yang, C., Liu, B., Lin, L., Sarker, S., Nahar, L., Yu, H., Cao, H., & Xiao, J. (2017). Bioactive compounds from marine macroalgae and their hypoglycemic benefits. *Trends in Food Science & Technology, 72*(4). https://doi.org/10.1016/j.tifs.2017.12.001

Zlatanos, S., Laskaridis, K., Feist, C., & Sagredos, A. (2006). Proximate composition, fatty acid analysis and protein digestibility-corrected amino acid score of three Mediterranean cephalopods. *Molecular Nutrition & Food Research, 50*(10), 967–970. https://doi.org/10.1002/mnfr.200600003

Zodape, G. V. (2018). Studies on the antibacterial activity of bioactive compounds of fish tetraodon fluviatilis of west coast of mumbai. *Biomedical and Pharmacology Journal, 11*(1). https://doi.org/10.13005/bpj/1401.

Zuo, W., & Kwok, H. F. (2021). Development of marine-derived compounds for cancer therapy. *Marine Drugs, 19*, 342. https://doi.org/10.3390/md19060342

Bioprospecting and Biodiversity Conservation

Laladhas Krishna Panicker, Arathy Sudhakaran, M. P. Vinod, K. Swathy, Achuthsankar S. Nair, and Oommen V. Oommen

Introduction

Bioprospecting is the process of scouring biodiversity for novel organic compounds with potential socioeconomic value. All living things and species, their genetic diversity, intricate community structures, and the variety of ecosystems that exist on Earth are all included in biodiversity. One of the topics that people discuss the most in the twenty-first century is biological diversity—both as a concept and in general. Human life depends directly on the diversity of nature around them. Even though ancient societies extensively studied and utilized the world's biological resources for human purposes, our understanding of biodiversity is still quite restricted. Plant variety is a worldwide resource that has enormous potential for the generation of more beneficial natural products but is currently little understood, insufficiently recorded, and frequently squandered.

The earth's life support system, which includes plant resources, is under threat from the growing human population as global change rates rise. It is now well acknowledged that scientific investigation of the enormous phytodiversity is likely to provide many more beneficial products for an incredibly broad range of human needs and interests. A vast range of food, medicines, cosmetics, fibers, and building materials are produced from flowering plants. These, however, make up a very small portion of all the species on Earth, and there is still a great deal of untapped resources.

L. Krishna Panicker (✉) · M. P. Vinod · K. Swathy · A. S. Nair · O. V. Oommen
Deptartment of Computational Biology & Bioinformatics, University of Kerala, Thiruvananthapuram, Kerala 695581, India
e-mail: laladhas@gmail.com

A. Sudhakaran
Department of Zoology, University of Kerala, Thiruvananthapuram, Kerala 695581, India

L. Krishna Panicker · O. V. Oommen
Center for Sustainable Environment-India Biodiversity Action, Thiruvananthapuram, Kerala, India

L. Krishna Panicker et al. (eds.), *Biodiversity and Business*,
https://doi.org/10.1007/978-3-031-71674-4_7

The Goteborg Resolution 1990, adopted at the International Society of Chemical Ecology meeting in Goteborg, Sweden, recognized the value of exploring biodiversity for novel products. Indeed, a wide range of pharmacological effects in different biological functions are ascribed to plants and plant compounds. Nowadays, it is considered that items generated from plants are safe and environmentally friendly, in contrast to synthetic materials that are thought to be incompatible for the environment and human health. This has made it necessary to explore botanical resources to investigate secure and efficient treatments for human illnesses. Industries employ a wide range of species in the bioprospecting of plants for medicinal compounds.

Indigenous knowledge is utilized by pharmaceutical corporations before screening for novel products. Indigenous people have a wealth of information about the characteristics of various native plants all across the world. Up until now, a plant species' resource values have been determined by evaluating the plant that yields the product or its derivatives, which can be used as models for imitation, modification, or other purposes. Thinking about potential natural evolutionary environments for the intended product is a common strategy for discovery. Next, a set of species or their habitats are recognized and investigated. Moreover, contemporary methods for drug discovery include combinatorial chemistry and rational drug design. The current belief is that in the future, natural products will probably supply the best lead compounds. We will find in this chapter how bioprospecting and ABS-biodiversity conservation may work together to preserve biodiversity sustainably for the future generations (Laladhas et al., 2015, 2016, 2017a, 2017b; Mithrambika, 2017; Nilayangode, 2016a, 2016b, 2016c, 2017; Oommen et al., 2022; Preetha et al., 2015).

Stopping Desertification and Enhancing Living Conditions

The annual plant hibiscus (*Hibiscus sabdariffa var. sabdariffa*) is indigenous to tropical and subtropical regions of the world. The industry that makes herbal teas and beverages use its swelling crimson calyces. This plant, which can thrive in arid and barren soil and is drought-tolerant, is crucial to the local farmers' livelihood in West Africa.

In Nigeria, where UEBT (Union for Ethical BioTrade) member REFEM (Industry) operates, hibiscus is frequently grown on formerly farmed land using labor-intensive, low-input multi-cropping techniques. The harvest season goes from November to April—the dry season—and every farmer must complete the harvest in a certain time range or risk losing the blossoms. This is because each bloom needs to be individually chosen, and there are no automated methods for this. In addition to exporting flowers, REFEM purchases raw materials from nearby growers and produces powders and extracts from them.

Market access is one of the largest problems local hibiscus growers have because of the intricacy of global rules that must be comprehended and followed. Due to wastage of crops, farmers lose out on vital revenue, and are occasionally compelled to sell below the cost of production or are taken advantage of by intermediaries in the supply chain, which also had an impact on productivity in certain seasons. Obtaining access to drying facilities that can eliminate contaminants and enhance the quality of the food is another difficulty. REFEM pays farmers a financial incentive to overcome these difficulties and meet quality criteria. This enhances farming practices and offers a reliable and immediate source of cash.

By building schools, awareness campaigns, women's empowerment initiatives, nutrition and health safety programs, and water boreholes to sustain 30,000 men, the corporation also assists local communities through non-monetary services. Additionally, REFEM is enhancing its biodiversity protection efforts in compliance with the UEBT Ethical Bio Trade Standard. This involves testing the fertility of the soil and restoring native plants to act as barriers against the occasional dry wind that blows in from the desert. This is essential to stop the loss of soil that causes desertification. For Hibiscus producers, the expenses of reforestation in terms of land are covered by crop resilience as a result of better soil quality. Due to its expanding international market, the business anticipates that hibiscus will continue to be significant in the food, beverage, and infusions sector. Notwithstanding the difficulties brought forth by climate change, REFEM hopes that Hibiscus will be a commodity that helps to both enhance the environment and farmers' standard of living (https://www.ethicalbiotrade.org/ingredient-stories/hibiscus).

Preservation of Shea Trees and Empowerment of Women

Shea (*Vitellaria paradoxa*) belonging to Sapotaceae family is native to the savannahs of West Africa, particularly Senegal, Ethiopia, and Uganda. Although these trees don't blossom until they are 15 to 20 years old, they continue to be more fruitful for 100 years. A Burkina Faso firm called Agrifaso hired women to harvest ripe nuts, which are highly sought after for use in skin care products, medications, and cooking oil. Demand drove up the price, providing rural women with a real source of income. The company purchases nuts from individuals and adds value before selling them to Europe. The business put conservation and sustainable usage policies into place. Long-term contracts with buyers and long-term conservation initiatives benefit families. (https://www.ethicalbiotrade.org/ingredient-stories/2019/6/19/shea-nuts).

Sustainable Access to the Marketing of "the Scent of the Vanishing Flora" in South Africa Derived from Clanwilliam Cedar

A native of South Africa, *Widdringtonia wallichii*, sometimes known as Clanwilliam cypress, *Widdringtonia cedar bergensis*, is a species of Widdringtonia. This tree is unique to the Western Cape Province's Cederberg Mountains, which are northeast of Cape Town. Due to excessive harvesting of its timber, which is the only source used to make the aroma, this priceless bioresource is in danger of being extinct. Through CapeNature, a public organization that manages the bioresource in the mountain responsibly, the majority of Heuningveli population was educated to protect and collect the bioresource.

The company that produces this bioresource's flavours and perfumes worldwide is Givaudan. Without taking any living samples from the mountain, the producers used cutting-edge technology to extract volatile compounds from the leaves and dead wood of Clanwilliam cedar and spread awareness of the value of protecting the endangered species. The producers created a synthetic perfume concentrate by using additional components to replicate the fragrance of Widdringtonia. This project serves as a model that will be on display to encourage the communities that help to preserve this endangered tree.

As per the South African benefit-sharing and access regulations, a benefit-sharing agreement was signed by the organization, Givaudan and The Body Shop together as users, and CapeNature as the provider. This instance is an intriguing example of the knowledge base connected with ABS equipment. The applicability of the ABS standards is not clear-cut in this case because headspace technology solely involves capturing volatile chemicals and does not need physical samples.

Prior informed consent has been acquired and mutually agreed upon agreements, including benefit sharing with resource provider CapeNature, have been authorized by the African national authorities. Scientifically carrying out its mandate, the public organization CapeNature was founded with a legislative duty for biodiversity protection in the Western Cape.

A benefit-sharing agreement was completed in October 2018, and to gain from the parameters that were mutually agreed upon, CapeNature also insisted on including the Heuningvlei community—a significant steward of the Clanwilliam cedar—in the process. Benefits that were monetary and non-monetary were considered. The process of creating and selling The Body Shop perfume as the finished product was taken into consideration while defining monetary advantages. These financial advantages have been deposited into the Bioprospecting Trust Fund in compliance with national ABS regulations. From there, CapeNature receives the funds for the conservation and livelihood programs that have been decided upon in the mutually agreed agreements between the supplier and users. The goals of these initiatives are to protect the Clanwilliam cedar and to uplift and sustain the way of life in the Heuningvlei neighborhood (Normand et al., 2021).

Integrating ABS Concepts into a Caucasus Asia Biodiversity Catalogue

A joint effort by Ilia State University in Georgia and the Alexander Koenig Zoological Research Museum developed an endeavour to investigate the species and genetic diversity of plants and animals in the Caucasus area. The funding for this worldwide hotspot for biodiversity was provided by the German Federal Ministry of Education and Research. The Caucasus Barcode of Life project (CaBOL) was started as a result of the initial pilot project's success. Its larger scope involved sophisticated research at both classical and molecular levels to barcode the species and create a database.

After the study is completed, the projects used the ABS framework to promote international cooperation to enhance science infrastructures, engage in teaching, and further investigate the biodiversity of the Caucasus. These initiatives promoted the creation of databases in Georgia and Armenia as well as the training of experts in molecular methods. Program partners were provided with cash for personnel and facilities, as well as assistance with curriculum development. The Caucasus would receive almost half of the funding for the Caucasus Barcode of Life initiative from the German Federal Ministry of Education and Research. The initiative offers funding for the creation of academic programs, training, jobs, infrastructure, open-access barcoding reference data, and scientific collaboration (Thorman et al., 2019).

Basic Study of Fijian and Asia–Pacific Marine Coral Reef Creatures

The goal of the study was to separate and extract natural compounds from microbiological specimens kept in a marine collection- an interdisciplinary collaboration between the University of the South Pacific, the Georgia Institute of Technology (GIT), the University of California at Davis's Scripps Institution of Oceanography (SIO), and the Fiji Ministry of Fisheries and Forests through the Institute of Applied Sciences (IAS). The idea of this project was to support reef conservation through education, training programs, systematic and scientifically planned management, and revenue production based on biodiversity. Utilizing cutting-edge scientific research techniques in conjunction with research and development activities on natural goods, pharmaceutical compounds are being developed to combat diseases including Alzheimer's, cancer, malaria, and TB (Aalbersberg et al., 1997).

The University of the South Pacific (USP), Fiji, and the US partners—the National Institutes of Health (NIH), the National Science Foundation (NSF), and the United States Agency for International Development (USAID)—share 50% of the earnings if a derivative is commercialized. Transfer of research results to a third party for commercialization will be with due permission and rewards. Local resource owners will get an equitable portion of all benefit-sharing at the local level, which will be divided by USP and the government. In terms of intellectual property rights, the partnership agreement states that sole inventors are entitled to patents for research conducted in their lab; filing and maintenance expenses are covered by the lab; in this case, the lab is GIT in the USA, which is one of the collaborative partners that have discovered new molecules. The accessor will cover monitoring support for research grants for junior researchers at USP (Government of Fiji, 2007).

If monetary value is obtained through licensing or royalty from the commercialization of a molecule, USP as a partner would be included in the filing for a patent or licensing where appropriate and serves as a focal point to coordinate the distribution of benefits between developers and the local communities. Additional financial assistance was given to Fiji in the form of a contribution to its initial investment, which totaled about FJD 1 million during its first term from 2006 to 2009. Building capacity, actively conserving marine resources, raising public awareness, and enabling stakeholders to collect and manage resources sustainably are examples of non-monetary benefits. One notable instance is the Navakavu project, which developed a locally administered marine conservation area for the Navakavu maritime area to support the site's conservation activities. As part of non-monetary advantages, more students were awarded fellowships and better, more advanced degrees (Feussner et al., 2012).

Reacting to the Emergence of Ash Dieback

Volunteers for conservation noticed signs of a strange illness in mid-September 2012 in historic woodland in Norfolk, United Kingdom. Ash trees are the third most prevalent tree in the country. DNA sequencing revealed after examination of the contaminated material that the infection was caused by a fungal pathogen that was proliferating throughout Europe. Several research organizations organized an ad hoc emergency reaction committee of scientists, realizing that Norfolk was probably at the edge of the pandemic moving from Europe, and they began sequencing the genomes of the virus and affected trees. They committed to providing instant open access to all data in the hopes of fostering a prompt reaction.

To make data open and accessible to biologists, genomic scientists, and the general public, they developed websites and software tools. For the latter, they created a Facebook game that had over 63,000 plays in its first year, allowing anyone to participate in the quick assembly of genomes. The "Open Dieback Project" led to an unparalleled rate of discoveries. The consortium found the genetic markers for trees that are less susceptible to the illness and the principles needed to choose and develop plants that are more resistant to it in a matter of years (Pautasso et al., 2013).

Research on the Microbiome of Soil and Water: The EcoBiomics Initiative

An interdepartmental project led by Agriculture and Agri-Food Canada (AAFC) aims to characterize aquatic and soil microbiomes and invertebrate zoo biomes in detail and test hypotheses to improve environmental monitoring, assessment, and remediation efforts. This will generate new knowledge to improve soil health and water quality. Additionally, it seeks to generate baselines of biodiversity and to evaluate potential future changes to soil and water biodiversity through significant long-term environmental monitoring sites across Canada. The goal of biological variety conservation is directly supported by the EcoBiomics initiative. This study profiles the microbial and invertebrate communities in different water- and soil-related environments, such as lakes, rivers, and pristine natural regions, using metagenomics techniques.

The findings of the sampling conducted in these areas using standardized DNA techniques will characterize the biodiversity in terms of taxonomy, ecology, and function. The capacity to identify the species in the sample and maybe their function in the ecosystem under study depends entirely on known sequencing information available from resources like GenBank. This initiative has the potential to publicly exchange knowledge and aid in the discovery and naming of species that science is currently unaware of.

To identify species and communities that could require preservation and, consequently, possible regulatory actions, monitoring is crucial. Maintaining variety is usually linked to healthy systems and thus production, as well as monitoring for invasives and functional changes that might harm productivity is necessary. As such, there is also a strong economic link to the production systems involved. Since EcoBiomics' production systems depend on sustainable ecosystem services derived from biodiversity for their productivity, the organization's goals also directly promote the sustainable use of biodiversity components. Metrics from the biomonitoring will be used to evaluate sustainability. It would also specify what kind of remediation—such as changing the composition of the soil or the water—is required to enable ecosystems to resume their "normal" functions (Edge et al., 2020).

Taxol's Sustainable Bioproduction for Chemotherapeutic Uses

An excellent illustration of how digital sequence information might assist sustainable usage is the endeavor to create an affordable, sustainable source of paclitaxel. Paclitaxel is a chemotherapeutic drug that is now marketed under the name Taxol and is used to treat several cancer types. Found in the bark of the Pacific Yew (*Taxus brevifolia*) tree, it was discovered in 1971 and licensed for use in medicine in 1993. Today, the World Health Organisation counts it among the safest and most effective medications required in a healthcare system. Thousands of trees had to be cut down for clinical studies in the middle of the 1980s, which made the source trees an endangered species. Since then, plant cell culture and semi-synthetic chemistry have been used to manufacture paclitaxel.

Because of these developments, the Pacific Yew is under less stress, which lessens the risk of extinction. However, availability and cost differ, and the medication is often listed on the American Society of Health-System Pharmacists' Drug Shortages List. Now that the biosynthetic pathway from the *T. brevifolia* genome has been extracted, many international research groups can use the digital sequence information to work towards the extremely challenging task of reprogramming species (like bacteria and yeast) that are suitable for large-scale, low-cost production with the large and complex paclitaxel biosynthetic pathway. This is made possible by the fact that these data are shared in public databases.

This approach offers the potential to enable the bioproduction of innovative chemical derivatives of the natural product, such as those with fewer side effects, in addition to offering the possibility of addressing global demands through low-cost, sustainable generation.

(https://www.researchgate.net/publication/352152547_The_contribution_of_Access_and_Benefit-Sharing_ABS_to_the_Sustainable_Development_Goals_-_Lessons_learned_and_best_practices).

The Potential for Biofortifying Maize Provitamin A Through the Identification of Spontaneous Genetic Variation

Researchers from CIMMYT (International Maize and Wheat Improvement Center) and several other educational and research institutions have been working on developing maize crops with higher levels of the precursors to vitamin A as part of a project funded by the US Department of Agriculture, Harvest Plus, and the Borlaug Fellowship. In sub-Saharan Africa, where vitamin A deficiency is frequent and can cause blindness and increased susceptibility to diseases, maize is an essential subsistence crop.

Researchers have been using expression analysis, association analysis, linkage mapping, and mutagenesis to find naturally occurring variations in the quantity of carotenoids generated in maize kernels. It has been demonstrated that natural variation at a lycopene cyclase locus influences the flow along certain carotenoid pathways, hence influencing the formation of provitamin A molecules. Breeders will be able to create maize grain with increased vitamin A levels by selecting the alleles with molecular markers found through sequencing data. This might have a transformative effect, improving the nutritional situation of millions of people in sub-Saharan Africa (Babu et al., 2013).

Conclusion

Overall, ABS is essential for the preservation of biodiversity, the advancement of sustainable development, and the fair distribution of the advantages resulting from genetic resources. ABS promotes responsible innovation and economic progress while assisting in the protection of biological variety via empowering communities, encouraging cooperation, and offering legal clarity.

Acknowledgements The author expresses gratitude to the Department of Computational Biology & Bioinformatics at the University of Kerala.

Conflict of Interest The authors state no conflict of interest.

References

Aalbersberg, W. G., Korovulavula, I., Parks, J. E., & Russell, D. (1997). The role of a Fijian community in a bioprospecting project. *Suva: Biodiversity Conservation Network, South Pacific Action Committee for Human Ecology and Environment*. https://pdf.usaid.gov/pdf_docs/PNACG917.pdf

Babu, R., Palacios, N., & Prasanna, B. M. (2013). Biofortified maize—A genetic avenue for nutritional security. *Translational Genomics for Crop Breeding: Abiotic Stress, Yield and Quality, 2*, 161–176. 10.1002/9781118728482.ch10

Edge, T. A., Baird, D. J., Bilodeau, G., Gagné, N., Greer, C., Konkin, D., Newton, G., Séguin, A., Beaudette, L., Bilkhu, S., & Bush, A. (2020). The Ecobiomics project: Advancing metagenomics assessment of soil health and freshwater quality in Canada. *Science of the Total Environment, 710*, 135906. DOI:10.1016/j.scitotenv.2019.135906. PMID: 31926407.

Feussner, K. D., Ragini, K., Kumar, R., Soapi, K. M., Aalbersberg, W. G., Harper, M. K., Carte, B., & Ireland, C. M. (2012). Investigations of the marine flora and fauna of the Fiji Islands. *Natural Product Reports, 29*(12), 1424–1462. https://doi.org/10.1039/C2NP20055D

Government of Fiji. (2007). *The Fiji national biodiversity strategy and action plan*. Republic of Fiji. https://www.cbd.int/doc/nbsap/nbsapcbw-pac-01/nbsapcbw-pac-01-fj-01-en.pdf

https://www.ethicalbiotrade.org/ingredient-stories/2019/6/19/shea-nuts

https://www.ethicalbiotrade.org/ingredient-stories/hibiscus

https://www.researchgate.net/publication/352152547_The_contribution_of_Access_and_Benefit-Sharing_ABS_to_the_Sustainable_Development_Goals_-_Lessons_learned_and_best_practices

Laladhas, K. P., Gopalan, S. V., Nilayangode, P., & Oommen, O. V. (2016). Conservation through payment for ecosystem services. *Biodiversity for sustainable development* (pp. 193–223). Springer International Publishing. https://doi.org/10.1007/978-3-319-42162-9_13

Laladhas, K. P., Nilayangode, P., & Oommen, O. V. (Eds.). (2017). *Biodiversity for sustainable development*. Springer International Publishing. https://doi.org/10.1007/978-3-319-42162-9_14

Laladhas, K. P., Nilayangode, P., Gopalan, S. V., Divya, L., & Oommen, O. V. (2017). Community conservation of Marine Turtles and sustainable ecosystems. *Biodiversity for Sustainable Development*, 177–184. https://doi.org/10.1007/978-3-319-42162-9_11

Laladhas, K. P., Preetha, N., Baijulal, B., & Oommen, V. (2015). *Conservation of medicinal plant resources through community born biodiversity management committee, Kerala* (p. 27). Biodiversity Conservation-Challenges for the Future. DOI:10.2174/97816810802151150101, eISBN: 978-1-68108-021-5, 2015 , ISBN: 978-1-68108-022-2

Mithrambika, N. B., Laladhas, K. P., Nilayangode, P., & Oommen, O. V. (2017). Decentralized governance for sustainable development. *Biodiversity for Sustainable Development*, 165–176. https://doi.org/10.1007/978-3-319-42162-9_10

Nilayangode, P., Cheruvat, D., Laladhas, K. P., & Oommen, O. V. (2016). ABS-Unlocking the opportunities. *Journal of Traditional and Folk Practices (JTFP), 4*(2). https://jtfp.jntbgri.res.in/index.php/jtfp/article/view/64

Nilayangode, P., Laladhas, K. P., & Oommen, O. V. (2016b). Biodiversity and sustainable development: A retrospect. *Biodiversity for sustainable development* (pp. 315–322). Springer International Publishing. https://doi.org/10.1007/978-3-319-42162-9_23

Nilayangode, P., Laladhas, K. P., Nair, G. U., Mathai, A., Baijulal, B., & Oommen, O. V. (2016c). Grassroots initiatives for environmental sustainability. *Biodiversity for sustainable development* (pp. 185–192). Springer International Publishing. DOI:10.1007/978-3-319-42162-9_12Corpus ID: 157892278

Nilayangode, P., Nair, G. U., Laladhas, K. P., & Oommen, O. V. (2017). Geographical indications and sustainable livelihood. *Biodiversity for Sustainable Development*, 225–234. https://doi.org/10.1007/978-3-319-42162-9_14

Normand, V., Oliva, M. J., Müller, S., al-Janabi, S., Zohren, Y., Nöske, N., Zedda, L., & Paulsch, A. (2021). The contribution of access and benefit-sharing (ABS) to the sustainable development goals. *Lessons Learned and Best Practices.* https://www.abs-biotrade.info/fileadmin/Media/Resources/Compendium-The-contribution-of-ABS-to-SDGs-Lessons-learned-best-practices-2021.pdf

Oommen, O. V., Laladhas, K. P., Nelliyat, P., & Pisupati, B. (Eds.). (2022). Biodiversity conservation through access and benefit sharing (ABS): Himalayas and Indian sub-continent. Springer. https://link.springer.com/book/10.1007/978-3-031-16186-5

Pautasso, M., Aas, G., Queloz, V., & Holdenrieder, O. (2013). European ash (Fraxinus excelsior) dieback–A conservation biology challenge. *Biological Conservation, 158*, 37–49. https://doi.org/10.1016/j.biocon.2012.08.026

Preetha, N., Laladhas, K. P., & Oommen, V. O. (2015). Biodiversity conservation-challenges for future and way forward. *Biodiversity and Conservation, 1*, 249.

Thormann, J., Ahrens, D., Anderson, C., Astrin, J., Mumladze, L., Rulik, B., Tarkhnishvili, D., & Espeland, M. (2019). A prelude to the caucasus barcode of life platform (CaBOL): Biodiversity days in Georgia in 2018 and 2019. *Bonn Zoological Bulletin, 68*(2), 275–296. http://dx.doi.org/10.20363/BZB-2019.68.2.275

Bioprospecting for Cosmetics

Laladhas Krishna Panicker, V. S. Soumya, Arathy Sudhakaran, M. P. Vinod, Neenu Mohan, K. Shahina, K. Sherin, and Oommen V. Oommen

Introduction

Investigating natural sources including plants, marine life, and microbes in search of novel bioactive compounds for cosmetic formulas is known as bioprospecting in the cosmetics industry. These bioactive substances frequently have anti-aging, moisturizing, anti-inflammatory, and antioxidant qualities that are advantageous for health care. Usually, the procedure begins with identifying natural sources, and the researchers locating potentially beneficial bioresources for cosmetic purposes. These might be specialized extracts with proven medicinal qualities- of animals, algae, fungi, or even plants. To isolate the necessary components, samples are taken from the sources that have been identified, and bioactive compunds are extracted utilizing a variety of procedures. Subsequently, in vitro and occasionally in vivo experiments are practiced to evaluate the extracted chemicals and the extracts for their cosmetic qualities. This entails stability, safety, and effectiveness testing. Cosmetic formulations are created once promising chemicals have been found. Creams, serums, lotions, and masks are some examples of these formulae. If successful, clinical trials are conducted on human volunteers to evaluate the safety and effectiveness of the formulations'. Regulatory clearance depending on the jurisdiction and the precise claims made about the cosmetic items, can be necessary. By doing this, safety of

L. Krishna Panicker (✉) · M. P. Vinod · N. Mohan · K. Shahina · K. Sherin · O. V. Oommen
Department of Computational Biology & Bioinformatics, University of Kerala, Thiruvananthapuram, Kerala 695581, India
e-mail: laladhas@gmail.com

L. Krishna Panicker · O. V. Oommen
Center for Sustainable Environment-India Biodiversity Action, Thiruvananthapuram, Kerala, India

V. S. Soumya
St. Stephens College, Kollam, Kerala, India

A. Sudhakaran
Department of Zoology, University of Kerala, Thiruvananthapuram, Kerala 695581, India

L. Krishna Panicker et al. (eds.), *Biodiversity and Business*,
https://doi.org/10.1007/978-3-031-71674-4_8

the items' for consumer usage is ensured. Following regulatory clearance, consumer marketing and sales of cosmetic goods can begin.

For the cosmetics business, bioprospecting has various benefits; It may lead to discovery of new chemicals, satisfy customer desire for sustainable and natural goods, and access traditional knowledge of indigenous populations. To protect biodiversity and uphold the rights of local communities and indigenous peoples, bioprospecting must be carried out ethically and sustainably. Sustainable sourcing methods are also necessary to avoid overexploitation and to protect biodiversity. Certain firms allocate their resources toward conservation programs, sustainable harvesting practices, and community development activities in the areas where these resources are located have been reported (Laladhas et al., 2015, 2016, 2017a, 2017b; Mithrambika, 2017, Nilayangode 2016a, 2016b, 2016c, 2017; Oommen et al., 2022; Preetha et al., 2015). All things considered, ABS-bioprospecting in the cosmetics industry presents chances for creativity and the creation of superior goods while supporting biodiversity conservation and upholding the rights of indigenous groups. This chapter covers bioprospecting in the cosmetics industry as well as access benefit sharing.

Buchu: Benefit Sharing with Traditional Knowledge Holders in South Africa

Buchu (*Agathosma betulina*) is a little blooming shrub native to the Western Cape of South Africa. These days, food, cosmetics, and medicines employ its essential oil and leaves, which have an aroma similar to peppermint. The indigenous Khoi and San people have been using it for millennia, for medicinal purposes, as an antiseptic, and in cosmetic applications when combined with animal fat. The Khoi and San people refer to buchu as the "holy plant" because of its significance.

The rights of indigenous people over their traditional knowledge related to bioresources were guaranteed by the CBD and Nagoya Protocol, which gave them the authority to establish regulations on biodiversity in South Africa. Discussions on the usage of local Buchu were held with businesses that utilize indigenous biological resources for product development, market research, and commercialization. The indigenous people would be benefited by such a study. Local laws and regulations also support the conservation of these resources to ensure sustainable harvest.

Buchu is a significant source of income for the local population in the area. The plant may be grown on hills where other crops do not fare as well since it grows well in dry environments. Provital, a Barcelona-based firm and a member of UEBT, creates botanical extracts and natural active ingredients for the cosmetics market. Provital became aware of Buch qualities through the study of the literature, and they started acting following both their ethical sourcing pledges and national laws governing access and benefit sharing (ABS).

To ensure that traditional knowledge holders would profit from the sales of Buchu, the first stage was developing a benefit-sharing agreement and negotiating with the

Khoi and San peoples with the assistance of local partners. Additionally, the providers and the holders of traditional knowledge came to a benefit-sharing arrangement amongst themselves. The businesses recognize the San and Khoi people as South Africa's legitimate original occupants and their traditional knowledge of the Buchu plant as part of these agreements. Provital obtained the necessary ABS permissions from the South African Department of Environment, Forestry, and Fisheries through negotiations and talks.

Today, Provital utilize Buchu to create "Affipore," a novel natural active component of a variety of cosmetic products. As part of its sourcing philosophy, which respects both people and the environment, it collaborates with local people to cultivate Buchu using organic methods, pays farmers fairly, and incorporates biodiversity principles. Additionally, it has ABS agreements in place with other nations for other components (https://www.ethicalbiotrade.org/ingredient-stories/buchu).

Carnauba

Copernicia prunifera, or carnauba a palm tree is significant for biodiversity and for supporting certain Brazilians who are economically disadvantaged. Lara is the President of Pontes Indústria de Cera, her family's fourth-generation business that manufactures wax and sells it to firms in the food, pharmaceutical, and cosmetics industries.

Carnauba wax, also known as queen of waxes, has hard and shiny properties, used as a gelling agent in a number of products. It is derived from wild trees (that are not cultivated by people), is used in a wide range of items, including cosmetics, sweets, and shoe polish. The leaves of carnauba palm is a native to the northeastern Brazilian states of Ceara, Piaui, Paraiba etc. The leaves are trimmed by local farmers and then sun-dried, a method that preserves the ecosystem and the integrity of the tree. After that, the leaves are ground into a fine powder and delivered to nearby refiners. When it rains and there are limited possibilities to work in agriculture, the crop provide much-needed jobs for the surrounding populations. The tree is so revered in the area that handicrafts and regional architecture frequently include its straw. Regional flags even display their pictures.

In recent years however, after investigating, the Brazilian government discovered unfavorable working conditions in a number of the nation's rural industries, including the production of carnauba wax. This is one of the reasons Lara is devoting even more time and energy to making sure her business sources ethically, including cooperating with UEBT to make changes.

The collaboration of businesses with UEBT was a crucial tool in redefining approaches to guaranteeing respect for people and nature. To assist in identifying subpar working conditions, the firm is now expanding on its prior efforts from 2016 to provide the government with access to all of its audit reports and supply chain data. A team of fifteen local experts, including solicitors and health and safety engineers, are currently in charge of a sizable educational initiative that was just introduced in the Carnauba fields. Pontes has given a variety of heat bottles, resting tents, bundles of personal protection equipment, and portable toilets in the area.

Employees are also taught proper gathering techniques and how to preserve Carnauba trees for the purpose of sustaining the region's biodiversity during the training sessions. The collectors of Carnauba operate as independent contractors and are not obliged to sell the powder to the Pontes firm in exchange for any equipment or training. During the harvest season, the program provides free over-the-phone legal counsel to workers seven days a week, empowering them to understand their rights.

The firm was evaluated as part of its engagement with UEBT, and after submitting an improvement plan, it was accepted as the organization's first Carnauba wax maker member. There are now four UEBT-member carnauba wax producers. New members pledge to strive for ethical sourcing conformity with global standards and must gradually demonstrate progress in this regard. The member company's work plan is approved by UEBT once it has verified the commitment.

For now, Pontes is the only company selling some UEBT-certified carnauba. After independent audits confirm that the regions where carnauba wax is obtained conform with the Ethical BioTrade Standard (UEBT standard), certification is given. These audits aid in ensuring compliance with the standards for local development, fair and equitable benefit sharing, worker rights, and the sustainable use of biodiversity. This involves audits to verify adherence to global laws governing labour standards, minimum age requirements for employment, freedom of association and collective bargaining, and occupational health and safety. It also involves controls on salaries and working conditions. To encourage sector reform, the Initiative for Responsible Carnauba (IRC) was established. The IRC is an alliance of businesses (manufacturers, distributors, and end-users of carnauba wax), non-profits (UEBT included), and governmental organisations that collaborate to guarantee conscientious manufacturing. Manufacturers in the IRC who sell carnauba wax must demonstrate how they plan to comply with the UEBT Ethical BioTrade Standard (https://www.ethicalbiotrade.org/ingredient-stories/2018/10/16/carnauba).

"The Scent of the Endangered Flora," Sourced from Clanwilliam Cedar and Distributed in South Africa in a Sustainable Manner

Widdringtonia wallichii is a native plant of South Africa, sometimes known as Clanwilliam cypress or Clanwilliam cedar. This tree is unique to the Western Cape Province's Cederberg Mountains, which are northeast of Cape Town. Due to excessive harvesting of its timber, which is the only source used to make the aroma, this priceless bioresource is in danger of being extinct. Through CapeNature, a public organisation that manages the bioresource in the mountain responsibly, the majority Heuningveli population was educated to protect and collect the bioresource.

The company Givaudan produces worldwide, flavours and perfumes from this bioresource. Without taking any live samples from the mountain, the producers used cutting-edge technology to extract volatile compounds from the leaves and dead wood of Clanwilliam cedar and spread awareness of the value of protecting the endangered species. By creating a synthetic perfume concentration with additional chemicals, the makers were able to replicate the fragrance of Widdringtonia. This project serves as a model that will be on display to encourage the communities to preserve this endangered tree.

Following South African benefit-sharing and access regulations, a benefit-sharing agreement was signed by the organisation, Givaudan and The Body Shop together as users, and CapeNature as the provider. This instance is an intriguing example of the knowledge base connected with ABS equipment. The applicability of the ABS standards is not clear-cut in this case because headspace technology solely involves capturing volatile chemicals and does not need physical samples.

Prior informed consent has been acquired and mutually agreed upon agreements, including benefit sharing with resource provider CapeNature, have been authorised by the African national authorities. Scientifically carrying out its mandate, the public organisation CapeNature was founded with legislative duty for biodiversity protection in the Western Cape.

A benefit-sharing agreement was completed in October 2018, and for the Heuningvlei community—a significant steward of the Clanwilliam cedar—to gain from the parameters that were mutually agreed upon, CapeNature insisted on including them in the process. Benefits that were monetary and non-monetary were taken into account. The process of creating and selling The Body Shop perfume as the finished product was taken into consideration while defining monetary advantages.

The financial benefits have been deposited into the Bioprospecting Trust Fund in compliance with national ABS regulations. From there, CapeNature receives the funds for the conservation and livelihood programs that have been decided upon in the mutually agreed agreements between the supplier and users. The goals of these initiatives are to protect the Clanwilliam cedar and to uplift and sustain the way of life in the Heuningvlei neighbourhood. (Normand et al., 2021).

An ABS Agreement by Bhutan, Asia for the Usage of Orchids

Bhutan is a nation rich in biological variety located in the ecologically significant Himalayan region, known for its exceptional conservation policies. A great variety of orchid species, many of which are indigenous, have been identified and preserved. In Bhutan 469 orchid species are known to exist. A few orchid species are used as food items, but many species are used in traditional medicine, which is a vital component of healthcare in rural communities.

The management of biodiversity protection and sustainable use in Bhutan is overseen by the National Biodiversity Centre (NBC). The commercialization of wild orchids, such as *Cymbidium erythraeum*, holds substantial value. A business called

Quantum Pharmaceuticals Limited (QPL) applied for a licence to study and utilise the Cymbidium erythraeum species for commercial purposes to the Ministry of Agriculture and Forests. By removing the cellulose from orchids, cosmetic lotions have been created. This marked the beginning of the large-scale cultivation of Cymbidium erythraeum species (http://www.nbc.gov.bt/wp-content/uploads/2010/06/ABS-Toolkit-final.pdf. Accessed 28.4.2020).

The goal of the study was to determine whether the cellulose found in orchid cells might be utilised to make a cosmetic cream. The demand for orchids expanded as a result of the commercial possibilities. Previously, the Cymbidium erythraeum species was mostly taken from the wild rather than being grown. NBC and QPL signed a Memorandum of Agreement to commercialise orchids sustainably and agreed ABS terms with the local inhabitants following the legislation that Bhutan has implemented. The Biodiversity Centre designated a group (Community group) of eighteen houses in Samtengang village, Nisho Gewog, Wangduephodrang, to cultivate and propagate C. erythraeum to extract its constituent elements.

QPL as the user and the NBC, who was acting on behalf of the providers, signed a memorandum of understanding. The agreement anticipates royalties on the sales of the cosmetic product as well as monetary advantages, whereby each family is paid a premium price for their accessed genetic resource. The money received from the propagation house's orchid flower production is administered through a communal fund under the recently created bylaw. a direct transfer of funds to the Bhutan Access and Benefit Sharing (BABS) Fund, which aims to maintain livelihoods while also conserving Bhutan's bioresources.

QPL decided to improve the infrastructure for the orchid propagation house, provide training to the community on orchid propagation, and use their household collections as the mother stock for the propagation house. In addition, the community was trained in orchid propagation. Benefits will be discussed and decided upon by both parties. QPL Company has successfully implemented research and development in Bhutan and has obtained patents in a few other countries. The creation of formulas for aesthetic skin treatments was the subject of the patent that was obtained. The therapy addresses skin wrinkling, elasticity, and firmness loss, and enhanced pigmentation associated with ageing (Susanne, 2017).

Latin America's Creation of Cosmetics Based on the Bioresources of the Amazon

Brazilian biodiversity is promoted economically by Natura Cosmetics, which goes by the trade name Natura Ekos. Local communities in Amazon have established sustainable cooperation to ensure the continued gathering and supply of raw materials needed to make the product. These indigenous groups, known as "extractive communities," are smallholders with authorization to harvest natural bioresources from the Amazon, such as Andiroba and Murumuru, in the Medio Jurua area.

The native palm tree of the Amazon, Murumuru, yields butter from its seed which is rich in fatty acids and lauric acid, making it an intriguing option for use as an emollient in cosmetic products for dry skin and hair. Andiroba tree is another tree endemic to Brazil and is known for its restorative and emollient properties for skin when prepared raw as oil. To provide a consistent and sustainable supply of Andiroba oil and Murumuru butter, Natura Cosmetics has engaged and forged relationships with the communities in Media Jurua. Natura Cosmetics takes into account not only the technical aspects of these supply chains but also makes sure that they positively impact the well-being and standard of living of the community.

The agreement was reached between Natura, the entity tasked with managing the region where the resources were used, and the representatives of three local community organisations, and the Chico Mendes Institute for Biodiversity Conservation (ICMBIO). It was thought to be a valuable investment since it is consistent with Natura's larger sustainability goal and allowed for the formation of a strong ABS agreement in terms of benefit-sharing mechanisms, even if the involvement of several players contributed to a lengthy process for its ABS agreement. Learnings from this experience have also been beneficial in promoting the creation of new supply chains by preserving and enhancing the region's bioresources and forming solid and reliable alliances with Amazonian people.

Financial benefits are linked to the profits made by Natura products made with Andiroba oil and Murumuru butter. These profits are allocated to a fund that supports conservation, sustainable bioresource use, and local livelihoods. It was agreed that 0.75% of product revenue during the commercialization period would go towards the conservation projects and that this percentage would be paid annually. For example, Natura shared benefits totalling about € 133,899.42 in 2017, supporting projects that involved four local community organisations that served 485 families from 42 communities. In 2018, benefits totalling € 167,602.82 were distributed to nine organisations.

The establishment of the Medio Jurua Fund, which is cooperatively administered by all stakeholders and facilitates transparent benefit sharing, was a crucial component of the Mutually Agreed Terms. It guarantees positive contributions through socio-environmental initiatives that benefit the whole community. An executive secretary and a committee made up of members from Natura, ICMBio, and the three community organisations oversee the Fund's administrative structure. The executive secretary is in charge of overseeing and managing the projects' execution as well as the committee's administrative operations. The fund works on several conservation initiatives, such as protecting forest regions, pirarucu fish, and endangered turtles.

Benefits that are not monetary include strengthening capacity through environmental education initiatives, such as youth-focused species conservation programs and management training. The construction of upgraded infrastructure for three community stores to provide more economical access to necessities including food, fishing tackle, and fuel. Increasing the production of bioresources and improving the quality of Andiroba oil would improve supply chains and empower the local population in decision-making (https://www.premiumbeautynews.com/en/murumuru-seeds-a-cosmetic,8860).

The Manufacturing of Cosmetics from Uchuva Calix in Colombia, Latin America

Commonly referred to as golden berries or cape berries, *Physalis peruviana* is a plant that is widely found in South America and is native to Ecuador, Peru, and Columbia. It is also known by its country of origin, uchuva. Its fruit is consumed, and in both Europe and the United States of America there is a high demand for it.

However, there is a limited market for uchuva calix, which is a by-product that providers often dump. Cosmo's research and development on thc uchuva calyx revealed a strong natural component with beauty benefits. A lotion for eye contouring and anti-aging was created using this novel component.

The first international business to secure a contract for access to Colombia's genetic resources was Cosmo International, a firm that develops natural components for cosmetics. In 2017, Cosmo International and delegates from the Colombian Ministry of Environment and Sustainable Development—the relevant government authority—signed a ten-year contract. The "Universidad Agraria de Colombia" was also involved in the benefit negotiations to offer scientific support and guidance on non-monetary advantages. OCATI, the supplier, was tasked with facilitating the connection between the farmers who produced the raw materials and the buyers.

A one-time payment of around sixty dollars for each species that is researched and developed was agreed; in this case, "*Physalis peruviana*" is the species utilised. An agreed proportion of the net margin—which is the total sales of ingredients less the expenses of purchasing raw materials, processing, administration, and taxes—is paid annually. The research findings will be published in a scientific publication, and OCATI and Uchuva will provide yearly workshops for farmers to discuss the research findings and provide training on topics about the sustainable use of biodiversity and commerce. It was decided to fund and assist two environmentally friendly company ventures started by farmers that provide OCATI with sustainable farming methods (UEBT Conference, 2019).

Aroma and Taste of Kurinji Flower Nectar

This honey is collected from the Western Ghats region during the flowering season of the shrub known as Kurinji or Neelakurinji (*Strobilanthes kunthiana*), which grows in the Western Ghats of Kerala and Tamil Nadu's Shola forests, India. These Neelakurinji (blue flower) blossoms are most likely the source of the name Nilgiris Hills, which translates to "Blue Mountains." Since Kurinji blossoms once every twelve years, so this honey is among the rarest biological resources available. This honey smells powerful and tastes very sweet. Honey from Kurunji Hill is extracted using natural bee colonies.

In the Shola jungles of Tamil Nadu, the Paliyans, an indigenous group, collect wild honey from the Kurunji flower. A Kodaikanal-based small business called "Hoopoe on a Hill" collaborates with the local indigenous tribes to promote their wild honey. The Indian supplier, "Jasmine Concrete," bought industrial honey jars from "Hoopoe on a Hill" to ship to Firmenich Grasse, a French fragrance and flavour manufacturer.

Firmenich identified airborne chemicals coming from the honey (not from the bloom itself) through research and development employing cutting-edge analytical technologies. The method just requires access to a sample of commercial honey; neither the blossom nor the honey is physically extracted throughout the process. Firmenich uses additional components from their fragrance and flavour palette, together with their technological know-how, to recreate the aroma of Kurinji honey.

The NBA's standards for upfront compensation for research initiatives are followed in terms of monetary advantages. If commercialization proceeds, further benefit-sharing would have to be agreed upon by current legislation. The communities concerned and the governmental bodies in charge of protecting these priceless treasures will split the rewards. The money obtained as per the requirements will be deposited into the National Biodiversity Fund, a legally mandated fund intended for livelihood and conservation. (https://biodiv.de/fileadmin/user_upload/PDF/Projekteaktuell/Handout_The_contribution_of_ABS_to_the_SDGs_final.pdf).

Determine Cosmetics in Vietnam—Technological Evaluation

The French company BASF Beauty Care Solutions France SAS (BASF) accessed *Salvia miltiorrhiza*, a perennial plant known locally under the names red sage, tan shen, and danshen belonging to the *Lamiaceae* family, from local sellers in 2017. The plant's root is prized greatly for its ancient medicinal uses. It grows on slopes and along stream banks at elevations of 90 to 1,200 m. It is native to Vietnam, China, and Japan. To assist in the discovery and screening of the plants' cosmetic constituents, BASF and the Centre for Rural Economy Development agreed to a partnership agreement after the Ministry of Natural Resources and Environment confirmed the registration in February 2018. In April 2018, the cosmetics firm and the local supplier inked an ABS deal after negotiations. In May 2018, the People's committee at the commune level certified the deal. Access to the bioresource and benefit sharing for commercial application was completed in April 2020 following national rules.

Financial gains amount to 1% of the yearly revenue from ingredients used in cosmetics that are directly developed from access to biological resources. Of this, 50% is set aside for the provider to support community development initiatives aimed at enhancing standard of living of the local community or conserving biodiversity, and the remaining 50% is paid to the state budget. The advantages to the provider will be utilized towards capacity-building initiatives including educational programs on sustainable harvesting practices, farming methods, and environmental preservation and conservation. The projects aimed at conserving biodiversity will be funded by state funds (Ministry of Natural Resources and Environment, Vietnam Environment Administration).

A Cosmetic Derived from the Active Components of Costa Rican Green Coffee

Coffea arabica, also referred to as Arabian coffee and a member of the *Rubiaceae* family, is a major cultivar species and is said to have been the first coffee species to be grown, accounting for roughly 60% of the world's supply. Raw, unroasted coffee beans are known as green coffee beans. Some of the beneficial, organic compounds in thc beans appear to be destroyed during the roasting process. *C. arabica* has biochemical constituents that have been found and isolated that have strong anti-aging properties that revitalise and support healthy skin.

The French cosmetics business Chanel Parfums Beauté (Chanel) began studying green coffee molecules from Costa Rica's Nicoya Peninsula in 2014 to create a cosmetic formulation. Two compounds, Kahweol and Cafestol, which are present in this coffee and have incredibly strong antioxidant and anti-inflammatory qualities, were identified by their investigation. Chanel collaborated with a nearby oil seed producer, who created a novel method for removing the oil from the dry beans, because such molecules are only found in the oleaginous form of the green coffee.

In compliance with Costa Rica's ABS regulations, the study plan was examined under the direction of Comisión Nacional para la Gestión de la Biodiversidad (CONAGEBIO), the national body in charge of allowing access to genetic resources. A license for the commercial use of Nicoya green coffee's genetic and biochemical resources was given by CONAGEBIO in 2016.

As non-financial benefits, Chanel signed a three-year renewable framework agreement for sustainably grown plant-based raw materials, that provides financial security to the raw material providers; additionally, Chanel financed a new coffee-roasting facility to encourage the production and sale of locally roasted coffee, increasing the purchase price of the coffee and supporting the livelihood capacity (https://bit.ly/32NbdeL).

The Potential Use of Marine Microalgae and Cyanobacteria in Cosmetics

Components produced by cyanobacteria and microalgae might be useful to the cosmetics (personal care goods) sector. Cosmetic and cosmeceutical products now contain active compounds isolated from certain microalgae; nevertheless, further investigation into their mechanisms of action is required. Cosmetics are products that work with excipients and active chemicals tailored to certain skin types (normal, oily, combination, sensitive, etc.) to improve the structure, morphology, and look of skin. The term "cosmetic products with biologically active ingredients purporting to have medical or drug-like benefits" is known as "cosmeceutics," although the term does not have formal recognition. The fact that active compounds made from microalgae may be utilised to treat seborrhea, prevent blemishes, heal damaged skin, and stop inflammation is significant.

Polysaccharides, which are also derived from macroalgae and utilised as gelling agents, thickeners, and moisturizers in various cosmetic formulae, are among the active substances from microalgae that have potential use in cosmetics (for example, the genus Chlorella). Furthermore, some of these polysaccharides—particularly ß-1,3-glucan—are strong immunostimulators and effective free-radical collectors, which makes them suitable candidates for use in skincare cosmetics, particularly in the prevention of external ageing, which is brought on by free radical-induced oxidation and environmental factors.

Furthermore, since ß-1,3-glucan from fungi and plants seems to have anti-inflammatory properties, it stands to reason that microalgal glucans would likewise have these properties. This would make microalgal glucan research intriguing for the cosmetics industry, which targets sensitive and reactive skin types. *Porphyridium, Nostoc flegelliforme, Skeletonema diatom*, and the *Chlorella* genus are species that are high in ß-glucans. Vitamin E is a common ingredient in cosmetic compositions and is also regarded as an efficient antioxidant. Sulfated PolySaccharide (sPS), which is produced and secreted by marine microalgae, has been shown by (Raposo et al. 2013) to have the ability to inhibit the accumulation and activity of free radicals. These findings are relevant to skin ageing supplements and cosmeceuticals. Extracts from *Chlorella vulgaris* have also been reported to support collagen repair mechanisms. Consequently, sPS may function as defence mechanism against these agents of oxidative and radical stress.

Additional examples include skin-tightening treatments made from extracts of the marine microalgae *Nannochloropsis oculata*, which Pentapharm (Basel, Switzerland) recently introduced. Furthermore, *D. salina* algal extract Blue Retinol™ promotes the development and multiplication of skin cells. Another product, SILIDINE® by Greentech USA, was created using the metabolic induction technique from the purple-red alga *Porphyridium cruentum*. Its benefits include improved vascular tonicity, a reduction in heavy leg syndrome, an improvement in the appearance of the skin, and a reduction in the redness and effects of rosacea. Additionally, this product is suggested for increase-of body tonicity, sensitive skin, redness prevention, vascular tonicity, and uniformity and brightness of complexion. Furthermore, a protein-rich Arthrospira extract is said to tighten skin, prevent stria, and heal indications of ageing (Protulines, Exsymol S.A.M., Monaco).

Several new patents have been filed recently for compositions suitable for topical administration that include microalgal exopolysaccharide particles (e.g., the exopolysaccharides produced by microalgae of the genus Parachlorella) for enhancing the health and appearance of skin by Solazyme Inc. (San Francisco, CA, USA) and others that include biomass from *Chlorella protothecoides* (Solazyme Inc.) or microalgae biomass combined with microalgal oil by Terravia Holdings, Inc. (San Francisco, CA, USA), which recently introduced GoldenChlorellaTM and AlgaPürTM Algae Oils, both of which are claimed to have significant cosmetic benefits for skin and hair. Various bioactive compounds could be considered potential candidates as active ingredients in cosmetics and cosmeceuticals for skin care (Mourelle et al., 2017).

Conclusion

To sum up, bioprospecting in the cosmetics industry presents a viable way to find novel, sustainable chemicals sourced from biodiversity. Through innovation and the creation of products with distinct skincare advantages, cosmetic enterprises may leverage the abundance of natural resources present in plants, sea organisms, and microbes. However, bioprospecting must be carried out ethically and responsibly, taking into account the rights and conservation activities of the indigenous groups'. To ensure just and equitable relationships between biotechnology companies, cosmetic makers, and custodial communities or nations, access and benefit-sharing agreements are essential. ABS in bioprospecting has the potential to significantly boost innovation in the cosmetics sector, assist livelihoods of the local people', and promote biodiversity conservation as the demand for natural and eco-friendly cosmetic goods rises. The cosmetics business may support social responsibility and environmental stewardship by adopting sustainable practices and upholding the concepts of Access and Benefit-Sharing.

Acknowledgements The authors express gratitude to the Department of Computational Biology & Bioinformatics at the University of Kerala and the Council of Scientific and Industrial Research, New Delhi, India, for the Senior Research Fellowship to RS (08/666(0003)/2019-EMR-I).

Conflict of Interest The authors state no conflict of interest.

References

"Cosmo's experience of ABS process in Colombia—example of the uchuva extract", presentation by Cosmo International Fragrances, Natural Ingredients Division (then Cosmo International Ingredients), UEBT Conference, May 2019.

de Jesus Raposo, M. F., De Morais, R. M. S. C., & de Morais, A. M. M. B. (2013). Bioactivity and applications of sulphated polysaccharides from marine microalgae. *Marine Drugs, 11*(1), 233–252. https://doi.org/10.3390/md11010233

Kum, K. Y. (2020). Herbal product's quality control using ^{1}H-NMR metabolomics approach–An example from Danshen (Salvia miltiorrhiza Bunge.) (Doctoral dissertation, UCL (University College London)).

GIZ: ABS - Practical examples from Central America and the Dominican Republic. https://bit.ly/32NbdeL

https://biodiv.de/fileadmin/user_upload/PDF/Projekteaktuell/Handout_The_contribution_of_ABS_to_the_SDGs_final.pdf

https://www.ethicalbiotrade.org/ingredient-stories/2018/10/16/carnauba

https://www.ethicalbiotrade.org/ingredient-stories/buchu

https://www.premiumbeautynews.com/en/murumuru-seeds-a-cosmetic,8860

Laladhas, K. P., Gopalan, S. V., Nilayangode, P., & Oommen, O. V. (2016). Conservation through payment for ecosystem services. *Biodiversity for sustainable development* (pp. 193–223). Springer International Publishing. https://doi.org/10.1007/978-3-319-42162-9_13

Laladhas, K. P., Nilayangode, P., Gopalan, S. V., Divya, L., & Oommen, O. V. (2017). Community conservation of Marine Turtles and sustainable ecosystems. *Biodiversity for Sustainable Development*, 177–184. https://doi.org/10.1007/978-3-319-42162-9_11

Laladhas, K. P., Preetha, N., Baijulal, B., & Oommen, V. (2015). *Conservation of medicinal plant resources through community born biodiversity management committee, Kerala* (p. 27). Biodiversity Conservation-Challenges for the Future. DOI:10.2174/97816810802151150100050

Mithrambika, N. B., Laladhas, K. P., Nilayangode, P., & Oommen, O. V. (2017). Decentralized governance for sustainable development. *Biodiversity for Sustainable Development*, 165–176. https://doi.org/10.1007/978-3-319-42162-9_10

Mourelle, M. L., Gómez, C. P., & Legido, J. L. (2017). The potential use of marine microalgae and cyanobacteria in cosmetics and thalassotherapy. *Cosmetics, 4*(4), 46. https://doi.org/10.3390/cosmetics4040046

NBC. (2018). Access and benefit-sharing toolkit for the management of genetic resources and associated traditional knowledge in Bhutan. http://www.nbc.gov.bt/wp-content/uploads/2010/06/ABS-Toolkit-final.pdf. Accessed April 28, 2020.

Nilayangode, P., Cheruvat, D., Laladhas, K. P., & Oommen, O. V. (2016a). ABS-unlocking the opportunities. *Journal of Traditional and Folk Practices (JTFP), 4*(2). https://jtfp.jntbgri.res.in/index.php/jtfp/article/view/64

Nilayangode, P., Laladhas, K. P., & Oommen, O. V. (2016b). Biodiversity and sustainable development: A retrospect. *Biodiversity for sustainable development* (pp. 315–322). Springer International Publishing. https://doi.org/10.1007/978-3-319-42162-9_23

Nilayangode, P., Laladhas, K. P., Nair, G. U., Mathai, A., Baijulal, B., & Oommen, O. V. (2016c). Grassroots initiatives for environmental sustainability. *Biodiversity for sustainable development* (pp. 185–192). Springer International Publishing. https://doi.org/10.1007/978-3-319-42162-9_12

Nilayangode, P., Laladhas, K.P., Oommen, O.V. (2017). Biodiversity and sustainable development: A retrospect. In K. Laladhas, P. Nilayangode, & O. V. Oommen (Eds.), *Biodiversity for sustainable development. Environmental challenges and solutions* (Vol. 3). Springer. https://doi.org/10.1007/978-3-319-42162-9_23

Nilayangode, P., Nair, G. U., Laladhas, K. P., & Oommen, O. V. (2017). Geographical indications and sustainable livelihood. *Biodiversity for Sustainable Development*, 225–234. https://doi.org/10.1007/978-3-319-42162-9_14

Normand, V., Oliva, M. J., Müller, S., al-Janabi, S., Zohren, Y., Nöske, N., Zedda, L., & Paulsch, A. (2021). The Contribution of access and benefit-sharing (ABS) to the sustainable development goals. *Lessons Learned and Best Practices*. https://www.abs-biotrade.info/fileadmin/Media/Resources/Compendium-The-contribution-of-ABS-to-SDGs-Lessons-learned-best-practices-2021.pdf

Oommen, O. V., Laladhas, K. P., Nelliyat, P., & Pisupati, B. (Eds.). (2022). Biodiversity conservation through access and benefit sharing (ABS): Himalayas and indian sub-continent. Springer. https://link.springer.com/book/10.1007/978-3-031-16186-5

Preetha, N., Laladhas, K. P., & Oommen, V. O. (2015). Biodiversity conservation-challenges for future and way forward. *Biodiversity and Conservation, 1*, 249. https://link.springer.com/book/10.1007/978-3-031-16186-5

Susanne, M. (2017). *Orchids of Bhutan*. http://susannemasters.com/wp-content/uploads/2017/01/Orchids-of-Bhutan.-June.-The-Orchid-Review..pdf. Accessed April 28, 202.

Benefit-Sharing and Bioprospecting for Cosmeceuticals

Pravi Prasad and Sreejith Parameswara Panicker

Introduction

Cosmeceuticals are a category of biological products that lie between cosmetics and pharmaceuticals, and possess medicinal and therapeutic properties. It provides an extensive selection of hair and skin-protective options. However, these products often fall into a regulatory gray zone between pharmaceuticals and cosmetics, each subject to distinct regulations. Hence, the effectiveness of cosmeceuticals differs. In order to choose products that are suitable for their specific requirements and skin type, clients should thoroughly examine product labels and consult dermatologists or skincare experts. Various aspects interconnect cosmeceuticals and biodiversity, including the impact of biodiversity, ethical sourcing, biodiversity conservation, and the use of biodiversity in product marketing. Bioprospecting and cosmeceuticals are two distinct fields that involve exploring and utilizing natural resources for various purposes, such as developing medications and skincare products. Bioprospecting, also known as biodiversity prospecting, is the search and analysis of natural resources, primarily derived from microorganisms, plants, and animals, in order to discover new pharmaceuticals, biotechnology products, and other useful chemicals. Researchers conducting bioprospecting may uncover plant extracts, marine species, or microbes with potential skincare benefits, enabling the creation of innovative cosmeceuticals. This procedure emphasizes the significance of using sustainable and ethical methods to obtain natural components. The goal is to protect biodiversity and ensure that indigenous groups receive fair benefits when appropriate. The global cosmetics industry is expanding due to changing lifestyles. In contemporary times, makeup has become an indispensable element of every individual's daily life. Furthermore,

P. Prasad · S. P. Panicker (✉)
Department of Zoology, University of Kerala, Thiruvananthapuram, Kerala, India
e-mail: p.sreejith@gmail.com; psreejith@keralauniversity.ac.in

Advanced Centre for Regenerative Medicine and Stem Cell Research in Cutaneous Biology (AcREM-STEM), University of Kerala, Thiruvananthapuram, Kerala 695581, India

L. Krishna Panicker et al. (eds.), *Biodiversity and Business*,
https://doi.org/10.1007/978-3-031-71674-4_9

a significant factor influencing the global market's use of cosmetics is the growing awareness of an individual's inner intelligence and external beauty. By approximately 2030, it is anticipated that the cosmeceutical sector will experience significant expansion, with the Asia Pacific region predicted to dominate the worldwide market share.

In 1984, Dr. Albert Kligman from the University of Pennsylvania introduced the word "cosmeceuticals" to describe a category of products combining cosmetics and medications features (Pandey et al., 2019). It demonstrates a pharmacological therapeutic advantage (Arora et al., 2012). Cosmeceuticals are a type of skincare and cosmetic product with medical and cosmetic properties (Draelos, 2008; Mahesh et al., 2019). These items have been developed to aid the skin in both cosmetic and therapeutic ways (Ganesan & Choi, 2016; Shanbhag et al., 2019). Cosmeceutical products offer several advantages, such as combating the effects of aging, moisturizing the skin, enhancing skin radiance, treating acne, providing sun protection, and promoting personal cleanliness (Couteau & Coiffard, 2016; Datta & Paramesh, 2010; Draelos, 2021; Juncan et al., 2021; Levy & Emer 2012). Active substances such as antioxidants, peptides, vitamins, and other components found in cosmetics have been scientifically proven to have favorable effects on the skin (Hoang et al., 2021; Michalak 2022; Michalak et al., 2021; Souto et al., 2020). The U.S. Food and Drug Administration (FDA), a regulatory body, does not officially recognize the term "cosmeceutical" (Newburger, 2009). These commodities sometimes fall within the regulatory ambiguity between pharmaceuticals and cosmetics, subject to distinct regulations (Baki, 2022). Therefore, the effectiveness of cosmeceuticals differs clients should thoroughly examine product labels and consult dermatologists or skincare experts to choose products suitable for their specific requirements and skin conditions.

Cosmeceuticals substantially increased in popularity during the 1980s and 1990s (Jones, 2011). Nevertheless, individuals who were part of the upper echelon of society and had substantial financial resources could obtain those commodities. During that period, white supremacy was dominant. The process of industrialization has significantly increased the popularity of cosmeceuticals and made them more readily available to consumers. Not predominantly white countries also began producing goods tailored to their needs, applications and wants.

Additionally, there was an increase in trading at that period. Ultimately, it precipitates the inception of the cosmeceuticals industry (Hunt et al., 2011). Cosmeceuticals can be categorized according to the origin of their constituents, and certain products may belong to many categories depending on their source and mechanism. Cosmeceuticals can be classified into three main groups based on their purpose: skin care, hair care, and other applications. Based on the active component and its purpose, it is classified as retinoids, sunscreens, moisturizers, depigmenting agents, exfoliants, antioxidants, proteins/peptides, and growth factors (Alamgir & Alamgir, 2017; Draelos, 2019). The primary hair disorders that affect the population include alopecia (hair loss), premature greying, dandruff, and related problems. Shampoos, conditioners, hair straighteners, hair colors, henna, hair growth stimulants, and hair serums are often used in hair care cosmeceuticals (Fernanda & Dias, 2015). The primary concerns in skin care include skin aging, pigmentation, sun tanning, burning, and cancer (Geller et al., 2002; Gloster et al., 2006; Lin & Fisher, 2007; Reichrath et al., 2014; Tobin, 2017).

The most often utilized cosmeceuticals for skin care are vitamins, hydroxy acids, peptides, growth factors, and botanicals. The topical applications of vitamins include using vitamin A or retinoids to diminish the appearance of fine lines and wrinkles. Vitamin E, or alpha-tocopherol, is utilized for sun protection. Vitamin B3, also known as nicotinamide or niacinamide, is a potent hydrating and depigmentation agent that helps maintain smooth skin texture (Bissett et al., 2005; Dreher et al., 1998; Kang & Voorhees, 1998). Vitamin C is commonly applied topically because of its extensive use as an antioxidant and anti-inflammatory agent (Pinnell et al., 2001). The next category of skin cosmeceuticals consists of hydroxy acids, including α-hydroxy acids, β-hydroxy acids, polyhydroxy acids, and bionic acids. These acids provide many cosmetic advantages, such as anti-aging effects and depigmentation, particularly for acne-prone skin (Green et al., 2009). The peptide pal-KTTKS, or Matrixyl, stimulates collagen synthesis and enhances wound healing. Cu-GHK also promotes collagen synthesis, while Ac-EEMQRR, better known as Argireline, reduces wrinkle lines (Mas-Chamberlin et al., 2002; Robinson et al., 2002). Growth factors are regulatory proteins that play a role in the signaling pathway of wound healing and are also employed in skin care cosmeceuticals. Kinetin, sometimes called N-6 furfuryladenine, is a widely recognized growth factor that has been extensively researched in human skin fibroblast cultures. It has anti-aging and antioxidant effects and can improve skin texture, reduce hyperpigmentation, and act as a potent moisturizing agent (Levin & Momin, 2010; Martin & Glaser, 2011). In summary, it can be inferred that cosmeceuticals offer comprehensive skin and hair advantages.

However, ongoing debates exist about whether cosmeceuticals are classified as medications or cosmetics. There are misconceptions about the ingredients, the product's efficacy, and side effects. These doubts and questions contribute to skepticism about the validity of such goods. Although there are differences, cosmetics, nutraceuticals, and pharmaceuticals are similar. Nutraceuticals deliver nutritional benefits through diet. The term "cosmeceuticals" refers to products that combine elements of cosmetics and pharmaceuticals, primarily focusing on their medicinal properties (Rahul & Sreedhar, 2022). Nutraceuticals primarily establish a connection between food and health based on the correlation between well-being and dietary patterns. The nutraceuticals mostly contain active phytochemicals, particularly polyphenols (Espín et al., 2007). Pharmaceuticals, in contrast, refer specifically to drugs or medicines extensively utilized to prevent and treat disorders (Greene, 2007; Schmidt et al., 2007).

Cosmeceuticals; Biodiversity and Business

Biodiversity encompasses all living forms on Earth, including plants, animals, microbes, and the ecosystems they inhabit. Biodiversity primarily serves the purpose of sustaining and harmonizing life (Eldredge, 2000; Haahtela, 2019; Myers, 1996). Biodiversity has a crucial role in the generation of biological products. Cosmeceuticals are a significant category of biological chemicals that frequently exhibit beneficial skin and hair maintenance and improvement qualities. Natural substances with biological functions form the basis of Cosmeceuticals. Herbal components are renowned for enhancing skin and hair appearance (Dwivedi & Tripathi, 2014; Joshi & Pawar, 2015). Cosmetic manufacturers have transitioned to using herbal compounds as the primary source of components due to their minimal adverse effects and significant biological activity (Bent & Ko, 2004; Setty & Sigal, 2005; Verma & Singh, 2008). Essential oils, such as coconut, almond, rosemary, olive, sesame, and castor, are a prominent category of cosmeceuticals derived from plants. These oils contain natural compounds that have biological properties (Al-Waili, 2003; Asokan et al., 2009; Deen et al., 2021; Soler et al., 1988; Tan et al., 2009).

Beeswax derived from *Apis mellifera* possesses antioxidant, anti-inflammatory, and antibacterial characteristics, making it suitable for skin moisturizers and softeners (Fratini et al., 2016). Carnauba Wax derived from Copernicia prunifera and C. Cerifera is utilized in the production of lipstick and lip balm (Chowdhury, 2002). Cosmetics utilize gums like Tragacanth, derived from the stems and branches of *Astragalus gummifer*, as emollients. Agar, obtained from Gracilaria, is a cosmetics gelling agent (Glicksman, 2019; Rathod & Mehta, 2015). *Rosa damascene* and *Rosa centifolia* yield rose oils, while Lavandula limon officinalis yields lavender oil. These oils possess anti-inflammatory and antibacterial qualities (Łyczko et al., 2019; Pourabbas et al., 2005). The saffron obtained from *Crocus sativus*, curcumin from *Curcuma longa*, and sandalwood obtained from the genus Santalum are evergreen natural colorants (Cooksey, 2013; Rodriguez-Amaya, 2016). Antioxidants, such as beta-carotene found in carrots and tomatoes, protect against the harmful effects of UV radiation. Resveratrol, derived from berries and grapes, promotes a longer cell lifespan. Green tea extracts help prevent skin aging and the development of skin malignancies (Hsu, 2005; Larsson et al., 2007; Rahman et al., 2020). Several skin-protecting compounds utilized in cosmeceuticals include Aloe derived from *Aloe vera*, which offers sun protection against UVB and UVA radiation. Jojoba Oil, derived from the *Simmondsia chinensis* plant, is a moisturizer that diminishes wrinkles, eliminates suntans, and treats acne (Boudreau & Beland, 2006; Gentry, 1958). Hair cosmeceuticals are popular biological products used for personal care and appearance enhancement. Coconut oil derived from *Cocos nucifera* is a vital biological element that effectively protects hair from damage, prevents dandruff, conditions the hair, and enhances its general health (Goyal et al., 2022). Castor oil derived from *Ricinus communis* nourishes the scalp, prevents dandruff, promotes the health of hair follicles, and eventually stimulates hair growth (Trak & Chauhan, 2022). The plant *Lawsonia inermis*, commonly known as henna, yields the most commonly used hair dye. In addition, it promotes hair growth, reduces dandruff, conditions the hair, and protects against an itchy scalp (Semwal et al., 2014). People commonly use neem, derived from the *Azadirachta indica* plant, as an anti-dandruff treatment (Baby et al., 2022). These are some of the main natural substances that nature abundantly offers for the benefit of human well-being.

Natural biological products have an authentic source, few adverse effects, achieve advertised results, and exhibit superior overall efficacy compared to synthetic products. There has been a significant and sudden change in the production and promotion of biological chemicals, specifically cosmeceuticals. By approximately 2030, it is anticipated that the cosmeceutical sector will experience significant expansion, with the Asia Pacific Region predicted to dominate the worldwide market share. We expect organic and natural components to dominate the cosmeceuticals market due to their numerous advantages over synthetic substances. Their tagline "go back to roots" substantiates the key advantages of being chemical-free and less damaging. Another crucial determinant for business expansion is the constantly evolving social landscape. The current lifestyle demonstrates an unwavering commitment to appear and feel nice to the fullest extent. It will improve one's physical look, social standing, and appreciation, enhancing emotional well-being and overall confidence. It enables individuals to achieve excellence in both their personal and professional lives. Therefore, this psychological factor also contributes to the emergence and dominance of the cosmeceutical sector. Lastly, the impact of social media and new marketing methods undoubtedly contributes to the gradual growth of the cosmeceutical business. Social media is a highly influential and convenient instrument. Utilizing a sponsored promotion strategy to market products, particularly with the assistance of a well-known influencer, significantly enhances product sales. The reviews and ratings contribute to brand awareness and stimulate market expansion (72, 73, 74, 75).

Bioprospecting and Business Sharing

Bioprospecting is the methodical and scientific exploration of biodiversity to discover potential micro- and macro compounds, nanoparticles, components, and genetic information that can be used in the production and promotion of commercial industries, ultimately benefiting humanity and contributing to economic prosperity. The biodiversity study is contingent on the product's societal and economic importance. Industries such as pharmaceuticals, manufacturing, agriculture, aquaculture, bioremediation, biomining, biomimetic engineering, nanotechnology, and others mostly utilize biological resources. We conduct the search, identification, isolation, extraction, and utilization of potential chemicals, genes, metabolic pathways, structures, materials, and behaviors of biological components. Bioprospecting operates effectively by relying on the principles of conserving biodiversity, sustainably managing natural resources, and promoting economic growth (Beattie et al., 2011; Dobrowolski et al., 2017; Mateo et al., 2001; Nunes & van den Bergh, 2001; Rowland et al., 2018). Exploring biological resources, particularly plants, animals, and microorganisms, for their potential use in obtaining unique biological properties can contribute to developing innovative treatments for serious diseases like malaria, tuberculosis, HIV/AIDS, and cancer. It promotes and enhances the mutual reliance between the natural world and human beings (Castree, 2003; Demunshi & Chugh, 2010; Ngo et al., 2013; Shiva, 2007; Strobel & Daisy, 2003).

The potential economic benefits of bioprospecting lead to the implementation of strict laws and regulations to prevent fraudulent and unethical acts during the process. Nature bestows fairness and equality on all individuals. Therefore, it is not owned by anyone, but rather possesses ownership over everything. Therefore, it is crucial to distribute the benefits of bioprospecting, whether monetary or non-monetary, fairly and equitably. Biopiracy is a multifaceted problem that occurs when natural resources are used without adequately compensating or sharing benefits with the communities that have traditionally managed these resources. Powerful entities, such as multinational corporations, extremely wealthy individuals, government officials, and socially influential individuals, frequently take advantage of disadvantaged, traditional, and indigenous communities to exploit the natural resources they have access to. Financial stability and technology-driven processes are the means by which individuals can generate profits from biological variety. Developed countries often exploit emerging and underdeveloped countries, which possess enormous natural resources but lack biological diversity. Furthermore, individuals who prioritize power only appreciate the ultimate advantage, then neglect it without further attention. The prevention of cultural appropriation, the loss of traditional knowledge, and unjust economic practices should be achieved through legal, ethical, and equitable means (Bhattacharya, 2014; Chen, 2006; De Werra, 2009; Efferth et al., 2016; Nordin et al., 2012).

Biodiversity is consistently an authentic supply of elements for humanity's well-being. Natural resources play a significant role in the manufacturing of biological goods. Cosmeceuticals are a category of biological chemicals that provide a diverse range of advantages in terms of safeguarding hair and skin. The hair care cosmeceuticals include shampoos, conditioners, hair straightening products, hair colors, henna, hair growth products, and hair serums. On the other hand, skin care products comprise vitamins, hydroxy acids, peptides, growth factors, and botanicals. Despite the various biological qualities of cosmeceuticals, the regulatory association does not recognize their legitimacy and credibility due to their misplacement between cosmetics and drugs. Current lifestyle trends, improved job prospects, and the unwavering desire to enhance one's appearance and well-being are driving the projected rapid growth of the cosmeceuticals business by 2030. Since the product comes from nature, ensuring a fair distribution of financial and non-financial benefits in society is crucial.

Combining biodiversity and business *via* bioprospecting and benefit sharing is pivotal in developing cosmeceuticals. Companies can utilize diverse natural resources to develop innovative products that satisfy consumer needs and simultaneously encourage sustainability. Efficient benefit-sharing agreements guarantee that the communities contributing these resources receive equitable remuneration, promoting conservation endeavors and bolstering local economies. In essence, the interdependent connection between biodiversity and business not only stimulates economic expansion but also aids in the conservation of our planet's natural resources.

References

Alamgir, A. N. M., & Alamgir, A. N. M. (2017). Classification of drugs, nutraceuticals, functional food, and cosmeceuticals; proteins, peptides, and enzymes as drugs. In *Therapeutic use of medicinal plants and their extracts: Volume 1: Pharmacognosy* (pp. 125–175). Springer International Publishing.

Al-Waili, N. S. (2003). Topical application of natural honey, beeswax and olive oil mixture for atopic dermatitis or psoriasis: Partially controlled, single-blinded study. *Complementary Therapies in Medicine, 11*(4), 226–234.

Arora, N., Agarwal, S., & Murthy, R. S. R. (2012). Latest technology advances in cosmaceuticals. *International Journal of Pharmaceutical Sciences and Drug Research, 4*(3), 168–182.

Asokan, S., Emmadi, P., & Chamundeswari, R. (2009). Effect of oil pulling on plaque induced gingivitis: A randomized, controlled, triple-blind study. *Indian Journal of Dental Research, 20*(1), 47–51.

Baby, A. R., Freire, T. B., Marques, G. D. A., Rijo, P., Lima, F. V., Carvalho, J. C. M. D., ... & Morocho-Jácome, A. L. (2022). Azadirachta indica (Neem) as a potential natural active for dermocosmetic and topical products: A narrative review. *Cosmetics, 9*(3), 58.

Baki, G. (2022). *Introduction to cosmetic formulation and technology*. Wiley.

Beattie, A. J., Hay, M., Magnusson, B., de Nys, R., Smeathers, J., & Vincent, J. F. (2011). Ecology and bioprospecting. *Austral Ecology, 36*(3), 341–356.

Bent, S., & Ko, R. (2004). Commonly used herbal medicines in the United States: A review. *The American Journal of Medicine, 116*(7), 478–485.

Bhattacharya, S. (2014). Bioprospecting, biopiracy and food security in India: The emerging sides of neoliberalism. *International Letters of Social and Humanistic Sciences, 12*, 49–56.

Bissett, D. L., Oblong, J. E., & Berge, C. A. (2005). Niacinamide: AB vitamin that improves aging facial skin appearance. *Dermatologic Surgery, 31*, 860–866.

Boudreau, M. D., & Beland, F. A. (2006). An evaluation of the biological and toxicological properties of Aloe barbadensis (miller), Aloe vera. *Journal of Environmental Science and Health Part C, 24*(1), 103–154.

Castree, N. (2003). Bioprospecting: From theory to practice (and back again). *Transactions of the Institute of British Geographers, 28*(1), 35–55.

Chen, J. (2006). There's no such thing as biopiracy… and it's a good thing too. *McGeorge Law Review, 37*, 1.

Chowdhury, M. M. U. (2002). Allergic contact dermatitis from prime yellow carnauba wax and coathylene in mascara. *Contact Dermatitis, 46*(4), 244–244.

Cooksey, C. (2013). Tyrian purple: The first four thousand years. *Science Progress, 96*(2), 171–186.

Cosmetics Antioxidant Market. (2023). Cosmetic Antioxidants Market by Source (Natural, Synthetic), Type (Vitamins, Enzymes, Polyphenols), Function (anti-aging, Hair Conditioning, UV Protection), and Application (skin Care, Hair care, Makeup)—Global Forecast to 2028. CH6364. Markets and Markets. https://www.marketsandmarkets.com/Market-Reports/cosmetic-antioxidant-market-144119844.html

Cosmetics Market. (2022). Cosmetics Market Size by Product Type (Hair Care, Fragrance, Skin Care, Makeup, Others), End-Use (Women, Men), Regions, Global Industry Analysis, Share, Growth, Trends, and Forecast 2022 to 2030. TBI-12994. The Brainy Insights. https://www.thebrainyinsights.com/report/cosmetics-market-12994

Cosmetics Market Analysis. (2023). Cosmetics Market, By Product Type (Skin Cosmetics, Hair Cosmetics, Nail Cosmetics, Eye Cosmetics, Others), By Category (Organic & Natural, Conventional), By Distribution Channel (Store-Based, Non-Store-Based), and By Geography (North America, Europe, Asia Pacific, Latin America.

Couteau, C., & Coiffard, L. (2016). Overview of skin whitening agents: Drugs and cosmetic products. *Cosmetics, 3*(3), 27.

Datta, H. S., & Paramesh, R. (2010). Trends in aging and skin care: Ayurvedic concepts. *Journal of Ayurveda and Integrative Medicine, 1*(2), 110.

De Werra, J. (2009). Fighting against biopiracy: Does the obligation to disclose in patent applications truly help. *Vanderbilt Journal of Transnational Law, 42*, 143.
Deen, A., Visvanathan, R., Wickramarachchi, D., Marikkar, N., Nammi, S., Jayawardana, B. C., & Liyanage, R. (2021). Chemical composition and health benefits of coconut oil: An overview. *Journal of the Science of Food and Agriculture, 101*(6), 2182–2193.
Demunshi, Y., & Chugh, A. (2010). Role of traditional knowledge in marine bioprospecting. *Biodiversity and Conservation, 19*, 3015–3033.
Dobrowolski, J. W., Bedla, D., Czech, T., Gambuś, F., Górecka, K., Kiszczak, W., ... & Zabochnicka-Świątek, M. (2017). Integrated innovative biotechnology for optimization of environmental bioprocesses and a green economy. In *Optimization and applicability of bioprocesses* (pp. 27–71).
Draelos, Z. D. (2008). The cosmeceutical realm. *Clinics in Dermatology, 26*(6), 627–632.
Draelos, Z. D. (2019). Cosmeceuticals. In *Evidence-based procedural dermatology* (pp. 479–497). Springer.
Draelos, Z. D. (Ed.). (2021). *Cosmetic dermatology: products and procedures*. Wiley.
Dreher, F., Gabard, B., Schwindt, D. A., & Maibach, H. I. (1998). Topical melatonin in combination with vitamins E and C protects skin from ultraviolet-induced erythema: A human study in vivo. *British Journal of Dermatology, 139*(2), 332–339.
Dwivedi, V., & Tripathi, S. (2014). Review study on potential activity of Piper betle. *Journal of Pharmacognosy and Phytochemistry, 3*(4), 93–98.
Efferth, T., Banerjee, M., Paul, N. W., Abdelfatah, S., Arend, J., Elhassan, G., ... & Titinchi, S. J. (2016). Biopiracy of natural products and good bioprospecting practice. *Phytomedicine, 23*(2), 166-173.
Eldredge, N. (2000). *Life in the balance: Humanity and the biodiversity crisis*. Princeton University Press.
Espín, J. C., García-Conesa, M. T., & Tomás-Barberán, F. A. (2007). Nutraceuticals: Facts and fiction. *Phytochemistry, 68*(22–24), 2986–3008.
Fernanda, M., & Dias, G. (2015). Hair cosmetics: An overview. *International Journal of Trichology, 7*(1), 2–2.
Fratini, F., Cilia, G., Turchi, B., & Felicioli, A. (2016). Beeswax: A minireview of its antimicrobial activity and its application in medicine. *Asian Pacific Journal of Tropical Medicine, 9*(9), 839–843.
Ganesan, P., & Choi, D. K. (2016). Current application of phytocompound-based nanocosmeceuticals for beauty and skin therapy. *International journal of nanomedicine*, 1987–2007.
Geller, A. C., Colditz, G., Oliveria, S., Emmons, K., Jorgensen, C., Aweh, G. N., & Frazier, A. L. (2002). Use of sunscreen, sunburning rates, and tanning bed use among more than 10 000 US children and adolescents. *Pediatrics, 109*(6), 1009–1014.
Gentry, H. S. (1958). The natural history of jojoba (Simmondsia chinensis) and its cultural aspects. *Economic Botany, 12*(3), 261–295.
Glicksman, M. (2019). Gum tragacanth. In *Food hydrocolloids* (pp. 49–60). CRC Press.
Gloster, H. M., Jr., & Neal, K. (2006). Skin cancer in skin of color. *Journal of the American Academy of Dermatology, 55*(5), 741–760.
Goyal, A., Sharma, A., Kaur, J., Kumari, S., Garg, M., Sindhu, R. K., ... & Abdel-Daim, M. M. (2022). Bioactive-based cosmeceuticals: An update on emerging trends. *Molecules, 27*(3), 828.
Green, B. A., Ruey, J. Y., & Van Scott, E. J. (2009). Clinical and cosmeceutical uses of hydroxyacids. *Clinics in Dermatology, 27*(5), 495–501.
Greene, J. A. (2007). *Prescribing by numbers: Drugs and the definition of disease*. JHU Press.
Haahtela, T. (2019). A biodiversity hypothesis. *Allergy, 74*(8), 1445–1456.
Hoang, H. T., Moon, J. Y., & Lee, Y. C. (2021). Natural antioxidants from plant extracts in skincare cosmetics: Recent applications, challenges and perspectives. *Cosmetics, 8*(4), 106.
Hsu, S. (2005). Green tea and the skin. *Journal of the American Academy of Dermatology, 52*(6), 1049–1059.

Hunt, K. A., Fate, J., & Dodds, B. (2011). Cultural and social influences on the perception of beauty: A case analysis of the cosmetics industry. *Journal of Business Case Studies (JBCS), 7*(1).

Jones, G. (2011). Globalization and beauty: A historical and firm perspective. *EurAmerica, 41*(4).

Joshi, L. S., & Pawar, H. A. (2015). Herbal cosmetics and cosmeceuticals: An overview. *Natural Products Chemistry & Research, 3*(2), 170.

Juncan, A. M., Moisă, D. G., Santini, A., Morgovan, C., Rus, L. L., Vonica-Ţincu, A. L., & Loghin, F. (2021). Advantages of hyaluronic acid and its combination with other bioactive ingredients in cosmeceuticals. *Molecules, 26*(15), 4429.

Kang, S., & Voorhees, J. J. (1998). Photoaging therapy with topical tretinoin: An evidence-based analysis. *Journal of the American Academy of Dermatology, 39*(2), S55–S61.

Larsson, S. C., Bergkvist, L., Näslund, I., Rutegård, J., & Wolk, A. (2007). Vitamin A, retinol, and carotenoids and the risk of gastric cancer: A prospective cohort study. *The American Journal of Clinical Nutrition, 85*(2), 497–503.

Levin, J., & Momin, S. B. (2010). How much do we really know about our favorite cosmeceutical ingredients? *The Journal of Clinical and Aesthetic Dermatology, 3*(2), 22.

Levy, L. L., & Emer, J. J. (2012). Emotional benefit of cosmetic camouflage in the treatment of facial skin conditions: Personal experience and review. In *Clinical, cosmetic and investigational dermatology* (pp. 173–182). Springer.

Lin, J. Y., & Fisher, D. E. (2007). Melanocyte biology and skin pigmentation. *Nature, 445*(7130), 843–850.

Łyczko, J., Jałoszyński, K., Surma, M., García-Garví, J. M., Carbonell-Barrachina, Á. A., & Szumny, A. (2019). Determination of various drying methods' impact on odour quality of true lavender (Lavandula angustifolia Mill.) flowers. *Molecules, 24*(16), 2900.

Mahesh, S. K., Fathima, J., & Veena, V. G. (2019). Cosmetic potential of natural products: industrial applications. In *Natural bio-active compounds: Volume 2: chemistry, pharmacology and health care practices* (pp. 215–250). Springer.

Martin, K. I., & Glaser, D. A. (2011). Cosmeceuticals: The new medicine of beauty. *Missouri Medicine, 108*(1), 60.

Mas-Chamberlin, C., Lintner, K., Basset, L., Adhoute, H., & Revuz, J. (2002). Relevance of anti-wrinkle treatment of a peptide: 4 months clinical double blind study vs excipient. *Annals of Dermatol Venereol, 129*, 1S456.

Mateo, N., Nader, W., & Tamayo, G. (2001). Bioprospecting. *Encyclopedia of Biodiversity, 1*, 471–487.

Michalak, M. (2022). Plant-derived antioxidants: Significance in skin health and the ageing process. *International Journal of Molecular Sciences, 23*(2), 585.

Michalak, M., Pierzak, M., Kręcisz, B., & Suliga, E. (2021). Bioactive compounds for skin health: A review. *Nutrients, 13*(1), 203.

Middle East & Africa. CMI5987. *Coherent market insights*. https://www.coherentmarketinsights.com/market-insight/cosmetics-market-5987

Myers, N. (1996). Environmental services of biodiversity. *Proceedings of the National Academy of Sciences, 93*(7), 2764–2769.

Newburger, A. E. (2009). Cosmeceuticals: Myths and misconceptions. *Clinics in Dermatology, 27*(5), 446–452.

Ngo, L. T., Okogun, J. I., & Folk, W. R. (2013). 21st century natural product research and drug development and traditional medicines. *Natural Product Reports, 30*(4), 584–592.

Nordin, R., Hassan, K. H., & Zainol, Z. A. (2012). Traditional knowledge documentation: Preventing or promoting biopiracy. *Pertanika Journal of Social Sciences & Humanities, 20*, 11–22.

Nunes, P. A., & van den Bergh, J. C. (2001). Economic valuation of biodiversity: Sense or nonsense? *Ecological Economics, 39*(2), 203–222.

Pandey, A., Jatana, G. K., & Sonthalia, S. (2019). *Cosmeceuticals*.

Pharmaceuticals. (2023). Pharmaceuticals. Cosmeceuticals Market Size, Share & COVID-19 Impact Analysis, By Product Type [Skin Care (Anti-Ageing, Skin Whitening, Sun Protection, Professional Skincare, Anti-Acne, and Others), Hair Care (Hair Growth, Anti-Dandruff, and

Others), Injectable (Botulinum Toxins, Dermal Fillers, and Other Injectable), and Others (Lip Care, Tooth Whitening, Anti-Cellulite/Body Slimming Preparations, Oral Care)], By Distribution Channel [Pharmacy/Drug Stores, Online Stores, and Others], and Regional Forecast, 2023–2030. FBI102521. Fortune Business Insights. https://www.fortunebusinessinsights.com/pharmaceutical-industry

Pinnell, S. R., Yang, H., Omar, M., Riviere, N. M., Debuys, H. V., Walker, L. C., … & Levine, M. (2001). Topical L-ascorbic acid: percutaneous absorption studies. *Dermatologic Surgery, 27*(2), 137–142.

Pourabbas, R., Delazar, A., & Chitsazi, M. T. (2005). *The effect of German chamomile mouthwash on dental plaque and gingival inflammation.*

Rahman, M. H., Akter, R., Bhattacharya, T., Abdel-Daim, M. M., Alkahtani, S., Arafah, M. W., … & Mittal, V. (2020). Resveratrol and neuroprotection: Impact and its therapeutic potential in Alzheimer's disease. *Frontiers in Pharmacology, 11*, 619024.

Rahul, C., & Sreedhar, D. (2022). Nutraceuticals and Cosmeceuticals–A brief overview. *Pharma Times, 54*(6), 12–14.

Rathod, H. J., & Mehta, D. P. (2015). A review on pharmaceutical gel. *International Journal of Pharmaceutical Sciences, 1*(1), 33–47.

Reichrath, J., Leiter, U., Eigentler, T., & Garbe, C. (2014). Epidemiology of skin cancer. In *Sunlight, vitamin D and skin cancer* (pp. 120–140).

Robinson, L., Fitzgerald, N., & Berge, C. (2002). Pentapeptide offers improvement in human photoaged facial skin. *Annals of Dermatol Venereol, 129*, 1S105.

Rodriguez-Amaya, D. B. (2016). Natural food pigments and colorants. *Current Opinion in Food Science, 7*, 20–26.

Rowland, I., Gibson, G., Heinken, A., Scott, K., Swann, J., Thiele, I., & Tuohy, K. (2018). Gut microbiota functions: Metabolism of nutrients and other food components. *European Journal of Nutrition, 57*, 1–24.

Schmidt, B. M., Ribnicky, D. M., Lipsky, P. E., & Raskin, I. (2007). Revisiting the ancient concept of botanical therapeutics. *Nature Chemical Biology, 3*(7), 360–366.

Semwal, R. B., Semwal, D. K., Combrinck, S., Cartwright-Jones, C., & Viljoen, A. (2014). Lawsonia inermis L.(henna): Ethnobotanical, phytochemical and pharmacological aspects. *Journal of Ethnopharmacology, 155*(1), 80–103.

Setty, A. R., & Sigal, L. H. (2005, June). Herbal medications commonly used in the practice of rheumatology: mechanisms of action, efficacy, and side effects. In *Seminars in arthritis and rheumatism* (Vol. 34, No. 6, pp. 773–784). WB Saunders.

Shanbhag, S., Nayak, A., Narayan, R., & Nayak, U. Y. (2019). Anti-aging and sunscreens: Paradigm shift in cosmetics. *Advanced Pharmaceutical Bulletin, 9*(3), 348.

Shiva, V. (2007). Bioprospecting as sophisticated biopiracy. *Signs: Journal of Women in Culture and Society, 32*(2), 307–313.

Soler, L., Canellas, J., & Saura-Calixto, F. (1988). Oil content and fatty acid composition of developing almond seeds. *Journal of Agricultural and Food Chemistry, 36*(4), 695–697.

Souto, E. B., Fernandes, A. R., Martins-Gomes, C., Coutinho, T. E., Durazzo, A., Lucarini, M., … & Santini, A. (2020). Nanomaterials for skin delivery of cosmeceuticals and pharmaceuticals. *Applied Sciences, 10*(5), 1594.

Strobel, G., & Daisy, B. (2003). Bioprospecting for microbial endophytes and their natural products. *Microbiology and Molecular Biology Reviews, 67*(4), 491–502.

Tan, Q. G., Cai, X. H., Du, Z. Z., & Luo, X. D. (2009). Three terpenoids and a tocopherol-related compound from Ricinus communis. *Helvetica Chimica Acta, 92*(12), 2762–2768.

Tobin, D. J. (2017). Introduction to skin aging. *Journal of Tissue Viability, 26*(1), 37–46.

Trak, N. H. D. T. H., & Chauhan, M. F. H. D. D. (2022). *Skin care secrets from herbal world*. Notion Press.

Verma, S., & Singh, S. P. (2008). Current and future status of herbal medicines. *Veterinary World, 1*(11), 347.

Bioprospecting for Food and Nutrition

A. S. Vijayasree, Laladhas Krishna Panicker, Junaida Ibrahim, Neenu Mohan, A. Shanitha, Achuthsankar S. Nair, P. R. Sudhakaran, and Oommen V. Oommen

Introduction

In the last few years, the food industry's rapid expansion has coincided with a rise in public awareness of the depletion of natural resources. Achieving high crop yields is necessary for the agricultural industry to meet the growing demand for food, which has a detrimental impact on sustainable production targets (Tyagi et al., 2022). Within the field of biodiversity, bioprospecting for food and nutrition is an important and fascinating activity. It entails investigating the vast diversity of life on Earth to recognize and make use of species—especially plants—for their nutritional worth and possible advantages to human health. Despite the worldwide pandemic in 2020, efforts are on to approve a global biodiversity framework at the 15th Conference of the Parties to the Convention on Biological Diversity (CBD), which is currently set for 2021. Furthermore, October 29, 2020, was the tenth anniversary of the Nagoya Protocol's adoption, which governed access to genetic resources and the just and equitable distribution of benefits resulting from their use. In an era where people are looking to nature more and more for solutions about health and wellness, bioprospecting is an intriguing area of study.

A. S. Vijayasree
PG and Research Department of Zoology, Fatima Mata National College (Autonomous), Kollam, Kerala, India

L. Krishna Panicker (✉) · J. Ibrahim · N. Mohan · A. Shanitha · A. S. Nair · P. R. Sudhakaran · O. V. Oommen
Department of Computational Biology and Bioinformatics, University of Kerala, Kariavattom, Thiruvananthapuram, Kerala 695581, India
e-mail: laladhas@gmail.com

L. Krishna Panicker · O. V. Oommen
CSEIBA: Center for, Sustainable Environment-India Biodiversity Action, Thiruvananthapuram, Kerala, India

L. Krishna Panicker et al. (eds.), *Biodiversity and Business*,
https://doi.org/10.1007/978-3-031-71674-4_10

Searching for nature's hidden elements takes you to some of the most isolated and exotic places on Earth, from impenetrable jungles to the lowest ocean depths, much like a treasure hunt. These marvels of nature have the potential to revolutionize our diet and supplementation, making them formidable friends in our quest for longevity and health. Unicellular or colonial photosynthetic organisms, microalgae are primarily found in natural aquatic settings such as coastal sea areas and inland waters. According to Ratha and Prasanna (2012), there are between 200,000 and 800,000 species of microalgae in the world. These algae have a lot of potential applications in biotechnology, biorefinery, and bioprospection (Brasilet al., 2017). Fundamentally, bioprospecting can be thought of as the scientific equivalent of a treasure hunt. Finding plants, animals, and microorganisms that may hold the key to the next major development in nutrition and health is the main goal of researching the natural world. Scientists searching for novel chemicals that may become the next big thing in the nutraceutical and functional food industries venture into unexplored territory with the utmost respect for nature and cutting-edge technology.

Biodiversity provides important opportunities for sustainable lives and is necessary for nutrition and food security has been reported elsewhere (Laladhas et al., 2015, 2016, 2017a, 2017b; Mithrambika, 2017; Oommen et al., 2022; Preetha et al., 2015; Nilayangode 2016a, 2016b, 2016c, 2017). Maintaining and creating opportunities that promote human well-being requires environmental integrity. The current body of knowledge demands that the sustainable use of biodiversity in food security and nutrition programs be promoted right away to help meet the Millennium Development Goals.

—from 'Main conclusions of the consultation on the CBD cross-cutting initiative on biodiversity for food and nutrition'

Approximately 7000 plant species and several thousand animal species have been used for human nourishment since agriculture started about 12,000 years ago. While some indigenous and traditional societies still use 200 or more species in their diets today, there has been a global tendency toward diet simplification, which has had detrimental effects on human health, nutritional balance, and food security. Under the Convention, a cross-cutting initiative on biodiversity for food and nutrition aims to advance the sustainable use of biodiversity in programs that promote improved human nutrition and food security.

Natural Medicine at Our Fingertips: Bioprospecting

Biodiversity refers to the variety of life forms on Earth, including plants, animals, fungi, and microorganisms. Bioprospecting is understood as the use of biodiversity in the provision of resources for the discovery, classification, investigation, and/or formulation of new sources of chemical compounds, genes, proteins, and other components with potential economic and biotechnological value (Marques et al., 2013). Bioprospecting, an interdisciplinary field at the intersection of biology, ecology, and applied sciences, is the systematic search for valuable compounds, genes, organisms, or materials derived from living organisms. Bioprospecting, also referred to as biodiversity prospecting, is the methodical and sustainable study of genetic and biochemical resources found in nature with the goal of turning them into useful goods. It is the process of searching natural sources for small and macromolecules, as well as biochemical and genetic information that may be used to create commercially valuable products for the pharmaceutical, agricultural, aquaculture, bioremediation, cosmetic, or nanotechnology sectors. Bioprospecting is understood as the use of biodiversity in the provision of resources for the discovery, classification, investigation, and/or formulation of new sources of chemical compounds, genes, proteins, and other components with potential economic and biotechnological value (Berlinck, 2012; Marques et al., 2013; Sacarro Junior, 2011). The term "bioprospecting" describes the application of scientific knowledge to explore unexplored natural resources to generate new business prospects. It can be described as the skillful discovery of new, economically viable organisms, especially undiscovered species, which increases the value of bioresources and encourages the preservation of biodiversity. Bioprospecting, often referred to as bio exploration, is the process of investigating biological variety to find and use novel substances, species, or genetic resources for a range of purposes. These resources can include anything from industrial enzymes and pharmaceuticals to dietary supplements and crops. The scope of bioprospecting encompasses diverse ecosystems such as forests, oceans, deserts, and even urban environments, where unique organisms may harbour valuable traits or biochemical compounds.

Many previous bioprospecting initiatives have concentrated on terrestrial plants, fungi, and actinobacteria; however, as a way to discover new chemicals with unique biological functions, interest in less studied habitats and animals, such as seas and oceans, is developing. Species can be arbitrarily chosen and screened for bioactivity using ecological, ethnobiological, ethnomedical, historical, or genetic data, or they can be screened at random. Biopiracy is the term for when biological resources or indigenous knowledge from a place are unethically taken or economically exploited without giving due compensation. Numerous international agreements have been created to give nations legal protection against biopiracy and to give business entities investment assurance. These consist of the Nagoya Protocol and the UN Convention on Biological Diversity. More treaties are presently being negotiated by the WIPO to fill in gaps in this area. Legislation has been created to fight the concerns of overharvesting particular species and environmental degradation that are linked with bioprospecting. Instances comprise domestic legislation like the United States Marine Mammal Protection Act and the United States Endangered Species Act, as well as global accords like the United Nations Convention on Biological Diversity, the United Nations Convention on the Law of the Sea, and the Antarctic Treaty. The scope of bioprospecting for food and nutrition is vast and encompasses various dimensions, including scientific research, economic development, environmental sustainability, and societal well-being.

Using Nature's Richness to Create Nutritious Gold

It is quite amazing how a molecule found in the environment may become a product on the shelves of your local supermarket. It combines cutting-edge technology, rigorous science, and conventional wisdom. By using genomic sequencing, scientists can unlock the genetic codes of microorganisms and plants, identifying the precise substances that have positive health effects.

Bioprospecting drives innovation in the food and nutrition industries by uncovering new sources of nutrients, flavors, and functional components. New food items, nutritional supplements, and nutraceuticals that satisfy customer preferences and health requirements are created as a result of this innovation. The utilization of plants as nutraceuticals represents a significant intersection between traditional knowledge, modern science, and healthcare. This emerging field harnesses the therapeutic potential of plant-derived compounds to promote health, prevent diseases, and enhance well-being. This chapter explores the diverse roles of plants as nutraceuticals, encompassing their biochemical constituents, physiological effects, and potential applications in dietary supplementation and functional foods.

Nutraceuticals

Nutraceuticals and bioprospecting represent two intertwined frontiers in the realm of food, health, and biodiversity. Nutraceuticals, a portmanteau of "nutrition" and "pharmaceuticals," refer to products derived from food sources with purported health benefits beyond basic nutrition. These can include dietary supplements, functional foods, and herbal products, among others. Bioprospecting, on the other hand, is the systematic exploration of biodiversity to discover novel compounds, genes, or organisms with potential commercial or medicinal value.

The convergence of nutraceuticals and bioprospecting reflects a growing interest in harnessing nature's bounty to address contemporary health challenges while also promoting conservation and sustainable use of biological resources. This synergy offers a promising avenue for discovering new bioactive compounds and ingredients that can support human health and well-being.

The introduction of nutraceuticals represents a paradigm shift in healthcare, emphasizing prevention and holistic approaches to wellness. Consumers are increasingly seeking natural alternatives to traditional pharmaceuticals, driving demand for products that offer both nutritional value and therapeutic benefits. This has spurred interest in exploring diverse ecosystems—from rainforests to coral reefs—to uncover nature's pharmacopeia and develop innovative health solutions.

Bioprospecting serves as the gateway to unlocking nature's treasure trove of bioactive compounds. By tapping into the rich biodiversity of our planet, scientists can discover novel molecules with potential applications in nutraceuticals, pharmaceuticals, cosmetics, and more. From antioxidant-rich fruits to microbiome-modulating microbes, the possibilities are vast and diverse.

However, the pursuit of nutraceuticals through bioprospecting also raises important ethical, legal, and environmental considerations. Indigenous knowledge, biodiversity conservation, and equitable benefit-sharing must be central to these endeavours to ensure that communities are not exploited, and ecosystems are not harmed in the quest for natural health solutions. In this introduction, we will explore deeper into the fascinating world of nutraceuticals and bioprospecting, exploring their significance, challenges, and potential impacts on human health, biodiversity conservation, and sustainable development. Through interdisciplinary collaboration and responsible stewardship of our natural resources, we can harness the power of nature to nourish both the body and the planet.

Bioactive Compounds in Plants

Plants are rich sources of bioactive compounds such as polyphenols, flavonoids, alkaloids, terpenoids, and other phytochemicals. These compounds possess various biological activities, including antioxidant, anti-inflammatory, antimicrobial, and anticancer properties. Through a comprehensive examination of these phytochemicals, this chapter elucidates their mechanisms of action and potential health benefits.

Physiological Effects and Health Benefits

The consumption of plant-based nutraceuticals has been associated with numerous health benefits, ranging from cardiovascular protection and immune modulation to cognitive enhancement and anti-aging effects. By exploring the physiological effects of specific plant-derived compounds, this chapter delineates their impact on human health and well-being, drawing upon both traditional wisdom and contemporary scientific research.

Applications in Dietary Supplementation and Functional Foods

Plants as nutraceuticals find wide-ranging applications in dietary supplementation and the development of functional foods and beverages. From herbal supplements and botanical extracts to fortified food products, the integration of plant-derived ingredients offers opportunities to enhance nutritional value, improve taste and texture, and meet consumer demands for natural and functional ingredients. Through case studies and market analyses, this chapter examines the growing trends and commercial potential of plant-based nutraceuticals in the food and beverage industry.

Addressing Benefit Sharing and Traditional Knowledge Protection

This chapter explores the intricate details of benefit sharing and the protection of traditional knowledge, particularly in the commercialization of products derived from indigenous plants or practices. Through case studies such as Allanblackia oil, avocado farming in Peru, Bergamot essential oil production in Italy, and Carnauba wax extraction in Brazil, it underscores the importance of fair and equitable distribution of benefits to local communities while preserving their cultural heritage and traditional practices.

Allanblackia

Allanblackia, a genus of tropical trees native to Africa, has garnered attention in recent years for its potential as a nutraceutical ingredient. Extracts from the seeds of Allanblackia species have been studied for their health-promoting properties, ranging from cardiovascular benefits to anti-inflammatory effects. The biochemical composition, physiological effects, and potential applications of Allanblackia as a nutraceutical ingredient shed light on its role in promoting human health and well-being.

The *Clusiaceae* or *Guttiferae* family includes the genus Allanblackia. They are widespread throughout Africa's rainforest regions, primarily in Uganda, Cameroon, Eastern Nigeria, and Southern and Northern Angola (Azebazeet al., 2009; Crockett, 2015). The taxonomic group Allanblackia presents a compelling study subject for phytochemical analysis due to its remarkable structural variability as well as the varied biological properties of its secondary metabolites, which include triterpenoids, xanthones, benzophenones, and bioflavonoids (Mountessouet al., 2018; Nganouet al., 2019). The genus name honours Allan Black. It contains the following species:

- *Allanblackia floribunda* (Nigeria to DR Congo and Angola)
- *Allanblackiagabonensis*
- *Allanblackiakimbiliensis*
- *Allanblackiakisonghi*
- *Allanblackiamarienii*
- *Allanblackia parviflora* (Upper Guinea, from Ghana westwards)

- *Allanblackiastaneriana*
- *Allanblackiastuhlmannii* (Eastern arc mountains—Usambara Tanzania)
- *Allanblackiaulugurensis* (Eastern arc mountains—Uluguru Tanzania)

Uses: The diverse applications of Allanblackia species have led to their reputation as adaptable species in nature. Indeed, Allanblackia species are used by local communities in various African countries for traditional herbal medicine, in addition to their traditional use as a means of extracting edible oil from its seeds. It is possible to extract Allanblackia oil from Allanblackia. Local populations from the uplands of Tanzania's Eastern Arc Mountains to the lowlands of Sierra Leone, West Africa, have been gathering the oil-rich seeds of the Allanblackia tree for generations, for food, to make cooking oil, and more recently, for soap. The possibility for increased pay for African farmers exists due to the domestication of this tree and the growing demand for its oil, which was formerly primarily collected from the wild. The nine Allanblackia species are limited to the humid forests of West, Central, and East Africa, and they are found in areas with significant biodiversity and habitat loss. The International Union for Conservation of Nature (IUCN) Red List has three of the species listed as "vulnerable". Tropical rainforests are its preferred habitat, yet it also thrives in areas with cultivated agriculture.

There is a significant demand for the local forest resources in many Allanblackia locations, which highlights the need to invest in this tree to meet conservation, forest restoration, and poverty reduction objectives. These sites are also thought to be rich in biodiversity and high in poverty. Aside from its high content of stearic and oleic acids, the oil's precise melting point—roughly 340C—and ability to stay solid at room temperature make it attractive from a commercial standpoint. Allanblackia oil requires little further modification throughout the manufacturing process than conventional palm oil, including fractionation, hardness, etc. This is the main benefit. The Novella Partnership was founded in 2002 to accelerate Allanblackia oil production in Ghana, Tanzania, and Nigeria while also encouraging sustainable enterprise, lowering poverty, and advancing the preservation of biodiversity in Africa. The World Agroforestry Centre (ICRAF) is leading research on domesticating Allanblackia to raise harvest levels to a point where they are commercially viable. IUCN works to facilitate the integration of forest landscape restoration principles into the various models for increased production of Allanblackia, and supports the development of a strategy for market differentiation for Allanblackia oil in collaboration with the Union for Ethical Biotrade. For financial stability, the Allanblackia harvest season—which spans from January to April and coincides with other cash crops—is especially important. The optimal traits for domesticating wild female Allanblackia trees—only female plants bear fruit—have been identified with considerable progress. By vegetative propagation, trees with both male and female flowers can also be produced, and they can start producing fruit as soon as four or five years after planting. These improvements are necessary to reduce the amount of planting.

Biochemical Composition of Allanblackia

Allanblackia seeds are rich in bioactive compounds such as fatty acids, particularly stearic acid, oleic acid, and linoleic acid. These fatty acids contribute to the nutritional value of Allanblackia oil and are associated with various health benefits, including cholesterol modulation and anti-inflammatory effects. By elucidating the biochemical composition of Allanblackia seeds and their derivatives, this chapter provides insights into the potential mechanisms underlying their nutraceutical properties.

Physiological Effects and Health Benefits

Studies have suggested that consumption of Allanblackia oil may have beneficial effects on cardiovascular health, including the modulation of lipid profiles and improvement of endothelial function. Additionally, Allanblackia extracts have demonstrated anti-inflammatory and antioxidant activities, which may contribute to their potential in preventing chronic diseases and supporting overall well-being. Through a review of preclinical and clinical studies, this chapter explores the physiological effects and health benefits associated with Allanblackia as a nutraceutical ingredient.

Allanblackia oil and extracts hold promise as ingredients in functional foods, dietary supplements, and nutraceutical formulations. Their high content of unsaturated fatty acids, coupled with their potential health-promoting properties, makes them suitable for incorporation into products targeting cardiovascular health, inflammation, and oxidative stress. This chapter examines the current landscape of Allanblackia-based products and explores opportunities for innovation in the development of functional foods and dietary supplements enriched with Allanblackia.

Avocado

Avocado, scientifically known as *Persea americana*, has gained significant recognition as a versatile fruit with various health-promoting properties. Beyond its culinary appeal, avocado is increasingly being explored for its potential as a nutraceutical ingredient. Rich in beneficial nutrients and bioactive compounds, avocado offers a range of health benefits, making it a promising candidate for inclusion in functional foods, dietary supplements, and nutraceutical formulations. Agriculture-related investment and development are occurring in sub-Saharan African nations in the twenty-first century (Scoones et al., 2018). According to Arshad (2022), this is anticipated to have a significant impact on smallholder farmers' income and standard of living. Furthermore, horticultural items including fruits, vegetables, and cut flowers have become popular cash crops, replacing many traditional export crops like cotton, pyrethrum, sisal, and tobacco as the main export crops.

According to Repoa (2018), Avocado (*Perseaamericana*) is one of these crops. *Persea americana*, known in Latin America as "avocado" or "agauacate," is a type of warm-climate tree whose fruit is used both for direct consumption and the manufacturing of a variety of goods, such as oil and cosmetics (Schaffer et al., 2013). Current estimates suggest that approximately 5 million tonnes are consumed annually worldwide. A portion of the rising craze for avocados in developed Asia, North America, and Europe in recent years can be attributed to lifestyle shifts and certain health advantages.

For thousands of years, Peruvian civilizations such as the Moche, Chimú, and Incas have grown avocados. Nonetheless, throughout the past ten or so years, the crop has grown astronomically, and both domestic and foreign investors have developed sizable new plantations with an eye toward export. The increased demand for avocados around the world has led to a recent rise in the popularity of avocado farming. But to ensure that avocado farming is sustainable over the long term, as with any other agricultural activity, sustainability must be given top priority. According to Tanzanialnvest (2023), the Tanzania Horticultural Association (TAHA) projected that the nation's avocado exports would total 11,237 tonnes, or 510 containers, valued at USD 33 million. Tanzania was expected to produce USD 45 million in foreign exchange and export 15,000 tonnes of avocados in 2023 (Tanzanialnvest, 2023). However, improved global marketing strategies and other socioeconomic shifts are also responsible for the high annual consumption of avocados, which is estimated to be 3.5 and 1.05 kg per capita in the USA and Europe (Netherlands: 2.28 kg), respectively (CBI, https://www.cbi.eu/). Thus, throughout the nation's producing regions, avocado cultivation has grown to be a significant component of livelihood diversification. It is noteworthy that, particularly in developing nations, livelihood diversification dramatically boosts economic growth and lowers rural poverty (Kadigi et al., 2022).

Therefore, diversification is essential for increasing smallholder farmers' portfolios of sources of income and enhancing rural incomes, which in turn raises the standard of life for the rural poor (Pritchard et al., 2019). Mexico, the fruit's home country, is the primary provider of avocados to North America throughout the year. In the winter, Chile, Mexico, Israel, and Spain supply avocados to Europe, while Peru, South Africa, and Kenya supply avocados during the summer. Persistence Market Research projects that the avocado market will expand at a compound annual rate of 6.2% from 2017 to 2027. Moreover, avocados' very stable and reliable supply has kept retail prices in the EU relatively stable (https://www.cbi.eu/). In 2018 the United States, the Netherlands, France, Italy, the United Kingdom, and Spain were the top importers of avocados. Even though avocados can be grown in a variety of soil types, they cannot live in wet conditions. They are also not very tolerant of salinization of the soil. Avocados thrive best in soil pH values between 5 and 7. The optimum soils for avocado farming are those that are slightly acidic (pH 6–7), while alkaline soils can still support avocado production if they drain properly. An abundant supply of organic matter is necessary for the successful cultivation of avocados because it retains moisture in the soil and provides the plant with essential nutrients like potassium, phosphate, and nitrogen. The addition of aged manure or compost can help improve poor soils. Enhancing root growth and oxygenation through the addition of organic matter to compacted soil mitigates the adverse consequences of excessive watering.

Spring is the ideal season to grow avocado plants. This allows the tree plenty of time to set roots before the arrival of winter. Furthermore, young trees are more susceptible to sun damage if planted in the summer since they are less able to absorb water.

When establishing an avocado tree, it is important to handle its delicate root structure with caution. When planting a tree, make sure the pit is larger than the tree's roots. It is advised that the depth of the pit match the height of the root ball. A tree may wither if it is planted too deeply or too shallowly, which will restrict the root system's access to nutrients and water. Avocados have traditionally been cultivated in monoculture on plantations as the main method of export-based production. Because this kind of agriculture heavily depends on irrigation infrastructure and management techniques that deteriorate soil quality and, consequently, reduce its capacity to hold water, it is linked to excessive water use. Additionally, because monoculture farming employs an excessive number of herbicides and inorganic fertilizers, it has an adverse effect on the environment. As a result, considering growing avocados organically in places like California, Peru, and Spain is necessary because there is less degradation of the water quality than in traditional agriculture.

Avocado's nutritional profile and health-promoting properties make it an attractive ingredient for incorporation into functional foods, nutraceutical supplements, and health products. From avocado oil and avocado powder to avocado-based snacks and beverages, the versatility of avocado enables its integration into diverse formulations targeting specific health outcomes. This chapter explores the current market landscape of avocado-derived products and identifies opportunities for innovation in the development of avocado-based nutraceuticals.

The Peruvian avocado sector has been successful, but it still faces several obstacles, including weather variations, competition in the market, and the need for upgraded infrastructure. Drought dangers persist in Peru in 2023, despite expected improvements in avocado production quantities and exports to the United States. The Peruvian Association of Hass Avocado Producers is optimistic despite the upcoming weather conditions. There are many of prospects for growth and expansion due to the rising demand for avocados worldwide. Peru's avocado business is well-positioned to continue growing in the international market thanks to investments in R&D, sustainable agricultural practices, and grower engagement.

Biochemical Composition of Avocado

Avocado is renowned for its high content of monounsaturated fatty acids, particularly oleic acid, which contributes to its favourable lipid profile. Additionally, avocados are rich in dietary fibre, vitamins (such as vitamin E, vitamin K, and B vitamins), minerals (including potassium, magnesium, and copper), and phytochemicals like carotenoids and phytosterols. These bioactive compounds play key roles in supporting various physiological functions and conferring health benefits.

Physiological Effects and Health Benefits

Consumption of avocado has been associated with numerous health benefits, including cardiovascular protection, weight management, and improved metabolic health. The monounsaturated fats in avocado are linked to reduced levels of LDL cholesterol and improved lipid profiles, while its fibre content supports digestive health and satiety. Moreover, the presence of antioxidants like vitamin E and carotenoids contributes to the fruit's anti-inflammatory and antioxidant properties, which may help mitigate oxidative stress and chronic disease risk.

Bergamot

Bergamot, derived from the citrus fruit *Citrus bergamia*, has gained attention not only for its aromatic properties but also for its potential health benefits. Beyond its traditional use in perfumery and flavouring, bergamot is increasingly recognized for its nutraceutical properties. Rich in bioactive compounds such as flavonoids, polyphenols, and essential oils, bergamot offers a range of health-promoting effects that make it a promising candidate for inclusion in functional foods, dietary supplements, and nutraceutical formulations.

The citrus fruit known as bergamot (*Citrus bergamia* Risso &Poit., syn.: *Citrus aurantium* L. subsp. C. *bergamia* (Risso et Poit.) Wight et Arn. Engl.) is aromatic and bears a resemblance to orange in size and colour, with a hint of lemon flavour. It is a member of the Citrus genus Rutaceae family. Ferrari initially described it as an "aurantium stellatum et roseum" in his 1646 monograph *Hesperides sive de malorumaureorumcultura et usu.,* It was not until 1708 that Volkamer gave it a detailed description in his Nurnbergische Hesperides, referring to it as "gloria limonumfructtus inter omnes nobilissimus," describing the unique and desirable qualities of its essential oil. In the Mediterranean, bergamot oil has been valued for its unique qualities for several centuries. As early as 1750, its benefits were acknowledged. Both the name's significance and its centre of origin are unclear as of yet. The essence of bergamot is a volatile, naturally occurring, limpid complex compound with a strong scent. It is produced by a variety of aromatic plants as a secondary metabolite and is typically extracted by steam or hydro-distillation. Every essential oil is a very complex mixture of molecules with 20–70 low molecular-weight components present in varying concentrations.

One of the most essential components of perfumes is bergamot oil, which blends into practically any perfume composition to give it a pleasant, refreshing aroma. As a result, no perfume is made today without bergamot essential oil (BEO). It gives the scent weight and substance while also adding delicate top notes. Essential oils, which are phytochemicals, have been deemed appealing due to their diverse range of biological actions. The clear, yellow-green liquid known as bergamot essence is made from cold pressing the fresh fruit's epicarp and partially its mesocarp. It consists of a volatile (93–96% of total) and a nonvolatile (4–7% of total) fraction; the former contains monoterpene and sesquiterpene hydrocarbons such as limonene (40%) and oxygenated derivatives such as linalool (8%) and linalyl acetate (28%). The flavour is marked by oxygenated molecules, rather than the hydrocarbon portion, which is not fundamentally responsible for defining the essence's olfactory character (Sakurada et al., 2011).

Biochemical Composition of Bergamot

Bergamot contains a plethora of bioactive compounds, including flavonoids (such as hesperidin and naringin), polyphenols (such as neoeriocitrin and brutieridin), and essential oils (notably limonene, linalool, and bergapten). These compounds contribute to bergamot's antioxidant, anti-inflammatory, and cholesterol-lowering properties. By elucidating the biochemical composition of bergamot, this chapter provides insights into its potential mechanisms of action and health-promoting effects.

Physiological Effects and Health Benefits

Research suggests that bergamot consumption may confer various health benefits, including cardiovascular protection, metabolic support, and mood enhancement. The flavonoids and polyphenols in bergamot exert antioxidant effects, which may help mitigate oxidative stress and inflammation associated with chronic diseases. Moreover, bergamot essential oil components have been investigated for their potential to improve lipid profiles, reduce cholesterol levels, and enhance mood and cognitive function.

Bergamot's unique flavour profile and nutraceutical properties make it a versatile ingredient for incorporation into functional foods, dietary supplements, and nutraceutical formulations. From bergamot-infused teas and juices to bergamot extracts and supplements, there is growing interest in leveraging bergamot's health-promoting effects for consumer products targeting cardiovascular health, mood support, and overall well-being. This chapter explores the current market landscape of bergamot-derived products and identifies opportunities for innovation in the development of bergamot-based nutraceuticals.

The majority of the world's supply of this citrus essential oil is produced in the Italian city of Bergamo, which is where Bergamot oil originated. Although the fruit itself is inedible, the plant is cultivated in small amounts in Mauritius, where the inhabitants drink the juice. Additionally, fragrances and cosmetics include it. Bergamot oil has long been thought to have anti-infective qualities, although little research has been done on its potential medical benefits (Pizzimenti et al., 1998). It is widely used in developed nations to treat anxiety, depression, and schizophrenia, as well as certain dementia symptoms, acute and chronic pain, and cancer (Perry & Perry, 2006). In the culinary arts and the confectionery sector, bergamot fruit juice is also utilised as a flavouring agent in the creation of fresh and dry pastries, candies, candied peel, ice cream, and sorbets, in addition to soft drinks, cocktails, and liqueurs (Montemurro, 2015). Furthermore, another bergamot product that is useful for improving soil is "pastazzo," which is made of the fruit's pulp and peel and is obtained after the essence is extracted (Foti, 2015). Furthermore, a great deal of interest has been shown by the scientific community due to the richness of nutraceutical chemicals found in bergamot pulp and juice. The antioxidant capacity of phenolic compounds originating from plants is actually of great interest due to its positive impact on human health. Many bergamot formulations that enhance inadequate gastrointestinal absorption have been developed. The fruit is rich in polyphenols, which are present in these formulations (Mollace et al., 2019). Moreover, combinations of bergamot extracts with extracts from other foods, like artichokes (Cicero et al., 2019), wild cardoon (Ferro et al., 2020), or olives and with pure compounds, like vitamin K2 (Bonfigli et al., 2020), flavonoids, and pectins (Capomollaet al., 2019), encourage favourable interactions that may increase the range of illnesses or risk conditions that can be successfully treated.

The pitted fruit of the bergamot tree has a pear-like form and can reach a height of 15 feet. Columbus brought the Italian city of Bergamo to Europe, which is where the citrus essential oil originated. This oil was used as a folk treatment for worms and fever in Italy for a long time. It has a moderate risk of phototoxicity, although its chemical composition is unclear. Businesses should thoroughly investigate the safety of the oil for these reasons. A significant pillar of the region of Calabria's prosperous economy is the bergamot plantation. It requires protection as the study of its biological activity is still in its early stages and, in recent decades, there has been progress towards the methodical and thorough investigation of possible phototherapeutics.

The company is supplied by Bergamot farmers in Southern Italy, whose UEBT member Capua 1880 has obtained UEBT natural ingredient certification. The farmers' modest groves span fewer than 10 hectares. Approximately 1,200 farmers, grouped into cooperatives, work two or three jobs in addition to growing crops for their consumption on their citrus farms. Some work as full-time farmers and grow other citrus fruits like olives in addition to bergamot. Even the first processing of essential oils and other substances based on bergamot for the food industry is done by a few farmers. Growers that supply Bergamot to Capua are collaborating with a fifth-generation family company that supplies the international fragrance and flavour market. The company uses a cold method (pelatrice) to extract the oil from the fruit. Capua has just purchased extra property in hectares to establish a didactic farm that will "cascade" gains in biodiversity and sound agricultural practices along the entire supply chain.

A few biodiversity-related initiatives include designating specific places on farms to draw pollinators, lowering or switching to natural pesticide substitutes, planting a variety of citrus fruits to boost agricultural resilience, and conserving water by utilising highly effective irrigation techniques. Consequently, CO2 emissions were reduced by 90% and water usage was reduced by 40% on plantations that were part of the UEBT effort. As the market grows, Capua is pleased that the UEBT designation is becoming more well-known. However, to maintain transparency and combat green washing, they believe it is essential that an impartial non-profit organisation evaluate and disseminate its initiatives.

Capua in the 1880s is one instance. This 80-person family business specialises in the extraction and processing of all Italian citrus fruits, including essential bergamot oils. Capua 1880 pays weekly fruit prices and establishes multi-year contracts with all of the major cooperatives to support farmers. To expand or establish themselves, farmers gain from steady, predictable earnings and systems that support long-term, sustainable farming.

Marula Oil

Marula oil, extracted from the kernels of the marula fruit (*Sclerocaryabirrea*), has garnered attention for its potential as a nutraceutical ingredient. Traditionally used in Africa for its medicinal and cosmetic properties, marula oil is rich in beneficial compounds such as fatty acids, antioxidants, and vitamins. In certain regions of southern Africa, the marula fruit tree (*Sclerocarya birrea*) is indigenous. Though they were originally uncommon, the wild trees are now grown. The marula tree was once thought to be sacred and associated with happily ever after and fertility. It was also believed that elephants may get intoxicated by the fermented fruit of the marula tree, as they appear to enjoy its delectable flavour just as much as humans. Throughout Africa, the marula tree is utilized in a variety of ways for culinary and medicinal purposes. A hard, brown nut with malleable, white kernels at its core is found inside each marula fruit. Although the nut's outer husk can also be used to extract marula oil, the kernels are the main source of the oil. Marula oil works well as a skin and hair treatment since it is easily absorbed and high in protein. Marula oil is a base note found in body lotions, soaps, and perfumes. It smells delicious and fragrant with a hint of warmth and nutty undertones. Many different types of cosmetic products contain marula oil as a component. It is also available for purchase as an essential oil.

It's beneficial components include:

1. L-arginine and glutamic acid are amino acids with moisturizing and anti-aging qualities.

2. Fatty acids, have moisturizing and emollient properties, such as myristic, oleic, stearic, and palmitic acids.
3. Antioxidants, such as vitamins E and C and phenolic compounds, combat free radicals and may prevent pollution and the sun's UV rays from damaging skin.
4. Research from 2018 Trusted Source suggests that marula oil may help to strengthen and moisturize dry, brittle, or damaged hair.
5. The same 2018 study claims that the oil provides moisturizing and hydrating qualities for extremely dry skin. It may function rapidly because the skin absorbs it well.
6. Marula oil can be applied to the cuticles and nail beds to hydrate them. This might lessen the appearance of hangnails and broken nail beds. However, more investigation is required into how marula oil affects nails.

To verify all of its qualities and how they could improve someone's health and well-being, more research is required. It is easiest to think of the Marula oil case as an early instance of ABS. There was no governing legislative structure in place when Marula oil began to be commercialized; in fact, negotiations were still ongoing for the AU Model Law and the Bonn Guidelines. The long-term success of the commercialization attempt was thus recognized to be significantly impacted by compliance with the ABS provisions of the Convention on Biological Diversity. In Africa, marula and related traditional knowledge (TK) are widely used. Early on, stakeholders in Namibia realized that other TK holders and resource owners in the region needed to be treated fairly and equally. The Marula oil case complies with the government's "Namibia Vision 2030," which was established in 2004 and states that natural capital in Namibia, particularly resources derived from forests, can be used in a participatory and sustainable manner to promote both economic and environmental well-being.

Biochemical Composition of Marula Oil

Marula oil is renowned for its high content of monounsaturated fatty acids, particularly oleic acid and palmitic acid, which contribute to its emollient and moisturizing properties. Additionally, marula oil contains tocopherols (vitamin E), sterols, and phenolic compounds, which possess antioxidant and anti-inflammatory effects. By elucidating the biochemical composition of marula oil, this chapter provides insights into its potential mechanisms of action and health-promoting effects.

Physiological Effects and Health Benefits

Research suggests that marula oil consumption and topical application may confer various health benefits, including skin hydration, anti-aging effects, and protection against oxidative stress. The fatty acids and antioxidants in marula oil help maintain skin barrier function, improve skin elasticity, and reduce the appearance of wrinkles and fine lines. Moreover, marula oil's anti-inflammatory properties may benefit individuals with inflammatory skin conditions such as acne and eczema.

Uchuva (*Physalis Peruviana*)

Many medicinally useful and preventative components have been found in plants, which have been used to treat a wide range of illnesses. Over time, there has been a consistent rise in the utilization of herbal remedies. The centre of genetic diversity for *Physalis peruviana* is the Andes mountains of Ecuador, Chile, Colombia, and Peru. It grows in forests, forest edges, and riparian areas. It grows at high elevations of 500–3,000 m (1,600–9,800 ft) in its native region, but may also be found at sea level in Oceania and Pacific islands where it occurs widely in subtropical and warm, temperate conditions.

Uchuva, also known as *Physalis peruviana* or Cape gooseberry, is a small, golden-colored fruit native to the Andes region of South America. Renowned for its unique flavor and nutritional richness, Uchuva has gained recognition as a potential nutraceutical ingredient. Rich in vitamins, minerals, antioxidants, and bioactive compounds, Uchuva offers a range of health-promoting properties that make it a promising candidate for incorporation into functional foods, dietary supplements, and nutraceutical formulations. This chapter explores the biochemical composition, physiological effects, and potential applications of Uchuva as a nutraceutical, shedding light on its role in enhancing human health and well-being.

Biochemical Composition of Uchuva

Uchuva is packed with essential nutrients and bioactive compounds, including vitamin C, vitamin A, vitamin B complex, iron, phosphorus, and flavonoids such as quercetin and kaempferol. These compounds contribute to uchuva's antioxidant, anti-inflammatory, and immune-modulating properties. By elucidating the biochemical composition of uchuva, this chapter provides insights into its potential mechanisms of action and health-promoting effects.

Physiological Effects and Health Benefits

Consumption of Uchuva has been associated with various health benefits, including immune support, eye health, and skin protection. The high vitamin C content of Uchuva strengthens the immune system and enhances collagen synthesis, promoting wound healing and skin elasticity. Additionally, the presence of carotenoids like beta-carotene and zeaxanthin supports eye health and protects against age-related macular degeneration. Moreover, the antioxidant and anti-inflammatory properties of Uchuva may help reduce the risk of chronic diseases and support overall well-being.

Microalgae

The creation of research techniques for microalgal bioprospecting and its application in supplementing animal feed mark significant advancements in the sustainable growth and enhancement of aquaculture, particularly as a substitute for the conventional dietary approaches used and valuing the use of products derived from these microorganisms (Arialdoet al., 2019).

Microalgae are gaining attention as potential sources of nutraceuticals due to their rich nutritional composition and various health benefits. Nutraceuticals are food or food products that provide health benefits beyond basic nutrition, often due to their bioactive compounds.

Following are the reasons why microalgae are being explored for nutraceutical purposes:

1. **High Nutritional Value**: Microalgae such as spirulina, chlorella, and astaxanthin are rich in essential nutrients including proteins, vitamins, minerals, and essential fatty acids.
2. **Rich in Antioxidants**: Many microalgae contain high levels of antioxidants such as carotenoids, chlorophyll, and phycocyanin. These antioxidants help neutralize free radicals in the body, reducing oxidative stress and lowering the risk of chronic diseases.
3. **Omega-3 Fatty Acids**: Some microalgae, particularly certain species of marine microalgae like *Schizochytrium* and *Nannochloropsis*, are natural sources of omega-3 fatty acids such as EPA and DHA, which are beneficial for heart health, brain function, and inflammation reduction.
4. **Phytonutrients**: Microalgae contain various phytonutrients such as phycocyanin, which has anti-inflammatory and immune-boosting properties, and chlorophyll, which may aid in detoxification and promote overall health.
5. **Sustainable and Environmentally Friendly**: Microalgae cultivation requires minimal land and water compared to traditional agriculture, making it a sustainable and environmentally friendly option for producing nutraceuticals. Additionally, microalgae can be grown in controlled environments, reducing the risk of contamination.
6. **Potential Health Benefits**: Consuming microalgae-based nutraceuticals has been associated with various health benefits, including immune support, improved cardiovascular health, enhanced cognitive function, and potential anti-cancer effects.

Versatile Applications: Microalgae-derived nutraceuticals can be incorporated into various food and beverage products, dietary supplements, and even skincare products, offering a wide range of applications for consumers.

Despite their potential, it's important to note that further research is needed to fully understand the efficacy and safety of microalgae-derived nutraceuticals, as well as to optimize cultivation and extraction processes to ensure consistency and quality in production.

Carnauba Wax

Carnauba wax, also referred to as Brazil wax or palm wax, is a natural wax derived from the leaves of the *Copernicia prunifera*, also known as the carnauba palm. It is predominantly grown in the Brazilian states of Ceará, Piau, and Rio Grande do Norte in the northeast. The palm tree species known as carnauba (*Copernicia prunifera*) is native to the semi-arid region of northeastern Brazil. It grows wild in the Cerrado and Caatinga environments, two highly diverse and endangered ecosystems. The carnauba trees provide vital ecosystem services, such as buffering watersheds, preventing soil erosion, and providing a home for local species, which help to preserve these special environments. For the residents in the impoverished northeastern region, the tree is a vital source of revenue.

It is typically found as hard, yellow–brown flakes when pure. To obtain it, leaves are gathered and the leaves dried, then beaten to release the wax before bleaching and refining it. During the dry season, when there aren't many alternative occupations in agriculture, carnauba powder is extracted for use in wax manufacture. Carnauba wax, sometimes referred to as the "Queen of wax," is a valuable export that has special qualities that make it an essential component of the food, cosmetics, pharmaceutical, and polishing sectors.

Composition and Uses: The main components of carnauba are fatty alcohols (12 wt%), ω-hydroxycarboxylic acids (13.0 wt%), diesters of 4-hydroxycinnamic acid (21.0 wt%), and aliphatic esters (40 wt%). Most of the compounds are produced from alcohols and acids in the C26–C30 range. It stands out due to its high diester content and methoxycinnamic acid level.

- Carnauba wax is used to make automotive waxes, shoe polishes, dental floss, culinary goods (like candy), musical instrument polishes, floor and furniture waxes, and more. It produces a glossy finish when combined with turpentine and beeswax.
- Carnauba wax is used as a thickening in cosmetics like lipstick, eyeliner, mascara, eye shadow, foundation, deodorant, and skincare and sun care preparations because of its hypoallergenic, emollient, and glossing qualities.
- Carnauba wax is too brittle to be used alone, so it's usually mixed with other waxes (often beeswax) to waterproof and cure leather goods. This gives the leather a high-gloss sheen while boosting its hardness and endurance.

- Carnauba wax can be applied as a mould releaser in the production of polymers reinforced with fibres.
- The hand-harvested carnauba wax is processed to make a product that may be applied to automobiles.

Of the 22,409 tonnes of carnauba wax produced in Brazil in 2006, 14% was solid and 86% was powdered. Carnauba wax is exported by 20 to 25 companies in Brazil, who first refine the wax before selling it to other countries or purchasing it straight from farmers. According to the Brazilian Ministry of Development, Industry and Foreign Trade, the major destinations for exported carnauba wax are: United States (25%), Japan (15–25%), Germany (10–15%), Netherlands (5%). Italy (5%), other countries (18%).

To preserve the carnauba palm tree and its ability to produce wax, only its leaves are taken during the carnauba wax harvesting process. Significant contributions to the ecological and biodiversity of the area are also made by carnauba palm palms. Because of its unique properties and versatility, carnauba wax is highly sought after as an ingredient in a wide range of products. Its vast use in a multitude of industries can be attributed to its natural origin, sustainability, and versatility of uses.

The labour-intensive process of extracting carnauba wax from palm leaves requires the employment of trained labourers to maintain the integrity of the carnauba palm tree population and sustainable harvesting methods.

1. **Intangible cultural heritage offers real-world illustrations of instructional materials and approaches.**

Communities have consistently discovered methods to organise and pass on to future generations their knowledge, life skills, and competencies about their natural and social environments. All children should have access to high-quality education without excluding the next generation from this abundant resource that is so closely linked to their cultural identity. To fully utilise the wealth that intangible cultural heritage offers, quality education must, on the one hand, fully integrate it into the curriculum of all pertinent disciplines and, on the other hand, look for ways to maximise the educational potential of traditional modes and methods of transmitting intangible cultural heritage within educational systems.

2. **The preservation of intangible cultural property can promote social inclusion and cohesiveness.**

Communities and groups can be strengthened inclusively by implementing social practices, rituals, and celebratory events that give structure to their existence. Communities have always looked for methods to organise and pass on to the next generation their knowledge, skills, and competencies in life, particularly about their natural and social environments.

3. **Gender roles and identities are formed and passed down through intangible cultural legacy, which makes it essential to gender equality.** Communities transmit their gender-related beliefs, conventions, and expectations through intangible cultural legacy, which also shapes the gender identities of their people.

Furthermore, these gender norms frequently dictate who can access and participate in particular expressions of heritage. For instance, the creation of traditional crafts frequently depends on a particular division of labour related to gender, whereas performing arts are a privileged arena for the open expression of gender norms and expectations.

Environmental Sustainability

Maintaining a stable climate, managing natural resources responsibly, and preserving biodiversity is all necessary for environmental sustainability. Reducing the human, societal, and financial consequences of natural disasters and climate change requires strengthening the resilience of vulnerable groups. Ancient customs, beliefs, and knowledge that have been passed down through the ages and are preserved as intangible cultural heritage have influenced human societies' relationships with the environment for millennia. In many areas today, including biodiversity protection, sustainable resource management, and natural disaster planning and response, the value of intangible cultural heritage to environmental sustainability is acknowledged.

The year 2020 holds great significance for biodiversity. The process of adopting a global biodiversity framework at the 15th Conference of the Parties to the CBD, which has been rescheduled for 2021, is still moving forward despite the global epidemic. Furthermore, October 29, 2020, marked a decade since the Nagoya Protocol on access and benefit-sharing (ABS) was adopted. The Nagoya Protocol has been ratified by 129 CBD Parties as of right now. Nonetheless, a lot of nations are either still creating ABS regulations or are just beginning to put them into practice. Given the additional responsibilities under the Nagoya Protocol, there is so comparatively little experience with implementing ABS in practice. Because of this, it is still difficult to find examples that are far enough along to allow for the assessment of contributions to the Sustainable Development Goals and the drawing of lessons.

Indigenous and local groups frequently hold important traditional knowledge regarding the cultivation and uses of baobab trees. Genetic resources can be accessed and used in a variety of ways. Access and benefit-sharing describe how the advantages of this use are distributed among the individuals or nations who use the resources (users) and the individuals or nations that supply them (providers). By utilising mechanisms like Access and Benefit Sharing (ABS) agreements, it is ensured that local communities maintain their intellectual property rights while profiting from the commercial usage of baobab resources. One of the most significant native fruit trees in Africa is the baobab. Fruits from this plant are very nutrient-dense and can help fight hunger and food poverty. Baobab trees have great cultural and traditional significance and are recognised as distinctive symbols in many civilizations. Local communities' identities and traditions are preserved to some extent through the preservation of these trees and the information that goes along with them regarding their functions and cultural value.

The goal of the 2016–2019 BAOFOOD project was to advance the use, processing, and market development of baobab to enhance rural livelihoods, food security, and nutrition. The German Federal Ministry of Food and Agriculture provided the funding (BMEL). Research projects were carried out by a variety of partners and covered every aspect of the value chain, from taking stock of baobab trees to creating a pilot processing facility for the benefit of the surrounding communities. To improve food security, nutrition, and rural livelihoods in Kenya and Sudan, the BAOFOOD project (2016–2019) intends to increase the domestication, production, market development, processing, and consumption of baobab (*Adansonia digitata* L.) German project partners were responsible for conducting a portion of the nutritional analyses of baobab fruit pulp because Kenya lacked the necessary laboratory equipment (HPLC–DAD, high-throughput HPTLC).

Conclusion

In compliance with the Nagoya agreement, the export of baobab samples and sample use by German project partners required prior authorization. One of the first cases of this type that Kenyan authorities had to deal with was the BAOFOOD scheme. It took about two years from the start of the negotiations to the obtaining of the necessary agreements. While acknowledging the non-commercial nature of the research initiative, Kenya Wildlife Service's first statements projected substantial financial benefits to be expected from the partnerships.

The legal frameworks known as Agreements on Benefit Sharing and Access (ABS) control the distribution of advantages arising from the utilisation of biological resources, such as baobab trees. Through these agreements, local communities will be fairly compensated for the use of their resources, sustainable practices will be supported, and scientific research and collaboration will be encouraged. To guarantee both legal access and the fair and equitable sharing goal of the Protocol, structures for compliance and transparency that complement national ABS measures must be established. To provide direction to both local administration and scholars, successful reference cases are thus desperately needed.

In conclusion, the exploration of bioprospecting for food and nutrition underscores the intricate interplay between economic development, environmental conservation, and cultural preservation. The examples provided illustrate the importance of addressing issues such as benefit sharing, protection of traditional knowledge, and sustainable resource management when commercializing products derived from indigenous plants and practices.

From the commercialization of Allanblackia oil to avocado farming in Peru, and bergamot production in Italy to carnauba wax extraction in Brazil, each case study exemplifies a unique approach to balancing economic opportunities with environmental stewardship and cultural heritage preservation. These endeavours not only offer economic benefits to local communities but also contribute to biodiversity conservation, sustainable agriculture, and the preservation of traditional knowledge systems.

The concept of Access and Benefit Sharing (ABS) agreements, as seen in the case of baobab resources, emphasizes fairness, equity, and responsible resource management, ensuring that the utilization of natural resources benefits local communities while supporting sustainable practices and facilitating scientific research and cooperation.

Moreover, the chapter underscores the increasing attention given to the ethical considerations surrounding bioprospecting and the commercialization of natural resources. By promoting fair trade practices, ethical sourcing, and adherence to international sustainability standards, stakeholders can ensure that bioprospecting efforts enhance food security and nutritional quality while respecting the rights and well-being of indigenous communities and preserving valuable ecosystems.

In essence, the multifaceted dynamics of benefit sharing, intellectual property rights, and sustainable management practices in bioprospecting for food and nutrition highlight the need for holistic approaches that integrate economic, environmental, and cultural considerations. Only through such comprehensive strategies can we harness the potential of bioprospecting to address global challenges while safeguarding our natural and cultural heritage for future generations. Overall, the commercialization of natural resources through bioprospecting can offer promising opportunities for scientific discovery and economic development, but it requires careful consideration of ethical, environmental, and legal issues to ensure that it is carried out in a socially responsible and sustainable manner.

Conflict of Interest The authors state no conflict of interest.

References

Arialdo, M. S., Silvia Maria, M., & Alan, C. C. (2019). Bioprospection of bio compounds and dietary supplements of microalgae with immunostimulating activity: A comprehensive review. *PeerJ, 7*, e7685. https://doi.org/10.7717/peerj.7685

Arshad, A. (2022). Impact of financial inclusion on food security: Evidence from developing countries. *International Journal of Social Economics, 49*, 336–355. ISSN: 0306-8293- https://doi.org/10.1108/IJSE-08-2021-0462

Azebaze, A. G. B., Menasria, F., Noumi, L. G., Nguemfo, E. L., Tchamfo, M. F., Nkengfack, A. E., Kolb, J. P., & Meyer, M. (2009). Xanthones from the seeds of *Allanblackiamonticola* and their apoptotic and antiproliferative activities. *Planta Medica, 75*, 243–248. https://doi.org/10.1055/s-0028-1088375

Berlinck, R. G. S. (2012). Bioprospecção no Brasil: Um breve histórico. *Ciência e Cultura, 64*, 27–30. https://doi.org/10.21800/S0009-67252012000300010

Bonfigli, A. R., Protic, O., Olivieri, F., Montesanto, A., Malatesta, G., di Pillo, R., & Antonicelli, R. (2020). Effects of a novel nutraceutical combination (BruMeCholTM) in subjects with mild hypercholesterolemia: Study proto-col of a randomized, double-blind, controlled trial. *Trials, 21*(1), 616. https://doi.org/10.1186/s13063-020-04551-4

Brasil, B. S. A., Silva, F. C. P., & Siqueira, F. G. (2017). Microalgae biorefineries: The Brazilian scenario in perspective. *New Biotechnology, 39*, 90–98. https://doi.org/10.1016/j.nbt.2016.04.007

Capomolla, A. S., Janda, E., Paone, S., Parafati, M., Sawicki, T., Mollace, R., Ragusa, S., & Mollace, V. (2019). Atherogenic index reduction and weight loss in metabolic syndrome patients treated with a novel pectin-enriched formulation of bergamot polyphenols. *Nutrients, 11*(6), 1271. https://doi.org/10.3390/nu11061271

Cicero, A. F. G., Fogacci, F., Bove, M., Giovannini, M., & Borghi, C. (2019). Three-arm, placebo-controlled, randomized clinical trial evaluating the metabolic effect of acombined nutraceutical containing a bergamot standardized flavonoid extractindyslipidemic overweight subjects. *Phytotherapy Research, 33*(8), 2094–2101. https://doi.org/10.1002/ptr.6402

Crockett, L. S. (2015). Allanblackia oil: Phytochemistry and use as functional food. *International Journal of Molecular Sciences, 16*, 22333–22349. doi: 10.3390/ijms160922333. PMID: 26389891; PMCID: PMC4613311

Ferro, Y., Montalcini, T., Mazza, E., Foti, D., Angotti, E., Gliozzi, M., Nucera, S., Paone, S., Bombardelli, E., & Aversa, I. (2020). Randomized clinical trial: Bergamot Citrus and wild cardoon reduce liver steatosis and body weight in non-diabetic individuals aged over 50 years. *Frontiers in Endocrinology, 11*, 494. https://doi.org/10.3389/fendo.2020.00494

Foti, F. (2015). *UtilizzazioneZootecnica del Pastazzo Fresco di Bergamotto* (pp. 177–186). Proceedings of Congress "Storie di bergamotto" Artemis Ed., Reggio Calabria, Italy.

Kadigi, R. M. J., Robinson, E., Szabo, S., Kangile, J., Mgeni, C. P., de Maria, M., Tsusaka, T., & Nhau, B. (2022). Revisiting the Solow-swan model of income convergence in the context of coffee producing and re-exporting countries in the world. *Sustainable Futures, 4*, Article 100082. https://doi.org/10.1016/j.sftr.2022.100082

Laladhas, K. P., Gopalan, S. V., Nilayangode, P., & Oommen, O. V. (2016). Conservation through payment for ecosystem services. *Biodiversity for sustainable development* (pp. 193–223). Springer International Publishing. https://doi.org/10.1007/978-3-319-42162-9_13

Laladhas, K. P., Nilayangode, P., & Oommen, O. V. (Eds.). (2017a). *Biodiversity for sustainable development*. Springer International Publishing. https://link.springer.com/book/10.1007/978-3-319-42162-9

Laladhas, K. P., Nilayangode, P., Gopalan, S. V., Divya, L., & Oommen, O. V. (2017b). Community conservation of marine turtles and sustainable ecosystems. *Biodiversity for Sustainable Development*, 177–184. https://link.springer.com/book/10.1007/978-3-319-42162-9

Laladhas, K. P., Preetha, N., Baijulal, B., & Oommen, V. (2015). *Conservation of medicinal plant resources through community born biodiversity management committee, Kerala, India* (p. 27). Biodiversity Conservation-Challenges for the Future.

Marques, L. G., Santos, M. R., Raffo, J., & Pessoa, C. (2013). Redes de Bioprospecção no Brasil: Cooperação para o desenvolvimentotecnológico. *Revista de DesenvolvimentoEconômico, 15*, 164–172.

Mithrambika, N. B., Laladhas, K. P., Nilayangode, P., & Oommen, O. V. (2017). Decentralized governance for sustainable development. In *Biodiversity for sustainable development* (pp. 165–176). Springer International Publishing. https://doi.org/10.1007/978-3-319-42162-9_10

Mollace, V., Scicchitano, M., Paone, S., Casale, F., Calandruccio, C., GliozziM, M. V., Carresi, C., Maiuolo, J., & Nucera, S. (2019). Hypoglycemicand hypolipemic effects of a new lecithin formulation of bergamot poly-phenolic fraction: A double blind, randomized, placebo-controlled study. *EndocrMetab Immune Disord Drug Targets, 19*(2), 136–143. https://doi.org/10.2174/1871530319666181203151513

Montemurro, V. (2015). *Il Bergamotto: Bene Prezioso a Tutela Della Salute* (Artemis Ed.). Reggio Calabria.

Mountessou, B. Y. G., Tchamgoue, J., Paul, D. J., Tchuenguem, R. T., Surup, F., & Choudhary, M. I. (2018). Two xanthones and two rotameric (3⟶8) biflavonoids from the Cameroonian medicinal plant *Allanblackiafloribunda Oliv.* (Guttiferae). *Tetrahedron Letters, 59*, 4545–4550. https://doi.org/10.1016/j.tetlet.2018.11.035

Nganou, K. B., Konga, S. I., Fangam, G. A., Bitchagno, M. G., Sonfack, G., Nayim, P., Celik, I., Koyuturk, S., Kuete, V., & Tane, P. (2019). Guttiferone BL with antibacterial activity from the fruits of *Allanblackiagabonensis*. https://doi.org/10.1080/14786419.2018.1465424

Nilayangode, P., Cheruvat, D., Laladhas, K. P., & Oommen, O. V. (2016a). ABS-Unlocking the opportunities. *Journal of Traditional and Folk Practices (JTFP), 4*(2). https://jtfp.jntbgri.res.in/index.php/jtfp/article/view/64

Nilayangode, P., Laladhas, K. P., & Oommen, O. V. (2016b). Biodiversity and sustainable development: A retrospect. *Biodiversity for sustainable development* (pp. 315–322). Springer International Publishing. https://doi.org/10.1007/978-3-319-42162-9_23

Nilayangode, P., Laladhas, K. P., Nair, G. U., Mathai, A., Baijulal, B., & Oommen, O. V. (2016c). Grassroots initiatives for environmental sustainability. *Biodiversity for sustainable development* (pp. 185–192). Springer International Publishing. https://doi.org/10.1007/978-3-319-42162-9_12

Nilayangode, P., Nair, G. U., Laladhas, K. P., & Oommen, O. V. (2017). Geographical indications and sustainable livelihood. In *Biodiversity for sustainable development* (pp. 225–234). Springer International Publishing. https://doi.org/10.1007/978-3-319-42162-9_14

Oommen, O. V., Laladhas, K. P., Nelliyat, P., & Pisupati, B. (Eds.). (2022). *Biodiversity conservation through access and benefit sharing (ABS): Himalayas and Indian sub-continent*. Springer. https://link.springer.com/book/10.1007/978-3-031-16186-5

Perry, N., & Perry, E. (2006). Aromatherapy in the management of psychiatric disorders clinical and neuropharmacological perspectives. *CNS Drugs, 20*, 257–280.

Pizzimenti, F., Tulino, G., & Marino, A. (1998). Antimicrobial and antifungal activity of bergamot oil. In *Program and Abstracts of the International Congress: 'Bergamotto 98. Stato dell'arte', Reggio Calabria, Italy, 1998. Abstract no. 38.* Laruffa, Reggio Calabria, Italy.

Preetha, N., Laladhas, K. P., & Oommen, V. O. (2015). Biodiversity conservation-challenges for future and way forward. *Biodiversity and Conservation, 1*, 249.

Pritchard, B., Rammohan, A., & Vicol, M. (2019). The Importance of non-farm livelihoods for household food security and dietary diversity in Rural Myanmar. *Journal of Rural Studies, 67*, 89–100. https://doi.org/10.1016/j.jrurstud.2019.02.017

Ratha, S. K., & Prasanna, R. (2012). Bioprospecting microalgae as potential sources of Green Energy—challenges and perspectives(Review). *Applied Biochemistry and Microbiology, 48*, 109–125. PMID: 22586907

REPOA. (2018). *Improving Tanzania's competitiveness of avocado ('green gold') value chain and exports: A case for targeted regulatory, policy and institutional reforms* (Policy Brief No. PB 1/2018). https://www.repoa.or.tz/wp-content/uploads/2020/07/AVOCADO_value_chain.pdf

Sacarro Junior, N. L. (2011). *Desafios da bioprospecção no Brasil. Publicações da secretaria de assuntosestratégicos da presidência da república.* Instituto de Pesquisa EconômiaAplicada; Brasília, Distrito Federal. https://repositorio.ipea.gov.br/bitstream/11058/1568/1/TD_1569.pdf

Sakurada, T., Mizoguchi, H., Kuwahata, H., Katsuyama, S., Komatsu, T., & Morrone, L. A. (2011). Intraplantar injection of bergamot essential oil induces peripheral antinociception mediated by opioid mechanism. *Pharmacology, Biochemistry and Behavior, 97*, 436–443. https://doi.org/10.1016/j.pbb.2010.09.020

Schaffer, B., Wolstenholme, B. N., & Whiley, A. W. (2013). *The avocado: Botany, production and uses*. An online publication by CABI. https://www.cabi.org/cabebooks/ebook/20133051561

Scoones, I., Mavedzenge, B., & Murimbarimba, F. (2018). Medium-scale commercial farms in Africa: The experience of the 'native purchase areas' in Zimbabwe. *Africa, 88*, 597–619. https://doi.org/10.1017/S0001972018000244

TanzaniaInvest. (2023). *Avocados*. https://www.tanzaniainvest.com/avocado

Tyagi, J., Ahmad, S., & Malik, M. (2022). Nitrogenous fertilizers: Impact on environment sustainability, mitigation strategies, and challenges. *International Journal of Environmental Science and Technology, 19*, 11649–11672. https://doi.org/10.1007/s13762-022-04027-9

Bioprospecting and Access and Benefit Sharing (ABS) from Protected Areas (PAs): Scope and Challenges

Prakash Nelliyat

Background

For preserving the natural habitants in ecologically sensitive areas, their declaration as Protected Areas (PAs) has become a universally acceptable strategy. Generally, the PAs of a country are the ecologically sensitive and biological or genetic resources-rich spots with unique flora and fauna. Countries consider that preserving their rich biodiversity spots will facilitate their future utilisation and also offer 'Existence Value'. PAs not only include the landscapes (especially forests) and inland water bodies (wetlands) but ocean or marine PAs also. Besides, there is trans-boundary PAs (where areas overlap in multiple nations).

In recent years, global biodiversity is facing a huge threat and the reasons include habitant loss due to over extraction of resources, indiscriminate discharge of pollutants (air, liquid and solid), spread of Invasive Alien Species (IAS), lack of ecosystem integrity, fragmentation of ecosystems, and climate change induced adverse impacts like frequent floods and droughts (CBD, 2010). Broadly, the direct impacts of biodiversity degradations are more for the most vulnerable communities in developing countries. Further, it adversely affects in achieving most of the Millennium Development Goals (MDGs) related to biodiversity, such as: poverty alleviation, water and food security, and environmental sustainability (Lopoukhine et al., 2012).

Considering all these points, one can strongly advocate that biodiversity conservation is needed at the priority level, in which the concept of PAs is extremely important. PAs are of several kinds and with different magnitudes of regulations, based on the prevailing legal and socio-political norms in a country. In each country, the Government has come up with different legal norms (rules and regulations) as well as institutional structures for the management of PAs. Generally, the public

P. Nelliyat (✉)
Centre for Biodiversity Policy and Law, National Biodiversity Authority (MoEFCC, Govt. of India), Taramani, Chennai 600113, India
e-mail: prakashnelliyat@gmail.com

L. Krishna Panicker et al. (eds.), *Biodiversity and Business*,
https://doi.org/10.1007/978-3-031-71674-4_11

accessibility to collect the resources from the PAs is nil or limited in all nations (Nelliyat & Meenakumari, 2018).

On the other hand, PAs are buffer zones of unique biological resources containing genetic materials and their utilization for bio-prospecting is having a huge scope and this should be seriously debated against the backdrop of the CBD objectives and the Nagoya Protocol principles. Further, the ABS amount generated through the commercial utilization of the biological resources from PAs might be an effective and continuous or sustainable source of revenue for PAs' management. However, PA management through ABS is not much debated in various forums, including the CBD's Conference of Parties (COP).

In this context, this chapter examines the scope of ABS from the PAs and the legal and institutional measures required for its operationalization. Since the available literature linking the ABS and PAs is limited, the author collected the relevant information from the existing literature which examines the ABS and PA separately. Besides, the author conducted a series of interviews with ABS' specialists and PA managers and synergised their views.

PAs and Their Ecological and Economic Significance

Different agencies defined PAs in a diversified manner. The International Union of Conservation of Nature (IUCN) stated that a PA is a clearly demarcated geographical area of a nation managed through stringent legal norms with the intention of long term conservation of nature with cultural values and ecosystem services (Dudley, 2008). In PAs one can identify different types of key biodiversity spots, that include: important plant diversity, endemic bird zones, indigenous communities' settlements, and zero extinction sites. According to Nagulendran (2013) PAs provide, innumerable goods and services, including those of cultural and aesthetic value and reservoirs of carbon stocks.

PAs' contribution to the conservation of the natural and cultural resources of the world is significant. PAs are the buffer zones of unique natural resources and their contribution in ensuring ecosystem services and livelihood options for nature dependent communities is significant. The large number of rural communities, that lives on the fringes (surroundings) of PAs, primarily depends on PAs' resources for their food and medicines or livelihoods. In this manner PAs play a significant role in generating employment and income, thereby reducing their poverty. Most of the PAs are the hotspots for eco-tourism and generate income for the Government and the local communities. Further, they sustain the heritage and cultural values of the indigenous and local communities who live in PAs and their surroundings. According to a study in 2010, worldwide, there were 1,61,000 PAs which cover around 17 million sq. km. or 11.5% of the worlds' land surface areas. Further, there are 6,800 marine PAs designated covering 1.2% of the worlds' oceans (Nelliyat & Meenakumari, 2018).

PAs' Management Under the CBD

According to the CBD, PAs is the major international legal strategy for safeguarding biodiversity. Article 2 of the CBD mentioned that; PA is the geographically defined area designated, regulated and managed for achieving targeted conservation objectives. Article 8 of the Convention also emphasised the establishment and management of PAs through "In Situ" conservation (CBD, 2008). Based on it, one can easily come to the conclusion that, for fulfilling the objectives of the CBD, particularly the conservation of biodiversity or biological resources containing genetic materials, the role of PAs is significant. Further, PAs are the settlements of the indigenous and local community, which has a thorough and extensive knowledge of the biological resources surrounding it (Laird et al., 2003).

However, the PAs' concept with restriction on genetic or biological resources' access for commercial utilization or bio-prospecting through stringent legal norms and the possible scope on imposing the benefit sharing principle should be the concern for further debate and action. What is needed is a universally acceptable and clear strategy for accessing the genetic or biological resources from the PA and the benefit sharing guidelines on it.

The CBD' COP VII (Decision VII/28) adapted the Programme of Work on Protected Areas (PoWPA), which discussed the various elements including: (a) direct action, planning, choosing, forming, strengthening and controlling the PAs, (b) ways and options for improving governance, stakeholder participation and principle of equity, (c) enabling programs and activities with respect to the PA, and (d) fixation of standards, assessment and monitoring criteria for PAs.

CBD stated that PoWPA can contribute to achieve 2010 targets and reduce the present biodiversity loss at the regional, national and international levels. Further, it can contribute to reduction in poverty sustainable development and achieve the Millennium Development Goals (MDGs). Goal 2.1(element 2) of PoWPA itself emphasises on the 'promotion of equity and benefit sharing'. It proposed that the costs and benefits arising from the establishment and management of PAs should be shared equitably and the required mechanism should be developed (Nelliyat & Meenakumari, 2018).

Realising the significance of PAs in biodiversity conservation at various levels (global, national and local) the COP10 of the CBD formulated a Global Target (Target No.11) to manage the PAs. It stated that 17% of the terrestrial and inland water and 10% of the coastal and marine areas should be conserved. In this way, the PAs' ecosystem services should be generated not only for the present generation but for future generations also.

In brief, it is very clear that the CBD also provides adequate consideration of the PAs as a viable option for biodiversity conservation. But the overall scope of ABS from PA is not discussed much even in the CBD forums. No doubt that, the CBD is taking the initiative for the conservation of biodiversity through both the Nagoya Protocol and the PoWPA. Ideologically these two initiatives have huge inter linkages. However, in different activities of the convention, they were observed to be separate. Hence, on a priority basis, there is an urgent need for the synergies between the Nagoya Protocol on ABS and PoWPA within the CBD framework itself.

Bio-Prospecting from PAs and Their ABS

After the advent of biotechnology in the developed nations of the world, bio-prospectors seriously target the rich biodiversity spots located in the developing nations and indiscriminately extracted their bio-wealth (biological and genetic resources) to a considerable extent. Genetic resources were valued significantly and incorporated in research and development (R&D) by different biological or genetic resources-based industrial sectors and derived different consumer products ranging from life saving medicines, to health care products, nautrasuiticals, botanicals, cosmetics, etc.

PAs (both terrestrial and marine) are treasure boxes of biodiversity with rich flora and fauna. This wealth attracts bio-prospectors who search for active compounds and genetic materials from plants and animals. Besides, PAs and their surroundings are the epic spots of the prevalence of local and indigenous communities, who have innumerable biological resources—associated traditional knowledge. The knowledge holders are also the centre of attraction to the bio-prospectors as they make it easy for them to bio-survey and do research in identifying a particular compound they are looking for. In this way bio-prospectors can save huge amounts of money and time, which are crucial in modern comparative business.

In recent decades, bio-prospecting from the PAs has contributed to developing valuable commercial products. Hence, further scope of bio-prospecting from the PAs is huge and the genetic materials may be the source for product development in many sectors, including pharmaceuticals. Despite all these, the need for a systematic approach in accessing the genetic materials from the PAs and their ABS possibilities is not thought about or sensitised to the PAs managers, who are the 'gate keepers' of the PAs.

PAs and their genetic materials may be considered as the 'option value' of ecosystems. Option value is the premium placed on maintaining ecosystems and their services for future use. Hence, PAs are buffer stocks of genetic materials and play a significant role in bio-prospecting in the future. Presently, CBD has Universal membership (196 Parties) and 142 countries have ratified the Nagoya Protocol on ABS. In this context, the ABS possibilities from the PA become much more significant. However, in different parts of the world the PAs' managers enforce strict legal norms for restricting the access to resources. Recently, countries are designating their PAs as "Community Conserved Areas" with entitlement over the resources to the community. This is revealing the changes in the mindset of PA managers and a move towards joint management of PAs, both by the community and the Governments. In this way, the benefits will be channelized for resources' management. There are evidences for bio-prospecting from the PAs' ends with valuable consumer products with high commercial significance and value and some of the examples are indicated below:

- Pharmaceutical Sandiummum (cyclosporine) was the top selling drug in 2000. Through its sale, the marketing company (Novarties) earned US$ 1.2 billion. A researcher at Sandoz (which became Novartis after a 1996 merger with Ciba Geigy) collected soil samples from Hardanger Vidda National Park in Norway during 1969. By 1972 the immunosuppressant properties of cyclosporine were identified in the Soil sample and in 1983 Sandoz introduced Sandimmum and marketed it.
- During 1966, the thermophile thermos aquaticus was collected from Yellow Stone National Park in US by scientists. In 1984, a DNA polymerase enzyme was isolated and used in a series of biotechnological applications with an annual sales amount of around US $ 200 million (Nelliyat & Meenakumari, 2018).

However, it is doubtful to what extent the PA managers realised these facts. In brief, all these developments throw lights towards the legal accessing of resources from the PAs and their ABS. Otherwise; these are huge possibilities in misusing the valuable genetic resources available in the PAs.

PAs and the ABS: Indian Experience

India is one of the mega bio-diverse countries in the world. Around 27% of India's geographical areas have been designated as PAs through different legal frameworks protecting its biodiversity:

The Wildlife Protection Act 1972 declared areas of adequate ecological, floral, faunal, natural and geo-morphological significance as National Parks, Wildlife Sanctuaries, Conservation Reserves and Common Reserves for protecting and developing wildlife and its environment. India's PAs include: 25 marine PAs in peninsular India and 106 PAs in the Islands, 28 elephant reserves and 39 tiger reserves. Table 1 gives the detailed picture of four different types of PAs in India.

Out of the 1014 PAs, 5 were designated as World Heritage sites by the UNESCO. India paid special attention to conserve trans-boundary PAs through coordinated conservation efforts with bilateral and multilateral co-operation. India is also a party

Table 1 Status of protected area in India (November, 2023)

National parks			Wildlife sanctuaries			Conservation reserves			Community reserves		
Number	Area (km^2)	% of State Area	Number	Area (km^2)	% of State Area	Number	Area (km^2)	% of State Area	Number	Area (km^2)	% of State Area
106	44,403	1.35	573	123,762	3.76	115	5549	0.17	220	1455	0.043

Source Wildlife Institute of India (2024)

in the major international conventions: Convention on International Trade in Endangered Species (CITES) of wild flora and fauna, International Union for the Convention of Nature (IUCN), International Convention for the Regulation of Whaling, and the UNESCO World Heritage Committee and Convention on Migratory Species.

Under the Indian Forest Act (1927) the National Forest Policy initiated the conversion of natural forests having rich biodiversity, by declaring them as Reserve Forests and Protected Forests, thus prohibiting the clearing of such areas, felling of trees and mining which damage the green cover. India's wetlands are also legally protected under the Wetland Conservation and Management Rules 2017 notified under the Environmental (protection) Act of 1986. There are 26 Ramsar Wetland sites in India spread over an area of 12,119 sq. kms. The Biological Diversity Act 2002 (Section 37) empowers the state Government to declare areas with significant biological diversity as 'Biodiversity Heritage Sites' in consultation with the local Government (Nelliyat & Meenakumari, 2018).

The UNDP and MoEFCC (2018) estimated the total area under different categories of conservation policy, through different Acts and legal norms in India. Accordingly, the total area under conservation is 9,14,074 sq. kms (27% of India's geographical area). Out of it 1,62,072 sq. kms. (18%) comes under the Wildlife Protection Act—1972; 6,05,347 sq. kms. (66%) under the Forest Act—1927; 1,45,714sq. kms (16%) are designated as wetlands under the Environmental (Protection) Act—1986; and 941 sq. kms. as Biodiversity Heritage Sites under the Biological Diversity Act—2002.

No doubt that, the local communities which have the customary rights on the resources are collecting the required resources from these PAs and a major share is going for commercial utilisation. Besides, illegal extraction (bio-piracy) may also occur. Therefore, mainstreaming the ABS in PAs is the need of the hour and in this regard, the Government should take appropriate policy decisions at the earliest.

Scope for Bio-Prospecting and ABS from the PAs

The above discussion clearly reveals that commercial utilization of genetic and biological resources from the PAs is progressing both in a formal and informal manner during the last few decades. However, there are huge uncertainties among the PA managers and the ABS enforcement agencies regarding this. The bio-prospecting partnership in Yellowstone National Park in the US offers maximum revenue to the park authority and has become an effective monetory option for park management. But, in South Africa, park managers refused commercial collections of biological resources from the PAs and are waiting for finalization of the national legislation. This situation prevails in many other countries also. On the other hand, park managements in Uganda, Cameroon and Cote d Ivoire take pro-active interim measures to deal with ABS instead of waiting for the finalization of the national legislations (Nelliyat & Meenakumari, 2018).

In brief, a range of policy options for enforcing ABS for PAs are progressing at Community, national and inter-governmental levels. From the Inter-governmental perspective, the International Treaty on Plant Genetic Resources for Food and Agriculture (ITPGRFA) and the CBD regularised principles such as Prior Informed Consent, Mutually Agreed Terms, and Benefits Sharing in connection with bio-prospecting or access, facilitate the use and exchange of biological and genetic resources.

As a party to the CBD, India has also taken effective legal, institutional and administrative measures to implement the ABS (Nelliyat, 2022) and it is reflected in the resources from the PAs too. In India, large quantities of biological resources collected by the tribal and local communities are being used for commercial purposes. From the business angle, industries from different sectors prefer biological resources collected from the wild due to their superior quality.

In India's Nagoya Protocol implantation trajectory, 95% of the ABS amount collected by the Competent National Authority (National Biodiversity Authority—NBA) is from red sanders (*pterocarpus santalinus*), an endemic species, existing in the Seshachalam forests, a Biosphere reserve (designated through the UNESCO's Man and Biosphere programme by the Government of India) in Andhra Pradesh. Out of the ABS amount a portion is used by the Andhra Pradesh Forest Department for the conservation of red sanders (Nelliyat, 2022).

In brief, it is clear from the above discussion that different PAs and ABS models are progressing and experimenting in the world with different success ranges. One can see all these models are on a trial basis at this stage and we need to formulate the most appropriate strategy to utilise the valuable genetic materials or biological resources from the PAs for bio-prospects and reap its benefits.

Conclusion and Way Forward

Globally, countries are taking proactive measures for biodiversity conservation and in which: (a) the declaration of ecosystem or biodiversity hotspots are PAs and (b) the conservation and sustainable use of biodiversity through ABS are predominant strategies. No doubt that both these strategies aim to sustain the ecosystem services of nature and maintain the stock of biodiversity. Ideally, when the resources in PAs are conserved through stringent legal norms with maximum restriction of their access or extraction, the ABS philosophy encouraged to sustainably use the resources with legal permits as well as access and benefit sharing. This dichotomy (contradictory approach to achieve the same goal) needs to be resolved at the CBD level itself, as PoWPA and Nagoya Protocol on ABS are functions in the CBD. As most of the countries are ratifying the Nagoya Protocol on ABS and seriously working out the required legal and institutional measures, the conservation of biodiversity though an age old strategy like PAs with limited or no access to biological or genetic resources needs to be re-examined.

Genetic or biological resources are renewable natural resources. Hence, their extraction within their regeneration for bio-prospecting and achieving health and food security and enhancing human welfare, through manufacturing of divergent consumer products, is a biologically and socio-economically acceptable venture. Besides, resources' conservation and their sustainable use through benefit sharing can achieve the principle of equity.

What is needed is a collaborative attempt from parties of the CBD to synergise the philosophy of PoWPA and the Nagoya Protocol on ABS and work for maintaining and enhancing the richness and stock of biodiversity in the universe. ABS should also be viewed as an innovative internal financial mechanism, as the required fund for biodiversity conservation can be mobilised from the biodiversity sector itself, which is the need of the hour.

Acknowledgements This paper is the abstract of the research the Author carried out at the Centre for Biodiversity Policy and Law (CEBPOL) with the support of the Norwegian Government. I am extremely thankful to Dr. S. K. Khanduri IFS (WL), Ministry of Environment Forest and Climate Change for his intensive review of the first draft and his valuable suggestions on linking the concept of Protected Areas and the ABS.

References

CBD. (2008). *Protected areas in today's world: Their values and benefits for the welfare of the planet.* Montreal, Technical Series no. 36, i–vii + 96 pages. Convention on Biological Diversity.

CBD. (2010). *Global biodiversity outlook 3.* Montreal: Secretariat of the Convention on Biological Diversity.

Dudley, N. (Ed.). (2008). *Guidelines for applying protected area management categories.* Gland.

Laird, S., Johnston, S., Wynberg, R., Lisinge, E., & Lohan, D. (2003). *Biodiversity access and benefit sharing policies for protected areas; An introduction.* UNU-IAS, Yokohama

Lopoukhine, N., Crawhall, N., Dudley, N., Figgis, P., Karibuhoye, C., Laffoley, D., Miranda Londoño, J., MacKinnon, K., & Sandwith, T. (2012). *Protected areas: Providing natural solutions to 21st Century challenges.* S.A.P.I.EN.S (Surveys and Perspectives Integrating Environment and Society) Paper. Vol. 5, Issue 2. IUCN Commissions.

Nagulendran, K. (2013). *ABS interventions to strengthen protected area management.* Paper presented in 1st Asia Parks Congress 13–17 November, Sendai, Japan.

Nelliyat, P., (2022). Implementation of the Nagoya protocol in India: Experience, challenges and learning. In O. V. Oomen, K. P. Laladhas, P. Nelliyat, & B. Pisupati (Eds.), *Biodiversity conservation through access and benefit sharing.* Springer International Publishing.

Nelliyat, P., & Meenakumari, B. (2018). *Protected areas and access and benefit sharing (ABS): A review.* Centre for Biodiversity Policy and Law, National Biodiversity Authority.

UNDP & MoEFCC. (2018). *Achievement of Aichi Biodiversity targets 11 and 16: Success stories from India.* Ministry of Environment Forests and Climate Change, Government of India.

Wildlife Institute of India. (2024). *Protected areas of India: Database.* National Wildlife Database Cell, Wildlife Institute of India. National Wildlife Database | Wildlife Institute of India, an Autonomous Institute of MoEF, Govt. of India (wii.gov.in).

Corporate/Business Responsibilities in the Application of Life

ABS-Bioprospecting Research and Benefit Sharing

Soumya V. S., Laladhas Krishna Panicker, Vijalakshmi B., Aswathy T. R., Prasad G., Achuthsankar S. Nair, and Oommen V. Oommen

Introduction

One of the most crucial areas of biodiversity law, developing recently is ABS. A system known as "access and benefit sharing" ensures that everyone has a fair and equal share of the advantages associated with using biological resources for profit. The concept of "access and benefit-sharing" (ABS) explains how genetic resources can be acquired and how the benefits of their use are shared between the people or countries that utilise the resources (users) and the people or countries who offer them (providers). The sources of genetic resources include governments or civil society organisations, which may also include local communities and private landowners within a country. These groups have the authority to permit access to genetic resources and to divide the benefits of their utilisation.

The access and benefit-sharing provisions of the Convention on Biological Diversity (CBD) are designed to ensure that physical access to genetic resources is facilitated and that the advantages of their utilisation are equitably distributed to the providers. This on occasion includes crucial conventional knowledge about genetic resources acquired from ILCs. The benefits to be shared might be both financial and non-financial, such as the sharing of royalties when the resources are used to create a product for sale or the advancement of research skills and knowledge. It is crucial that institutional structures like those established by the CBD and in the

S. V. S.
St. Stephen's College, Kollam 689695, India

L. Krishna Panicker (✉) · V. B. · A. T. R. · P. G. · A. S. Nair · O. V. Oommen
Department of Computational Biology & Bioinformatics, University of Kerala, Thiruvananthapuram 695581, Kerala, India
e-mail: laladhas@gmail.com

L. Krishna Panicker · O. V. Oommen
CSEIBA: Center Sustainable Environment-India Biodiversity Action, Thiruvananthapuram, Kerala, India

L. Krishna Panicker et al. (eds.), *Biodiversity and Business*,
https://doi.org/10.1007/978-3-031-71674-4_12

Bonn, Guidelines be understood and respected by both consumers and providers. These assist governments in creating their national frameworks that guarantee fair and equitable access and benefit sharing.

Prior informed consent (PIC), which is provided by a provider to a user, as well as agreements between the two parties to establish mutually agreed terms (MAT), ensure fair and equitable distribution of genetic resources and the benefits associated with them. Prior informed consent (PIC) is the authorization granted to a user by the responsible national authority of a country that is the source of genetic resources prior to accessing those resources, following the necessary national institutional and legal framework.

Mutually Agreed Terms (MAT) refers to an understanding made between the owners of genetic resources and their users regarding the terms under which the resources may be accessed and used, as well as the rewards to be distributed equally. The standards for these conditions are laid forth in Article 15 of the CBD, which was passed in 1992, creates a global framework for access to genetic resources as well as the fair and equitable distribution of the benefits deriving from their use.

Genetic Resource Providers: State-owned natural resources fall within their sovereign authority. They must create circumstances that make it easier for people to utilise these resources for uses that don't harm the environment. To grant access and distribute benefits fairly, resource providers agree to agreements that include PIC and MAT. Other parties, such as indigenous and local communities (ILCs), may have the right to negotiate access and benefit-sharing arrangements under the laws of the providers' jurisdiction. The involvement of ILCs is required when traditional knowledge connected to genetic resources is accessed. Users must give suppliers credit for any advantages they receive from using genetic resources. For a variety of reasons, ranging from fundamental research to the creation of new products, they seek access to genetic resources. They are a varied group that includes botanical gardens, institutions of higher learning, collectors, and the pharmaceutical, agricultural, and cosmetic industries.

To facilitate access, users need a clear and transparent approach that outlines whom to contact and the conditions and practises in provider countries. National Focal Points are in charge of disseminating this data. Competent National Authorities (CNAs): CNAs are organisations that governments create to represent providers on a local or national level and to allow access to consumers of their genetic resources.

Biodiversity, genetic/biological resources, and associated traditional knowledge were once considered common property (a worldwide public good), but Access and Benefit Sharing (ABS), which has acquired universal recognition, has transformed this into a state's sovereign right. For the ABS to be operationalized as a cutting-edge financial and incentive mechanism for biodiversity conservation and its sustainable use, with the support of the local communities who are the custodians of resources, the countries, including India, were compelled to establish the institutional, administrative, and legal frameworks required. Some of the established models from Indian scenario have been reported by different authors (Laladhas et al., 2013, 2017a, 2017b, 2023a, 2023b, 2023c; Nilayangod et al., 2023; Nelliyat et al., 2023).

Even if ABS has a large potential to protect biodiversity from its current crisis, its implementation comes with significant difficulties that should be taken into consideration with proactive measures. This chapter discusses the ABS bioprospecting incentives for in situ knowledge generation through biodiscovery from the perspective of a dynamic approach to the economic institutions of contracts and property rights and optimal resource allocation under conditions of perfect rationality, but it also focuses on issues of dynamic efficiency, such as knowledge acquisition and incentives for preserving future possibilities of use under uncertain circumstances. We aimed to demonstrate the significance of examining the importance of bioprospecting using ABS by applying this dynamic method to the process of discoveries which are enlisted below (CBD, 2011).

Jeevani and ABS to Kani Tribes

The use and merits of a biological resource and the way by which the local people use the resources for their regular livelihood activities and the products derived therefrom are different in different areas. We have several models successfully practiced by local communities for the benefit of their livelihood.

An Example: Jeevani from tribes of Kerala

Jeevani, a drug, was developed by the researchers of JNTBGRI from the knowledge associated with bioresources from the Kani Tribe from Kerala, India. As demanded by the tribes, this herbal remedy is well known for its anti-fatigue, immunity-enhancing, and liver-protecting properties. Research from the plant *Trichopus zeylanicus*, vernacularly named Arogyapaacha are used by the Kanitribals in their traditional medicine. 'Plathis' are the traditional healers of the Kani tribe and the customary right of the medicine is vested with these people and also to transfer the right.

The knowledge was exposed to the scientists by three Kani tribal members and they isolated 12 active compounds from *T. zeylanicus*, and developed the drug "Jevaani". The technology was then licensed to Arya Vaidya Pharmacy, Ltd., an Indian pharmaceutical manufacturer for its commercialization. Based on the MAT a Trust Fund was established to transfer the economic benefits arising from the commercialization of the TK-based drug "Jevaani". This model offers very important and interesting lessons on the role of intellectual property rights in benefit-sharing, traditional medicinal knowledge associated with genetic resources, the operations of the Fund from benefit-sharing with the involvement of all relevant stakeholders, as well as the sustainable harvesting of the arogyapaacha plant and its conservation. (https://web.worldbank.org/archive/website01219/WEB/IMAGES/KANI.PDF).

Marula Products and Benefit Sharing in Namibia

Namibia's Eudafano Women's Cooperative (EWC) and the Marula tree *Sclerocarya birrea* form an integral part of their diet, tradition, and culture of rural communities in southern Africa and also is prime commodity to various commercial initiatives. In Namibia, marula-commercialization is undertaken through a collaborative project between primary producers, a cooperative, contract processors, and a local NGO. The Marula tree (*Sclerocarya birrea*) is the prime source of marula oil which is produced by women in rural Namibian communities using traditional techniques – and sold within communities. Marula oil is used as a meat preserve, skin moisturizer, and ingredient for popular foods such as jam and alcoholic beverages (local marula beer).

From local use to Eudafano Women's Cooperative: EWC becomes an early member of PhytoTrade Africa, a non-profit trade association of the natural products industry in southern Africa. EWC receives fair prices for the marula products it produces for international markets. PhytoTrade Africa ensured fair benefit sharing for the use of their traditional knowledge.

A joint partnership has been developed between the rural Namibian marula producers, EWC, and PhytoTrade in collaboration with Aldivia S.A, an international R&D company specializing in producing natural and organic ingredients for cosmetic manufacturers. Consequently, the active chemical ingredient in the Marula Oil—named Maruline was isolated, which is 100% natural marula oil with enhanced antioxidant properties. Process for making Maruline co-owned by Aldivia and the primary African producers represented by the Southern African Natural Products Trade Association.

Maruline was a credible internationally viable product that gained patent and trade market. The demand for Marula shot up to twenty tons of oil worth over US $20 million in 2008. Retail prices for cosmetics with marula sell for four times more than products without it Marula producers were receiving over US $60,000 annually and by 2010 were receiving US $2.35 per kilogram of marula. Production of Maruline was co-owned by Aldivia and the primary African producers represented by the Southern African Natural Products Trade Association(RRB, Leakey, 2002).

Chamomile, Star of Medicinal Herb

Chamomile (*Matricaria chamomilla L.*) is one of the world's most important medicinal herbs, often found near roads, landfills, or in cultivated fields as a weed. This herb is native to southern and eastern Europe and is also grown in Germany, Hungary, France, Russia, Croatia, and Argentina.

Chamomile is an important source of income in Croatia, used traditionally by farming families where UEBT member Jan-Spider works with 40 cooperatives. These families make a good part of their livelihood from the cultivation of Chamomile. For local people, Chamomile is well-known for its use as an infusion for intestinal pain, cold, and better sleep.

Chamomile is collected by special flower harvester machines that look like big combs with knives. Once collected, Jan-Spider processes the Chamomile from its whole flowers or cuts it for tea filter bags. These Chamomile infusion products are sold mostly to the food sector for herbal tea products, and also to cosmetic and pharmaceutical companies.

Jan-Spider works to help family growers thrive by providing knowledge and advise on Chamomile and good farming practices, providing seeds, helping farmers obtain finance, and paying fair prices. The company also takes actions to support the local community as a whole including sponsoring school and childcare activities, and local hospitals. Spider Group a socially responsible company is actively involved in the cooperative farming of Chamomile, and sustainable harvesting to maintain the natural environment for people and wildlife in that region (https://www.ethicalbiotrade.org/ingredient-stories/chamomile).

An Extended ABS Cooperation for a Medication to Treat Bronchitis with Indigenous Populations

Pelargonium sidoides is a plant from the Geraniaceae family and is endemic to Lesotho and different regions of South Africa and is used for traditional medicinal uses. This has been used to treat respiratory tract infections. In 1974, a doctoral study commissioned by the German pharmaceutical company ISO-Arzneimittel analyzed Pelargonium plant ingredients and identified active constituents effective for treating bronchial conditions. In 1987, the company developed a product called Umckaloabo for the symptomatic treatment of bronchitis. This happened a long time before South Africa became a Party to the CBD. Later in 2008, Schwabe with its local partner Parceval requested an integrated bioprospecting and export permit with benefit-sharing agreements in South Africa through the King Sandile Development Trust and other traditional councils that are harvesting *Pelargonium* (Lewu et al., 2007).

The Benefit Sharing Agreement was finally signed and then endorsed by the Minister of Forestry, Fisheries, and the Environment before it becomes a legally binding document. Monetary benefits were negotiated with the traditional council based on the quantities of raw bioresource collected in their areas as per the agreed rate. This project is a typical model for rural people to have a sustainable harvest of bioresources, research, and development and facilitated livelihood (Chennells & Roger, 2013). Non-monetary benefits endorsed by this agreement include the infrastructure upgradation and renovation of the Early Childhood Development Center to nourish the children. Awareness and motivation programs and skill developments in teaching were providedand educational buildings were conducted. A community cultivation program for food security in the rural areas along with the cultivation *of Pelargonium* for sustainable harvest was also promoted (Morris & Christopher, 2016).

Sustainable Access to the Marketing of "the Scent of the Vanishing Flora" in South Africa Derived from Clanwilliam Cedar

Widdringtonia wallichii, Clanwilliam cedar or Clanwilliam cypress, previously named *Widdringtonia cedar bergensis* is a species of Widdringtonia native to South Africa. This tree is endemic to the Cederberg Mountains, northeast of Cape Town in Western Cape Province. This precious bioresource is on the verge of extinction due to overharvest of its timber which is the solitary source for the making of the fragrance. The predominant community called Heuningveli, who live the mountain was edified to conserve and harvest the bioresource via CapeNature, a public institution responsibly managing the bioresource in the mountain. Givaudan is the enterprise globally manufacturing the flavors and fragrances from this bioresource. With the help of advanced technologies, the manufacturers extracted volatile molecules from leaves and dead woods of Clanwilliam cedar without removing any live samples from the mountain and raised awareness of the importance of conserving the threatened taxa. The manufacturers were able to recreate the scent of Widdringtonia—that is, they developed a perfume concentrate using other ingredients via synthetic mode. This project is a model to be displayed to support the conservation of this endangered tree and the communities that contribute to its preservation.

A benefit-sharing agreement was signed under South African rules on access and benefit-sharing, involving the institution, CapeNature as the provider, and Givaudan and The Body Shop jointly as users. This case is an interesting display of how a knowledge associated with a bioresource is enters into ABS machinery. Here the relevance of ABS requirements is not straightforward, as the involvement of headspace technology does not need physical samples and only captures volatile molecules. The CapeNature, the public institution established with statutory responsibility for biodiversity conservation in the Western Cape scientifically carried out its responsibility.

In October 2018, a benefit-sharing agreement was signed and CapeNature also insisted on involving the Heuningvlei community in this process, as an important steward of the Clanwilliam cedar, so that the community could benefit from the mutually agreed terms. Both monetary and non-monetary benefits were considered. Monetary benefits were defined taking into account the making and turnover of The Body Shop perfume as the final product. Following national ABS requirements, these monetary benefits have been paid into the Bioprospecting Trust Fund, which then transfers them to CapeNature for the conservation and livelihood projects agreed upon in the mutually agreed terms between providers and users. These projects are aimed at preserving the Clanwilliam cedar and supporting and empowering the livelihoods in the Heuningvlei community (Normand et al., 2020).

Access and Benefit-Sharing of Rooibos

Rooibos (*Aspalathus linearis*) or "red bush" is a broom-like plant from the family Fabaceae endemic to Cederberg Mountains of South Africa and is used to treat dermatitis, allergies, and digestive problems. Rooibos is produced in huge quantities and generally used in tea, food inclusive, flavours, extracts, and cosmetics. Production of rooibos is an important industry for South Africa with an estimated worth of around US$ 22.2 million per year, employing around 5,000 people. The San and Khoikhoi communities played a key role in the development of the rooibos industry. The San are represented by the South African San Council. The Khoikhoi people are represented by the National Khoi & San Council (NKSC).

The South African Rooibos Council (SARC), is active in bioprospecting and bio trade along with Nestlé, for new beverage products and cosmetics. The first ABS agreement was made in 2014 between Nestlé Professional (SA) and the San Council and NKSC. The 3% of net sales is shared among the two council 50% each. A biannual payment for bioprospecting trust fund was managed by DEFF. Community programmes, bursaries, employment, and programmes specific to the needs of the community were met from non-monetary funds (https://rooibos.abs-biotrade.info/#/abs-implications).

An Agreement Between ABS and the Dominican Republic of Latin America for the Commercial Usage of Scorpion Venom

Local communities from Neiba, Bahoruco province, traditionally use scorpion venom from the species *Rhopalurus princeps* also called blue scorpion and *R. abudi* for treating bone aliments effectively. Medolife S.R.L., an international pharmaceutical company housed in the USA was interested in the potential bio-capability of the scorpion venom and was granted a research permit in 2018. The Ministry of Environment and Natural Resources of the Dominican Republic was able to develop the traditional knowledge associated with the blue scorpion, which led to the development of the biocultural protocol for the commercial use of scorpion venom. Based on the research and development from the scorpion venom, the U S company developed a medicine with anti-cancer properties. ABS agreement leading to monitary benefits includes the contribution of 5% of the annual net profits from the commercialisation of the developed drug to the Ministry of Environment. Non-monetary benefits constitute the construction of a research center -Biotherium in which the *Rhopalurus princeps* specimens will be reared, training for students on sample collection, laboratory work, and transfer of the standardized technique used in the extraction of scorpion venom. R & D products will be labelled specifying their geographical origin and 30% of the direct jobs generated will be given to the people from the indigenous community, especially women (Al-Janabi et al., 2019).

Historical Information About the Hoodia Plant

The succulent species known as Hoodia is native to Southern Africa. The indigenous San people have used it for millennia to ward off hunger and thirst during extended hunting expeditions and times when food is in short supply. Generations of San people have passed along traditional knowledge of hoodia. The active Hoodia chemicals, which were shown to decrease appetite, were patented by the Council for Scientific and Industrial Research (CSIR), a South African organization, in 1996. To develop and market a Hoodia-based product, CSIR entered into licensing arrangements with certain major pharmaceutical companies due to the commercial potential of Hoodia as an appetite suppressant in the anti-obesity sector.

However, these actions were initially performed without the San people's consent. As a result of media coverage of the potentially lucrative agreements made to develop the plant's properties and a protest by a South African NGO, steps were taken to start talks between CSIR and the San people. As a result, the San Hoodia Benefit-Sharing Trust and a benefit-sharing contract with financial and non-financial incentives were created. If a product was successfully commercialized, and the contract required royalty payments in addition to milestone payments during the product development period.

The money will go towards the improvement of research and the preservation of the San people's traditional knowledge and legacy as well as the growth, education, and training of the San community. Larger sums of money are anticipated to flow from the agreement in the future, although some payments have already been made to the trust fund and are being used to fortify the San people's institutional foundation throughout Southern Africa. (CBD, 2011 https://www.cbd.int › info kit › revised › web › all-files-en).

Colonial Bioprospecting of Devil's Claw: The Case of Harpago Tea

When colonial scientists in South-West Africa, now Namibia, first collected and described the plant, the first scientific paper on the medical applications of Devil's claw was published in 1822. However, the German imperialist G.H. Mehnert would not claim the 'discovery' of the therapeutic benefits of Devil's claw in modern-day

Namibia until 1907, during the Scramble for Africa, after tapping into the rich mine of Traditional Knowledge(TK) of the Khoi and San peoples.

The dried Devil's claw tubers were sold by this early colonialist turned bioprospector to Germany, where B. Zorn at the University of Jena first investigated the material in the 1950s. Western medicine later recognized the Devil's Claw's analgesic and anti-inflammatory effects. The German company Erwin Hagen Naturheilmittel GmbH began receiving considerable quantities of Devil's claw tubers in 1962 from the Namibian company Harpago (Pty) Ltd (Windhoek, South Africa). In the 1960s, dried Devil's claw tubers were used in Germany as a general blood cleanser, including a bitter tea is made from it intended to alleviate dyspepsia.

The Khoi and San communities, the rightful owners of the TK connected to the therapeutic benefits of Devil's Claw, would have received compensation from Harpago (Pyt) Ltd. had Harpago tea been produced after the CBD went into effect in 1993. Unfortunately, the indigenous people of Namibia were not benefited financially from their knowledge of the medical uses of the Devil's claw, which is used to make and sell Harpago tea. Instead, the intermediaries and businesses continue to profit greatly from the Devil's claw. Additionally, manufacturers continue to protect their Devil's claw-based herbal drugs through trade secrets and brand names, and the San and Khoi peoples' rights to their ethnobotanical knowledge of the Devil's claw's medicinal properties are not recognized. This is because extraction patents. processing techniques were made available to commercial businesses in Germany and the UK (Cole et al., 2001).

Other Indigenous Plant-Based Product Trading—the Case of Namibian Myrrh

Namibian myrrh, or *Commiphora wildii*, is primarily found in isolated areas of the Kunene Region where it is known as Omumbiri. The Himba people traditionally use the resin that the plant generates to create a body scent. Resin is typically produced by a single *C. wildii* plant. Conservancy members gathered a total of five tonnes of *C. wildii* resin in 2007, marking the start of commercial production. Between 2007 and 2008, harvesters made a total of N$250,000. Currently, five conservancies—Puros, Orupembe, Marienfluss, Sanitatas, and Okondjombo—harvest about 50 tonnes of resin annually. The Opuwo Processing Facility (OPF), which separates the essential oil from *C. wildii* resin, is owned by the Indigenous Natural Products Trust. OPF offers the essential oil to businesses in Namibia, South Africa, France, and Germany, for N$5000 per kg. Samples of the essential oil and resin were brought to the In-Cosmetics Trade Fair in Paris in 2008.

The indigenous knowledge utilized to produce the essential oil from *C. wildii* resin belongs to the Himba, who are the proper custodians of it. Afriplex Ltd. (Paarl, South Africa), a South African enterprise, and the local populace entered into the first-ever ABS agreement in April 2010 under the auspices of the Kunene Commiphora

Conservancies Association, a coalition of five conservancies. Each harvester made between N$1000 and N$5000 (on average N$1500) over the three to four months of the harvesting season, paying N$50 per kilogram of gum.

The Case of Latent HIV Activator Prostratin; Associated Knowledge from Western Samoa

Making use of age-old wisdom, Samoan traditional healers have used the plant to cure liver diseases whose symptoms resembled those of yellow fever and hepatitis. The study of Samoan medicinal plants began in 1984 thanks to the cooperation of traditional healers from the Western Samoan hamlet of Falealupo and Dr. Paul Alan Cox, who was then attending Brigham Young University in Utah. The indigenous people used (*Homalanthus nutans*) to cure a liver ailment that seemed to be yellow fever but turned out to be hepatitis. Prostaglandin, which is presently a potential medication for eradicating latent HIV infection from T cells in patients getting highly aggressive anti-retroviral therapy (HAART), was discovered by Cox and his colleagues during their ongoing investigation of this tree.

The extract exhibited substantial anti-HIV activity in a cell-based screen. Through extract fractionation guided by bioassays, chemists generated a 12-deoxy phorbol known as prostratin. Researchers from the NCI did not find that prostratin promoted the growth of tumors, in contrast to other phorbol esters. Furthermore, prostratin was a potent promoter of HIV expression in T-cell lines that were latently infected, according to research done by a team at the University of California, Los Angeles (UCLA) AIDS Institute.

Prostratin's surprising discovery first disqualified it from usage as an anti-HIV drug, however, it was soon realized that prostratin's role as a viral activator may be highly advantageous in HIV therapy. The medication may be used to extract latent HIV from lymph nodes, and then HAART may be administered to eradicate the virus. Prostratin research and development are expected to progress with the partnership with ARA, a community-based non-profit research organization that is collaborating with labs throughout the world and conducting a multifaceted campaign against AIDS. ARA is funding clinical trials for prostratin in part.

Under the agreement, the community has been given approximately $480,000 to build walkways, schools, clinics, water supplies, and an aerial walkway above a rainforest canopy. Additionally, an endowment for the rainforest has been established. After drug screens at NCI demonstrated the substance's therapeutic potential (sourced from a Samoan medicinal tree), NIH and ARA entered into a licensing arrangement in 2001 for the substance's further research and development (Gupta et al., 2005).

Reverse Pharmacology

It is defined as a rigorous scientific process that integrates experiential observations and documented clinical experiences to generate leads through transdisciplinary exploratory studies. Subsequently, these leads are developed into drug candidates or formulations through extensive preclinical and clinical research (Vaidya et al., 2007).In this approach, "safety" is still the most crucial first step, and effectiveness is a question of confirmation. The new aspect of this strategy is how it combines current traditional knowledge, like that of Ayurveda, with the use of cutting-edge procedures and technology to provide better and safer results. The foundation for reverse pharmacology of medicinal plants was laid by Sir Ram Nath Chopra and Gananath Sen by concentrating on the effects of scientifically validated Ayurvedic drugs (Ramalingaswami et al., 1982). *Rauwolfia serpentine* Benth was the significant finding made possible by this technology. Sen and Bose were able to demonstrate the plant's sedative and hypotensive qualities in addition to its peculiar side effects, which included depression, extrapyramidal syndromes, and gynecomastia (Sen, 1931). The development of innovative antidepressants, Parkinson's drugs, and prolactin-reducers underwent a sea change as a result of this work (Svensson, 1980). Reserpine is one Rauwolfia alkaloid that is antihypertensive and was made available in this way.

About two or three decades ago, some promising work was produced as a consequence of a joint drug research effort between the Indian Council of Medical Research (ICMR) and the Council for Scientific and Industrial Research (CSIR) of the Government of India. Under the direction of Ayurveda, Commiphoramukul was utilized to make the drug guggulipid, which lowers cholesterol (Satyavati, 1988). The Drug Controller General of India (DCGI) approved the drug's marketing in 1986. The business that manufactures and distributes guggulipid is Cipla Pharmaceuticals Limited, but the lack of legitimate raw ingredients has continued to prevent the drug from being widely used. A memory enhancer derived from *Bacopa monnieri* has also been licensed and marketed by the Central Drug Research Institute, located in Lucknow. Unfortunately, not enough research has been done on the huge potential of these approaches to develop goods that can compete on a global scale.

Psoriasis

One of the most widespread dermatological conditions, psoriasis has no preventative or curative treatments and affects around 2% of the global population. Lupin Limited in India pursued the development of a single plant-based oral formulation using a reverse pharmacology technique as part of the NMITLI project (Satyanarayana, 2007) The herbal medication (Desoris) is an extract from just one plant that successfully alters cellular activity, improving psoriatic lesions (Arora et al., 2003). The application for an investigational new drug (IND) was submitted in India after extensive research. This product has been created following DCGI standards for novel medication development and FDA regulations for botanical drug products.

National Bioprospecting Strategy

India's commitment to implementing the CBD is centered on its National Biodiversity Act and Rules. The Act's scope is much greater than what is described in this article. However, to make the Act relevant and helpful for conservation and development, its implementation necessitates the strengthening of human resource, institutional, and financial capacity as well as a much-needed rise in local public awareness. To ensure that the implementation experience of India may be adequately reflected in the international negotiations on the establishment of the international system, efforts are now being made. In this context, we appreciate the chance to speak with anyone interested in our experiences with the creation and application of the Act and Rules. Although national activities are based on national priorities, they have a significant global impact.

Our ability to harness biological diversity sustainably is essential to our continued existence on this planet. The earth's biological diversity is what makes it home to life. The commercialization of seed production, monoculture, and the protection of novel plant varieties, microbes, and genetically modified organisms are all encouraged by IPR. But our abundant biodiversity is in jeopardy.

Although developed nations lack abundant biological resources, they are well developed in terms of infrastructure for R & D. They make use of the biological resources imported from underdeveloped nations. Without any protection from patents, plant breeder's rights, or other forms of intellectual property (IP), genetic information is transferred from developing to developed nations. A balance needs to be struck between IPR and environmentally responsible exploitation of biological diversity. The need for sustainable use of biological resources is urgent; damage to human existence on this planet will result from misuse, overuse, or abuse of our natural and biological resources.

Benefit sharing is a component of the larger Access and Benefit Sharing (ABS) concept found in international agreements like the 2001 International Treaty on Plant Genetic Resources for Food and Agriculture (ITPGRFA), the 2010 Nagoya Protocol (NP), and the Convention on Biological Diversity (CBD). Benefit sharing creates legal safeguards to make sure that people can only utilize genetic resources with permission and under mutually agreed-upon conditions. India has been a leader in the push for successful benefit-sharing agreements at the global level and was one of the first nations to develop benefit-sharing legislation.

Although the emphasis has been on generating money, domestic implementation has remained inadequate. A few benefit-sharing agreements with local communities have been reached, and despite this, very little actual revenue has resulted from these agreements. Both the allocation of actual benefits to communities under benefit sharing in India and the development impacts on neighborhood communities are still unknown. This article investigates India's benefit-sharing program to shed light on why it has been so challenging to put into practice. At the domestic level, poor execution has been largely attributed to the absence of a clearly stated purpose. The existence of a regime complex has an impact on policymaking and decision-making processes on a global scale. Domestic constraints mixed with the complexity of the global regime lead to poor benefit-sharing implementation in India.

Conclusion

This work discusses the usefulness of bioprospecting for traditional knowledge (TK) and genetic resources (GR) as input for industry and research. The monetary worth of plant genetic information from the perspective of traditional knowledge holders may be evaluated using a contingent valuation modeling technique, as demonstrated by a special "access and benefit sharing" (ABS) bioprospecting contract. The study highlights some of the most significant shortcomings in the assessment of traditional knowledge related to ABS instances from the standpoint of evolving institutions. It emphasizes the need to go beyond conventional market methods for valuing GR/TK economically and for protecting local people's traditional knowledge acceptably.

Acknowledgements The author expresses gratitude to the Department of Computational Biology & Bioinformatics at the University of Kerala and VSS thanks the Council of Scientific and Industrial Research, New Delhi, India, for the Senior Research Fellowship (08/666(0003)/2019-EMR-I).

Conflict of Interest The authors state no conflict of interest.

References

Al-Janabi, S., Feit, U., Fenster, E., Greiber, T., & Schauerte, P. (Eds.). (2019). *Vilm ABS Dialogue 2018-Informing about Domestic Measures for Access to Genetic Resources.* Bundesamt für Naturschutz. https://www.bfn.de/sites/default/files/BfN/service/Dokumente/skripten/skript524.pdf

Arora, S., Gupta, L., Srivastava, V., Sanganabhatla, N., & Saraf, D., Lupin Ltd, (2003). *Herbal composition for treating various disorders including psoriasis, a process for preparation thereof and method for treatment of such disorders.* U.S. Patent Application 10/340,195. PAT: EP1467745

CBD (2011) Introduction to access and benefit sharing. https://www.cbd.int › infokit › revised › web › all-files-en. https://www.cbd.int/abs/infokit/revised/web/all-files-en.pdf

Chennells, R. (2013). Traditional knowledge and benefit sharing after the nagoya protocol: Three cases from South Africa. *9/2 Law, Environment and Development Journal*, 2013, 163. http://www.lead-journal.org/content/13163.pdf

Cole, D., & Du Plessis, P. (2001). Namibian Devil's Claw (Harpagophytum spp.): a case study on benefit-sharing arrangements. *Unpublished report prepared for the Namibian Ministry of Environment and Tourism, Directorate of Environmental Affairs, Namibian National Biodiversity Programme.*

Gupta, R., Gabrielsen, B., & Ferguson, S. M. (2005). Nature's medicines: traditional knowledge and intellectual property management. *Case studies from the National Institutes of Health (NIH), USA. Current drug discovery technologies*, 2(4), 203–219. https://bfn.bsz-bw.de/frontdoor/deliver/index/docId/145/file/Skript_524.pdf

https://rooibos.abs-biotrade.info/#/abs-implications

https://web.worldbank.org/archive/website01219/WEB/IMAGES/KANI.PDF

https://www.ethicalbiotrade.org/ingredient-stories/chamomile

Kanikaram Satyanarayana, K. S. (2007). *Current IP management issues for health and agriculture in India.* Centre for the Management of Intellectual Property in Health Research and Development (MIHR). (pp. 1605–1620).

Laladhas, K. P., Nilayangode, P., & Oommen, O. V. (Eds.). (2017a). *Biodiversity for sustainable development.* Springer International Publishing. https://link.springer.com/book/10.1007/978-3-319-42162-9

Laladhas, K. P., Nilayangode, P., Gopalan, S. V., Divya, L., & Oommen, O. V. (2017b). *Community conservation of marine turtles and sustainable ecosystems.* Biodiversity for Sustainable Development, 177–184. https://link.springer.com/book/10.1007/978-3-319-42162-9

Laladhas, K. P., Preetha, N., BaijuLal, B., & Oommen, O. V. (2013). Region-specific biodiversity education-the role of People's Biodiversity Register. *Current Science (00113891), 105*(12).

Laladhas, K. P., Rani, J. R., & Dionys ius, J. M., Nair, A. S., Sudhakaran, P. R., & Oommen, O. V. (2023a). Achievements in India's ABS Mechanism. *Biodiversity Conservation Through Access and Benefit Sharing (ABS) Himalayas and Indian Sub-Continent* (pp. 37–53). Springer International Publishing. https://doi.org/10.1007/978-3-031-16186-5_3

Laladhas, K. P., Rani, J. R., Vijaya Sree, A. S., Nair, A. S., & Oommen, O. V. (2023b). Statutory and obligatory responsibilities of state biodiversity boards for the conservation of indigenous biodiversity and ABS. *Biodiversity conservation through Access and Benefit Sharing (ABS) Himalayas and Indian Sub-Continent* (pp. 55–77). Springer International Publishing. https://link.springer.com/book/10.1007/978-3-031-16186-5

Laladhas, K. P., Fila, M., Nair, A. S., & Oommen, O. V. (2023c). Joint Biodiversity Management Committees: A Mechanism to Manage and Conserve Eco-Regions. *Advances in Environmental and Engineering Research, 4*(3), 1–25. https://doi.org/10.21926/aeer.2303042

Leakey, R. R. B. (2002). A summary of knowledge on marula. *Southern African Forestry Journal., 196*, 67–77.

Lewu, F. B., Adebola, P. O., & Afolayan, A. J. (2007). Commercial harvesting of Pelargonium sidoides in the Eastern Cape, South Africa: Striking a balance between resource conservation and rural livelihoods. *Journal of Arid Environments, 70*(2), 380–388. https://api.semanticscholar.org/CorpusID:85329296

Morris, C. (2016). Bioprospecting, rights, and traditional authority in South Africa. In *American ethnologist*, *43*(3), 525–539. https://doi.org/10.1111/amet.12344

Nelliyat, P., Pisupati, B., Laladhas, K. P., & Oommen, O. V. (2023). Convention on biological diversity, access and benefit sharing: A glance. *Biodiversity conservation through Access and Benefit Sharing (ABS) Himalayas and Indian Sub-Continent* (pp. 3–16). Springer International Publishing. https://doi.org/10.1007/978-3-031-16186-5_1

Nilayangode, P., Laladhas, K. P., & Oommen, O. V. (2023). Strengthening ABS Implementation in India. *Biodiversity conservation through Access and Benefit Sharing (ABS) Himalayas and Indian Sub-Continent* (pp. 249–261). Springer International Publishing. https://link.springer.com/book/10.1007/978-3-031-16186-5

Normand, V., Oliva, M. J., Müller, S., al-Janabi, S., Nöske, N., Zedda, L., & Paulsch, A. (2020). *The contribution of access and benefit-sharing (ABS) to the sustainable development goals: Lessons learned and best practices.*

Ramalingaswami, V., & Satyavati, G. V. (1982). Sir Ram Nath Chopra. *The Indian Journal of Medical Research, 76*, V–VI. PMID: 6764453

Satyavati, G. V. (1988). Gum guggul (Commiphoramukul)–the success story of an ancient insight leading to a modern discovery. *The Indian Journal of Medical Research, 87*, 327–335. PMID: 3049326

Sen, G. (1931). Rauwolfia serpentina, a new Indian drug for insanity and high blood pressure. *Ind Med World, 2*, 194–201. https://doi.org/10.4103/ijmr.ijmr_1674_21. PMID: 34782547; PMCID: PMC8715689.

Svensson, T. H. (1980). Effect of chronic treatment with tricyclic antidepressant drugs on identified brain noradrenergic and serotonergic neurons. *Acta PsychiatricaScandinavica, 61*(S280), 121–131. https://doi.org/10.1111/acps.1980.61.s280.121

Vaidya, A. D., & Devasagayam, T. P. (2007). Current status of herbal drugs in India: An overview. *Journal of Clinical Biochemistry and Nutrition, 41*(1), 1–11. https://doi.org/10.3164/jcbn.2007001

Industrial Sector Wise Analysis of Bio-Prospecting in Access and Benefit Sharing (ABS) Perspective

Prakash Nelliyat

Background

After the emergence of biotechnology in western developed nations, the dependency on the genetic resources for bio-prospecting particularly from the biodiversity-rich developing nations accelerated. Bio-prospecting is the exploration of biodiversity for economically and commercially valuable bio-chemicals and genetic materials. (Bavikatte & Morten, 2015). Many biological resources -based industries (Pharmaceuticals, botanicals, Agricultural seeds, ornamental horticulture, crop protection products or bio-fertilizers and pesticides, health and personal care products, cosmetics, food and beverages, textiles, leather, paper and pulp, jute, etc.) are using bio-chemical and genetic materials as inputs or raw-materials and manufacturing different consumer products and earning huge benefits or profit. These industries' contribution to global Gross Domestic Product (GDP) is significant. The annual market, during 2010–11, of various kinds of products originated from genetic resources in different sectors was estimated by the CBD (2013c) as: Pharmaceuticals—US$ 955 billion, Cosmetics—US$ 426 billion, botanicals—US$ 84 billion, food and beverages—US$ 11.6 trillion, seeds—US$ 45 billion, crop protection products—US$ 40 billion and industrial biotechnology—US$ 71 billion.

After globalization, the biological resources- based industrial products' markets accelerated and spread all over the world at the cost of third world biodiversity. Most of the biodiversity spots in developing countries were experiencing the over extraction of resources and species extinction. The ultimate impact is the degradation of biodiversity and the reduction in ecosystem goods and services and its negative impact is for the entire global community and even for business as it leads to raw-material insecurity. In this peculiar situation, the CBD's principle of Access and

P. Nelliyat (✉)
Centre for Biodiversity Policy and Law, National Biodiversity Authority (MoEFCC, Govt. of India), Taramani, Chennai, India
e-mail: prakashnelliyat@gmail.com

L. Krishna Panicker et al. (eds.), *Biodiversity and Business*,
https://doi.org/10.1007/978-3-031-71674-4_13

Benefit Sharing (ABS) is extremely significant where the synergy between business and biodiversity prevails (Nelliyat & Meenakumari, 2018).

For the effective enforcement of the ABS, the cooperation of the business group is important. A thorough understanding about different industries link with biodiversity and bio-prosperity is a pre-condition for various ABS stakeholders, including enforcement agencies, who act as mediators between the providers (local and indigenous communities) and users (bio-prospectors) of resources, and all the biological or genetic resources based industries, who are supposed to come forward for benefit sharing.

In this context, this chapter examines the nature and prospects of bio-prospecting in different industrial sectors and their present initiatives on ABS and its future scope as an innovative financial mechanism for biodiversity management. The relevant data has been collected from the available literature as well as through interviews with ABS experts and industrialists who represent different sectors. Further, the knowledge acquired by the author from visits to different biological resources based industries and discussions with bio-prospectors in different states in India and the insights obtained through the "ABS Expert Committee Meetings" of the National Biodiversity Authority (NBA), where ABS applications are being scrutinised and approval granted with benefit sharing norms as well as negotiations with applicants are taking place, has also been incorporated.

Why is Sectoral Specific Approach to ABS Significant?

Genetic or biological resources based manufacturing companies are using biological components including genetic materials from plants, animals, microorganism etc. from different ecosystems. However, among these industries the nature of bio-persecution and the volume of biological or genetic resources usage vary considerably. Further, the application of modern technology and the degree of Research and Development (R&D) also varies considerably in different sectors as per their needs.

In the case of pharmaceuticals (modern drugs) and agriculture seeds industries, the volume of collection of biological resources from nature is limited. These industries will engage in R&D in high magnitudes for identifying the bio-compounds or genetic materials. Sometimes these companies also have in-house facilities for culturing or multiplying certain biological resources. When we approached a leading vaccine manufacturing company in Chennai, India, the scientists told us "*we take only a drop of bio-ingredient from nature and multiply it in our laboratory. Hence, we are not disturbing the biodiversity, and hence require the ABS exemption*". Similarly, in a seed manufacturing company (located in Hyderabad, India) the scientist informed us "*I take germplasam from the field and after rigorous R&D process come up with genetically modified seeds. Hence, as the efforts of our company are significant in product development, sharing benefits is difficult*".

On the other hand, in the botanical and food processing industries' case, there is an extremely different scenario. These companies required huge volumes of biological resources as raw-materials. Generally, the application of modern biotechnology or R&D is less in this sector than in the pharmaceutical and seed processing sectors. In botanical and food processing, typical production as input to output transformation is taking place. When we visited a major botanical company in Bangalore, we realized they are collecting herbs and aromatic plants from all over India and trucks of the raw-materials in the company godown / warehouse. After the huge laborious process valuable extracts are prepared and exported to different countries, where they are used for manufacturing various botanical end products including nutraceuticals.

It is clear from the above discussion that, there is a huge variation in the production and input requirements of different biological resources based sectors. Sometimes, a particular biological resource may generate different values in different sectors. All these aspects create challenges in the valuation of biological resources for ABS and fixing the benefit sharing norms / amount by the enforcement agencies.

The ABS and its acceptance in bio-prospecting and business model is a new process which emerged during the post-CBD era. Sara and Rachel (2012) stated that ABS related laws, policies and strategies should be promoted within the dynamism of bio-economy and bio-science. Further, the business sectors should realise their responsibilities in CBD and Nagoya Protocol and follow the fundamental principles of ABS. But, the experience from the parties of the CBD reveal that industries always followed a lenient and careless approach to ABS without realizing their obligations. Even if 142 countries ratified the Nagoya Protocol on ABS, most of them are yet to establish the legal measures and institutional arrangements for enforcing ABS in their country. Further, a clear understanding of the underlying principles of ABS, especially conservation and sustainable use of biodiversity will ensure the raw-material security for their businesses forever, but this has not been imbibed in the business community. Further, maintaining the rich biodiversity stock is a pre-requisite for future bio-surveys and research and identifying new compounds, which may further enhance business.

Studies reveal that industries' engagement with ABS is progressing and varying considerably over the last 20 years. In the earlier days of CBD, debates associated with ABS were primarily confined to the pharmaceutical and agriculture seed sectors. But, recently the discussions spread over other biological resources dependent sectors too. The mass public demand for natural products in all sectors and the application of biotechnology are the factors that have influenced this. Presently, all the sectors are engaging the application of biotechnology and involving in research and developing genetic and bio-chemical components to some degree. Besides, their ABS awareness is also slowly progressing.

However, smaller entrepreneurs and academic institutions' awareness of CBD and Nagoya Protocol obligation is inconsistent. Many smaller companies' major complaint is that they do not have the "band width" or internal staff including legal experts to facilitate ABS norms and agreements (Sara & Rachel, 2012). Therefore, these companies are shying away from collecting biological or genetic resources and ABS obligations. In this inconsistent and divergent scenario, irrespective of the sectors and size of the units, the universal enforcement of ABS and operationalization is a challenge, and it should be seriously taken into account in the CBD forums.

India's ABS: A Sector Specific Analysis

India is one of the pioneering countries in the world, who took appropriate measures (legal, institutional and administrative) for enforcing the ABS mechanism. India enacted the 'Biologist Diversity Act—2002' and 'Rules – 2004', prior to the emergence of the Nagoya Protocol on ABS, 2010. However, India's legal framework is able to fulfil the requirements of the Protocol's criteria. Subsequently, India came up with ABS Guidelines (2014) and amended the Biological Diversity Act in 2023, based on their past two decades of implementation experience. Since India is a federation, in its different States, specific 'Biodiversity Rules' also developed.

For enforcing the ABS a three tier institutional structure (National Biodiversity Authority—NBA at the national level, State Biodiversity Boards—SBBs at the provincial level and Biodiversity Management Committee at the local body level) is introduced. At present (April, 2024) NBA approved 5259 ABS agreements. Besides, SBBs are also signing ABS agreements. With the ABS money, various biodiversity conservation efforts are also progressing.

Gayathri and Rana (2018) conducted a sector specific ABS study considering 1853 ABS agreements approved by the NBA and its findings are given below:

- ABS agreements are classified under six heads: pharmaceuticals, research, nutraceuticals, export, environmental bio-remediation and cosmetics. There is a considerable variation in the numbers of agreements under each head.
- The Pharmaceutical sector is the major sector with 1010 (out of 1853) agreements followed by cosmetics (351 agreements).
- Environmental bio-remediation (with 64 agreements) and cosmetics (with 48 agreements) are less significant sectors.
- India's ABS collection, through the export of biological resources like red sanders, cattle embryos, sea weeds and *neem* shows a considerable rise in recent years and has mobilised huge amounts.

The percentage of sector specific ABS agreement in India is given in the following Fig. 1.

Bio-Prospecting: A Sectoral Analysis

Commercial utilization of different genetic or biological resources in different industrial sectors and their ABS associated issues and challenges are summarised below:

A. **Pharmaceuticals**

For achieving the health security, the role of the pharmaceutical sector is critical, especially in recent decades as humanity is experiencing pandemics like Covid-19. This Sector is flourishing all over the world in different degrees for attacking various

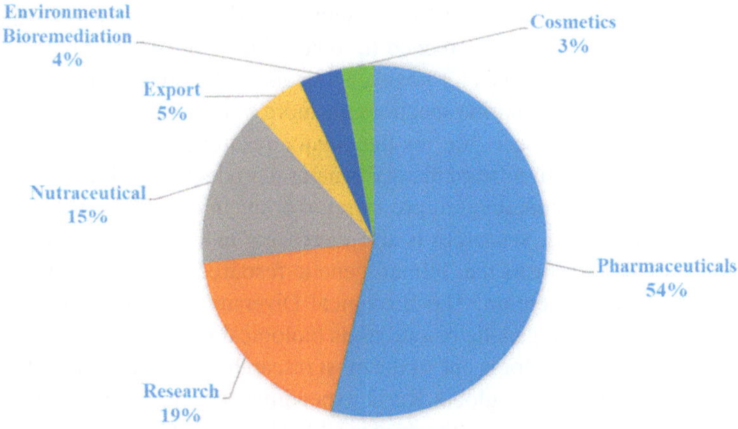

Fig. 1 Percentage share of sector specific ABS in India

diseases. One peculiarity of this sector is; the discovery of medicine in one country (one part of the world) may benefit all the countries (the entire world). Among the total revenue (sales) generated in the pharmaceutical sector, the top 10 companies' contribution is US$ 353 billion or 60% of the total revenue. In recent years, a significant structural transformation is being experienced by the pharmaceutical sector. The growth and marketing of pharmaceuticals in developed nations such as Japan, Europe and US has slowed down considerably, with high growth in emerging economies like China, India and Brazil (CBD, 2013).

In research and development along with the private sector, the Government and academic organisations also play a critical role in the pharmaceutical sector. Recently, most of the natural products- based drug discovery is initiated by the smaller companies and the promising products are licensed to big companies for development. Hence, the fresh demand for genetic resources is primarily from the smaller companies and not from the larger ones. This is a serious concern in enforcing the ABS in this sector. Traditionally, pharmaceutical companies are vertically integrated. However, nowadays, many companies manufacture and sell a combination of products that include: devices, nutraceuticals, medical or laboratory, cosmetics and beauty care products, agricultural and animal health care items, flavours and fragrances, and vitamins.

A range of natural ingredients including plants, fungi, insects, marine organisms microorganisms, animal and human genetic resources are contributing to drug manufacturing. Even if plants are prevalent in pharmacopoeia, only 5–15% (around 250,000 to 500,000) plant species have been investigated for examining the bio-active compounds so far. This indicates that the potential of biodiversity for discovering new medicines is tremendous in future. Nowadays, a majority of the research in the pharmaceutical sector focuses on microorganisms, especially from the sea. Microorganisms are a solid source for bioactive metabolites, penicillins and different antibiotics and many other promising products.

Marine organisms are another prominent source used by the pharmaceutical sector, but only a small component is screened till date. Marine organisms are the richest sources of bioactive compounds for novel chemicals not found in terrestrial sources. However, the collection of marine specimens is more costly and complex than terrestrial plant ingredients. Insects are another major resource in this sector. Animal genetic materials have a significant history in natural product drug discovery. Around 23% of all compounds contained in prescription drugs in the US are from animals. The use of human genetic resources is also increasing in the pharmaceutical sector. However, the CBD reaffirms that human genetic resources are not included in the frameworks of the Convention. The Biological Diversity Act (2002) of India also does not consider human genetic resources as biological resources.

The Pharmaceutical sector is one of the most research-intensive industrial sectors. An estimate reveals that the global R&D expenditure during 1998 was US$ 21.1 billion and the cost of developing the drug is US$ 231–500 million. In different parts of the world the local and the indigenous committees' biological resources associated TK and experience are guides to drug discovery. Out of the 120 products derived from plants, during 1985, 75% were discovered through the basis of their traditional medical use.

B. **Seed Industries**

The UN's 'Development of Economics and Social Affairs' (DESA) recent data on 'World population prospects' indicates that the world population is likely to peak this century at around 10.3 billion. For achieving food security, especially in the era of population explosion, a proportionate growth in agriculture and allied activities is required. Food production needs to be enhanced with modern agricultural practices. The application of genetic resources (from traditional varieties) in plant breeding and developing high yielding varieties of seeds is a pre-requisite in enhancing crop productivity. As seed production is a multi-billion business and generates huge profit/benefits to the companies, the scope of ABS in the seed sector needs to be explored. The CBD (2013a) stated that, the Nagoya Protocol on ABS and the Treaty on Plant Genetic Resources for Food and Agriculture (ITIGRFA) needs to create an opportunity and platform to develop ABS solutions in the seed sector.

Kate and Sarah (2000) stated that countries do not depend on their food on Plant Genetic Resources for Food and Agriculture (PGRFA) once it obtained from elsewhere. But, the questions are: (a) whether countries are self sufficient for food production or germplasm used for plant breeding and (b) does our dependencies on PGFRA continue? A study, covering 11 countries indicates that breeders still regularly use germplasm from other countries to develop new plant varieties.

In brief, both improving the existing varieties of crops and developing new crops is a continuous process and requires access to genetic resources. The Seed sector is very much engaged in R&D. The seed sector's usage of wild species requires substantial investment to convert into commercially viable products; besides it is highly risky. However, still many companies are coming forward to invest in cropping wild relatives primarily due to: consumer choice or demand, and presence of genes for stress resistance and productivity enhancement (CBD, 2013a). The market share of the top 10 companies in the global commercial seed market was around US $ 20 billion in 2009, which comes to 59% of the total value. The Available data reveals that the US and China have the highest valued domestic seed markets in this sector.

C. **Botanical Industries**

Botanical or herbal medicines are from plants or plant extracts; they may be eaten or applied to the skin. From ancient times herbal medicines are used to treat illness and to improve body functions. Recently, global botanical medicine markets have grown considerably with a wide range of products. Nowadays, apart from medicines, botanicals are widely used for promoting human wellbeing. Hence, in the botanical sector, a wide range of industries, products, markets, R&D and regulations has emerged. Recently, the scope of herbal products manufacturing has expanded beyond plants as raw-materials. The companies also focus on fungi, minerals, shells, bee products and certain animal parts. The annual growth rate of botanical medicine industries is between 10–20%.

In India, botanical medicines are called on 'ayurvedic medicines' with complex formulations of more than 30 ingredients. These companies have significant differences in their size and production processes. Botanical industries include small and marginal family-run companies that produce traditional medicines and large companies that engage in extensive R&D and manufacture sizeable pharmaceuticals. In many countries a smaller group of very extensive number of units dominate in this sector (CBD, 2013b). In the botanical industries' supply chain, a large number of divergent actors are involved, including; growers, harvesters, brokers, traders, exporters, processing companies, manufacturers, distributors, retailers, wholesalers etc.

In recent years, botanical processing has transformed into cheap labour centred countries like India and China. Raw-materials or herbs harvested around the world are shipped to be processed in these countries for extraction. Afterwards, the intermediate or final products are shipped to the home countries to manufacture the end products (Sarah & Rachel, 2013). Generally, accessing the bulk raw-materials, which is further trading of commodities, does not fall within the scope of ABS or the Nagoya protocol. But, from the botanical industries' perspective, sourcing raw-materials is crucial, where ABS policy makers' attention is important. Further, the long chain of intermediaries' stand in ABS also needs to be taken into account (CBD, 2013b). The market of the botanical industries increased from US$ 84 billion (2010) to US$ 107 billion (2017). Nowadays, companies are coming forward and taking stronger position in terms of Intellectual Property Right (IPR) and applying a scientific approach to production and patenting more. TK is the backbone of the botanical industries; however, identifying its contribution for assigning ABS is a herculean task. Large companies are undertaking advanced research with the support of the Academic and Research Institutions and the Government. Besides, companies are also using pharmaceutical R&D in developing new botanical products.

D. **Cosmetics**

Cosmetics are one of the fast growing industries in the world. As their demands from different segments of communities are used in different applications ranging from hair care, nutricosmetics or beauty supplements, cosmeceuticals typically include bioactive compounds, perfumes and fragrances related to beauty and personal care. Over the past decades, due to the growing consumer interest in health and well-being and fair trade products, the trend in using natural or botanical ingredients in the cosmetics sector has grown considerably.

One can see, in the cosmetic sector there is an enormous variation in investments and approaches in R&D, application of technology, and patenting (Rachel & Sarah, 2013). Recently the demand for purer "natural cosmetics"—containing a wide range of plant-based materials including fats, oils and waxes, essential oils and oleoresins, plant extracts and colourants has increased significantly. The global sales for the "natural cosmetics" during 2011 were US$ 26.3 billion. Europe and the United States account for around 40% of the global natural cosmetics market. Within Europe, France and Germany represent the strongest markets. However, in the near future, Eastern or Central Europe and Brazil will create more opportunities for high growth in the natural cosmetics market.

Different consumer products are being manufactured in the cosmetics sector. Among them, skin care products played a significant role in the sales (42%), followed by Fragrances (10%), Oral care (10%), Make-up (8%), and other toiletries (15%). Like other sectors, in cosmetics also, a few companies are playing a significant role in the market. In 2010, nearly 50% of the total market sales of personal care products using natural ingredients were manufactured by ten companies. Among these, Johnson & Johnson, L'Oréal and Colgate-Palmolive play a major role (Rachel & Sarah, 2013). Along with large companies, a significant number of small and medium enterprises (SMEs) are also functioning. For example in Europe, out of the 4,000 cosmetics companies two-thirds are SMEs. Nowadays, major companies are paying attention to biodiversity conservation; this is indicated in their reports. Out of the twenty largest beauty care companies, 80% mentioned biodiversity in their corporate sustainability reports. The usage and trade of raw-materials (biological resources) in the cosmetics sector is similar to that of the botanical and food sectors. A majority of these sectors is using the same plants or intermediaries in their production.

For mobilizing the biological resources, companies are following different models. One common model, which is followed by many companies, is mobilizing through local dealers, who source plants from the local growers or collectors. Sometimes, the resources may pass through a number of different local traders or cooperatives before going for export. Resources will be stored in the large warehouses of international trading companies, particularly in Tokyo, New York, Hong Kong, and Hamburg and they play a major role in quality control and pricing, and also act as clearing-houses for the raw-materials' trade (Rachel & Sarah, 2013). Another emerging model is the outsourcing of extraction and processing to low-cost labour-centres in China and India.

The companies spend huge amounts for R&D and an estimate revealed that around US$ 9 billion is spent each year in the cosmetic sector. Generally, cosmetics companies are benefitting (science and technology) from the other sectors including pharmaceuticals (cosmeceuticals), food (nutricosmetics), and biotechnology. The dependence on TK in this sector is considerable. However, certain companies are hesitant to incorporate TK into their R&D as this might adversely impact on the patent. Since, the cosmetic sector is intensively using different and unique biological resources; the scope of ABS is significant in this sector.

E. **Crop Protection Products**

The growing demand for organic agricultural products is the base for the emergence of production and marketing of crop protection products at the global level. Both in developed and developing countries this sector's growth is significant, where agricultural research and extension institutions, farmers, NGOs and traders role is critical. In the agriculture sector a considerable quantity of chemical fertilizers and pesticides has been used for enhancing the production, which adversely affects the environment, especially the soil and water quality. Further, the risks in food quality are also high when bio-magnification possibilities exist. This realisation enhanced the scope of organic farming globally. Now-a-days the demand for organic products is increasing all over the world.

In bio-fertilizer and pesticides manufacturing microorganisms play a significant role and the research in this line is progressing all over the world. In the markets of global bio-pesticides, 10 companies control 82% of the market. The crop protection products' sale has increased from US$ 25 billion (1990) to US$ 40 billion (2010).

In agriculture, the continuous focus on herbicides for insect resistance is needed, which is the challenge of crop protection product companies. In this sector, large agro-chemical companies are working with smaller companies for collecting soil samples, segregation and testing of microorganisms and screen the DNA from the microbes. Generally, the dependency of TK in this sector is insignificant. Big companies are aware about the ABS and their responsibilities to follow the Nagoya Protocol on ABS, but their active participation in the process is yet to start.

F. **Horticulture**

Horticulture is a major section of agriculture. The main activities of the horticulture sector include: facilitations of ornamental horticulture, fruits culture, flower culture etc. and its production increased considerably in recent periods, particularly after the life style changes of the people. Ornamental horticulture is still largely under the control of small and medium size companies with the application of limited level technologies predominating the conventional breeding practices. For first-time species breeding, companies depend on Public Institutions and Universities because of the high costs involved in this (Raichal Waynberg, 2013).

After1980, a steady growth has been observed in the global horticulture industries, with a shift in the production from the developed to the developing countries. The total value of global horticultural products manufactured in 2011 is US$ 19 billion.

Historically, the Netherlands is the centre of flower production. The use of R&D in the horticulture sector is a growing and a dynamic process and highly related to agriculture. Generally, development will first take place in the agricultural crop fields, then vegetables and finally trickle down to the ornamental horticulture crops. Application of DNA technologies is too expensive in this sector. The role of R&D in this sector is determined by the size of the unit and its location. Now-a-days, in the ornamental horticulture sector the dependency on genetic resources from the wild are limited. They are using the already available resources with them or other commonly available stocks. Further, most of the companies acquired thc genetic resources, before the emergence of ABS regulations. Horticulture companies may require new wild genetic resources only for developing completely new species and to develop new colours, trails and characteristics.

G. **Industrial Biotechnology**

The application of bio-technology to eco-efficient production and processing of materials and chemicals and bio-energy is broadly considered by the industrial biotechnology sector. It utilises the capability of microorganism, and enzymes to manufacture products in food and feed, paper and pulps, textiles, etc. (CBD, 2013d). Advancement in science and technology, concerns on energy security, climate change, and green manufacturing have influenced the growth of this sector worldwide. Now-a-days small and large size industrial biotechnology companies are coming forward for producing bio-chemicals, bio-fuels and bio-plastics.

The global reserve of goods produced using Industrial bio-technologies was about US$ 65–78 billion during 2010. The largest industrial biotech companies are located in Europe and US. In many countries this sector is flourishing with the support of Government subsidy as this sector's contribution in safeguarding the environment and biodiversity is significant. The R&D in this sector is progressing over periods at a significant level. However, the dependency on TK is negligible in this sector.

H. **Nutraceuticals**

The term "nutraceutical" is derived from two different words namely: "*nutrient*" (a nourishing food component) and "*pharmaceutical*" (a medical drug). The name was coined in 1989 by Stephen De Felice, Founder and Chairman of the 'Foundation for Innovation in Medicine', an American organization located in New Jersey. Nutraceuticals may be any food ingredient or product consumed for its health and medical benefits, including the prevention of disease and or for treating a particular disease. Nutraceutical products include: entire diets (macrobiotic): dietary supplements, isolated nutrients, functional or medical foods, biotechnology enhanced foods, and fortified processed foods such as soups, cereals, and beverages. Through nutraceuticals' consumption extra health benefits may be obtained in addition to the basic nutritional value found in foods (Kate & Sarah, 2000).

Nutraceuticals are grouped in four major categories: dietary supplements, medicinal food, functional food, and farmaceuticals. A dietary supplement is a product that contains nutrients derived from food products, and is concentrated in liquid, powder, capsule or pill form. Medical food is formulated to be consumed internally, under the supervision of a qualified physician. Functional food is a category of items which includes whole foods and fortified, enriched dietary components that may reduce the risk of chronic disease and provide health-benefits. Farmaceuticals are medically valuable components produced from modified agricultural crops or animals.

Generally, nutraceuticals may improve health, prevent chronic diseases, postpone the aging process and ultimately increase the life expectancy. Besides, they support the functions and integrity of the body. They can prevent life threatening diseases such as diabetes, gastrointestinal disorders, as well as different infections. The global nutraceuticals' sale value during 2013 was UD$ 300 billion. The use of TK in this sector is limited and R&D is a continuous process.

Summary and Conclusion

It is very clear from the above analysis that a wide range of companies are using genetic materials or biological resources in their business model in different quantities based on their requirement. However, the companies and other stakeholders are not clear about the resources' access process and the benefit sharing formalities under the ambit of ABS. Similarly, the providers of biological or genetic resources are also not aware about their rights and responsibilities under the ABS mechanism. Even if CBD is working with its parties on these issues, most of the users of the biological resources (business communities) are not very keen on ABS, as they generally assume that ABS is an additional financial burden for their business. This is a major concern as well as a challenge for the success of CBD and the ABS regime, which should be take in to account seriously.

However, the sector or industry wise analysis carried out in this chapter clearly reveals the significant differences in the ABS track (Nelliyat & Meenakumari, 2018). The pharmaceutical sector is one of the predominant industrial sectors, which uses a wide range of genetic resources for manufacturing different drugs. The R&D investments in this sector are relatively high when compared to other sectors and the dependency on the biological resources associated TK is also transparent, but at a diminishing rate. Seed companies collect genetic materials / germplasm from the repositories (national or international) or from the wild in small quantities. After rigorous research, companies have come up with genetically modified high yielding varieties of crops. As in this sector, the R&D costs are substantial in the overall production costs of the genetically modified seeds, companies are hesitant in complying with the ABS.

The botanical sectors normally use wide ranges of herbs as inputs (raw-materials) in bulk quantities for manufacturing different products having huge markets, both at the domestic and international levels. In recent decades, botanical industries are also involved in intensive R&D. Historically, the role of TK in product development and manufacturing is significant. Considering the growing importance of the botanical products in human health and welfare, the scope of ABS in all segments of botanicals is important. The TK element in this sector may be clarified, and not become a constraint in the ABS process.

With the changing life style and the modernization trend, the demand for cosmetic items increased in developed as well as developing nations. Industries are using different biological resources. R&D is predominant in the cosmetic sector. As many highly biological resources based value added products are manufactured in this sector, the scope of ABS is extremely important. Food and beverage industries flourishing all over the world also use a wide range of biological resources (both plants and animals) and manufacture different consumer products. Broadly, ABS is a new phenomenon in this sector and requires much more clarity on ITPGRFA and the exemptions on agricultural products. The R&D in the food and beverage sector is progressing and coming up with modern dietary products, coping with the demand in accordance with the changing life style of the people with the help of different biological resources.

Considering the biological resource predominant dependencies in the crop protection products' manufacturing sector and its emerging markets with organic farming as its base, ABS has to be initiated in this sector. The horticulture sector (particularly ornamental horticulture) is under transformation from the wild natural species to the newly adopted ones with changing human tastes. As this sector uses genetic resources for its already available stocks and the newly identified wild ones, the scope of ABS is enormous. Industrial biotechnology is another emerging area, with emphasis on more environment friendly product manufacturing. This sector levels up further with the advent of synthetic biology. Nutraceuticals are the growing sector in which more attention to ABS is required.

In brief, even if industries have made some progress in complying with the domestic legislation pertaining to the ABS during the last decade, much more effort (cooperation and support) is required for the success of ABS to a great extent. It is also clear that understanding ABS and the level of its acceptance varies between the different industrial sectors. Strict enforcement of the national legislation on ABS is required, but in the absence of a strong institutional and administrative network it becomes a difficult task. The voluntary acceptance of ABS by all the genetic or biological resources- based industrial units is a prerequisite for the success of ABS. We realize that industries are far ahead in operationalizing the Corporate Social Responsibility. Similarly, industries should come forward and amicably understand the concept and the principles of the CBD and the Nagoya Protocol on ABS and comply with the legal norms pertaining to it. They should realise the fact that, the success of ABS is their success, as it ensures raw-material (genetic/biological resources) security forever to sustain their businesses.

Acknowledgements This Chapter is the outcome of the research the Author done at the Centre for Biodiversity Policy and Law (CEBPOL) with the support of the Norwegian Government. I am extremely thankful to Dr. L Pushpa Kumar, Professor, Faculty of Law, University of Delhi for his review of the draft and his valuable suggestions.

References

Bavikatte Kabir Sanjay, & Morten Walløe Tvedt. (2015). Beyond the thumbrule approach: Regulatory innovations for bioprospecting in India. 11/1 *Law, Environment and Development Journal*. http://www.lead-journal.org

CBD (2013). *Bioscience at a crossroads: The pharmaceutical industry and the Nagoya protocol*. Secretariat of the Convention on Biological Diversity, Montreal.

CBD (2013a). *Bioscience at a Crossroads: Agriculture and the Nagoya Protocol*. Secretariat of the Convention on Biological Diversity, Montreal.

CBD (2013b). *Bioscience at a crossroads: Botanicals and the Nagoya protocol*. Fact Sheet, Secretariat of the Convention on Biological Diversity, Montreal.

CBD (2013c). *Bioscience at a crossroads: Implementing the Nagoya protocol in a time of scientific, technological and industry change*. Fact Sheet, Secretariat of the Convention on Biological Diversity, Montreal.

CBD (2013d). *Bioscience at a crossroads: Industrial biotechnology and the Nagoya protocol*. Fact Sheet, Secretariat of the Convention on Biological Diversity, Montreal.

Gayathri Shanbhag, N., & Rana, R. S. (2018). *Regulation of access to biological resources and benefit sharing in India: An analytical study*. National Biodiversity Authority, Chennai.

Kerry, K., & ten and Sarah A Laird,. (2000). *Commercial use of biodiversity: Access to genetic resources and benefit sharing*. Earthscan Publications Ltd.

Nelliyat and Meenakumari. (2018). *Compliance of Access and Benefit Sharing (ABS): A sector specific review*. Published by the Centre for Biodiversity Policy and Law.

Rachel, W., & Sarah, L. (2013). *Bioscience at a crossroads: Access and benefit sharing in a time of scientific, technological and industry change: The cosmetics sector*. Secretariat of the convention on biological diversity.

Sarah, L., & Rachel, W. (2012). *Bioscience at a crossroads: Implementing the Nagoya protocol on access and benefit sharing in a time of scientific, technological and industry change*. Convention on Biological Diversity.

Sarah, L., & Rachel, W. (2013). *Bioscience at a crossroads: Access and benefit sharing in a time of scientific, technological and industry change: Botanicals*. secretariat of the convention on biological diversity.

Emerging Paradigms in Natural Products-Based Drug Discovery

S. J. Soumya, K. R. Arya, C. S. Abhinand, Anuroopa G. Nadh, J. R. Rani, Oommen V. Oommen, and P. R. Sudhakaran

Introduction

A rich biodiversity of various life forms exists on Earth comprising more than 50 million species of plants and animals. Recent advances in Science and Technology make use of the immense potential of various life forms, particularly plant biodiversity. Despite the abundance of plant species on Earth, only a limited number have been utilised for human purposes. Plant biodiversity is rich in several bioactive phytocompounds that have been used to cure and treat various diseases since ancient times. The Traditional medicine systems of countries like India and China rely on herbal medicines [H-son-Mou & Paul, 1986; Kapoor, 1990]. These herbal medicines mostly originate from tropical regions and native folklore. In addition to this, herbal compounds have been widely used in the pharmaceutical industry for developing drugs to treat various diseases.

Tropical plants have served as sources of medicinal compounds for centuries. Over 80,000 plant species have been documented for their medicinal properties. The book 'Hortus Malabaricus' documents the diversities and medicinal properties of the flora

S. J. Soumya · K. R. Arya · C. S. Abhinand · A. G. Nadh · O. V. Oommen · P. R. Sudhakaran (✉)
Department of Computational Biology and Bioinformatics, University of Kerala, Thiruvananthapuram, Kerala 695581, India
e-mail: prslab@gmail.com

C. S. Abhinand
Department of Virus and Genomics, Institute of Advanced Virology, Thiruvananthapuram 695317, India

A. G. Nadh
Kerala University of Digital Sciences, Innovation and Technology (Digital University Kerala), Thiruvananthapuram, Kerala 695317, India

J. R. Rani
School of Biotechnology & Pharmaceutical Sciences, Vignan's Foundation for Science, Technology & Research, Guntur, Andra Pradesh, India

L. Krishna Panicker et al. (eds.), *Biodiversity and Business*,
https://doi.org/10.1007/978-3-031-71674-4_14

on the Malabar coast. It contains botanical descriptions, medicinal uses, and ethnobotanical knowledge of plants. It also mentions traditional healing practices using indigenous plant collections of the Malabar region. About 780 species of important plants of the Malabar region are described in the book including their medicinal properties, methods of preparation, and application of the medicine [Manilal, 1984]. Approximately more than 25% of the medicines are derived from plants with 70% of drugs either directly derived from plants or modified phytocompounds [Jantan et al., 2015]. Most phytocompounds are secondary metabolites produced by plants as part of their defense mechanisms. These compounds exhibit structural variations with promising biological activities due to the rich biodiversity, which is scarce in synthetic compounds. Tropical regions, particularly rain forests are rich reservoirs of plant biodiversity and most of them are not explored in terms of their biological potential. Research is still going on to explore bioactive compounds from plants for drug discovery and development by integrating traditional knowledge with modern healthcare systems.

Initially, plant extracts were used to treat diseases until the identification of bioactive compounds from plants. Phytochemicals isolated from plants were directly used as drugs and morphine from the seed pods of poppy plants was the first herbal drug identified in 1804. [Dias et al., 2012]. This discovery marked a turning point in the drug discovery process leading to the isolation and purification of several bioactive compounds from plant extracts. Later on, the elucidation of their structures led to the synthesis of phytocompounds and structural modifications for the development of new drugs [Kotadiya, 2023]. Examples of other herbal drugs include cardiac glycosides such as digoxin and digitoxin, as well as anti-cancer drugs such as vincristine and vinblastine. This chapter focuses on various approaches used to identify pharmacologically active phytocompounds for drug discovery and development along with the importance of conservation of plant biodiversity for the sustainability of traditional herbal medicines.

Conventional Approaches in Herbal Drug Discovery

As the phytokingdom and its phytochemical diversity are vast, scientists depend on traditional knowledge-based herbal medicines to reduce the chemical space during the initial biological screening in the drug discovery process. Here, the selection of plants is primarily based on indigeneous knowledge, which provides information about the biological activities of plants based on traditional practices. This ethnobotanical approach to plant selection has proven to be very successful in identifying potential compounds with pharmacological activity from medicinal plants.

Conventional approaches in herbal drug discovery involve several sequential steps beginning with plant collection followed by phytochemical extraction, fractionation, analysing bioactivity, characterisation, lead optimisation, preclinical and clinical trials, and ultimately the potential lead compounds being marketed as herbal drugs. After the plant has been selected, the plant material is collected from the natural habitat with proper authentication. The next step involves extracting phytocompounds from the dried/fresh plant material using different solvents and extraction techniques. Commonly used solvents for extracting phytocompounds include water, methanol, chloroform, hexane, petroleum ether, ethyl acetate, etc. Soxhlet extraction, hydro distillation, cold extraction, and maceration are among the most commonly used extraction techniques. The crude extract is then subjected to fractionation using chromatographic techniques like column chromatography and High-pressure liqud chromatography (HPLC) to isolate individual phytocompounds/groups of phytocompounds. The biological activity of these compounds can be examined using various assays either *in vitro* cell culture studies or biochemical assays. The phytocompounds possessing bioactivity are further characterised by nuclear magnetic resonance (NMR) and mass spectrometry (MS) such as Liquid chromatography-mass spectrometry (LC–MS) and Gas chromatography-mass spectrometry (GC–MS) techniques [Atanasov et al., 2021]. The lead compounds showing potential bioactivity are further optimised through structural modifications to enhance their pharmacokinetic properties. Selected lead compounds then undergo preclinical trials using animal models, followed by clinical trials in humans to examine their safety and efficacy. The compounds that are successful in clinical trials will get regulatory approval for commercialisation and can be marketed for therapeutic use.

Polyherbal formulations which are generally prescribed in traditional medical practice such as Ayurveda are made of multiple medicinal plants, containing numerous phytocompounds ranging from 100–1000 or more. So it is very challenging to identify the potential lead compounds among them by screening manually [Dagur et al., 2023]. Moreover, the conventional approach to drug discovery is not only ineffective but also tedious and time-consuming often taking 15–20 years for completion. Furthermore, it is not sure that lead compounds will be successful in clinical trials. In this context, computational approaches in herbal drug discovery have gained much attention recently because they help reduce the cost and time taken by the conventional drug discovery process.

Omics Approaches in Herbal Drug Discovery

The new era of herbal drug discovery is being revolutionised by the application of multi-omics technologies in medicinal plant research, leading to the identification of genes, metabolites, and enzymes involved in the synthesis of novel phytocompounds. Transcriptomics, proteomics, and metabolomics help explore novel pharmacologically active phytocompounds, their chemical diversity, and, their utilisation for drug discovery. High-throughput Genome sequencing provides information on the genetic composition of medicinal plants, while transcriptomics helps identify gene expression patterns and provides information on the genes involved in the synthesis of phytochemicals. Proteomics helps in identifying proteins involved in the synthesis of phytochemicals and associated metabolic pathways. Metabolomics provides the complete set of metabolites in the plants. The combination of these multi-omics approaches helps elucidate the molecular mechanisms in phytochemical synthetic pathways. This useful information is provided by various bioinformatics tools through the analysis of omics data. These high-throughput technologies together with multi-omics strategies enable a deeper understanding of the genetic and biochemical mechanisms underlying the production of pharmacologically active compounds in medicinal plants, paving the way for enhanced natural drug discovery [Zhang et al., 2023].

Phytochemical genomics is a new field that focuses on identifying the genomic basis of phytochemicals. The availability of the genome of medicinal plants has supported the functional characterisation of genes involved in the synthesis of secondary metabolites. Sequencing the chloroplast and mitochondrial genome leads to the development of molecular markers for identifying the authenticity of medicinal plants. Traditional genomics provides genetic information about an organism whereas phytochemical genomics correlates genetic information with metabolomics to understand the association between genes and phytochemicals [Saito, 2013]. Next-generation sequencing (NGS)—based gene expression profiling helps elucidate the molecular mechanisms underlying plant metabolism. Mass spectrometry, particularly LC–MS and NMR assists in identifying phytochemical composition and metabolite profiling. These techniques enable the identification of metabolic pathways and their crosstalk in the synthesis and function of metabolites. Co-expression analysis by integrating gene expression profiles with metabolomics data provides information about the functional association between genes and metabolite synthesis [Yonekura-Sakakibara et al., 2012]. The co-expression analysis of transcriptome data from *Catharanthus roseus* identified genes involved in alkaloid biosynthesis [Giddings et al., 2011]. Similarly, genes involved in the synthesis of sweetener glycyrrhizin in licorice were identified enabling the microbial production of glycyrrhizin [Seki et al., 2011]. The genes involved in the biosynthesis of morphine in the opium poppy plant have been identified through omics approaches [Gesell et al., 2009].

Metabolomic data analysis helps in understanding plant metabolomes, discovering novel phytochemicals, and elucidating the biochemical pathways involved in plant metabolism [Turi et al., 2015]. Targeted metabolomics, untargeted metabolomics, and a hybrid of targeted-untargeted metabolomics are different approaches used to identify phytocompounds. In targeted metabolomics, known phytochemicals in a sample are identified using reference molecules. Untargeted metabolomics involves the analysis of all metabolites present in a sample without using reference compounds allowing for the discovery of novel compounds and metabolic pathways. The hybrid approach combines both targeted and untargeted metabolomics by integrating the targeted analysis of known compounds with the untargeted analysis, offering a balanced strategy for discovering phytocompounds. Indeed, metabolomics strategies contribute to the development of new bioactive compounds with potential applications in pharmaceuticals.

One of the major challenges in phytochemical genomics is annotating gene expression data and metabolomics data. In metabolomics, it is very difficult to identify metabolites corresponding to the peaks in mass spectrometry and NMR data as well as to identify and characterise unknown metabolites. Sophisticated computational tools help addressing these challenges to some extent. Bioinformatics tools analyse large-scale data sets on genomics and metabolomics to extract meaningful information. Advanced computational tools and databases help in the identification and annotation of unknown metabolites. In addition to this, systems biology approaches to data integration of different '-omics' data help understand the interaction between genes and metabolites and elucidate plant metabolic pathway maps.

Computational Approaches in Herbal Drug Discovery

Computational approaches help in accelerating the identification of lead phytocompounds and herbal drug development. Computational tools enable the screening of a large library of phytochemicals against target proteins relevant to a particular disease condition. This virtual screening increases the chances of identifying lead compounds from herbs more easily compared to the conventional approach. In this approach, molecular docking and simulation techniques are mostly employed to predict the interaction between phytocompounds and target proteins ultimately to identify the bioactive lead compounds [Hemaiswarya et al., 2022].

Computer-Aided Drug Discovery (CADD)

As mentioned earlier, Computer Aided Drug Discovery (CADD) is a computational method used in the drug discovery process to identify potential lead compounds using molecular docking, quantitative structure–activity relationship (QSAR) analysis, and pharmacophore modelling. CADD plays a crucial role in filtering large libraries of compounds, designing multi-target drugs and predicting pharmacokinetics and pharmacodynamics properties. CADD is categorised into Structure-based drug designing (SBDD) and Ligand-based drug designing (LBDD). The schematic representation of general work flow of SBDD and LBDD is given in Fig. 1.

Structure-Based Drug Designing (SBDD)

Structure-based drug designing (SBDD) is an approach that utilises the three-dimensional (3D) structure of the target protein to design a molecule that binds to the target. For this, the 3D structure of the protein determined by X-ray crystallography and NMR are selected. Once the protein structure is known, SBDD specifically identifies the molecules which bind to specific sites on the protein by molecular docking. The interaction between the protein and the molecule is evaluated based on bonding interactions, binding energy, and shape complementarity.

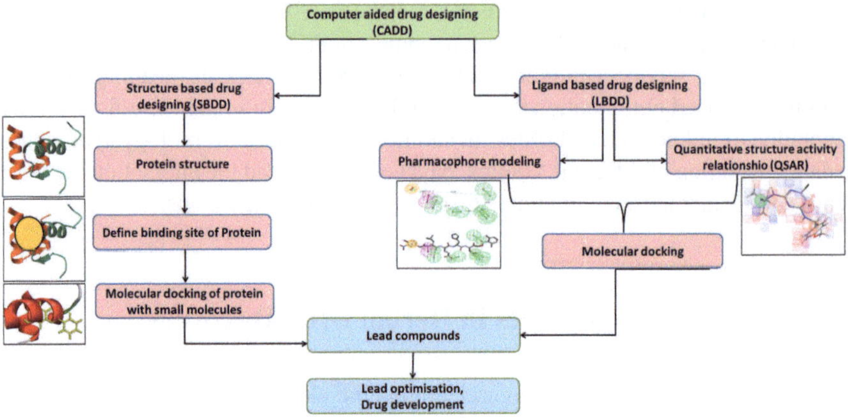

Fig. 1 Schematic representation of CADD workflow. The general workflow of SBDD and LBDD leading to lead compound identification is given

Important steps in SBDD include the identification of the target, screening of ligands, identification of hit compounds, identification of lead compounds, and experimental validation. Identification of target—Target protein involved in disease is selected and its 3D structure is either downloaded from available databases or modelled computationally, Screening of ligands—The library of molecules is employed in molecular docking with the protein, their interactions and binding affinity are analysed, Identification of hit compounds—Molecules that interact with the protein are identified as potential hits, Identification of lead compounds—Hit compounds are optimised through structural modification to get lead compounds with favourable pharmacokinetic properties, Experimental validation—The predicted bioactivity of lead compounds is experimentally validated.

SBDD typically relies on the protein structure and function to design molecules that interact specifically with the active site of the protein, ultimately leading to the development of potential drug candidates. SBDD is a rapidly evolving area with many successful case studies reported in drug discovery. SBDD leads to the development of small molecules such as Captopril, Dorzolamide, Saquinavir, Zanamivir, Oseltamivir, Aliskiren, Boceprevir, Nolatrexed, TMI-005, LY-517717, Rupintrivir and NVP-AUY922 that are in clinical trials or have been approved for therapeutic use [Junaida et al., 2023; Talele et al., 2010]. The design of captopril, an antihypertensive drug and reversible inhibitor of angiotensin-converting enzyme (ACE), stands as one of the early successes of SBDD. Dorzolamide, an inhibitor of carbonic anhydrase, is approved for topical use in the treatment of glaucoma; Saquinavir, HIV-protease inhibitor; Oseltamivir and Zanamivir, inhibitors of neuraminidase enzyme, are used for the treatment of influenza; Aliskiren, an inhibitor of renin, is used for the treatment of hypertension; Boceprevir, an inhibitor of serine protease involved in hepatitis C virus (HCV) replication, possesses anti-HCV activity; Nolatrexed dihydrochloride, an inhibitor of thymidylate synthase, is used as anti-cancer chemotherapeutic drug; TMI-005, an inhibitor of Tumor necrosis factor- α converting enzyme, is used for the treatment of inflammatory diseases; LY-517717, an inhibitor of factor Xa serine protease, is used for regulating blood coagulation; Rupintrivir, an inhibitor of human rhino virus (HRV) 3C protease, possess anti-HRV activity; NVP-AUY922, an inhibitor of heat shock protein 90 (HSP90), is used for anti-cancer therapy, all of these drugs were designed using SBDD.

SBDD approach employing different molecular docking tools has also been used to predict in silico phytocompounds against potent target proteins in various pathological processes. Based on molecular docking studies, phytocompounds that target the main protease of SARA-CoV-2 have been predicted to show anti-viral effect [Hossain et al., 2023]. Our studies, as discussed in detail in the following section, reported identification of lead compounds from medicinal plants using SBDD. The anti-angiogenic effect of Triphala churna was identified using molecular docking studies. Triphala churna phytochemicals such as punicalagin and chebulagic acid inhibit angiogenesis by targeting multiple signalling proteins of VEGF/VEGFR2 signaling pathway [Abhinand et al., 2020]. Similarly, potential lead compounds targeting BACE1 protein were identified from phytochemicals of medicinal plants and this is relevant for managing Alzheimer's disease [Nadh et al., 2022].

Ligand-Based Drug Designing (LBDD)

Ligand-based drug designing (LBDD) uses computational methods to identify lead compounds based on the properties of known ligands of the target protein to design compounds with the same pharmacological properties. This approach focuses on understanding the structure–activity relationships (SAR) of ligands and employs techniques such as pharmacophore modeling, similarity searching, and quantitative structure–activity relationship (QSAR) analysis. By using the knowledge of the properties of the ligands and their interactions with the target protein, LBDD enables the discovery of drug candidates that exhibit desired pharmacological effects.

LBDD involves the following steps: (a) Ligand Selection: Known ligands that bind to the target protein are selected from experimental data or databases. (b) Pharmacophore Modeling: Pharmacophore modeling involves identifying key pharmacophore features of the known ligands such as hydrogen bond donors, acceptors, aromatic rings, etc. which are essential for their biological activity. (c) Virtual Screening: Computational methods are used to search chemical databases for compounds that match the identified pharmacophore features of known ligands by similarity searching, pharmacophore-based screening, or QSAR analysis. (d) Hit Identification: Compounds that match the pharmacophore features are selected as potential hits. (e) Lead Optimization: Hit compounds are optimized through chemical modification to enhance their pharmacokinetic properties. (f) Experimental Validation: The lead compounds are tested experimentally to evaluate their biological activity. LBDD helps design compounds without the knowledge of 3D structure of the target protein.

The identification of camptothecin as a human RAD9 protein inhibitor, relevant to anti-cancer therapy was achieved using ligand-based pharmacophore modelling approach [Prasad et al., 2013]. 4-Oxothiazolidines and 5-Arylidines were identified as inhibitors of 5-LOX based on QSAR analysis of LBDD [Choudhary et al., 2010]. Novel antimicrobial compounds against *Staphylococcus aureus* were identified by LBDD approach [Ye et al., 2022]. Kaempferol from *Clitoria ternatea* has been identified as a potent inhibitor of snake venom phospholipase A2 (PLA2) through pharmacophore modeling [Suveena et al., 2022].

Computational Approaches to Identify Multiple Ligands Against Multiple Targets

Conventional drug discovery approaches focus on the treatment of diseases by mono ligand—monotarget concept. In this approach, the drug specifically targets a single biological molecule, particularly a protein relevant to the disease. However, several diseases are caused by multiple factors, particularly the dysregulation of multiple proteins in various signaling pathways. Therapeutics using single-target drugs may not be effective in these conditions. Moreover, the development of drug resistance is due to the formation of compensatory mechanisms, such as the existence of alternative or bypass signaling pathways. Multiple proteins contribute to disease progression and drug resistance, and it is crucial to identify ligands against multiple targets. Recent drug discovery approaches are focusing on multi-ligand, multi-target concepts other than mono-target to increase the efficacy of treatment strategies in multifactorial diseases and drug-resistant conditions. In this regard, natural product-based drug discovery can be made attractive by leveraging traditional knowledge so that the phytochemical space for screening by computational approaches can be reduced. A few examples in this regard are illustrated below.

Mono-ligand mono-target concept: There are several studies that employ CADD approach to identify phytocompounds from medicinal plants that target proteins involved in pathology of several diseases. In these studies, the phytoprofile of the medicinal plants and the traditional knowledge of their pharmacological effect, provide the basic information. In Ayurveda, a group of medicinal plants termed 'Medhya rasayana' is specifically used for managing neurological disorders. It is known to be effective in improving cognitive and intellect. It is evident from many scientific investigations that all the medicinal plants of Medhya Rasayana individually exhibit neuroprotective effects against various neurological disorders. An *in silico* study coupled with *in vitro* techniques analyzed the pharmacological activity of bio-active compounds present in the medicinal plants of Medhya rasayana against Alzheimer's disease target beta-site amyloid precursor protein cleaving enzyme 1 (BACE1). From structure-based docking approach, it was observed that

11 compounds out of 2562 phytochemicals from these plants were identified as potential hits, based on docking scores, binding energies, and interactions with the critical amino acid residues of BACE1 compared to the reference drug. Also, the possible neurological activities of these compounds were predicted. Out of the 11 compounds screened, two compounds, 'Convolidine' from the plant *Convolvulus pleuricaulis* Choisy and 'N-(4-hydroxybutyl) phthalimide' from *Glycyrrhiza glabra* satisfied the pharmacological parameters of drug likeness and ADMET (Absorption, Distribution, Metabolism, Excretion, and Toxicity). The binding stability of these compounds against BACE1 was confirmed by molecular dynamic simulation and post-dynamic MM/GBSA calculations. The interactions of these compounds with the critical amino acids in the active site of BACE1 revealed the possible inhibitory potential. The predicted antagonistic activity of the compound 'Convolidine' against BACE1 was validated using an enzymatic assay. Considering the narrow therapeutic window of available BACE1 inhibitors, the identified compound with optimal pharmacological profiles is a step forward in identifying better compounds with favorable therapeutic properties from traditionally employed medicinal plants [Nadh et al., 2022].

Multi-ligand multi-target concept: The progression of the disease in several pathological conditions involves dysregulation of multiple physiological processes. For instance, in diabetes, which is one of the highly prevalent non-communicable diseases globally, endothelial dysfunction, dysregulation of metabolism, insulin resistance, and inflammation do occur and contribute to the progression and complication of the disease. In such conditions monotarget approach may not yield desired effects and require targeting multiple such dysregulated pathways. The potential of medicinal plant-based formulations made of multiple medicinal plants can be examined by employing computational tools, as briefly indicated here. Considering the importance of targeting angiogenesis for treating numerous pathological conditions, a study was carried out to identify the anti-angiogenic compounds. Triphala churna (THL), a polyherbal formulation encompassing multiple bioactive constituents, has long been utilized in traditional medicine for diverse diseases. Employing molecular docking-based approaches, a study was performed to elucidate the bioactive compounds within THL that show inhibitory effects on key proteins involved in the VEGF/VEGFR2-mediated angiogenesis pathway. The signature metabolites of the constituent herbs in THL were identified from the literature and molecular docking calculations were done using AutoDock vina. The binding affinities of each compound against each of the targets were analyzed to assess the potential efficacy of individual phytochemicals. Subsequently, the prediction efficacy of a phytocompound for angiogenesis was estimated as the sum of its binding affinity values with all the target proteins obtained by molecular docking. Further, the prediction efficacy of the polyherbal formulation was elucidated as the sum of the prediction efficacy values of all the bioactive compounds selected for docking. To further confirm the affinity of the compounds, molecular docking was carried out against crucial target proteins using the LIBDOCK protocol in Discovery Studio 2022. The findings revealed that THL exhibits significant anti-angiogenic activity, with punicalagin and chebulagic acid identified as top compounds to its efficacy. This *in silico* approach provides valuable insights into the therapeutic potential of herbal formulations, offering a promising strategy for evaluating their efficacy in treating various diseases [Abhinand et al., 2020].

Multi-ligand multi-target concept in drug resistance: Glioblastoma (GBM) is a tumor condition that showed resistance to anti-angiogenic therapies particularly anti-VEGF resistance. Considering the complex and dynamic nature of Glioblastoma, understanding pathways and mechanisms associated with drug resistance is relevant for therapeutics. It appears that drug resistance is due to the existence of alternative signalling pathways other than VEGF/VEGFR signalling. Most of the studies on anti-cancer effects were focused on identifying a single compound against a single molecular target or certain biological process such as apoptosis, rather than a systemic approach aimed to target multiple components of multiple pathways that may contribute to tumor angiogenesis and growth. Recent technologies particularly transcriptome analysis would help identify differentially expressed genes in anti-VEGF resistant glioblastoma. A detailed bioinformatics analysis of transcriptome data from bevacizumab resistant glioblastoma xenografts identified multiple differentially expressed genes (DEGs) involved in angiogenesis [Arya et al., 2021]. Then traditional knowledge-inspired identification of medicinal plants with neuroprotective effects and computational prediction of phytocompounds from these plants that target these products of the key DEGs through CADD has been carried out. Multiple phytochemicals of Medhya rasayana plants targeting multiple proteins involved in drug resistance were identified. So instead of using mono-drug monotherapy, a multi-drug multi-target therapy can be employed for effectively targeting drug resistance in GBM. However experimental validation of these *in silico* predictions is required before proceeding further to explore its therapeutic potential.

AI-Based Drug Discovery

AI-based drug discovery refers to the application of artificial intelligence (AI) techniques and machine learning algorithms to accelerate and enhance drug discovery and development. This approach needs large datasets, computational modeling, and predictive analytics to identify potential drug candidates [Kate et al., 2023].

AI-based drug discovery includes the following methodologies: (a) Virtual Screening: AI algorithms are used to screen large databases of molecules and predict the possibility of their binding to target proteins. This approach helps prioritize compounds for experimental validation based on their predicted activity and pharmacological properties. (b) De Novo Drug Design: AI algorithms can generate novel molecules with desired pharmacological properties by exploring vast chemical space and optimizing molecular properties using generative models and reinforcement learning techniques. (c) Predictive Modeling: Machine learning algorithms are trained on different datasets of chemical compounds, biological assays, and clinical data to predict various properties relevant to drug discovery, such as toxicity, bioactivity, and pharmacokinetics. (d) Target Identification and Validation: AI algorithms analyze biological data to identify potential drug targets implicated in disease pathways and predict their druggability. (e) Lead Optimization: AI techniques are used to optimize lead compounds by predicting their binding affinity and pharmacokinetic properties. (f) Drug Repurposing: AI algorithms analyze large-scale omics data and literature to identify existing drugs with other potential therapeutic applications.

Computational approaches use several phytochemical datasets to accelerate *in silico* drug discovery. The availability of these datasets holds potential for the discovery of new drugs from natural products [Pathania et al., 2015]. These datasets compile accessible information about indigenous phytochemicals and their derivatives, which can be an input source for AI-ML algorithms for virtual screening and drug discovery. In Ayurveda, herbal formulations are based on empirical knowledge rather than understanding active ingredients. In such a scenario, leveraging traditional knowledge can complement modern drug discovery approaches. The integration of a phytochemical database with an Ayurvedic pharmacology database helps in drug discovery based on polyherbal formulations.

Deep learning approach in drug discovery: The discovery of antibiotics using a deep learning approach was reported by Strokes et al., 2020. In this approach, a trained deep neural network is capable of predicting molecules with antibiotic activity in silico from a collection of over 107 million compounds. Halicin was identified as the broad-spectrum antibiotic in the study. AI-based drug discovery improved prediction accuracy through the deep learning algorithm. In addition to this, machine learning (ML) approaches help in predicting narrow-spectrum antibiotics, which act against specific bacterial strains, thus reducing antibiotic resistance.

Deep learning approaches include the following strategies: (a) Training deep neural networks with available data on a diverse set of molecules, here in the abovementioned study, data on molecules causing bacterial growth inhibition, enables the prediction of potential lead compounds with specific pharmacological activity. (b) Developing modern neural molecular representations to improve prediction accuracy. (c) Developing algorithms for predicting molecular properties using ML techniques, can perform in silico screening of huge chemical space for drug discovery compared to other approaches. (d) Deep neural network model constructed molecular graphs based on specific properties using a directed message-passing approach. (e) Hyperparameter optimisation and ensembling techniques improve the performance of deep neural network model.

Nowadays, AI-based drug discovery is getting more attention due to its advantages over CADD. AI algorithms can analyse vast amounts of data compared to other approaches, thereby speeding up drug discovery. In just 8 months, Evotec and Exscientia companies identified an anti-cancer molecule using AI tools [Savage, 2021]. AI, together with ML, offers efficiency, accuracy, cost reduction, and novel innovations in drug discovery and development [Blanco-González et al., 2023]. The general workflow of the deep learning approach in drug discovery is given in Fig. 2.

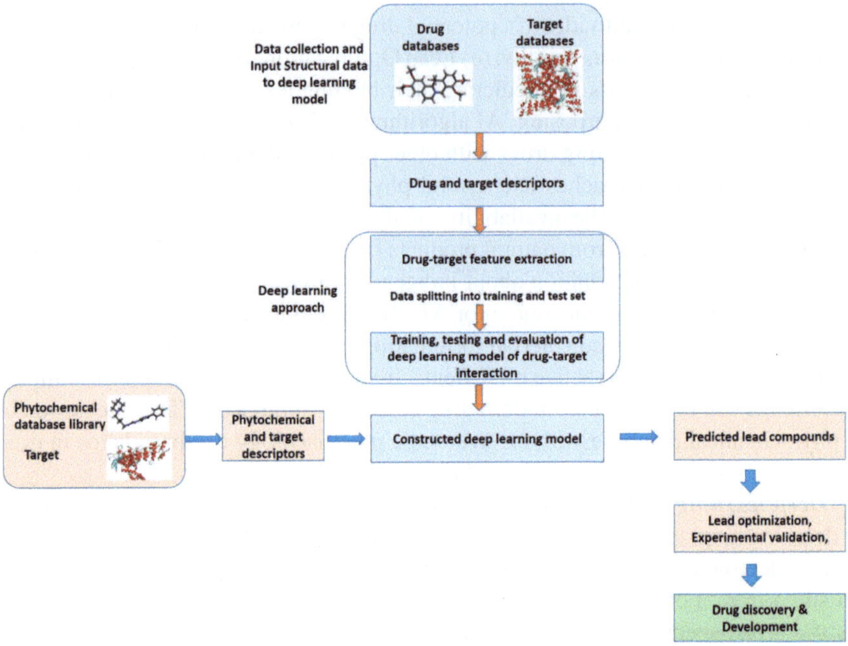

Fig. 2 Schematic representation of deep learning approach in drug discovery

Significance of Computational Drug Discovery

In recent times, with the development of cutting-edge technologies, it has become possible to overcome the challenges in improving the efficacy of drug discovery, and identifying the precise targets of natural products, particularly using computational approaches. Computational methods facilitate the identification and validation of molecular targets for drug development. By analysing biological pathways and protein structures, it is possible to identify key targets involved in disease processes and design compounds to modulate their activity.

As mentioned earlier, computational drug discovery aids in the identification of new drugs by screening vast libraries of natural compounds to target relevant proteins, thereby reducing the time and costs associated with conventional approaches. In addition to this, another interesting strategy with the natural compounds is their potential for multi-target action. Many plant-based products consist of polyherbal formulations, containing multiple phytocompounds. These phytocompounds modulate the activity of multiple target proteins synergistically, enhancing their overall therapeutic efficacy. The synergistic effects of multiple phytocompounds on multiple targets can be identified using computational drug discovery approaches. Computational approaches also play a crucial role in identifying therapeutics during drug resistance conditions. A high-throughput approach is employed to identify alternate pathways contributing to drug resistance in disease conditions by identifying differentially expressed genes (DEGs) in drug-resistant conditions. After identifying DEGs relevant to drug resistance, traditional knowledge-driven computational drug discovery approach can be utilised to target the proteins coded by DEGs, thereby combating drug resistance. Multiple phytocompounds from the polyherbal formulation interact with multiple targets involved in drug resistance in Glioblastoma.

Thus, computational approaches ease the identification of new herbal drugs by screening a large library of compounds, the multi-ligand multi-target effect of polyherbal formulations, and the identification of proteins and alternate pathways contributing to drug resistance. Then, these computational methods enable the identification of phytocompounds targeting the proteins involved in drug resistance. Subsequently, the interaction between potential lead compounds and proteins predicted by computational methods can be validated using computational molecular simulation studies to assess the stability of protein—ligand complex as well as experimentally to examine the bioactivity. Considering conventional methods, manual screening of a large number of compounds, identification of multi ligand-multi target concept, and examination of proteins causing drug resistance experimentally are laborious and time-consuming. In that scenario, computational approaches, excelled by filtering potential compounds with pharmacological activity, favorable drug-likeness, and pharmacokinetic parameters from the vast space of phytochemicals, acting just like a sieve.

Preserving Plant Biodiversity for the Sustainability of Natural Sources

For tens of thousands of years, humans have been using natural resources, particularly plants, for various purposes, including food, agriculture, and therapeutics. Furthermore, modern pharmaceuticals also utilise natural resources based on ancient knowledge for drug discovery. Therefore, preserving nature is very important as it serves as a vital source of bioactive compounds with medicinal value. Preserving plant biodiversity is crucial to expanding the diversity of phytocompounds necessary for drug discovery and development [Sharma & Sarkar, 2013]. Loss of natural habitat of medicinal plants due to deforestation, urbanisation, and climate change influence the availability of herbal medicines. Ensuring the quality and authenticity of medicinal plants is also essential. Changes in the agro-climatic conditions, harvesting, and processing methods can affect the efficiency of herbal medicines. Therefore, protecting tropical ecosystems ensures a sustainable resource of potential medicines [Bhujun et al., 2017].

The over harvesting of medicinal plants leads to the depletion and extinction of plant species and the loss of biodiversity. To overcome this, responsible utilisation of natural resources is to be practiced. Another limitation is the genetic variability of medicinal plants in different regions, resulting in variations in phytochemical composition. To address this issue, it is important to promote the cultivation of medicinal plants under optimal growth conditions such as a favorable climate, good soil quality, sufficient water availability, and proper pest control measures. Moreover, conserving medicinal plants with the desired phytochemical profile can be achieved through the maintenance of preferred medicinal plants using plant tissue culture methods. A multidisciplinary approach is needed to ensure the sustainable production and utilization of medicinal plants by combining sustainable cultivation practices and conservation efforts.

Conclusion

In general, computational drug discovery approaches reduce the time and cost involved in drug discovery compared to conventional approaches. Molecular docking allows the evaluation of ligand–protein interactions to identify potential binding modes and lead compounds. They also predict the drug-likeness and Absorption, Digestion, Metabolism, Excretion and Toxicity (ADMET) descriptors of the molecules thereby prioritize compounds for experimental validation. They facilitate the high throughput screening of vast natural product libraries to identify novel drug candidates from herbal sources. This minimizes the need for laborious experimental screening. Computational approaches complement experimental techniques in drug discovery by providing information on potential lead compounds and targets. The experimental validation of the *in silico* prediction is necessary for confirming its pharmacological potential. Therefore, it is possible to accelerate herbal drug discovery by integrating computational and experimental methods.

Information about the diversity of plants is often dispersed across various sources. Therefore, it is crucial to link biodiversity knowledge with pharmacological resources by utilising both traditional knowledge and modern drug discovery approaches. The combination of traditional knowledge and computational approaches holds immense potential for advancing drug discovery from natural sources. By integrating ancient wisdom with modern science, it is possible to explore new therapeutic possibilities based on medicinal plants.

References

Abhinand, C. S., Athira, P. A., Soumya, S. J., & Sudhakaran, P. R. (2020). Multiple targets directed multiple ligands: An in silico and in vitro approach to evaluating the effect of Triphala on Angiogenesis. *Biomolecules, 10*(2), 177. https://doi.org/10.3390/biom10020177

Arya, K. R., Bharath Chand, R. P., Abhinand, C. S., Nair, A. S., Oommen, O. V., & Sudhakaran, P. R. (2021). Identification of hub genes and key pathways associated with Anti-*VEGF* resistant glioblastoma using gene expression data analysis. *Biomolecules, 11*(3), 403. https://doi.org/10.3390/biom11030403

Atanasov, A. G., Zotchev, S. B., Dirsch, V. M., et al. (2021). Natural products in drug discovery: Advances and opportunities. *Nature Reviews. Drug Discovery, 20*, 200–216. https://doi.org/10.1038/s41573-020-00114-z

Blanco-González, A., Cabezón, A., Seco-González, A., Conde-Torres, D., Antelo-Riveiro, P., Piñeiro, Á., & Garcia-Fandino, R. (2023). The role of AI in drug discovery: Challenges, Opportunities, and Strategies. *Pharmaceuticals (Basel), 16*(6), 891. https://doi.org/10.3390/ph16060891

Choudhary, A. N., Kumar, A., & Juyal, V. (2010). Quantitative structure activity relationship (QSAR) analysis of substituted 4-oxothiazolidines and 5-arylidines as lipoxygenase inhibitors. Mini reviews in medicinal chemistry, 10(8), 705–714. https://doi.org/10.2174/138955710791572505

Dagur, P., Shreya, Ghosh, R., Rakshit, G., Biswas, A., Ghosh, M. (2023). Computational Methods in natural products-based drug discovery. In M. Rudrapal & Khan, J. (Eds.), *CADD and Informatics in Drug Discovery. Interdisciplinary Biotechnological Advances*. Springer. https://doi.org/10.1007/978-981-99-1316-9_5

Dias, D. A., Urban, S., & Roessner, U. (2012). A historical overview of natural products in drug discovery. Metabolites, 2(2), 303–336. https://doi.org/10.3390/metabo2020303

Gesell, A., Rolf, M., Ziegler, J., Díaz Chávez, M. L., Huang, F. C., & Kutchan, T. M. (2009). CYP719B1 is salutaridine synthase, the C-C phenol-coupling enzyme of morphine biosynthesis in opium poppy. *Journal of Biological Chemistry, 284*(36), 24432–24442. https://doi.org/10.1074/jbc.M109.033373

Giddings, L. A., Liscombe, D. K., Hamilton, J. P., Childs, K. L., DellaPenna, D., Buell, C. R., & O'Connor, S. E. (2011). A stereoselective hydroxylation step of alkaloid biosynthesis by a unique cytochrome P450 in Catharanthus roseus. *Journal of Biological Chemistry, 286*(19), 16751–16757. https://doi.org/10.1074/jbc.M111.225383

Hemaiswarya, S., Prabhakar, P. K., Doble, M. (2022). Computational methods to study herb-drug interactions. In *Herb-drug combinations*. Springer. https://doi.org/10.1007/978-981-19-5125-1_15

Hossain, A., Rahman, M. E., Rahman, M. S., Nasirujjaman, K., Matin, M. N., Faruqe, M. O., & Rabbee, M. F. (2023). Identification of medicinal plant-based phytochemicals as a potential inhibitor for SARS-CoV-2 main protease (M^{pro}) using molecular docking and deep learning methods. *Computers in Biology and Medicine, 157*, 106785. https://doi.org/10.1016/j.compbiomed.2023.106785

Hson-Mou Chang and Paul Pui-Hay But (1986). Pharmacology and Applications of Chinese Materia Medica, 1, 740. https://doi.org/10.1142/0284

Jantan, I., Bukhari, S. N. A., Mohamed, M. A. S., Wai, L. K., Mesaik, M. A. (2015). The evolving role of natural products from the tropical rainforests as a replenishable source of new drug leads. *Drug discovery and development—from molecules to medicine*. InTech. https://doi.org/10.5772/59603

Kapoor, L. D. (1990). *CRC handbook of ayurvedic medicinal plants*. CRC Press. https://www.scirp.org/reference/referencespapers?referenceid=666449

Kate, A., Seth, E., Singh, A., Chakole, C. M., Chauhan, M. K., Singh, R. K., Maddalwar, S., & Mishra, M. (2023). Artificial intelligence for computer-aided drug discovery. *Drug Res, 73*, 369–377. https://doi.org/10.1055/a-2076-3359 ISSN 2194–9379.

Kotadiya, M. (2023). Drug repurposing: scopes in herbal/natural products-based drug discovery and role of in silico techniques [Internet]. *Drug Repurposing—Advances, Scopes and Opportunities in Drug Discovery*. Intech Open. https://doi.org/10.5772/intechopen.109821

Manilal K.S.(1984). Hortus malabaricus and the ethnoiatrical knowledge of ancient malabar. Anc Sci Life, 4(2),96-9. PMID: 22557457; PMCID: PMC3331504.

Prasad, N., Kanakaveti, V., Siddhartha Eadlapalli, Ramakrishna Vadde, Angamba Potshangbam Meetei, Vindal, V. (2013). Ligand-Based pharmacophore modeling and virtual screening of RAD9 Inhibitors. *Journal of Chemistry*, 679459. https://doi.org/10.1155/2013/679459

Nadh, A. G., Revikumar, A., Sudhakaran, P. R., & Nair, A. S. (2022). Identification of potential lead compounds against BACE1 through in-silico screening of phytochemicals of Medhya rasayana plants for Alzheimer's disease management. *Computers in Biology and Medicine, 145*, 105422. https://doi.org/10.1016/j.compbiomed.2022.105422

Neergheen-Bhujun, V., Awan, A. T., Baran, Y., Bunnefeld, N., Chan, K., Dela Cruz, T. E., Egamberdieva, D., Elsässer, S., Johnson, M. V., Komai, S., Konevega, A. L., Malone, J. H., Mason, P., Nguon, R., Piper, R., Shrestha, U. B., Pešić, M., & Kagansky, A. (2017). Biodiversity, drug discovery, and the future of global health: Introducing the biodiversity to biomedicine consortium, a call to action. *Journal of Global Health, 7*(2), 020304. https://doi.org/10.7189/jogh.07.020304

Ondetti, M. A., Rubin, B., & Cushman, D. W. (1977). Design of specific inhibitors of angiotensin-converting enzyme: New class of orally active antihypertensive agents. *Science, 196*(4288), 441–444. https://doi.org/10.1126/science.191908

Saito, K. (2013). Phytochemical genomics–a new trend. *Current Opinion in Plant Biology, 16*(3), 373–380. https://doi.org/10.1016/j.pbi.2013.04.001

Savage, N. (2021). Tapping into the drug discovery potential of AI. *Nature*. https://doi.org/10.1038/d43747-021-00045-7

Seki, H., Sawai, S., Ohyama, K., Mizutani, M., Ohnishi, T., Sudo, H., Fukushima, E. O., Akashi, T., Aoki, T., Saito, K., & Muranaka, T. (2011). Triterpene functional genomics in licorice for identification of CYP72A154 involved in the biosynthesis of glycyrrhizin. *The Plant Cell, 23*(11), 4112–4123. https://doi.org/10.1105/tpc.110.082685

Sharma, V., & Sarkar, I. N. (2013). Leveraging biodiversity knowledge for potential phytotherapeutic applications. *Journal of the American Medical Informatics Association, 20*(4), 668–679. https://doi.org/10.1136/amiajnl-2012-001445

Shivalika Pathania, Sai Mukund Ramakrishnan, Ganesh Bagler. (2015). Phytochemica: a platform to explore phytochemicals of medicinal plants, *Database*, bav075. https://doi.org/10.1093/database/bav075

Stokes, J. M., Yang, K., Swanson, K., Jin, W., Cubillos-Ruiz, A., Donghia, N. M., MacNair, C. R., French, S., Carfrae, L. A., Bloom-Ackermann, Z., Tran, V. M., Chiappino-Pepe, A., Badran, A. H., Andrews, I. W., Chory, E. J., Church, G. M., Brown, E. D., Jaakkola, T. S., Barzilay, R., & Collins, J. J. (2020). A deep learning approach to antibiotic discovery. *Cell, 180*(4), 688–702.e13. https://doi.org/10.1016/j.cell.2020.01.021. Erratum in: Cell. 2020 Apr 16; 181(2): 475–483.

Suveena, S., Saraswathy, V., Junaida, M. I., Vinod, M. P., Laladhas, K. P. ORCID Icon, Achuthsankar, S., Nair, Sudhakaran P. R., Oommen V. Oommen (2020). In silico screening of the phytochemicals present in Clitoria ternatea L. as the inhibitors of snake venom phospholipase A_2 (PLA_2). *Journal of Biomolecular Structure and Dynamics, 41*(16): 7874–7883. https://doi.org/10.1080/07391102.2022.2126889

Talele, T. T., Khedkar, S. A., & Rigby, A. C. (2010). Successful applications of computer aided drug discovery: Moving drugs from concept to the clinic. *Current Topics in Medicinal Chemistry, 10*(1), 127–141. https://doi.org/10.2174/156802610790232251

Turi, C. E., Finley, J., Shipley, P. R., Murch, S. J., & Brown, P. N. (2015). Metabolomics for phytochemical discovery: Development of statistical approaches using a cranberry model system. *Journal of Natural Products, 78*(4), 953–966. https://doi.org/10.1021/np500667z

Ye, J., Yang, X., & Ma, C. (2022). Ligand-Based drug design of novel antimicrobials against staphylococcus aureus by targeting bacterial transcription. *International Journal of Molecular Sciences, 24*(1), 339. https://doi.org/10.3390/ijms24010339

Yonekura-Sakakibara, K., Fukushima, A., & Saito, K. (2013). Transcriptome data modeling for targeted plant metabolic engineering. *Current Opinion in Biotechnology, 24*(2), 285–290. https://doi.org/10.1016/j.copbio.2012.10.018

Zhang, W., Zeng, Y., Jiao, M., Ye, C., Li, Y., Liu, C., & Wang, J. (2023). Integration of high-throughput omics technologies in medicinal plant research: The new era of natural drug discovery. *Frontiers in Plant Science, 14*, 1073848. https://doi.org/10.3389/fpls.2023.1073848

Bioprospecting from Medicinal Plants and Benefit Sharing

Nilayangode Preetha, V. Balakrishnan, and George Thomas

Introduction

Bioprospecting is defined as 'the search for valuable genetic and biochemical information found in wild animals, plants or microbial organisms' for product development as a purely scientific and commercial endeavour (Priya & Kurian, 2018). In an ASEAN Workshop on TRIPS Agreement, 2001, Bioprospecting has been defined as the systematic search for and development of new sources of chemical compounds, genes, micro-organisms, macro-organisms, and other valuable products from nature. In brief, the methodical and structured search for valuable products from biological resources, such as plants, microorganisms, animals, etc., is known as bioprospecting. Numerous sectors, the most well-known of which is the pharmaceutical sector, as well as other agricultural sectors, are involved in it. This involves creating pharmaceuticals or agroproducts either directly from natural products or by developing them based on traditional knowledge of bioresources, followed by in-vitro biological activity screening and field trials. This includes creating industrial products like paints, glues, dyes, and cosmetics from natural products; creating natural products to supplement the agri-food industry by using them directly as food, food additives, dietary supplements, fodder, and fertilizer; creating natural products for biofuel, drug development and biodiversity exploration for wild genes, and subsequent development of commercial goods or genetically modified organisms.

The diversity of life on Earth is thought to number over 10 million plant, microbial, and animal species. Over the previous 3.5 billion years, natural selection has changed the genetic and metabolic makeup of these organisms. Worldwide, between 50,000 and 80,000 plant species are utilized for therapeutic purposes. Almost 80% of people in developing and underdeveloped nations rely entirely on herbal medications for primary healthcare, while more than 25% of recommended medicines in

N. Preetha (✉) · V. Balakrishnan · G. Thomas
Kerala State Biodiversity Board, Thiruvananthapuram, Kerala, India
e-mail: preetha.nilayangod@gmail.com

L. Krishna Panicker et al. (eds.), *Biodiversity and Business*,
https://doi.org/10.1007/978-3-031-71674-4_15

industrialized countries are derived from wild plant species. In Africa, 80% of people use traditional medications, while 65% of people in India only have access to traditional medical systems. In India, plants are used in approximately 90% of ayurvedic remedies with many ayurvedic remedies being polyherbal, involving a combination of 3–30 plants. An estimate states that various higher plants provide about 25% of the drugs now in use (WHO, 2002).

Depending on their source, therapeutic substances can be classified as natural, synthetic, or semi-synthetic. Natural therapeutic agents are made from substances that are present in nature and comprise extracts of active ingredients derived from a variety of sources, such as microorganisms, minerals, plants, and animals. Natural products have been crucial in the development of novel chemical entities (NCEs), between 1981 and 2002, natural products or products derived from them accounted for about 28% of NCEs (Newman et al., 2003). An additional 20% of NCEs were classified as natural product mimics, which indicates that the synthetic molecule came from research on natural products. Only between 5 and 15% of the 250,000–500,000 higher plant species have been investigated for active compounds (https://www.cbd.int/financial/bensharing/g-tleclair.doc). Current calculations indicate that just one in 250,000 natural samples would be a commercially viable drug, and the development would take decades to complete, one of the major factors affecting the acceptability of bioprospecting by the pharmaceutical industry. Despite this pharmaceutical industry is probably the sector relying the most on bioprospecting as 11 of the 25 best-selling drugs in the world are based on natural products and represent 42% of sales.

The biotechnology industry in India is divided into four sectors: biopharmaceuticals, bioagriculture, biotechnology, and bioservices.

Biopharmaceuticals: One of the world's largest exporters of inexpensive medications and vaccinations is India. India has the greatest number of biosimilars authorized for the domestic market.

Bioagriculture: India is the world's fifth-largest producer of organic produce and one of the world's top producers of Bt cotton, with roughly 55% of its land dedicated to agriculture and related activities. BioAgri, has the potential to almost quadruple its contribution to the BioEconomy from $10.5 to $20 billion by 2025.

Bioindustrial: Across the nation, waste disposal and manufacturing are changing as a result of the integration of biotechnology into industrial processes.

Bio IT & Services: India is home to the majority of US FDA-approved factories outside of the US and has excellent capabilities in contract manufacturing, research, and clinical trials. (www.investindia.gov.in).

Bioprospecting and Bioeconomy

It is estimated that there are between 2.5 and 3.5 million different plant species in the world. Plants are the most common source of natural medicine because of their diversity in composition, structure, and chemistry. It is estimated that 35,000 different plant species are used to treat various diseases in different parts of the world

(Kong et al., 2003). Natural product research keeps coming up with new ways for the pharmaceutical sector to find lead chemicals for drugs. Approximately 25% of drugs that are currently in use come from various higher plants, according to estimates. In addition, during the past ten years, there has been a massive surge in interest in plant-based goods, particularly in cosmetics and phytotherapeutic supplements (nutraceuticals).

Globally, almost 40% of the world's economy is directly derived from biodiversity. Bioprospecting has been widely recognized as a means of creating value from global biodiversity resources, with the generated revenue being able to support conservation programs in developing countries.

Phytopharmaceuticals play a crucial role in the global economy, being widely popular among people of all ages. According to a study conducted by a global news wire, the global herbal medicine market is projected to reach a staggering value of 550 billion dollars by 2030. A significant increase from 83 billion dollars in 2019 and in 2005, the worldwide herbal industry for plant-derived medications was valued at 18 billion dollars. Leading the global herb market are China and India, with China exporting 120,000 tonnes of herbal remedies annually, followed by India with 32,000 tonnes. Europe, on the other hand, stands as the largest importer of medicinal plants, bringing in over 400,000 tonnes each year to meet local demand (Briskin, 2000; Khan, 2014). It is worth noting that researchers have estimated that only around 15% of medicinal plants have undergone phytochemical analysis, and a mere 6% have undergone biological screening (Newman et al., 2002). This leaves a vast potential for the development of new and more effective medicinal agents. The United States has witnessed a noticeable shift towards the increased use of herbal products, with the overall market for herbal goods estimated to be worth 230 billion dollars in 2021 and projected to reach 430 billion dollars in 2028 (Briskin, 2000).

One crucial aspect to consider is the valuation of biological resources and how to measure it. In 2019, India's Bio-economy was estimated at $62.5 billion, with BioPharma contributing $36 billion (58% of the total Bioeconomy), Bio agriculture $12.0 billion (19%), bioindustrial $5.0 billion (8%), and Research $9.8 billion. Enzymes and biofuels sectors are reported to make up 46 and 54% of the total BioIndustrial segment, respectively (The India BioEconomy Report, 2020).

India is recognized as a major source of bio-prospecting, as demonstrated by the origin of 80% of medicinal plant patents in the U.S. and the fact that 50% of drugs listed in the British Pharmacopoeia are derived from medicinal plants found in the Western Himalayas (CSIR News, 2001) India, being a major player, exports medicinal plants worth Rs. 1200 million annually. India alone boasts over 8000 different types of medicinal plants, with nearly half of them being higher flowering plant species.

Bioprospecting and Traditional Knowledge

The idea of bioprospecting stems from the understanding of the significance of natural product discovery for the creation of novel crops and medications, which are frequently grounded in traditional knowledge (Robinson, 2012).

TK-based bio-prospecting may significantly cut the costs of pharmaceutical R&D, as compared to the conventional system of screening millions of synthesized chemical formulations.

Many bioprospecting companies and research groups target traditional and folk medicines for samples instead of randomly screening for the potent biological activity of natural products. This is because there is a high probability that these ethnobotanical remedies contain biologically active compounds that have been selected through hundreds of years of trial and error in folk medicine. As a result, the pharmaceutical and biotech industries are paying more attention to drugs and other goods developed locally in more countries like China, India, and Africa.

The first evidence of the usage of medicinal plants for the creation of herbal drugs dates back 5000 years, to a Sumerian clay slab found in Nagpur, India (Kelly, 2009). About 365 plant products were listed in the oldest Chinese herbal book, "Pen T'Sao," which was penned by Emperor Shen Nung somewhere around 2500 BC, for the treatment of various human maladies (Petrovska, 2012). Approximately in 800 BC, Homer's epic poem The Iliad and Odysseys listed 63 plant species that might be used medicinally (Toplak & Galle, 2005). The usage of several spices, including clove, pepper, and nutmeg, by Arya Monks, was mentioned in the sacred legendary texts of India, the "Vedas" for its potential to cure ailments (Agarwal, 2013).

A New Era in Bioprospecting

Isolation of morphine, the first natural and pure plant-derived compound, from *Papaver somniferum* in 1803 marked the beginning of a new era of drug discovery. The anticancer medications vinblastine and vincristine have been developed later on. Scientists from the University of Western Ontario in Canada and the American pharmaceutical giant Eli Lilly made these discoveries about medications in the 1960s, using the native Madagascar plant known as Rosy Periwinkle. The plant was used to cure diabetes in Madagascar's traditional folk medicine until the discovery of these alkaloids. These vinca alkaloids were isolated by the bio prospectors, and they subsequently proved to be highly profitable chemotherapy medications. But none of the money made was shared with Madagascar by the American business or the Canadian university. Another instance is the application made by the American pharmaceutical company WR Grace and Company for strong anti-fungal medications made from neem trees. In actuality, neem tree extracts were traditionally utilized in Ayurveda in India for a wide range of ailments, and several organizations were against that patent application. The matter was finally settled in 2005 when

WR Grace and Company lost their last attempt to overturn the patent's denial. The cinchona tree yields quinine, which is used in Peruvian folk medicine. Artemisinin and its derivatives were extracted from ancient Chinese remedies, and the opium poppy yields the alkaloids morphine and codine, which are utilized in a variety of indigenous medicines across the globe.

The anticancer medication etoposide was created using the chemical podophyllotoxin, which was extracted from the herb *Podophyllum peltatum*, which was utilized by Native American tribes in North America to treat warts. Herbal formulations are frequently developed into polymolecular botanical supplements, functional foods, and neutraceuticals to get around several challenges encountered during the rigorous drug development and formal approval processes. One such formulation is "BGR-34," which is allegedly antidiabetic and was developed by CSIR, India. The US National Cancer Institute's extensive natural product screening program for anticancer properties is a well-known example. Over almost 20 years, the program tested extracts from over 70,000 different plant species, and it produced just one drug: paclitaxel, a blockbuster medication and commercially successful chemotherapy treatment. On the other hand, a striking illustration comes from the time a Swiss scientist carried a sample of soil back to his lab in Switzerland while on a family vacation in Norway. The commercially effective immunosuppressant cyclosporine-A was developed as a result of additional studies. An estimated USD 1 billion is made annually from sales of this popular medication. Since Norway and Switzerland had no agreements for fair trade, Norway lost its whole profit share from this popular medication to Switzerland (Nasim et al., 2022).

Of the 252 medications that the World Health Organization has designated as essential and primary for the treatment of human health, 11% are wholly derived from plants. Quinine and quinidine from Cinchona species, atropine from *Atropa belladonna*, digoxin from Digitalis species, vincristrine and vinblastine from *Catharanthusroseus*, morphine, taxol from *Taxusbaccata*, curcumin from *Curcuma longa*, azadirachtin from *Azadirachta indica*, reserpine from *Rauwolfia serpentina*, and codeine from *Papaver somniferum* are some significant plant-produced bioactive molecules linked to human health. Remarkably, just one lead phytochemical out of five thousand has been evaluated to be effective in clinical studies and authorized for use as a medicine. For instance, the cholesterol-lowering medications lovastatin and mevastatin (compactin), were isolated from the Penicillin species. Ivermectins, were isolated from the Streptomycetes species and developed into anthelmintic and antiparasitic medications. The antiasthmatic medications salbutamol and salmetrol (beta agonist) were synthesized using ephedrine, which was extracted from *Ephedra sinca*.

About 70,000 plant species have been found to have medicinal qualities and are used in folk medicine worldwide (Alamgir, 2017). Around 25% of medications recommended globally for the treatment of diseases are plant-based. Presently, the majority of the world's species are found in 17 countries that have been designated as mega-biodiversity centers, including India. About 10,000 different plant species are utilized medicinally in China (He, 1998).

More than 5000 plant species are utilized medicinally in Africa (Luyt et al., 1999). In a recent investigation, the World Conservation Union (IUCN) discovered that 72,000 species of higher plants—or 17% of all plants—are utilized globally to extract medicinal compounds. According to past reports from the World Health Organization (WHO), 80% of the world's population still receives their basic medical treatment from traditional medicines (IUCN, 2006). Details on the conventional methods of harvesting, preparing, storing, and administering the organisms and their natural products offer important insights into the properties of the active ingredients.

Approximately seventy thousand herbal plants have been utilized medicinally, mostly in Asian medicine. These therapeutic plants are a treasure trove of limitless ethnobotanical chemicals that are currently being used in a variety of medication delivery systems.

Seventy-seven percent of the 119 plant-based medications that were made available in the United States between 1959 and 1980 were discovered through chemical research aimed at separating the active ingredients from the plants that are used in traditional medicine. The quest for novel pharmaceuticals persists, primarily employing the ethno-directed bio-rational methodology. More than 88,000 isolates were listed in the NAPRALERT database in 1999, and many of them served as the basis for well-known medications that are still on the market today (Table 1).

Name of plant	Biological name	Patent granted
Neem	Azadirachta indica	European Patent
Basmati Rice	Oryzasativa	US patent 5663484
Amla	Phyllanthus emblica	USPTO has granted five patents Four patents filed by Japanese Patent Office
Kala Jeera	Cuminumcyminum	US 5653981
Kumari	Aloe barbadensis	US 5652265
Amaltas	Cassia fistula	US 5411733
Pomegranate	Punicagranatum	US 541173325
Harad	Terminalia chebula	US 5529778
Aswagandha	Withaniasomnifera	US 5466452
Turmeric	Curcuma longa	US patent 5401504

Bioutilization of Medicinal Plants—India

The Indian system of medicine includes six recognized systems of medicine-Ayurveda, Siddha, Unani and Yoga, Naturopathy and Homoeopathy. In India, institutional mechanisms and programs directly related to the use of medicinal plants are under the Ministry of Ayurveda, Yoga and Natural Therapy, Unani, Siddha and Homeopathy (AYUSH) (Table 2).

Table 1 Important natural products derived from plant sources

Botanical source	Ethnobotanical compounds	Therapeutic application
Hyperianperfotum	Hypericin	Immunogenic cell death inducer
Lithospermumerythrorhizon	Shikonin	Immunogenic cell death inducer
Scutellariabaicalensis	Wogonine	Immunogenic cell death inducer
Piper nigrum	Piperine	Nanotheranostic agent for cancer treatment
Mucunapruriens	L-Dopa	Parkinson's disease
Taxusbrevifolia	Taxol	Cancer
Catharanthusroseus	Vinblastine & Vincristine	Cancer
Camptothecaacuminata	Camptothecin	Cancer
Aloe species	Aloe	Treatment of burns and wounds
Willow (Salix species)	Aspirin	Pain relief, promotion of heart- health, blood thinning
Digitalis purpurea	Digitalis	Heart failure treatment
Cinnamomumcamphora	Camphor	Rheumatic pain relief
Oenotherabiennis	Gamma-linolenic acid	Eczema, cancer, and nerve-damage treatment
Hoodia gordonii	Hoodia	Weight loss
Camptothecaacuminata	Irinotecan	Colon cancer treatment
Combretumcaffrum	Combretastatin	Cancer
Podophyllumpeltatum	Podophyllotoxin	Antitumor
Galanthusworonowii	Galantamine	Alzheimer's disease
Gingko biloba	Gingkolides	Alzheimer's disease
Artemisia annua	Artemisinin	Malaria
Atropabelladona	Tiotropium	Pulmonary disease
Rauwolfia serpentina	Reserpine	Antipsychotic, antihypertensive
Berberis vulgaris L	Berberine, Jatrorrhizine, Palmatine, Ceptisine	Antidiabetic, Anticancer, Antibacterial, Analgesic, Antiinflammatory, and Cardiovascular
Cinchona spp	Quinine	Antimalarial drugs
Ephedra sinica	Pseudoephedrine	Bronchodilation, nasal decongestion, allergies

(continued)

Table 1 (continued)

Botanical source	Ethnobotanical compounds	Therapeutic application
Pilocarpus species	China Pilocarpine	Glaucoma treatment
Datura and Solananceae species	Scopolamine	Sedation, motionsickness reduction
Illiciumverum	Americas Shikimic acid	Cure of viral diseases Laurance, 2005 (influenza)
Artemisia annua	Artemisinin	Type I diabetes and cancer
Salvia divinorum	Salvinorin A	Neuro-psycho pharmacotherapeutic plant-based drugs
Cleome	Pinocembrin, Kaempferol, Kaempferitrin	Anti-cancer
Silybummarianum	Silymarin	Hepatoprotective activities
Coleus forskohlii	Forskolin	Antiglaucoma drug
Curcuma longa L. (Turmeric)	Curcumin	Antioxidant, anti-inflammatory, arthritis, metabolic syndrome and pain
Galanthus nivalis	Galantamine	Alzhemer
Capsicum annuum	Capsaicin	Pain relievers
Artemisia glabella	Arglabin	Anti-tumor
Genista tinctoria L	Genistein	Anticancer, Alzheimer's disease
Vitis vinifera L	Resveratrol	Chemotherapeutic, antidiabetic, antioxidant
Azadirachta indica A. Juss (Neem)	Azadirachtin	Insecticidal and antimicrobial
Panax ginseng (Black ginseng)	Extract	Anticancer; anti-inflammatory
Althaea officinalis	Extract	Anti-inflammatory
Punica granatum (Pomegranate)	Extract	Antidiarrheal activity
Averrhoa carambola	Extract	CNS depressant and hypnotic properties
Ganoderma lucidum, Glycyrrhizauralensis and *Sophoraflavescens*	Extracts	Anti-asthma
Trigonella foenum-graceum L	Trigonelline, Diaszhenin	Antidiabetic, Anti-conception
Capsicum annuum	Capsaicin	Antilithogenic effect, Antiinflammatory

Compiled from Bast (2019)

Table 2 Medicinal plants used in the Indian system of medicine

Sl. No	Name of the system	Total number of plants mentioned
1	[a]Ayurveda	1200
2	[a]Siddha	800
3	[a]Unani	700
4	Folk medicine/Tribal medicine	10,000
5	Modern medicine	90

[a]Roughly about 50% of medicinal plants mentioned in this system of medicine are common to all four systems

The market size of India's beauty, cosmetic, and grooming market is expected to reach US$20,000 million (INR 140,000 crores) by 2025. This offers a huge scope for the herbal cosmetics market, which is growing at the rate of 12% per annum. As per the CII survey of Ayurveda Industries, the size of the Indian market was US$3000 million (INR 21,000 crores) in 2016 and INR 30,000 crores by the end of 2018. Ayurveda product's share in the market was 75% and the share of services was 25%. There are 8000 registered pharmacies and 400,000 registered practitioners in India. Over 30,000 branded and 1500 traditional AYUSH products are available in the market.

(*Compiled from CII, CII-PwC Report on Ayurveda, Annual Report of Ministry of AYUSH, ASSOCHAM Report, Media reports*). As per the latest official data, 794 herbal and ayurvedic industries are registered and 144 cosmetic industries are listed in Kerala. It is reported that 12% of the nation's formal ayurvedic industrial units are established in Kerala. These companies have an annual turn around of about Rs. 650 crores.

Kerala State Biodiversity Board had done an assessment of the Tradable Bioresources of Kerala and their Economic potential with financial support from the Rebuild Kerala Initiative. The total annual value of the bioresources at their origin is Rs. 66,883.11 crore, which indicates the magnitude of the contribution of biodiversity (in the form of bioresources) in the State. The annual export value of bioresources as well as bioresources-based products from Kerala through its major sea ports is (Rs. 21,760 crore) far higher than the annual bioresources as well as bioresources based products import value (Rs. 4414 crore). The balance of trade of Rs. 17,347 crore indicated the richness of Kerala's bio-wealth (Nelliyat & Preetha, 2022).

As part of this, a survey of the Ayurvedic industries of Kerala and the raw materials utilized were done. The highest consumed plant parts are fruits 20.64% and root, root tubers & rhizomes 21.86%. The total projected quantity of medicinal plants estimated is 25,516.6 MT based on data collected from 2015 to 2020. Out of top 100 with regard to quantity 60 species are consumed above 100 MT annually. The rest of the 40 species is between 95 and 60 MT consumed by the manufacturing units. In Kerala, there are two broad categories of supply chains in the medicinal plant trade—one for fresh plant parts and the other for dried plant parts. Starting from the primary collectors, there are multiple players in the medicinal plant supply chain and the price of these plants varies when they pass from one hand to another. Major players in the supply chain consist of primary collectors/cultivators, private traders, Commission agents, wholesalers, and final consumers (Rajasekaran et al., 2022) (Table 3).

Medicinal Plants at Risk

Medicinal plants are now under great pressure due to their increased demand, habitat loss, narrow range of distribution, introduction of invasive species, climatic changes, etc. On a global basis, IUCN has estimated that about 13.49% of the world's vascular plants (3,00,000 species), totalling about 40,468 species are under varying degrees of threat. Many plant species are becoming extinct (123 are extinct and 37 are extinct in the wild) and under threatened categories CR (3325), EN (6063), VU (7072), or vulnerable to extinction due to various reasons like population loss, loss of pollinators, loss of reproduction and/or seed germination capacity, habitat destruction (both natural and human-induced), over-exploitation and loss of genetic variability (IUCN, 2020). About 11.53% of vascular plants, totalling about 2142 species are red-listed in India. Out of these 8 are extinct, 432 species are threatened (CR, EN and VU) and nearly 54 species are grouped under near threatened (IUCN, 2020). Of 48,655 plant species documented (including viruses, bacteria, algae, fungi, and lichens) 9500 species have ethnobotanical importance and 7500 species are in medicinal use for indigenous health practices as well as modern systems of medicines (Kumar et al., 2013).

In Kerala, with financial support from the Kerala State Biodiversity Board, a Red list of threatened plants of Kerala has been prepared by Jawaharlal Nehru Tropical Research Institute, using the criterion of IUICN. A total of 1274 taxa, belonging to various plant groups such as Bryophytes, Pteridophytes, Gymnosperms, and Angiosperms are coming under various categories such as Extinct in the Wild (EW), critically endangered (CR), Endangered (EN), Vulnerable (VU), Near Threatened (NT), Least Concern (LC) and Data Deficient (DD). 264 species of flowering plants (angiosperms) in the State are facing the threat of extinction in varying degrees. They have been categorised into 44 Critically Endangered (CR) species, 113 Endangered (EN) species, and 107 Vulnerable (VU) species. Besides these, four gymnosperms, three pteridophytes, and one bryophyte also have been listed as 'threatened' in the State.

An analysis of the 255 species commercially utilized by the Ayush industry in Kerala indicates that 90 species are threatened and are being destructively harvested (Table 4).

Table 3 Major bioresources used by Ayurveda industries in Kerala

Sl. No	Botanical name	ApproxQuantity (MT)
1	*Citrus limon*	1545.8
2	*Cocos nucifera*	1207.9
3	*Phyllanthus emblica*	981.6
4	*Terminalia chebula*	913.7
5	*Tamarindus indica*	834.8
6	*Asparagus racemosus*	777.8
7	*Zingiber officinale*	702
8	*Sida cordifolia*	642.1
9	*Tinospora cordifolia*	632
10	*Aloe vera*	561.3
11	*Momordi cadioica*	533.6
12	*Aegle marmelos*	513.3
13	*Curcuma longa*	423.5
14	*Withania somnifera*	414.1
15	*Bacopa monnieri*	410.3
16	*Eclipta prostrata*	398.8
17	*Sidar hombifolia*	387.4
18	*Piper longum*	331.8
19	*Phyllanthus emblica*	315.5
20	*Vitis vinifera*	303.2
21	*Cedrus deodara*	302.4
22	*Azadirachta indica*	298.4
23	*Justicia beddomei*	288.1
24	*Solanum melongena*	285.7
25	*Boerhavia diffusa*	281.8
26	*Cyperus rotundus*	272.2
27	*Tribulus terrestris*	263.1
28	*Moringa oleifera*	256.9
29	*Ricinus communis*	254.8
30	*Nilgirianthus ciliatus*	236.9
31	*Cuminum cyminum*	229.2
32	*Terminalia bellericca*	225.5
33	*Cynodon dactylon*	219.3
34	*Premna serratifolia*	186.4
35	*Woodfordia fruticosa*	186.2
36	*Pseudarthria viscida*	183.6
37	*Crocus sativus*	174.3
38	*Pongamia pinnata*	174.3

(continued)

Table 3 (continued)

Sl. No	Botanical name	ApproxQuantity (MT)
39	*Tragiain volucrata*	168.9
40	*Hordeum vulgare*	164.5
41	*Gmelina arborea*	163.2
42	*Desmodium gangeticum*	162
43	*Allium cepa*	153.7
44	*Oroxy lumindicum*	152
45	*Stereospermum chelonoides*	152
46	*Aerva lanata*	151.3
47	*Wrightia tinctoria*	151
48	*Piper betle*	148.7
49	*Spermacoce hispida*	145.5
50	*Alpinia officinarum*	142.7
51	*Oldenlandia corymbosa*	140.3
52	*Piper nigrum*	138.1
53	*Acorus calamus*	137.9
54	*Operculina turpethum*	133.7
55	*Glycyrrhiza glabra*	133.4
56	*Coriandrum sativum*	131.2
57	*Indigofera tinctoria*	124.2
58	*Hedyotis corymbosa*	111
59	*Trichosanthes lobata*	108.6
60	*Chrysopogon zizanioides*	100.8

International and National Legislations for Access and Benefit-Sharing

The value of bioresources was acknowledged when the pharmaceutical giant Merck gained access to chemical extracts from Costa Rica's National Biodiversity Institute (INBio) in exchange for some $1.1 million and an unspecified percentage of future royalties (Blum, 2010) Another well-known case of bio-prospecting involves the case of rosy periwinkle of the island of Madagascar the source of alkaloids that helped in the cure of Hodgkin's disease and acute lymphoycticleukemia. This was used by Eli Lilly to produce two life-saving anti-cancer drugs, Vincristine and Vinblastine. In the absence of an adequate regulatory mechanism, although the company utilized the traditional knowledge benifitted from the sale of drugs, but it neither entered into nor honoured any benefit-sharing agreement nor did it share royalties arising from it with the Malagasy people (Torrance, 2000).

The Convention on Biological Diversity, an international treaty signed by 196 countries as an outcome of the Rio Erath Summit of 1992 made it compulsory to

Table 4 Threatened species of which roots, whole plant, rhizome, or all of them are extracted

Sl No	Botanical name	Part used	Threat status
1	*Aconitum ferox*	Tuberous root	EN
2	*Aconitum heterophyllum*	Tuberous root	CR
3	*Acorus calamus*	Rhizome	EN
4	*Justicia beddomei*	Roots, Leaves	CR
5	*Andrographis paniculata*	Whole plant	VU
6	*Aristolochia bracteolata*	Roots	VU
7	*Aristolochia indica*	Roots	VU
8	*Asparagus racemosus*	Tuberous root	EN
9	*Bacopa monnieri*	Whole plant	VU
10	*Chonemorpha fragrans*	Roots	EN
11	*Citrullus colocynthis*	Roots	VU
12	*Clerodendrum serratum*	Roots	EN
13	*Coptis teeta*	Tuberous root	EN
14	*Costus speciosus*	Tuberous root	VU
15	*Curcuma aromatica*	Rhizomes	VU
16	*Holostemma ada-kodien*	Tuberous root	EN
17	*Nardostachys jatamansi*	Roots	CR
18	*Nilgirianthus ciliatus*	Whole plant, Roots	VU
19	*Nymphaea nouchali*	Rhizomes	EN
20	*Ocimum gratissimum*	Roots	VU
21	*Operculina turpethum*	Roots	EN
22	*Oroxylum indicum*	Roots	EN
23	*Picrorhiza kurroa*	Roots	EN
24	*Plumbago zeylanica*	Tuberous root	VU
25	*Pongamia pinnata*	Roots, Kernel, Bark, Fruits	EN
26	*Pseudarthria viscida*	Roots	VU
27	*Rauwolfia serpentina*	Roots	EN
28	*Salacia oblonga*	Roots	VU
29	*Saussurea costus*	Roots	CR
30	*Valeriana wallichii*	Roots	VU

obtain Prior Informed Consents (PICs) from the party providing biological resources, to protect indigenous and traditional knowledge of the communities involved. The treaty was later modified to include transparent legal frameworks for equitable benefit sharing in the meeting held at Nagoya, Japan in 2010, a supplementary agreement commonly known as Nagoya Protocol. The protocol entered into force in 2014 and has been ratified by 96 parties. These protocols made it necessary to have Prior

Informed Consent (PIC) from the provider of bioresource before collecting the biological materials.

The Nagoya Protocol details the rights and obligations of countries of origin of genetic resources and associated traditional knowledge (provider countries) and countries in whose jurisdiction these genetic resources and associated traditional knowledge are used (user countries). The Nagoya Protocol also recognizes the rights of indigenous and local communities over genetic resources and associated traditional knowledge following the domestic laws of the provider countries. It makes it obligatory for parties to ensure that users of genetic resources and associated traditional knowledge (bioprospectors) comply with the domestic ABS regulatory requirements of the provider countries. This national obligation on bioprospectors requires them to not only comply with domestic ABS frameworks of provider countries but where required by such laws enter into ABS agreements with legitimate providers of such resources and knowledge.

Article 8(j) of the CBD establishes a crucial connection between the preservation of biological variety and the traditional knowledge, innovations, and practices of indigenous peoples and local communities. States are required to safeguard communities' rights to consent to non-traditional uses of their knowledge and to share in any benefits that may result from the research to preserve and market this type of traditional knowledge.

The Convention on Biological Diversity recognise that nations have ownership of, and can control access to, species ("genetic resources") within their boundaries; and mandates equitable sharing of the benefits derived from biodiversity. The convention specifies that research on the uses of biodiversity should be facilitated, with all contracting parties providing appropriate access to biodiversity (article 15) and for the transfer of technology (article 16). Nations "shall endeavour to create conditions to facilitate access to genetic resources for environmentally sound uses" and to "develop and carry out scientific research based on genetic resources" (article 15). The Kunming-Montreal Global Biodiversity Framework has 23 action-oriented global targets to be achieved over the decade to 2030. GOAL C has the vision to Share Benefits Fairly—The monetary and non-monetary benefits from the utilization of genetic resources and digital sequence information on genetic resources, and of traditional knowledge associated with genetic resources, as applicable, are shared fairly and equitably, including, as appropriate with indigenous peoples and local communities, and substantially increased by 2050, while ensuring traditional knowledge associated with genetic resources is appropriately protected, thereby contributing to the conservation and sustainable use of biodiversity, following internationally agreed access and benefit-sharing instruments.

The legal issues related to bio-prospecting have been brought to the forefront as developing nations rich in biological resources and traditional knowledge, have become the epicenter of research by the developed nations for obtaining patents and establishing a monopoly over these resources/knowledge. The World Intellectual Property Organization (WIPO) in 2001 established the Traditional Knowledge Task Force, whose report indicates that India is an important source country for bioprospecting. The Task Force looked up patents pertaining to classical knowledge systems in worldwide patent databases. Over 5000 patent references pertaining to 90 medicinal plants were found in the US Patent and Trademark Office alone as a result of the search. 80% of the patents relating to these 90 medicinal plants came from seven plants, all seven of which were of Indian origin (CSIR, 2001).

The Biological Diversity Act, of 2002 was enacted by the Indian parliament in response to India's commitments under the Convention on Biological Diversity (CBD) of 1992. The Act provides for the conservation of biodiversity, sustainable use of bioresources, and fair and equitable sharing of benefits of commercial utilization of bioresources through a three-tier structure consisting of National Biodiversity Authority (NBA), State Biodiversity Boards (SBB) and Biodiversity Management Committee (BMC) at the local level. As per the provisions of the Biological Diversity Act 2002 for access to traditional knowledge and Medicinal plants by companies and traders and for commercial utilization prior approval from the National Biodiversity Authority/State Biodiversity Boards, as the case may be has to be obtained. Before approvals are granted National Biodiversity Authority (NBA)/State Biodiversity Board (SBB) shall consult the Biodiversity Management Committee to seek their consent for access to bioresources and/or traditional knowledge from the local people. Further, the National Biodiversity Authority (NBA)/State Biodiversity Board (SBB) would grant approval in the form of an agreement that comprises the terms and conditions including the benefit sharing amount payable to NBA by the applicant. The benefit shared by the users of biodiversity shall be ploughed back to the beneficiaries viz individual/communities/BMCs, etc. who provided the resources/ traditional knowledge. The Guidelines on Access and Benefit Sharing propose that Indian entities commercially utilizing bioresources have to pay between 3 and 5% on the extracted bioresources or between 0.01 and 0.05% on the annual gross ex-factory sales as part of access benefit sharing. Traders and manufacturers of herbs and herbal products must pay either between 1 and 5% of the purchase price or between 0.1 and 0.5% of the sale price, depending on the scale of their commercial operations.

In India, the recent amendments to the Biological Diversity Act 2002 bring in issues relating to the applicability of ABS to medicinal plant resources. The amendments have great significance as they influence research across sectors based on bioresources such as pharmaceuticals, agriculture, cosmetics, and traditional medicine, etc. The process for obtaining Intellectual Property Rights (IPR) on biological resources and associated traditional knowledge has been described as per Section 6. Foreign entities must obtain prior approval to secure IPR on Indian biological resources, whereas Indian entities need only register with the National Biodiversity Authority (NBA). NBA's prior approval is needed, if these entities intend to commercialise the IPR, All except Indian citizens and entities residing in India will have to seek prior approval from the NBA to access biological materials of Indian origin deposited in the repositories before the grant of a patent. This may prevent biopiracy and safeguard our bioresources both within and outside the country whereas a new section 36A authorises the NBA to monitor and regulate the access and utilisation of biological resources obtained from foreign countries.

As per Section 7 of the Act, commercial utilization of bioresources shall be with prior approval of SBB and is subject to terms and conditions as set by SBB including sharing of benefits with the local community. The Amendments—Section 7 has exempted codified traditional knowledge, cultivated medicinal plants and their products, local people and communities of the area, vaidyas, and registered AYUSH practitioners practicing indigenous medicines, including Indian systems of medicine from the purview of the Act. In the case of cultivated medicinal plants, the exemption shall be available only if a certificate of origin is obtained from the Biodiversity Management Committee. The codified traditional knowledge" has been defined as the knowledge derived from authoritative books specified in the First Schedule to the Drugs and Cosmetics Act, 1940. The first schedule comprises nearly 100 books listed under Ayurveda, Siddha, and Unani systems. There are speculations that all traditional knowledge and practices associated thereto, which are either in the digital format available in the Traditional Knowledge Digital Library (TKDL) or in the Peoples Biodiversity Register (PBR), will now be exempted from the purview of ABS. But as per the Amendments 2023, only the codified traditional knowledge is exempted under section 7(1) and not the traditional knowledge per se. It has been a long demand of the Ayush companies of Kerala to exempt classical medicines and bring only patented products under the purview of the Act. Hence this amendment may enable to bring the Ayush companies under the purview of the Act. Value-added products and derivatives of biological resources have been exempted (Section 2(3)). India is a major destination for spice industries, oleoresins, and marine products such as fish oil, squalene, etc. Section 40 has expanded the exemptions for Normally Traded as Commodities (NTAC) to include derivatives of NTAC bioresources, agricultural wastes, and cultivated medicinal plants, but this is limited to Indian entities. Section 55 of the Principal Act was amended and the Bill aimed to decriminalize offenses under the Act and replace imprisonment with penalties ranging from Rs. 100,000 to Rs. 5 million. For continuous violations, an additional penalty of up to ₹10 million could apply.

Challenges in Bioprospecting

The loss of natural product materials used in traditional medicines, the generational loss of traditional medicine expertise, and the absence of a legal framework to conduct bioprospecting are the three biggest hazards associated with adopting ethnopharmacology in the drug discovery process. When a species disappears, traditional medicine practitioners can look for replacements, as more than half of the therapeutic plants used are harvested from the wild. However, using alternative natural products obscures a loss that is not immediately apparent in terms of the identification of new biologically active compounds—no evidence suggests that the alternative natural product used in traditional medicine applications contains the same biologically active compound. It has been estimated that the drug discovery process takes more than 10 years (Reichert, 2003) and costs more than 800 million dollars (Dickson & Gagnon, 2004). It has been estimated that only one out of every 5000 lead compounds progress through clinical studies and receive approval for usage. Small amounts of naturally occurring products are usually isolated, which is insufficient for lead development, lead optimization, and clinical trials.

Bioprospecting is also affected by the sustainability of medicinal plant resources, intricate trade networks, problems with traceability, the focus of academic research and publications, lack of harmonized system codes (HS Codes) for the export of Indian medicinal plants, etc. Import and export restrictions for specific products in India are established via the Indian Trade Classification (ITC) developed in accordance with the internationally standardized tariff nomenclature, known as the Harmonized System of Coding (HS), developed and maintained by the World Customs Organization (WCO), an independent intergovernmental organizaon based in Brussels, Belgium, with over 200 member countries. The export of Ayurvedic products takes place under the Indian Trade Classification based on Harmonized System of Coding (ITCHS) through two HS codes-30039011 for medicants and 30049011 for medicaments. The export of products of Unani, Siddha, and Homoeopathic systems takes place through HS codes-30039012, 30039013, and 30039014 (mendicants) and 30049012, 30049013, and 30049014 (medicaments) respectively (Ayush Export Promotion Council). Limitations exists of the ITC (HS) coding system in providing species-wise detail in foreign trade. Most of the products of Ayurveda, Homoeopathic, Siddha, and Unani systems, Herbal Products, and Medicinal Plant products are not identified under specific HS Codes (International Harmonised Commodity Description and Coding System).

The existing trade classification system tends to club many medicinal plants in trade under the head 'Others'. As such, this 8-digit ITC (HS) Code is limited in its applicability to 400-odd medicinal plant species that are under foreign trade to and from India. This inability of the existing ITC (HS) Codes to provide medicinal plant-wise data on exports/imports make it difficult to develop strategies for the development and management of medicinal plant resources in the country. It also brings in problems related to "intellectual property rights." Starting from the sourcing of raw materials, sharing along the chain of custody, the working conditions that exist between producers and harvesters, and the fairness of supply agreements. Medicinal plant trade raises issues of equity, the way benefits are distributed along the supply chain. The supply chain extends to 3–4 tiers without much value addition but with an increase in sale price at each level.

Prospects

India has a 3–5% market share in the global biotechnology industry and it ranks 12th as reported by the India Bioeconomy Report 2023, by the Association of Biotechnology Led Enterprises (ABLE). The paper states that in 2022, India's bioeconomy grew significantly, rising by 29% to a significant value of $137.24 billion. With a 59% contribution to the bioeconomy, the bio-industrial sector has been the leading player, followed by biopharma with a 49% share.

Most academic research results are at the stage of laboratory prototype or preclinical trials and require substantial further research and development before they can be turned into a commercially viable product. In India, not more than 5% of patents reach the market, and two-thirds of the research originates from Government labs. Experiences elsewhere have shown that the model has worked well when the research community is engaged with startups to commercialize the product. The report also notes that a total of 13,470 startups are anticipated for 2025 and 35,000 cumulative startups for 2030. In 2022, India's entrepreneurial scene experienced rapid expansion, with startups becoming an important force behind economic growth. India is ranked third in Asia-Pacific and 12th globally in terms of biotechnology, with a 3–5% market share in the biotech sector. Several initiatives like Make in India, Start-up India, Digital India, and Skill India have promoted technological innovations in all fields.

In the next five years, the Anusandhan National Research Foundation (NRF), will promote research and development and establish India as a leader in the field worldwide. A thorough inventory of India's biodiversity, with the possibility of cataloguing and mapping all living forms there along with related cultural and traditional practices, will be part of the National Mission for Biodiversity and Human Well-Being. Other objectives include evaluating the distribution and conservation status of India's biodiversity and establishing a thriving biodiversity-based economy. The natural services associated with biodiversity provided by India's forests alone are valued at ₹128 trillion/year. The Mission proposes a national effort to link biodiversity with peoples' economic prosperity and the effective use of bio-resources for renewable and sustainable livelihoods for people.

Since natural selection has been influencing natural products for billions of years, nature has produced structurally diverse chemicals that are nearly hard to create by synthetic chemistry methods. However, as structural chemistry has advanced, more complicated molecules—like taxol and cyclic depsipeptide—are currently being created in labs across the globe. One strategy that combines elements of combinatorial chemistry and bioprospecting is to build a systematic, annotated, and curated computational database of the structures of purified natural products, rather than screening the natural product extracts and purified compounds for different bioactivities. There are many natural product libraries available, which offer powerful resources for online natural product screening procedures. One of the earliest works is Hortus Malabaricus commissioned by Hendrik van Rheede, ex-Governor of Dutch Malabar, a compilation of around 700 indigenous plants with medicinal properties in Kerala. One such database is InterBioscreen (IBS, https://www.ibscreen.com/), perhaps the most extensive database of natural products with more than 50,000 accessions freely available to anyone. Other databases of natural products include Super Natural (http://bioinformatics.charite.de/supernatural/) and the Drug Discovery Portal of the University of Strathclyde, UK (www.ddp.strath.ac.uk). These databases can be searched to find potential structural matches, for the biological pathways—a process known as virtual or in silico screening through molecular docking. For example, using the IBS, investors Singh and Bast, 2015 were able to identify natural products Myricetin and Epigallocatechol- two compounds from green tea- as potential inhibitors for a crucial pathway in the progression of many cancers.

In 2000, the TKDL was set up by the Government of India to digitalise traditional knowledge such as Ayurveda, Unani, Siddha, and Yoga. The database curates details about the formulations, the plants, their quantities (composition is outlined in terms of grams or milliliters), their botanical identities, their use, the method of preparation, and the sources that documented the formulation. The Council of Scientific & Industrial Research has created the Traditional Knowledge Digital Library (http://www.tkdl.res.in/) with details of approximately 250,000 formulations for global patent offices to use as a reference tool for determining prior art. The importance of this resource has increased due to the growing utilization of traditional medicines and natural products in drug discovery. Additionally, the Indian Council of Medical Research has curated resources on the historical use of medicinal plants and sought to verify the effectiveness of numerous traditional Indian medicines. The production of synthetic libraries can benefit greatly from the use of natural materials as templates. Research in this area is typically informed by pre-existing indigenous knowledge that may be publicly accessible.

ABS agreements incorporating milestone payments or royalties, in the field of bioprospecting is difficult due to the extremely low success rate for drug development. In addition, the time lag for royalties is lengthy, taking anywhere from 10 to 12 years from discovery to benefits being received. Thus when the source country provides biological materials and the developed country undertakes the R&D, very little benefits are provided to the local community. Exporting raw or dry bioresources adds little value to the resources.

Conclusions

Bioprospecting can provide benefits only if research and development of novel products occur in the source country whereby resources, having low value are converted to products with the potential to yield much higher value to the source country. The nonmonetary benefits include jobs, training, investment in infrastructure, and the likelihood that scientific discoveries will lead to new investments in research and development. Developing countries like India should build their technical capability, add value to the resources, and create opportunities for benefit sharing with local communities, who have conserved these resources.

It is necessary to design a future ABS strategy that recognizes the results of research and promotes commercialization of the same at the same time ensuring sustainability of the resources. The "ABS potential" of the country needs to be studied and understood in a better manner.

Conflict of Interest The authors have no conflict of interest to declare.

References

Agarwal, M. K. (2013). The Vedic core of human history: And truth will be the Savior. ISBN:491715952, 9781491715956iUniverse (Publishers).

Alamgir, A. (2017). Therapeutic use of medicinal plants and their extracts: Volume 1. Pharmacognosy. https://doi.org/10.1007/978-3-319-63862-1

Bast, F. (2019). Bioprospecting. In Biodiversity and conservation by Dr Sunil Mittal. Published by INFLIBNET CENTER.

Blum, E. (2010). *Making biodiversity conservation profitable: A case study of the Merck/INBio Agreement*. https://doi.org/10.1080/00139157.1993.9929092

Briskin, D. P. (2000) Medicinal plants and phytomedicines. Linking plant biochemistry and physiology to human health. *Plant Physiology,* 124, 507–514. https://doi.org/10.1104/pp.124.2.507, PMID 1102770123.

Council of Scientific & Industrial Research. (2001). CSIR News 51 (5&6: 1–3).

Dickson, M., & Gagnon, J. P. (2004). Key factors in the rising cost of new drug discovery and development. *Nature Reviews Drug Discovery, 3*(5), 417–429.

India BioEconomy Report. (2020). Biotechnology Industry Research Assistance Council, New Delhi.

He, S.-A. (1998). Utilization and conservation of medicinal plants in China with special reference to Atractylodeslancea. In *Medicinal plants: Their role in health and biodiversity by Timothy R. Tomlinson and OlayiwolaAkerele* (pp. 161–168). https://www.jstor.org/stable/j.ctt15vt8s1

IUCN. (2006). International Union for Conservation of Nature, Natural Resources. Ecosystems, & Livelihoods Group. Conserving medicinal species: securing a healthy future.

The IUCN red list of threatened species. Version 2020–1. https://www.iucnredlist.org

Kelly, K. (2009). *History of medicine* (pp. 29–50). Facts on File.

Khan, H. (2014). Medicinal plants in light of history: Recognized therapeutic modality. *Journal of Evidence-Based Complementary & Alternative Medicine, 19*, 216–219. https://doi.org/10.1177/2156587214533346, PMID 24789912

Kong, J. M., Goh, N. K., Chia, L. S., & Chia, T. F. (2003). Recent advances in traditional plant drugs and orchids. *Acta Pharmacologica Sinica, 24*, 7–21. PMID 12511224.

Kumar, A., Pandey, V. C., Singh, A. G., & Tewari, D. D. (2013). Traditional uses of medicinal plants for dermatological healthcare management practices by the Tharu tribal community of Uttar Pradesh, India. *Genetic Resources and Crop Evolution, 60*, 203–224. https://doi.org/10.1007/s10722-012-9826-6

Luyt, R. P., Jäger, A. K., & Van Staden, J. (1999). The rational usage of DrimiarobustaBak. in traditional medicine. South African Journal of Botany, *65*(4), 291–294.

Nasim, N., Sandeep, I. S, & Mohanty, S. (2022). Plant-derived natural products for drug discovery: Current approaches and prospects. *Nucleus (Calcutta), 65*(3), 399–411. https://doi.org/10.1007/s13237-022-00405-3. Epub Oct 18. PMID: 36276225; PMCID: PMC9579558.

Newman, D. J., Cragg, G. M, & Snader, K. M. (2002). The influence of natural products upon drug discovery. *Natural Product Reports, 17*, 215–234. https://doi.org/10.1039/a902202c, PMID 10888010

Newman, D. J., Cragg, G. M., & Snader, K. M. (2003). Natural products as sources of new drugs over the period 1981–2002. *Journal of Natural Products, 66*(7), 1022–1037.

Nelliyat, P., & Preetha, N. (2022). *Tradable/commercially potential bioresources of Kerala and economic valuation in Kerala.* Kerala State Biodiversity Board

Petrovska, B. B. (2012). Historical study of medicinal plants' usage. *Pharmacognosy Reviews, 6*(11), 1.

Priya, R., & Kurian, C. M. (2018). Regulating access and protecting traditional health knowledge through intellectual property rights? Issues from a holistic health systems perspective. *Science, Technology and Society, 23*(3), 504–529.

Rajasekaran, S., Murugan, K., & Preetha, N. (2022). *Tradable/commercially potential floral bioresources of Kerala.* Kerala State Biodiversity Board.

Reichert, J. M. (2003). Trends in development and approval times for new therapeutics in the United States. *Nature Reviews Drug Discovery, 2*(9), 695–702.
Roberson, E. (2008). *Nature's pharmacy, our treasure chest: Why we must conserve our natural heritage* (A Native Plant Conservation Campaign Report). 16 p. https://www.biologicaldiversity.org/publications/papers/Medicinal_Plants_042008_lores.pdf
Robinson, D. F. (2012). Biopiracy and the innovations of Indigenous peoples and local communities. In *Indigenous people's innovation: Intellectual property pathways to development* (pp. 77–93).
Saklani, A., & Kutty, S. K. (2008). Plant-derived compounds in clinical trials. *Drug Discovery Today, 13*, 161–171. https://doi.org/10.1016/j.drudis.2007.10.010, PMID1827591421.
Torrance, A. W. (2000). Bioprospecting and the convention on biological diversity. *Harvard Law School.* https://dash.harvard.edu/bitstream/handle/1/8965586/Torrance,_Andrew_00.html?sequence=2.
Toplak and Galle. K. (2005) *Herbal remedies used in domestic medicine* (pp. 60–61). Mozaic Book.
Willison, K. D., & Andrews, G. J. (2004). Complementary medicine and older people: Past research and future directions. *Complementary Therapies in Nursing and Midwifery, 10*, 80–91. https://doi.org/10.1016/S1353-6117(03)00106-9, PMID 15135760.
WHO Essential Medicines and Health Products Information Portal, Trips, CBD and Traditional Medicines: Concepts and Questions, Report of an ASEAN Workshop on the TRIPS Agreement and Traditional Medicine (Jakarta, February 2001).
WHO Traditional Medicine Strategy 2002–2005 (WHO). (2002); Timmermans, K. (2003). Intellectual property rights and traditional medicine: Policy dilemmas at the interface. *Social Science and Medicine, 57*(4), 745.

Bioprospecting Nutraceuticals for Human Welfare

Mangala Hegde, Anamika Mishra, Ajaikumar B. Kunnumakkara, and Bharat B. Aggarwal

Introduction

The term "nutraceutical" is derived from the fusion of "nutrient" and "pharmaceuticals," initially coined by Stephen De Felice, Founder of the Foundation for Innovation in Medicine, in 1989 (Brower, 1998; Maddi et al., 2007; Vishvakarma et al., 2023). De Felice defined nutraceuticals as "foods or food components that provide medical or health benefits, including the prevention and treatment of diseases" (Bull et al., 2018; Kaur et al., 2019; Press C, 2002; Vishvakarma et al., 2023). Nutraceuticals represent dietary supplements containing bioactive substances or delivering bioactive agents, often surpassing normal food dosages, thereby enhancing health (Espin et al., 2007; Sauer & Plauth, 2017). Originating from various sectors such as the food industry, herbal and dietary supplement market, pharmaceutical industry, and agribusiness/nutrition conglomerates, nutraceuticals are subject to distinct regulatory frameworks by the Food and Drug Administration (FDA) compared to conventional medicines (Espin et al., 2007). The distinction between nutraceuticals and functional foods is not always clear-cut (Espin et al., 2007). The principal categories of nutraceuticals encompass herbal and natural products, dietary supplements, and functional foods (Das et al., 2012; Espin et al., 2007; Shinde et al., 2014). Globally, there is a growing demand for nutraceuticals and herbal extracts, with prominent markets in the USA, UK, Japan, and India (Das et al., 2012; Shinde et al., 2014). Examples of widely utilized and clinically successful nutraceuticals include green tea for weight

M. Hegde · A. Mishra · A. B. Kunnumakkara (✉)
Cancer Biology Laboratory, Department of Biosciences and Bioengineering,
Indian Institute of Technology Guwahati (IITG), Guwahati 781039, Assam, India
e-mail: kunnumakkara@iitg.ac.in

B. B. Aggarwal (✉)
Inflammation Research Center, San Diego, CA, USA
e-mail: bbaggarwal@gmail.com

L. Krishna Panicker et al. (eds.), *Biodiversity and Business*,
https://doi.org/10.1007/978-3-031-71674-4_16

loss and managing lifestyle-related disorders, turmeric and curcumin-based therapies for preventing and treating chronic diseases, as well as nutraceuticals targeted at combating Coronavirus infections and enhancing cognitive functions, such as *Ginkgo biloba* (Banik et al., 2021; Chen IJ et al., 2016; DeKosky et al., 2008; Jurgens et al., 2012; Hegde et al., 2023a; Kunnumakkara et al., 2023; Li et al., 2017; Shinde et al., 2014; Snitz et al., 2009). Additionally, there is increasing global consumption of vitamins, minerals, and glucosamine for arthritis management (Juturu & Kremer, 2010; Shinde et al., 2014). Nutraceuticals are available in various forms such as pills, tablets, and extracts, containing common phytochemicals like anthocyanins, coumarins, ellagic acid (EA), flavonols, and stilbenes (Espin et al., 2007; Sauer & Plauth, 2017). Nutraceuticals exhibit therapeutic efficacy across a spectrum of human ailments including arthritis, autoimmune diseases, cardiovascular diseases (CVDs), diabetes, neurological diseases, and obesity, owing to their natural origins and recognized safety for human consumption (Das et al., 2012; Carrizzo et al., 2020; Dutta et al., 2018; Garodia et al., 2023; Hegde et al., 2023a, 2023b; Kunnumakkara et al., 2023; Muralimanoharan et al., 2009; Sajeev et al., 2022; Sailo et al., 2024a, 2024b; Unnikrishnan et al., 2024). Extensive preclinical and clinical studies have demonstrated the safety and efficacy of nutraceuticals, emphasizing their potential incorporation into foods and diets for health promotion and disease prevention and treatment (AlAli et al., 2021; Gupta et al., 2021; Hegde et al., 2023a; Kunnumakkara et al., 2023).

Nutraceuticals: General Concept and Categories

Nutraceuticals, traditionally defined as health-beneficial natural products encompassing non-nutrient secondary metabolites, such as polyphenols and flavonoids derived from plants (Shinde et al., 2014). Presently, nutraceuticals are understood as any nature-derived compounds conferring health benefits and suitable for inclusion in the diet. This expanded definition encompasses various categories:

1. Micronutrients: Vitamins and minerals
2. Dietary supplements, including botanicals: Examples include Coenzyme Q, carnitine, Gingsing, *Gingko biloba*, Saint John's Wort, and Saw Palmetto.
3. Functional foods: Probiotics, prebiotics, foods fortified with stanols, oats, and omega-3 milk.
4. Secondary metabolites and other products: Polyphenols, flavonoids, terpenoids, glucosilonates, etc.
5. Medicinal foods: Health bars, lactoferrin for immune enhancement, transgenic plants for oral vaccination against infectious diseases (Shinde et al., 2014) (Fig. 1).

In the pharmaceutical industry, nutraceuticals are utilized to enhance physiological responses and are formulated into capsules, gels, granulates, liquids, pills, powders, ointments, solutions, and syrups (Sauer & Plauth, 2017). These formulations often

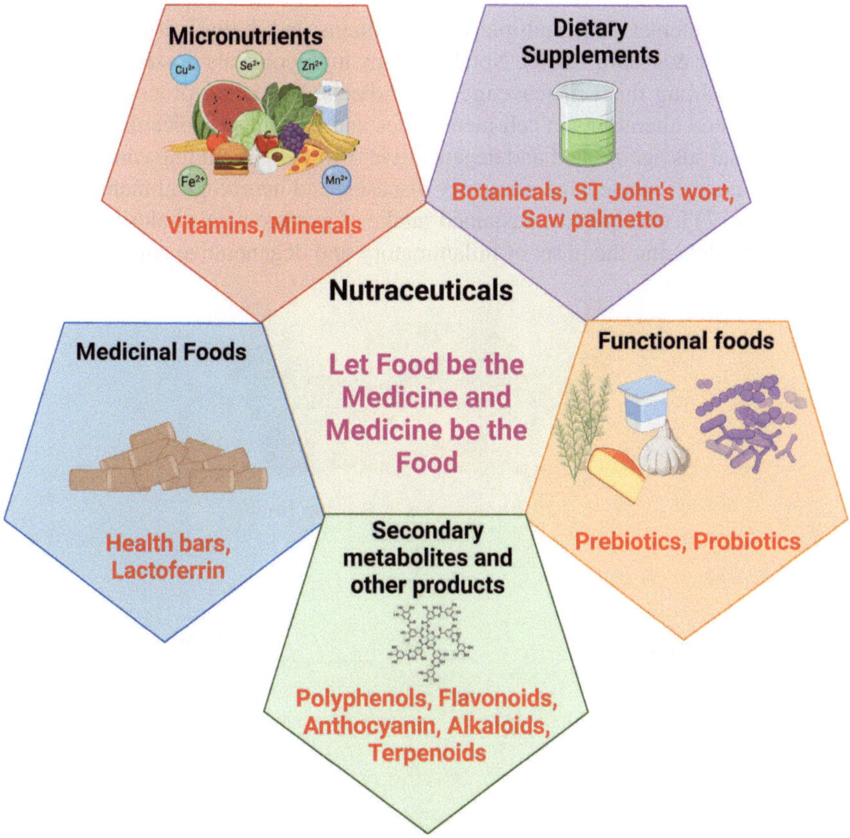

Fig. 1 Nutraceuticals belong to various categories that are consumed as food and considered generally safe

contain extracts, enriched substances, or isolated compounds sourced from natural reservoirs. For instance, while curcumin naturally occurs in turmeric, commercially available capsules like Meriva, Theracurmin, and Curcuwin have demonstrated superior bioavailability compared to pure curcumin (Hegde et al., 2023a). Similarly, resveratrol, typically found in red wine in limited quantities, is commonly provided in 500 mg capsules derived from *Vitis vinifera* extract (Augustin et al., 2013; Munin & Edwards-Levy, 2011, Espin et al., 2007).

Despite the diversity of diets, the secondary metabolites or other molecules in food generally do not exhibit adverse effects. Most dietary nutraceuticals are reported to be safe even at higher concentrations (Das et al., 2012; Hegde et al., 2023a; Kunnumakkara et al., 2023). Regulatory bodies such as the food and drug regulatory agencies in the USA classify several nutraceuticals as "generally recognized as safe (GRAS)" and establish "no-observed-adverse-effect levels (NOAELs)" (Das et al., 2012; Sauer & Plauth, 2017). Upon distribution within the human body, bioavailable nutraceuticals interact with multiple target proteins, exhibiting various biological activities (Sauer & Plauth, 2017). Notably, many nutraceuticals possess antioxidant properties, enabling them to scavenge free radicals. Oxidation processes and free radical-mediated alterations in cell membranes and biological molecules are implicated in numerous age-related and degenerative disorders, including cardiovascular diseases, cancers, diabetes, inflammatory diseases, and neurological manifestations (Table 1, Fig. 2) Consequently, sustained intake of nutraceuticals holds promise for mitigating or delaying the onset of inflammatory and degenerative conditions.

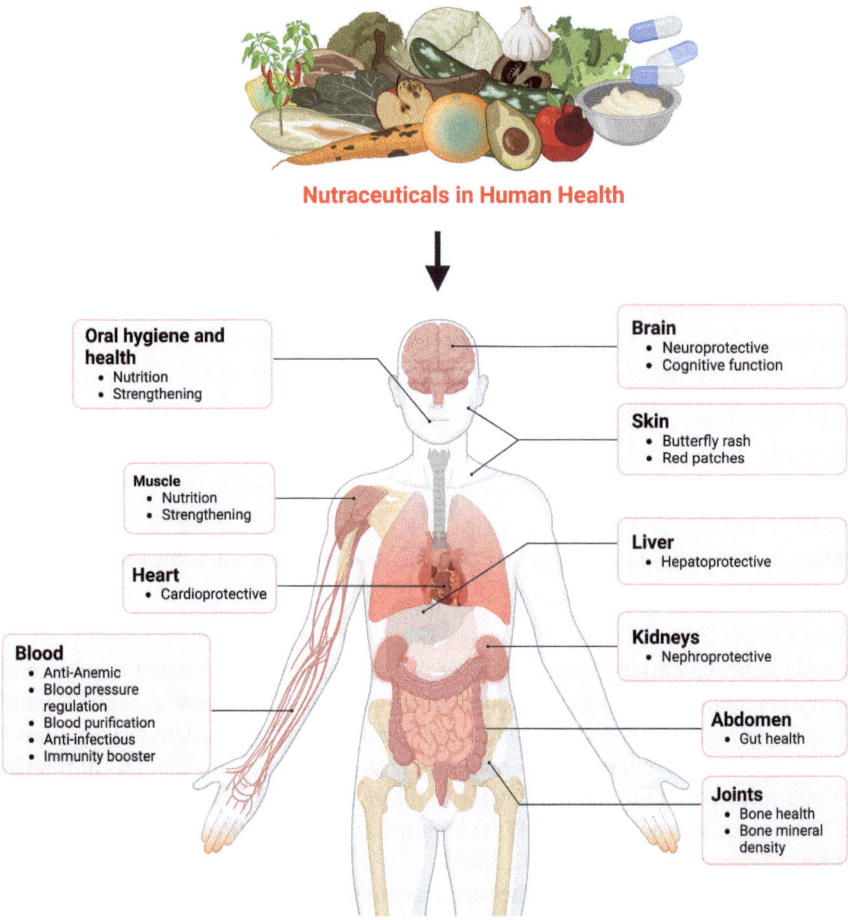

Fig. 2 Nutraceuticals and their health benefits and organ protective effects are shown

Table 1 Major nutraceuticals and their therapeutic targets: Observations from clinical trials

Nutraceuticals	Effects	Targets	References
Curcumin, Curcuminoids and formulations	Antioxidant, anti-inflammatory, anticancer, antiproliferative, antiaging, proapoptotic, epigenetic modulation, hepatoprotective, neuroprotective, nephroprotective, cardioprotective	NF-κB, TNF-α, p53, GSH, VEGF, COX-2, MDA, hs-CRP, TLR4, IL-6, IL-8, CK, IL-10, IL-2, IL-17, RORγt, TGF-β, FOXP3, Bcl-2, Bax, ALDH, IL-1, STAT3, PSA, CA-19-9, CEA, MCP-1, LDL, HDL, MMP-2, MMP-9, SOD, GPx, IFN, AST, ALT, GGT, Nrf-2, CAT, Ki-67, IAPP, GSK-3β, miR-16, miR-17–92, miR-21, miR-27, miR-29b, miR-126, miR-128, miR-132, miR-141, miR-146a, miR-155, miR-200, miR-326, miR-550, miR-340, miR-15a, miR-19b, miR-106b, miR-320a, miR-363, miR-31, miR-181c, miR-150, miR-599	Alwi et al. (2008), Dhillon et al. (2008), Dominiak et al. (2010), He et al. (2011), Panahi et al. (2014), James et al. (2015), Dolati et al. (2018), Falgiano et al. (2018), Greil et al. (2018), McAllister et al. (2018), Hajialilo et al. (2019), Hatab et al. (2019), Jazayeri-Tehrani et al. (2019), Rajuddin et al. (2019), Tanabe et al. (2019), Ahmadi et al. (2020), Mogharrabi et al. (2020), Thota et al. (2020), Amalraj et al. (2021), Briata et al. (2021), Farshbaf-Khalili et al. (2021), Helli et al. (2021), Silalahi et al. (2021), Tahmasebi et al. (2021), Farshbaf-Khalili et al. (2022), Karimi et al. (2022), Miranda-Castro et al. (2022) and Suryawati et al., (2022)
Resveratrol	Antioxidant, anti-inflammatory, anticancer, antiproliferative, modulation of lipid metabolism	NF-κB, TNF-α, MMP-2, MMP-9, MMP-10, IL-6 IL-4, FGF-2, IL-12P40, IL-12P70, RANTES, Aβ42, IL-1β, IL-18, miR-34a, miR-375, miR-21, miR-192, miR-126, miR-132, ATF4, ATF6, CHOP1, GRP78, XBP1, LH, FSH, CK, AMPK, SIRT1, PGC-1α, osteocalcin, BAP, TRACP-5b, IL-8, galectin 3, xanthine oxidase, procollagen type I C peptide, SIRT3, SOD, CRP, ACE-2	Chen J et al. (2016), Moussa et al. (2017), Beijers et al. (2020), Brenjian et al. (2020), Kawamura et al. (2020), de Ligt et al. (2021), Mansour et al. (2021), Mahjabeen et al. (2022), Corbi et al. (2023), Garcia-Martinez et al. (2023), Hassan et al. (2023), Nikniaz et al. (2023), Zheng et al. (2023) and Zhou et al. (2023)
Quercetin	Antioxidant, anti-inflammatory, anticancer, epigenetic modulation	TNF-α, IL-6, IGF-1, IGF-2, LH	Sgro et al., (2021) and Vaez et al. (2023)

(continued)

Table 1 (continued)

Nutraceuticals	Effects	Targets	References
Epigallocatechin 3-O-gallate	Antioxidant, proapoptotic, antiproliferative, anti-inflammatory, anticancer, epigenetic modulation	NF-κB, FOXO1, SIRT1, CAT, FABP1, GSTA2, ACSL1, CPT2, ACC1, FAS, SEB-1	Yoon et al. (2013) and Yuan et al. (2020)
PUFA	Antioxidant, anti-inflammatory	NF-κB, TNF-α, IL-6, IL-1β, IL-10, CD68, hs-CRP, calgranulin, BDNF, MCP-1, IL-1RA, SAA, IFN-γ, IL-8, CD14, Adiponectin, leptin, adipoR2, ICAM-1, IL-2R, MMP-9	Dangardt et al. (2010), Kiecolt-Glaser et al. (2011), Guebre-Egziabher et al. (2013), Ferguson et al. (2014), Shah et al. (2017), Smesny et al. (2017), Rodriguez-Cruz et al. (2018), Awoyemi et al. (2019), Castro Dos Santos et al. (2020), Tingo et al. (2022) and Chang et al. (2023)
Genistein	Antioxidant, anti-inflammatory	IL-6, TNF-α, c-Terminal telopeptide, osteocalcin, bone-ALP, HCF-2, MEK, MMP-2, sclerostin, adiponectin	Squadrito et al. (2013), Bilir et al. (2017), Amanat et al. (2018), Zhang et al. (2019) and Squadrito et al. (2023)
Ellagic acid	Antioxidant, anti-inflammatory	TNF-α, IL-6, CRP, BDNF, serotonin, IFN-γ, IDO	Ghadimi et al. (2021), Kazemi et al. (2021) and Hajiluian et al. (2023)
Vitamin supplementation	Antioxidant, anti-inflammatory, anti-neurodegenerative, anti-cancer	CRP, thrombomodulin, 15-HPGD, COX-2, IL-18, TNF-α, IL-6, IFN-γ, CXCL-10, CXCL-12, CCL-2, CCL-5, CCL-11, PF-4, Arginase I	Fowler et al. (2019), Ostadmohammadi et al. (2019), Gibbs et al. (2021), Johny et al. (2022), Chen H et al. (2016), Scott et al. (2017) and Abiri et al. (2022)
Spirulina	Anti-inflammatory	AST, ALT, MST, FST, IGF-1, CRP, pentraxin-3	Bagheri et al. (2022), Tamtaji et al. (2023) and Moradi et al. (2024)

Nutraceuticals for Human Welfare: Evidence from Clinical Trials

Nutraceuticals have been reported to show beneficial effects on humans in treating various chronic ailments. This section will summarize the various nutraceuticals tested clinically and shown beneficial effects (Fig. 3). Tables 1 and 2 summarize the clinical trials and reported targets of multiple nutraceuticals (Abiri et al., 2022; Ahmadi et al., 2020; Alwi et al., 2008; Amalraj et al., 2021; Amanat et al., 2018; Awoyemi et al., 2019; Bagheri et al., 2022; Beijers et al., 2020; Bilir et al., 2017; Brenjian et al., 2020; Briata et al., 2021; Castro Dos Santos et al., 2020; Chang et al., 2023; Chen H et al., 2016; Chen J et al., 2016; Corbi et al., 2023; Dangardt et al., 2010; de Ligt et al., 2021; Dhillon et al., 2008; Dolati et al., 2018; Dominiak et al., 2010; Falgiano et al., 2018; Farshbaf-Khalili et al., 2021; Farshbaf-Khalili et al., 2022; Ferguson et al., 2014; Fowler et al., 2019; Garcia-Martinez et al., 2023; Ghadimi et al., 2021; Gibbs et al., 2021; Greil et al., 2018; Guebre-Egziabher et al., 2013; Hajialilo et al., 2019; Hajiluian et al., 2023; Hassan et al., 2023; Hatab et al., 2019; He et al., 2011; Helli et al., 2021; James et al., 2015; Jazayeri-Tehrani et al., 2019; Johny et al., 2022; Karimi et al., 2022; Kawamura et al., 2020; Kazemi et al., 2021; Kiecolt-Glaser et al., 2011; Mahjabeen et al., 2022; Mansour et al., 2021; McAllister et al., 2018; Miranda-Castro et al., 2022; Mogharrabi et al., 2020; Moradi et al., 2024; Moussa et al., 2017; Nikniaz et al., 2023; Ostadmohammadi et al., 2019; Panahi et al., 2014; Rajuddin et al., 2019; Rodriguez-Cruz et al., 2018; Scott et al., 2017; Sgro et al., 2021; Shah et al., 2017; Silalahi et al., 2021; Smesny et al., 2017; Squadrito et al., 2013; Squadrito et al., 2023; Suryawati et al., 2022; Tahmasebi et al., 2021; Tamtaji et al., 2023; Tanabe et al., 2019; Thota et al., 2020; Tingo et al., 2022; Vaez et al., 2023; Yoon et al., 2013; Yuan et al., 2020; Zhang et al., 2019; Zheng et al., 2023; Zhou et al., 2023).

Nutraceuticals in the Prevention and Treatment of Infectious Diseases

Nutraceuticals demonstrate promise in preventing and alleviating infectious diseases, effectively reducing their impact and providing therapeutic benefits. This section outlines the potential of nutraceuticals in addressing diverse bacterial, viral, and other infections.

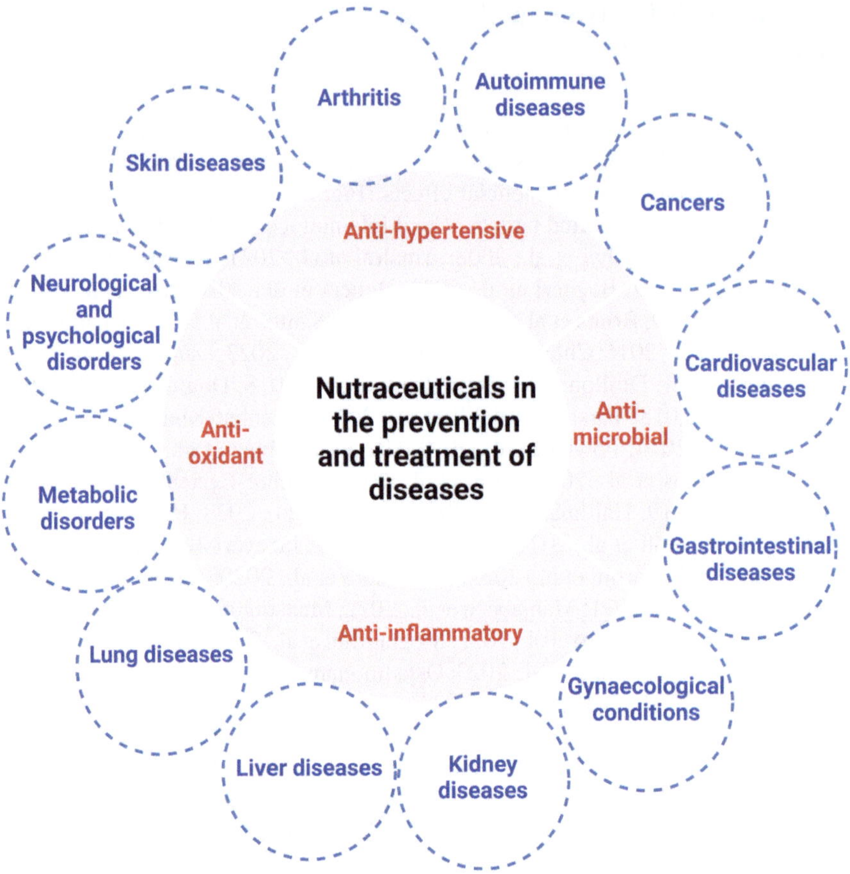

Fig. 3 Nutraceuticals are beneficial in treating various human ailments

Nutraceuticals in the Prevention and Treatment of Viral Diseases

In the absence of specific pharmacological interventions for emerging viral diseases, bolstering the immune system through nutritional interventions assumes paramount importance (Costagliola et al., 2021; Singh et al., 2021). Given the COVID-19 pandemic and the dearth of targeted antiviral medications, symptomatic support remains the primary therapeutic approach (Costagliola et al., 2021; Singh et al., 2021). Nutraceuticals, comprising compounds like resveratrol, quercetin, curcumin, EGCG, N-acetyl cysteine (NAC), and palmitoylethanolamide (PEA), exhibit antiviral, anti-inflammatory, and immunomodulatory properties (Costagliola et al., 2021; Singh et al., 2021). These attributes are particularly relevant in combatting viral infections, including coronaviruse. The antiviral efficacy

Table 2 Registered clinical trials in NIH (https://clinicaltrials.gov/) on nutraceuticals (as accessed on 11th April 2024)

Diseases	Status	Phase	Study design	Reg. No
Aging	Completed	II	Randomized	NCT01963767
	Completed	I	Randomized	NCT01179789
Aging/Osteoporosis/ Bonemetabolism	Recruiting	NA	Randomized	NCT04669782
Allergic Rhinoconjunctivitis	Completed	NA	Randomized	NCT03365648
Alzheimer's disease	Completed	I/II	Randomized	NCT00580931
	Completed	II	Randomized, Crossover	NCT01320527
	Suspended	II	Randomized, Crossover	NCT02033941
Appetite|Post-prandial Glucose response	Withdrawn	NA	Randomized, Crossover	NCT02111486
Atherosclerosis/ Inflammation/ Hypercholesterolemia	Completed	IV	Randomized	NCT02422927
Hypercholesterolemia in HIV-infected patients	Completed	IV	Randomized, Crossover	NCT03470376
Blood pressure	Completed	NA	Randomized	NCT01956994
Body weight	Unknown	NA	Randomized	NCT05283525
Cancer	Withdrawn	III	Non-Randomized	NCT01032512
	Unknown	III	Randomized	NCT02468882
Cancer of breast	Terminated	II	Randomized	NCT02795572
Cancer of breast/lung/ prostate	Completed	II	Randomized	NCT05137067
Cancer of pancreas	Terminated	NA	Randomized	NCT02681601
Cancer of prostate	Active not recruiting	NA	Randomized	NCT01653925
Colorectal cancer	Completed	I/II	Randomized	NCT01916239
Cognitive impairment	Recruiting	NA	Randomized	NCT04705220
Coronary artery disease	Unknown	IV	Randomized	NCT01490229
Coronary heart disease	Completed	II	Randomized, Crossover	NCT00924391
CVDs	Completed	II	Randomized	NCT01449110
CVDs/T2D/ Hypercholesterolemia	Withdrawn	NA	Randomized	NCT01954472
CVDs	Completed	NA	Randomized, Crossover	NCT01163773
CVDs	Completed	II	Randomized, Crossover	NCT00839774

(continued)

Table 2 (continued)

Diseases	Status	Phase	Study design	Reg. No
CVDs/ Hypercholesterolemia/ T2D/Metabolic syndrome	Completed	NA	Randomized	NCT02481466
T2D/Metabolic syndrome/ Obesity/CVDs/ Hypercholesterolemia	Wthdrawn	NA	Randomized	NCT02078635
CVDs/Hyperlipidemia	Completed	NA	Randomized	NCT00438425
Diabetes	Completed	II	Randomized, Crossover	NCT00877968
Diabetes prevention or Obesity prevention	Completed	NA	Randomized, Crossover	NCT02366572
Diabetes risk/ Insulin resisatnce	Completed	NA	Randomized	NCT03512665
Diarrhea/Mortality/ Hospitalization	Completed	NA	Randomized	NCT00278746
Dry eye disease	Completed	NA	Randomized	NCT03141931
Dyslipidaemias	Completed	NA	Randomized	NCT04186780
	Recruiting	NA	Randomized	NCT06340308
	Completed	NA	Randomized	NCT03019263
Endometriosis	Terminated	NA	Randomized	NCT04091191
Fatigue	Completed	NA	Randomized	NCT04261881
Fertility (Male infertility)	Not yet recruiting	II	Randomized	NCT06091969
Fragile X syndrome	Completed	II	Randomized	NCT05030129
Glaucoma/Glaucoma eye/ Glaucoma open-angle	Withdrawn	IV	Randomized	NCT04676126
Hair thinning	Completed	NA	Randomized	NCT03206567
	Unknown	NA	Randomized	NCT04048031
	Terminated	NA	Randomized, Crossover	NCT03709563
Heart failure with preserved ejection fraction	Completed	II	Randomized, Crossover	NCT04633460
Hypercholesterolemia	Completed	IV	Randomized	NCT02078167
	Completed	NA	Randomized	NCT03065491
	Completed	NA	Randomized, Crossover	NCT01478789
	Completed	II	Randomized	NCT00981981
	Completed	NA	Observational	NCT04433429
	Completed	NA	Observational	NCT05000541

(continued)

Table 2 (continued)

Diseases	Status	Phase	Study design	Reg. No
	Completed	NA	Randomized	NCT03739242
	Completed	NA	Randomized	NCT04677335
Hyperlipidaemia	Completed	NA	Randomized	NCT01152073
	Unknown	IV	Randomized	NCT01649986
Hyperlipidaemia/ Inflammation/Oxidative stress	Completed	I	Randomized, Crossover	NCT01047280
Hyperuricemia	Completed	IV	Randomized	NCT04161872
Insomnia	Completed	NA	Randomized	NCT05609890
Insulin resistance/Obesity/ IBD	Completed	NA	Randomized	NCT04810572
Ischemic heart disease	Completed	NA	Randomized	NCT03411369
Joint pain	Completed	NA	Randomized	NCT03209895
Keratoconus	Completed	NA	Randomized	NCT05756062
Metabolic syndrome	Completed	NA	Randomized, Crossover	NCT01351012
	Completed	NA	Observational	NCT01293162
	Completed	NA	Randomized, Crossover	NCT02106208
	Not yet recruiting	NA	NA	NCT06240273
	Completed	NA	Randomized	NCT02295176
Muscle development	Completed	NA	Randomized	NCT01580189
Myalgic encephalomyelitis	Completed	NA	NA	NCT05128292
NAFLD	Completed	NA	Randomized	NCT02369536
	Recruiting	IV	Randomized	NCT04475276
	Completed	NA	Randomized	NCT03614039
Neurofibromatosis 1	Recruiting	I	NA	NCT05363267
Obesity	Completed	NA	Randomized	NCT01656681
	Completed	II	Randomized	NCT01272739
Osteoarthritis	Completed	NA	NA	NCT02614274
	Completed	NA	Randomized	NCT03612986
Other eating disorders	Completed	NA	Randomized	NCT01535846
Periodontitis	Terminated	NA	Randomized	NCT04692246
	Recruiting	NA	Randomized	NCT05482373
	Recruiting	NA	Non-Randomized	NCT05864768
PLDS	Completed	II/III	Randomized	NCT04141969

(continued)

Table 2 (continued)

Diseases	Status	Phase	Study design	Reg. No
Psychological disease/ Burn out	Completed	NA	Randomized	NCT04532892
Retinal disease	Completed	NA	Randomized	NCT03712670
Schizophrenia and related disorders	Active not recruiting	NA	Randomized	NCT04366401
Senile purpura	Unknown	NA	Randomized	NCT01183910
Sickle cell disease	Active, Not Recruiting	NA	Randomized	NCT05791591
Skin Acne Vulgaris|Acne	Withdrawn	NA	Randomized	NCT04534140
T2D	Completed	NA	Randomized	NCT02693392
	Withdrawn	II	Randomized	NCT01275027
T2D with nephropathy	Unknown	II	Randomized	NCT03622762
Vascular resistance|Hypertension	Terminated	I/II	Randomized	NCT01842399
Vasomotor symptoms in post menopausal women	Completed	NA	Randomized	NCT05813067

Abbreviation
CVDs, Cardiovascular diseases; HIV, Human immunodeficiency virus; IBD, Inflammatory bowel disease; NA, Not applicable; NAFLD, Non-alcoholic fatty liver disease; PLDS, Post-Lyme Disease Syndrome; T2D, Type 2 diabetes mellitus

of these nutraceuticals primarily stems from their ability to mitigate inflammation by inhibiting NLRP3 inflammasome-mediated IL-beta production and pro-inflammatory cytokines. Furthermore, they demonstrate viral replication inhibition by regulating COVID-19 main protease (Mpro), thus modulating viral pathogenesis and providing symptomatic relief (Costagliola et al., 2021; Singh et al., 2021). Resveratrol is a polyphenolic stilbene compound found in fermented products like red wine, grapes, mulberries, and peanuts. It possesses antiviral properties, inhibiting virus protein production and gene expression, and shows promise against a range of viruses including influenza, hepatitis C, and herpes simplex. Quercetin, a bioflavonoid, demonstrates broad-spectrum antiviral activity against both DNA and RNA viruses (Costagliola et al., 2021; Singh et al., 2021). Curcumin, derived from turmeric, exhibits antiviral effects by blocking virus entry and replication (Ahmadi et al., 2020; Asadirad et al., 2022; Askari et al., 2022; Cristina Da Silva et al., 2020; da Silva et al., 2020; Hassaniazad et al., 2021; Hellou et al., 2022; Honarkar Shafie et al., 2022; Karimi et al., 2020; Karimi et al., 2022; Naeini et al., 2022; Saber-Moghaddam et al., 2021; Silva et al., 2019; Suryawati et al., 2022; Tahmasebi et al., 2021; Valizadeh et al., 2020). EGCG, found in tea, acts as a zinc ionophore and may inhibit viral replication, while NAC functions as an antioxidant and anti-inflammatory, aiding in viral defense. PEA, derived from plants and animals, modulates immune responses and reduces respiratory viral infection incidences (Costagliola et al., 2021; Singh et al., 2021). The antiviral activities of these nutraceuticals have been clinically tested and significant positive effects in reducing the symptoms and viral burden have been observed (Costagliola et al., 2021; Singh et al., 2021). Moreover, a nutrient-rich diet, comprising vitamins (A, D, C, E, B6, B12), minerals (calcium, magnesium), and trace elements (zinc, copper, selenium), supports immunity and mitigates viral infections. Vitamins bolster immune defenses: C and E strengthen barriers; C, D, and E enhance cellular immunity; A regulates humoral and cell-mediated responses. Vitamin D modulates respiratory infections and inflammation, while vitamin C aids recovery and reduces morbidity from viral illnesses. Vitamin E offers protection against upper respiratory infections (Costagliola et al., 2021; Singh et al., 2021). B complex vitamins improve respiratory function and immune responses, with B12 potentially reducing COVID-19 severity (Derosa et al., 2021; Singh et al., 2021). Dietary strategies rich in these nutrients promote overall health and resilience against viral infections. Further, various nutrients play crucial roles in antiviral immunity. Zinc modulates immune responses and can reduce the duration and severity of viral infections, including COVID-19 (Derosa et al., 2021). Copper aids in immune cell development and has demonstrated antiviral properties. Selenium supplementation enhances antiviral responses against influenza strains. Magnesium regulates immune functions and may improve the immune response against viral infections (Costagliola et al., 2021; Singh et al., 2021). Multi-nutrient supplements, including trace elements and vitamins, have shown favorable effects on antiviral immunity. Additionally, milk proteins and peptides, functional foods, such as honey and berries, and probiotics can enhance antiviral immunity and mitigate viral infections (Costagliola et al., 2021; Derosa et al., 2021; Singh et al., 2021). Further research is warranted to explore their efficacy in antiviral therapy.

In conclusion, nutraceuticals hold promise in combating viral infections by modulating inflammation, viral replication, and immune responses. Their integration into therapeutic strategies warrants rigorous clinical evaluation to ensure safety and efficacy, indicating the importance of continued research in this domain.

Nutraceuticals in the Prevention and Treatment of Bacterial Diseases

Bacterial infections give rise to diseases that pose significant challenges to human health and impose a substantial impact on global well-being by claiming many lives. Adding to this difficulty is the growing threat of antibiotic resistance, which worsens the severity of bacterial infections (Maglangit et al., 2021). Some of the most notorious ones are *Mycobacterium tuberculosis*, which is known for causing multidrug-resistant Tuberculosis (MDR-TB) (Joean et al., 2020), as well as other challenging forms such as *Enterococcus faecium*, *Staphylococcus aureus*, *Klebsiella pneumoniae*, *Acinetobacter baumannii*, *Pseudomonas aeruginosa*, and different species of

Enterobacter (Mancuso et al., 2021). As antibiotic resistance instances continue to increase, the need for alternate treatment methods grows increasingly urgent.

Helicobacter pylori (*H. pylori*) infection is a significant risk factor for gastric inflammation, ulcers, and the development of gastric cancer. While eradication of *H. pylori* is essential, it alone may not suffice in preventing gastric cancer, as the infection primarily promotes carcinogenesis rather than acting as a direct carcinogen. Thus, targeting the chronic inflammatory processes induced by *H. pylori* is crucial for cancer prevention. Beyond antimicrobial approaches, non-antimicrobial strategies are being explored to prevent *H. pylori*-associated gastric cancer. Phytoceuticals, derived from plants, show promise in this regard. Products like Korean red ginseng, green tea, red wine, flavonoids, broccoli sprouts, garlic, and probiotics have been found to inhibit *H. pylori* colonization, reduce gastric inflammation by inhibiting cytokine release, and suppress precancerous changes by various mechanisms, including inducing apoptosis and inhibiting mutagenesis. Recently, Bae et al. identified β-carotene as a potential treatment to counteract *H. pylori*-induced inflammation (Abbas et al., 2017; Bae et al., 2021; Ooi & Pak, 2021). This compound, found abundantly in fungi, plants, and fruits, demonstrated inhibitory effects on key pathways involved in inflammation and cell invasion induced by *H. pylori* (Bae et al., 2021). Additionally, β-carotene promoted the expression of protective factors, such as peroxisome proliferator-activated receptor-gamma and catalase, reducing oxidative stress within infected cells. Although challenges remain before phytoceuticals can be widely adopted as standard treatments for *H. pylori* infection, they hold potential as potent tools for suppressing or modulating the disease-associated effects of *H. pylori* infection. Further research is needed to elucidate their mechanisms of action and establish their efficacy in clinical settings. By targeting the inflammatory and carcinogenic processes associated with *H. pylori* infection, these non-antimicrobial approaches offer new avenues for the prevention of *H. pylori*-associated gastric diseases, including cancer.

Nutraceuticals have been extensively explored for their potential activity against bacterial diseases. The most considerable example is the role of probiotics against bacterial infection (Heineman et al., 2012; Rueda-Robles et al., 2022). The antibacterial potential of probiotics against bacterial pathogens including *H.pylori*, *Clostridium difficile*, and Vancomycin-Resistant *Enterococci* has been tested (Heineman et al., 2012). In a clinical trial conducted by Thomas et al. the efficacy of VE303 was determined against *Clostridioides difficile* infection (CDI) (Louie et al., 2023). VE303 is a specific combination of 8 strains of commensal Clostridia bacteria. They included 79 patients to conduct their study and study group includes the high-dose group (n = 30), low-dose group (n = 27), and placebo (n = 22) (Louie et al., 2023). The researchers noted that high-dose VE303 effectively prevented recurrent *Clostridioides difficile* infection (CDI) in people who had laboratory-confirmed CDI with one or more previous episodes in the past 6 months, as well as in those with initial CDI who were at high risk for recurrence, this was found in comparison to a placebo (Louie et al., 2023). Considering the significant benefit of probiotics against bacterial infections, researchers are involved in modulating the probiotics in order to enhance their efficacy (Louie et al., 2023). In another study by Karimi et al. they designed various gel formulations containing *Lactobacillus rhamnosus* to evaluate the stability of the formulations (Karimi et al., 2023). Further, they tested the formulations against methicillin-resistant *Staphylococcus* (MRSA) aureus and vancomycin-resistant Enterococcus (VRE). They found their formulations to be effective against both MRSA and VRE (Karimi et al., 2023).

Apart from probiotics, various herbal extracts have also been screened for potential antibacterial activity. The polar extracts of Sumac, a spice with medicinal properties from the Mediterranean region and the Middle East possess antibacterial activity against *H. pylori* (Martinelli et al., 2022). All the polar extracts of Sumac show activity against *H. pylori* with a minimum inhibitory concentration (MIC) value of 100 μg/mL, except the aqueous extract (Martinelli et al., 2022). In a similar study, the methanolic extract of *Quararibea funebris* flowers was screened for antibacterial potential against *Staphylococcus aureus*, *Enterococcus faecalis*, *Escherichia coli*, and *Salmonella typhi* (*S.typhi*) (Jimenez-Fernandez et al., 2023). The extract shows the highest activity against *S. typhi*. The extracts of herbal churnas from the Indian Ayurveda system have also shown antibacterial activity against various bacterial strain with significant activity (Tambekar & Dahikar, 2011).

In addition to the crude extracts of plant samples, the phytochemicals have also been investigated for their antibacterial potential. Resveratrol, a polyphenol present in several foods including blueberries, peanuts, grapes, and red wine, is well known for its therapeutic applications (Meng et al., 2020). Resveratrol has been evaluated for antibacterial activity against various pathogenic gram-positive and gram-negative bacteria namely Staphylococcus aureus, Enterococcus faecalis, Mycobacterium tuberculosis, Escherichia coli, Pseudomonas aeruginosa, etc. (Vestergaard & Ingmer, 2019). Resveratrol exhibits growth inhibition in many bacterial strains at a concentration of even less than 100 μg/mL, including Helicobacter pylori with a MIC/MBC of 25–100 μg/mL (Mahady & Pendland, 2000; Martini et al., 2011, Paulo et al., 2011), and Vibrio cholerae with a MIC of 60 μg/mL (Augustine et al., 2014). Another phytochemical Genistein which is found in soybeans is reported to have antibacterial activity against *Staphylococcus aureus* and *Bacillus anthracis* at 100 μM concentration of Genistein (Hong et al., 2006).

Studies are available for analyzing the impact of combination therapy of nutraceuticals along with antibiotics for the effective management of bacterial diseases. Phytochemicals such as Carnosic acid, Carnosol, Epigallocatechin-gallate, and Tea catechin, are used in combination with antibiotics Tetracycline, Erythromycin, Ampicillin, etc. to combat the resistance towards antibiotics (Hemaiswarya et al., 2008).

While there is existing research supporting the advantages of using nutraceuticals to control bacterial infections, more comprehensive investigations are required to explore their potential as a medication candidate for treating such diseases.

Nutraceuticals in the Prevention and Treatment of Non-infectious Diseases

Nutraceuticals in the Prevention and Treatment of Arthritis

Arthritis, a chronic musculoskeletal disorder characterized by joint pain, swelling, and stiffness, encompasses over 100 types, including osteoarthritis, psoriatic arthritis, rheumatoid arthritis (RA), and septic arthritis (Alam et al., 2014; Grover & Samson, 2016). While RA is the second most common form of arthritis, its management can benefit from dietary interventions, such as the incorporation of flavonoids like apigenin found in fruits and vegetables, which have demonstrated inhibition of collagenase activity and TNF-related apoptosis-inducing ligand effects on RA fibroblast-like synoviocytes (Alam et al., 2014; Firestein 1997). Additionally, compounds like arctigenin and curcumin have shown anti-inflammatory effects through the modulation of NF-κB activation and suppression of inflammatory enzymes, suggesting potential therapeutic roles in RA and osteoarthritis management (Alam et al., 2014; Li et al., 2023; Thomas et al., 2021). Supplementation of nutraceuticals including curcumin as application of herbal ointments and various oils are reported to be beneficial in prevention and management of pain and symptoms in clinical settings. Curcumin supplementation in various forms showed reductions in inflammatory markers and disease activity scores in patients with RA and osteoarthritis (Amalraj et al., 2017; Anusha et al., 2019; Atabaki et al., 2020; Atabaki et al., 2022; Baqer & Mohammed, 2004; Bayet-Robert et al., 2010; Belcaro et al., 2010; Belcaro et al., 2014; Deodhar et al., 1980; Gupte et al., 2019; Haroyan et al., 2018; Hashemzadeh et al., 2020; Hemmati et al., 2016; Henrotin et al., 2014; Jamali et al., 2020; Jamali et al., 2021; Javadi et al., 2019; Khanna et al., 2020; Liu et al., 2021; Nakagawa et al., 2014; Nivetha et al., 2022; Panahi et al., 2014b, 2016; Pourhabibi-Zarandi et al., 2022; Rahimnia et al., 2015; Shep et al., 2019; Sterzi et al., 2016; Thomas et al., 2021, Chandran & Goel, 2012; Henrotin et al., 2019; Javadi et al., 2019; Pourhabibi-Zarandi et al., 2022). Other nutraceuticals, including EGCG, gallic acid, genistein, kaempferol, quercetin, and sodium ferulate, exhibit promising anti-inflammatory and antioxidant properties, presenting new avenues for alleviating RA symptoms and improving the quality of life for affected individuals (Alam et al., 2014).

Nutraceuticals in the Prevention and Treatment of Autoimmune Diseases

In recent years, advancements in autoimmune disease management have been significant, yet the absence of a definitive cure has led to increased utilization of complementary and alternative medicines (Bright, 2004). While not curative, herbal medicines and dietary supplements have shown promise in alleviating symptoms of organ-specific autoimmune diseases, being effective, affordable, and relatively safe (Bright, 2004). Numerous herbs, spices, and plant extracts have demonstrated utility in treating conditions such as rheumatoid arthritis, type 1 diabetes, systemic lupus erythematosus, and myocarditis (Hussan et al., 2018; Kumar et al., 2021; Kuo, 1986; Patavino & Brady, 2001; Taibi & Bourguignon, 2003; Winiarska-Mieczan et al., 2022). Studies highlight the immunomodulatory actions of substances like 1,25-dihydroxyvitamin D3, *Ginkgo biloba* extracts, cannabinoids from *Cannabis sativa*, and Padma 28 in experiment models of autoimmune encephalomyelitis (EAE) and multiple sclerosis (MS) (Badmaev et al., 1999; Bowling & Stewart, 2003; Pryce et al., 2003). Research into the protective effects of curcumin in EAE reveals a decrease in disease severity and incidence in mice. This effect was associated with the inhibition of interleukin-12 (IL-12) and Th1 differentiation, suggesting the IL-12/Th1 axis is a novel therapeutic target for autoimmune diseases treated with nutraceuticals (Natarajan & Bright, 2002a; 2002b). IL-12, a key cytokine, triggers signal transduction pathways involving Janus kinases (JAKs) and signal transducer and activator of transcription (STAT) proteins, crucial for Th1 cell-mediated autoimmune diseases including MS. Nutraceuticals have been found to modulate these pathways, inhibiting IL-12-induced differentiation of Th1 cells and ameliorating EAE (Hayakawa et al., 1998; Furuya et al., 2002; Kang et al., 1999a; 1999b, 2001; Nakada et al., 2002; Natarajan & Bright, 2002a). In addition, nano-curcumin supplementation at 80 mg/day reduced serum IL-17, RORγt, miR-200, miR-141, and miR-155 levels in ankylosing spondylitis patients (Ahmadi et al., 2020; Hajialilo et al., 2019). Compounds like 15-deoxy-Δ12,14-prostaglandin J2 and ciglitazone have shown efficacy in blocking IL-12 signaling via the JAK-STAT pathway, thus mitigating autoimmune disease progression (Natarajan & Bright, 2002b). Curcumin, in particular, also exhibits dose-dependent inhibition of IL-12-induced JAK-STAT pathway activation, further highlighting its potential as a therapeutic agent in autoimmune diseases mediated by Th1 cells (Natarajan & Bright, 2002b). Therefore, understanding the molecular mechanisms underlying IL-12 signaling regulation by nutraceuticals is crucial for developing novel therapeutic strategies for Th1 cell-mediated autoimmune diseases, offering hope for improved patient outcomes and quality of life.

Nutraceuticals in the Prevention and Treatment of Cancers

Cancer is a lifestyle disorder characterized by uncontrolled division of cells and is regarded as one of the deadliest diseases worldwide. The global incidence of cancer burden is alarmingly rising and the inefficiency and adverse side effects of established therapies including chemo, radio, and immunotherapies necessitate the development of novel therapeutics for this malady (Bray et al., 2024). Recently, the use of nutraceuticals for the prevention, treatment and management of cancer has gained significant attention due to their pleiotropic effects and non-toxic nature. These nutraceuticals are structurally and functionally diverse bioactive molecules derived from plants, animals and microbes (Das et al., 2012; Dutta et al., 2018; Carrizzo et al., 2020; Sajeev et al., 2022; Garodia et al., 2023; Hegde et al., 2023a, 2023b; Kunnumakkara et al., 2023; Muralimanoharan et al., 2009; Sailo et al., 2024a, 2024b; Unnikrishnan et al., 2024). These compounds prevent and reverse cancers through various mechanisms such as inhibition of proliferation and cell cycle progression, inhibition of immune evasion, triggering apoptosis and autophagy, inhibiting invasion, migration and metastasis etc. Nutraceuticals are also known to resensitize the therapeutic resistant cancers to therapies through the inhibition of efflux transporters such as breast cancer resistance protein and multi-drug resistance protein (Kuppusamy et al., 2014). They also reduce the toxicity of chemotherapeutic agents and thereby decreasing the required dosage of these drugs. Besides, nutraceuticals are easy to extract and hence, are less expensive compared to conventional pharmaceuticals. Moreover, potential chemo-preventive and radio-preventive effects of nutraceuticals are also been reported in clinical trials (Kuppusamy et al., 2014). Over the years, numerous studies have highlighted the potential anti-cancer properties of turmeric and its key component curcumin across various cancer types. Clinical trials have demonstrated improvements in antioxidant activity, quality of life, and survival rates in cancer patients consuming turmeric or curcumin supplements. From breast and bladder cancer to pancreatic and prostate cancer, curcumin has shown promising results in reducing tumor markers, enhancing therapeutic effects, and alleviating treatment-related symptoms. Moreover, its efficacy extends to conditions like Barrett's esophagus, chronic myelogenous leukemia, and head and neck squamous cell carcinoma (Bayet-Robert et al., 2010; Chaitanya et al., 2020; Cheng et al., 2001; Choi YH et al., 2019; Cruz-Correa et al., 2006; Delavarian et al., 2019; Dhillon et al., 2008; Dützmann et al., 2016; Epelbaum et al., 2010; Esposito et al., 2017; Farhadi et al., 2018; Garcea et al., 2005; Gbolahan et al., 2022; Golombick et al., 2012; Greil et al., 2018; Hatab et al., 2019; He et al., 2011; Hidayat et al., 2021; Howells et al., 2019; Irving et al., 2013; James et al., 2015; Kanai et al., 2012; Kanai et al., 2013; Kia et al., 2021; Ledda et al., 2012; Lu et al., 2013; Mahammedi et al., 2016; Martínez et al., 2019; Molana 2013; Neetha et al., 2020; O'Rawe et al., 2022; Panahi et al., 2014; Parsons et al., 2016; Passildas-Jahanmohan et al., 2021; Pastorelli et al., 2018; Pinheiro et al., 2019; Qiao et al., 2021; Rawat et al., 2012; Redorta et al., 2021; Ryan et al., 2013; Ryan Wolf et al., 2018, 2020; Saadipoor et al., 2019; Saghatelyan et al., 2020; Sandoughdaran et al., 2020; Santosa et al., 2022; Shah et al., 2020; Sharma et al., 2004; Stancioiu et al., 2019; Talakesh et al., 2022). Apart from turmeric and curcumin, a huge number of nutraceuticals are potential agents in preventing the occurrence of cancers. For instance, the consumption of lycopene among men resulted in a 31% reduction in the risk of developing pancreatic cancer. A mounting body of evidence indicates that nutraceuticals possess the capacity to impede the progression of cancer by cellular signaling pathways, modulating microRNAs (miRNAs), and the epigenome. Besides, these nutraceuticals reported to be acting on multiple targets and major signaling molecules and pathways that are affected by these compounds are epidermal growth factor receptor (EGFR) family receptors, insulin-like growth factor receptor (IGFR), JAK/STAT pathway, MAPK/ERK pathway, PI3K/Akt/mTOR pathway, Ras/Raf signaling, Sonic hedgehog signaling, Wnt/β-catenin signaling pathway etc. (Kuppusamy et al., 2014).

Moreover, nutraceuticals modulate the pharmacokinetics and pharmacodynamics of chemotherapeutic agents by regulating ATP-binding cassette proteins, BCRP, p-gp and MRP. Numerous nutraceuticals, including 3,3′-diindolylmethane, [6]-gingerol, [6]-paradol, ajoene, allicin, allyl isothiocyanate, anethole, apigenin, butein, cardamomin, carnosol, caryophyllene, capsaicin, cinnamaldehyde, chalcone, diallyl sulfide, EGCG, eugenol, fisetin, humulene, indole-3-carbinol, lycopene, kaempferol, limonene, myrcene, perillyl alcohol, phytic acid, piperine, quercetin, resveratrol, silibinin, sulforaphane, ursolic acid, withaferin A, xanthohumol, and zingerone have been observed to downregulate pivotal signaling pathways such as PI3K, NF-kB, and Akt, among others, which are imperative for the oncogenic processes underlying cancer development (Kuppusamy et al., 2014). Numerous clinical trials both randomized and non-randomized trials have reported the efficacy, safety and molecular targets of these nutraceuticals (Kuppusamy et al., 2014). Initial studies have shown the poor solubility and bioavailability of certain nutraceuticals preventing their use as therapeutics. However, recent research has evidenced the efficacy of nutraceuticals in treating cancer including curcumin even when undetectable levels are present in serum and the development of novel formulations that enhanced the bioavailability of these compounds further to treat cancer for example, various generations of curcumin formulations are found to be having excellent bioavailability compared pure curcumin alone without compromising its efficacy (Hegde et al., 2023a; Kunnumakkara et al., 2023). Therefore, the multifaceted properties of nutraceuticals, demonstrated through their ability to target inflammation and pivotal signaling pathways involved in cancer development, signify their potential as integral components of future cancer treatments. Their natural origins and diverse mechanisms of action offer a promising avenue for the development of personalized and complementary therapeutic approaches in oncology. Thus, nutraceuticals represent a burgeoning field in cancer medicine with the capacity to augment traditional treatments and improve patient outcomes.

Nutraceuticals in the Prevention and Treatment of Cardiovascular Diseases

Cardiovascular diseases (CVDs), the term is used for diverse conditions namely coronary artery disease (CAD) and cerebrovascular disease, peripheral artery disease (PAD), and aortic atherosclerosis (Lopez et al., 2023). CVDs continues to be the primary cause of mortality globally and significantly contributes to the deterioration of health of the individuals (Vaduganathan et al., 2022). Common risk factors for CVDs include high blood pressure, dietary hazards, smoking, air pollution, dyslipidemia, high body mass index, and kidney dysfunction (Vaduganathan et al., 2022). Nutraceuticals having therapeutic effects against various diseases have also been explored for the treatment of CVDs (Rivellese et al., 2019). Neutraceuticals serve as an important agent for the management of CVDs in diverse ways such as managing high blood pressure, preventing atherosclerosis, reducing sugar and cholesterol levels, etc. (Cicero et al., 2015, Carrizzo et al., 2020).

High blood pressure or hypertension is one of the most significant risk factors for CVDs (Fuchs & Whelton, 2020). Resveratrol a potent antioxidant that is found in rich amounts in grapes, is reported to possess antihypertensive effects in animal studies (Zordoky et al., 2015). Wong et al. conducted a study to assess the antihypertensive effects of resveratrol by measuring the flow-mediated dilatation of the brachial artery (FMD). They have seen a substantial rise in FMD compared to a placebo after administering three different doses of resveratrol (Wong et al., 2011). Quercetin, a compound found in several plant-based foods such as apples, onions, citrus fruits, berries, broccoli, tea, etc. (Bhagwat et al., 2011), has shown antihypertensive properties in multiple research investigations (Larson AJ et al., 2012). Quercetin has been shown to have a statistically significant impact in reducing high blood pressure (Serban et al., 2016). In addition to quercetin and resveratrol, other compounds such as epicatechin, curcumin, and berberine are also recognized for their antihypertensive properties (Carrizzo et al., 2020). These bioactive compounds are not only effective for hypertension rather they have also shown effectiveness against diabetes, atherosclerosis, and heart failure (Carrizzo et al., 2020).

Recent research indicates that probiotics, prebiotics, and synbiotics (a combination of both probiotics and prebiotics) may play a crucial role as dietary components in the prevention and treatment of CVDs (Olas, 2020). Probiotics and prebiotics have a positive effect on managing CVDs by functioning as antioxidants, anti-inflammatory agents, and antihypertensives, and by lowering cholesterol levels (Oniszczuk et al., 2021). Similarly, CVDs preventive properties of omega-3 fatty acids (FAs), namely eicosapentaenoic (EPA) and docosahexaenoic (DHA), are well recognized (Watanabe & Tatsuna, 2020). Khan et al., conducted a comprehensive analysis to assess the impact of omega-3 FAs on CVDs. Their findings indicate that omega-3 FAs effectively decrease cardiovascular associated mortality and enhance cardiovascular outcomes (Khan et al., 2021).

These studies highlight the importance of nutraceutical compounds in the treatment of CVDs. Further clinical studies are necessary to create efficacious pharmaceuticals derived from nutraceuticals for the treatment of CVDs.

Nutraceuticals in the Prevention and Treatment of Gastrointestinal Diseases

Digestive disorders are on the rise, affecting millions worldwide. Conditions like gastroesophageal reflux disease (GERD), irritable bowel syndrome, celiac disease, food allergies, and inflammatory bowel diseases are becoming increasingly prevalent (Cencic & Chingwaru, 2010). The GI tract functions as a complex ecosystem, with over 400 bacterial species influencing various physiological processes (Cencic & Chingwaru, 2010; Luo et al., 2022). These bacteria produce essential nutrients and metabolites, contribute to mucosal protection, and educate the immune system. The intricate interplay between the mucosal immune system and gut microbiota maintains immune homeostasis, crucial for overall health (Cencic & Chingwaru, 2010; O'Sullivan et al., 2005). Constipation, characterized by infrequent or difficult bowel movements, is a prevalent gastrointestinal issue often necessitating medical intervention due to its prolonged persistence (Chatoor & Emnauel, 2009). Laxatives, substances facilitating stool passage, are commonly recommended for occasional constipation, with botanical products historically employed for this purpose (Gao et al., 2020). Senna, an anthraquinone derivative, is a well-utilized laxative known for its efficacy in altering colonic motility and increasing fluid volume and some anthraquinone derivatives are approved by FDA (Cirillo & Capasso, 2015; Gao et al., 2020). Cascara, another anthranoid-containing laxative, has demonstrated efficacy in improving stool frequency and consistency in clinical trials (Gao et al., 2020). Psyllium, a bulk laxative, is frequently used to alleviate constipation but may lead to appetite loss and delayed gastric emptying (Erdogan et al., 2016; Jalanka et al., 2019). Irritable Bowel Syndrome (IBS), characterized by abdominal discomfort and altered bowel habits, often coexists with psychological conditions (Gao et al., 2020). Probiotics, like *Saccharomyces cerevisiae*, have been explored for their potential in managing IBS symptoms, although caution is advised due to the risk of invasive yeast infections, especially in immunocompromised individuals (Fadhel et al., 2019; Fakruddin et al., 2017; Munoz et al., 2005; Spiller et al., 2016). Botanicals like *Andrographis paniculata* and *Boswellia serrata* exhibit anti-inflammatory properties, potentially alleviating IBS symptoms (Belcaro et al., 2017; Gao et al., 2020). Structural gastrointestinal disorders, including inflammatory bowel disease (IBD), gastrointestinal neoplasia, and diverticular disease, pose significant health challenges (Gao et al., 2020). *Curcuma longa*, containing curcumin, has demonstrated efficacy in improving IBS symptoms and ulcerative colitis in clinical trials (Belcaro et al., 2017; Hanai et al., 2006; Holt et al., 2005; Lang et al., 2015; Portincasa et al., 2016; Ng et al., 2018; Singla et al., 2014; Suskind et al., 2013). Also, anthocyanins have shown promise in managing inflammation associated with IBD (Farzaei et al., 2018). Moreover, GABA and resveratrol exhibit potential in suppressing tumor growth in gastrointestinal neoplasia (Gao et al., 2020). Omega-3 FAs and folate-rich foods like dark green leafy vegetables and fruits show promise in reducing the risk of colorectal cancer by modulating inflammation and DNA repair mechanisms (Gao et al., 2020).

In conclusion, a multifaceted approach incorporating dietary modifications, botanical remedies, and targeted interventions holds promise in managing a spectrum of gastrointestinal disorders, from functional bowel issues to structural abnormalities and neoplastic conditions. Further research is needed to elucidate the efficacy and safety profiles of these interventions and optimize their clinical utility.

Nutraceuticals in the Prevention and Treatment of Gynecological Conditions

Throughout a woman's life, she may experience various disorders, including dysmenorrhea during young adulthood, polycystic ovarian syndrome (PCOS), premenstrual syndrome, mastalgia, infertility in the reproductive age group, and menopausal symptoms during menopause (Chauhan et al., 2022). While standard drugs are recommended for these disorders, the scientific literature supports the evidence-based use of nutraceuticals (Chauhan et al., 2022).

Prolactin levels and hormonal imbalances significantly affect ovulatory cycles. Lowering prolactin levels and restoring hormonal balance positively impact ovulation and PCOS treatment by reducing cyst formation and improving ovulatory cycles (Chauhan et al., 2022). Vitex and turmeric are two herbs beneficial for PCOS (Kar et al., 2024; Kamal et al., 2021; Shah & Shrivastava, 2022). Vitex agnus-castus, also known as chaste berry, contains compounds like viteagnusin and flavonoids that suppress prolactin levels and increase ovulation (Feyzollahi et al., 2021; Kakadia et al., 2019; Kar et al., 2024; Westphal et al., 2006). *Curcuma longa*, or turmeric, rich in curcuminoids, enhances ovulation by improving follicular sheath formation and reducing serum progesterone levels and inflammation in PCOS (Asan et al., 2020; Heshmati et al., 2020, 2021; Jamilian et al., 2020; Lakshmi et al., 2023; Sohaei et al., 2019). Anti-androgenic herbs like *Glycyrrhiza glabra, Linum usitatissimum, Mentha spicata, Cocos nucifera*, and *Punica granatum* can also manage PCOS by regulating androgen levels (Lakshmi et al., 2023; Manouchehri et al., 2023). *Glycyrrhiza glabra* contains glycyrrhizin, which acts as a potent anti-androgen, while *Linum usitatissimum* reduces testosterone levels and symptoms associated with PCOS (Lakshmi et al., 2023). *Mentha spicata* improves ovarian cysts by regulating hormone levels, while *Cocos nucifera* modulates sex hormone levels and reduces cyst formation. *Punica granatum* contains phenolic compounds that normalize testosterone levels in women with PCOS (Lakshmi et al., 2023). *Aloe vera* regulates steroidogenic activity, suppresses androgen secretion, and restores metabolic balance, making it beneficial for PCOS management (Desai et al., 2012; Lakshmi et al., 2023). In addition, resveratrol was also shown promising results in treating PCOS wherein it improved menstrual cyclicity and reduced inflammation and ER stress through the downregulation of NF-κB (Brenjian et al., 2020; Hassan et al., 2023; Mansour et al., 2021). For female infertility, nutraceutical interventions have shown promise (Chauhan et al., 2022). CoQ10 supplementation increases clinical pregnancy rates in females undergoing assisted reproductive technology for infertility (Ben-Meir et al., 2015; Chauhan et al., 2022; Florou et al., 2020). Lycopene supplementation reduces the time to conception in cases of primary infertility, while omega-3 FAs positively influence oocyte quality (Barsky & Blesson, 2020; Chauhan et al., 2022; Ruder et al., 2014). Curcumin reduces inflammation, oxidative stress, and angiogenesis, aiding in infertility resolution, particularly in patients with endometriosis (Arablou et al., 2018; Vallee & Lecarpentier, 2020). Evening primrose oil and Shatavari, an ayurvedic herb, offer relief from menopausal symptoms and support postmenopausal bone and muscle health, respectively (Chauhan et al., 2022; O'Leary et al., 2021).

A healthy and varied diet supplying essential macronutrients and micronutrients is necessary for optimal cellular and tissue functioning. Supplements containing amino acids support cellular processes, reproduction, and fetal growth during pregnancy (Chauhan et al., 2022; Gao H, 2020; Manta-Vogli et al., 2020). Adequate nutrition during pregnancy and lactation ensures proper fetal development and neonatal health. Addressing nutritional needs from menarche to menopause is vital for women's health across all life stages, including adolescence, childbearing years, and menopause. Awareness of dietary choices and nutrition's impact on health is essential for promoting well-being and preventing non-communicable diseases in future generations (Gao H, 2020; Chauhan et al., 2022; Manta-Vogli et al., 2020).

In conclusion, women experience a spectrum of disorders throughout life, from dysmenorrhea in youth to menopausal symptoms in later years. While standard drugs are recommended, nutraceutical interventions offer evidence-based support, particularly in managing conditions like PCOS and infertility, where they can positively impact hormonal balance and reproductive health. Ensuring adequate nutrition, including supplementation with amino acids, from adolescence through menopause is crucial for women's overall health and the prevention of non-communicable diseases.

Nutraceuticals in the Prevention and Treatment of Kidney Diseases

Chronic kidney disease (CKD) represents a global health challenge with its prevalence steadily increasing worldwide. Estimates suggest that CKD affects 8–16% of the global population (Coresh et al., 2007; Rysz et al., 2021). Patients afflicted with CKD face a substantially increased risk of CVDs, with mortality rates from CVD tenfold higher compared to age-matched counterparts, even before reaching the stage necessitating dialysis or kidney transplantation (Rysz et al., 2021; Keith et al., 2004). Encouragingly, clinical trials have shown that curcumin treatment can lead to a significant decrease in plasma TGF-β levels and attenuate proteinuria and inflammation in CKD patients, thereby indicating its potential therapeutic utility (Jimenez-Osorio et al., 2016; Moreillon et al., 2013; Pivari et al., 2022; Rysz et al., 2021). Additionally, herbal supplements containing curcumin have demonstrated efficacy in reducing systemic inflammation markers in non-dialysis CKD patients. Curcumin's antioxidant properties have been validated in various studies, showcasing its ability to mitigate oxidative stress in CKD patients, potentially offering protective effects against disease progression (Aggarwal & Sung, 2009; Jimenez-Osorio et al., 2016; Rysz et al., 2021). In the realm of renal transplantation, curcumin has shown promise in improving early graft function and reducing acute rejection rates, possibly through the induction of heme oxygenase-1, an enzyme involved in anti-inflammatory responses and apoptosis modulation (Hill-Kapturczak et al., 2001; Rysz et al., 2021; Shoskes et al., 2005). In conclusion, while the literature emphasizes curcumin's multifaceted therapeutic properties in CKD and related comorbidities, further research, particularly large-scale clinical trials, is warranted to elucidate its full therapeutic potential in CKD management and transplantation. Additionally, exploring synergistic combinations of curcumin with other nutraceuticals or conventional therapies may hold promise in optimizing treatment outcomes while minimizing adverse effects.

The prolonged presence of polyphenols in the bloodstream potentially extends their effects on vascular health, although further investigation is essential to ensure safety and mitigate distant adverse effects (Rysz et al., 2021). Human trials demonstrate positive outcomes with catechin supplementation, notably a daily intake of 455 mg, equivalent to four cups of decaffeinated green tea. This regimen significantly reduces reactive oxygen species production, plasma hypochlorous acid and hydrogen peroxide activities, as well as levels of inflammatory markers (Hsu et al., 2007). While green tea consumption correlates with improved endothelial function in CKD patients and kidney failure patients, its effects on clinical parameters remain inconclusive (Park et al., 2010; Rysz et al., 2021). Systematic review and Meta-analyses suggest a potential role for green tea in managing obesity (Hursel & Viechtbauer, 2009; Lin et al., 2020). Vitamin D deficiency exacerbates cardiovascular risks in CKD patients, emphasizing the importance of supplementation in ameliorating endothelial dysfunction (Dusso, 2011; Navaneethan et al., 2011). Omega-3 FAs exhibit promise in modulating cardiovascular risk factors and endothelial functions in chronic kidney failure cases, although further research is warranted for conclusive evidence (Rysz et al., 2021; Zanetti et al., 2017). Vitamin K supplementation shows potential in reducing vascular calcification, especially in CKD patients (Fusaro et al., 2011; Kaesler et al., 2021). Combining nutraceuticals with complementary mechanisms of action may enhance efficacy while minimizing adverse effects, presenting a promising avenue for holistic therapeutic approaches.

Nutraceuticals in the Prevention and Treatment of Liver Diseases

NAFLD is the prevalent chronic liver disease linked to metabolic syndrome, with a complex etiology (Del Ben et al., 2017). Therapeutic strategies include lifestyle management, nutraceuticals, target oxidative stress and lipid metabolism (Del Ben et al., 2014, 2017). Antioxidants like vitamin E show promise in NASH treatment, even in the advanced stage (Del Ben et al., 2014, 2017). Polyphenols also inhibit liver fat accumulation. Silymarin demonstrates favorable effects, while vitamin D and carnitine outcomes remain uncertain (Del Ben et al., 2017). Vitamin D, derived from skin exposure to ultraviolet light, regulates mineral metabolism, immune function, and insulin secretion. Studies suggest a link between vitamin D deficiency and NAFLD, with potential impacts on disease severity (Barchetta et al., 2011; Del Ben et al., 2017; Eliades & Spyrou, 2015; Pinelli et al., 2010; Sharifi et al., 2014; Targher et al., 2007). L-carnitine aids in FA transport, stabilizes cell membranes, and reduces serum lipid levels (Al-Dhuayan, 2023; Flanagan et al., 2010; Virmani & Cirulli, 2022). Clinical trials demonstrate its effectiveness in improving liver function, insulin resistance, and inflammatory markers in NASH and NAFLD patients. Combined with lifestyle changes, L-carnitine supplementation shows promise in NASH treatment, although more research is warranted (Hazzan et al., 2022; Malaguarnera et al., 2010; Saneian et al., 2021; Somi et al., 2014). Oxidative stress plays a crucial role in NAFLD progression, with increased reactive oxygen species contributing to inflammation and fibrosis (Del Ben et al., 2017). Antioxidants like vitamin E have shown potential in improving liver histology and mitigate inflammation (Kim et al., 2015a; Mosca et al., 2021; Sanyal et al., 2010; Sebastiani et al., 2020). Similarly, polyphenols, particularly curcumin, silymarin and resveratrol, exhibit hepatoprotective effects, while omega-3 FAs modulate lipid metabolism and reduce hepatic fat content (Bakker et al., 2019; Cerletti et al., 2020; Chashmniam et al., 2019; Chen S et al., 2015; Faghihzadeh et al., 2014; Feher et al., 1989; Green et al., 2020; Hariri et al., 2020; Jazayeri-Tehrani et al., 2019; Mirhafez et al., 2019a, 2019b; Mirhafez et al., 2021; Mirnezami et al., 2020; Moradi Kelardeh et al., 2020; Navarro et al., 2019; Nouri-Vaskeh et al., 2020a, 2020b; Oscarsson et al., 2018; Panahi et al., 2016b, 2017, 2018; Rahmani et al., 2016; Saadati et al., 2018, 2019a, 2019b; Saberi-Karimian et al., 2020; Smid et al., 2022; Wah Kheong et al., 2017). Anthocyanins, betaine, and other compounds offer additional avenues for NAFLD treatment, showing promising results in reducing liver enzymes and improving insulin resistance (Del Ben et al., 2017). Further, Phyllanthus extracts have been shown to mitigate steatosis, hepatic fibrosis and NAFLD (Abu Hassan et al., 2023; Al Zarzour et al., 2017; Huang et al., 2017; Luo et al., 2022; Wong et al., 2013; Zarzour et al., 2018). However, comprehensive clinical trials with histological assessments are necessary to establish their efficacy conclusively.

In summary, nutraceuticals hold potential in NAFLD management, targeting various pathways involved in disease progression. While some compounds show favorable outcomes in clinical trials, others require further investigation. A holistic approach, combining nutraceuticals with lifestyle interventions, may offer comprehensive treatment for NAFLD, mitigating its impact on global healthcare systems.

Nutraceuticals in the Prevention and Treatment of Lung Diseases

Lung diseases affect people of all age groups and cause significant health issues, financial burdens, morbidity, and death worldwide due to their remoteness, prolonged hospitalization, and inaccessible healthcare facilities (Chan et al., 2023; Gulati et al., 2016; Rahman et al., 2022). The most common severe forms of lung diseases include bronchial asthma, chronic obstructive pulmonary disease (COPD), and lung infections such as pulmonary tuberculosis and pneumonia (Chan et al., 2023; Rahman et al., 2022). As per the findings of the World Health Organization (WHO), there exists an approximate populace of 235 million individuals afflicted with mild to moderate manifestations of asthma. Of note, COPD singularly contributes to the demise of over three million individuals annually, representing approximately 6% of total global mortality (Chan et al., 2023). A diverse array of pharmacological interventions are commonly employed in managing these conditions, encompassing anti-inflammatory agents, bronchodilators, antibiotics, antiallergics, mucolytics, and antineoplastic agents (Gulati et al., 2016). Given the escalating concerns regarding the diminishing efficacy and safety profile of conventional pharmacotherapy, there emerges an urgent imperative for the exploration and development of safer and potentially efficacious alternative or adjunctive therapeutic modalities (Chan et al., 2023; Gulati et al., 2016; Rahman et al., 2022). The significance of nutrition in maintaining health and the value of natural substances and nutritional supplements in preventive healthcare are progressively gaining recognition. Natural products including crude extracts, vegetables, fruits and metabolites extracted from different plants are tested and showed remarkable potential in treating lung diseases (Cao et al., 2019; Gulati et al., 2016; Wang et al., 2015). Plant-derived compounds as monotherapy in treating Coronavirus infections and lung cancer is gaining increasing attention due to their effectiveness and low or no adverse side effects. Plant-derivatives including baicalein, curcumin, ginkogolic acid, resveratrol, and shiraiachrome A for COVID-19 treatment and daurisoline, deguelin, erianin and graveospene A, for the treatment of lung cancer are widely explored and showed significant efficacy (Rahman et al., 2022). A thorough investigation into five significant categories, namely triterpenoids, glycosides, flavonoids, alkaloids, and polyphenols, along with various combinations, notably including triptolide, has revealed remarkable efficacy in attenuating inflammatory cytokine expression among patients diagnosed with bronchial asthma. Notably, supplementation with omega-3 polyunsaturated FAs, butyrate, beta-alanine, melatonin, vitamin D, magnesium, dietary nitrate, probiotics, curcumin and its formulations, and quercetin are clinically beneficial in the treatment and management of patients with COPD (Alasmari et al., 2024; Baggs et al., 2023; De Brandt et al., 2022; Engelen et al., 2022; Karim et al., 2022; Pavitt et al., 2020, 2022; Qaisar et al., 2024; Rafiq et al., 2022; Viana et al., 2023; Zanforlini et al., 2022; Zare'i et al., 2024). The most prevalent and widely accepted natural medicinal extracts studied for lung diseases are derived from *Aesculus hippocastanum, Aglaia foveolata, Azadirachta indica, Baccharis retusa, Calophyllum brasiliense, Camellia sinensis, Curcuma longa, Dendrobium pulchellum, Dracocephalum rupestre, Ginkgo biloba, Glycyrrhiza glabra, Herba Paederiae, Lycoris radiate, Myrciaria cauliflora, Nelumbo nucifera Gaertn, Nigella sativa, Picrorhiza scrophulariiflora, Polygonum cuspidatum, Psoralea corylifolia, Rheum officinale, Scutellaria baicalensis, Scrophularia scorodonia, Silybum marianum, Tanacetum parthenium, Tripterygium regelii, Tripterygium wilfordii, Uncaria tomentosa, Vanilla planifolia* and *Ziziphus jujube* plant species (Rahman et al., 2022). However, further investigation is warranted to elucidate the underlying mechanisms of action and to ascertain the efficacy and safety profiles of natural substances before their clinical implementation in patients with lung diseases.

Nutraceuticals in the Prevention and Treatment of Metabolic Disorders

Metabolic syndrome encompasses a cluster of metabolic risk factors such as insulin resistance, hyperinsulinemia, impaired glucose tolerance, type 2 diabetes mellitus, visceral obesity, atherogenic dyslipidemia, and high blood pressure, collectively elevating the risk for CVDs (Sirtori et al., 2017). Prevalent in nearly 30% of adults in Western societies, particularly where obesity and unhealthy dietary habits prevail, metabolic syndrome poses a significant public health concern (Sirtori et al., 2017).

Numerous nutraceuticals exhibit anti-obesity properties and hold promise in managing excess weight in obese individuals. Compounds such as psyllium fibers, Momordica, conjugated linoleic acid, and capsaicin demonstrate anti-obesity activities (Sirtori et al., 2017; Cicero & Colletti, 2016). In addition, vitamin C, sylvester G, fenugreek, glucomannan, and chitosan present potential for reducing fat accumulation in obese patients (Cicero & Colletti, 2016). Dietary supplementation with these nutraceuticals offers a viable approach to addressing obesity, even without exercise, which is often hindered by psychological complexities (Cicero & Colletti, 2016).

In diabetes mellitus, characterized by abnormal blood glucose levels due to inadequate insulin synthesis. Nutraceutical interventions play a crucial role in the management of diabetes and associated complications (Cicero & Colletti, 2016). Omega-3 FAs, particularly DHA, serve as insulin-resistant modulators, while lipoic acid exhibits antioxidative properties effective in diabetes treatment. Magnesium supplementation improves insulin sensitivity, and vitamins D, calcium, and chromium picolinate aid in diabetes management. Additionally, plants like cinnamon and bitter melon show effectiveness in diabetes treatment (Cicero & Colletti, 2016).

Flavonoids, abundant in various fruits and vegetables, play a significant role in cardiovascular health (Cicero & Colletti, 2016). Flavonoid derivatives such as flavone glycosides, ginkgo, quercetin, and hesperidins possess antiplatelet and vasoprotective properties, preventing CVDs. These compounds mitigate joint disorders commonly associated with obesity by enhancing vessel strength and microcirculation (Cicero & Colletti, 2016).

Hence, nutraceuticals show promise in managing metabolic disorders, but further research is needed to understand their mechanisms and clinical implications fully.

Nutraceuticals in the Prevention and Treatment of Neurological Disorders

Neurological diseases are a major global health problem that places a substantial burden on human well-being. Neurological disorders include various conditions namely Alzheimer's disease and other dementias, headaches, idiopathic epilepsy, motor neuron disease, multiple sclerosis, Parkinson's disease (PD), muscular dystrophy, and Huntington's disease (Steinmetz, 2021). Neurological disorders accounted for 3% of the total global burden of disease (Thakur et al., 2016). Although the percent occurrence of neurological disorders appears small disorders, still they stand as the top-ranked contributor to global nervous disability-adjusted life years (DALYs) and years of life lost (YLLs) in Global Burden of Disease 2021 (Steinmetz, 2021).

Nutraceutical agents known for their diverse medicinal benefits, have also been known for their potential role against neurodegenerative disorders as well. Nutraceutical agents show their potential effect by reducing inflammation in the neurons and neuro-oxidative stress, regulating energy metabolism, and enhancing neurogenesis through a variety of signaling pathways (Lohani et al., 2024). Various studies have been conducted to evaluate the benefits of nutraceuticals for neurodegenerative studies. Thota et al. has conducted a clinical trial to determine the potential of curcumin against Alzheimer's disease (Thota et al., 2020). Curcumin is well known for its effectiveness against neurological disorders (Garodia et al., 2023). Thota et al. conducted their study with 29 participants, 15 allocated to the placebo group and 14 allocated to the curcumin treatment group. At the end of the treatment, they determined the level of Islet amyloid polypeptide (IAPP), and they observed a reduction in levels of IAPP, indicating the potential benefit of curcumin treatment for the management of Alzheimer's disease. Apart from curcumin, quercetin, EGCG, resveratrol, morin, and luteolin are some other notable examples of naturally occurring compounds that are effective for the management of Alzheimer's disease (Calfio et al., 2020; Moussa et al., 2017).

PD is a neurological disorder characterized by a drastic reduction of dopamine levels; nutraceuticals have also shown a considerable potential in management of PD (Makkar et al., 2020). Curcumin, quercetin, fish oil, probiotics, licorice, NAC, and vitamins are some of the nutraceuticals that are effective against PD (Borzabadi et al., 2018; Bytowska et al., 2023; Coe et al., 2022; da Silva et al., 2008; de et al., 2006; Du et al., 2022; Fan et al., 2018; Monti et al., 2016; Pantzaris et al., 2021; Pathak-Gandhi & Vaidya, 2017; Petramfar et al., 2020).

Neurological disorders are a serious concern in the field of medicine, and in the current scenario due to lack of specific efficient therapy, there is a need to explore other options. Nutraceuticals, due to their effectiveness and safety, are considered viable candidates for drug development targeting neurological illnesses.

Nutraceuticals in the Prevention and Treatment of Oral Diseases

Nutraceuticals play a pivotal role in preventing and managing various dental conditions, including dental caries, periodontal diseases, orthodontic complications, oral surgery, and implantology (Cenzato et al., 2023; Seminario-Amez et al., 2017; Inchingolo et al., 2023). Dental caries, driven primarily by microbial factors, can be mitigated through the use of prebiotics and probiotics, which modulate the oral microbiota and inhibit the growth of cariogenic pathogens like *S. mutans* (Cenzato et al., 2023; Zaura & Twetman, 2019). Similarly, in periodontal diseases, probiotics have shown significant improvements in clinical parameters such as bleeding on probing and gingival index, highlighting their potential as adjunctive therapy (Iniesta et al., 2012; Seminario-Amez et al., 2017; Slawik et al., 2011). Moreover, the integration of long-chain n3 PUFAs has demonstrated anti-inflammatory and antimicrobial effects, contributing to periodontal health (Azuma et al., 2022; Chee 2015). In orthodontics, nutraceuticals offer potential solutions to plaque accumulation and microbial activity associated with fixed orthodontic appliances. While probiotics may not directly impact gingival inflammation, they exhibit antimicrobial activity against oral pathogens. Probiotics also show promise in managing peri-implant mucositis, although their efficacy in improving probing depths requires further investigation (Cenzato et al., 2023). In oral surgery and implantology, vitamin D supplementation has emerged as a critical factor influencing osseointegration, with deficiencies linked to reduced implant success rates (Alsulaimani et al., 2022; Makke, 2022; Kuhad 2023; Werny et al., 2022).

Overall, nutraceutical interventions present viable options for enhancing oral health outcomes across various dental disciplines, although their precise mechanisms and clinical effectiveness warrant ongoing research and validation.

Nutraceuticals in the Prevention and Treatment of Psychological Disorders

Psychopathological conditions have a negative impact on an individual's social well-being. Psychological disorders include several conditions such as anxiety, depression, post-traumatic stress disorder (PTSD), and stress-related psychiatric illnesses (Agrawal et al., 2023). These diseases often manifest in individuals experiencing symptoms such as hallucinations, cognitive impairment, difficulty concentrating, indecisiveness, delusions, and sleeplessness. Nutraceuticals have been recognized for their role in the treatment of psychotic disorders such as schizophrenia and bipolar disorder (Makkar et al., 2020). Nutraceuticals significantly enhance the therapeutic effectiveness of psychiatric medications by demonstrating their activity in a synergistic way (Makkar et al., 2020).

Omega-3 Fatty Acids naturally occurring in fish oil have been investigated for their beneficial role in the management of psychological disorders (Bozzatello et al., 2016). There is plenty of clinical research indicating the potential of Omega-3 Fatty acids in combating schizophrenia. In a trial conducted by Jamilian et al., where they examined the effects of 1 g/day of unspecified omega-3 compared to a placebo in 60 individuals with schizophrenia (Jamilian et al., 2014). The individuals in their study group were already on a typical antipsychotic drug. It was noted that Omega-3 performed much better than the placebo, as shown by the PANSS score (Jamilian et al., 2014).

Depression is a prevalent and intricate psychiatric condition that has a significant impact on a broad global population (Jach et al., 2023; Smith & Torres, 2014). Recent research has provided evidence supporting the connection between gut microbes and the brain (Appleton 2018), indicating that a healthy gut microbiome promotes mental well-being. As Probiotics are known to promote good bacteria, they are being evaluated for their potential to treat depression (Jach et al., 2023), several in vivo and human clinical investigations have been undertaken to support the antidepressant benefits of probiotics (Dey & Mookherjee, 2021). In clinical research, Messaoudi et al., used a Probiotic formulation of L. helveticus R0052 and B. longus R0175, which is commercially known as Probio'Stick (Lallemand Health Solutions Inc., Montreal, QC, Canada). A study conducted on both rats and humans ($N = 55$) discovered that it effectively reduces depression and related stress-induced disorders (Messaoudi et al., 2011).

The natural compounds are renowned for their efficacy in managing psychological disorders. The cannabinoids present in Cannabis sativa are well recognized for their efficacy in treating mood disorders, anxiety, and depression (Abrams 2018; Ferber et al., 2020). Likewise, the use of essential oils derived from aromatic plants is well recognized for its effectiveness in treating anxiety and sadness (Cao et al., 2023). Hossein Ebrahimi conducted a clinical investigation that demonstrated the beneficial benefits of Lavender and Chamomile aromatherapy (Ebrahimi et al., 2022).

Nutraceuticals have garnered significant interest due to their therapeutic advantages. Extensive research has been conducted on their efficacy in treating psychological disorders. Multiple preliminary investigations have shown their effectiveness for psychiatric conditions.

Nutraceuticals in the Prevention and Treatment of Skin Diseases

Nutraceuticals play a pivotal role in maintaining healthy skin, offering potential preventive and therapeutic benefits for various dermatologic conditions and offer protection against ultraviolet radiation, chemicals and microbes (Perez-Sanchez et al., 2018; Xie et al., 2022). Probiotics, phenolics, and vitamins are among the key nutraceuticals implicated in skin health maintenance (Nwanodi, 2018). Probiotics, notably *Lactobacillus* species, along with vitamin E and green tea catechins, exhibit a broad spectrum of skin protective mechanisms, with probiotics demonstrating particularly promising clinical efficacy (Heinrich et al., 2011; Nwanodi, 2018; Perez-Sanchez et al., 2018). Recent research highlights the efficacy of probiotics, such as *Lactobacillus rhamnosus* Goldin and Gorbach, in ameliorating atopic dermatitis and preventing opportunistic skin infections, including those associated with burns (da Costa Baptista et al., 2013). Phenolics, though less extensively studied for dermatologic conditions, show promise in photoprotection, with compounds like rutin demonstrating significant ultraviolet A and B radiation filtering and reactive oxygen species scavenging capabilities (Chaiprasongsuk et al., 2016; Charoenchon et al., 2022; Choi SJ et al., 2016; Farrar et al., 2015; Heinrich et al., 2011; Rhodes et al., 2013; Megow et al., 2017; Yoon et al., 2013). Optimized bioavailability, dosage, and formulation are crucial for maximizing the efficacy of nutraceuticals in skin health. For instance, orally consumed *Lactobacillus plantarum* K8 lysates have been shown to improve skin hydration, reduce horny layer thickness, and decrease transepidermal water loss (Kim et al., 2015b). These findings suggest a potential role for *L. plantarum* K8 lysates in reducing the reliance on topical corticosteroids and emollients for atopic dermatitis treatment (Kim et al., 2015b). Probiotics, when administered as live cultures or lysates, exhibit biological plausibility in skin health, with live cultures demonstrating superior efficacy over heat-killed preparations (da Costa Baptista et al., 2013; Nwanodi, 2018; Prince et al., 2012; Okada et al., 2010). Live cultures competitively deplete pathogens' nutrients and exhibit stronger receptor binding capabilities, making them more effective in preventing skin colonization by pathogenic microbes. Consequently, probiotics or lysates are recommended components of barrier creams and soaps to reduce pathogenic microbe skin colonization and prevent opportunistic infections (da Costa Baptista et al., 2013; Okada et al., 2010; Nwanodi 2018; Prince et al., 2012). Polyphenols, including green tea catechins and resveratrol, hold promise in photoprotection and anti-photocarcinogenesis (Fabbrocini et al., 2011; Farris et al., 2014; Ferzli et al., 2013; Heinrich et al., 2011; Hong et al., 2013; Katiyar, 2008; Ndiaye et al., 2011; Wisuitiprot et al., 2022). However, further human trials are warranted to fully elucidate their efficacy and mechanisms of action. Similarly, comprehensive studies on tocotrienols, quercetin, silymarin, and grape seed proanthocyanidins are needed to validate their potential benefits for skin health (Altaei, 2012; Amer et al., 2022; Becker-Schiebe et al., 2011; Berardesca et al., 2008; Jowkar et al., 2020; Pedrelli et al., 2012; Weber et al., 2003; Xie et al., 2022; Yamakoshi et al., 2004). Despite the growing market demand for nutraceuticals that improve skin health, robust scientific evidence supporting their efficacy remains limited. Challenges include the variability in doses, formulations, and study designs, as well as the need for standardized human trials with adequate statistical power. Future research efforts should focus on addressing these challenges to enhance the credibility and effectiveness of nutraceutical interventions for skin health.

Conclusion

Numerous clinical trials have established the safety and efficacy of nutraceuticals in the prevention, management and treatment of various chronic diseases where conventional drugs have failed to treat effectively. The non-toxic nature and comparatively lower price levels has made these agents more in demand. The food being medicine or medicine being food has also attracted the contemporary generation to avoid diseases instead of curing after getting the diseases. The recent tailored clinical trials and new technologies have also proved that the bioavailability of a few of these nutraceuticals is not an issue anymore. Unlike many pharmaceutical agents, nutraceuticals exhibit a favorable toxicity profile, making them well-tolerated and suitable for long-term use. Most of the nutraceuticals have been recognized as safe by FDA under the GRAS category. Traditional medicine across the world including Ayurveda and Traditional Chinese Medicine which have been used since centuries have shown the efficacy and safety of these compounds. The mechanism of action of several of these agents is well-established both in animal and clinical settings. Additionally, the widespread availability and accessibility of nutraceuticals align with endeavours to broaden healthcare access. Advancements in tailored clinical trial methodologies and technological innovations have addressed historical concerns regarding the bioavailability of certain nutraceutical compounds, bolstering confidence in their therapeutic utility. Through cutting-edge research approaches and innovative delivery systems, scientists have surmounted previous limitations associated with the bioavailability of nutraceuticals including solubility, absorption and tissue distribution, leading to improved efficacy and clinical outcomes. These advancements highlight the evolving landscape of nutraceutical research and its growing significance in contemporary healthcare paradigms. Crucially, the mechanistic foundations of nutraceutical action have been elucidated through rigorous experimentation in both preclinical animal models and human clinical trials. These investigations have shed light on the intricate molecular pathways through which nutraceuticals exert their therapeutic effects, providing valuable insights into their mechanisms of action and therapeutic potential. Therefore, nutraceuticals emerge as a promising frontier in the management of chronic diseases, offering a prospective avenue for therapeutic interventions.

In conclusion, nutraceuticals represent a promising frontier in the prevention, management, and treatment of chronic diseases, providing a holistic and cost-effective approach to healthcare. With a growing body of evidence supporting their safety, efficacy, and mechanistic functions, nutraceuticals hold considerable promise as integral components of future therapeutic strategies aimed at addressing the global burden of chronic illness.

Nutraceuticals, classically delineated as biologically active compounds found in natural sources, primarily comprising non-nutrient secondary metabolites like polyphenols and flavonoids originating from plants, have recently undergone a redefined conceptual framework (Al-Dhuayan, 2023). Presently, nutraceuticals are recognized as any naturally occurring compounds that confer health advantages and are suitable for integration into the human diet. This expanded delineation encompasses several categories: micronutrients, dietary supplements, inclusive of botanicals, functional foods, secondary metabolites and miscellaneous products and medicinal foods comprising health-oriented food products like nutrition bars, lactoferrin for immune enhancement, and transgenic plants engineered for oral vaccination against infectious diseases.

Nutraceuticals and their health benefits and organ protective effects are shown. Nutraceuticals, encompassing a diverse array of natural compounds, offer multifaceted health benefits and organ protective effects. Derived from plants, these bioactive substances like polyphenols and flavonoids exhibit antioxidant, anti-inflammatory, and anti-carcinogenic properties, aiding in disease prevention and management. Micronutrients like vitamins and minerals contribute to overall well-being, while dietary supplements such as Coenzyme Q and omega-3 fatty acids support cardiovascular health. Functional foods fortified with probiotics and prebiotics promote gut health, enhancing immune function. With medicinal foods like lactoferrin bolstering immunity and transgenic plants offering innovative avenues for vaccination, nutraceuticals stand at the forefront of holistic health promotion and organ protection.

Nutraceuticals play a vital role in preventing and treating various chronic diseases. These natural compounds, derived from food sources, offer therapeutic benefits without the adverse effects often associated with pharmaceuticals. Clinical studies demonstrate their efficacy in managing conditions like cardiovascular diseases, diabetes, and inflammatory disorders. Nutraceuticals, rich in antioxidants and bioactive substances, combat oxidative stress and inflammation, key contributors to chronic illnesses. Their growing popularity stems from their safety, affordability, and compatibility with holistic health approaches. Embracing nutraceuticals as adjuncts to conventional therapies holds promise for improving health outcomes and reducing the global burden of chronic diseases.

Acknowledgement This project was supported by BT/556/NE/U-Excel/2016 grant awarded to ABK by Department of Biotechnology (DBT), Government of India. Mangala Hegde acknowledges Research Associate fellowship (DBT-RA/2023/July/NE/4835) from the Department of Biotechnology (DBT), Government of India. All the figures were created using https://www.biorender.com/. The authors acknowledge the use of AI tools for improving the language and readability.

Author Contribution MH contributed to the literature survey, table preparation, initial drafting of the manuscript, figure preparation, and overall editing; AM-contributed to initial drafting of the manuscript, and overall editing; ABK and BBA contributed to conceptualization, funding, overall supervision, and supported review development, overall editing and critical overall manuscript revision.

Competing Interests The authors declare that they have no known competing financial interests or personal relationships that could have appeared to influence the work reported in this paper.

References

Abbas, S. H., Abdulridha, M. K., & Najeb, A. A. (2017). Potential benefit of curcumin adjuvant therapy to the standard Helicobacter pylori eradication therapy in patients with peptic ulcer disease. *Asian Journal of Pharmaceutical and Clinical Research, 10*(5), 313–317.

Abiri, B., Sarbakhsh, P., & Vafa, M. (2022). Randomized study of the effects of vitamin D and/or magnesium supplementation on mood, serum levels of BDNF, inflammation, and SIRT1 in obese women with mild to moderate depressive symptoms. *Nutritional Neuroscience, 25*(10), 2123–2135.

Abrams, D. I. (2018). The therapeutic effects of Cannabis and cannabinoids: An update from the National Academies of Sciences, Engineering and Medicine report. *European Journal of Internal Medicine, 49*, 7–11.

Abu Hassan, M. R., Said, R. H. M., Zainuddin, Z., Omar, H., Ali, S. M. M., Aris, S. A., & Chan, H. K. (2023). Effects of one-year supplementation with *Phyllanthus niruri* on fibrosis score and metabolic markers in patients with non-alcoholic fatty liver disease: A randomized, double-blind, placebo-controlled trial. Heliyon, *9*(6), e16652.

Aggarwal, B. B., & Sung, B. (2009). Pharmacological basis for the role of curcumin in chronic diseases: An age-old spice with modern targets. *Trends in Pharmacological Sciences, 30*(2), 85–94.

Agrawal, K., Chakraborty, P., Dewanjee, S., Arfin, S., Das, S. S., Dey, A., Moustafa, M., Mishra, P. C., Jafari, S. M., Jha, N. K., & Jha, S. K. (2023). Neuropharmacological interventions of quercetin and its derivatives in neurological and psychological disorders. *Neuroscience & Biobehavioral Reviews, 144*, 104955.

Ahmadi, M., Hajialilo, M., Dolati, S., Eghbal-Fard, S., Heydarlou, H., Ghaebi, M., Ghassembaglou, A., Aghebati-Maleki, L., Samadi Kafil, H., Kamrani, A., & Rahnama, B. (2020). The effects of nanocurcumin on Treg cell responses and treatment of ankylosing spondylitis patients: A randomized, double-blind, placebo-controlled clinical trial. *Journal of Cellular Biochemistry, 121*(1), 103–110.

Al Zarzour, R. H., Ahmad, M., Asmawi, M. Z., Kaur, G., Saeed, M. A. A., Al-Mansoub, M. A., Saghir, S. A. M., Usman, N. S., Al-Dulaimi, D. W., & Yam, M. F. (2017). *Phyllanthus niruri* standardized extract alleviates the progression of non-alcoholic fatty liver disease and decreases atherosclerotic risk in Sprague-Dawley Rats. *Nutrients, 9*(7).

Al-Dhuayan, I. S. (2023). Biomedical role of L-carnitine in several organ systems, cellular tissues, and COVID-19. *Brazilian Journal of Biology, 82*, e267633.

AlAli, M., Alqubaisy, M., Aljaafari, M. N., et al. (2021). Nutraceuticals: Transformation of conventional foods into health promoters/disease preventers and safety considerations. *Molecules, 26*(9).

Alam, M. N., Rahman, M. M., & Khalil, M. I. (2014). Nutraceuticals in arthritis management: A contemporary prospect of dietary phytochemicals. *The Open Nutraceuticals Journal, 7*(1).

Alasmari, A. M., Alsulayyim, A. S., Alghamdi, S. M., et al. (2024). Oral nitrate supplementation improves cardiovascular risk markers in COPD: ON-BC, a randomised controlled trial. *European Respiratory Journal, 63*(2).

Alsulaimani, L., Alqarni, A., Almarghlani, A., et al. (2022). The relationship between low serum vitamin D level and early dental implant failure: A systematic review. *Cureus, 14*(1), e21264.

Altaei, T. (2012). The treatment of melasma by silymarin cream. *BMC Dermatology, 12*, 18.

Alwi, I., Santoso, T., Suyono, S., et al. (2008). The effect of curcumin on lipid level in patients with acute coronary syndrome. *Acta Medica Indonesiana, 40*(4), 201–210.

Amalraj, A., Varma, K., Jacob, J., et al. (2017). A novel highly bioavailable curcumin formulation improves symptoms and diagnostic indicators in rheumatoid arthritis patients: A randomized, double-blind, placebo-controlled, two-dose, three-arm, and parallel-group study. *Journal of Medicinal Food, 20*(10), 1022–1030.

Amalraj, A., Varma, K., Jacob, J., et al. (2021). Efficacy and safety of a gut health product (Actbiome) prepared by incorporation of asafoetida-curcumin complex onto the turmeric dietary fiber in the management of gut health and intestinal microflora in healthy subjects: A randomized, double-blind, placebo controlled study. *Bioactive Carbohydrates and Dietary Fibre, 26*, 100280.

Amanat, S., Eftekhari, M. H., Fararouei, M., et al. (2018). Genistein supplementation improves insulin resistance and inflammatory state in non-alcoholic fatty liver patients: A randomized, controlled trial. *Clinical Nutrition, 37*(4), 1210–1215.

Amer, S. S., Mamdouh, W., Nasr, M., et al. (2022). Quercetin loaded cosm-nutraceutical electrospun composite nanofibers for acne alleviation: Preparation, characterization and experimental clinical appraisal. *International Journal of Pharmaceutics, 612*, 121309.

Anusha, D., Chaly, P. E., Junaid, M., et al. (2019). Efficacy of a mouthwash containing essential oils and curcumin as an adjunct to nonsurgical periodontal therapy among rheumatoid arthritis patients with chronic periodontitis: A randomized controlled trial. *Indian Journal of Dental Research, 30*(4), 506–511.

Appleton, J. (2018). The gut-brain axis: Influence of microbiota on mood and mental health. *Integrative Medicine (Encinitas), 17*(4), 28–32.

Arablou, T., & Kolahdouz-Mohammadi, R. (2018). Curcumin and endometriosis: Review on potential roles and molecular mechanisms. *Biomedicine & Pharmacotherapy, 97*, 91–97.

Asadirad, A., Nashibi, R., Khodadadi, A., et al. (2022). Antiinflammatory potential of nano-curcumin as an alternative therapeutic agent for the treatment of mild-to-moderate hospitalized COVID-19 patients in a placebo-controlled clinical trial. *Phytotherapy Research, 36*(2), 1023–1031.

Asan, S. A., Bas, M., Eren, B., et al. (2020). The effects of curcumin supplementation added to diet on anthropometric and biochemical status in women with polycystic ovary syndrome: A randomized, placebo-controlled trial. *Progress in Nutrition, 22*(4).

Askari, G., Sahebkar, A., Soleimani, D., et al. (2022). The efficacy of curcumin-piperine co-supplementation on clinical symptoms, duration, severity, and inflammatory factors in COVID-19 outpatients: A randomized double-blind, placebo-controlled trial. *Trials, 23*(1), 472.

Atabaki, M., Shariati-Sarabi, Z., Tavakkol-Afshari, J., et al. (2022). Curcumin as an effective suppressor of miRNA expression in patients with knee osteoarthritis. *Avicenna Journal of Phytomedicine, 12*(4), 346–356.

Atabaki, M., Shariati-Sarabi, Z., Tavakkol-Afshari, J., et al. (2020). Significant immunomodulatory properties of curcumin in patients with osteoarthritis; a successful clinical trial in Iran. *International Immunopharmacology, 85*, 106607.

Augustin, M. A., Sanguansri, L., & Lockett, T. (2013). Nano- and micro-encapsulated systems for enhancing the delivery of resveratrol. *Annals of the New York Academy of Sciences, 1290*, 107–112.

Augustine, N., Goel, A. K., Sivakumar, K. C., et al. (2014). Resveratrol–a potential inhibitor of biofilm formation in Vibrio cholerae. *Phytomedicine, 21*(3), 286–289.

Awoyemi, A., Troseid, M., Arnesen, H., et al. (2019). Effects of dietary intervention and n-3 PUFA supplementation on markers of gut-related inflammation and their association with cardiovascular events in a high-risk population. *Atherosclerosis, 286*, 53–59.

Azuma, M. M., Cardoso, C. B. M., da Silva, C. C., et al. (2022). The use of omega-3 fatty acids in the treatment of oral diseases. *Oral Diseases, 28*(2), 264–274.

Badmaev, V., Kozlowski, P. B., Schuller-Levis, G. B., et al. (1999). The therapeutic effect of an herbal formula Badmaev 28 (padma 28) on experimental allergic encephalomyelitis (EAE) in SJL/J mice. *Phytotherapy Research, 13*(3), 218–221.

Bae, S., Lim, J. W., & Kim, H. (2021). beta-carotene inhibits expression of matrix metalloproteinase-10 and invasion in helicobacter pylori-infected gastric epithelial cells. *Molecules, 26*(6).

Baggs, G. E., Middleton, C., Nelson, J. L., et al. (2023). Impact of a specialized oral nutritional supplement on quality of life in older adults following hospitalization: Post-hoc analysis of the NOURISH trial. *Clinical Nutrition, 42*(11), 2116–2123.

Bagheri, R., Negaresh, R., Motevalli, M. S., et al. (2022). Spirulina supplementation during gradual weight loss in competitive wrestlers. *British Journal of Nutrition, 127*(2), 248–256.

Bakker, N., van den Helder, R. S., Geenen, R. W. F., et al. (2019). Four weeks of preoperative omega-3 fatty acids reduce liver volume: A randomised controlled trial. *Obesity Surgery, 29*(7), 2037–2044.

Banik, K., Khatoon, E., Hegde, M., et al. (2021). A novel bioavailable curcumin-galactomannan complex modulates the genes responsible for the development of chronic diseases in mice: A RNA sequence analysis. *Life Sciences, 15*(287), 120074.

Baqer, M. S., & Mohammed, M. M. (2004). Evaluation of the anti-inflammatory effect of curcumin as an additive therapy to meloxicam in management of knee osteoarthritis.

Barchetta, I., Angelico, F., Del Ben, M., et al. (2011). Strong association between non alcoholic fatty liver disease (NAFLD) and low 25(OH) vitamin D levels in an adult population with normal serum liver enzymes. *BMC Medicine, 9*, 85.
Barsky, M., & Blesson, C. S. (2020). Oocytes, obesity, and omega-3 fatty acids. *Fertility and Sterility, 113*(1), 71–72.
Bayet-Robert, M., Kwiatowski, F., Leheurteur, M, et al. (2010). Phase I dose escalation trial of docetaxel plus curcumin in patients with advanced and metastatic breast cancer. *Cancer Biology & Therapy, 9*(1), 8–14.
Becker-Schiebe, M., Mengs, U., Schaefer, M., et al. (2011). Topical use of a silymarin-based preparation to prevent radiodermatitis: Results of a prospective study in breast cancer patients. *Strahlentherapie und Onkologie, 187*(8), 485–491.
Beijers, R. J., Gosker, H. R., Sanders, K. J., et al. (2020). Resveratrol and metabolic health in COPD: A proof-of-concept randomized controlled trial. *Clinical Nutrition, 39*(10), 2989–2997.
Belcaro, G., Cesarone, M., Dugall, M., et al. (2010). Product-evaluation registry of Meriva®, a curcumin-phosphatidylcholine complex, for the complementary management of osteoarthritis. *Panminerva Medica, 52*(2 Suppl 1), 55–62.
Belcaro, G., Dugall, M., Luzzi, R., et al. (2014). Meriva(R)+Glucosamine versus Condroitin+Glucosamine in patients with knee osteoarthritis: An observational study. *European Review for Medical and Pharmacological Sciences, 18*(24), 3959–3963.
Belcaro, G., Gizzi, G., Pellegrini, L., et al. (2017). Supplementation with a lecithin-based delivery form of Boswellia serrata extract (Casperome(R)) controls symptoms of mild irritable bowel syndrome. *European Review for Medical and Pharmacological Sciences, 21*(9), 2249–2254.
Ben-Meir, A., Burstein, E., Borrego-Alvarez, A., et al. (2015). Coenzyme Q10 restores oocyte mitochondrial function and fertility during reproductive aging. *Aging Cell, 14*(5), 887–895.
Berardesca, E., Cameli, N., Cavallotti, C., et al. (2008). Combined effects of silymarin and methyl-sulfonylmethane in the management of rosacea: Clinical and instrumental evaluation. *Journal of Cosmetic Dermatology, 7*(1), 8–14.
Bhagwat, S., Haytowitz, D. B., & Holden, J. M. (2011). *USDA database for the flavonoid content of selected foods, release 3* (US Department of Agriculture: Beltsville, MD, USA. 159).
Bilir, B., Sharma, N. V., Lee, J., et al. (2017). Effects of genistein supplementation on genome-wide DNA methylation and gene expression in patients with localized prostate cancer. *International Journal of Oncology, 51*(1), 223–234.
Borzabadi, S., Oryan, S., Eidi A, et al. (2018). The Effects of probiotic supplementation on gene expression related to inflammation, insulin and lipid in patients with Parkinson's disease: A randomized, double-blind, placebocontrolled Trial. *Archives of Iranian Medicine, 21*(7), 289–295.
Bowling, A. C., & Stewart, T. M. (2003). Current complementary and alternative therapies for multiple sclerosis. *Current Treatment Options in Neurology, 5*(1), 55–68.
Bozzatello, P., Brignolo, E., De Grandi, E., et al. (2016). Supplementation with omega-3 fatty acids in psychiatric disorders: A review of literature data. *Journal of Clinical Medicine, 5*(8), 67.
Bray, F., Laversanne, M., Sung, H., et al. (2024). Global cancer statistics 2022: GLOBOCAN estimates of incidence and mortality worldwide for 36 cancers in 185 countries. *CA: A Cancer Journal for Clinicians*.
Brenjian, S., Moini, A., Yamini, N., et al. (2020). Resveratrol treatment in patients with poly-cystic ovary syndrome decreased pro-inflammatory and endoplasmic reticulum stress markers. *American Journal of Reproductive Immunology, 83*(1), e13186.
Briata, I. M., Paleari, L., Rutigliani, M., et al. (2021). A presurgical study of curcumin combined with anthocyanin supplements in patients with colorectal adenomatous polyps. *International Journal of Molecular Sciences, 22*(20).
Bright, J. J. (2004). Targeting autoimmune diseases through nutraceuticals. *Nutrition, 20*(1), 39–43.
Brower, V. (1998). Nutraceuticals: Poised for a healthy slice of the healthcare market? *Nature Biotechnology, 16*(8), 728–731.
Bull, E., Rapport, L., & Lockwood B. (2018). 1. What is a nutraceutical. *Stroke, 13*, 57.

Bytowska, Z. K., Korewo-Labelle, D., Berezka, P., et al. (2023). Effect of 12-Week BMI-based vitamin D(3) supplementation in Parkinson's disease with deep brain stimulation on physical performance, inflammation, and vitamin D metabolites. *International Journal of Molecular Sciences, 24*(12).

Calfio, C., Gonzalez, A., Singh, S. K., et al. (2020). The emerging role of nutraceuticals and phytochemicals in the prevention and treatment of Alzheimer's disease. *Journal of Alzheimer's Disease, 77*(1), 33–51.

Cao, S.-Y., Li, Y., Meng, X., et al. (2019). Dietary natural products and lung cancer: Effects and mechanisms of action. *Journal of Functional Foods, 52*, 316–331.

Cao, X., Zhou, J., Liu, J., et al. (2023). Aromatherapy in anxiety, depression, and insomnia: A bibliometric study and visualization analysis. *Heliyon, 9*(7), e18380.

Carrizzo, A., Izzo, C., Forte, M., et al. (2020). A novel promising frontier for human health: The beneficial effects of nutraceuticals in cardiovascular diseases. *International Journal of Molecular Sciences, 21*(22).

Castro Dos Santos, N. C., Andere, N., Araujo, C. F., et al. (2020). Omega-3 PUFA and aspirin as adjuncts to periodontal debridement in patients with periodontitis and type 2 diabetes mellitus: Randomized clinical trial. *Journal of Periodontology, 91*(10), 1318–1327.

Cencic, A., & Chingwaru, W. (2010). The role of functional foods, nutraceuticals, and food supplements in intestinal health. *Nutrients, 2*(6), 611–625.

Cenzato, N., Khijmatgar, S., Carloni, P., et al. (2023). What is the use of nutraceuticals in dentistry? A scoping review. *European Review for Medical and Pharmacological Sciences, 27*(11), 4899–4913.

Cerletti, C., Colucci, M., Storto, M., et al. (2020). Randomised trial of chronic supplementation with a nutraceutical mixture in subjects with non-alcoholic fatty liver disease. *British Journal of Nutrition, 123*(2), 190–197.

Chaiprasongsuk, A., Onkoksoong, T., Pluemsamran, T., et al. (2016). Photoprotection by dietary phenolics against melanogenesis induced by UVA through Nrf2-dependent antioxidant responses. *Redox Biology, 8*, 79–90.

Chaitanya, N. C., Badam, R., Aryasri, A. S., et al. (2020). Efficacy of improvised topical zinc (1%) ora-base on oral mucositis during cancer chemo-radiation—A randomized study. *Journal of Nutritional Science and Vitaminology, 66*(2), 93–97.

Chan, Y., Raju Allam, V. S. R., Paudel, K. R., et al. (2023). Nutraceuticals: Unlocking newer paradigms in the mitigation of inflammatory lung diseases. *Critical Reviews in Food Science and Nutrition, 63*(19), 3302–3332.

Chandran, B., & Goel, A. (2012). A randomized, pilot study to assess the efficacy and safety of curcumin in patients with active rheumatoid arthritis. *Phytotherapy Research, 26*(11), 1719–1725.

Chang, J. P., Chang, S. S., Chen, H. T., et al. (2023). Omega-3 polyunsaturated fatty acids (n-3 PUFAs), somatic and fatigue symptoms in cardiovascular diseases comorbid major depressive disorder (MDD): A randomized controlled trial. *Brain, Behavior, and Immunity, 112*, 125–131.

Charoenchon, N., Rhodes, L. E., Nicolaou, A., et al. (2022). Ultraviolet radiation-induced degradation of dermal extracellular matrix and protection by green tea catechins: A randomized controlled trial. *Clinical and Experimental Dermatology, 47*(7), 1314–1323.

Chashmniam, S., Mirhafez, S. R., Dehabeh, M., et al. (2019). A pilot study of the effect of phospholipid curcumin on serum metabolomic profile in patients with non-alcoholic fatty liver disease: A randomized, double-blind, placebo-controlled trial. *European Journal of Clinical Nutrition, 73*(9), 1224–1235.

Chatoor, D., & Emmnauel, A. (2009). Constipation and evacuation disorders. *Best Practice & Research Clinical Gastroenterology, 23*(4), 517–530.

Chauhan, A., Thacker, H., Gandhi, A. B., et al. (2022). Expert consensus on the role of nutraceuticals in women's health: Menarche to menopause. *International Journal of Clinical Obstetrics and Gynaecology, 6*(5), 24–31.

Chee, B. (2015). Long chain omega-3 fatty acids as an adjunct to non-surgical periodontal therapy: A randomised double-blind placebo controlled trial 2015.

Chen, H., Liu, S., Ji, L., et al. (2016). Folic acid supplementation mitigates Alzheimer's disease by reducing inflammation: A randomized controlled trial. *Mediators of Inflammation*, 5912146.

Chen, I. J., Liu, C. Y., Chiu, J. P., et al. (2016). Therapeutic effect of high-dose green tea extract on weight reduction: A randomized, double-blind, placebo-controlled clinical trial. *Clinical Nutrition, 35*(3), 592–599.

Chen, J., Bai, Q., Zhao, Z., et al. (2016). Resveratrol improves delayed r-tPA treatment outcome by reducing MMPs. *Acta Neurologica Scandinavica, 134*(1), 54–60.

Chen, S., Zhao, X., Ran, L., et al. (2015). Resveratrol improves insulin resistance, glucose and lipid metabolism in patients with non-alcoholic fatty liver disease: A randomized controlled trial. *Digestive and Liver Disease, 47*(3), 226–232.

Cheng, A. L., Hsu, C. H., Lin, J. K., et al. (2001). Phase I clinical trial of curcumin, a chemopreventive agent, in patients with high-risk or pre-malignant lesions. *Anticancer Res, 21*(4B, 2895), e2900.

Choi, S. J., Lee, S. N., Kim, K., et al. (2016). Biological effects of rutin on skin aging. *International Journal of Molecular Medicine, 38*(1), 357–363.

Choi, Y. H., Han, D. H., Kim, S. W., et al. (2019). A randomized, double-blind, placebo-controlled trial to evaluate the role of curcumin in prostate cancer patients with intermittent androgen deprivation. *Prostate, 79*(6), 614–621.

Cicero, A. F., & Colletti, A. (2016). Role of phytochemicals in the management of metabolic syndrome. *Phytomedicine, 23*(11), 1134–1144.

Cicero, A. F., Parini, A., & Rosticci, M. (2015). Nutraceuticals and cholesterol-lowering action. *IJC Metabolic & Endocrine, 6*, 1–4.

Cirillo, C., & Capasso, R. (2015). Constipation and botanical medicines: An overview. *Phytotherapy Research, 29*(10), 1488–1493.

Coe, S., Andreoli, D., George, M., et al. (2022). A feasibility study to determine whether the daily consumption of flavonoid-rich pure cocoa has the potential to reduce fatigue and fatigability in people with Parkinson's (pwP). *Clinical Nutrition ESPEN, 48*, 68–73.

Collaborators GBDNSD. (2024). Global, regional, and national burden of disorders affecting the nervous system, 1990–2021: A systematic analysis for the Global Burden of Disease Study 2021. *Lancet Neurology, 23*(4), 344–381.

Corbi, G., Nobile, V., Conti, V., et al. (2023). Equol and resveratrol improve bone turnover biomarkers in postmenopausal women: A clinical trial. *International Journal of Molecular Sciences, 24*(15).

Coresh, J., Selvin, E., Stevens, L. A., et al. (2007). Prevalence of chronic kidney disease in the United States. *JAMA, 298*(17), 2038–2047.

Costagliola, G., Nuzzi, G., Spada, E., et al. (2021). Nutraceuticals in viral infections: An overview of the immunomodulating properties. *Nutrients, 13*(7).

Cristina Da Silva, F., Rosa, L. P., Santos, G. P. O., et al. (2020). Total mouth photodynamic therapy mediated by blue led and curcumin in individuals with AIDS. *Expert Review of Anti-infective Therapy, 18*(7), 689–696.

Cruz–Correa, M., Shoskes, D. A., Sanchez, P., et al. (2006). Combination treatment with curcumin and quercetin of adenomas in familial adenomatous polyposis. *Clinical Gastroenterology and Hepatology, 4*(8), 1035–1038.

da Costa Baptista, I. P., Accioly, E., & de Carvalho Padilha, P. (2013). Effect of the use of probiotics in the treatment of children with atopic dermatitis; a literature review. *Nutrición Hospitalaria, 28*(1), 16–26.

da Silva, T. A. L., de Medeiros, D. C., da Silva Cunha de Medeiros, R. C., et al. (2020). Influence of curcumin on glycemic profile, inflammatory markers, and oxidative stress in HIV-infected individuals: A randomized controlled trial. *Phytotherapy Research, 34*(9), 2323–2330.

da Silva, T. M., Munhoz, R. P., Alvarez, C., et al. (2008). Depression in Parkinson's disease: A double-blind, randomized, placebo-controlled pilot study of omega-3 fatty-acid supplementation. *Journal of Affective Disorders, 111*(2–3), 351–359.

Dangardt, F., Osika, W., Chen, Y., et al. (2010). Omega-3 fatty acid supplementation improves vascular function and reduces inflammation in obese adolescents. *Atherosclerosis, 212*(2), 580–585.

Das, L., Bhaumik, E., Raychaudhuri, U., et al. (2012). Role of nutraceuticals in human health. *Journal of Food Science and Technology, 49*(2), 173–183.

De Brandt, J., Derave, W., Vandenabeele, F., et al. (2022). Efficacy of 12 weeks oral beta-alanine supplementation in patients with chronic obstructive pulmonary disease: A double-blind, randomized, placebo-controlled trial. *Journal of Cachexia, Sarcopenia and Muscle, 13*(5), 2361–2372.

de Lau, L. M., Koudstaal, P. J., Witteman, J. C., et al. (2006). Dietary folate, vitamin B12, and vitamin B6 and the risk of Parkinson disease. *Neurology, 67*(2), 315–318.

de Ligt, M., Hesselink, M. K. C., Jorgensen, J., et al. (2021). Resveratrol supplementation reduces ACE2 expression in human adipose tissue. *Adipocyte, 10*(1), 408–411.

DeKosky, S. T., Williamson, J. D., Fitzpatrick, A. L., et al. (2008). Ginkgo biloba for prevention of dementia: A randomized controlled trial. *JAMA, 300*(19), 2253–2262.

Del Ben, M., Polimeni, L., Baratta, F., et al. (2014). Modern approach to the clinical management of non-alcoholic fatty liver disease. *World Journal of Gastroenterology, 20*(26), 8341–8350.

Del Ben, M., Polimeni, L., Baratta, F., et al. (2017). The role of nutraceuticals for the treatment of non-alcoholic fatty liver disease. *British Journal of Clinical Pharmacology, 83*(1), 88–95.

Delavarian, Z., Pakfetrat, A., Ghazi, A., et al. (2019). Oral administration of nanomicelle curcumin in the prevention of radiotherapy-induced mucositis in head and neck cancers. *Special Care in Dentistry, 39*(2), 166–172.

Deodhar, S. D., Sethi, R., & Srimal, R. C. (1980). Preliminary study on antirheumatic activity of curcumin (diferuloyl methane). *Indian Journal of Medical Research, 71*, 632–634.

Derosa, G., Maffioli, P., D'Angelo, A., et al. (2021). Nutraceutical approach to preventing coronavirus disease 2019 and related complications. *Frontiers in Immunology, 12*, 582556.

Desai, B. N., Maharjan, R. H., & Nampoothiri, L. P. (2012). Aloe barbadensis Mill. formulation restores lipid profile to normal in a letrozole-induced polycystic ovarian syndrome rat model. *Pharmacognosy Research, 4*(2), 109–15.

Dey, G., & Mookherjee, S. (2021). Probiotics-targeting new milestones from gut health to mental health. *FEMS Microbiology Letters, 368*(15).

Dhillon, N., Aggarwal, B. B., Newman, R. A., et al. (2008). Phase II trial of curcumin in patients with advanced pancreatic cancer. *Clinical Cancer Research, 14*(14), 4491–4499.

Dolati, S., Aghebati-Maleki, L., Ahmadi, M., et al. (2018). Nanocurcumin restores aberrant miRNA expression profile in multiple sclerosis, randomized, double-blind, placebo-controlled trial. *Journal of Cellular Physiology, 233*(7), 5222–5230.

Dominiak, K., McKinney, J., Heilbrun, L. K., et al. (2010). Critical need for clinical trials: An example of a pilot human intervention trial of a mixture of natural agents protecting lymphocytes against TNF-alpha induced activation of NF-kappaB. *Pharmaceutical Research, 27*(6), 1061–1065.

Du, Y., Li, Y., Xu, X., et al. (2022). Probiotics for constipation and gut microbiota in Parkinson's disease. *Parkinsonism & Related Disorders, 103*, 92–97.

Dusso, A. S. (2011). Kidney disease and vitamin D levels: 25-hydroxyvitamin D, 1,25-dihydroxyvitamin D, and VDR activation. *Kidney International Supplements, 1*(4), 136–141.

Dutta, S., Ali, K. M., Dash, S. K., et al. (2018). Role of nutraceuticals on health promotion and disease prevention: A review. *Journal of Drug Delivery and Therapeutics, 8*(4), 42–47.

Dützmann, S., Schiborr, C., Kocher, A., et al. (2016). Intratumoral concentrations and effects of orally administered micellar curcuminoids in glioblastoma patients. *Nutrition and Cancer, 68*(6), 943–948.

Ebrahimi, H., Mardani, A., Basirinezhad, M. H., et al. (2022). The effects of Lavender and Chamomile essential oil inhalation aromatherapy on depression, anxiety and stress in older community-dwelling people: A randomized controlled trial. *Explore (NY), 18*(3), 272–278.

Eliades, M., & Spyrou, E. (2015). Vitamin D: A new player in non-alcoholic fatty liver disease? *World Journal of Gastroenterology, 21*(6), 1718–1727.

Engelen, M., Jonker, R., Sulaiman, H., et al. (2022). Omega-3 polyunsaturated fatty acid supplementation improves postabsorptive and prandial protein metabolism in patients with chronic obstructive pulmonary disease: A randomized clinical trial. *American Journal of Clinical Nutrition, 116*(3), 686–698.

Epelbaum, R., Schaffer, M., Vizel, B., et al. (2010). Curcumin and gemcitabine in patients with advanced pancreatic cancer. *Nutrition and Cancer, 62*(8), 1137–1141.

Erdogan, A., Rao, S. S., Thiruvaiyaru, D., et al. (2016). Randomised clinical trial: mixed soluble/insoluble fibre vs. psyllium for chronic constipation. *Alimentary Pharmacology & Therapeutics, 44*(1), 35–44.

Espin, J. C., Garcia-Conesa, M. T., & Tomas-Barberan, F. A. (2007). Nutraceuticals: facts and fiction. *Phytochemistry, 68*(22–24), 2986–3008.

Esposito, T., Schettino, C., Polverino, P., et al. (2017). Synergistic interplay between curcumin and polyphenol-rich foods in the mediterranean diet: Therapeutic prospects for neurofibromatosis 1 patients. *Nutrients, 9*(7), 783.

Fabbrocini, G., Staibano, S., De Rosa, G., et al. (2011). Resveratrol-containing gel for the treatment of acne vulgaris: A single-blind, vehicle-controlled, pilot study. *American Journal of Clinical Dermatology, 12*(2), 133–141.

Fadhel, M., Patel, S., Liu, E., et al. (2019). Saccharomyces cerevisiae fungemia in a critically ill patient with acute cholangitis and long term probiotic use. *Medical Mycology Case Reports, 23*, 23–25.

Faghihzadeh, F., Adibi, P., Rafiei, R., et al. (2014). Resveratrol supplementation improves inflammatory biomarkers in patients with nonalcoholic fatty liver disease. *Nutrition Research, 34*(10), 837–843.

Fakruddin, M., Hossain, M. N., & Ahmed, M. M. (2017). Antimicrobial and antioxidant activities of Saccharomyces cerevisiae IFST062013, a potential probiotic. *BMC Complementary and Alternative Medicine, 17*(1), 64.

Falgiano, P. A., Gillum, T. L., Schall, Z. J., et al. (2018). Dietary curcumin supplementation does not alter peripheral blood mononuclear cell responses to exertional heat stress. *European Journal of Applied Physiology, 118*(12), 2707–2717.

Fan, D., Alamri, Y., Liu, K., et al. (2018). Supplementation of Blackcurrant anthocyanins increased cyclic glycine-proline in the cerebrospinal fluid of Parkinson patients: Potential treatment to improve insulin-like growth factor-1 function. *Nutrients, 10*(6).

Farhadi, M., Bakhshandeh, M., Shafiei, B., et al. (2018). The radioprotective effects of nano-curcumin against genotoxicity induced by iodine-131 in patients with differentiated thyroid carcinoma (DTC) by micronucleus assay. *International Journal of Cancer Management, 11*(6).

Farrar, M. D., Nicolaou, A., Clarke, K. A., et al. (2015). A randomized controlled trial of green tea catechins in protection against ultraviolet radiation-induced cutaneous inflammation. *American Journal of Clinical Nutrition, 102*(3), 608–615.

Farris, P., Yatskayer, M., Chen, N., et al. (2014). Evaluation of efficacy and tolerance of a night-time topical antioxidant containing resveratrol, baicalin, and vitamin e for treatment of mild to moderately photodamaged skin. *Journal of Drugs in Dermatology, 13*(12), 1467–1472.

Farshbaf-Khalili, A., Farajnia, S., Pourzeinali, S., et al. (2021). The effect of nanomicelle curcumin supplementation and Nigella sativa oil on the expression level of miRNA-21, miRNA-422a, and miRNA-503 gene in postmenopausal women with low bone mass density: A randomized, triple-blind, placebo-controlled clinical trial with factorial design. *Phytotherapy Research, 35*(11), 6216–6227.

Farshbaf-Khalili, A., Ostadrahimi, A., Mirghafourvand, M., et al. (2022). Clinical efficacy of curcumin and vitamin E on inflammatory-oxidative stress biomarkers and primary symptoms of menopause in healthy postmenopausal women: A triple-blind randomized controlled trial. *Journal of Nutrition and Metabolism*, 6339715.

Farzaei, M. H., El-Senduny, F. F., Momtaz, S., et al. (2018). An update on dietary consideration in inflammatory bowel disease: Anthocyanins and more. *Expert Review of Gastroenterology & Hepatology, 12*(10), 1007–1024.

Feher, J., Deak, G., Muzes, G., et al. (1989). Liver-protective action of silymarin therapy in chronic alcoholic liver diseases. *Orvosi Hetilap, 130*(51), 2723–2727.

Ferber, S. G., Namdar, D., Hen-Shoval, D., et al. (2020). The "entourage effect": Terpenes coupled with cannabinoids for the treatment of mood disorders and anxiety disorders. *Current Neuropharmacology, 18*(2), 87–96.

Ferguson, J. F., Mulvey, C. K., Patel, P. N., et al. (2014). Omega-3 PUFA supplementation and the response to evoked endotoxemia in healthy volunteers. *Molecular Nutrition & Food Research, 58*(3), 601–613.

Ferzli, G., Patel, M., Phrsai, N., et al. (2013). Reduction of facial redness with resveratrol added to topical product containing green tea polyphenols and caffeine. *Journal of Drugs in Dermatology, 12*(7), 770–774.

Feyzollahi, Z., Mohseni Kouchesfehani, H., Jalali, H., et al. (2021). Effect of Vitex agnus-castus ethanolic extract on hypothalamic KISS-1 gene expression in a rat model of polycystic ovary syndrome. *Avicenna Journal of Phytomedicine, 11*(3), 292–301.

Firestein, G. S. (1997). Etiology and pathogenesis of rheumatoid arthritis. *Textbook of Rheumatology, 851.*

Flanagan, J. L., Simmons, P. A., Vehige, J., et al. (2010). Role of carnitine in disease. *Nutrition & Metabolism (London), 16*(7), 30.

Florou, P., Anagnostis, P., Theocharis, P., et al. (2020). Does coenzyme Q(10) supplementation improve fertility outcomes in women undergoing assisted reproductive technology procedures? A systematic review and meta-analysis of randomized-controlled trials. *Journal of Assisted Reproduction and Genetics, 37*(10), 2377–2387.

Fowler, A. A., Truwit, J. D., Hite, R. D., et al. (2019). Effect of vitamin C infusion on organ failure and biomarkers of inflammation and vascular injury in patients with sepsis and severe acute respiratory failure: The CITRIS-ALI randomized clinical trial. *JAMA, 322*(13), 1261–1270.

Fuchs, F. D., & Whelton, P. K. (2020). High blood pressure and cardiovascular disease. *Hypertension, 75*(2), 285–292.

Furuya, Y., Kawakita, T., & Nomoto, K. (2002). Participation of contrasting changes in IL-10 and IL-12 production in the reduction of Th1-predominance by Hachimi-jio-gan in autoimmune MRL/MP-lpr/lpr mice. *Immunopharmacology and Immunotoxicology, 24*(1), 17–29.

Fusaro, M., Crepaldi, G., Maggi, S., et al. (2011). Vitamin K, bone fractures, and vascular calcifications in chronic kidney disease: An important but poorly studied relationship. *Journal of Endocrinological Investigation, 34*(4), 317–323.

Gao, H. (2020). Amino acids in reproductive nutrition and health. *Advances in Experimental Medicine and Biology, 1265*, 111–131.

Gao, X., Liu, J., Li, L., et al. (2020). A brief review of nutraceutical ingredients in gastrointestinal disorders: Evidence and suggestions. *International Journal of Molecular Sciences, 21*(5).

Garcea, G., Berry, D. P., Jones, D. J., et al. (2005). Consumption of the putative chemopreventive agent curcumin by cancer patients: Assessment of curcumin levels in the colorectum and their pharmacodynamic consequences. *Cancer Epidemiology Biomarkers & Prevention, 14*(1), 120–125.

Garcia-Martinez, B. I., Ruiz-Ramos, M., Pedraza-Chaverri, J., et al. (2023). Effect of resveratrol on markers of oxidative stress and sirtuin 1 in elderly adults with type 2 diabetes. *International Journal of Molecular Sciences, 24*(8).

Garodia, P., Hegde, M., Kunnumakkara, A. B., et al. (2023). Curcumin, inflammation, and neurological disorders: How are they linked? *Integrative Medicine Research, 12*(3), 100968.

Gbolahan, O. B., O'Neil, B. H., McRee, A. J., et al. (2022). A phase I evaluation of the effect of curcumin on dose-limiting toxicity and pharmacokinetics of irinotecan in participants with solid tumors. *Clinical and Translational Science, 15*(5), 1304–1315.

Ghadimi, M., Foroughi, F., Hashemipour, S., et al. (2021). Randomized double-blind clinical trial examining the Ellagic acid effects on glycemic status, insulin resistance, antioxidant, and inflammatory factors in patients with type 2 diabetes. *Phytotherapy Research, 35*(2), 1023–1032.

Gibbs, D. C., Fedirko, V., Baron, J. A., et al. (2021). Inflammation modulation by vitamin D and calcium in the morphologically normal colorectal mucosa of patients with colorectal adenoma in a clinical trial. *Cancer Prevention Research (Philadelphia, PA), 14*(1), 65–76.

Golombick, T., Diamond, T. H., Manoharan, A., et al. (2012) Monoclonal gammopathy of undetermined significance, smoldering multiple myeloma, and curcumin: A randomized, double-blind placebo-controlled cross-over 4g study and an open-label 8g extension study. *American Journal of Hematology, 87*(5), 455–460.

Green, C. J., Pramfalk, C., Charlton, C. A., et al. (2020). Hepatic de novo lipogenesis is suppressed and fat oxidation is increased by omega-3 fatty acids at the expense of glucose metabolism. *BMJ Open Diabetes Research Care, 8*(1).

Greil, R., Greil-Ressler, S., Weiss, L., et al. (2018). A phase 1 dose-escalation study on the safety, tolerability and activity of liposomal curcumin (Lipocurc()) in patients with locally advanced or metastatic cancer. *Cancer Chemotherapy and Pharmacology, 82*(4), 695–706.

Grover, A. K., & Samson, S. E. (2016). Benefits of antioxidant supplements for knee osteoarthritis: Rationale and reality. *Nutrition Journal, 5*(15), 1.

Guebre-Egziabher, F., Debard, C., Drai, J., et al. (2013). Differential dose effect of fish oil on inflammation and adipose tissue gene expression in chronic kidney disease patients. *Nutrition, 29*(5), 730–736.

Gulati, K., Rai, N., Chaudhary, S., et al. (2016). *Nutraceuticals in respiratory disorders* (pp. 75–86). Elsevier.

Gupta, R. C., Lall, R., & Srivastava, A. (2021). *Nutraceuticals: Efficacy, safety and toxicity.* Academic Press.

Gupte, P. A., Giramkar, S. A., Harke, S. M., et al. (2019). Evaluation of the efficacy and safety of Capsule Longvida((R)) Optimized Curcumin (solid lipid curcumin particles) in knee osteoarthritis: A pilot clinical study. *Journal of Inflammation Research, 12*, 145–152.

Hajialilo, M., Dolati, S., Abdolmohammadi-Vahid, S., et al. (2019). Nanocurcumin: A novel strategy in treating ankylosing spondylitis by modulating Th17 cells frequency and function. *Journal of Cellular Biochemistry, 120*(7), 12027–12038.

Hajiluian, G., Karegar, S. J., Shidfar, F., et al. (2023). The effects of Ellagic acid supplementation on neurotrophic, inflammation, and oxidative stress factors, and indoleamine 2, 3-dioxygenase gene expression in multiple sclerosis patients with mild to moderate depressive symptoms: A randomized, triple-blind, placebo-controlled trial. *Phytomedicine, 121*, 155094.

Hanai, H., Iida, T., Takeuchi, K., et al. (2006). Curcumin maintenance therapy for ulcerative colitis: randomized, multicenter, double-blind, placebo-controlled trial. *Clinical Gastroenterology and Hepatology, 4*(12), 1502–1506.

Hariri, M., Gholami, A., Mirhafez, S. R., et al. (2020). A pilot study of the effect of curcumin on epigenetic changes and DNA damage among patients with non-alcoholic fatty liver disease: A randomized, double-blind, placebo-controlled, clinical trial. *Complementary Therapies in Medicine, 51*, 102447.

Haroyan, A., Mukuchyan, V., Mkrtchyan, N., et al. (2018). Efficacy and safety of curcumin and its combination with boswellic acid in osteoarthritis: a comparative, randomized, double-blind, placebo-controlled study. *BMC Complementary and Alternative Medicine, 18*(1), 1–16.

Hashemzadeh, K., Davoudian, N., Jaafari, M. R., et al. (2020). The effect of nanocurcumin in improvement of knee osteoarthritis: A randomized clinical trial. *Current Rheumatology Reviews, 16*(2), 158–164.

Hassan, S., Shah, M., Malik, M. O., et al. (2023). Treatment with combined resveratrol and myoinositol ameliorates endocrine, metabolic alterations and perceived stress response in women with PCOS: A double-blind randomized clinical trial. *Endocrine, 79*(1), 208–220.

Hassaniazad, M., Eftekhar, E., Inchehsablagh, B. R., et al. (2021). A triple-blind, placebo-controlled, randomized clinical trial to evaluate the effect of curcumin-containing nanomicelles on cellular immune responses subtypes and clinical outcome in COVID-19 patients. *Phytotherapy Research, 35*(11), 6417–6427.

Hatab, H. M., Hamid, F. F. A., Soliman, A. F., et al. (2019). A combined treatment of curcumin, piperine, and taurine alters the circulating levels of IL-10 and miR-21 in hepatocellular carcinoma patients: A pilot study. *Journal of Gastrointestinal Oncology, 10*(4), 766.

Hayakawa, Y., Fujii, H., Hase, K., et al. (1998). Anti-metastatic and immunomodulating properties of the water extract from Celosia argentea seeds. *Biological and Pharmaceutical Bulletin, 21*(11), 1154–1159.

Hazzan, R., Abu Ahmad, N., Slim, W., et al. (2022). Hepatoprotective effect of combination of L-carnitine and magnesium-hydroxide in nonalcoholic fatty liver disease patients: A double-blinded randomized controlled pilot study. *European Review for Medical and Pharmacological Sciences, 26*(20), 7522–7532.

He, Z. Y., Shi, C. B., Wen, H., et al. (2011). Upregulation of p53 expression in patients with colorectal cancer by administration of curcumin. *Cancer Investigation, 29*(3), 208–213.

Hegde, M., Girisa, S., BharathwajChetty, B., et al. (2023a). Curcumin formulations for better bioavailability: What we learned from clinical trials thus far? *ACS Omega, 8*(12), 10713–10746.

Hegde, M., Girisa, S., Naliyadhara, N., et al. (2023b). Natural compounds targeting nuclear receptors for effective cancer therapy. *Cancer and Metastasis Reviews, 42*(3), 765–822.

Heineman, J., Bubenik, S., McClave, S., et al. (2012). Fighting fire with fire: Is it time to use probiotics to manage pathogenic bacterial diseases? *Current Gastroenterology Reports, 14*(4), 343–348.

Heinrich, U., Moore, C. E., De Spirt, S., et al. (2011). Green tea polyphenols provide photoprotection, increase microcirculation, and modulate skin properties of women. *Journal of Nutrition, 141*(6), 1202–1208.

Helli, B., Gerami, H., Kavianpour, M., et al. (2021). Curcumin nanomicelle improves lipid profile, stress oxidative factors and inflammatory markers in patients undergoing coronary elective angioplasty; a randomized clinical trial. *Endocrine, Metabolic & Immune Disorders Drug Targets.*

Hellou, E., Mohsin, J., Elemy, A., et al. (2022). Effect of ArtemiC in patients with COVID-19: A Phase II prospective study. *Journal of Cellular and Molecular Medicine, 26*(11), 3281–3289.

Hemaiswarya, S., Kruthiventi, A. K., & Doble, M. (2008). Synergism between natural products and antibiotics against infectious diseases. *Phytomedicine, 15*(8), 639–652.

Hemmati, A. A., Rajaee, E., Houshmand, G., et al. (2016). Study the effects of anti-inflammatory curcumex capsules containing three plants (ginger, curcumin and black pepper) in patients with active rheumatoid arthritis. *The IIOAB Journal, 7*(Suppl 5), 389–392.

Henrotin, Y., Gharbi, M., Dierckxsens, Y., et al. (2014). Decrease of a specific biomarker of collagen degradation in osteoarthritis, Coll2-1, by treatment with highly bioavailable curcumin during an exploratory clinical trial. *BMC Complementary and Alternative Medicine, 14*(1), 1–7.

Henrotin, Y., Malaise, M., Wittoek, R., et al. (2019). Bio-optimized Curcuma longa extract is efficient on knee osteoarthritis pain: A double-blind multicenter randomized placebo controlled three-arm study. *Arthritis Research & Therapy, 21*(1), 179.

Heshmati, J., Golab, F., Morvaridzadeh, M., et al. (2020). The effects of curcumin supplementation on oxidative stress, Sirtuin-1 and peroxisome proliferator activated receptor γ coactivator 1α gene expression in polycystic ovarian syndrome (PCOS) patients: A randomized placebo-controlled clinical trial. *Diabetes & Metabolic Syndrome, 14*(2), 77–82.

Heshmati, J., Moini, A., Sepidarkish, M., et al. (2021). Effects of curcumin supplementation on blood glucose, insulin resistance and androgens in patients with polycystic ovary syndrome: A randomized double-blind placebo-controlled clinical trial. *Phytomedicine, 80*, 153395.

Hidayat, Y. M., Wagey, F., Suardi, D., et al. (2021). Analysis of curcumin as a radiosensitizer in cancer therapy with serum survivin examination: Randomised control trial. *Asian Pacific Journal of Cancer Prevention, 22*(1), 139–143.

Hill-Kapturczak, N., Thamilselvan, V., Liu, F., et al. (2001). Mechanism of heme oxygenase-1 gene induction by curcumin in human renal proximal tubule cells. *American Journal of Physiology. Renal Physiology, 281*(5), F851–F859.

Holt, P. R., Katz, S., Kirshoff, R. J. D., et al. (2005). Curcumin therapy in inflammatory bowel disease: A pilot study. *Digestive Diseases and Sciences, 50*(11), 2191–2193.

Honarkar Shafie, E., Taheri, F., Alijani, N., et al. (2022). Effect of nanocurcumin supplementation on the severity of symptoms and length of hospital stay in patients with COVID-19: A randomized double-blind placebo-controlled trial. *Phytotherapy Research, 36*(2), 1013–1022.

Hong, H., Landauer, M. R., Foriska, M. A., et al. (2006). Antibacterial activity of the soy isoflavone genistein. *Journal of Basic Microbiology, 46*(4), 329–335.

Hong, Y. H., Jung, E. Y., Shin, K. S., et al. (2013). Tannase-converted green tea catechins and their anti-wrinkle activity in humans. *Journal of Cosmetic Dermatology, 12*(2), 137–143.

Howells, L. M., Iwuji, C. O. O., Irving, G. R. B., et al. (2019). Curcumin combined with FOLFOX chemotherapy is safe and tolerable in patients with metastatic colorectal cancer in a randomized phase iia trial. *Journal of Nutrition, 149*(7), 1133–1139.

Hsu, S. P., Wu, M. S., Yang, C. C., et al. (2007). Chronic green tea extract supplementation reduces hemodialysis-enhanced production of hydrogen peroxide and hypochlorous acid, atherosclerotic factors, and proinflammatory cytokines. *American Journal of Clinical Nutrition, 86*(5), 1539–1547.

Huang, C. Z., Tung, Y. T., Hsia, S. M., et al. (2017). The hepatoprotective effect of Phyllanthus emblica L. fruit on high fat diet-induced non-alcoholic fatty liver disease (NAFLD) in SD rats. Food and Function, *8*(2), 842–850.

Hursel, R., Viechtbauer, W., & Westerterp-Plantenga, M. S. (2009). The effects of green tea on weight loss and weight maintenance: A meta-analysis. *International Journal of Obesity, 33*(9), 956–961. https://doi.org/10.1038/ijo.2009.135

Hussan, F., Yahaya, M. F., Teoh, S. L., et al. (2018). Herbs for effective treatment of diabetes mellitus wounds: Medicinal chemistry and future therapeutic options. *Mini Reviews in Medicinal Chemistry, 18*(8), 697–710.

Inchingolo, F., Inchingolo, A. M., Malcangi, G., et al. (2023). The benefits of probiotics on oral health: Systematic review of the literature. *Pharmaceuticals (Basel), 16*(9).

Iniesta, M., Herrera, D., Montero, E., et al. (2012). Probiotic effects of orally administered Lactobacillus reuteri-containing tablets on the subgingival and salivary microbiota in patients with gingivitis. A randomized clinical trial. *Journal of Clinical Periodontology, 39*(8), 736–744.

Irving, G. R., Howells, L. M., Sale, S., et al. (2013). Prolonged biologically active colonic tissue levels of curcumin achieved after oral administration—A clinical pilot study including assessment of patient acceptability. *Cancer Prevention Research (Philadelphia, PA), 6*(2), 119–128.

Jach, M. E., Serefko, A., Szopa, A., et al. (2023). The role of probiotics and their metabolites in the treatment of depression. *Molecules, 28*(7), 3213.

Jalanka, J., Major, G., Murray, K., et al. (2019). The effect of psyllium husk on intestinal microbiota in constipated patients and healthy controls. *International Journal of Molecular Sciences, 20*(2).

Jamali, N., Adib-Hajbaghery, M., & Soleimani, A. (2020). The effect of curcumin ointment on knee pain in older adults with osteoarthritis: A randomized placebo trial. *BMC Complement Medicine and Therapies, 20*(1), 305.

Jamali, N., Adib-Hajbaghery, M., & Soleimani, A. (2021). The effect of curcumin ointment on the quality of life of older adults with knee osteoarthritis: A randomized placebo clinical trial. *Nursing and Midwifery Studies, 10*(2), 79.

James, M. I., Iwuji, C., Irving, G., et al. (2015). Curcumin inhibits cancer stem cell phenotypes in ex vivo models of colorectal liver metastases, and is clinically safe and tolerable in combination with FOLFOX chemotherapy. *Cancer Letters, 364*(2), 135–141.

Jamilian, H., Solhi, H., & Jamilian, M. (2014). Randomized, placebo-controlled clinical trial of omega-3 as supplemental treatment in schizophrenia. Global Journal of Health Science, *6*(7 Spec No), 103–108.

Jamilian, M., Foroozanfard, F., Kavossian, E., et al. (2020). Effects of curcumin on body weight, glycemic control and serum lipids in women with polycystic ovary syndrome: A randomized, double-blind, placebo-controlled trial. *Clinical Nutrition ESPEN, 36*, 128–133.

Javadi, M., Khadem Haghighian, H., Goodarzy, S., et al. (2019). Effect of curcumin nanomicelle on the clinical symptoms of patients with rheumatoid arthritis: A randomized, double-blind, controlled trial. *International Journal of Rheumatic Diseases, 22*(10), 1857–1862.

Jazayeri-Tehrani, S. A., Rezayat, S. M., Mansouri, S., et al. (2019). Nano-curcumin improves glucose indices, lipids, inflammation, and Nesfatin in overweight and obese patients with non-alcoholic fatty liver disease (NAFLD): A double-blind randomized placebo-controlled clinical trial. *Nutrition & Metabolism (London), 16*, 8.

Jimenez-Fernandez, M., Juarez-Trujillo, N., Mendoza-Lopez, M. R., et al. (2023). Nutraceutical potential, and antioxidant and antibacterial properties of *Quararibea funebris* flowers. *Food Chemistry, 15*(411), 135529.

Jimenez-Osorio, A. S., Garcia-Nino, W. R., Gonzalez-Reyes, S., et al. (2016). The effect of dietary supplementation with curcumin on redox status and Nrf2 activation in patients with nondiabetic or diabetic proteinuric chronic kidney disease: A pilot study. *Journal of Renal Nutrition, 26*(4), 237–244.

Joean, O., Thiele, T., Schutz, K., et al. (2020). Multidrug-resistant *Mycobacterium tuberculosis*: A report of cosmopolitan microbial migration and an analysis of best management practices. *BMC Infectious Diseases, 20*(1), 678.

Johny, E., Jala, A., Nath, B., et al. (2022). Vitamin D supplementation modulates platelet-mediated inflammation in subjects with type 2 diabetes: A randomized, double-blind, placebo-controlled trial. *Frontiers in Immunology, 13*, 869591.

Jowkar, F., Godarzi, H., & Parvizi, M. M. (2020). Can we consider silymarin as a treatment option for vitiligo? A double-blind controlled randomized clinical trial of phototherapy plus oral Silybum marianum product versus phototherapy alone. *The Journal of Dermatological Treatment, 31*(3), 256–260.

Jurgens, T. M., Whelan, A. M., Killian, L., et al. (2012). Green tea for weight loss and weight maintenance in overweight or obese adults. *Cochrane Database of Systematic Review, 12*(12), CD008650.

Juturu, V., & Kremer, J. (2010). Dietary approaches and alternative therapies for rheumatoid arthritis. *Current Nutrition & Food Science, 6*(4), 240–255.

Kaesler, N., Schurgers, L. J., & Floege, J. (2021). Vitamin K and cardiovascular complications in chronic kidney disease patients. *Kidney International, 100*(5), 1023–1036.

Kakadia, N., Patel, P., Deshpande, S, et al. (2019). Effect of Vitex negundo L. seeds in letrozole induced polycystic ovarian syndrome. *Journal of Traditional and Complementary Medicine, 9*(4), 336–345.

Kamal, D. A. M., Salamt, N., Yusuf, A. N. M., et al. (2021). Potential health benefits of curcumin on female reproductive disorders: A review. *Nutrients, 13*(9).

Kanai, M., Imaizumi, A., & Otsuka, Y., et al. (2012). Dose-escalation and pharmacokinetic study of nanoparticle curcumin, a potential anticancer agent with improved bioavailability, in healthy human volunteers. *Cancer Chemotherapy and Pharmacology, 69*(1), 65–70.

Kanai, M., Otsuka, Y., Otsuka, K., et al. (2013). A phase I study investigating the safety and pharmacokinetics of highly bioavailable curcumin (Theracurmin®) in cancer patients. *Cancer Chemotherapy and Pharmacology, 71*(6), 1521–1530.

Kang, B. Y., Chung, S. W., Chung, W., et al. (1999a). Inhibition of interleukin-12 production in lipopolysaccharide-activated macrophages by curcumin. *European Journal of Pharmacology, 384*(2–3), 191–195.

Kang, B. Y., Chung, S. W., & Kim, T. S. (2001). Inhibition of interleukin-12 production in lipopolysaccharide-activated mouse macrophages by hpyericin, an active component of Hypericum perforatum. *Planta Medica, 67*(4), 364–366.

Kang, B. Y., Song, Y. J., Kim, K. M., et al. (1999b). Curcumin inhibits Th1 cytokine profile in CD4+ T cells by suppressing interleukin-12 production in macrophages. *British Journal of Pharmacology, 128*(2), 380–384.

Kar, T. K., Sil, S., Ghosh, A., et al. (2024). Mitigation of letrozole induced polycystic ovarian syndrome associated inflammatory response and endocrinal dysfunction by Vitex negundo seeds. *Journal of Ovarian Research, 17*(1), 76.

Karim, A., Muhammad, T., Shahid Iqbal, M., et al. (2022). A multistrain probiotic improves handgrip strength and functional capacity in patients with COPD: A randomized controlled trial. *Archives of Gerontology and Geriatrics, 102*, 104721.

Karimi, A., Mahmoodpoor, A., Kooshki, F., et al. (2020). Effects of nanocurcumin on inflammatory factors and clinical outcomes in critically ill patients with sepsis: A pilot randomized clinical trial. *European Journal of Integrative Medicine, 36*, 101122.

Karimi, A., Naeini, F., Niazkar, H. R., et al. (2022). Nano-curcumin supplementation in critically ill patients with sepsis: A randomized clinical trial investigating the inflammatory biomarkers, oxidative stress indices, endothelial function, clinical outcomes and nutritional status. *Food & Function, 13*(12), 6596–6612.

Karimi, F., Azadi, A., Omidifar, N., et al. (2023). Pharmacotechnical aspects of a stable probiotic formulation toward multidrug-resistance antibacterial activity: Design and quality control. *BMC Complement Medicine and Therapies, 23*(1), 391.

Katiyar, S. K. (2008). Grape seed proanthocyanidines and skin cancer prevention: Inhibition of oxidative stress and protection of immune system. *Molecular Nutrition & Food Research, 52* (Suppl 1), S71–S76.

Kaur, S., Kumar, M., Pandit, K., et al. (2019). Potential health benefits of nutraceuticals for human health. In: *Environmental contaminants and natural products* (pp. 193–208). Bentham Science Publishers.

Kawamura, K., Fukumura, S., Nikaido, K., et al. (2020). Resveratrol improves motor function in patients with muscular dystrophies: An open-label, single-arm, phase IIa study. *Science and Reports, 10*(1), 20585.

Kazemi, M., Lalooha, F., Nooshabadi, M. R., et al. (2021). Randomized double blind clinical trial evaluating the Ellagic acid effects on insulin resistance, oxidative stress and sex hormones levels in women with polycystic ovarian syndrome. *Journal of Ovarian Research, 14*(1), 100.

Keith, D. S., Nichols, G. A., Gullion, C. M., et al. (2004). Longitudinal follow-up and outcomes among a population with chronic kidney disease in a large managed care organization. *Archives of Internal Medicine, 164*(6), 659–663.

Khan, S. U., Lone, A. N., Khan, M. S., et al. (2021). Effect of omega-3 fatty acids on cardiovascular outcomes: A systematic review and meta-analysis. *EClinicalMedicine, 38*.

Khanna, A., Das, S. S., Smina, T. P., et al. (2020). Curcumagalactomannoside/glucosamine combination improved joint health among osteoarthritic subjects as compared to chondroitin sulfate/glucosamine: Double-blinded, randomized controlled study. *Journal of Alternative and Complementary Medicine, 26*(10), 945–955.

Kia, S. J., Basirat, M., Saedi, H. S., et al. (2021). Effects of nanomicelle curcumin capsules on prevention and treatment of oral mucosits in patients under chemotherapy with or without head and neck radiotherapy: A randomized clinical trial. *BMC Complement Medicine and Therapies, 21*(1), 232.

Kiecolt-Glaser, J. K., Belury, M. A., Andridge, R., et al. (2011). Omega-3 supplementation lowers inflammation and anxiety in medical students: A randomized controlled trial. *Brain, Behavior, and Immunity, 25*(8), 1725–1734.

Kim, G. H., Chung, J. W., Lee, J. H., et al. (2015a). Effect of vitamin E in nonalcoholic fatty liver disease with metabolic syndrome: A propensity score-matched cohort study. *Clinical and Molecular Hepatology, 21*(4), 379–386.

Kim, H., Kim, H. R., Jeong, B. J., et al. (2015b). Effects of oral intake of kimchi-derived Lactobacillus plantarum K8 lysates on skin moisturizing. *Journal of Microbiology and Biotechnology, 25*(1), 74–80.

Kuhad, A. (2023). The role of vitamin D levels in early dental implant failure. *Journal of Long-Term Effects of Medical Implants, 33*(4), 1–8.

Kumar, S., Mittal, A., Babu, D., et al. (2021). Herbal medicines for diabetes management and its secondary complications. *Current Diabetes Review, 17*(4), 437–456.

Kunnumakkara, A. B., Hegde, M., Parama, D., et al. (2023). Role of turmeric and curcumin in prevention and treatment of chronic diseases: Lessons learned from clinical trials. *ACS Pharmacology and Translational Science, 6*(4), 447–518.

Kuo, C. (1986). Successful treatment of complete left bundle branch block complicating acute viral myocarditis employing Chinese herbs. *American Journal of Chinese Medicine, 14*(3–4), 124–130.

Kuppusamy, P., Yusoff, M. M., Maniam, G. P., et al. (2014). Nutraceuticals as potential therapeutic agents for colon cancer: A review. *Acta Pharmaceutica Sinica B, 4*(3), 173–181.

Lakshmi, J. N., Babu, A. N., Kiran, S. S. M., et al. (2023). Herbs as a source for the treatment of polycystic ovarian syndrome: A systematic review. *BioTech (Basel), 12*(1).

Lang, A., Salomon, N., Wu, J. C., et al. (2015.) Curcumin in combination with mesalamine induces remission in patients with mild-to-moderate ulcerative colitis in a randomized controlled trial. *Clinical Gastroenterology and Hepatology, 13*(8), 1444–1449. e1.

Larson A. J., Symons J. D., Jalili T. (2010) Quercetin: A Treatment for Hypertension?-A Review of Efficacy and Mechanisms. *Pharmaceuticals (Basel), 3*(1), 237–250. https://doi.org/10.3390/ph3010237. PMID: 27713250; PMCID: PMC3991028.

Ledda, A., Belcaro, G., Dugall, M., et al. (2012). Meriva(R), a lecithinized curcumin delivery system, in the control of benign prostatic hyperplasia: A pilot, product evaluation registry study. *Panminerva Medica, 54*(1 Suppl 4), 17–22.

Li, S., Zhang, X., Fang, Q., et al. (2017). Ginkgo biloba extract improved cognitive and neurological functions of acute ischaemic stroke: A randomised controlled trial. *Stroke and Vascular Neurology, 2*(4), 189–197.

Li, W., Yan, W., Liu, Y., et al. (2023). Treatment of rheumatoid arthritis with curcumin analog 3,5-bis(arylidene)-4-piperidone. *Future Medicinal Chemistry, 15*(22), 2051–2064.

Lin, Y., Shi, D., Su, B., et al. (2020). The effect of green tea supplementation on obesity: A systematic review and dose-response meta-analysis of randomized controlled trials. *Phytotherapy Research, 34*(10), 2459–2470.

Liu, X., Robbins, S., Eyles, J., et al. (2021). Efficacy and safety of a supplement combination on hand pain among people with symptomatic hand osteoarthritis an internet-based, randomised clinical trial the RADIANT study. *Osteoarthritis Cartilage, 29*(5), 667–677.

Lohani, A., Maurya, M., Kaur, R., et al. (2024). *Nutraceutical's potentials in neurodegenerative disease* (pp. 199–213). Nutraceutical Fruits and Foods for Neurodegenerative Disorders: Elsevier.

Lopez, E. O., Ballard, B. D., & Jan, A. (2023). *Cardiovascular disease.* StatPearls [Internet]: StatPearls Publishing.

Louie, T., Golan, Y., Khanna, S., et al. (2023). VE303, a defined bacterial consortium, for prevention of recurrent clostridioides difficile infection: A randomized clinical trial. *JAMA, 329*(16), 1356–1366.

Lu, Q., Jiang, F., Xu, R., et al. (2013). A pilot study on intravesical administration of curcumin for cystitis glandularis. *Evidence-Based Complementary and Alternative Medicine.*

Luo, X., Zhang, B., Pan, Y., et al. (2022). Phyllanthus emblica aqueous extract retards hepatic steatosis and fibrosis in NAFLD mice in association with the reshaping of intestinal microecology. *Frontiers in Pharmacology, 13*, 893561.

Maddi, V., Aragade, P., Digge, V., et al. (2007). Phcog Rev.: Short review importance of nutraceuticals in health management. *Pharmacognosy Reviews, 1*(2).

Maglangit, F., Yu, Y., & Deng, H. (2021). Bacterial pathogens: Threat or treat (a review on bioactive natural products from bacterial pathogens). *Natural Product Reports, 38*(4), 782–821.

Mahady, G. B., & Pendland, S. L. (2000). Resveratrol inhibits the growth of helicobacter pylori in vitro. *Official Journal of the American College of Gastroenterology| ACG, 95*(7), 1849.

Mahammedi, H., Planchat, E., Pouget, M., et al. (2016). The new combination docetaxel, prednisone and curcumin in patients with castration-resistant prostate cancer: A pilot phase II study. *Oncology, 90*(2), 69–78.

Mahjabeen, W., Khan, D. A., & Mirza, S. A. (2022). Role of resveratrol supplementation in regulation of glucose hemostasis, inflammation and oxidative stress in patients with diabetes mellitus type 2: A randomized, placebo-controlled trial. *Complementary Therapies in Medicine, 66*, 102819.

Makkar, R., Behl, T., Bungau, S., et al. (2020). Nutraceuticals in neurological disorders. *International Journal of Molecular Sciences, 21*(12), 4424.

Makke, A. (2022). Vitamin D supplementation for prevention of dental implant failure: A systematic review. International Journal of Dentistry, 2845902.

Malaguarnera, M., Gargante, M. P., Russo, C., et al. (2010). L-carnitine supplementation to diet: A new tool in treatment of nonalcoholic steatohepatitis–a randomized and controlled clinical trial. *American Journal of Gastroenterology, 105*(6), 1338–1345.

Mancuso, G., Midiri, A., Gerace, E., et al. (2021). Bacterial antibiotic resistance: The most critical pathogens. *Pathogens, 10*(10), 1310.

Manouchehri, A., Abbaszadeh, S., Ahmadi, M., et al. (2023). Polycystic ovaries and herbal remedies: A systematic review. *JBRA Assisted Reproduction, 27*(1), 85–91.

Mansour, A., Samadi, M., Sanginabadi, M., et al. (2021). Effect of resveratrol on menstrual cyclicity, hyperandrogenism and metabolic profile in women with PCOS. *Clinical Nutrition, 40*(6), 4106–4112.

Manta-Vogli, P. D., Schulpis, K. H., Dotsikas, Y., et al. (2020). The significant role of amino acids during pregnancy: Nutritional support. *The Journal of Maternal-Fetal & Neonatal Medicine, 33*(2), 334–340.

Martinelli, G., Angarano, M., Piazza, S., et al. (2022). The nutraceutical properties of sumac (Rhus coriaria L.) against gastritis: Antibacterial and anti-inflammatory activities in gastric epithelial cells infected with H. pylori. *Nutrients, 14*(9).

Martínez, N., Herrera, M., Frías, L., et al. (2019). A combination of hydroxytyrosol, omega-3 fatty acids and curcumin improves pain and inflammation among early stage breast cancer patients receiving adjuvant hormonal therapy: Results of a pilot study. *Clinical and Translational Oncology, 21*(4), 489–498.

Martini, S., Bonechi, C., Rossi, C., et al. (2011). Increased susceptibility to resveratrol of Helicobacter pylori strains isolated from patients with gastric carcinoma. *Journal of Natural Products, 74*(10), 2257–2260.

McAllister, M. J., Basham, S. A., Smith, J. W., et al. (2018). Effects of environmental heat and antioxidant ingestion on blood markers of oxidative stress in professional firefighters performing structural fire exercises. *Journal of Occupational and Environmental Medicine, 60*(11), e595–e601.

Megow, I., Darvin, M. E., Meinke, M. C., et al. (2017). A Randomized controlled trial of green tea beverages on the in vivo radical scavenging activity in human skin. *Skin Pharmacol Physiol., 30*(5), 225–233.

Meng, X., Zhou, J., & Zhao, C. N., et al. (2020). Health benefits and molecular mechanisms of resveratrol: A narrative review. *Foods, 9*(3).

Messaoudi, M., Lalonde, R., Violle, N., et al. (2011). Assessment of psychotropic-like properties of a probiotic formulation (Lactobacillus helveticus R0052 and Bifidobacterium longum R0175) in rats and human subjects. *British Journal of Nutrition, 105*(5), 755–764.

Miranda-Castro, S., Aidar, F. J., de Moura, S. S., et al. (2022). The curcumin supplementation with piperine can influence the acute elevation of exercise-induced cytokines: Double-blind crossover study. *Biology (Basel), 11*(4).

Mirhafez, S. R., Farimani, A. R., Dehhabe, M., et al. (2019a). Effect of phytosomal curcumin on circulating levels of adiponectin and leptin in patients with non-alcoholic fatty liver disease: A randomized, double-blind, placebo-controlled clinical trial. *Journal of Gastrointestinal and Liver Diseases, 1*(28), 183–189.

Mirhafez, S. R., Farimani, A. R., Gholami, A., et al. (2019b). The effect of curcumin with piperine supplementation on pro-oxidant and antioxidant balance in patients with non-alcoholic fatty liver disease: A randomized, double-blind, placebo-controlled trial. *Drug Metabolism and Personalized Therapy, 34*(2).

Mirhafez, S. R., Rezai, A., Dehabeh, M., et al. (2021). Efficacy of phytosomal curcumin among patients with non-alcoholic fatty liver disease. *International Journal for Vitamin and Nutrition Research, 91*(3–4), 278–286.

Mirnezami, M., Jafarimanesh, H., Rezagholizamenjany, M., et al. (2020). The effect of silymarin on liver enzymes in patients taking isotretinoin: A randomized clinical trial. *Dermatologic Therapy, 33*(2), e13236.

Mogharrabi, M., Rahimi, H. R., Hasanzadeh, S., et al. (2020). The effects of nanomicelle of curcumin on the matrix metalloproteinase (MMP-2, 9) activity and expression in patients with coronary artery disease (CAD): A randomized controlled clinical trial. *ARYA Atheroscler, 16*(3), 136–145.

Molana, S.-H. (2013). A pilot clinical trial of radioprotective effects of curcumin supplementation in patients with prostate cancer.

Monti, D. A., Zabrecky, G., Kremens, D., et al. (2016). N-acetyl cysteine may support dopamine neurons in Parkinson's disease: Preliminary clinical and cell line data. *PLoS ONE, 11*(6), e0157602.

Moradi Kelardeh, B., Rahmati-Ahmadabad, S., Farzanegi, P., et al. (2020). Effects of non-linear resistance training and curcumin supplementation on the liver biochemical markers levels and structure in older women with non-alcoholic fatty liver disease. *Journal of Bodywork and Movement Therapies, 24*(3), 154–160.

Moradi, S., Bagheri, R., Amirian, P., et al. (2024). Effects of Spirulina supplementation in patients with ulcerative colitis: A double-blind, placebo-controlled randomized trial. *BMC Complement Medicine and Therapies, 24*(1), 109.

Moreillon, J. J., Bowden, R. G., Deike, E., et al. (2013). The use of an anti-inflammatory supplement in patients with chronic kidney disease. *Journal of Complementary and Integrative Medicine, 1*, 10.

Mosca, A., Crudele, A., Smeriglio, A., et al. (2021). Antioxidant activity of hydroxytyrosol and vitamin E reduces systemic inflammation in children with paediatric NAFLD. *Digestive and Liver Disease, 53*(9), 1154–1158.

Moussa, C., Hebron, M., Huang, X., et al. (2017). Resveratrol regulates neuro-inflammation and induces adaptive immunity in Alzheimer's disease. *Journal of Neuroinflammation, 14*(1), 1.

Munin, A., & Edwards-Levy, F. (2011). Encapsulation of natural polyphenolic compounds; a review. *Pharmaceutics, 3*(4), 793–829.

Munoz, P., Bouza, E., Cuenca-Estrella, M., et al. (2005). Saccharomyces cerevisiae fungemia: An emerging infectious disease. *Clinical Infectious Diseases, 40*(11), 1625–1634.

Muralimanoharan, S. B., Kunnumakkara, A. B., Shylesh, B., et al. (2009). Butanol fraction containing berberine or related compound from nexrutine inhibits NFkappaB signaling and induces apoptosis in prostate cancer cells. *Prostate, 69*(5), 494–504.

Naeini, F., Tutunchi, H., Razmi, H., et al. (2022). Does nano-curcumin supplementation improve hematological indices in critically ill patients with sepsis? A randomized controlled clinical trial. *Journal of Food Biochemistry, 46*(5), e14093.

Nakada, T., Watanabe, K., Jin, G. B., et al. (2002). Effect of ninjin-youei-to on Th1/Th2 type cytokine production in different mouse strains. *American Journal of Chinese Medicine, 30*(2–3), 215–223.

Nakagawa, Y., Mukai, S., Yamada, S., et al. (2014). Short-term effects of highly-bioavailable curcumin for treating knee osteoarthritis: A randomized, double-blind, placebo-controlled prospective study. *Journal of Orthopaedic Science, 19*(6), 933–939.

Natarajan, C., & Bright, J. J. (2002a). Curcumin inhibits experimental allergic encephalomyelitis by blocking IL-12 signaling through Janus kinase-STAT pathway in T lymphocytes. *The Journal of Immunology, 168*(12), 6506–6513.

Natarajan, C., & Bright, J. J. (2002b). Peroxisome proliferator-activated receptor-gamma agonists inhibit experimental allergic encephalomyelitis by blocking IL-12 production, IL-12 signaling and Th1 differentiation. *Genes and Immunity, 3*(2), 59–70.

Navaneethan, S. D., Schold, J. D., Arrigain, S., et al. (2011). Low 25-hydroxyvitamin D levels and mortality in non-dialysis-dependent CKD. *American Journal of Kidney Diseases, 58*(4), 536–543.

Navarro, V. J., Belle, S. H., D'Amato, M., et al. (2019). Silymarin in non-cirrhotics with non-alcoholic steatohepatitis: A randomized, double-blind, placebo controlled trial. *PLoS ONE, 14*(9), e0221683.

Ndiaye, M., Philippe, C., Mukhtar, H., et al. (2011). The grape antioxidant resveratrol for skin disorders: Promise, prospects, and challenges. *Archives of Biochemistry and Biophysics, 508*(2), 164–170.

Neetha, M. C., Panchaksharappa, M. G., Pattabhiramasastry, S., et al. (2020). Chemopreventive synergism between green tea extract and curcumin in patients with potentially malignant oral disorders: A double-blind, randomized preliminary study. *The Journal of Contemporary Dental Practice, 21*(5), 521–531.

Ng, Q. X., Soh, A. Y. S., Loke, W., et al. (2018). A meta-analysis of the clinical use of curcumin for irritable bowel syndrome (IBS). *Journal of Clinical Medicine, 7*(10).

Nikniaz, S., Vaziri, F., & Mansouri, R. (2023). Impact of resveratrol supplementation on clinical parameters and inflammatory markers in patients with chronic periodontitis: A randomized clinical trail. *BMC Oral Health, 23*(1), 177.

Nivetha, B., Rahmathunisha, A., Lokeshwari, K., et al. (2022). Efficacy of nanocurcumin with application of iontophoresis on inflammatory arthritis patients. *Research Journal of Pharmacy and Technology, 15*(2), 825–829.

Nouri-Vaskeh, M., Afshan, H., Malek Mahdavi, A., et al. (2020). Curcumin ameliorates health-related quality of life in patients with liver cirrhosis: A randomized, double-blind placebo-controlled trial. *Complementary Therapies in Medicine, 49*, 102351.

Nouri-Vaskeh, M., Malek Mahdavi, A., Afshan, H., et al. (2020). Effect of curcumin supplementation on disease severity in patients with liver cirrhosis: A randomized controlled trial. *Phytotherapy Research, 34*(6), 1446–1454.

Nwanodi, O. (2018). Skin protective nutraceuticals: The current evidence in brief. *Healthcare (Basel), 6*(2).

Okada, H., Kuhn, C., Feillet, H., et al. (2010). The 'hygiene hypothesis' for autoimmune and allergic diseases: An update. *Clinical and Experimental Immunology, 160*(1), 1–9.

Olas, B. (2020). Probiotics, prebiotics and synbiotics-a promising strategy in prevention and treatment of cardiovascular diseases? *International Journal of Molecular Sciences, 21*(24).

O'Leary, M. F., Jackman, S. R., Sabou, V. R., et al. (2021). Shatavari supplementation in postmenopausal women improves handgrip strength and increases vastus lateralis myosin regulatory light chain phosphorylation but does not alter markers of bone turnover. *Nutrients, 13*(12).

Oniszczuk, A., Oniszczuk, T., Gancarz, M., et al. (2021). Role of gut microbiota, probiotics and prebiotics in the cardiovascular diseases. *Molecules, 26*(4), 1172.

Ooi, S. L., & Pak, S. C. (2021). Nutraceuticals in immune function. *Molecules, 26*(17).

O'Rawe, M., Wickremesekera, A. C., Pandey, R., et al. (2022). Treatment of glioblastoma with re-purposed renin-angiotensin system modulators: Results of a phase I clinical trial. *Journal of Clinical Neuroscience, 95*, 48–54.

O'Sullivan, G. C., Kelly, P., O'Halloran, S., et al. (2005). Probiotics: An emerging therapy. *Current Pharmaceutical Design, 11*(1), 3–10.

Oscarsson, J., Onnerhag, K., Riserus, U., et al. (2018). Effects of free omega-3 carboxylic acids and fenofibrate on liver fat content in patients with hypertriglyceridemia and non-alcoholic fatty liver disease: A double-blind, randomized, placebo-controlled study. *Journal of Clinical Lipidology, 12*(6), 1390–1403 e4.

Ostadmohammadi, V., Jamilian, M., Bahmani, F., et al. (2019). Vitamin D and probiotic co-supplementation affects mental health, hormonal, inflammatory and oxidative stress parameters in women with polycystic ovary syndrome. *Journal of Ovarian Research, 12*(1), 5.

Panahi, Y., Alishiri, G. H., Parvin, S., et al. (2016a). Mitigation of systemic oxidative stress by curcuminoids in osteoarthritis: Results of a randomized controlled trial. *Journal of Dietary Supplements, 13*(2), 209–220.

Panahi, Y., Kianpour, P., Mohtashami, R., et al. (2016b). Curcumin lowers serum lipids and uric acid in subjects with nonalcoholic fatty liver disease: A randomized controlled trial. *Journal of Cardiovascular Pharmacology, 68*(3), 223–229.

Panahi, Y., Kianpour, P., Mohtashami, R., et al. (2017). Efficacy and safety of phytosomal curcumin in non-alcoholic fatty liver disease: a randomized controlled trial. *Drug Research, 67*(4), 244–251.

Panahi, Y., Kianpour, P., Mohtashami, R., et al. (2018). Efficacy of phospholipidated curcumin in nonalcoholic fatty liver disease: A clinical study. *Journal of Asian Natural Products Research.*

Panahi, Y., Rahimnia, A. R., Sharafi, M., et al. (2014). Curcuminoid treatment for knee osteoarthritis: A randomized double-blind placebo-controlled trial. *Phytotherapy Research, 28*(11), 1625–1631.

Panahi, Y., Saadat, A., Beiraghdar, F., et al. (2014). Adjuvant therapy with bioavailability-boosted curcuminoids suppresses systemic inflammation and improves quality of life in patients with solid tumors: A randomized double-blind placebo-controlled trial. *Phytotherapy Research, 28*(10), 1461–1467.

Pantzaris, M., Loukaides, G., Paraskevis, D., et al. (2021). Neuroaspis PLP10, a nutritional formula rich in omega-3 and omega-6 fatty acids with antioxidant vitamins including gamma-tocopherol in early Parkinson's disease: A randomized, double-blind, placebo-controlled trial. *Clinical Neurology and Neurosurgery, 210*, 106954.

Park, C. S., Kim, W., Woo, J. S., et al. (2010). Green tea consumption improves endothelial function but not circulating endothelial progenitor cells in patients with chronic renal failure. *International Journal of Cardiology, 145*(2), 261–262.

Parsons, H. A., Baracos, V. E., Hong, D. S., et al. (2016). The effects of curcumin (diferuloylmethane) on body composition of patients with advanced pancreatic cancer. *Oncotarget, 7*(15), 20293–20304.

Passildas-Jahanmohan, J., Eymard, J. C., Pouget, M., et al. (2021). Multicenter randomized phase II study comparing docetaxel plus curcumin versus docetaxel plus placebo in first-line treatment of metastatic castration-resistant prostate cancer. *Cancer Medicine, 10*(7), 2332–2340.

Pastorelli, D., Fabricio, A. S. C., Giovanis, P., et al. (2018). Phytosome complex of curcumin as complementary therapy of advanced pancreatic cancer improves safety and efficacy of gemcitabine: Results of a prospective phase II trial. *Pharmacological Research, 132*, 72–79.

Patavino, T., & Brady, D. M. (2001). Natural medicine and nutritional therapy as an alternative treatment in systemic lupus erythematosus. *Alternative Medicine Review, 6*(5), 460–471.

Pathak-Gandhi, N., & Vaidya, A. D. (2017). Management of Parkinson's disease in Ayurveda: Medicinal plants and adjuvant measures. *Journal of Ethnopharmacology, 2*(197), 46–51.

Paulo, L., Oleastro, M., Gallardo, E., et al. (2011). Anti-Helicobacter pylori and urease inhibitory activities of resveratrol and red wine. *Food Research International, 44*(4), 964–969.

Pavitt, M. J., Lewis, A., Buttery, S. C., et al. (2022). Dietary nitrate supplementation to enhance exercise capacity in hypoxic COPD: EDEN-OX, a double-blind, placebo-controlled, randomised cross-over study. *Thorax, 77*(10), 968–975.

Pavitt, M. J., Tanner, R. J., Lewis, A., et al. (2020). Oral nitrate supplementation to enhance pulmonary rehabilitation in COPD: ON-EPIC a multicentre, double-blind, placebo-controlled, randomised parallel group study. *Thorax, 75*(7), 547–555.

Pedrelli, V. F., Lauriola, M. M., & Pigatto, P. D. (2012). Clinical evaluation of photoprotective effect by a topical antioxidants combination (tocopherols and tocotrienols). *Journal of the European Academy of Dermatology and Venereology, 26*(11), 1449–1453.

Perez-Sanchez, A., Barrajon-Catalan, E., Herranz-Lopez, M., et al. (2018). Nutraceuticals for skin care: A comprehensive review of human clinical studies. *Nutrients, 10*(4).
Petramfar, P., Hajari, F., Yousefi, G., et al. (2020). Efficacy of oral administration of licorice as an adjunct therapy on improving the symptoms of patients with Parkinson's disease, a randomized double blinded clinical trial. *Journal of Ethnopharmacology, 30*(247), 112226.
Pinelli, N. R., Jaber, L. A., Brown, M. B., et al. (2010). Serum 25-hydroxy vitamin D and insulin resistance, metabolic syndrome, and glucose intolerance among Arab Americans. *Diabetes Care, 33*(6), 1373–1375.
Pinheiro, S. L., Bonadiman, A. C., Borges Lemos, A., et al. (2019). Photobiomodulation therapy in cancer patients with mucositis: A clinical evaluation. *Photobiomodul Photomed Laser and Surgery, 37*(3), 142–150.
Pivari, F., Mingione, A., Piazzini, G., et al. (2022). Curcumin supplementation (Meriva((R))) modulates inflammation, lipid peroxidation and gut microbiota composition in chronic kidney disease. *Nutrients, 14*(1).
Portincasa, P., Bonfrate, L., Scribano, M. L., et al. (2016). Curcumin and fennel essential oil improve symptoms and quality of life in patients with irritable bowel syndrome. *Journal of Gastrointestinal and Liver Diseases, 25*(2), 151–157.
Pourhabibi-Zarandi, F., Rafraf, M., Zayeni, H., et al. (2022). Effects of curcumin supplementation on metabolic parameters, inflammatory factors and obesity values in women with rheumatoid arthritis: A randomized, double-blind, placebo-controlled clinical trial. *Phytotherapy Research, 36*(4), 1797–1806.
Press C. (2002). *Handbook of nutraceuticals and functional foods*. CRC Press.
Prince, T., McBain, A. J., & O'Neill, C. A. (2012). Lactobacillus reuteri protects epidermal keratinocytes from Staphylococcus aureus-induced cell death by competitive exclusion. *Applied and Environment Microbiology, 78*(15), 5119–5126.
Pryce, G., Ahmed, Z., Hankey, D. J., et al. (2003). Cannabinoids inhibit neurodegeneration in models of multiple sclerosis. *Brain, 126*(Pt 10), 2191–2202.
Qaisar, R., Karim, A., Muhammad, T., et al. (2024). Butyrate supplementation reduces sarcopenia by repairing neuromuscular junction in patients with chronic obstructive pulmonary disease. *Respiratory Medicine, 222*, 107510.
Qiao, J., Gan, Y., Gong, Y., et al. (2021). Combination therapy with curcumin plus tamsulosin and finasteride in the treatment of men with benign prostatic hyperplasia: A single center, randomized control study. *Translational Andrology and Urology, 10*(8), 3432–3439.
Rafiq, R., Aleva, F. E., Schrumpf, J. A., et al. (2022). Vitamin D supplementation in chronic obstructive pulmonary disease patients with low serum vitamin D: A randomized controlled trial. *American Journal of Clinical Nutrition, 116*(2), 491–499.
Rahimnia, A.-R., Panahi, Y., Alishiri, G, et al. (2015). Impact of supplementation with curcuminoids on systemic inflammation in patients with knee osteoarthritis: Findings from a randomized double-blind placebo-controlled trial. *Drug Research, 65*(10), 521–525.
Rahman, M. M., Bibi, S., Rahaman, M. S., et al. (2022). Natural therapeutics and nutraceuticals for lung diseases: Traditional significance, phytochemistry, and pharmacology. *Biomedicine & Pharmacotherapy, 150*, 113041.
Rahmani, S., Asgary, S., Askari, G., et al. (2016). Treatment of non-alcoholic fatty liver disease with curcumin: A randomized placebo-controlled trial. *Phytotherapy Research, 30*(9), 1540–1548.
Rajuddin, R., Wiweko, B., Nugroho, L. (2019). The effects of curcumin administration on expression patterns of VEGF and COX-2 in fertile endometrium: A randomised clinical trial. *International Journal of Applied Pharmaceutics, 11*(Special Issue 6), 149–152.
Rawat, N., Alhamdani, A., McAdam, E, et al. (2012). Curcumin abrogates bile-induced NF-κB activity and DNA damage in vitro and suppresses NF-κB activity whilst promoting apoptosis in vivo, suggesting chemopreventative potential in Barrett's oesophagus. *Clinical and Translational Oncology, 14*(4), 302–311.

Redorta, J. P., Sanguedolce, F., Pardo, G. S., et al. (2021). Multicentre international study for the prevention with iAluRil of radio-induced cystitis (MISTIC): A randomised controlled study. *European Urology Open Science, 26*, 45–54.

Rhodes, L. E., Darby, G., Massey, K. A., et al. (2013). Oral green tea catechin metabolites are incorporated into human skin and protect against UV radiation-induced cutaneous inflammation in association with reduced production of pro-inflammatory eicosanoid 12-hydroxyeicosatetraenoic acid. *British Journal of Nutrition, 110*(5), 891–900.

Rivellese, A. A., Ciciola, P., Costabile, G., et al. (2019). The possible role of nutraceuticals in the prevention of cardiovascular disease. *High Blood Pressure & Cardiovascular Prevention, 26*, 101–111.

Rodriguez-Cruz, M., Cruz-Guzman, O. D. R., Almeida-Becerril, T., et al. (2018). Potential therapeutic impact of omega-3 long chain-polyunsaturated fatty acids on inflammation markers in Duchenne muscular dystrophy: A double-blind, controlled randomized trial. *Clinical Nutrition, 37*(6 Pt A), 1840–1851.

Ruder, E. H., Hartman, T. J., Reindollar, R. H., et al. (2014). Female dietary antioxidant intake and time to pregnancy among couples treated for unexplained infertility. *Fertility and Sterility, 101*(3), 759–766.

Rueda-Robles, A., Rodríguez-Lara, A., Meyers, M. S., et al. (2022). Effect of probiotics on host-microbiota in bacterial infections. *Pathogens, 11*(9), 986.

Ryan, J. L., Heckler, C. E, Ling, M., et al. (2013). Curcumin for radiation dermatitis: a randomized, double-blind, placebo-controlled clinical trial of thirty breast cancer patients. *Radiation Research, 180*(1), 34–43.

Ryan Wolf, J., Gewandter, J. S., Bautista, J., et al. (2020). Utility of topical agents for radiation dermatitis and pain: A randomized clinical trial. *Supportive Care in Cancer, 28*(7), 3303–3311.

Ryan Wolf, J., Heckler, C. E., Guido, J. J., et al. (2018). Oral curcumin for radiation dermatitis: A URCC NCORP study of 686 breast cancer patients. *Supportive Care in Cancer, 26*(5), 1543–1552.

Rysz, J., Franczyk, B., Kujawski, K., et al. (2021). Are nutraceuticals beneficial in chronic kidney disease? *Pharmaceutics, 13*(2).

Saadati, S., Hatami, B., Yari, Z., et al. (2019a). The effects of curcumin supplementation on liver enzymes, lipid profile, glucose homeostasis, and hepatic steatosis and fibrosis in patients with non-alcoholic fatty liver disease. *European Journal of Clinical Nutrition, 73*(3), 441–449.

Saadati, S., Hekmatdoost, A., Hatami, B., et al. (2018). Comparing different non-invasive methods in assessment of the effects of curcumin on hepatic fibrosis in patients with non-alcoholic fatty liver disease. *Gastroenterology and Hepatology from Bed to Bench, 11*(Suppl 1), S8–S13.

Saadati, S., Sadeghi, A., Mansour, A., et al. (2019b). Curcumin and inflammation in non-alcoholic fatty liver disease: A randomized, placebo controlled clinical trial. *BMC Gastroenterology, 19*(1), 133.

Saadipoor, A., Razzaghdoust, A., Simforoosh, N., et al. (2019). Randomized, double-blind, placebo-controlled phase II trial of nanocurcumin in prostate cancer patients undergoing radiotherapy. *Phytotherapy Research, 33*(2), 370–378.

Saber-Moghaddam, N., Salari, S., Hejazi, S., et al. (2021). Oral nano-curcumin formulation efficacy in management of mild to moderate hospitalized coronavirus disease-19 patients: An open label nonrandomized clinical trial. *Phytotherapy Research, 35*(5), 2616–2623.

Saberi-Karimian, M., Keshvari, M., Ghayour-Mobarhan, M., et al. (2020). Effects of curcuminoids on inflammatory status in patients with non-alcoholic fatty liver disease: A randomized controlled trial. *Complementary Therapies in Medicine, 49*, 102322.

Saghatelyan, T., Tananyan, A., Janoyan, N., et al. (2020). Efficacy and safety of curcumin in combination with paclitaxel in patients with advanced, metastatic breast cancer: A comparative, randomized, double-blind, placebo-controlled clinical trial. *Phytomedicine, 70*, 153218.

Sailo, B. L., Chauhan, S., Hegde, M., et al. (2024a). Therapeutic potential of tocotrienols as chemosensitizers in cancer therapy. *Phytotherapy Research.*

Sailo, B. L., Liu, L., Chauhan, S., et al. (2024b). Harnessing sulforaphane potential as a chemosensitizing agent: A comprehensive review. *Cancers (Basel), 16*(2).

Sajeev, A., Hegde, M., Girisa, S., et al. (2022). Oroxylin A: A promising flavonoid for prevention and treatment of chronic diseases. *Biomolecules, 12*(9).

Sandoughdaran, S., Razzaghdoust, A., Tabibi, A., et al. (2020). Randomized, double-blind pilot study of nanocurcumin in bladder cancer patients receiving induction chemotherapy. *Urology Journal, 18*(3), 295–300.

Saneian, H., Khalilian, L., Heidari-Beni, M., et al. (2021). Effect of l-carnitine supplementation on children and adolescents with nonalcoholic fatty liver disease (NAFLD): A randomized, triple-blind, placebo-controlled clinical trial. *Journal of Pediatric Endocrinology and Metabolism, 34*(7), 897–904.

Santosa, D., Suharti, C., Riwanto, I., et al. (2022). Curcumin as adjuvant therapy to improve remission in myeloma patients: A pilot randomized clinical trial. *Caspian Journal of Internal Medicine, 13*(2), 9.

Sanyal, A. J., Chalasani, N., Kowdley, K. V., et al. (2010). Pioglitazone, vitamin E, or placebo for nonalcoholic steatohepatitis. *New England Journal of Medicine, 362*(18), 1675–1685.

Sauer, S., & Plauth, A. (2017). Health-beneficial nutraceuticals-myth or reality? *Applied Microbiology and Biotechnology, 101*(3), 951–961.

Scott, J. F., Das, L. M., Ahsanuddin, S., et al. (2017). Oral vitamin D rapidly attenuates inflammation from sunburn: An interventional study. *The Journal of Investigative Dermatology, 137*(10), 2078–2086.

Sebastiani, G., Saeed, S., Lebouche, B., et al. (2020). Vitamin E is an effective treatment for nonalcoholic steatohepatitis in HIV mono-infected patients. *AIDS, 34*(2), 237–244.

Seminario-Amez, M., Lopez-Lopez, J., Estrugo-Devesa, A., et al. (2017). Probiotics and oral health: A systematic review. *Medicina Oral, Patología Oral y Cirugía Bucal, 22*(3), e282–e288.

Serban, M. C., Sahebkar, A., Zanchetti, A., et al. (2016). Effects of quercetin on blood pressure: A systematic review and meta-analysis of randomized controlled trials. *Journal of the American Heart Association, 5*(7), e002713.

Sgro, P., Ceci, R., Lista, M., et al. (2021). Quercetin modulates IGF-I and IGF-II levels after eccentric exercise-induced muscle-damage: A placebo-controlled study. *Front Endocrinol (Lausanne), 12*, 745959.

Shah, M., & Shrivastava, V. K. (2022). Turmeric extract alleviates endocrine-metabolic disturbances in letrozole-induced PCOS by increasing adiponectin circulation: A comparison with Metformin. *Metabolism Open, 13*, 100160.

Shah, R. D., Xue, C., Zhang, H., et al. (2017). Expression of calgranulin genes S100A8, S100A9 and S100A12 is modulated by n-3 PUFA during inflammation in adipose tissue and mononuclear cells. *PLoS ONE, 12*(1), e0169614.

Shah, S., Rath, H., Sharma, G., et al. (2020). Effectiveness of curcumin mouthwash on radiation-induced oral mucositis among head and neck cancer patients: A triple-blind, pilot randomised controlled trial. *Indian Journal of Dental Research, 31*(5), 718–727.

Sharifi, N., Amani, R., Hajiani, E., et al. (2014). Does vitamin D improve liver enzymes, oxidative stress, and inflammatory biomarkers in adults with non-alcoholic fatty liver disease? A randomized clinical trial. *Endocrine, 47*(1), 70–80.

Sharma, R. A., Euden, S. A., Platton, S. L., et al. (2004). Phase I clinical trial of oral curcumin: Biomarkers of systemic activity and compliance. *Clinical Cancer Research, 10*(20), 6847–6854.

Shep, D., Khanwelkar, C., Gade, P., et al. (2019). Safety and efficacy of curcumin versus diclofenac in knee osteoarthritis: A randomized open-label parallel-arm study. *Trials, 20*(1), 214.

Shinde, N., Bangar, B., Deshmukh, S., et al. (2014). Nutraceuticals: A review on current status. *Research Journal of Pharmacy and Technology, 7*(1), 110–113.

Shoskes, D., Lapierre, C., Cruz-Correa, M., et al. (2005). Beneficial effects of the bioflavonoids curcumin and quercetin on early function in cadaveric renal transplantation: A randomized placebo controlled trial. *Transplantation, 80*(11), 1556–1559.

Silalahi, T., Alwi, I., Suyatna, F., et al. (2021). Curcumin's effect on inflammatory response following percutaneous coronary intervention in adult patients with stable coronary heart disease. *International Journal of Angiology, 30*(2), 132–138.

Silva, T. A. L., Medeiros, D. C., Medeiros, G., et al. (2019). Influence of curcumin supplementation on metabolic and lipid parameters of people living with HIV/AIDS: A randomized controlled trial. *BMC Complementary and Alternative Medicine, 19*(1), 202.

Singh, S., Kola, P., Kaur, D., et al. (2021). Therapeutic potential of nutraceuticals and dietary supplements in the prevention of viral diseases: A review. *Frontiers in Nutrition, 8*, 679312.

Singla, V., Pratap Mouli, V., Garg, S. K., et al. (2014). Induction with NCB-02 (curcumin) enema for mild-to-moderate distal ulcerative colitis—A randomized, placebo-controlled, pilot study. *Journal of Crohn's and Colitis, 8*(3), 208–214.

Sirtori, C. R., Pavanello, C., Calabresi, L., et al. (2017). Nutraceutical approaches to metabolic syndrome. *Annals of Medicine, 49*(8), 678–697.

Slawik, S., Staufenbiel, I., Schilke, R., et al. (2011). Probiotics affect the clinical inflammatory parameters of experimental gingivitis in humans. *European Journal of Clinical Nutrition, 65*(7), 857–863.

Smesny, S., Milleit, B., Schaefer, M. R., et al. (2017). Effects of omega-3 PUFA on immune markers in adolescent individuals at ultra-high risk for psychosis—Results of the randomized controlled Vienna omega-3 study. *Schizophrenia Research, 188*, 110–117.

Smid, V., Dvorak, K., Sedivy, P., et al. (2022). Effect of omega-3 polyunsaturated fatty acids on lipid metabolism in patients with metabolic syndrome and NAFLD. *Hepatology Communications, 6*(6), 1336–1349.

Smith, K., & De Torres, I. (2014). A world of depression. *Nature, 515*(181), 10–1038.

Snitz, B. E., O'Meara, E. S., Carlson, M. C., et al. (2009). Ginkgo biloba for preventing cognitive decline in older adults: A randomized trial. *JAMA, 302*(24), 2663–2670.

Sohaei, S., Amani, R., Tarrahi, M. J., et al. (2019). The effects of curcumin supplementation on glycemic status, lipid profile and hs-CRP levels in overweight/obese women with polycystic ovary syndrome: A randomized, double-blind, placebo-controlled clinical trial. *Complementary Therapies in Medicine, 47*, 102201.

Somi, M. H., Fatahi, E., Panahi, J., et al. (2014). Data from a randomized and controlled trial of LCarnitine prescription for the treatment for Non-Alcoholic Fatty Liver Disease. *Bioinformation, 10*(9), 575–579.

Spiller, R., Pelerin, F., Cayzeele Decherf, A., et al. (2016). Randomized double blind placebo-controlled trial of *Saccharomyces cerevisiae* CNCM I-3856 in irritable bowel syndrome: Improvement in abdominal pain and bloating in those with predominant constipation. *United European Gastroenterology Journal, 4*(3), 353–362.

Squadrito, F., Imbalzano, E., Rottura, M., et al. (2023). Effects of genistein aglycone in glucocorticoid induced osteoporosis: A randomized clinical trial in comparison with alendronate. *Biomedicine & Pharmacotherapy, 163*, 114821.

Squadrito, F., Marini, H., Bitto, A., et al. (2013). Genistein in the metabolic syndrome: Results of a randomized clinical trial. *Journal of Clinical Endocrinology and Metabolism, 98*(8), 3366–3374.

Stancioiu, F., Mihai, D., Papadakis, G. Z., et al. (2019). Treatment for benign thyroid nodules with a combination of natural extracts. *Molecular Medicine Reports, 20*(3), 2332–2338.

Steinmetz J. D., Seeher K. M., Schiess N., Nichols E., Cao B., Servili C., Cavallera V., Cousin E., Hagins H., Moberg M. E., Mehlman M. L. (2024). Global, regional, and national burden of disorders affecting the nervous system, 1990–2021: A systematic analysis for the Global Burden of Disease Study 2021. *The Lancet Neurology, 23*(4), 344–81.

Sterzi, S., Giordani, L., Morrone, M., et al. (2016). The efficacy and safety of a combination of glucosamine hydrochloride, chondroitin sulfate and bio-curcumin with exercise in the treatment of knee osteoarthritis: A randomized, double-blind, placebo-controlled study. *European Journal of Physical and Rehabilitation Medicine, 52*(3), 321–330.

Suryawati, B., Sari, Y., Avicena, A., et al. (2022). The effect of curcumin and virgin coconut oil towards cytokines levels in COVID-19 patients at Universitas Sebelas Maret Hospital, Surakarta, Indonesia. *Pharmacognosy Journal, 14*(1).

Suskind, D. L., Wahbeh, G., Burpee, T., et al. (2013). Tolerability of curcumin in pediatric inflammatory bowel disease: A forced dose titration study. *Journal of Pediatric Gastroenterology and Nutrition, 56*(3), 277.

Tahmasebi, S., Saeed, B. Q., Temirgalieva, E., et al. (2021). Nanocurcumin improves Treg cell responses in patients with mild and severe SARS-CoV2. *Life Sciences, 1*(276), 119437.

Taibi, D. M., & Bourguignon, C. (2003). The role of complementary and alternative therapies in managing rheumatoid arthritis. *Family and Community Health, 26*(1), 41–52.

Talakesh, T., Tabatabaee, N., Atoof, F., et al. (2022). Effect of nano-curcumin on radiotherapy-induced skin reaction in breast cancer patients: A randomized, triple-blind, placebo-controlled trial. *Current Radiopharmaceuticals.*

Tambekar, D., & Dahikar, S. (2011). Antibacterial activity of some Indian Ayurvedic preparations against enteric bacterial pathogens. *Journal of Advanced Pharmaceutical Technology & Research, 2*(1), 24–29.

Tamtaji, O. R., Heidari-Soureshjani, R., Asemi, Z., et al. (2023). The effects of spirulina intake on clinical and metabolic parameters in Alzheimer's disease: A randomized, double-blind, controlled trial. *Phytotherapy Research, 37*(7), 2957–2964.

Tanabe, Y., Chino, K., Ohnishi, T., et al. (2019). Effects of oral curcumin ingested before or after eccentric exercise on markers of muscle damage and inflammation. *Scandinavian Journal of Medicine and Science in Sports, 29*(4), 524–534.

Targher, G., Bertolini, L., Scala, L., et al. (2007). Associations between serum 25-hydroxyvitamin D3 concentrations and liver histology in patients with non-alcoholic fatty liver disease. *Nutrition, Metabolism, and Cardiovascular Diseases, 17*(7), 517–524.

Thakur, K. T., Albanese, E., Giannakopoulos, P., et al. (2016). Neurological disorders. *Disease Control Priorities,* 4, 87–107.

Thomas, J. V., Smina, T. P., Khanna, A., et al. (2021). Influence of a low-dose supplementation of curcumagalactomannoside complex (CurQfen) in knee osteoarthritis: A randomized, open-labeled, active-controlled clinical trial. *Phytotherapy Research, 35*(3), 1443–1455.

Thota, R. N., Rosato, J. I., Dias, C. B., et al. (2020). Dietary supplementation with curcumin reduce circulating levels of glycogen synthase kinase-3beta and islet amyloid polypeptide in adults with high risk of type 2 diabetes and Alzheimer's disease. *Nutrients, 12*(4).

Tingo, L., Hutchinson, A. N., Bergh, C., et al. (2022). Potential modulation of inflammation by probiotic and omega-3 supplementation in elderly with chronic low-grade inflammation—A randomized, placebo-controlled trial. *Nutrients, 14*(19).

Unnikrishnan, J., Hegde, M., Girisa, S., et al. (2024). Ellagic acid-gold nanoparticle complex as anticancer agent in Triple-negative breast cancer cells. *Nano LIFE.*

Vaduganathan, M., Mensah, G. A., Turco, J. V., et al. (2022). The global burden of cardiovascular diseases and risk: A compass for future health. *Journal of the American College of Cardiology, 80*(25), 2361–2371.

Vaez, S., Parivr, K., Amidi, F., et al. (2023Mar). Quercetin and polycystic ovary syndrome; inflammation, hormonal parameters and pregnancy outcome: A randomized clinical trial. *American Journal of Reproductive Immunology, 89*(3), e13644.

Valizadeh, H., Abdolmohammadi-Vahid, S., Danshina, S., et al. (2020). Nano-curcumin therapy, a promising method in modulating inflammatory cytokines in COVID-19 patients. *International Immunopharmacology, 89*(Pt B), 107088.

Vallee, A., & Lecarpentier, Y. (2020). Curcumin and endometriosis. *International Journal of Molecular Sciences, 21*(7).

Vestergaard, M., & Ingmer, H. (2019). Antibacterial and antifungal properties of resveratrol. *International Journal of Antimicrobial Agents, 53*(6), 716–723.

Viana, S., de Bruin, V. M. S., Vasconcelos, R. S., et al. (2023). Melatonin supplementation enhances pulmonary rehabilitation outcomes in COPD: A randomized, double-blind, placebo-controlled study. *Respiratory Medicine, 220*, 107441.

Virmani, M. A., & Cirulli, M. (2022). The role of L-carnitine in mitochondria, prevention of metabolic inflexibility and disease initiation. *International Journal of Molecular Sciences, 23*(5).

Vishvakarma, P., Mandal, S., & Verma, A. (2023). A review on current aspects of nutraceuticals and dietary supplements. *International Journal of Pharma Professional's Research (IJPPR), 14*(1), 78–91.

Wah Kheong, C., Nik Mustapha, N. R., & Mahadeva, S. (2017). A randomized trial of silymarin for the treatment of nonalcoholic steatohepatitis. *Clinical Gastroenterology and Hepatology, 15*(12), 1940–1949 e8.

Wang, M., Qin, S., Zhang, T., et al. (2015). The effect of fruit and vegetable intake on the development of lung cancer: A meta-analysis of 32 publications and 20,414 cases. *European Journal of Clinical Nutrition, 69*(11), 1184–1192.

Watanabe, Y., & Tatsuno, I. (2020). Prevention of cardiovascular events with omega-3 polyunsaturated fatty acids and the mechanism involved. *Journal of Atherosclerosis and Thrombosis, 27*(3), 183–198.

Weber, S. U., Thiele, J. J., Han, N., et al. (2003). Topical alpha-tocotrienol supplementation inhibits lipid peroxidation but fails to mitigate increased transepidermal water loss after benzoyl peroxide treatment of human skin. *Free Radical Biology & Medicine, 34*(2), 170–176.

Werny, J. G., Sagheb, K., Diaz, L., et al. (2022). Does vitamin D have an effect on osseointegration of dental implants? A systematic review. *International Journal of Implant Dentistry, 8*(1), 16.

Westphal, L. M., Polan, M. L., & Trant, A. S. (2006). Double-blind, placebo-controlled study of Fertilityblend: A nutritional supplement for improving fertility in women. *Clinical and Experimental Obstetrics and Gynecology, 33*(4), 205–208.

Winiarska-Mieczan, A., Tomaszewska, E., Donaldson, J., et al. (2022) The role of nutritional factors in the modulation of the composition of the gut microbiota in people with autoimmune diabetes. *Nutrients, 14*(12).

Wisuitiprot, W., Ingkaninan, K., Jones, S., et al. (2022). Effect of green tea extract loaded chitosan microparticles on facial skin: A split-face, double-blind, randomized placebo-controlled study. *Journal of Cosmetic Dermatology, 21*(9), 4001–4008.

Wong, R. H., Howe, P. R., Buckley, J. D., et al. (2011). Acute resveratrol supplementation improves flow-mediated dilatation in overweight/obese individuals with mildly elevated blood pressure. *Nutrition, Metabolism, and Cardiovascular Diseases, 21*(11), 851–856.

Wong, V. W., Wong, G. L., Chan, A. W., et al. (2013). Treatment of non-alcoholic steatohepatitis with *Phyllanthus urinaria*: A randomized trial. *Journal of Gastroenterology and Hepatology, 28*(1), 57–62.

Xie, Y., Zhu, G., Yi, J., et al. (2022). A new product of multi-plant extracts improved skin photoaging: An oral intake in vivo study. *Journal of Cosmetic Dermatology, 21*(8), 3406–3415.

Yamakoshi, J., Sano, A., Tokutake, S., et al. (2004). Oral intake of proanthocyanidin-rich extract from grape seeds improves chloasma. *Phytotherapy Research, 18*(11), 895–899.

Yoon, J. Y., Kwon, H. H., Min, S. U., et al. (2013). Epigallocatechin-3-gallate improves acne in humans by modulating intracellular molecular targets and inhibiting P. acnes. *Journal of Investigative Dermatology, 133*(2), 429–440.

Yuan, H., Li, Y., Ling, F., et al. (2020). The phytochemical epigallocatechin gallate prolongs the lifespan by improving lipid metabolism, reducing inflammation and oxidative stress in high-fat diet-fed obese rats. *Aging Cell, 19*(9), e13199.

Zanetti, M., Gortan Cappellari, G., Barbetta, D., et al. (2017). Omega 3 polyunsaturated fatty acids improve endothelial dysfunction in chronic renal failure: Role of eNOS activation and of oxidative stress. *Nutrients, 9*(8).

Zanforlini, B. M., Ceolin, C., Trevisan, C., et al. (2022). Clinical trial on the effects of oral magnesium supplementation in stable-phase COPD patients. *Aging Clinical and Experimental Research, 34*(1), 167–174.

Zare'i, M., Rabieepour, M., Ghareaghaji, R., et al. (2024). Nanocurcumin supplementation improves pulmonary function in severe COPD patients: A randomized, double blind, and placebo-controlled clinical trial. *Phytotherapy Research, 38*(3), 1224–1234.

Zarzour, R. H. A., Alshawsh, M. A., Asif, M., et al. (2018). Adipocytokine regulation and antiangiogenic activity underlie the molecular mechanisms of therapeutic effects of *Phyllanthus niruri* against non-alcoholic fatty liver disease. *Nutrients, 10*(8).

Zaura, E., & Twetman, S. (2019). Critical appraisal of oral pre- and probiotics for caries prevention and care. *Caries Research, 53*(5), 514–526.

Zhang, H., Gordon, R., Li, W., et al. (2019). Genistein treatment duration effects biomarkers of cell motility in human prostate. *PLoS ONE, 14*(3), e0214078.

Zheng, X., Hai, J., Yang, Y., et al. (2023). Effects of resveratrol supplementation on cardiac remodeling in hypertensive patients: A randomized controlled clinical trial. *Hypertension Research, 46*(6), 1493–1503.

Zhou, Y., Zeng, Y., Pan, Z., et al. (2023). A randomized trial on resveratrol supplement affecting lipid profile and other metabolic markers in subjects with dyslipidemia. *Nutrients, 15*(3).

Zordoky, B. N., Robertson, I. M., & Dyck, J. R. (2015). Preclinical and clinical evidence for the role of resveratrol in the treatment of cardiovascular diseases. *Biochimica et Biophysica Acta (BBA)-Molecular Basis of Disease, 1852*(6), 1155–1177.

Fostering Partnership in Bioprospecting

Access and Benefit Sharing (ABS) Potential Estimation: Methodology and Empirical Results

Prakash Nelliyat

Introduction

Access and Benefits Sharing (ABS), one of the key objectives of the Convention on Biological Diversity (CBD) is getting universal acceptability and the Parties of the Convention have been establishing the required legal and institutional frameworks for its operationalization. However, the practicability of the ABS (obtaining permission when accessing the genetic/biological resources for commercial purposes, benefit sharing by bio-prospectus to the community, conversation and sustainable use of biodiversity with the participation of the Indigenous and Local Communities—ILCs) is a slow process and countries are facing several challenges. Even if India made some achievements in this regard, there is still a long way to go.

ABS also can emerge as an innovative financial mechanism for biodiversity conservation. Globally, countries are facing difficulties in mobilizing the required money for biodiversity management. Presently, in many countries, ABS is in its infant stage and the Government does not have any ideas about the ABS potential. But genetic/biological resources are being used by the bio-prospectors and biological resources based industries and business sectors at an alarming rate and they are making huge benefits out of it. Generally, the industries are not coming forward and complying with the legal provisions of ABS and benefit sharing. Further, policymakers do not have any idea about the ABS potential in their jurisdiction. The study attempts to assess the ABS potential of Kerala, one of the richest biodiversity states in India.

The biodiversity of Kerala provides multiple ecosystem services as well as a source of income and livelihood options for millions of poor people. The biodiversity goods (genetic/biological resources) of Kerala are raw materials for manufacturing

P. Nelliyat (✉)
Centre for Biodiversity Policy and Law, National Biodiversity Authority, Taramani, Chennai 600113, India
e-mail: prakashnelliyat@gmail.com

L. Krishna Panicker et al. (eds.), *Biodiversity and Business*,
https://doi.org/10.1007/978-3-031-71674-4_17

different consumer products or the benefit/profits options for a large number of bio-entrepreneurs not only in Kerala but in other parts of the country and abroad. In this context, the management and sustainable use of biodiversity is significant. However, mobilizing the required funds for biodiversity conservation is a challenge, where the State can promote ABS as an effective financial option on the grounds of its unique biodiversity or rich biological resources.

Kerala's forest resources, including Timber and Non-Timber Forest Products (NTFPs), marine resources, agricultural crops, and livestock resources have high commercial significance. Understanding the commercial use of these resources and the value generated at different levels (inputs and final product manufacturing) as well as the possible ABS out of it, helps the Government to take appropriate policy measures towards biodiversity conservation and sustainable use. This exercise has been done with the support of the available secondary data collected from various Government Departments. Broadly, the industries' cooperation, in sharing reliable information, was limited.

The first part of the chapter focused on the ABS philosophy in the context of biodiversity conservation and India's achievements in operationalizing the ABS. The second part is about Kerala's biodiversity and its commercial significance and the need for escalation of the ABS. The third section discussed the methodology, database for ABS estimation, and practical exercise with the empirical results of Kerala. The last section discusses the challenges in ABS estimation and the need for further research in this area.

Access and Benefit Sharing (ABS) an Emerging Option for Biodiversity Conservation

During the last few decades, global biodiversity has been under serious pressure primarily due to anthropogenic reasons. The rapid growth of population, expansion of production, trade, and market (particularly after globalization) induced natural resources over-exploitation and pollution, rigorous land use change, hampering ecosystem stability, etc. are the major human-induced reasons for the degradation of biodiversity. Besides, climate change-induced frequent rainfall variations (flood and drought) and widespread growth of invasive alien species also adversely affect biodiversity. The recent report of the Intergovernmental Science-Policy Platform on Biodiversity and Ecosystem Services (IPBES) Global Assessment presents a warning picture of the deterioration of the health of ecosystems (United Nations, 2019). But, the fact is that biodiversity once lost is lost forever as it is irreversible (Nelliyat, 2017).

At the same time, we are also realising that the conservation of biodiversity is needed for global harmony and human existence. After this realization at the international level, serious initiatives emerged under the CBD; among them, ABS is a landmark. ABS refers to how genetic resources can be accessed and how users and providers reach agreements on the fair and equitable sharing of the benefits that might result from their use (CBD, 2011a).

Users of the genetic resources include research institutes or companies seeking access to basic scientific research or product development. Providers include States (including local communities) that have sovereign rights over biological resources. The Provider country may engage the indigenous and local communities (ILCs) to negotiate the terms and conditions associated with ABS. The participation of ILCs is necessary when traditional knowledge (TK) associated with genetic resources is being accessed.

Article 15 of the CBD sets out rules that govern ABS. Under these rules, the governments of countries have two key responsibilities: (a) to put in place a system that facilitates access to genetic resources for environmentally sound purposes. (b) to ensure that the benefits resulting from their use are shared fairly and equitably between the users and the providers. Access to genetic resources can lead to benefits for both users and providers. The fact is that through the ABS mechanism, the genetic resources were accessed and used to maximize the benefits for users and providers and the ecology and communities where they are found (CBD, 2011b).

For developing countries, granting access to genetic resources in exchange for a share of monetary and non-monetary benefits could contribute significantly to poverty alleviation and sustainable development. In certain cases, access to genetic resources may depend on using the traditional knowledge of ILCs. ABS rules recognize the value of this knowledge by requiring users to obtain permission to use it and to share any benefits that result from its use with the communities who own it (CBD, 2011b).

Both users and providers must understand and respect the legal, administrative, and policy frameworks at the national and local levels, as well as those outlined in the Convention of Biological Diversity (CBD) and the Nagoya Protocol on ABS. The ABS is based on prior informed consent (PIC) being granted by a provider to a user, and negotiations between both parties that result in mutually agreed terms (MAT). Presently, 196 countries are Parties in the CBD. Out of this 142 countries ratified the Nagoya Protocol on ABS (CBD, 2024). It implies that a substantial number of countries have accepted biodiversity conservation through the ABS mechanism.

In brief, ABS is emerging as an innovative approach and an incentive mechanism in biodiversity conservation and its sustainable utilization. The ABS framework provides formal guidance for how biological or genetic resources are accessed, and for the people or countries that provide them. Broadly, the ABS philosophy proposes that providers of biological resources are entitled to receive fair benefits from the users. In this context, ABS balances the rights of the users of biological resources with the rights of the providers of such resources. Further, the ABS can manage biodiversity as a community asset and support biodiversity-based businesses effectively and sustainably (Nelliyat & Pisupati, 2013).

India's Achievements in ABS Initiatives

As a party to the CBD, India's initiatives in fulfilling its objectives were significant. India ratified the Nagoya Protocol on ABS in 2014 and implemented the Protocol through various legal, administrative, policy, and institutional measures. Before the ratification of the Nagoya Protocol, India enacted the Biological Diversity Act in 2002 and the Biological Diversity Rules in 2004 (concurrent with the CBD objectives). Significant changes were made to the Biological Diversity Act, 2002 by way of amendments in 2023. Further, India came up with the Guidelines on Access to Biological Resources and Associated Knowledge and Benefit Sharing Regulations, 2014, as well as several Notifications and State level legislations. The BD Act is implemented through a three-tier institutional mechanism: the National Biodiversity Authority (NBA) at the national level; State Biodiversity Boards (SBBs) at the State Government or provincial level; and the Biodiversity Management Committees (BMCs) set up by the elected bodies at the local level. Broadly, the major steps taken by the country in implementing the Nagoya Protocol on ABS are appreciable at the individual level. India's status and achievements in the implementation of the Nagoya Protocol are submitted to the CBD in the form of the "Interim National Report on the Implementation of the Nagoya Protocol" (CBD, 2017).

The BD Act, 2002 proposes the ABS provisions when users access biological resources and associated traditional knowledge with commercial intent. Section 3 of the BD Act insists that 'certain persons do not undertake biodiversity-related activities without the approval of the NBA'. Section 4 mentioned that the 'results of research should not be transferred to certain persons without the approval of the NBA'. Section 6 pointed out that 'applications of IPR are not to be made without the approval of the NBA'. Similarly, as per Section 7, 'prior intimation of SBBs for obtaining biological resources for certain purposes' is required (NBA, 2010). ABS Guidelines, 2014 specified the benefit-sharing criteria when utilizing biological resources and associated traditional knowledge for different purposes (MoEFCC, 2014). ABS agreements under the BD Act are divided into four categories, and necessitate the signing of legally binding arguments through various forms. During the past few years, the NBA put a robust and responsive ABS in place. India is the only contrasting party in the CBD that has entered into more than 5000 ABS agreements so far. This is a huge step in operationalizing the ABS provision within the Act as well as implementing the provisions under Article 15 of the CBD (NBA, 2024). As of April 2024 NBA received 8252 ABS applications in different categories. Out of it, 5259 applications were approved and ABS agreements were signed (Table 1).

The table reveals that there is a steady growth in ABS agreements in the country over a period. A majority of the applications (81%) are for Form 3, for approval for obtaining the IPR followed by Form 1—access to biological resources for research or commercial purposes—(14%). Industries that use genetic/biological resources for commercial purposes and research results transfer and IPR application should follow the BD Act provisions. But, in reality, a majority of the industries do not comply with the Act or share the benefits acquired through bio-prospecting with

Table 1 Status of ABS approval granted by NBA (up to April 2024)

Year	Form I (access to bio-resources for research/commercial purpose)	Form II (transfer of research results)	Form III (approval for obtaining IPR)	Form IV (third party transfer)	Form B	Total
2006–2007	4	1	0	2	0	7
2007–2008	5	3	12	6	0	26
2008–2009	4	4	21	6	0	35
2009–2010	2	1	9	1	0	13
2010–2011	3	1	4	1	0	9
2011–2012	1	2	6	0	0	9
2012–2013	1	0	8	7	0	16
2013–2014	1	0	14	2	0	17
2014–2015	19	0	22	1	0	42
2015–2016	31	1	51	2	7	92
2016–2017	36	4	127	0	15	182
2017–2018	36	2	246	1	31	316
2018–2019	65	1	208	1	17	292
2019–2020	30	1	256	0	39	326
2020–2021	76	9	968	0	22	1075
2021–2022	171	3	714	4	19	911
2022–2023	132	6	755	0	25	918
2023–2024	121	5	827	0	14	967
2024–2025	1	1	2	0	0	4
Total	739	45	4252	34	189	5259

Source NBA (2024)

the local community (through NBA/SBBs) for biodiversity conservation. Hence, a significant portion of finance (ABS amount) which is supposed to be assigned for biodiversity conservation is being lost for the country. This is a concern not only for the management of biodiversity but to sustain the livelihood of millions of poor people, who depend upon the biodiversity of the country for their livelihood.

These non-compliance (BD Act) issues of biological resources-based industries and other stakeholders were discussed and debated in various forums, but the results are sketchy. Different actors raised their concerns in complying with the Act. Policy-makers and enforcement agencies are completely unaware of the spectrum of ABS in the country. In India, NBA and SBBs mobilized an amount of Rs. 1600 Million so far through the ABS. However, we don't have any idea about the overall ABS potential of India: how much we have collected, and how much we have missed. As the methodology and the estimation are cumbersome and complex tasks, considering the range of uncertainties many researchers have not attempted the ABS potential estimation. In this context, the estimation of the ABS potential will enhance its scope.

The present study is a pioneering attempt in the estimation of the ABS potential of one of the biodiversity-rich States (Kerala) in India.

Biodiversity of Kerala and Its Economic Significance

Kerala, one of the richest biodiversity coverage geographical areas in the country, is located in the southern part of India (Fig. 1). The state's eastern part consists of the Western Ghats, which is also the catchment of several perennial rivers and consists of large numbers of unique flora and fauna. This special agroecological zone in the Western Ghats region is famous for different varieties of medicinal and aromatic plants and species which have high demand at national and international levels.

The mountain landscape vegetation of these areas consists of sholas, grasslands, dry mixed and moist deciduous forests, forest plantations (especially eucalyptuses wattle, pine, teak, and sandal) commercial plantations (tea, coffee, and spices), horticultural fields and mixed farms. Many timber species in Kerala's forests have good economic value. There are 117 species of trees used for softwood purposes, 146 species for the extraction of tannin, 29 species for the extraction of latex, 111 species for the extraction of gum, 65 species for the extraction of resin, and 284 species for the extraction of oil. Further, 1170 species are used medicinally. The traditional and local communities who are empowered with biological resources based TK in the Western Ghats landscape use the locally available medicinal and aromatic plants and different agricultural products for their day-to-day life (Nelliyat, 2020).

The western part of the State is the Arabian Sea coast. This low-lying coastal area consists of several estuaries and backwaters and is rich in different and unique aquatic flora and fauna. Broadly, the wetland ecosystem of the State primarily consists of freshwater lakes, ponds, streams, rivers, and brackish water estuaries and backwaters. These water bodies are rich in aquatic flora and fauna having significant market potential and economic value. The brackish water areas in the coastal belt of the State are famous for aquaculture.

Kerala is a major producer of various agricultural produce that includes: coconuts, spices, cashews, different varieties of grains, vegetables, and fruits. The agro-climatic conditions in the hilly areas of the State are conducive for the growth of highly diversified floral species which are ecologically and economically vital for the day-to-day livelihood of the tribal community. These areas are also famous for commercial plantations such as tea, cardamom, coffee, and home gardens. Despite its limited land mass, Kerala accounts for nearly 20 percent of the country's total food exports (Nelliyat, 2020).

The unique landscape and the agro-climatic conditions in Kerala enrich different ecosystems, which are the buffer zones of biodiversity and consist of a large number of biological resources having significant economic potential and commercial values. Biological resources-related sectors (food processing, spices, rubber, ayurveda, handlooms, textiles includes cotton-based apparel and garments, coir, traditional wood carvings, etc.) that obtain their raw materials from the biodiversity, are playing a

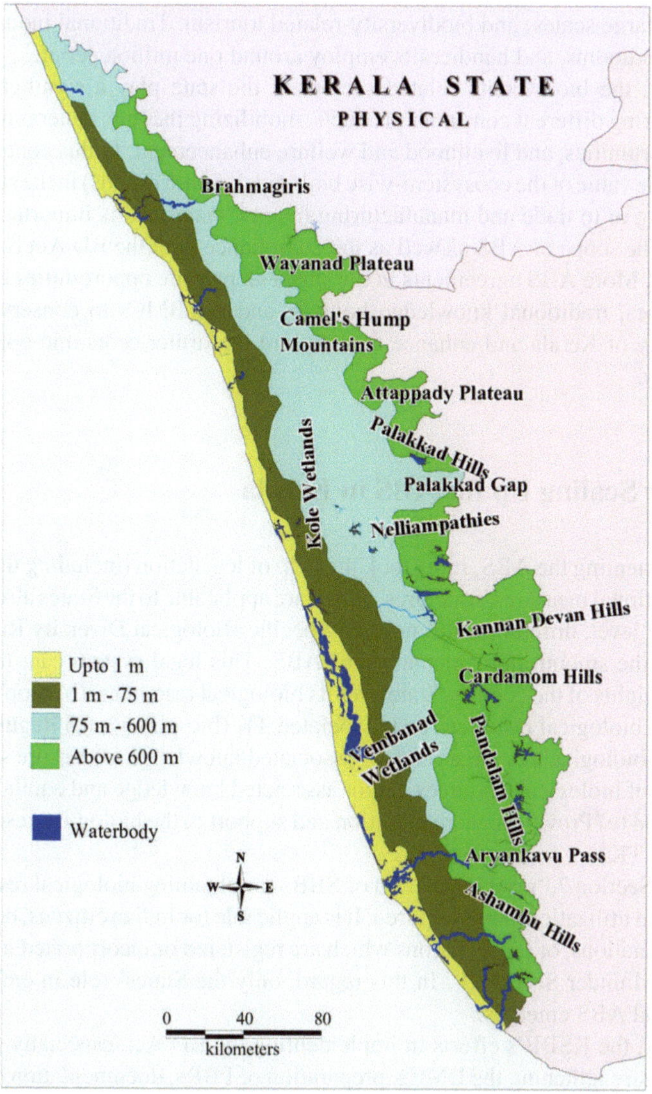

Fig. 1 Map of Kerala. *Source* Kerala geography map—Search Images (bing.com)

significant role in building Kerala's economy. Kerala is one of the leading producers of natural rubber, coir, coconut, cashew, coffee, and spices in India. In the allied agriculture sector, livestock is the fastest-growing sector. Around 10% of the State's population is directly dependent on biodiversity for their livelihood. In addition, a considerable number of people are indirectly dependent on biological resources-based trade and value addition (manufacturing of different consumer products on small and large scales) and biodiversity-related tourism. Traditional industries such as coir, handlooms, and handicrafts employ around one million people.

In brief, the biodiversity related sectors of the state play a significant role in manufacturing different consumer products, mobilizing income, generating employment opportunities, and livelihood and welfare enhancement. In this context, understanding the value of the ecosystem-wise biological wealth (goods) in the state, which are entering in to trade and manufacturing (as raw materials) is important. Further, assessing the scope of ABS as well as the compliance with the BD Act is extremely significant. More ABS agreements in the state can provide opportunities to the local communities, traditional knowledge holders, and the BMCs to conserve the rich biodiversity of Kerala and enhance and explore its further economic opportunities sustainably.

Need for Scaling Up the ABS in Kerala

For implementing the ABS, India took the help of legislation (including the BD Act) and institutional measures/structures, which are applicable to the States also. Besides, at the state level, different states notified Specific Biological Diversity Rules, introduced for the smooth implementation of ABS. This legal system can: (a) Provide sovereign rights of the country/State over its biological resources, (b) Stop misappropriation of biological resources and associated TK (bio-piracy), (c) Regulate access and use of biological resources and /or associated knowledge (d) Ensure sustainable utilisation of biological resources and/or associated knowledge and equitable benefit sharing, and (e) Provide legal recognition and support to the biological resources and associated TK.

As per Section 7, 'prior intimation of SBBs for obtaining biological resources for commercial utilization, are is required. It is applicable for Indian citizens, body corporates, associations, or organisations which are registered or incorporated in India and not covered under Section 3'. In this regard, only the States' role in enforcing the BD Act and ABS emerges.

Broadly, the KSBB's efforts in implementing the BD Act, especially for formulating and strengthening the BMCs, preparation of PBRs, documentation of tradable biological resources, capacity building, and awareness generation on various provisions of BD Act to different stakeholders are commendable and serve as a model to other SBBs in the country. However, KSBB signed only four ABS agreements and mobilized Rs. 88,310/- so far.

ABS Potential Estimation: Process, Methodology and Database

As per the "ABS Guidelines – 2014", ABS can be availed of from the biological resources-based industries either:

(a) Based on the biological resources' purchased price by the industries, based on the raw-material cost, prescribed as 3–5%.
  or
(b) Based on the ex-factory sale value of the product minus government taxes, where biological resources are involved in production fully or partially (0.1–0.5%).

ABS Potential Estimation Based on Biological Resources' Purchased Price by the Industries

For the successful implementation of this approach, we need (a) a comprehensive list of biological resources-based industries that come under the ambit of ABS and (b) the database on industries' use or consumption of biological resources (type, volume, purchase price, etc.). Unfortunately, we are not in a position to get this information (each industry-wise) either through the secondary data or through the primary sources. In Kerala, neither the KSBB nor any other agency list this. When we approach some industries randomly, they are unwilling to share it (business secrets). Most industries are also unaware of the origin of the biological resources (from where and when it is collected or cultivated). They are getting it from the wholesalers or traders. In this situation, the ABS amount to be collected by the Board from the bio-prospecting industries may not be distributed to the provider. Further, industries in Kerala obtaining their biological resources from other states of India or abroad also complicates the benefit-sharing process.

Ecosystem Goods Based: Approach, Process and Database

Due to all these constraints, we searched for possible alternative approaches and realised that the ecosystem goods-based approach has been followed with certain specific assumptions; this has been discussed with the ABS experts and fine-tuned. The base for this assessment is the ecosystem-wise biological resources (quantity and their market value) estimation and ABS share derivation with certain assumptions. In this regard, the major processes are:

1. Classified the state's ecosystems into different categories, such as forests, marine and inland, agriculture, livestock, and tourism.
2. List out the major goods having market and commercial value and the quantities coming from each ecosystem.
3. Collected the value or market price of the resources.

4. Identified the ratio of biological resources that go for commercial utilization from the total volume of the resources.
5. Estimate the ABS potential considering the norms prescribed in the ABS Guidelines.

Biological resources, such as timber and non-timber forest products, fish, agricultural produce, and livestock produce were considered for valuation and ABS potential estimation (the value of tourism from natural ecosystems was calculated with respect to its recreational use value). The biological resources (goods from ecosystems) that have market value have been captured in the study. In other words, the biological resource value at the ecosystem stage (from which it is extracted) is the criterion for the assessment. District-wise details were collected for five years (2015–2020), and the cumulative average value was considered for analysis to avoid the error in extreme cases/years.

The key sources of data are the published documents by various Government Departments: State Planning Board, Department of Statistics; Departments of Forests, Fisheries, Agriculture, Animal Husbandry, Livestock, and Tourism; Kerala State Biodiversity Board, Directorate of Industries and Commerce, District Industries Centre, etc. Besides, available secondary data was also collected from academic and research institutions, different industries and business enterprises, private institutions, and NGOs. Primary data was collected through household surveys/FGD among the target groups (farmers, fishermen, tribes, etc.), and interviews with biological resource suppliers/traders (timber depot officials, bidders of NTFP, agriculture produce and fish) and users (industries).

The Forest sectors' database includes species-wise quantity and value of different forest products (Timber and NTFPs). The quantity and value of industrial wood sold /auctioned at the depots are collected from the Forest Department. Besides, there is a considerable production of wood from social forests and other plantations in forests and outside forests or the forest fringe land. These details were collected from the Kerala Forest Development Cooperation (KFDC). Data on Marayoor sandalwood was also collected. Species-wise quantity and values of different NTFPs were collected from the Kerala State SC/ST Federation. The data on wood outside forests was compiled from a 'Report on Saw-mills of Kerala' prepared by the Directorate of Economics and Statistics, Government of Kerala.

In the fishery sector, species-wise quantity and value of fish from the ocean, coastal and offshore waters, and inland waters were considered. The Fisheries Department is the principal source of data for the fishing sector. For estimating the species-wise quantity and value, the average annual auction prices of marine and inland fish collected at the landing centers were used.

Agriculture sector data were primarily collected from Agricultural Statistics 2018–2019 and Price Statistics 2017. This database includes crop-wise area under cultivation, crop production, and productivity, the farm price of produce, and the cost of cultivation. The major crops evaluated include:: Paddy, Spices, Sugar crops, Fresh fruits, Dry fruits, Tubers, Vegetables, Oil seeds, Fibre, Drugs, Narcotics Plantation Crops, Fodder, Medicinal plants, etc.

Livestock analysis focuses on the production of milk, meat, and eggs and its quantity and value. Indigenous cattle buffalo crossbred cattle and goats are the main sources of milk production. Cattle, buffalo, pig, goat, poultry, fowl, and duck meat were considered. Data were collected from the Livestock Statistics and Economic Review (2019), published by the Kerala State Planning Board.

Biodiversity and eco-tourism-related data were compiled from published reports of Kerala Tourism Statistics, (2019) and Kerala Forest Department to understand the income generated from the tourism sector. However, we have not done any ABS calculations.

The valuation of biological resources originated from different ecosystems of the State and its ABS potential is explained below.

Forest Sector/Ecosystem

High-Value Bio-Resources (Sandalwood and Other Timbers)

Kerala's forest ecosystem is rich in timber (wood) and NTFPs. *Marayoor* sandalwood has a high demand even in international markets. In the case of NBA, out of the total ABS amount collected so far, around 95% has been obtained from red sanders. The ABS potential of high-value biological resources will be estimated based on the norms prescribed in the ABS Guidelines as well as the Guidelines issued by the NBA for the red sanders' ABS estimation. The following table (Table 2) provides an overall picture (quantity and value) of the timber sector (both in the forest and outside the forest) in Kerala, based on auction value.

As per the estimates, the total value generated through the timber sector in Kerala is Rs. 21,910 million of which Rs. 2097 Million is from forests and Rs. 19,813 Million through timber outside the forests. The timber auctioned by the timber depots and KFDC is primarily industrial wood and it may be used in different sectors including the domestic. Hence, only 50% of the timber value is considered for the ABS estimation. The ABS Guidelines prescribed that 3–5% of biological resources' purchase price by the industry might be the ABS share. However, we considered its average value as 4% for the ABS estimation. In the case of Marayoor sandalwoods, as a certain portion is going to the temples for ritual purposes, 50% is assumed to go to industry and considered for the ABS estimation. Timber outside the forests is from private lands including the rubber garden or from Government (public) lands, in which a considerable share may go for own (domestic) use. We are assuming that 20% of the total quantity may go to the industrial sector for value addition and 4% is considered as ABS amount. The estimated ABS Potential of timber both from the forests and outside it is Rs. 200 million per year (Table 2).

Table 2 ABS potential of timber (based on auction value)

Timber source	Mode of estimation	Quantity (M^3/ Kg)	Value (Rs. Millions)		Possible ABS potential value			
			Value	%	% of value considered for ABS estimation with assumption (value consider for ABS)	ABS Ratio	ABS amount (Rs. Millions)	%
Timber from forests								
27 timber depots	Cumulative annual average (2015–2020)	26422.07 (M^3)	1540	73	50% of the timber from the depots go for value addition in high-end industries (Rs. 770 Million)	4%	31	
KFDC	Cumulative annual average (2015–2020)	9684.30 (M^3)	59	3	50% of the timber from KFDC go for value addition in high-end industries (Rs.30 Million)		1	
Marayoor Sandalwood	Cumulative annual average (2015–2020)	72,991 (Kg)	498	24	50% go for value addition in high-end industries (Remaining go for ritual purpose in temples) (Rs. 249 Million)		10	
Forest (Total)			2097	100	(Rs. 1049 Million)		42	21

(continued)

Table 2 (continued)

Timber source	Mode of estimation	Quantity (M^3/ Kg)	Value (Rs. Millions)		Possible ABS potential value			
			Value	%	% of value considered for ABS estimation with assumption (value consider for ABS)	ABS Ratio	ABS amount (Rs. Millions)	%
Timber outside forest								
Outside forest (Total)	Annual total (2014–2015)	22,56,219 (M^3)	19,813		20% go for value addition in high-end industries (80% for landowners self use) (Rs. 3963 Million)	4%	158	79
Grand total			21,910		Rs. 5012 Million		200	100

Table 3 ABS potential of NTFPs in Kerala

Method/year	Quantity (Kg)	Total value (Rs. in Millions)	Value considered for ABS (50% of total value) (Rs. in Millions)	ABS potential ratio	ABS Potential (Rs. in Millions)
Cumulative averagc (2015–2019)	826,574	42	21	3%	0.6

Non Timber Forest Products (NTFPs)

Some of the NTFPs, particularly the medicinal plants and aromatic plants from Kerala's forests have high ABS potential. The cumulative annual average value of the various NTFPs is Rs. 42 Million. However, we assumed that around 50% of NTFPs go for domestic (non-commercial) purposes and traditional practices. As per the Biological Diversity Act, the biological resources used by the local people and communities, including growers and cultivators of biodiversity, and *vaids* and *hakims*, who have been practicing indigenous medicine, are exempted from the provisions of the Act. Hence, only 50% of the total value (Rs. 21 Million) is considered for the ABS potential estimation. The benefit sharing ratio considered is 3% (lowest range) of the value. Accordingly, the ABS potential of the NTFPs of Kerala is Rs. 0.6 Million per year (Table 3).

Fisheries Sector

Both the marine and inland fisheries in Kerala generated significant value and can also contribute substantially to the ABS. Some of the fish species available in the State have high export value. Fish processing industries located in Kerala as well as in neighboring states use the fish landed or cultured in Kerala as raw material and manufacture different value-added products (fish oil, fish meat/*surumi* products, etc.). The total value of fish is Rs. 121,574 Million, in which marine fisheries' contribution is Rs. 83,167 Million (68%) and the inland's is Rs. 38,407 Million (32%).

Through the discussion with experts, we realize that 10% of the total fish yield or the catch is going for value addition or used by fish processing industries or exports and considered for ABS estimation (around 90% of the total fish catch in the State is used for domestic/local or non-commercial consumption)., The benefit sharing ratio taken into account is 3% of the share value for both the marine and inland sectors. Accordingly, the total ABS potential of the fisheries sector in the State is Rs. 365 Million per year, which includes Rs. 250 Million from marine and Rs. 115 Million from the inland sectors (Table 4).

Table 4 ABS potential of fisheries

Method/year	Quantity (MT)	Total value (Rs. Millions)	Value considered for ABS (10% total value)	ABS potential (Ratio)	ABS potential (RS. Millions)	%
Marine fisheries						
Cumulative average (2015–2019)	518,783	83,167 (68%)	8317	(3% of the value)	250	68
Inland fisheries						
Cumulative average (2015–2019)	197,086	38,407 (32%)	3841	(3% of the value)	115	32
Total	715,869	121,574 (100%)	12158		365	100

Table 5 ABS potential of agriculture

Year	Production (MT)	Value (Rs. Millions)	Value considered for ABS (Rs. Millions) (10% of the total value)	ABS potential ratio	Potential ABS value (Rs. Millions)
2018–2019	5,213,126.3	236,141	23,614	3% of the value	708

Agriculture Sector

For the agriculture sector, 26 major crops items of Kerala were considered for estimating the net value which comes to Rs. 236,141 Million. Out of this, only 10% (Rs. 23,614 Million) is considered for ABS estimation, as most of the crop items are directly used for household purposes or domestic consumption. Further, considerable volumes of agricultural produce are under NTC (Normally Traded as Commodities). The benefit-sharing ratio considered for estimating the ABS is 3% of the value. In brief, the ABS potential of Kerala's agriculture sector is Rs. 708 Million per year (Table 5).

Livestock Sector

The total annual value generated from the livestock sector through milk, egg, and meat production is Rs. 289,168 Million. As most of the livestock products are used directly in the domestic sectors, we considered a limited percentage of the value for ABS estimation (10% for milk and 5% for egg and meat). Accordingly, the total

Table 6 ABS potential of livestock

Livestock product	Production	Value (Rs. Millions)	Value considered for ABS (Rs. Millions)	ABS potential ratio	Potential ABS value (Rs. Millions)	%
Milk	2456 (Million Litres)	124,791	12,479 (10% of the total value)	(3% of the value)	374	60
Egg	2180 (Million numbers)	13,095	655 (5% of the total value)		20	3
Meat	469 (Million Kg)	151,282	7564 (5% of the total value)		227	37
Total		289,168	20,698		621	100

value eligible for ABS is Rs. 20,698 Million. The benefit-sharing ratio considered in the livestock sector is 3%. Accordingly, the projected ABS in the livestock sector in Kerala is Rs. 621 Million per year, which includes Rs. 374 Million from milk; Rs. 20 Million from eggs, and Rs. 227 Million from meat (Table 6).

The total value of biological resources of Kerala and its ABS potential: In brief, the total value of the goods derived from different ecosystems of the State is Rs.668,835 Million. The estimated biological resources (purchase price of industry) based ABS potential of Kerala would be Rs. 1895 Million (Table 7).

In the assessment, we are assuming that from the total volume of biological resources (goods) after the domestic use the remaining will be used by the industries as raw materials or for commercial purposes. But, some percentage of it may go for direct export, which may not reach the industries in Kerala. However, during the consultation with the industries, we realized that industries in Kerala are using the biological resources imported from other countries also as raw materials. Hence, we assume that the volume of exports may equal the volume of imports.

ABS Potential Estimation Based on the Biological Resources Based Manufacturing Industries Ex-factory Sale Value (Turnover)

The biological resources-based industries in the State are classified into two categories: (a) Micro, Small, and Medium Enterprises (MSMEs) and (b) large factories. Relevant data (category-wise number of units, production/products, annual turnover, etc.) were collected from the Directorate of Industries and Commerce and Annual Survey of Industries. The analysis of data on MSMEs and large factories in Kerala was carried out to derive the sector-wise annual turnover. The ABS potential is estimated based on the norms prescribed in the ABS Guidelines (0.1–0.5% of ex-factory

Table 7 Bio-resources based ABS potential of Kerala

S. No	Sectors	Total value (Rs. Millions)	Value considered for ABS potential estimation (Rs. Millions)	ABS potential (Rs. Millions)	%
1	Forest				
	Timber (including outside forest)	21,910	5012	200	
	NTFPs	42	21	0.6	
Total		21,952	5033	200.6	11
2	Fisheries				
	Marine	83,167	8317	250	
	Inland	38,407	3841	115	
Total		121,574	12158	365	19
3	Agriculture	23,61,41	23,614	708	37
4	Livestock	289,168	20698	621	33
Total of ecosystem/ biodiversity goods		668,835	61503	1894.6	100
	Tourism (ecosystem/ biodiversity services)	391,970[a]			

Note [a]Revenue from biodiversity-attributed tourism, including eco-tourism

sale value). In the MSMEs case, we estimated ABS based on 0.1% of the turnover (minimum number), as its overall investment is relatively small compared to the large factories (Table 8). NBA also followed the minimum or medium range values (prescribed in the Guidelines) for ABS estimation.

As per the estimation, the ABS potential of all the biological resources based MSMEs in the State is Rs. 240 Million. In this, the Food processing industries play a significant role with Rs. 128 Million (53.62%) followed by Wood, Bamboo and Cane based industries (Rs. 24 Million—9.95%), Rubber based products (Rs. 23 Million—9.44%), and Marine products and Sea foods (Rs. 22 Million—9.24%). In the Other industrial units, the ABS potential is relatively small.

For the large factories, the ABS potential is estimated based on 0.2% of the turnover, as its overall investment is high (Table 9).

The estimation of the ABS potential of factories in the State emphasized (a) fully on biological resources-based factories and (b) partial biological resources-based factories. In the fully biological resources-based factories the estimated ABS potential is Rs. 877 Million. Out of this, the manufacture of food products plays a significant role with Rs. 705 Million (80.38%) followed by the manufacture of pharmaceuticals, medicinal, chemical, and botanical products with Rs. 50 Million

Table 8 ABS potential of biological resources based enterprises (MSMEs)

Sl No	Category	Number of enterprises		Annual turnover (Rs. Millions)	ABS potential (0.1% of turnover)	
		No	%		Amount (Rs. Millions)	%
1	Ayurveda and Herbal cosmetics	586	1.23	4365	4	1.90
2	Food Processing	15,927	33.50	128,761	128	53.62
3	Cashew products other than nuts	207	0.44	1055	1	0.44
4	Marine products and Sea foods	174	0.37	22,499	22	9.24
5	Textiles and Handlooms	9964	20.96	12,549	13	5.23
6	Coir products	1159	2.44	7668	8	3.19
7	Wood, Bamboo and Cane based industries	9629	20.25	23,890	24	9.95
8	Herbal Wellness Centres	3245	6.83	2085	2	0.87
9	Paper based products and printing	3317	6.98	8840	9	3.68
10	Rubber based products (Tyres, Foot wares etc.)	2023	4.25	22,664	23	9.44
11	Wax products	254	0.53	122	0.1	0.10
12	Others	1056	2.22	5621	6	2.34
	Total	47,541	100.00	240,119	240.1	100.00

(5.75%) and manufacture of leather and related products with Rs. 48 Million (5.44%). Other factories' ABS potential is relatively small.

In the partially biological resources-based factories' case (2-digit NIC classifications), along with biological resources the non-biological resources may also have been used as raw materials. Hence, only 50% of the total output value is considered for the ABS estimation. Accordingly, the partial biological resources-based factories' total ABS potential is Rs. 361 Million.

The following table (Table 10) provides a comprehensive picture about the ABS potential of the manufacturing industrial units in the state, and the amount is around Rs. 1478 Million per year.

Table 9 ABS potential of biological resources-based factories in Kerala (major factory group at 2-digit NIC 2008)

NIC code	Factory/ description	Factories		Total output/turnover		ABS potential (0.2% of turnover)	
		Number	%	Value (Rs. Millions)	%	Amount (Rs. Millions)	%
A. *Fully Biological Resources Based*							
10	Manufacture of food products	1624	45.88	352,568	80.38	705	80.38
11	Manufacture of beverages	63	1.78	12,230	2.79	24	2.79
12	Manufacture of tobacco products	500	14.12	1608	0.37	3	0.37
15	Manufacture of leather and related products	203	5.73	23,842	5.44	48	5.44
16	Manufacture of wood and products of wood and cork, except furniture; manufacture of articles of straw and plaiting materials	857	24.21	15,349	3.50	31	3.50
17	Manufacture of paper and paper products	134	3.79	7794	1.77	16	1.77
21	Manufacture of pharmaceuticals, medicinal chemical and botanical products	159	4.49	25,221	5.75	50	5.75
	Total	3540	100	438,612	100	877	100
B. *Partially Biological Resources Based*[a]							
13	Manufacture of textiles	423	32.36	42,691	11.81	21,346 (43[a])	11.81
14	Manufacture of wearing apparel	51	3.90	8112	2.25	4056 (8[a])	2.24
22	Manufacture of rubber and plastic products	629	48.13	105,428	29.17	52,714 (105[a])	29.17
31	Manufacture of furniture	130	9.95	4809	1.33	2405 (5[a])	1.33
32	Other manufacturing	74	5.66	200,372	55.44	100,186 (200[a])	55.44

(continued)

Table 9 (continued)

NIC code	Factory/ description	Factories		Total output/turnover		ABS potential (0.2% of turnover)	
		Number	%	Value (Rs. Millions)	%	Amount (Rs. Millions)	%
	Total	1307	100	361,412	100	180,706 (361[a])	100.00
	Grand total (A+B)	4847		800,024		1238	

Note [a]In Partially biological resources-based industries, only 50% of the total output value (turnover) is considered for ABS amount estimation
Source Estimated based on Annual Survey of Industries 2017–2018

Table 10 Consolidated ABS potential from biological resource-based industrial units in Kerala

S No	Biological resource-based manufacturing	ABS amount (Rs. Millions)	Percentage
1	MSME	240	16
2	Large factories	1238	84
	Total	1478	100

Conclusion

For ABS estimation two approaches were followed. The first approach focussed on the biological resources' purchased price by the industries, based on the industry's raw material cost. This assessment has been carried out by taking into account the biological resource value of the ecosystem, from which it is extracted and goes to the manufacturing sector as raw materials for value addition. For different sectors district-wise data were collected for five years (2015–2020) and data analyses were carried out. The cumulative average value was considered to avoid the error in extreme years and for the ABS potential differentiated weightage has been assigned based on assumptions. The biological resources-based ABS potential of Kerala is Rs. 1895 Million per year. The second approach is based on industries (MSMEs and Large Factories) turnover-based assessment. In the MSMEs' case, the ABS potential is Rs. 240 Million/year. The total amount of the large factories' ABS potential assessed for fully biological resources-based industries and partially biological resources-based industries is Rs. 1238 Million. In brief, the total ABS potential through the biological resources-based industries turnover in Kerala is Rs. 1478 Million/year.

The ABS estimation carried out is based on Section 7 of the BD Act (prior intimation for obtaining biological resources for commercial activities) only. However, beyond that, Kerala's biological resources are used nationally and internationally for obtaining the IPR, and this ABS amount is not captured in this study.

For operationalization of ABS in a vast country like India, sincere efforts from all the stakeholders are required. NBA and SBBs don't have any idea about the biological resources-based industries or traders in their respective states, which come under the ambit of the Biological Diversity Act and the ABS. The challenge includes:

- The majority of the biological resources-based industries and traders are not aware of the BD Act and the complaints provisions.
- Biological resources trade and its businesses are highly unorganized. There is no database on the biological resources' collection/cultivation, trade, and commercial use.
- Industries are not bothered about the biological resources (raw materials) stock or their sustainability.
- Biological resources come under the NTC list and the associated ABS issues are not clear.
- Presently, the SBBs are focusing on the biological resources collected from the wild for the ABS. However, biological resources other than wild items need to be thought about and need a policy on handling the cultivated biological resources and their ABS.
- Traditional practices, especially practicing indigenous medicines, are exempted from the BD Act. In this context, many village and household entrepreneurs are claiming that their manufacture belongs to indigenous practice, which should be resolved.
- Similarly, value-added products (products which may contain portions or extracts of plants and animals in unrecognizable and physically inseparable form) are also exempted from the BD Act. However, many manufacturers are using secondary bio-resources including extracts as raw materials and manufacturing final products. According to them, extracts are value-added products, and hence require exemption from the Act, but it is controversial.

In brief, for precise ABS estimation, a more reliable methodology, appropriate data on the commercial users of biological resources, and the support of key stakeholders, especially industries, are required. The present study is a pioneering attempt at ABS potential estimation with several constraints and limitations. Hence, the methodology and the assumptions assigned for the estimations must be critically evaluated and debated further and need to encourage more reliable ABS potential estimation. No doubt that it will offer good insights for the enforcement agencies and policymakers to leverage the ABS for biodiversity conservation and its sustainable use.

Acknowledgements The ABS potential estimation carried out in this chapter is the extension of the work done by the author through the RKI Project (Economic Valuation of Biological resources) initiated by the Kerala State Biodiversity Board (KSBB), during 2021–2022, where he served as the Subject Expert. I am very grateful to Dr. Usha Titus IAS, the then Principal Secretary, Environment Department, Government of Kerala, and Chairperson of KSBB (additional charge) who gave me the responsibility to initiate the study. I am thankful to the other Board officials for their encouragement and to the project team for their support in extensive data collection. Recently, the methodology and the ABS potential criteria have been further debated, and based on the expert guidance a re-estimate has been done, which is given in this chapter. I am also thankful to Dr. Narendran

Thiruthy, Assistant Professor, Rajiv Gandhi School of Intellectual Property Law, Indian Institute of Technology—Kharagpur for reviewing the draft of this chapter and his critical comments.

References

CBD. (2011a). Nagoya protocol on access to genetic resources and the fair and equitable sharing of benefits arising from their utilization to the convention on biological diversity. Convention on Biological Diversity, United Nations.

CBD. (2011b). Convention on biological diversity text and annexes. Secretariat of the Convention on Biological Diversity, Montreal.

CBD. (2017). *Interim National Report on the implementation of the Nagoya protocol.* https://absch.cbd.int/en/search?schema=absNationalReport&currentPage=3&rowsPerPage=25

CBD. (2024). *The Clearing-House Mechanism (CHM) of the Convention on biological diversity.* Available: Database | Clearing-House Mechanism | CHM | CBD

MoEFCC. (2014). Guidelines on access to biological resources and associated knowledge and benefit sharing regulations—2014. Notification, Ministry of Environment, Forests and Climate Change (National Biodiversity Authority), New Delhi.

NBA. (2010). Biological Diversity Act, 2002. National Biodiversity Authority (NBA), Chennai.

NBA. (2024). *Approval granted to the applicants.* National Biodiversity Authority (NBA), Chennai National Biodiversity Authority—Approval granted to the Applicants (nbaindia.org).

Nelliyat, P. (2017). Bio-resources valuation for ensuring equity in access and benefit sharing: Issues and challenges. In K. P. Laladhas, P. Nilayangode & O. V. Oomen (Eds.), *Biodiversity for sustainable development.* Springer International Publishing.

Nelliyat, P. (2020). Tradable bio-resources' documentation (Database) and identification of its ABS potential with supply chain: A manual. UNDP—India Sponsored High Range Mountain (Munnar) Landscape Project, Kerala State Biodiversity Board (KSBB).

Nelliyet, P., & Pisupati, B. (2013). *Biodiversity economics from access and benefit sharing perspective* (Dissemination Paper -1 in 'Valuation of Biodiversity'), National Biodiversity Authority, Chennai.

United Nation. (2019). UN report: Nature's dangerous decline 'unprecedented'; species extinction rates 'accelerating. Available: UN Report: Nature's Dangerous Decline 'Unprecedented'; Species Extinction Rates 'Accelerating'—United Nations Sustainable Development.

Bioprospecting of Plant Diversity in India: Trends from CBD to the Present and Prospects in Traditional Medicine-Based Drug Discovery

T. P. Ijinu, N. M. Krishnakumar, M. Navas, Vinodkumar T. G. Nair, S. Rajasekharan, and P. Pushpangadan

Introduction

Bioprospecting is a systematic search for biochemical compounds or genetic materials from natural sources, such as plants, animals, fungi, and microorganisms, with potential applications in various industries, including pharmaceuticals, agriculture, and biotechnology. This practice stems from the recognition of the immense biodiversity found in ecosystems worldwide, harbouring untapped resources that could address pressing societal needs. It encompasses three main areas: "chemical prospecting," "gene prospecting," and "bionic prospecting" (Ijinu

T. P. Ijinu
Naturæ Scientific, Kerala University Business Innovation and Incubation Centre, Karyavattom Campus, Thiruvananthapuram, Kerala 695581, India

T. P. Ijinu · S. Rajasekharan · P. Pushpangadan (✉)
The National Society of Ethnopharmacology, VRA 179, Mannamoola, Thiruvananthapuram, Kerala 695005, India
e-mail: palpupushpangadan@gmail.com

N. M. Krishnakumar
Department of Biosciences, Rajagiri College of Social Sciences (Autonomous), Ernakulam, Kerala 683104, India

Present Address:
M. Navas
State Medicinal Plants Board, Regional Office, Poojappura P.O., Thiruvananthapuram, Kerala 695012, India

M. Navas · V. T. G. Nair
Ethnomedicine and Ethnopharmacology Division, Jawaharlal Nehru Tropical Botanic Garden and Research Institute, Thiruvananthapuram, Kerala 695562, India

S. Rajasekharan
Kerala State Biodiversity Board, Thiruvananthapuram, Kerala 695003, India

L. Krishna Panicker et al. (eds.), *Biodiversity and Business*,
https://doi.org/10.1007/978-3-031-71674-4_18

et al., 2020; Laird & Ten Kate, 2002; Mateo et al., 2001; Pushpangadan et al., 2012b, 2017, 2018a, 2018f; Pushpangadan & Ijinu, 2017). The total annual global market for products derived from genetic resources is estimated to be between US$500 billion and US$800 billion (Ten Kate & Laird, 2021). Bioprospecting involves the collection, identification, and analysis of biological samples to uncover novel molecules or biological processes that possess valuable properties such as medicinal efficacy, pest resistance, and environmental remediation capabilities. Historically, indigenous communities have long engaged in bioprospecting and harnessing traditional knowledge (TK) to discover medicinal plants and natural remedies. However, modern bioprospecting integrates advanced scientific techniques, including genomics, proteomics, transcriptomics, metabolomics, and bioinformatics, to expedite the discovery process and maximise potential benefits. Despite its promise, bioprospecting raises ethical and legal considerations, particularly regarding the equitable sharing of benefits and protection of biodiversity and indigenous rights. International agreements, such as the Nagoya Protocol, aim to address these concerns by establishing guidelines for access and benefit-sharing (ABS) arrangements between resource-rich countries and entities conducting bioprospecting activities (Pushpangadan & Ijinu, 2017, 2019).

Plant diversity in India is of immense importance owing to its rich cultural heritage, ecological significance, and economic value. India ranks 8th in the world and 4th in Asia among mega-diverse countries, home to approximately 7–8% of the global species of plants, animals, fungi, and microorganisms within its diverse ecosystems, ranging from the Himalayas in the north to the Western Ghats and the Eastern Ghats in the south (NBA, 2018, 2022). The country's flora includes approximately 45,500 species, with over 5,285 endemic species (NBA, 2018). India's remarkable biodiversity is complemented by its rich cultural diversity and unique wealth of TK. This knowledge has been developed, preserved, and practised by millions of ethnic and indigenous people residing in rural and forested areas. Botanical richness is shaped by various factors, including the country's varied topography, climatic conditions, and historical influences. Plants have played a central role in India's traditions, medicine, and rituals for many millennia. Ayurveda, the ancient Indian system of medicine, relies heavily on plant-based remedies, with thousands of plant species being used for their medicinal properties. Many traditional herbal medicines have been scientifically validated, and continue to be important sources of pharmaceutical compounds (Che et al., 2024; Pushpangadan et al., 2013a). Some plants hold religious significance in India, with numerous species revered as sacred and integral to rituals and festivals. For example, the Tulsi plant (*Ocimum tenuiflorum*) is considered holy and is often grown in homes and temples (Ijinu et al., 2023a). The Banyan tree (*Ficus benghalensis*) is another species that symbolises longevity and wisdom in Hindu mythology (Chakraborty et al., 2022).

Ecologically, plant diversity in India sustains myriad ecosystems, ranging from tropical rainforests and mangroves to alpine meadows and desert scrublands. These ecosystems provide habitats and food for a diverse array of wildlife, including many endemic and endangered species. The preservation of plant diversity is crucial for maintaining ecosystem balance, resilience, and functioning, such as regulating climate, purifying water, and preventing soil erosion (De Deyn & Kooistra, 2021). Moreover, plants form the foundation of food chains and provide essential ecosystem services such as pollination and nutrient cycling, which are vital for human well-being. Economically, plant diversity in India contributes significantly to the agriculture, forestry, and pharmaceutical industries. India is a centre of crop diversity, with thousands of cultivated plant species, including rice, wheat, cotton, and spices, supporting the livelihoods of millions of farmers. Traditional agroecosystems, such as rice paddies and mixed-cropping systems, have evolved over centuries, harnessing the diversity of plant species to enhance resilience and productivity. Therefore, protecting and sustainably managing India's plant diversity is essential for ensuring food security, livelihoods, and economic prosperity for present and future generations.

The Convention on Biological Diversity

The Convention on Biological Diversity (CBD), adopted at the United Nations Conference on Environment and Development (UNCED) in Rio de Janeiro in 1992, is a landmark international treaty aimed at conserving biodiversity, promoting the sustainable use of biological resources, and ensuring the equitable sharing of benefits derived from genetic resources. With its three main objectives—conservation of biological diversity, sustainable use of its components, and fair and equitable sharing of benefits arising from genetic resources—the CBD represents a comprehensive framework for addressing the global challenges posed by biodiversity loss and ecosystem degradation (Pushpangadan & Ijinu, 2017; Pushpangadan et al., 2012a, 2017). Central to the CBD's approach is the recognition of biodiversity as essential for the well-being of present and future generations, acknowledging its intrinsic value, as well as its instrumental role in supporting ecosystem services, human health, and socio-economic development. Through the CBD, countries implement national strategies and action plans, establish protected areas, conserve endangered species, promote sustainable resource management practices, and foster international cooperation to address transboundary conservation issues. Moreover, the CBD includes provisions for technology transfer, capacity building, and financial assistance to support developing countries in their conservation efforts. As the primary international instrument for biodiversity conservation, the CBD continues to serve as a catalyst for global cooperation, scientific research, and policy development to safeguard Earth's biological heritage for generations to come. India became a party to the Convention on 19 May 1994 (Venkataraman & Latha, 2008).

Traditional Knowledge and India

TK holds profound importance in Indian culture, serving as a repository of wisdom passed down through generations through oral traditions, ancient texts, and experiential learning. Rooted in the rich tapestry of India's diverse cultures, languages, and belief systems, TK encompasses a wide array of domains including agriculture, medicine, spirituality, and governance. Indigenous knowledge reflects centuries of observation, experimentation, and adaptation to local environments, fostering harmony between human communities and the natural world. In Indian society, TK is revered as a source of resilience, sustainability, and cultural identity, shaping daily practices, rituals, and social norms. Moreover, TK systems embody holistic approaches to health, well-being, and resource management, offering valuable insights into alternative paradigms of development that prioritise harmony, equity, and interconnectedness. As India confronts contemporary challenges, such as environmental degradation, climate change, and health disparities, the revival and integration of TK hold immense potential to foster resilience, foster sustainable solutions, and promote cultural heritage conservation. Thus, TK remains an integral pillar of Indian culture, embodying the wisdom of the past and guiding aspirations for a more harmonious and sustainable future (Pushpangadan & Ijinu, 2017; Pushpangadan et al., 2017, 2018a).

Ayurveda, an ancient medical system that originated more than 5,000 years ago, is one of the most significant representations of TK in Indian culture. The term Ayurveda comes from Sanskrit and translates to "knowledge of life". This system promotes a holistic approach to health and wellness by treating the body, mind, and spirit as interconnected entities. A key aspect of Ayurveda is the assessment of an individual's constitution, known as *prakriti* (body type) (Pushpangadan et al., 2015a). Central to Ayurvedic philosophy is the concept of doshas: *Vata*, *Pitta*, and *Kapha* (VPK), which are elemental forces that regulate the body's constitution and physiological processes. *Prakriti* is determined by the balance of these three doshas, which are derived from a combination of five basic elements: space, air, fire, water, and earth. A *Vata*-dominant constitution, characterised by the dominance of air and space, is marked by traits such as dryness, lightness, hyperactivity, and reduced sleep. Those with a *Pitta*-dominant constitution, influenced by fire and water, tend to have a higher metabolic rate, increased appetite, moderate build, oily skin, good memory, and a quick temper. In contrast, individuals with a *Kapha*-dominant constitution, governed by water and earth, are prone to weight gain, lower metabolism, increased sleep, and inertia (Che et al., 2017, 2024). Drawing on the principles of herbalism, dietetics, yoga, and meditation, Ayurvedic practitioners prescribe personalised treatments to restore balance and harmony, addressing not only the symptoms of the disease but also its underlying causes (Kanherkar et al., 2017; Pushpangadan et al., 2018d). The efficacy of Ayurveda in promoting health and healing has been corroborated by modern scientific research, leading to a resurgence of interest in this traditional healing art, both within India and globally.

In addition to medicine, TK plays a pivotal role in agriculture, where ancient farming practices and indigenous crop varieties have sustained Indian communities for generations. Practices such as organic farming, crop rotation, and mixed cropping are deeply rooted in traditional agricultural wisdom, enabling farmers to cultivate diverse crops, while preserving soil fertility and biodiversity (Singh et al., 2010; Souravi & Rajasekharan, 2019). Traditional seed-saving techniques passed down through oral traditions and community networks, have safeguarded indigenous crop varieties adapted to local climates, pests, and soil conditions, contributing to food security and resilience in the face of environmental challenges. Moreover, TK systems such as agroforestry, permaculture, and watershed management offer innovative solutions to contemporary issues, such as soil erosion, water scarcity, and climate change adaptation, highlighting the relevance of indigenous wisdom in addressing pressing global challenges.

Beyond practical applications, TK also encompasses spiritual and cultural dimensions that shape the world view and social fabric of Indian society. From ancient scriptures such as the *Vedas*, *Upanishads*, and *Puranas* to folk tales, myths, and legends passed down through oral traditions, Indian culture has been replete with stories, symbols, and rituals that embody timeless truths and moral teachings. These narratives reflect universal themes of love, duty, sacrifice, and redemption, offering moral guidance and spiritual insight to individuals and communities. Moreover, traditional arts such as music, dance, painting, and sculpture serve as vehicles for transmitting cultural values, fostering creativity, and forging connections between the past, present, and future generations. Whether through classical dance forms like *Bharatanatyam* and *Kathak*, or folk traditions such as *Baul* music and *Warli* painting, traditional arts embody the essence of Indian culture, celebrating the diversity and unity of the human experience. Thus, TK occupies a central place in Indian culture, from the healing arts of Ayurveda to the sustainable practices of indigenous agriculture and from the spiritual teachings of ancient texts to the expressive forms of traditional arts. TK permeates every aspect of life in India, shaping values, beliefs, and behaviours in profound ways. As India navigates the complexities of modernity and globalisation, the preservation and revitalisation of TK are more crucial than ever, not only for preserving cultural heritage but also for fostering sustainable development, social justice, and spiritual well-being in an increasingly interconnected world.

Early Bioprospecting Efforts in India

Bioprospecting efforts in India are rooted in indigenous knowledge systems that recognise the medicinal, nutritional, and economic value of the country's diverse flora and fauna. Indigenous communities across India have long engaged in the systematic exploration of their natural surroundings, identifying and utilising a wide array of plant and animal species for various purposes. Traditional healers, known as *vaidyas*, *siddhas*, or *hakims*, play a pivotal role in this process, drawing upon centuries-old herbal remedies and folk medicines to treat a myriad of ailments and to promote well-being. Ancient Indian texts, such as the *Vedas*, the *Atharva Veda*, and the *Charaka Samhita*, contain detailed descriptions of medicinal plants and their therapeutic properties, providing a valuable source of knowledge and inspiration for early bioprospectors. Moreover, India's cultural and religious traditions, which venerate certain plants and animals as sacred, have further, motivated efforts to explore and conserve the country's biological diversity.

Colonial rule, spanning over two centuries, brought about significant changes in India's socio-economic, political, and cultural landscapes, profoundly influencing indigenous knowledge systems, resource management practices, and scientific enquiry. One of the most significant impacts of colonialism was the commodification and commercialisation of India's natural resources, including its rich botanical diversity. British colonial administrators and European naturalists viewed India's flora and fauna as valuable commodities that could be exploited for economic gain, leading to the establishment of botanical gardens, herbaria, and research institutions aimed at cataloguing, classifying, and extracting valuable plant species for export to Europe. The Royal Botanic Gardens at Kew, founded in the eighteenth century, played a central role in this process, serving as a hub for the collection, cultivation, and dissemination of exotic plant species across the British Empire, including India (Baber, 2016). The exploitation of India's natural resources during this period has led to the depletion of certain species, habitat destruction, and erosion of TK systems, highlighting the need for a more sustainable and inclusive approach to bioprospecting. Despite these challenges, colonialism has advanced scientific knowledge and technology in India, laying the foundation for modern bioprospecting and pharmaceutical research. Institutions such as the Indian Medical Service and the Calcutta Botanic Garden have facilitated scientific exchange and collaboration, integrating traditional Indian medicine with Western practices and documenting thousands of plant species and their medicinal properties (Kumar, 2017; Sikarwar, 2020).

In the post-independence era, India embarked on a journey of scientific exploration and conservation aimed at harnessing biodiversity for the benefit of its people. The establishment of institutions such as the Indian Council of Agricultural Research (ICAR), Council of Scientific and Industrial Research (CSIR), Indian Council of Medical Research (ICMR), Department of Biotechnology (DBT), and Ministry of AYUSH paved the way for systematic research and development in fields such as agriculture, pharmaceuticals, and biotechnology. Scientists and researchers have begun to systematically catalogue and study India's plant diversity and identify novel bioactive compounds with potential applications in medicine, agriculture, and industry. TK holders, including tribal communities and local healers, have been increasingly recognised as valuable partners in bioprospecting efforts, with initiatives such as the National Medicinal Plants Board (NMPB) and the Traditional Knowledge Digital Library (TKDL) seeking to document and protect indigenous knowledge systems. TKDL serves as a valuable resource for patent examiners and policymakers, facilitating the recognition and protection of India's TK systems (Pushpangadan & Ijinu, 2017; Pushpangadan et al., 2014b, 2016b). Several significant initiatives have been launched to advance bioprospecting and biotechnology in India. These include the

"Network Programme on Bioprospecting of Biological Wealth using Biotechnological Tools" during the 9th Plan and the Biotechnology Industry Research Assistance Council/Biotechnology Industry Partnership Programme (BIRAC/BIPP) by the Department of Biotechnology. The CSIR initiated the New Millennium Indian Technology Leadership Initiative (NMITLI) and Open-Source Drug Discovery (OSDD). The Golden Triangle Partnership (GTP), a collaborative effort by the Department of AYUSH (now the Ministry of AYUSH), ICMR, and CSIR, focuses on validating plant drugs and developing new ones of national and global importance (Pushpangadan & Ijinu, 2017; Pushpangadan et al., 2015a).

The evolution of bioprospecting regulations in India can be traced back to the country's commitment to international agreements such as the CBD, ratified in 1992. The CBD laid the foundation for the establishment of national frameworks for access to genetic resources and the fair and equitable sharing of benefits arising from their utilisation. In response to the CBD, India enacted the Biological Diversity Act (2002), which provides a comprehensive legal framework for conservation, sustainable use, and equitable sharing of benefits derived from biological resources. The Act establishes the National Biodiversity Authority (NBA) as the apex body responsible for regulating access to biological resources and ensuring compliance with ABS (ABS) provisions. The Biological Diversity Rules (2004) and subsequent amendments provide detailed guidelines for obtaining prior approval and benefit-sharing agreements for bioprospecting. As a party to the Nagoya Protocol, India has been actively involved in its implementation to safeguard its rich biodiversity and TK. The protocol supports India's efforts to regulate access to its genetic resources and ensure that benefits, such as technology transfer, research collaboration, and financial returns, are shared with local communities and indigenous peoples. These regulations require researchers, companies, and organisations to seek permission from the NBA before accessing biological resources for research, commercialisation, or any other purpose.

Collaborative initiatives among government agencies, research institutions, industry stakeholders, and local communities help bridge the gap between TK and modern science, fostering innovation, entrepreneurship, and sustainable development. Despite these regulatory advancements, challenges remain in the effective implementation and enforcement of bioprospecting regulations, particularly at the grassroots level. Issues such as inadequate awareness, capacity, and resources as well as the complex nature of TK documentation and benefit-sharing negotiations pose significant hurdles to the effective implementation of bioprospecting regulations in India. Moving forward, concerted efforts are needed to enhance stakeholder engagement, capacity building, and transparency in bioprospecting processes, ensuring that the benefits of India's biodiversity are equitably shared for the benefit of present and future generations while safeguarding the rights and interests of local communities and indigenous peoples.

International Agreements in Bioprospecting

The international agreements mentioned below provide a framework for addressing the complex issues surrounding bioprospecting and contribute to the development of ethical, equitable, and sustainable approaches to accessing and utilising genetic resources and TK (Pushpangadan & Ijinu, 2017, 2019; Pushpangadan et al., 2017).

1. Convention on Biological Diversity:

   - Adopted in 1992, it is the primary international treaty governing bioprospecting and biodiversity conservation.
   - Key objectives include conservation of biological diversity, sustainable use of its components, and fair and equitable sharing of benefits arising from the utilisation of genetic resources.
   - Promotes the establishment of national legal frameworks for ABS and encourages the recognition and protection of TK.

2. Nagoya Protocol on Access to Genetic Resources and the Fair and Equitable Sharing of Benefits Arising from their Utilization:

   - Adopted in 2010 as a supplementary agreement to the CBD, the Nagoya Protocol provides a legally binding ABS framework.
   - It establishes transparent and predictable rules for accessing genetic resources and ensuring fair and equitable sharing of benefits, including monetary and non-monetary benefits, with providers of genetic resources and TK holders.
   - The Nagoya Protocol emphasises the importance of prior informed consent (PIC) and mutually agreed terms (MAT) in bioprospecting agreements.

3. International Treaty on Plant Genetic Resources for Food and Agriculture:

   - The International Treaty on Plant Genetic Resources for Food and Agriculture (ITPGRFA), adopted in 2001 under the auspices of the Food and Agriculture Organization (FAO), specifically focuses on plant genetic resources for food and agriculture.
   - It aims to ensure the conservation and sustainable use of plant genetic resources and equitable benefit-sharing, particularly in the context of agriculture and food security.
   - The treaty promotes facilitated access to plant genetic resources for research, breeding, and training while recognising the rights of farmers and indigenous communities.

4. World Intellectual Property Organization (WIPO) Intergovernmental Committee (IGC) on Intellectual Property and Genetic Resources, Traditional Knowledge, and Folklore:

- The IGC explores ways to ensure the effective protection of TK and genetic resources against misappropriation and unfair commercialisation through the development of *sui generis* intellectual property regimes, disclosure requirements, and benefit-sharing mechanisms.

5. United Nations Declaration on the Rights of Indigenous Peoples (UNDRIP):

   - While not specifically focused on bioprospecting, the UNDRIP recognises the rights of indigenous peoples to maintain, control, protect, and develop their cultural heritage, TK, and genetic resources.
   - It calls for the full and effective participation of indigenous peoples in decision-making processes that affect them and emphasises the importance of obtaining their free, prior, and informed consent in matters related to resource utilisation and development.

National Legislations on Bioprospecting

Bioprospecting activities in India are governed by several national legislation and policies aimed at regulating access to biological resources, ensuring equitable benefit-sharing, protecting TK, and promoting biodiversity conservation. These national legislations and policies constitute the legal framework governing bioprospecting activities in India, ensuring that they are conducted in a transparent, sustainable, and socially responsible manner, while promoting biodiversity conservation and equitable benefit-sharing (Pushpangadan & Ijinu, 2017, 2019; Pushpangadan et al., 2017). The key legislation and policies are as follows.

1. Biological Diversity Act, 2002:

   - The Biological Diversity Act (BDA) provides a comprehensive legal framework for the conservation, sustainable use, and equitable sharing of benefits derived from biological resources.
   - The BDA established the National Biodiversity Authority (NBA) as the principal regulatory body responsible for implementing the provisions of the Act at the national level.
   - It mandates the establishment of State Biodiversity Boards (SBBs) and Biodiversity Management Committees (BMCs) at the local level to regulate access to biological resources and to oversee biodiversity conservation initiatives.

2. Biological Diversity Rules, 2004:

   - The Biological Diversity Rules (BDR) operationalises the provisions of the BDA and provides detailed guidelines for implementing its various provisions.
   - The rules specify procedures for obtaining prior approval and access permits for bioprospecting activities, including research, commercial utilisation, and IPR applications.

- They outlined the mechanisms for sharing benefits arising from the utilisation of biological resources, including monetary and non-monetary benefits, with stakeholders, including indigenous and local communities.

3. Guidelines on Access to Biological Resources and Associated Knowledge and Benefit-sharing Regulations, 2014:

   - The ABS guidelines provide a framework for regulating access to biological resources and associated TK in accordance with the provisions of the BDA and Nagoya Protocol.
   - They outlined the procedures for obtaining PIC and MAT from stakeholders, including indigenous and local communities, before undertaking bioprospecting activities.
   - The guidelines emphasise the importance of equitable benefit-sharing and the establishment of benefit-sharing agreements (BSAs) to ensure that the benefits derived from bioprospecting are shared fairly with providers of genetic resources and TK holders.

4. Protection of Plant Varieties and Farmers' Rights Act, 2001:

   - The Protection of Plant Varieties and Farmers' Rights (PPV&FR) Act provides legal protection for plant varieties and promotes the rights of both farmers and breeders.
   - While not specifically focused on bioprospecting, the Act includes provisions related to the conservation and sustainable use of plant genetic resources, and recognises the contributions of farmers and indigenous communities to the conservation and development of plant genetic resources.

Compliance Mechanisms and Enforcement of Regulations

In India, compliance with bioprospecting and ABS regulations is enforced by legal frameworks, regulatory authorities, monitoring mechanisms, and enforcement measures. Established under the Biological Diversity Act, of 2002, the NBA is the main regulatory body overseeing ABS at the national level, granting approval for access to biological resources, and overseeing benefit-sharing agreements. Each state has an SBB that implements ABS regulations and grants access approvals within its jurisdiction, whereas local BMCs handle permissions and compliance locally. Bioprospecting requires obtaining PIC from indigenous and local communities, along with MAT, which outlines access and benefit-sharing conditions. Access Permits and Benefit-sharing Agreements are issued by the NBA or SBBs after PIC and MAT are completed. These permits allow bioprospectors to use biological resources for research or commercial purposes, detailing both monetary and non-monetary benefit-sharing obligations. Regular reports on bioprospecting activities and compliance are required, and regulatory authorities should conduct site visits, audits, and inspections. Non-compliance can lead to penalties, fines, or the revocation of access permits, with legal action possible under the Biological Diversity Act, of 2002. Efforts to enhance stakeholder capacity include awareness-raising initiatives, workshops, training programs, and outreach activities to promote understanding and collaboration regarding ABS regulations.

Indian Case Studies

The following are notable case studies of bioprospecting initiatives and biopiracy disputes in India that highlight both the successes and challenges faced in implementing TK protection and ABS mechanisms.

TBGRI-Kani Benefit-Sharing Model

One of the most prominent bioprospecting initiatives in India is the development of the herbal formulation *Jeevani*, from the plant *Arogyapacha* (*Trichopus zeylanicus* Gaertn.), which possess anti-stress, anti-fatigue, and immune-enhancing properties (George et al., 2016; Ijinu et al., 2023b). The Kani tribe of Kerala (India) traditionally uses the fruit of this plant for its energising properties. Researchers from the Jawaharlal Nehru Tropical Botanic Garden and Research Institute (KSCSTE-JNTBGRI) in Kerala recognise the potential of this plant in collaboration with the Kani tribe to study and commercialise it. Scientists at CSIR-IIIM (RRL Jammu) and JNTBGRI discovered that the leaves of the plant had similar properties and used them to develop a polyherbal formulary drug, Jeevani, based on tribal knowledge, which gained national and international recognition. The technology to produce the formulation was licenced to Arya Vaidya Pharmacy (Coimbatore) Limited (1995) for the highest herbal drug licence fee in India's history at that time. In a landmark decision of the Executive Committee, TBGRI and the Kani tribe established a benefit-sharing agreement, ensuring that the tribe received 50% of profits from *Jeevani* sales. This agreement includes provisions for financial compensation and initiatives to improve the tribe's living conditions. The *Jeevani* case is often cited as a successful example of benefit-sharing, where TK holders receive tangible benefits from the commercialisation of their knowledge. Articles 8(j) and 15.7 of the CBD were directly applicable to this benefit-sharing model. Furthermore, this model distinguishes India from being the first country in the world to experiment with a benefit-sharing approach that implements Article 8(j) in both the letter and spirit. The benefit-sharing model established by this initiative has set a precedent for other bioprospecting projects (George et al., 2014; Pushpangadan & Ijinu, 2016, 2017; Pushpangadan et al., 2014a, 2014b, 2015a, 2016a, 2016b, 2017, 2018b, 2019; Rajasekharan et al., 2012). Dr. Pushpangadan, the senior author of this article, received the UNDP Equator Prize (2002) in the individual category for their work on developing the sustainable use of a wild plant (*T. zeylanicus*), demonstrating a successful model of equitable benefit-sharing, and recognising the IPR of the indigenous Kani community of the Western Ghats of Kerala, India. The senior authors of this article, PP and SR, are the major architects to this benefit-sharing model.

Many criticisms of the TBGRI-Kani model of benefit sharing exist in the public domain. Most critics are unaware of the story and base their critiques on previously published materials. The surprising thing is the comparison of this pre-CBD story with post-CBD developments, including the ABS protocol. The TBGRI-Kani story began in 1987 in the Agasthyamalai Hills of the Thiruvananthapuram district in Kerala, India, as part of the All-India Coordinated Research Project on Ethnobiology (AICRPE). This is a pre-CBD case and no prior model exists. Moreover, the Biological Diversity Act, the Rules, the National Biodiversity Authority, and any other stringent biodiversity policies did not exist in India at that time.

Second Model from JNTBGRI

The study was based on the TK of a medicinal plant (Code No. 222) disclosed by a traditional healer (*vaidyar*) after signing a PIC and non-disclosure agreement with KSCSTE-JNTBGRI. The traditional healer claimed plant-treated diabetes, liver disorders, jaundice, and fatigue, but no prior information was available. The preclinical study aimed to evaluate these claims, explore formulation possibilities, and conduct scientific validation following the WHO guidelines. The results showed that the plant had significant antidiabetic, hepatoprotective, antifatigue, and antioxidant properties (Krishnakumar et al., 2016a, 2016b, 2017a, 2017b, 2018a, 2018b, 2020; Latha et al., 2015). An Indian patent (no. 357405) for a polyherbal formulation was granted on 1 February 2021 including the traditional healer as an inventor. This case marks the first ABS study in which a traditional healer is listed as an inventor, ensuring their rights and royalty share upon commercialisation. The traditional healer will be eligible for all rights even later when the technology transfer is complete, and for a share of the royalty when the drug is commercialised. The authors of this article, SR, MN, NMK, and Dr. S. R. Suja of JNTBGRI are part of this invention.

Neem Tree and Biopiracy

The neem tree, referred to as "*sarva-roga nivarini*" in Sanskrit, which means "the healer of all ailments," has been a cornerstone of Indian traditional medicine for centuries. It has been documented in ancient Indian texts dating back over 2000 years and has been utilised in agriculture as a natural insect and pest repellent in both human and veterinary medicine and for its antifungal and antibacterial properties, as well as in the production of toiletries and cosmetics. In the 1990s, the U.S. Department of Agriculture (USDA) and W. R. Grace and Co. obtained a patent related to a fungicide (Winterbottom, 2021). Indian activists and scientists contested the patent, arguing that it constituted biopiracy since the use of neem was already well documented in Indian TK. The legal battle culminated in the European Patent Office (EPO), revoking the patent in 2000 and acknowledging prior art and TK associated with neem. This case underscores the need for robust mechanisms to protect TK and prevent biopiracy, leading to greater awareness and strengthening of India's ABS frameworks.

Turmeric Patent Dispute

The patent dispute involved the traditional use of turmeric (*haldi*, *Curcuma longa* L.) in India. In 1995, American researchers from the University of Mississippi Medical Center were granted a U.S. patent by the United States Patent and Trademark Office (USPTO) for the "use of turmeric in wound healing." Given the long-standing use of turmeric in Indian traditional medicine for the treatment of wounds and other ailments, the patent has sparked controversy. The CSIR challenged the patent, arguing that the use of turmeric for healing wounds was not a novel invention, but rather an existing traditional practice well-documented in ancient Sanskrit texts and Ayurvedic literature (Halliburton, 2017). The USPTO eventually revoked the patent in 1997, recognising TK as a prior art. This case highlights the importance of documenting TK to protect it from misappropriation and to ensure that bioprospecting activities respect existing knowledge systems. This dispute prompted countries, especially third-world countries, to bolster their legal frameworks to protect TK. Indian initiatives such as CSIR-TKDL were established to document TK systematically and make it accessible to patent examiners globally.

Basmati Rice Case

The case of basmati rice is a significant example of biopiracy and intellectual property rights disputes involving TK and agricultural products. It is renowned for its unique aroma, flavour, and long slender grains. Basmati has been cultivated and consumed for centuries, and its name itself is derived from the Sanskrit word "Vasmati," meaning "fragrant." In 1997, RiceTec Inc., a U.S.-based company, was granted a patent by the USPTO for certain strains of basmati rice and their production methods, sparking controversy and opposition from India and Pakistan. This patent was viewed as an appropriation of a traditional product, threatening the economic interests of regional farmers. The Indian government, alongside NGOs and farmer groups, challenged the patent on grounds of prior art and sought to protect basmati rice under the geographical indication (GI) framework, ensuring that only rice grown in specific regions could be marketed as "basmati" (Halliburton, 2017). In 2001, the USPTO invalidated several claims of the RiceTec patent, and in 2016 India secured GI status for basmati rice. This case also underscores the need for robust legal frameworks to protect TK, and demonstrates the effectiveness of international advocacy in combating biopiracy.

Divya Pharmacy v. Union of India and Others

This case is a landmark legal battle in the context of ABS regulations under the 2002 BDA in India. This case highlights the complexities and challenges of implementing ABS mechanisms, particularly in the context of TK and commercial bioprospecting. Divya Pharmacy, an Ayurvedic company associated with Patanjali Ayurved Ltd., was required to comply with the Act, which mandates benefit-sharing with local communities for the use of biological resources. When the Uttarakhand State Biodiversity Board demanded benefit-sharing fees, Divya Pharmacy challenged the notices, arguing that ABS provisions were primarily intended for foreign entities. The case raised key legal issues regarding the applicability of ABS provisions to Indian companies, the interpretation of the Act's sections on access and benefit sharing, and the constitutional validity of these provisions. In 2018, the Uttarakhand High Court upheld the applicability of ABS provisions to Indian companies, emphasising that the Act applies to all entities and that benefit-sharing is essential for conservation and equitable use. The court supported the regulatory roles of SBBs and the NBA. The case clarified the legal obligations of Indian companies, strengthened the regulatory framework, and set a precedent for future ABS-related legal disputes, underscoring the importance of compliance to ensure fair compensation for local communities and TK holders who are responsible for conserving and protecting such resources (Nomani, 2020). This case serves as a crucial reference for the ongoing development and enforcement of ABS regulations in India.

Prospects for Traditional Medicine-Based Drug Discovery

Owing to its rich biodiversity, India has emerged as a significant player in the global bioprospecting landscape. The country's diverse ecosystems, ranging from the Himalayan region in the north to the Western Ghats in the south, harbour a plethora of plant species with immense potential for pharmaceutical, agricultural, and industrial applications. The rich biodiversity of India is matched by equally rich cultural diversity and a unique wealth of TK systems developed, preserved, and practiced by millions of ethnic and indigenous people living in rural and forested areas (Pushpangadan, 1998; Pushpangadan et al., 2015b). In India, the vast amount of this type of information collected through AICRPE is locked in unattended reports due to a lack of appropriate resources. The project ran from 1982 to 1998 and spanned over 16 years. TK was collected from approximately 10,000 plants during this period. Of the 8,000 wild plant species traditionally employed by indigenous communities for medicinal purposes, approximately 950 have emerged as new claims that merit scientific investigation. Furthermore, over 3,900 wild plant species are used as supplementary foods and vegetables among these tribal populations. Of the 400 plant species designated for fodder, approximately 100 have been recommended for broad use. Three hundred wild species, such as piscicides or pesticides, are harnessed by tribal communities. Notably, at least 175 of these species have a great potential for further development as safe and effective biopesticides (Ijinu et al., 2016, 2022a, 2022b; Pushpangadan, 1998, 2002; Pushpangadan & Ijinu, 2016, 2019; Pushpangadan & Pradeep, 2008; Pushpangadan et al., 2014a, 2014c, 2015a, 2016a, 2016b, 2017, 2018a, 2018e; Sreedevi et al., 2013).

Recently, the Jawaharlal Nehru Tropical Botanic Garden and Research Institute (JNTBGRI) in Thiruvananthapuram, India, completed the systematic documentation of TK associated with plants used for food and medicine in Kerala State under a project funded by the Ministry of AYUSH, Government of India. From 2010 to 2014, JNTBGRI documented 26,419 pieces of information under the leadership of Dr. S. Rajasekharan, of which 3,392 (12.84%) were recorded from different tribal communities and 23,027 (87.16%) from non-tribal communities. In the second phase, from 2019 to 2022, 9,034 pieces of information were documented, with 675 records from different tribal communities and 8,359 from non-tribal communities. Since 1987, JNTBGRI has documented approximately 45,000 pieces of TK information across Kerala. In addition, the Amity Institute for Herbal and Biotech Products Development (AIHBPD) in Thiruvananthapuram, India documented 6,008 pieces of information under the leadership of Dr. P. Pushpangadan, based on TK related to plants used for food and medicine from 34 gramapanchayats in 12 districts of Kerala, India, under projects funded by the Department of Science and Technology and Council of Scientific and Industrial Research, Government of India (Chithra et al., 2020a, 2020b; Ijinu et al., 2011, 2022a, 2022b, 2022c). Scientists and industrialists consider TK a cost-effective method for discovering novel molecules and products with high economic value. This chapter particularly focuses on the trends toward plant-based drug discovery with special reference to plant diversity in India. Moreover, there are many other applications of plant diversity in various sectors, such as agriculture (crop yield, pest control, and soil fertility), food industries (nutrients, natural colours, flavours, and preservatives), and clothing and textiles (cotton, flax, hemp, and natural dyes).

The investigation of TK systems regarding the plant world is the main focus of the scientific field known as "ethnobotany." In modern drug discovery, ethnobotany is a cost-effective approach for identifying unique and valuable molecules with significant therapeutic potential (Pushpangadan et al., 2013b). It is widely believed that the chance of finding a bioactive molecule by random screening of plant samples is 1 in 10,000 and the chance of developing a marketable drug is 1 in 4. In comparison, the success rate of identifying a bioactive molecule by selective screening based on ethnobotanical leads is 1 in 100, and the discovery rate of a marketable drug is 1 in 2 (Pushpangadan & Ijinu, 2017; Pushpangadan & Nair, 2005; Pushpangadan et al., 2016a). Many modern drugs are derived from plants and were initially identified through ethnobotanical and ethnopharmacological research (George & Ijinu, 2011; Pushpangadan et al., 2014b, 2015a, 2016a, 2016b, 2018a, 2018e). For instance, aspirin (acetylsalicylic acid; analgesic, anti-inflammatory) is derived from *Salix* spp. (willow tree, Salicaceae); reserpine (antihypertensive, tranquilliser) is extracted from *Rauvolfia* spp. (sarpagandha, Apocynaceae); quinine (antimalarial) is sourced from *Cinchona* spp. (Rubiaceae); and eserine (physostigmine, acetylcholinesterase inhibitor) is obtained from *Physostigma venenosum* Balf. (Fabaceae). Other examples include vinblastine (anticancer) from *Catharanthus roseus* (L.) G.Don (Apocynaceae); diosgenin (cardio-tonic) from *Dioscorea* spp. (Dioscoreaceae); solasodine (anticancer, cholinergic) from *Solanum* spp. (Solanaceae); morphine (opioid analgesic) from *Papaver somniferum* L. (Papaveraceae); cocaine (anaesthetic, stimulant) from *Erythroxylum coca* Lam. (Erythroxylaceae); atropine (anticholinergic) from *Atropa bella-donna* L.(Solanaceae); digoxin (cardiotonic) from *Digitalis lanata* Ehrh. (Plantaginaceae); etoposide (anticancer) from *Podophyllum peltatum* L. (Berberidaceae); etc. (Balick & Cox, 2020; Che et al., 2017, 2024; Duke et al. 2000; Farnsworth et al., 1985; George & Ijinu, 2011; Heinrich, 2000; Kingston, 2011; Ijinu et al., 2020, 2022; Newman & Cragg, 2016, 2020; Soejarto et al., 2011; Süntar, 2020; Verpoorte et al., 2005). Furthermore, in 1971, artemisinin (Qinghaosu) was first isolated from *Artemisia annua* L., and Youyou Tu, the researcher responsible for its discovery was awarded the Nobel Prize in Physiology or Medicine in 2015 for its development as an effective antimalarial drug and elucidation of its mechanism of action (Tu, 2016).

Some important plants listed in the AYUSH system (Indian Systems of Medicine, ISM) that have been investigated in both in India and abroad are *Withania somnifera* (L.) Dunal (withanolides, withaferin; anti-inflammatory, immunomodulatory), *Boswellia serrata* Roxb. (boswellic acids; anti-inflammatory), *Berberis aristata* DC.(berberine; anti-inflammatory, antileishmanial), *Picrorhiza kurroa* Royle ex Benth. (kutkoside; hepatoprotective), *Curcuma longa* L.(curcumin; anti-inflammatory), *Commiphora wightii* (Arn.) Bhandari(syn. *C. mukul*; guggulsterone; anti-inflammatory, hypolipidaemic), *Embelia ribes* Burm.f.(embelin; anti-inflammatory, antifertility), *Rauvolfia serpentina* (L.) Benth. ex Kurz(reserpine; cardioprotective), *Coleus barbatus* var. *barbatus* (syn. *C. forskohlii*; forskolin; cardioprotective), *Momordica charantia* L.(charantin; antidiabetic), *Andrographis paniculata* (Burm.f.) Wall. ex Nees (andrographolide; antidiabetic, hepatoprotective), *Dysoxylum gotadhora* (Buch.-Ham.) Mabb.(syn. *D. binectariferum*; rohitukine; anticancer), *Podophyllum hexandrum* Royle (syn. *P. emodi*; podophyllotoxin; anticancer), *Holarrhena pubescens* Wall. ex G.Don (*H. antidysenterica*; conessine; antidiarrheal), *Justicia adhatoda* L. (*Adhatoda vasica*; vasicine; antiasthmatic,bronchodilator), *Azadirachta indica* A.Juss. (nimbolide, gedunin; antimalarial), *Bacopa monnieri* (L.) Wettst. (bacosides, memory enhancer, neuroprotective), *Centella asiatica* (L.) Urb. (madecassic acid, madecassosides, asiatic acid, asiaticoside; memory enhancers, neuroprotective), *Senna alexandrina* var. *alexandrina* (*Cassia angustifolia*; sennosides; constipation), *Vincetoxicum indicum* (Burm.f.) Mabb. (*Tylophora indica*; tylophorine; bronchodilator), *Asparagus racemosus* Willd. (shatavarin; tonic, galactagogue), *Glycyrrhiza glabra* L. (glycyrrhizin; antiulcer, antitussive), *Aloe vera* (L.) Burm.f. (aloin; demulcent, skin diseases), *Trigonella foenum-graecum* L. (trigonellin; antidiabetic, hypolipidaemic), *Pterocarpus marsupium* Roxb. (liquiritigenin, isoliquiritigenin; antidiabetic), *Boerhavia coccinea* Mill. (*B. diffusa*; boeravinones; hepatoprotective), *Achyranthes aspera* L. (achyranthine; diuretic), *Anethum graveolens* L. (anethole; digestive, carminative), *Cullen corylifolium* (L.) Medik. (*Psoralea corylifolia*; psoralen; vitiligo), *Garcinia cowa* var. *cowa* (*G. cambogia*; hydroxycitric acid; antiobesity), *Aegle marmelos* (L.) Corrêa (aegeline, marmelosin; bowel disease), *Tinospora cordifolia* (Willd.) Hook.f. & Thomson (tinosporic acid, cordifoliosides; immunomodulatory), etc. (Bhutani & Gohil, 2010; Ijinu et al., 2022b; Katiyar et al., 2012; Patwardhan et al., 2005; Pushpangadan et al., 2015a).

Traditional Knowledge Integration

Integrating TK, particularly the medicinal and economic properties of plant species, with modern bioprospecting efforts represents a promising pathway for sustainable development. This integration leverages centuries-old wisdom from indigenous and local communities and offers valuable insights that can expedite the discovery and development of new molecules and products while conserving biodiversity. Benefits include cultural preservation, economic opportunities through fair benefit-sharing agreements, and the promotion of sustainable harvesting practices. However, challenges such as protecting intellectual property rights (IPR), ensuring ethical practices, and scientifically validating traditional medicine (TM) in drug development must be addressed. Pathways to sustainable development include fostering collaborative research, capacity building, implementing strong policies, and adhering to sustainable practices. Balancing scientific innovation with ethical considerations and cultural respect can lead to new resource discoveries that benefit both biodiversity and local communities.

There has been growing interest in incorporating TM into mainstream healthcare systems (Che et al., 2017; Ijinu et al., 2022a, 2022b; Pushpangadan et al., 2017). Many studies have shown that TM can effectively treat certain diseases and often has fewer side effects than modern pharmaceuticals. Furthermore, TM is important for promoting community-based healthcare and supporting the local economies. For the reasons mentioned previously, true integration of both systems for major diseases is neither possible nor necessary. For diseases for which modern medicines yield true benefits, including drugs for any chronic ailments and lifestyle and lifesaving diseases such as cancer, cardiovascular diseases, and diabetes, modern medicines should be used. Since traditional medicines are plant-based and have few side effects, they may be added as adjuvant therapy, knowing fully well that they are not curative but possibly immune potentiators. However, challenges and concerns are associated with the use of TM in healthcare settings. For example, there is a lack of standardisation and regulation in the production and use of TM, which can pose risks to patient safety. Furthermore, issues around the IPR and commercialisation of TK/TM can lead to the exploitation and loss of cultural heritage. The role of TK/TM in healthcare is a complex and multifaceted issue that requires careful consideration and collaboration among traditional healers, healthcare professionals, policymakers, and communities.

Pharmaceutical Research

Drug discovery from medicinal plants typically involves the isolation and identification of active compounds, followed by preclinical screening, which includes both in vitro and in vivo experiments, to determine their potential as therapeutic agents. Once a promising compound has been identified, it can be further optimised through drug development, which involves modifications to improve its potency, selectivity, and safety. However, this method is time-consuming and expensive. On average, it requires 10 to 15 years and a financial investment ranging from USD 1.5 to 2.0 billion to introduce a new medication to the market successfully—the clinical trial phases of the drug development cycle account for approximately half of this duration and cost. The remaining 50% of R&D expenses are allocated to preclinical compound discovery, testing, and navigation through regulatory processes (Harrer et al., 2019). This transition from new chemical entities (NCE) to clinically accepted drugs involves integration of the activities of many disciplines. Only a few pharmacologically active compounds have advanced drug development, and an even smaller number have made it from development to the market. It is estimated that out of 5000 chemical entities synthesised or isolated from natural resources, only 250 will undergo animal testing (preclinical), five will undergo clinical testing in healthy volunteers or patients, and only one will ultimately enter the marketplace (Iwu, 2002).

Pharmaceutical research gained momentum when natural product chemists and pharmacologists began dissecting the chemistry underlying TM. Scientific advancements have led to the discovery of numerous novel compounds, resulting in new treatments for infections, cancers, ulcers, and heart diseases. While some breakthroughs came from random plant screenings, many others resulted from extensive scientific observations. The landscape of natural product research, specifically plant-based drug research, has transformed dramatically with the advent of high-throughput screening (HTS) methods, enabling the rapid screening of thousands of samples within 24 h for specific activities. Unlike synthetic chemists, who cannot match the volume of compound production, they have turned to combinatorial chemistry and a novel method for testing mixtures of compounds synthesised via solid-phase techniques. HTS has opened new possibilities for natural products by facilitating the rapid screening of extensive extract libraries and greatly enhancing bioassay-guided fractionation, a former bottleneck in plant extract studies. The success of natural product chemistry over the years has been bolstered by advanced spectroscopic methods and effective extraction techniques such as chromatography. The search for novel bioactive natural products involves meticulous extraction, isolation, purification, and structural elucidation, driving both chemical and biological investigations.

Drug discovery and development have recently shifted from focusing on single active components and specific targets to a "multi-component" "multi-target" approach. This shift reflects a deeper understanding of TMs, such as the AYUSH system and traditional Chinese medicine (TCM), which has long been recognised, the intricate interactions between multiple components of TM, and their effects on various targets in the body. These TM systems prioritise holistic healthcare, aiming to restore balance and harmony within the body. They acknowledged that therapeutic effects are not solely attributed to a single component acting on a single target but result from the synergistic interactions of multiple components on multiple targets. Network pharmacology, a relatively new field, merges principles from TM and modern pharmacology to study the complex interactions among drugs, targets, and diseases. It aims to understand the molecular processes underlying the therapeutic effects of natural products and to create novel approaches for drug discovery. Through the examination of interactions among components and targets within a network, researchers have gained insights into the intricate biological processes that drive diseases. This understanding will enable the design of more effective and targeted therapies, leading to advancements in treatment strategies. This transition in pharmacological thinking emphasises the significance of understanding the synergistic effects of multiple components and their interactions within a biological network, thereby providing a comprehensive approach to drug discovery and development. It amalgamates the wisdom and experiences of our ancient *Rishis* and practitioners encoded in ancient treatises such as *Charaka Samhita* and *Susruta Samhita*, enabling researchers and pharmaceutical companies to harness the power of synergy and combination therapies, thereby addressing complex diseases and improving patient outcomes.

Technological Advancements

Advances in biotechnology and computational biology have revolutionised the development of bioprospecting methodologies. HTS techniques, genomic sequencing, and computational modelling have expedited the discovery and characterisation of bioactive compounds from plant sources. These technological innovations have enhanced the efficiency and accuracy of the bioprospecting processes, thereby facilitating the identification of novel compounds with diverse applications. Recent advancements in the human genome project and "multi-omics"/systems biology have revolutionised the drug discovery process (George & Ijinu, 2011; Pushpangadan et al., 2013a, 2015b, 2018c). These advancements have enabled reverse pharmacology to identify potential drug candidates for existing diseases by using modern testing methods and advanced hyphenated instruments. Currently, this process involves screening hits, interactive medicinal chemistry, and optimising compounds to minimise potential side effects while improving their affinity and selectivity (Najmi et al., 2022; Nicolaou, 2014). During the drug development phase, efforts have been made to enhance efficacy, potency, metabolic stability, and oral bioavailability. Impressive developments in "multi-omics" have enabled an in-depth scientific approach to TM by integrating systems biology. Moreover, recent advancements have indicated that TM and Western medicine share more similarities than previously assumed. Systems biology is poised to be a valuable tool in TM research, offering the following benefits:

- Proof of concept, efficacy, and safety of TM preparations.
- Mechanistic studies and the development of biomarkers and surrogate endpoints.
- Fingerprinting of TM products to guide quality control and production.
- Discovery and validation of new drug leads and targets.

These deliverables highlight the potential of systems biology to advance TM research and its integration into mainstream medical practice.

Various hyphenated techniques have emerged as powerful tools in TM-based drug discovery to maximise efficiency and success rate. These methods include gas chromatography-tandem mass spectrometry (GC-MS/MS) (Chiu & Kuo, 2020; Rosado et al., 2018; Woźniak et al., 2018), liquid chromatography-tandem mass spectrometry (LC-MS/MS) (Jin et al., 2017; Wolfender et al., 2019; Zhu et al., 2020), thermogravimetry tandem gas chromatography-mass spectrometry (TGA/STA-GC-MS) (Awuchi et al., 2022), liquid chromatography-tandem nuclear magnetic resonance (HPLC-NMR-MS) (Gajula et al., 2021; Wolfender et al., 2019), gas chromatography-tandem infrared spectroscopy (GC-FTIR) (Salerno et al., 2020; Song et al., 2020), and capillary electrophoresis tandem mass spectrometry (CE-MS) (Cieslarova et al., 2017; Ozawa et al., 2020). By combining the separation capabilities of chromatography with the sensitivity and specificity of mass spectrometry and nuclear magnetic resonance, these analytical techniques facilitate the identification, quantification, and characterisation of drugs and their metabolites in complex biological matrices. Another powerful technique is liquid chromatography-mass spectrometry imaging (LC-MSI), which combines the spatial information provided by imaging techniques with the molecular characterisation capabilities of mass spectrometry. LC-MSI enables researchers to visualise the distribution of bioactive compounds within plant tissues, thereby providing valuable insight into their localisation and accumulation patterns. This information is crucial for understanding the biosynthesis of secondary metabolites and their metabolic pathways. LC-MSI also helps identify potential sources of bioactive compounds for drug discovery (Wu & Lu, 2022; Wu et al., 2020a, 2020b).

In addition, computational approaches have proven to be highly efficient and cost-effective for predicting drug-target interactions. Drug discovery has undergone a significant transformation, moving away from a linear approach that focuses on a single drug, target, and disease to a network-based approach driven by advancements in network pharmacology and systems biology. This network approach considers multiple drugs, targets, and diseases simultaneously. As a result, a single drug can now demonstrate activity against multiple targets simultaneously rather than solely targeting a single molecular target. The integration of artificial intelligence (AI) into TM-based drug discovery and development has ushered in a new era of scientific exploration and optimisation. AI algorithms provide researchers with powerful tools to analyse complex biological systems, unravel the mechanisms of action of TM, and identify novel drug targets. By merging multi-omics data, AI enables a thorough understanding of biological interactions and enhances the precision and customisation of medicines. Moreover, AI accelerates the discovery process and reduces the time and costs associated with drug development. However, validating AI-generated results through rigorous scientific methodologies and clinical trials is crucial (Blanco-González et al., 2023; Chu et al., 2022; Ng et al., 2024). Thus, combining the wisdom of TM with the power of AI can fully unlock the potential of TM, revolutionise drug development, and improve healthcare outcomes. Multidisciplinary research teams, international collaborations, and continued support for the scientific validation of TK are essential for realising the promise of this field and benefiting global health and well-being.

Need for Sustainable Bioprospecting Practices

Conservation-driven sustainable bioprospecting practices are critical for ensuring the long-term survival of biological resources and ecosystems while maximising socioeconomic benefits. These practices play a significant role in preserving biodiversity by advocating the responsible use and management of biological resources, minimising the impact on ecosystems, and safeguarding natural habitats for future generations. Specifically, sustainable bioprospecting focuses on practices that prioritise the conservation of plant species and their habitats. This involves extensive biodiversity assessments, implementation of ABS mechanisms, and the engagement of communities in conservation efforts. Moreover, sustainable bioprospecting helps maintain essential ecosystem services such as pollination, water purification, and climate regulation, which are vital for human well-being. Moreover, sustainable bioprospecting respects and protects the TK of indigenous communities and local people, who often possess invaluable insights into the use of medicinal plants and other natural resources. This approach preserves TK and ensures its transmission to future generations while providing equitable benefits to the communities that hold it. Economically, sustainable bioprospecting supports local development by offering a stable source of income and employment opportunities in the long term. Sustainable bioprospecting also enhances scientific research and innovation by providing access to diverse biological resources and allowing researchers to discover new species, novel biological compounds, and innovative solutions to challenges in medicine, agriculture, and biotechnology (Pushpangadan et al., 2012c). Culturally, sustainable bioprospecting fosters positive relationships and partnerships with indigenous peoples and local communities by respecting their cultural heritage and rights and promoting social equity, empowerment, and community development.

Conclusion and Future Prospects

Several future directions could shape bioprospecting in India. These include investing in R&D capabilities; integrating modern technologies such as genomics, bioinformatics, and HTS; creating an environment for entrepreneurship and innovation; recognising and harnessing TK from indigenous communities; strengthening human resource capacity; raising awareness; and international collaboration. India can harness its biodiversity, scientific expertise, and TK systems to realise its full potential for sustainable development, economic growth, and biodiversity conservation. The formulation of robust policy frameworks and regulatory mechanisms has been instrumental in governing bioprospecting activities in India. Initiatives such as the BDA (2002) and NBA have provided legal frameworks for access to biological resources and the equitable sharing of benefits arising from their utilisation. These policies aim to balance the objectives of biodiversity conservation, sustainable development, and socioeconomic equity. In conclusion, the bioprospecting of plant diversity in India has witnessed significant advancements and evolving trends since the establishment of CBD. Through a combination of TK integration, scientific research, conservation efforts, technological innovations, and policy interventions, India continues to harness its botanical wealth for societal benefits, while ensuring the sustainable management of its natural resources.

Acknowledgements NMK and MN gratefully acknowledge the financial support received from the Kerala State Biodiversity Board, Government of Kerala (no. 740/A13/2022/KSBB).

Conflict of Interest The authors have no conflict of interest to declare.

References

Awuchi, C., Godswill, H., & Twinomuhwezi, C. G. (2022). Hyphenated techniques. In C. Egbuna, K. C. Patrick-Iwuanyanwu, M. A. Shah, J. C. Ifemeje, & A. Rasul (Eds.), *Analytical techniques in Biosciences: From basics to applications* (1st ed., pp. 125–145). Academic Press, Elsevier. https://doi.org/10.1016/B978-0-12-822654-4.00015-4

Baber, Z. (2016). The plants of empire: Botanic gardens, colonial power and botanical knowledge. *Journal of Contemporary Asia, 46*, 659–679. https://doi.org/10.1080/00472336.2016.1185796

Balick, M. J., & Cox, P. A. (2020). *Plants, people, and culture: The science of ethnobotany* (2nd ed., pp. 228). Garland Science. ISBN: 978-0815345909

Bhutani, K. K., & Gohil, V. M. (2010). Natural products drug discovery research in India: Status and appraisal. *Indian Journal of Experimental Biology, 48*, 199–207.

Blanco-González, A., Cabezón, A., Seco-González, A., Conde-Torres, D., Antelo-Riveiro, P., Piñeiro, Á., & Garcia-Fandino, R. (2023). The role of AI in drug discovery: Challenges, opportunities, and strategies. *Pharmaceuticals (Basel), 16*, 891. https://doi.org/10.3390/ph16060891

Chakraborty, A., Mahajan, S., Bisht, M. S., & Sharma, V. K. (2022). Genome sequencing and comparative analysis of *Ficus benghalensis* and *Ficus religiosa* species reveal evolutionary mechanisms of longevity. *iScience, 25*, 105100. https://doi.org/10.1016/j.isci.2022.105100

Che, C.-T., George, V., Ijinu, T. P., Pushpangadan, P., & Andrae-Marobela, K. (2024). Traditional medicine. In S. B. McCreath & Y. N. Clement (Eds.), *Pharmacognosy: Fundamentals, applications and strategies* (2nd ed., pp. 11–28). Academic Press, Elsevier. https://doi.org/10.1016/B978-0-443-18657-8.00037-2. ISBN: 978-0443186578

Che, C.-T., George, V., Ijinu, T. P., Pushpangadan, P., & Andrae-Marobela, K. (2017). Traditional medicine. In S. Badal & R. Delgoda (Eds.), *Pharmacognosy: Fundamentals, applications and strategies* (1st ed., pp. 15–30). Academic Press, Elsevier. https://doi.org/10.1016/B978-0-12-802104-0.00002-0. ISBN: 978-0128021040

Chithra, M. A., Ijinu, T. P., Kharkwal, H., Sharma, R. K., Janardhanan, K. K., Pushpangadan, P., & George, V. (2020a). *Cocos nucifera* L. inflorescence extract: An effective hepatoprotective agent. *Indian Journal of Traditional Knowledge (IJTK), 19*, 128–136. https://doi.org/10.56042/ijtk.v19i1.30852

Chithra, M. A., Ijinu, T. P., Kharkwal, H., Sharma, R. K., Pushpangadan, P., & George, V. (2020b). Phenolic rich *Cocos nucifera* inflorescence extract ameliorates inflammatory responses in LPS-stimulated RAW264.7 macrophages and toxin-induced murine models. *Inflammopharmacology, 28*, 1073–1089. https://doi.org/10.1007/s10787-019-00620-6

Chiu, H.-H., & Kuo, C.-H. (2020). Gas chromatography-mass spectrometry-based analytical strategies for fatty acid analysis in biological samples. *Journal of Food and Drug Analysis, 28*, 60–73. https://doi.org/10.1016/j.jfda.2019.10.003

Chu, H., Moon, S., Park, J., Bak, S., Ko, Y., & Youn, B.-Y. (2022). The use of artificial intelligence in complementary and alternative medicine: A systematic scoping review. *Frontiers in Pharmacology, 13*, 826044. https://doi.org/10.3389/fphar.2022.826044

Cieslarova, Z., Lopes, F. S., do Lago, C. L., França, M. C. Jr, & Colnaghi Simionato, A. V. (2017). Capillary electrophoresis tandem mass spectrometry determination of glutamic acid and homocysteine's metabolites: Potential biomarkers of amyotrophic lateral sclerosis. *Talanta, 170*, 63–68. https://doi.org/10.1016/j.talanta.2017.03.103

De Deyn, G. B., & Kooistra, L. (2021). The role of soils in habitat creation, maintenance and restoration. *Philosophical Transactions of the Royal Society of London. Series B, Biological Sciences, 376*, 20200170. https://doi.org/10.1098/rstb.2020.0170

Duke, S. O., Rimando, A. M., Dayan, F. E., Canel, C., Wedge, D. E., Tellez, M. R., Schrader, K. K., Weston, L. A., Smillie, T. J., Paul, R. N., & Duke, M. V. (2000). Strategies for the discovery of bioactive phytochemicals. In W. R. Bidlack, S. T. Omaye, M. S. Meskin, & D. K. W. Topham (Eds.), *Phytochemicals as bioactive agents* (pp. 1–20). Technomic Publishing Co. ISBN: 978-0429181252

Farnsworth, N. L., Akerele, O., Bingel, A. S., Soejarto, D. D., & Guo, Z. (1985). Medicinal plants in therapy. *Bulletin of the World Health Organization, 63*, 965–981.

Gajula, R., Nageswara, S., & Nanjappan, S. (2021). Metabolomics: A Recent advanced omics technology in herbal medicine research. In T. Aftab & K. R. Hakeem (Eds.), *Medicinal and aromatic plants: Expanding their horizons through omics* (pp. 97–117). Academic Press. ISBN: 9780128195901

George, V., & Ijinu, T. P. (2011). Scientific validation of traditional knowledge and development of herbal drugs. In S. Rajasekharan & P. G. Latha (Eds.), *Traditional and folk practices: Contemporary relevance and future prospects* (pp. 56–63). Jawaharlal Nehru Tropical Botanic Garden and Research Institute.

George, V., Ijinu, T. P., Chithra, M. A., & Pushpangadan, P. (2016). Can local health traditions and tribal medicines strengthen Ayurveda?, Case study 2. *Trichopus zeylanicus* ssp. *travancoricus* Burkill ex Narayanan. *Journal Traditional and Folk Practices, 4*, 1–14. https://jtfp.jntbgri.res.in/index.php/jtfp/article/view/62/58

George, V., Rajasekharan, S., Ijinu, T. P., & Pushpangadan, P. (2014). Intellectual property rights (IPR), traditional medicine (TM) and traditional knowledge (TK), In V. N. R. Pillai (Ed.), *Compendium of invited papers—Traditional knowledge* (pp. 80–102). 26th Kerala Science Congress, Kerala State Council of Science, Technology and Environment. ISBN: 978-8192467412

Halliburton, M. (2017). *India and the patent wars: Pharmaceuticals in the new intellectual property regime* (p. 187). ILR Press, Cornell University Press. ISBN: 978-1501713972

Harrer, S., Shah, P., Antony, B., & Hu, J. (2019). Artificial intelligence for clinical trial design. *Trends in Pharmacological Sciences, 40*, 577–591. https://doi.org/10.1016/j.tips.2019.05.005

Heinrich, M. (2000). Ethnobotany and its role in drug development. *Phytotherapy Research, 14*, 479–488. https://doi.org/10.1002/1099-1573(200011)14:7%3c479::aid-ptr958%3e3.0.co;2-2

Ijinu, T. P., Anish, N., Shiju, H., George, V., & Pushpangadan, P. (2011). Home gardens for nutritional and primary health security of rural poor of south Kerala. *Indian Journal of Traditional Knowledge, 10*, 413–428.

Ijinu, T. P., Aswany, T., Chithra, M. A., Priya Rani, M., George, V., & Pushpangadan, P. (2023a). Phytochemical and pharmacological aspects of "Arogyapacha", *Trichopus zeylanicus* Gaertn. In T. Pullaiah (Ed.), *Phytochemistry and pharmacology of medicinal plants* (Vol. 1, pp. 433–441). Apple Academic Press. ISBN: 978-1774915622

Ijinu, T. P., George, V., & Pushpangadan, P. (2020). Bioprospecting of medicinal and aromatic plants. In K. V. Peter (Ed.), *Bioprospecting of underutilized horticulture crops* (pp. 1–23). Brillion Publishing. ISBN: 978-9389350104

Ijinu, T. P., George, V., & Pushpangadan, P. (2022a). History of research on medicinal plants in India, In A. Máthé & I. A. Khan (Eds.), *Medicinal and aromatic plants of India* (Vol. 1, pp. 35–61). Springer International Publishing. https://doi.org/10.1007/978-3-030-98701-5_2. ISBN: 978=3030987008

Ijinu, T. P., Jacob, J., Sasikumar, P., Varghese, T., Ragesh, R. N., & Pushpangadan, P. (2023b). Phytochemical constituents and pharmacology of *Ocimum tenuiflorum* L. In T. Pullaiah (Ed.), *Bioactives and pharmacology of Lamiaceae* (pp. 367–388). Apple Academic Press. ISBN: 978-1774912935

Ijinu, T. P., Latha, P. G., George, V., & Pushpangadan, P. (2016). Total phenolic and flavonoid contents and *in vitro* antioxidant potential of methanolic extract of *Blepharis maderaspatensis* (L.) B. Heyne ex Roth whole plant. *Annals of Phytomedicine, 5*, 85–90.

Ijinu, T. P., Manikantan, K., Aneeshkumar, A. L., Soorya, V., Latha, P. G., George, V., & Pushpangadan, P. (2022c). Ameliorative effect of the ethnomedicinal plant *Blepharis maderaspatensis* (L.) B.Heyne ex Roth on lipopolysaccharide-induced acute liver inflammation and oxidative stress in Wistar rat model. *Phytomedicine Plus, 2*, 100247. https://doi.org/10.1016/j.phyplu.2022.100247

Ijinu, T. P., Rajasekharan, S., George, V., & Pushpangadan, P. (2022b). Access and benefit sharing in indigenous knowledge stewardship and sustainable development. In O. V. Oommen, K. P. Laladhas, P. Nelliyat, & B. Pisupati (Eds.), *Biodiversity conservation through access and benefit sharing (ABS)* (pp. 163–1858). Springer International Publishing. https://doi.org/10.1007/978-3-031-16186-5_8. ISBN: 978-3031161858

Iwu, M. M. (2002). Ethnobotanical approach to pharmaceutical drug discovery: Strengths and limitations. In M. M. Iwu & J. C. Wootton (Eds.), *Ethnomedicine and drug discovery, Advances in Phytomedicine* (Vol. 1, pp. 309–320). Elsevier. https://doi.org/10.1016/S1572-557X(02)80034-4

Jin, Y., Zhang, J., Zhao, W., Zhang, W., Wang, L., Zhou, J., & Li, Y. (2017). Development and validation of a multiclass method for the quantification of veterinary drug residues in honey and royal jelly by liquid chromatography–tandem mass spectrometry. *Food Chemistry, 221*, 1298–1307. https://doi.org/10.1016/j.foodchem.2016.11.026

Kanherkar, R. R., Stair, S. E., Bhatia-Dey, N., Mills, P. J., Chopra, D., & Csoka, A. B. (2017). Epigenetic mechanisms of integrative medicine. *Evidence-Based Complementary and Alternative Medicine, 2017*, 1–19. https://doi.org/10.1155/2017/4365429

Katiyar, C., Gupta, A., Kanjilal, S., & Katiyar, S. (2012). Drug discovery from plant sources: An integrated approach. *AYU (An International Quarterly Journal of Research in Ayurveda), 33*, 10–19. https://doi.org/10.4103/0974-8520.100295

Kingston, D. G. (2011). Modern natural products drug discovery and its relevance to biodiversity conservation. *Journal of Natural Products, 74*, 496–511.

Krishnakumar, N. M., Latha, P. G., Rajasekharan, S., Suja, S. R., Dan, M., Sabulal, B., & Navas, M. (2016a). Preclinical studies of a medicinal plant (coded) based on traditional knowledge with special reference to access and benefit sharing, Part I. *Journal of Traditional and Folk Practices, 2*(3&4), 151–159.

Krishnakumar, N. M., Latha, P. G., Rajasekharan, S., Suja, S. R., Dan, M., Sabulal, B., & Navas, M. (2016b). Protective effect of coded plant (222) leaf extract against streptozotocin-induced diabetes in Wistar albino rats, Part II. *Journal of Traditional and Folk Practices, 4*, 37–47.

Krishnakumar, N. M., Latha, P. G., Rajasekharan, S., Suja, S. R., Dan, M., Sabulal, B., & Navas, M. (2017a). Evaluation of the protective effect of coded plant leaf ethanolic extract (222) against paracetamol-induced hepatotoxicity and oxidative stress in Wistar albino rats, Part III. *Journal of Traditional and Folk Practices, 5*, 8–15.

Krishnakumar, N. M., Latha, P. G., Rajasekharan, S., Suja, S. R., Dan, M., Sabulal, B., & Navas, M. (2017b). Assessment of hepatoprotective activity of coded plant leaf ethanolic extract (222) against carbon tetrachloride-induced hepatotoxicity in Wistar rats, Part IV. *Journal of Traditional and Folk Practices, 5*, 111–120.

Krishnakumar, N. M., Latha, P. G., Rajasekharan, S., Suja, S. R., Dan, M., Sabulal, B., & Navas, M. (2018a). Coded plant (222) leaf ethanolic extract ameliorates ethanol-induced liver damage and oxidative stress in Wistar albino rats, Part V. *Journal of Traditional and Folk Practices, 6*, 39–48.

Krishnakumar, N. M., Latha, P. G., Rajasekharan, S., Suja, S. R., Dan, M., Sabulal, B., & Navas, M. (2018b). Hepatoprotective activity of coded plant (222) leaf extract in d-galactosamine induced hepatotoxicity: Biochemical and histopathological analysis in experimental animals, Part VI. *Journal of Traditional and Folk Practices, 6*, 39–48.

Krishnakumar, N. M., Latha, P. G., Rajasekharan, S., Suja, S. R., Dan, M., Sabulal, B., & Navas, M. (2020). Phytochemical analysis and estimation of phytoconstituents in coded plant (222) leaves, Part VII. *Journal of Traditional and Folk Practices, 7&8*, 55–63.

Kumar, D. (2017). Unequal contenders, uneven ground: Medical encounters in British India, 1820–1920. In A. Cunningham & B. Andrews (Eds.), *Western medicine as contested knowledge* (pp. 172–211). Manchester University Press. https://doi.org/10.7765/9781526123572.00013

Laird, S. A., & Ten Kate, K. (2002). Linking biodiversity prospecting and forest conservation. In S. Pagiola, J. Bishop, & N. Landell-Mills (Eds.), *Selling forest environmental services* (pp. 151–172). Earthscan.

Latha, P. G., Rajasekharan, S., Suja, S. R., Dan, M., Sabulal, B., Navas, M., Krishnakumar, N. M., & Shahul Hameed Vaidyar, T. M. (2015). *Ethnopharmacological evaluation of antidiabetic, hepatoprotective, antioxidant, antifatigue potential of selected herbal coded drugs based on traditional knowledge (Preclinical studies)* (p. 115). Jawaharlal Nehru Tropical Botanic Garden and Research Institute. ISBN: 978-8192467443

Mateo, N., Nader, W., & Tamayo, G. (2001). Bioprospecting. In S. A. Levin (Ed.), *Encyclopaedia of biodiversity* (pp. 471–488). Academic Press.

Najmi, A., Javed, S. A., Al Bratty, M., & Alhazmi, H. A. (2022). Modern approaches in the discovery and development of plant-based natural products and their analogues as potential therapeutic agents. *Molecules, 27*, 349. https://doi.org/10.3390/molecules27020349

NBA. (2018). *National Biodiversity Authority, Annual report (2017–2018)* (p. 91). Ministry of Environment, Forest and Climate Change, Government of India.

NBA. (2022). *National Biodiversity Authority, Annual report (2021–2022)* (p. 94). Ministry of Environment, Forest and Climate Change, Government of India.

Newman, D. J., & Cragg, G. M. (2016). Natural products as sources of new drugs from 1981 to 2014. *Journal of Natural Products, 79*, 629–661. https://doi.org/10.1021/acs.jnatprod.5b01055

Newman, D. J., & Cragg, G. M. (2020). Natural products as sources of new drugs over the nearly four decades from 01/1981 to 09/2019. *Journal of Natural Products, 83*, 770–803. https://doi.org/10.1021/acs.jnatprod.9b01285

Ng, J. Y., Cramer, H., Lee, M. S., & Moher, D. (2024). Traditional, complementary, and integrative medicine and artificial intelligence: Novel opportunities in healthcare. *Integrative Medicine Research, 13*, 101024. https://doi.org/10.1016/j.imr.2024.101024

Nicolaou, K. C. (2014). Advancing the drug discovery and development process. *Angewandte Chemie (International Ed. in English), 53*, 9128–9140. https://doi.org/10.1002/anie.201404761

Nomani, Z. M. (2020). Divya pharmacy v. Union of India high court of Uttarakhand, India; 2018 SCC online utt. 1035: Writ petition (M/S) no. 3437 of 2016; Judgment reserved on September 7, 2018; Judgment delivered on December 21, 2018. *Biotechnology Law Report, 39*, 122–128. https://doi.org/10.1089/blr.2020.29161.zmn

Ozawa, H., Hirayama, A., Ishikawa, T., Kudo, R., Maruyama, M., Shoji, F., Doke, T., Ishimoto, T., Maruyama, S., Soga, T., & Tomita, M. (2020). Comprehensive dipeptide profiling and quantitation by capillary electrophoresis and liquid chromatography coupled with tandem mass spectrometry. *Analytical Chemistry, 92*, 9799–9806. https://doi.org/10.1021/acs.analchem.0c01258

Patwardhan, B., Warude, D., Pushpangadan, P., & Bhatt, N. (2005). Ayurveda and traditional Chinese medicine: A comparative overview. *Evidence-Based Complementary and Alternative Medicine, 2*, 465–473. https://doi.org/10.1093/ecam/neh140

Pushpangadan, P. (1998). *Ethnobiology in India: A status report on all-India co-ordinated research project on ethnobiology*. Ministry of Environment, Forests & Climate Change, Government of India.

Pushpangadan, P. (2002). Biodiversity and emerging benefit sharing arrangements challenges and opportunities for India. *Proceedings of the National Academy of Sciences, India Section B, 68*, 297–314.

Pushpangadan, P., Dan, V. M., Ijinu, T. P., & George, V. (2012a). New chemical entities emerging in Noni (*Morinda citrifolia*) from modern biotechnological intervention. In T. Marimuthu (Ed.), *Noni-The tool for wellness. Noni search 2012: Proceedings of seventh national symposium* (pp. 233–247). World Noni Research Foundation, International Society for Noni Science.

Pushpangadan, P., Dan, V. M., Ijinu, T. P., & George, V. (2013a). Phytochemical characterization of medicinal plant in herbal drugs. In I. P. Tripathi (Ed.), *Chemistry, biochemistry and ayurveda of Indian medicinal plants, Part 1, Identification and conservation of medicinal plants* (pp. 1–12). International E-Publication. ISBN: 978-8183520041

Pushpangadan, P., George, V., & Ijinu, T. P. (2015a). Traditional medicine and ethnopharmacology research in India. In A. K. Chauhan, P. Pushpangadan, & V. George (Eds.), *Natural products: Recent advances* (pp. 120–146). Write and Print Publications. ISBN: 978-8192970530

Pushpangadan, P., George, V., & Ijinu, T. P. (2016a). Ethnobotany, ethnopharmacology, bioprospecting and patents. In T. Pullaiah, K. V. Krishnamurthy, & B. Bahadur (Eds.), *Ethnobotany of India, vol 1: Eastern Ghats and Deccan* (pp. 461–482). Apple Academic Press. ISBN: 978-1771883382

Pushpangadan, P., George, V., & Ijinu, T. P. (2018a). Community based traditional knowledge and protection by *sui generis* and intellectual property rights. *Journal Traditional and Folk Practices, 6*, 3–12. https://doi.org/10.25173/jtfp.2018.6.2.100http:/www.jtfp.jntbgri.res.in

Pushpangadan, P., George, V., Ijinu, T. P., & Chithra, M. A. (2018b). Ethnomedicine, health food and nutraceuticals-traditional wisdom of maternal and child health in India. *Edelweiss Applied Science and Technology, 2*, 113–117. https://doi.org/10.33805/2576.8484.124

Pushpangadan, P., George, V., Ijinu, T. P., & Chithra, M. A. (2018c). Traditional knowledge related IPR and their bioprospecting for development of nutraceuticals and healthcare products. *Acta Scientific Medical Sciences, 2*, 43–48.

Pushpangadan, P., George, V., Ijinu, T. P., & Chithra, M. A. (2018d). Biodiversity, bioprospecting, traditional Knowledge, sustainable development and value-added products: A review. *Journal of Traditional Medicine and Clinical Naturopathy, 7*, 1000256. https://doi.org/10.4172/2573-4555.1000256

Pushpangadan, P., George, V., Ijinu, T. P., Chithra, M. A., & Govidarajan, R. (2018e). Resurgence of traditional medicine systems with special reference to Indian systems of medicine and modern scientific developments. *Journal of Traditional Medicine and Clinical Naturopathy, 7*, 1000276. https://doi.org/10.4172/2573-4555.1000275

Pushpangadan, P., George, V., Ijinu, T. P., & Rajasekharan, S. (2016b). Ethnobotany, ethnobiology, ethnopharmacology, bioprospecting of traditional knowledge and evolution of benefit sharing.

In A. K. Jain (Ed.), *Indian ethnobotany: Emerging trends* (pp. 1–23). Scientific Publishers. ISBN: 978-8172339609

Pushpangadan, P., George, V., Ijinu, T. P., & Rajasekharan, S. (2017). Ethnopharmacology, traditional knowledge and intellectual property rights. In K. P. Laladhas, P. Nilayangod, & O. V. Oommen (Eds.), *Biodiversity for sustainable development* (pp. 97–119). Springer International Publishing. https://doi.org/10.1007/978-3-319-42162-9_6

Pushpangadan, P., George, V., Sreedevi, P., Ijinu, T. P., & Ninawe, A. S. (2014a). Ethnozoological knowledge of Indian scheduled tribe, scheduled caste and rural communities. *Indian Journal of Traditional Knowledge, 13*, 735–741.

Pushpangadan, P., & Ijinu, T. P. (2016). Biodiversity, bio-prospecting, intellectual property rights, conservation and sustainable development: An Indian experiment, In N. N. C. Awasthi (Ed.), *Proceedings of the national seminar on biodiversity conservation and sustainable development* (pp. 43–55). Brahmanand College.

Pushpangadan, P., & Ijinu, T. P. (2017). Intellectual property rights on traditional knowledge and their significance in sustainable societal development. In B. P. R. Narasimharao, E. Wright, S. Prasad, & M. Joshi (Eds.), *Handbook of research on science education and university outreach as a tool for regional development* (pp. 157–185). IGI Global. https://doi.org/10.4018/978-1-5225-1880-8.ch013. ISBN: 978-1522518808

Pushpangadan, P., & Ijinu, T. P. (2019). Tribal upliftment, intellectual property rights and benefit sharing. In T. Basar (Ed.), *Traditional knowledge and traditional cultural expressions: National and community perspectives* (pp. 154–169). National Law University and Judicial Academy. ISBN: 978-8194108603

Pushpangadan, P., Ijinu, T. P., Chithra, M. A., & George, V. (2019). Tribal culture and traditional knowledge of India through ages. In S. K. Misra & S. Chaudhary (Eds.), *Traditional healing systems of India* (pp. 26–29). Indian Trust for Rural Heritage and Development.

Pushpangadan, P., Ijinu, T. P., & Dan, V. M. (2018f). Evolution of ethnopharmacology research in India and their bioprospecting strategies. In A. P. Das & S. Bera (Eds.), *Plant diversity in the Himalaya hotspot region* (Vol. 2, pp. 787–806). Dehradun. ISBN: 978-81-211-0985-7

Pushpangadan, P., Ijinu, T. P., Dan, V. M., & George, V. (2012b). The role of science and technology in biodiversity based sustainable development. In T. K. Sheng & D. Solijan (Eds.), *Science and technology policy for sustainable development* (pp. 125–135). Daya Publishing House. ISBN: 978-8170357858

Pushpangadan, P., Ijinu, T. P., Dan, V. M., & George, V. (2015b). Trends in bioprospecting of biodiversity in new drug design. *Pleione, 9*, 267–282.

Pushpangadan, P., Ijinu, T. P., Dan, V. M., Thomas, A., Avinash, S., & George, V. (2012c). Recent advances of agricultural biotechnology in the light of climate change. *Proceedings of the National Academy of Sciences, India Section B, 82*, 381–386. https://doi.org/10.1007/s40011-012-0101-6

Pushpangadan, P., Ijinu, T. P., & George, V. (2013b). Opportunities for ethnobotanical studies on medicinal and aromatic plants in India. *Ethnobotany, 25*, 22–32.

Pushpangadan, P., Ijinu, T. P., & George, V. (2014b). Origin, growth and development of ethnobiology/ethnobotany, human culture and civilization. In S. Kumar (Ed.), *Ethnobotanical studies in India* (pp. 1–11). Deep Publications. ISBN: 978-9380702063

Pushpangadan, P., Ijinu, T. P., & George, V. (2014c). Intellectual property rights (IPR) regime and the traditional knowledge/indigenous knowledge holders. In N. P. Chaubey & N. S. Rao (Eds.), *Proceedings of 38th Indian social science congress* (pp. 45–60). Indian Academy of Social Sciences.

Pushpangadan, P., & Nair, K. N. (2005). Value addition and commercialisation of biodiversity and associated traditional knowledge in the context of the intellectual property regime. *Journal of Intellectual Property Rights, 10*, 441–453.

Pushpangadan, P., & Pradeep, P. R. J. (2008). *A glimpse at tribal India: An ethnobiological enquiry* (p. 167). Amity Institute for Herbal and Biotech Products Development.

Rajasekharan, S., Pushpangadan, P., Latha, P. G., Vinodkumar, T., & Nair, G. (2012). *Learning from traditional knowledge, Intellectual property, Traditional knowledge: Community prospective on access and benefit sharing, ABS Kani model of access and benefit sharing* (p. 28). Jawaharlal Nehru Tropical Botanic Garden and Research Institute. ISBN: 978-8192009872

Rosado, T., Oppolzer, D., Cruz, B., Barroso, M., Varela, S., Oliveira, V., Leitão, C., & Gallardo, E. (2018). Development and validation of a gas chromatography/tandem mass spectrometry method for simultaneous quantitation of several antipsychotics in human plasma and oral fluid. *Rapid Communications in Mass Spectrometry, 32*, 2081–2095. https://doi.org/10.1002/rcm.8087

Salerno, T. M. G., Donato, P., Frison, G., Zamengo, L., & Mondello, L. (2020). Gas chromatography—Fourier transform infrared spectroscopy for unambiguous determination of illicit drugs: A proof of concept. *Frontiers in Chemistry, 8*, 624. https://doi.org/10.3389/fchem.2020.00624

Sikarwar, R. L. S. (2020). History and footprints of plant explorations in Indian subcontinent. *The Journal of the Indian Botanical Society, 100*, 1–17. ISSN: 0019-4468

Singh, R. K., Pretty, J., & Pilgrim, S. (2010). Traditional knowledge and biocultural diversity: Learning from tribal communities for sustainable development in northeast India. *Journal of Environmental Planning and Management, 53*, 511–533. https://doi.org/10.1080/09640561003722343

Soejarto, D. D., Gyllenhaal, C., Riley, M. C., & Zhang, H. (2011). Ethnobotany of natural products. In J. M. Pezzuto & M. J. Kato (Eds.), *Phytochemistry and pharmacognosy*. UNESCO-EOLSS Publishers.

Song, Y., Cong, Y., Wang, B., & Zhang, N. (2020). Applications of Fourier transform infrared spectroscopy to pharmaceutical preparations. *Expert Opinion on Drug Delivery, 17*, 551–571. https://doi.org/10.1080/17425247.2020.1737671

Souravi, K., & Rajasekharan, P. E. (2019). Traditional knowledge in horticulture. In P. Rajasekharan & V. Rao (Eds.), *Conservation and utilization of horticultural genetic resources* (pp. 621–636). Springer Singapore. https://doi.org/10.1007/978-981-13-3669-0_21

Sreedevi, P., Ijinu, T. P., Anzar, S., Bincy, A. J., George, V., Rajasekharan, S., & Pushpangadan, P. (2013). Ethnobiology, ethnobotany, ethnomedicine and traditional knowledge with special reference to India. *Annals of Phytomedicine, 2*, 4–12.

Süntar, I. (2020). Importance of ethnopharmacological studies in drug discovery: Role of medicinal plants. *Phytochemistry Reviews, 19*, 1199–1209. https://doi.org/10.1007/s11101-019-09629-9

Ten Kate, K., & Laird, S. A. (2021). *The commercial use of biodiversity: Access to genetic resources and benefit-sharing* (p. 414). Routledge library editions: Ecology. Taylor & Francis. ISBN: 978-0367357641

Tu, Y. (2016). Artemisinin—A gift from traditional Chinese medicine to the world (Nobel lecture). *Angewandte Chemie (International Ed. in English), 55*, 10210–10226. https://doi.org/10.1002/anie.201601967

Venkataraman, K., & Latha, S. S. (2008). Intellectual property rights, traditional knowledge and biodiversity of India. *Journal of Intellectual Property Rights, 13*, 326–335.

Verpoorte, R., Choi, Y. H., & Kim, H. K. (2005). Ethnopharmacology and systems biology: A perfect holistic match. *Journal of Ethnopharmacology, 100*, 53–56. https://doi.org/10.1016/j.jep.2005.05.033

Winterbottom, A. (2021). Becoming "traditional": A transnational history of neem and biopiracy discourse. *Osiris, 36*, 262–283. https://doi.org/10.1086/713897

Wolfender, J.-L., Nuzillard, J.-M., van der Hooft, J. J. J., Renault, J.-H., & Bertrand, S. (2019). Accelerating metabolite identification in natural product research: Toward an ideal combination of liquid chromatography–high-resolution tandem mass spectrometry and NMR profiling, in silico databases, and chemometrics. *Analytical Chemistry, 91*, 704–742. https://doi.org/10.1021/acs.analchem.8b05112

Woźniak, M. K., Wiergowski, M., Aszyk, J., Kubica, P., Namieśnik, J., & Biziuk, M. (2018). Application of gas chromatography–tandem mass spectrometry for the determination of amphetamine-type stimulants in blood and urine. *Journal of Pharmaceutical and Biomedical Analysis, 148*, 58–64. https://doi.org/10.1016/j.jpba.2017.09.020

Wu, Q., Huang, Z., Wang, Y., Zhang, Z., & Lu, H. (2020a). Absolute quantitative imaging of sphingolipids in brain tissue by exhaustive liquid microjunction surface sampling–liquid chromatography–mass spectrometry. *Journal of Chromatography A, 1609*, 460436. https://doi.org/10.1016/j.chroma.2019.460436

Wu, Q., Li, Y., Wang, Y., & Lu, H. (2020b). Quantitative mass spectrometry imaging of amino acids with isomer differentiation in brain tissue via exhaustive liquid microjunction surface sampling–tandem mass tags labeling–ultra performance liquid chromatography–mass spectrometry. *Journal of Chromatography A, 1621*, 461086. https://doi.org/10.1016/j.chroma.2020.461086

Wu, Q., & Lu, H. (2022). Quantitative mass spectrometry imaging with liquid microjunction surface sampling. *Methods in Molecular Biology, 2437*, 181–194. https://doi.org/10.1007/978-1-0716-2030-4_13

Zhu, S., Zheng, Z., Peng, H., Sun, J., Zhao, X.-E., & Liu, H. (2020). Quadruplex stable isotope derivatization strategy for the determination of panaxadiol and panaxatriol in foodstuffs and medicinal materials using ultra high-performance liquid chromatography tandem mass spectrometry. *Journal of Chromatography A, 1616*, 460794. https://doi.org/10.1016/j.chroma.2019.460794

Inequity, in Commerce over Biodiversity, to the Benefit Claimer!

M. K. Ramesh

Introduction

Bio-resources, in all their varieties, are viewed as "Green Gold". Their products, besides contributing to the well-being of humanity, have, been earning rich dividends for their producers. Exploring and further mining the vastness of its potential, in the marketplace, has been a subject of great interest to biotechnologists, food processors, healthcare specialists and the like. Food and Pharmaceutical Industries, for instance, have demonstrated, over time that they are real money-spinners. Of late, with the increasing application of science and technology in these sectors, the scope for an increase in the new and improved varieties and the demand for such products and the consequent economic benefits have been, on the ascendant. The inventors and innovators of such products and processes associated with them, have always been given due recognition and ensured a share in the profits of the investing entrepreneurs, for their intellectual labour, in the well-evolved law, on the subject. However, the plight of the practitioners, carriers, inheritors, and conservators of the traditional wisdom associated with these resources, in this regard, has been quite pathetic. They have been treated as a class apart, negatively. Let alone, getting duly rewarded for their "anonymous and silent" contribution to the conservation and environmentally sustainable use of bio-resources, the legal regime hardly recognised and conferred any right or entitlement, to them. Since traditional practices and knowledge systems are in the Public Domain, the Intellectual Property Law has no place or space for them. While one of the principles of law underlying any transaction is that there shall not be any, "Unjust Enrichment", the hard law, at the enforcement level, made little attempt to pay attention to the unfortunate plight of these vulnerable victims, who are the real custodians and conservators of the traditions and practices that have ensured the resources well-being and continued significance and value to humanity across

M. K. Ramesh (✉)
National Law School of India University, Bengaluru, India
e-mail: mkramesh@nls.ac.in

L. Krishna Panicker et al. (eds.), *Biodiversity and Business*,
https://doi.org/10.1007/978-3-031-71674-4_19

generations. Nor those, who made a business out of them and earned huge profits, hardly cared to share the benefits with the ones who provided the source, base and capital for such ventures. It is a system that has been both iniquitous and unjust.

Wheels of time began to turn, to highlight this sorry state of affairs, leading to a host of legal developments over the issue. International Law, arguably, for the first time, through a legal prescription, provided the conceptual frame creating a right and the mechanism for its realization. In terms of concretising the idea, the (CBD 1992),[1] was the major legal development. It, was indeed, a path-breaking one. It has been over three decades since such a conscious and legally crystallised effort was put on the anvil of application. Nonetheless, in the submission of this author, both the efforts in law and the actions that have followed it, instead of ensuring the system turning a new leaf, have been very late in coming into effect, with less than adequate safeguards, in the realization of the right in any significant manner. They are, almost inconsequential in effect and have hardly arrested the trend of perpetuation of the inequity and injustice to the resource, the community and their customary practices. The following is a little elaboration of this assertion.

In this effort, after briefly laying down the various elements of the Equity aspect of Sharing of Benefits, a brief examination of the evolution of the law and its working would be carried out. How the changes, over time, attested the contributions of the communities, their traditions associated with bioresources, provided legal recognition, by according a status of a right holder in law and developed mechanisms for a share in the benefits besides, providing safeguards against any mischief that would lead to its denial to the, would then be considered. Further, how and to what extent have these laws been able to ensure arresting the pernicious practice of "unjust enrichment" of those who least deserved it, and their limitations, would get underlined in the enquiry. The enquiry traces the developments of the law—both international and Indian—on the subject. The primary focus would be on CBD, with reference, in appropriate places, to the law on Protection of Plant Varieties and Farmers' Rights.[2]

Biodiversity Traditions, Knowledge Systems, Communities and the Rationale for the Right to a Fair and Equitable Sharing of Benefits

Life and life forms, variety their uses and the knowledge associated with them are as old as human appearance on earth. In different regions of the World, such knowledge evolved into traditions and practices of certain communities of people, over time.

---

[1] The Convention on Biological Diversity, 1992 (CBD), along with its two protocols: The Montreal Protocol, 2000 (on Biosafety) and Nagoya Protocol, 2014 (on Access and Sharing of Benefits—ABS) and the Indian law: The Biodiversity Act, 2002.

[2] International Treaty on Plant Genetic Resources for Food and Agriculture, 2001 (ITPGRFA) and The Protection of Plant Variety and Farmers Right Act, 2001 (PPVFRA).

Their utility and value having stood the test of time and transmitted across generations, continue to be conserved, protected and practised, in modern times as well. We have now, individuals and communities identified as Traditional Knowledge (TK) Holders, Carriers, Conservers and practitioners of the art and craft of managing and putting into different kinds of uses of bio-resources and the knowledge associated with them, for the good of humanity (wipo 2011).[3] Practice of TK, for livelihood and as a profession that would contribute to sustainable development (sustainable use of the resource, its conservation and evolution, besides service to society), has been acknowledged and respected, since time immemorial. The realisation has been that the continued existence of such TK holders and the practice of their art, besides securing the health and well-being of the society, ensured the conservation and protection of the resource as well. As such, enabling and facilitating the community, as an entitlement, to access, use, manage and put into application the TK, in relation to them, for a variety of purposes, as their Common Property, has been part of social ordering as well, in several countries.

With the advent of modern science and its application (especially, untested and unverified, biotechnological processes and the increased efforts of the Pharmaceutical and food industries in accessing and appropriating the bioresources and the TK, about them for commercial purposes), the industrial revolution led to an increase in mass production of goods (with bioresources as raw material), increasing consumerist practice (of use, overuse and abuse of biological resources), accelerated environmentally destructive developmental activities (leading to loss of flora and fauna) and the like, the world over, has, escalated into global concern (Common Concern of humanity) about this resource and the knowledge systems associated with it (Characterised as, Global Commons), among the Community of nations, in recent times[4] (Earth Commission).

There is a worrisome spinoff of the unregulated and ever-increasing application of biotechnological processes and Commercialisation of Bio-resources, knowledge and the practices associated with them, by the modern scientific community and the industrial houses. It is twin dangers of, over exploitation and loss of the resource base (loss of Common Property and the increase in Common Concern over its safety, security and integrity) and denial of any share in the benefits derivable from the commercial application of it, to the Community and the traditional Practitioners. This is iniquitous to both the resource, TK and the Community associated with them. Resource that is meant for the Common Good should not be allowed to be exploited by a few, who have no exclusive claim over it, to the detriment of the resource. Further, the holders, conservers and carriers of the good Bio-diverse traditions, as its custodians, have the first charge over them. They cannot be deprived of their just entitlement to derive the benefits resulting from their labour. It is not just a denial of the latter's right to share the benefits. The consequential impact of such a situation is more devastating: the Practitioner and the Community had acquired an identity and a status of their own, owing to their long-established umbilical relationship with the

[3] The systems of Ayurveda, Unani and Siddha, are examples of this tradition.

[4] See, the Preamble to CBD, supra, 1.

resource, over Centuries of proximity with it. The identity gets lost and the Status is dented.

Equity demands that the no contribution, that has resulted in benefit of some kind, should go unrewarded. There has to be a mechanism for setting markers for determination of the extent of contribution, based upon which the apportionment of benefits arising from the commercial application of the resource and the TK and the associated practices. Viewed from another angle, the one who puts TK and customary practices associated with the Bioresources to commercial use, should recognise and incentivise the ones, from whose ideas, traditions and practices he has enriched himself, by sharing a fair and just portion of it. The "doctrine of unjust enrichment", in law, requires that nobody shall enrich oneself at the cost of another. Such a beneficiary, has the obligation of restitution to the other by properly, suitably and fairly compensating the other. The entire edifice of the law of Contracts is built on this foundational principle. There is a contract, by implication (Indian Contract Act),[5] between the Commercial user and the TK Holder that the latter would get a fair share of the benefits of such an application. Thus, Environmental Equity (bio-resource conservation and biosafety) and Human Equity (a fair share in the benefits to the holder and practitioner of the bio-resource and TK over it, resulting from its commercial application), constitute, the core aspect of equity in Bioresource management.

Customary practices have their rightful place and the status of the holders and practitioners of TK are acknowledged in the socially organised life of the people. But, the ability to exercise a host of these rights and realising the benefits arising out of it, is possible only when there is a legal attestation of approval and accommodation of it. The legally recognised right would entitle one to challenge, in an appropriate Court of law, any violation of it, this includes, the right to demand adequate compensation for the denial of benefits from the one who has unjustly enriched himself. The other dimension of this right is that, in addition to claiming a share in the benefits, the act of denial would be tantamount to a criminal offence of piracy and scope exists for penal action for the piratical act of such wrongdoer. The path for the recognition and creation of a status, besides, derivation of a right therefrom, for these, in the legal ordering, especially those associated with bioresources, has been quite tortuous. Mustering legal support for what appears logical and just, ironically, has been subjected to a host of hurdles. This finds elaboration and the reasons for the same, in the following part of this work.

[5] See, Sections 68 to 72 of the Indian Contract Act, 1872.

Tryst of TK Concerning Bioresources (or, More Appropriately, ETK—Environmental Traditional Knowledge), Its Holders and Practitioners, with the Law

Social Mores, Value systems and practices that have shaped and given dynamism to our traditions, over time have, majorly contributed to the evolution of the human race. They have impacted every aspect of our existence, including, personal, familial, societal and property relations. But it is an irony that, as a general rule, the traditions and practices over public resources, hardly gets the respect, regard and recognition, due to them, in our legal order. Creation of rights in law, over public resources, in the community and the individuals comprising it and imposition of corresponding duties and obligations on others, based on the time-honoured customs and traditions of the former, outside personal and familial relationships, have hardly been in existence anywhere in the World (Langton 2004).[6] Traditions of communities of people concerning natural resources are glorified and cherished as, "living law", in all social formations (Customary law, is the other expression used for them, without giving them the force of law!). However, the Lex Scriptum (Statutes, handcrafted by the legislature), Precedential Law (Judge-made law) and, the practices and procedures of the bureaucracy, seemingly imprisoned by legislative prescriptive codes of conduct (Administrative Law), hardly take cognizance of them (Lewans, 2018).[7]

Even the Human Rights Jurisprudence, at the international level, offers little hope. It does elaborately catalogue a host of rights of individuals and communities of people (that include, protection of the community and the identity of the individual within it), ordains, Nation-State systems to provide an enabling environment to facilitate and ensure them to continue to nurture and practice their traditions and customary practices, without hindrance (subject, however, to reasonable restrictions of public health, order, morality, national security and the like).There are procedural and institutional mechanisms, created at the international, regional and even national legal systems to ensure the realisation of some of these rights! But these are available and confined to Political and Civil Rights (like, non-discrimination, equality, right to life, etc.) The ones that are the subject of scrutiny here (classified under, Economic, Social and Cultural Rights), let alone getting the protection, they find themselves left in the margins and at the mercy of legal systems of the member countries of these international legal instruments! National Legal Systems, put all their energy and efforts in taking care of the P& C Rights and have consistently postponed the realization of

[6] The general practice, the world over, is that the religious practices and traditions, associated with familial relationship, marriage, adoption, etc., are classed as Personal laws. Respective Religious dictates regulate them. State law—Public Law—as a general rule, has little to do with them.

[7] Customs and traditions, acquirepropriore vigor(the force of "law"), only when there is either a legislative recognition or judicial endorsement. As such, in the absence of such an approbation, they neither create right in one and obligation in another, under the law in force in the legal regime.

ESC Rights, in a "progressive way" over time—without setting any time limit![8] (De Schutter, 2019). At the national level, the obstacles are very many, real and almost insurmountable! The deterrent, in legal systems like India, is essentially rooted in the convoluted notion of, "sovereignty", as given to us by the British, during their time and even after!

Sovereignty, Common Property, Equity

True, that by Personal and customary laws, the traditions and practices do create a particular legal status and confer a clear identity to Communities of people. This is preserved and protected, by the creation of a special status for the protection of Customs and traditions of Tribal Communities, under the Constitution of a number of Countries, including India[9] (Sohoni, 2021). Right of protection to the tribal properties is also assured under the laws enacted by the States in India. However, the traditions and Practices, that include right over natural resources in the public domain, for communities of people, tribal or otherwise, are not assured under the legal systems. The customs and traditions define Communities of people and determine their distinctive identity in the social milieu. However, ironically, their customs and traditions like, practices of conservation, protection, maintenance, management and sustainable use of resources and the traditions associated with them, nurtured and kept alive by them, hardly get legal support, to make claims or to secure and realise them, for their benefit. Let alone a proprietary right, even the usufructuary right is not an assured one, as a matter of entitlement to them. This is the crux of the argument, here.

There are a variety of reasons for such non-recognition. Vagueness in their formulations; non-confinement of the Community to any single region and diversity in practices that are, at times, in conflict with each other, are some of them, that hinder the legal system to accord support to them.

In pure Anglo-Saxon jurisprudential terms—the system India adopted and continuing to follow, even after independence (a legacy of the colonial rule of the British in India)—natural resources and all that is there in the public domain is, State Property, over which, no single individual or community, can make any exclusive claim. The government, representing the State would alone, have the authority to decide on the manner of their use and application, guided by public interest. This agency of the State, determines what is in public interest and acts on it. This control over resources exercised by the State is, envisioned as the adjunct of Sovereign right of the

---

[8] See, Universal Declaration of Human Rights, 1948 (United Nations General Assembly), along with the twin international Human Rights Instruments: International Convention on Political and Civil Rights, 1956 and International Convention on Economic, Social and Cultural Rights,1966.

[9] Under the Indian Constitution, several provisions exist for their protection ((Art.19(5)) and designation of areas for the conservation and protection of their traditions, practices and political organisation (Arts. 244& 244-A, read with Scheduled Five and Six).

State. There is a bit of a problem here: The law recognises only two kinds of Property, namely, Public or State Property and Private Property. Since private property remains an exclusive property of one having proprietary right over it and the Public/State Property belongs to the State agency to deal with it, no legal space exists for a third category—Common Property or Community Property. Common Property is that property, over which the members of a community have a co-equal use right[10] (Jodha, 1986). It may be a village pasture or a Water Body or anything over which a determinate group of people, closest to it, having close bond of attachment to it, have rights, as determined by the custom or tradition prevalent and practised over a long period. This category of property and the rights of the community over it were part of Indian Jurisprudence, recognised and supported under both the public law and Personal law., in the past. Forests, for example, were under the control and authority of the communities of people living there, under the notion of Common Property Resources, before the advent of British in India[11] (Guha, 1999). This complexion of the property was completely changed by the British, when they started making policies and laws concerning Forests in Colonial India that culminated in the enactment of the Indian Forest Act, 1927. Communitarian claims over forests, got relegated to concessions and privileges, that too for the purpose of livelihood, at the whim of the administration[12] (Guha, 1983). The story is the same with all other natural resources, the laws in relation to which (land, Water, minerals, etc.,) were crafted by the British. They continue to operate, even now, retaining this feature and flavour!

Under personal laws, Common Property was, indeed, at the core of property relations! Under Hindu Law, for instance, in the Hindu Undivided Family, all the property (referred to as Coparcenary Property), belonged to the entire family. No individual member of the family could predicate any aspect of such property as exclusively belonging to him—owned by him—, till the family remained undivided. Property under reference here was quite broad as to cover both ancestral property and the property earned by any of its members by their own effort, as well![13] (Raghavachariar & Venkataraman, 1980) Notion of private property, with regard to anything held under Coparcenary, hardly existed, nor even scope for asserting individual intellectual property right, with in the family! This changed, almost a decade before the British left India. They enacted, The Hindu Gains of Learning Act, 1938, by which any property acquired from the earnings of one's education ("gains of learning"), through profession, employment, etc., became the exclusive private property of that member of the Hindu family[14] (Bhat, 1985). By a sleight of legal hand, the British succeeded in making inroads into traditional practices of certain communities of people and

[10] Jodha, N. S. (1986, July 5). Common property resources and rural poor in dry regions of India. *Economic and Political Weekly, 21*(27), 1169–1181.

[11] Guha, S. (1999). *Environment and Ethnicity in India, 1200–1991.* Combridge University Press.

[12] Guha, R. (1983, October 29). Forestry in British and post-British India: A historical analysis. *Economic and Political Weekly, 18*(44), 1882–1896.

[13] See N. R. Raghavachariar, *Hindu law: Principles and Precedents* (8th ed., thoroughly rev. & enl. by S. Venkataraman). Madras Law Journal Office, Nagpur (1987).

[14] s.2(b), Hindu Gains of Learning Act, 1930.

privatise that which, as part of Custom, was Common Property within the family! This continues to be the law, even to this day.

While, the "gains of learning" got the legal shield of protection, even under Hindu Personal Law, in this manner, the same logic was not applied in the case of TK over Bioresources (ETK) to extend such a protection. Even the intellectual Property Law, that evolved in India after its independence, had no place for TK. Starting with the Indian Patents Act, 1970, the legal regime established a monopolistic individual or corporate right over the process or product that resulted from innovation, invention, novelty (that was not obvious) and something of commercial value. The hindrance for its inability to extend the same protection to TK, was for two reasons: TK, by its very nature. was already in the public domain (quite obvious and not novel—a requirement for consideration as Intellectual Property, for protection!) and that as Property, it was, "Commons" (and not, private property—another requirement—a handicap!).

As such, claims for a share in the benefits of Commercial application of TK, by others, as a matter of right in law, did not figure in any significant way, till almost the end of the last and the very beginning of the present millennium!

Having waded through, some of the major hurdles for legal recognition for ETK, as above, it is the argument of this work that even after the evolution of the much needed legal prescriptive codes of conduct (that started at the cusp of the two millenia) and the policy decisions that have given some kind of recognition and protection, to such a claim, they are no more than acts of tokenism than Just and Equitous—both at the conceptual level and on the ground in action! For this, as indicated at the outset, two legal developments—one concerning Biological Diversity and another on Protecting Farmers' Rights—are presented here. They, perhaps, more than any other effort, present a fascinating account of the evolution and the current predicaments in the legal ordering on the subject.

International Legal Developments (CBD & ITPGFRA)

The Convention on Biological Diversity, 1992 (CBD) and the Nagoya Protocol, 2010: This, arguably, is the first international legal arrangement that recognised the contributions of traditions and customs of communities of people and their knowledge systems (TK/ETK), in the conservation, protection and sustainable use of biological resources. In a nutshell, anchored to the principle of equity, this was the major potent international legal prescription that conferred the legal status of Holders and Practitioners of ETK, to them and created a derivative right, in their favour, to a share in the benefits derivable from the commercial application of the ETK.

CBD, signed by over 150 Countries in the Earth Summit on Environment and Development, at Rio-de-Janeiro in 1992, came into force from 1993, with more than 190 Countries subscribing to it. It legally recognises, the close and traditional dependence of many indigenous and local communities embodying traditional lifestyles on biological resources, and the need for ensuring sharing equitably benefits arising from the use of traditional knowledge, innovations and practices relevant to the conservation of biological diversity and the sustainable use of its components[15] (CBD). This constitutes one of its three principal objectives of CBD15.

It requires the member countries, to provide effective legal frame and work out mechanisms for, (i) respecting, preserving and maintaining knowledge, innovations and practices of indigenous and local communities embodying traditional lifestyles relevant for the conservation and sustainable use of biological diversity; (ii) promoting their wider application with the approval and involvement of the holders of such knowledge, innovations and practices and (iii) to facilitate, the equitable sharing of the benefits arising from the utilization of such knowledge, innovations and practices.[16] By applying the International Law Principle of Sovereign Rights of States[17] (Weissberg, 1975), each and every Country, under the Convention, had the power of regulation of access to, nature and form utilization of the resource and evolve the modalities of sharing the benefits.

CBD, as the framework Convention, laid out the contours of the right. It was without doubt, the first major step in doing equity to the Community and the nation of origin, by creating a right in their favour, for a fair and equitable share in the benefits arising from the utilisation of the bio resource, knowledge systems and practices associated with it. By one stroke, the Convention ensured the observance of the principle of equity and prevention of piracy. The onus was on the member Countries, as the Sovereign Right holders, to develop a robust mechanism, besides Cooperating with each other, to realise the objective. This was, indeed, a laudatory effort and a real legal empowerment in breaking free from the historical experiences of deceit, inequity and injustice. But, being a Framework Convention, the details, to act as guidelines for action, to the States, were left to be developed separately through a Protocol. Monitoring Compliance with the Convention provisions too, had to be developed through the Protocol. As it so happened, it took 18 long years to develop the formula for ensuring equity in sharing of Benefits, through a Protocol (Nagoya Protocol, 2010). The delay displayed a lack of a sense of urgency in the International Community of nations, as the Parties to the Convention. The international effort was further weakened by the United State (as one of the biodiversity-rich Countries and one of the major actors in catalysing the evolution of CBD). It did not become a Party to this legal arrangement, causing a major dent to universal support to the Convention.

The Nagoya Protocol on Access to Genetic Resources and the Fair and Equitable Sharing of Benefits Arising from their Utilization to the Convention on Biological Diversity, 2010 (Nagoya Protocol on Access and Benefit sharing-ABS), covers TK/

[15] Preamble to CBD.

[16] Art.8(j).

[17] See Resolutions, 523 (VI) of 12 January 1952 and 626 (VII) of the General Assembly of United Nations,21 December 1952; this was reiterated and elaborately stated in, the resolution 1803 (XVII) of 14 December 1962, of the General Assembly, entitled, "Permanent sovereignty over natural resources".

ETK and the benefits arising from its utilization, as well. It establishes a framework for accessing Genetic resources and associated TK, and for sharing of benefits arising from their use. It stipulates the manner in which researchers and companies can obtain access and the sharing of benefits arising from their use. It also sets out obligations for Parties to provide that users of Genetic Resources within their jurisdiction respect the domestic regulatory framework of Parties from where resources have been accessed. Benefits to be shared may be monetary or non-monetary. The Annexure to the Protocol contains non-exhaustive list of benefits. It should note here that while the Protocol does not apply to pre-Protocol accessions, it is made clear that it is operable for the materials accessed after entry into force of CBD. This requires reiteration as in India and, perhaps in many other Countries as well, arguments have been put forth, quite successfully, by such accessors, to ensure the postponement of the operation of the law, till the time national legal instruments, incorporate and put them into application (such an escape route is allowed to be taken, in a number of instances in India, either on account of reluctance/ignorance on the part of the authorities or as fait accompli by the Judiciary!) Such, delayed action has contributed to the business as usual, unfair advantage for the accessor and continuing denial of the benefits for the TK holders.

As required under the Protocol, the Domestic level access measures should, create legal certainty, clarity and transparency; provide fair and non-arbitrary rules and procedures; establish clear rules and procedures for Prior Informed Consent (PIC) and Material Agreement Transfer (MAT) and publish permits on ABS Clearing House to constitute Internationally Recognised Certificate of Compliance (IRCC). It requires the Domestic level benefit sharing measures, should, provide for fair and equitable sharing of benefits arising from utilisation of GRs, as well as subsequent applications and commercialisation and ensures that sharing of benefits is subject to MAT. The Benefits may be monetary (such as access fees, licence fees, milestone payments, royalty), or non-monetary (such as transfer of technology, sharing results of research etc.). The Compliance obligations, constitute the core aspects of the Protocol and are quite a challenge for implementation. Under that rubric, the Parties are required: to ensure that the Genetic Resources, utilised within their jurisdiction, have been accessed in accordance with PIC and that MAT are put in place; to address situations of non-compliance and to cooperate in cases of alleged violation of domestic ABS requirements. Their Obligations for monitoring the utilisation of Genetic Resources, to ensure Compliance include designation of one or more effective check points for collection of information at any stage of R&D, innovation, pre-commercialisation or commercialisation. These could be Patent offices, Custom authorities, Market approval offices, Research funding agencies etc.

A range of tools and mechanisms for effective implementation are provided by the Nagoya Protocol, to assist the contracting parties. They include, Establishment of National Focal Points (NFPs) and Competent National Authorities (CNAs,) to serve as contact points for information, granting access, or to ensure compliance; an Access and Benefit-sharing Clearing-House to share information, such as domestic regulatory ABS requirements or information on NFPs and CNAs and Capacity-building to support key aspects of implementation.[18] (CBD-Nagoya Protocol) Having entered into force from 12th October, 2014, after ratification by 50 member countries, the Protocol has now over 130 ratifications. India was one of the earliest signatories to have developed the ABS mechanism domestically. It acted as a major catalyst for a large number of Countries to subscribe to the Protocol.

Implementation and Compliance with the Obligations of the Protocol, besides being quite challenging, requires a lot of preparation on the part of the Member Countries. In addition, to the delay in setting up the required institutional arrangements, there has been added flab to the existing bureaucracy, along with its inherent problems of lack of coordination, lack of expertise and experience and poor implementation, etc. The experience of the last fourteen years, across member Countries, has been such that there is very little to report on ABS benefits and the same reaching the TK holders, quite negligible and grossly inadequate. A number of reasons exist for this that include, Member Countries being still in the process of, (i) creating the required bureaucracy and institutional arrangements and (ii) developing guidelines for the implementation of the Nagoya frame for ABS. This is in addition to the coordination deficit, inherent among different agencies of State and the continued non-cooperation and non-compliance, with impunity, by the corporate entities either, by taking advantage of the vagueness in the formulations or by ingeniously carving out exceptions and making excuses and continue with their familiar, 'business as usual' approach (may be considered for critique, after examination of the Indian Law).

Protection of Plant Varieties and Farmers right Law: Although, the legal developments on Plant Varieties, was subsequent to CBD, at the international level since, in India, this law and the ABS mechanisms under it were evolved, before similar developments on Biodiversity law and the two legal regimes are currently operating in parallel (of course, not in conflict), it would be worthwhile to briefly reflect upon this law, with particular reference to ABS, now. This may also provide the base for comparison between the two laws, on the subject, as well.[19]

The Traditional farming Community in India, like elsewhere, are the carriers, conservers and custodians of agricultural practices (like, seed saving exchange and barter of agricultural products, seeds etc.). As a Customary practice, they enjoyed the freedom of experimentation with seeds, produce and products and carry out innovations in them, in an evolutionary way, over time, in their farmlands. For them, the farmland was their muse and the produce and the products, their music—an integral aspect of their culture! Good traditional practices, that were environmentally benign, continued and flourished and those that were harmful, discarded or avoided, from time to time. This was bound to change, through a legal regime under the World Trade Organisation. After joining the global trade body, in 1995, India became obligated to implement the Agreement on Trade-Related aspects of Intellectual Property Rights (TRIPS). After debating for a while over the introduction of intellectual property

[18] Gist of the relevant aspects of the Nagoya Protocol, 2010.

[19] Ch.6, Farmers Rights, Sections 39–46, PPVFRA, 2001.

rights in Indian agriculture, it was felt that that there were two choices with us: either to enact a law that protected the interests of farming communities or to accept the framework of plant breeders' rights given by the International Union for Protection of New Plant Varieties (UPOV Convention), a model perfectly suitable to the Agri-business model of the West. Since the latter, denied the farmers, the freedom to reuse farm-saved seeds and to exchange them with their neighbours, there was not much of an option, but to develop a sui generis system that, besides conforming to TRIPS, also protected the interests of the Farming Community. With this objective India enacted the Protection of Plant Varieties and Farmers' Rights Act, in 2001. In the Chapter on Farmers' Rights, under which, (i) farmers are recognised as plant breeders, having the right to register their varieties; (ii) farmers engaged in the conservation of genetic resources of landraces and wild relatives of economic plants and their improvement through selection and preservation are recognised and rewarded; and, (iii) the traditional practices of the farmers of saving seeds from one harvest and using the saved seeds, either for sowing for their next harvest or sharing them with their farm neighbours, got protected.

In the same year (2001) the International Treaty on Plant Genetic Resources for Food and Agriculture (ITPGRFA—also known as "Seed Treaty"), was adopted. India became a signatory to it. It is a comprehensive international agreement for ensuring food security through conservation, exchange and sustainable use of the world's Plant Genetic Resources for Food and Agriculture and requires the recipients to fairly and equitably share the benefits they derive from the use of these genetic materials with the countries where they have originated. It formally acknowledges the enormous contribution of indigenous people and small-holder farmers as traditional custodians of the world's food crops. The global arrangement is significant, in that, at the very outset, it formally acknowledges the enormous contribution of indigenous people and small-holder farmers, as traditional custodians of the world's food crop. It also recognises farmers' rights, subject to national laws.

Under the PPVFRA, referred above—the second and the third stipulations under it, in particular-, there is recognition of certain aspects the Framers' Traditions and even protection extended to them. But, in effect, they are not rights, but mere concessions and exceptions to the rule! The rule, stipulates that the traditional farmer can register as Breeders. Once, this is done, they will be subjected to a host of requirements to conform (like, the produce, seed quality, use of raw material for production, etc. need have to follow certain standard specifications and certification mechanisms), to be fit for getting protection in the market place, under the law and for availing the benefits under the ABS mechanism evolved under the law. But these are Breeders Right and not the same as to be in sync with the practices of the Traditional Farmers. Instead of giving the shield of protection to all the rights of the Traditional Farmers, including a fair and equitable share in the benefits, the PPVFRA has only it adopted the ABS Mechanism applicable to the Breeders. It is pretty clear that the regime is intended to protect the intellectual property of the Modern Breeder and not those of the Traditional Farmer. As such, under the PPVFRA, the benefits derived from the TK of the Traditional Farmer, by the recipient has little legal compulsion to share it with the Traditional Farmer.

The element of "equity", as observed under the Biodiversity Regime, through CBD, is remarkably broad enough to include, both environmental equity (bioresource conservation and biosafety, as the top priority, which shall not be compromised while accommodating sustainable use—covering commerce, as well) and Human equity (sharing of benefits, for benefit claimers). These are integral aspects of Indian Biodiversity Law, as well. PPVFRA covers only the commerce and the human equity aspects.

Biodiversity Act, 2002 (the Act), Rules & Notifications on ABS

The design of the Act is very much[20] in line with CBD, having the three principal objects, namely: (i) conservation of both biological[21] resources and knowledge systems associated with them, their safety and security; (ii) their sustainable use and (iii) ensuring equity in the sharing benefits derivable from their use.[22] The law is intended to be worked in such a manner as to ensure that all activities, either through research or commerce and all other kinds of uses, either for sustenance or livelihood, do not, by any stretch, affect, offend or compromise the realization of the said principal objectives.

The Provisions for the third Component of the Objects of the Act (ensuring equity in the sharing benefits derivable from their use), the subject matter of the enquiry here, are quite elaborately laid out. Interestingly, the details of sharing of benefits, in the form of rules, were made and brought into application, as early as 2004, long before the international design for such an arrangement was unveiled at the international level, in the form of Nagoya Protocol, 2010! In that sense, the Indian legal system was far ahead of the global community, in crystallising the idea, creating institutions, for action. Under the Act, once the access is permitted and the benefit claimers identified, fair and equitable sharing on benefits, included, grant of joint ownership of IPR, transfer of technology, location of production, research and development units in the area of access, and creation of an association of Indian scientists, benefit claimers and the local people with research and development in biological resources and bio survey and bio utilization. It is also indicated in the Act, for proposing direct financial mechanisms, as benefit sharing that include, setting up of venture capital funds and payment of monetary compensation to the benefit claimers as the NBA may deem fit. The NBA is also mandated under the law to frame guidelines to effect benefit sharing.[23] Following the Nagoya Protocol (and, to be in sync with it) the Government of India, firmed this up with the details of procedures and measures to give effect to

[20] The Biological Diversity Act, 2002 (18 of 2003)—hence forth, the Act.

[21] The Biological Diversity Rules, 2004—hence forth, the Rule/s.

[22] Preamble to the Act.

[23] s.21 of the Act, read withrule.20.

the ABS Mechanism, through a set of Guidelines by a notification.[24] But, as can be seen from a detailed reading and analysis of the notification, this legal instrument was more of an elaboration and detailing of whatever was provided under the extant Act and the Rules, than spelling out anything substantive in nature.

Although, the Act is neither an overarching nor an umbrella law on the subject, its sweep, import and impacts are quite significant and far reaching. As a matter of law and fact, the limitations on its operations are just a few, confined primarily concerning its relationship with Forest and wildlife laws. This law, it is made clear in the statute itself that it, should be understood, applied and interpreted in such a manner as not to come in conflict with the working of the laws on Forests and wildlife. As such, the scope and ambit of application of all these laws, both in letter and spirit, are complementary to each other.[25] This law, however, overarches the intellectual property law in the sense that any applicant, before approaching the concerned authorities, for grant of an intellectual property for any invention concerning a bioresource obtained from India, should have obtained approval in advance of such an application from the National Biodiversity Authority (NBA) under this law.[26] The limitation in this regard, as stipulated in the statute, is that such a prior approval from the NBA is not required when one makes an application for securing any right under the Protection of Plant Varieties Act, 2001.[27] This apart, there is little left for exclusion from operation of this law over anything concerning bio resources or the knowledge, traditions and practices associated with it in India. The statute even goes to the extent of empowering the NBA to act on behalf of the government of India for taking, "any measures that are necessary to oppose grant of intellectual property rights in any country outside India on any biological resource obtained from India or knowledge associated with such biological resource which is derived from India".[28]

Without any doubt, this is the first major instance of a law filling the legislative vacuum concerning intellectual property rights in ETK and providing for an equitable share in the benefits of its use, for the benefit claimers recognised under it. The stand out feature of this law is the clear formulation of rules and regulations, in according recognition to ETK, identifying the legal status of community/individual, to claim benefits, detailing the nature and manner of sharing benefits with the ETK holders and Practitioners, in the event of commercial application of the Bioresources, the knowledges and practices associated with them and the institutional arrangements for ensuring the dispensation. The Act, at the very outset, accords a legal status to the right holders as benefit claimers and describes them as, "the conservers of biological resources, thereby products, creators and holders of knowledge and information relating to the use of such biological resources, innovations and practices associated

[24] The Guidelines on Access to Biological Resources and Associated Knowledge and Benefit Sharing Regulations, 2014. This was further updated and amended in 2019.

[25] s.59 of the Act.

[26] s.6 of the Act.

[27] s.6(3) of the Act.

[28] s.18(4) of the Act.

with such use and application".[29] The Act, puts the status of the Benefit Claimer to be on par with the Central Government and an Authority constituted under it, by providing that in addition to these, the Benefit Claimer, has the right to bring in an action before a Court for an offence against his interest, thirty days after submitting a complaint, in that regard, with the authority concerned with his right.[30] This is very significant, in that it opens an avenue to the Benefit Claimer to get his grievance redressed judicially, when he is not satisfied with the action taken by the authorities do not satisfy him. It further states that the National Biodiversity Authority shall determine the benefit sharing or the payment of monetary compensation and other non-monetary benefits to the benefit claimers.[31] At the State level, the State Biodiversity Board would determine the manner of its realisation.[32] It also provides that the local people and communities of the area, including growers and cultivators of biodiversity are exempt from the regulatory control of the State. In addition, the Practitioners of TK (like the Vaids and Hakims) would similarly have the unfettered right to continue their practice.[33]

The status accorded to the third tier of governance in the biodiversity management scheme—the Biodiversity Management Committee, under this law deserves particular mention. The local government (under the Constitutional hierarchical scheme of governance, this is the primary unit of governance, below the Centre and the State Governments; in rural areas, it is in the form of a Panchayat and in the urban milieu, it assumes the nomenclature of a Municipality or a Municipal Corporation), brings this entity into existence. Except at the time of its birth and during the process of conduct of periodic elections to the post of its chairman. When the Local Body has all the authority, the Biodiversity Management Committee functions with full autonomy. Under the existing scheme of things, once constituted, it has a ring of permanence to its existence, so much so that it survives and outlives not only the expiry of the term of office of the very local body that was responsible for its birth but also all its successors that assume office, through periodic elections.[34] The whole range of functions that are required to be performed by it include, promotion of conservation, sustainable use and documentation of biological diversity, preservation of habitats, conservation of land races, folk varieties and cultivars, domesticated stocks and breeds of animals and microorganisms and chronicling of knowledge relating to biodiversity.[35] Even the upper layers of governance under the Act—the National Biodiversity Authority and the State Biodiversity Boards—are mandated to consult this entity while taking any decision relating to the use of biological resources and knowledge associated with such resources occurring within its territorial jurisdiction and in deciding on the

[29] s.2(a) of the Act.

[30] s.61(b) of the Act.

[31] s.21 #) of the Act.

[32] s.23(a) of the Act.

[33] Proviso to s.7 of the Act.

[34] s.41 of the Act, read with Rule 22. The only stipulation is for the tenure of the Chairman of BMC, which is , for three years.

[35] s.41 of the Act.

sharing of benefits, at the local and community level.[36] Further, this body has been given the authority to levy and collect an access fee, when any entity accesses and collects any biological resource for commercial purposes from areas falling within its territorial jurisdiction.[37] Under the Indian Constitutional Scheme and within any of the statutory formulation, one may dare say, there is hardly any parallel, either in composition, characterization or functioning, to compare with this institution at the grass root level of governance. As such, use of the expression, "Committee" to describe this body, that gives the impression of a temporarily functioning adhoc entity, is both misleading and does not render justice to its stature or status, envisaged under the Act Verily, the Biodiversity Management Committee is the linchpin of the whole system of biodiversity management in India. The uniqueness of the visualisation of an entity, in the entire scheme of governance, especially, in its role in identifying the Beneficiaries and the function of mandatory consultation by the upper tiers of governance, before the determination of the equitable aspect of the share in the benefits, in effect, has deepened the democratic spirit, as never before, in the Indian legal system.

Perpetuation of Unfairness and in Equity in Sharing Benefits

After highlighting the positive and salutary aspects of the legal prescriptions, as above, when one turns to the action and working of the legal scheme, there is total disillusionment and despair, with the system coming a cropper, in translating the Prescriptions into practice and in providing the institutional spine and support to the realisation of the right to fair and equitable share in the benefits to the Holders, Carriers and Practitioners of ETK. Some of the ailments and deficiencies, in support of the thread of thought under consideration here, are discussed in brief, as below:

1. Experience of poor and inadequate returns: As the saying goes, "the proof of pudding is in eating", the effectiveness of a legal regime, in the realisation of the objectives is, to a large extent, dependent on its realisation at the ground level. On this score, the experience has been quite dismal. There is a wealth of literature that is available, by both NBA[38] and other publications,[39] that refer to a large number of agreements signed (over 100 agreements, as early as 2016—with little evidence

[36] s.41(2) of the Act

[37] s.41(3) of the Act

[38] "Access and Benefit Sharing Experiences from India", NBA Portal: https//nabaindia.org.

[39] See, "CHASING BENEFITS": Issues on Access to Genetic Resources and Traditional Knowledge with reference to India's Biodiversity Regime—A post-Nagoya Protocol view on Access and Benefit Sharing—Kanchi Kohli and Shalini Bhutani, briefing paper for the NGO Alliance on CBD (India),supported through WWF-India's CSO initiative. Kalpavruksh; also see, "Biodiversity and Access and Benefit Sharing in India" National Law School of India University Portal: https//:abs.nls.ac.in.

as to the share of the benefit claimers, in the bargains!) and quite substantial contributions obtained, in specific instances (like, over Rs.200 crores, just in one deal concerning Red Sanders!—which is more an instance of the result of assertion of sovereign right than hardly anything to do with ETK or Practice of it or percolation of benefits to local community!). there is very little information made available as to how much of what was collected, was ploughed back to meet the interests of the, "benefit claimers" (there is, for example, an interesting narration about a deal with a Japanese firm about the Neem Leaves, a part of which of the benefits utilised for planting of Neem Saplings, in the area from where they were sourced!!). Obviously, there is little to report about this. Instead, there are reports made available about the difficulties experienced by the Authorities in realising this objective. Like, for instance, the NBA, as early as in 2009, revealed that the sharing some of the practical problems it faced in identifying the rightful "benefit claimers", in relation to the illegal trade in medicinal plants both across borders and within the country, found it impossible to trace the community or local healers that knew about these or were involved in growing them. With the result, it found, recovering "benefits" from the pharmaceutical company making the money from commercialisation became a challenge to it.[40] Although, this experience was long before the Nagoya Protocol and the subsequent legal developments in India, in this regard, the difficulty experienced and shared, which continues even now, is yet to be satisfactorily addressed.

2. Inordinate delay in Law-Making and putting the Law into action: The "common concern of humanity" that compelled the community of nations to evolve a strong legal fence, to arrest the alarming trend in Biodiversity loss, its unsustainable use, bio-piracy and inequity in the sharing of benefits, resulted in the making of CBD, in 1992. There was a sense of urgency for the member Countries to design their own laws, in conformity with the international frame and to get into action without any further loss of time. Time was of essence, as expeditious action alone gave hope to humanity for securing environmental and human equity. But National Laws were a long while in coming. India was no exception. The Indian statute, as we know it, had one of the longest gestation periods. With over half a dozen official drafts kept in circulation in the public domain, for nearly a decade, the law was finalised in the year 2002. It came into force at the national level, two years after it was passed by the Parliament, with the rules made under it getting the law-makers nod in 2004. It took roughly another 3 to 5 years for about a dozen States, to craft their own rules, as provided in the Statute itself, for making the law operable in each one of them.

Added to this delay, was the curious case of making of ABS Regulations and putting them into application. Although, the ABS Regulations were quite imaginatively integrated with the Act and the Rules, a separate Notification of Guidelines were issued as Regulations, in 2014, with the reasoning that it was required for strengthening the administration with the details to the ideas encapsulated in the superior legislative instruments. This was also, it was claimed, to ensure that the same were in conformity with the newly agreed international arrangement (justified,

[40] At the National Forum for Policy Dialogue on "Six Years of the Implementation of the Biological Diversity Act 2002" co-organised by Kalpavriksh and GRAIN at New Delhi on 3rd February 2009.

partly, for the efforts in filling in the gaps and a few additional aspects put in place, internalising the Nagoya stipulations). Justifications apart, precious time was lost, in which, with the business-as-usual approach of loot and scoot continued unabated, with both the resource, TK and the communitarian claims for a fair and equitable share continued to suffer, in an immeasurable way.

Further, imprecision in the legislative formulations, giving scope for varied interpretations and opened routes to escape from being brought under the law, to many an entrepreneur. For instance, the interpretations, to the expression, "Value added product", exacerbated the problem for implementation. The Act defined them as, "products which may contain portions or extracts of plants and animals in unrecognizable and physically inseparable form".[41] The expression simply meant that any product, by virtue of addition of a minute quantity of a bioresource in an unrecognizable manner (extract of a herb and the like), resulted in a new product, that enhanced the market value of the original product. Such a product, as provided in the law, was exempt from the application of ABS Regulations. Well, lack of clarity in explaining this, nor any effort in illustrating this in the Rules and guidelines, led to a lot of claims being made for exemptions of a number of Products. A product, Chyavanprash, was put in the list of exempted products, by the Authorities, without any explanation as to what was the "value added" to "which product" by this "new" product, to deserve to be in the exalted list! Such claims of exemptions, too, have contributed to dilution of the law and its enforcement. As a consequence, ABS, to that extent, remains an unrealised dream!

3. Administrative Architectural birth-pangs, deficiencies, distortions and other inadequacies: The test of any good law, is its well-coordinated, institutional structure sitting firmly on the foundations of competent personnel, getting into action, in all earnestness and promptly, in an efficient manner. This, unfortunately, has not been the case with what emerged under the law, over a long lapse of time.

The National Biodiversity Authority (NBA), the top layer of governance, was brought into existence, with in a couple of years after the Act and the Rules were made. The second layer of governance—the State Biodiversity Boards—in most of the States, could see the light of the day, after about 15 years of the making of the law. As a matter of fact, the formation of third tier—the Biodiversity Management Committees—in all the States in India, took a lot more time, with a large majority of the members in them, knowing very little of their functions, even to this day. Let alone strengthening grass root governance, as was intended, the grass is yet

[41] The first Committee comprised of four members, headed by, Amarjeet Ahuja, the Former Joint Secretary, MoEFCC. It had a mandate of doing a quick review of the extant law and suggest reforms, in four months time. About half a dozen recommendations were made. No information is made availableas to, any follow up measures! The Second Committee, comprising of over one dozen members, was headed by A. K. Goel, former Joint Secretary to MoEFCC. Its mandate was to carry out a thorough evaluation of the Act and to come with recommendations for reforms. After forty odd meetings, in a little over an years' time, it suggested 66 reforms to the government. Only six of them (that primarily dealt with procedural aspects and a little re-structuring of the NBA—not much of a significance, in substantive terms), were accommodated in the amendment! This author was a member of both the Committees!

to develop roots, even after two decades of making of the Law! As such, in terms of institutional arrangement and working, Biodiversity law in India, except at the highest and the middle levels, continues to be in the nascent stage of construction and evolution, with different aspects of it coming into effect and operation at different points of time, with little resonance and with meagre returns! A classic case of, "Closing the Stable door after the horse has bolted away"!

4. The Bungle of Biodiversity (Amendment Act), 2023: All the analysis carried out, so far, concern the law, as it existed on 31st March 2024. As is evident, the Act required amendments, not just for the reasons listed, as above, but for a host of other reasons, as well (like, clarity of language and prescriptions, strengthening institutional arrangements, classifying varieties of deviations and weighted remedial and punitive measures, and the like). As a matter of fact, two expert Committees were constituted, in the last five years,[39] to recommend reformative measures to the Government. These Committees went the whole hog of the law and its working and came up with a host of recommendations (over 70 recommendations, in all!), to strengthen the law, implementation, compliance measures and institutional reformative actions, etc., for the realisation of the avowed objectives of the Act. Unfortunately, most of the recommendations did not pass muster with the government. The amendments, effected and brought into application with effect from the first of April, 2024, are a far cry from those that were required. They are quite ill-conceived and not much supportive of the realisation of the purposes for which the law ids brought into application, in the first place. The brief, here in below, are confined to the aspects affecting the third objective—fair and equitable sharing of benefits of the use of bioresources, knowledge and traditions associated with them (and, more in particular, to the perpetuation of inequity to ETK holders, carriers and Practitioners):

(a) Sovereign Right given up and sovereign function, paralysed: The extant law provided for imposition of penal sanctions, for violations under the law and non-compliance with any of the Orders of the authorities. These included, imposition of fines, penalties, and criminal sanctions like, imprisonment, de-licensing and the like. Violations were non-negotiable and punishable under the law.[42] Resource and the knowledge systems, including ETK, were accorded the status of "national asset" and any infringement was viewed as a "crime against the State"—offending either the Sovereign right of the State or performance of sovereign functions of the State. Under this law, the highest authority (National Biodiversity Authority), was representing the Sovereign and performing sovereign functions. This constituted the core value of the law and was very necessary, to protect national interest, to deal with "biopiracy" and safeguard the sovereign right. International law, respected and recognised this right and this provision was the symbol of such an assertion.

By amending Sections 55 and 58, in one sweep, the criminal nature of the offences got transformed into civil "wrongs" and were regularised as non-offensive, through a "compensatory" mechanism. As explained to the Parliament, the idea was to decriminalise offences by charging a penalty between Rs. one lakh and Rs. 50 lakhs and to

---

[42] ss.55 & 58, under the Act.

impose additional penalty of up to one crore rupees for continuing contraventions. Decriminalisation has been so meticulously argued for that it has conveniently left untouched, all the instances of violations that have happened during the last 2 decades of the working of the law that invited penal action, either not getting reported or acted upon, to the advantage of the "invisible offender" (whose number is phenomenally high). In effect, it appears, to have given the offenders an amnesty for challenging the sovereign authority of the State(?!) A new entity—Adjudicating Officer—is created, by the insertion two provisions, with the responsibility of characterising the nature of deviant conduct and determining the amount of compensation payable for violations.[43] This literally re-writes Indian Criminal Justice Jurisprudence, by enabling an individual officer of the government, who need not be a Law Professional, to transform a criminal offence into a civil wrong, at his discretion.

"Decriminalisation" and characterising the violation as a "civil wrong" is, also an open invitation to challenge the Sovereign Right and get away with the loot. The end result is to enable one to violate the law and carry on, his "business as usual", after compensating the wrong. No justification exists for resorting to this measure as the facts point to the contrary the facts are,

- In the entire history of the working of this law, the application of the provision was rarest of the rare and that too, after giving a lot of leeway, before initiating action;
- There were inordinate delays in bringing the institutions of enforcement under the law and to that extent, initiation of the penal actions were long time in coming into fruition;
- Constant and relentless lobbying by a number of vested interest groups with various other Ministries and govt. agencies and successfully putting pressure on the Environment Ministry and the authorities to go slow on actions.

(b). Inequity to Practitioners of ETK and the Benefit Claimers: The existing law, by carving an exception to the application of s.7., conferred an exclusive right of exemption and exclusion of the Traditional Practitioners and the local communities from the obligations of seeking permissions and approvals of the authorities, for access to the resource and the requirement of sharing benefits derived from their application, as it was no more than continuation of their tradition and practice—an integral aspect of their identity and culture. This was, arguably, the first path-breaking legislative recognition, in the entire world, of conferring a clear legal status to Traditions and Practices of People, that can be enforced, to ensure that it, besides, mainstreamed Traditional Practices as an important strategy of Resource Conservation and Environmentally Sustainable Use (a very distinct, Indian Trait), legal safeguard exclusively existed for them, to retain their identity, Status and enjoy the fruits of their labour, all by themselves, without any, "unqualified" entity, getting entry into this exclusive club, with potential to disturb their identity and over time, overwhelm them by getting all the benefits and render the system iniquitous as to deny what is legitimately due, exclusively to the community. Notwithstanding the

[43] ss.55A & 55B.

relentless pressure and lobbying by the vested interests, expert legal opinion, that the government secured, from time to time, made sure that the legal provision was not diluted in any way, till now. The amendment, now includes, "the codified traditional knowledge, cultivated medicinal plants and its products, registered AYUSH practitioners…", under exemption. The justification is that the AYUSH Practitioners carry out their profession, trade and calling, as per the time-honoured traditions. Such Practice, keeps the tradition alive and that any commerce done there by is illustrative of sustainable use. Both, are perfectly in tune with the two principal objectives of the law—Conservation and Sustainable Use. But, while this may be true, this is not the issue. The issue is that of satisfying the third object: "equity in sharing benefits" and this condition, with the amendment of inclusion of the new players, does not get satisfied. All the three objects ought to be satisfied under this law. Since that did not happen, there does not exist a valid reason for this addition. The issue is that of, "practice of trade in the traditional way", which is not the same as, "the carrier of tradition, allowed to continue to carry on with the tradition". The difference between the two is that while, the new entrant is making the claim for exclusion for having, "acquired a skill", but the existing one who is under exemption has been, "retaining and continuing with one's identity and practice". The difference also is that while, the former, is trying to cash in, "on borrowed knowledge and skill", the latter is a "natural" who is encouraged through exemption to continue to develop, "the innate, inborn knowledge and naturally evolving practice and skill", unabated. To put it succinctly, it is the case of the imposter, getting the benefits of the knowledge, skill and its application, none of which are his or part of his tradition, all for himself, without the obligation to share it with the traditional practitioner and the local communities, in an equitous manner and hence unfair and unjust. The bigger alarm is the scope created thus, for corporate capture that would drive away, the "real benefit claimers" from getting what they are entitled for. It is true that there are still some traditional Practitioners' families carrying on the trade, in small scale and tiny sector. They do need support and encouragement. For this, the amendment should have included only these within the exemption; as such a thing would satisfy all the three objects of the law. This is not a very difficult thing to achieve. All that is required is to include this group in the definition of, "benefit claimers", clearly stipulate the obligation of the BMC, in identifying such "traditional entrepreneurs", in the relevant provision of the law and further stipulate the maximum limit of the turnover for such an enterprise. Such a reasoned distinction of this group from the other, "AYUSH Practitioners", should have happened. Since, this is part of the sui generis system evolved under this law, it would have also satisfy the requirements of the WTO requirements concerning Intellectual Property Rights. The aspect of working out the facilitative mechanisms for the "traditional entrepreneurs" can be detailed in the rules (like, organising them in to cooperative ventures, as to become models for the larger enterprises to emulate and adopt their good practices).

Further, the use of the expression, "codified traditional knowledge", is quite vague. There is a danger of this expression getting misused. There is a need for clarification, by inclusion in the definition clause as to what this expression conveys, to avoid possible mischievous interpretations. This cannot be left to the rules to be detailed, for the simple reason that this goes to the root, substance and realisation of the traditional intellectual property rights of communities of people in India.

(c) Contribution to Democratic Deficit: Under s.41, a proviso is introduced, to the provision of the mandatory obligation of every Local Self Government (LSG—to constitute a BMC in each local body). It states, "that the State Government may constitute Biodiversity Management Committees at the intermediate or district Panchayat level for achieving the objectives of the Act". The proviso, in effect, gives the alternative avenue to the State Government, to constitute BMCs at intermediate and District Panchayat level, according to its convenience. It is a clear case of playing into the designs of the one under an obligation to comply (the Local Self-Governments, functioning under the Ministries of Urban and Rural Development) and who is guilty of lethargy, neglect and utter disregard of the real intent of the law maker. By legitimising such an attitude and practice, the legislator expresses satisfaction both in this new formulation and the action of the State Govt. That it would suffice in, "achieving the objectives of the Act". There is no doubt that it is a huge task, to set up a BMC, at every level of the LSG. But this is a must to take democracy to the doorstep of the local communities and the "Benefit Claimers" Nothing short of it would be a mockery of democracy. In effect, the obligation of the LSG, gets shifted to the State Govt, in a diluted form, with no guarantee of a BMC, in every LSG and there by greatly curtailing the opportunity of deepening democracy at the grassroots.

Secondly, when the Functions of the BMCs are read along with the changes effected to s.7 (referred to earlier) and s.2, especially in the discharge of the primary function of preparing, maintaining and having the custody of the Peoples Biodiversity Registers (PBRs), a disconcerting and an undemocratic situation emerges. With the insertion in in s.7, "Codified Traditional Knowledge (CTK)", does not require observance of the requirements of s.7. This expression, gets described, under s.2(ae), as, "the knowledge derived from authoritative books specified in the First Schedule to the Drugs and Cosmetics Act, 1940". This, in effect means that BMCs will have nothing to do with CTK—they would neither prepare and document them, nor would they certify about certain aspects of the CTK. Through this amendment, to S.7, the Practitioners of CTK get exempted from the requirements of the provision of the obligation of sharing benefits from the commercial application of such TK. With one major crucial aspect of the entry in the PBRs getting out of the control of BMCs, let alone doing inequity to the real local "benefit claimers", the democratic exercise of power of recording and monitoring of the resources and their application is stunted. The object of empowering the local communities has given way to giving more strength to State Government, not by default, but by design and expediency of the latter, at the expense of LSGs and even BMCs!

There is another clear and present danger to even the preparation, documentation, maintenance of and control over PBRs, by the BMCs. An initiative, aggressively promoted by the Science and Technology Ministry, has led to the preparation of a draft of Guidelines for Handling and Working of Traditional Knowledges and Systems associated with them. This would come under the aegis of Traditional Knowledge

Digital Library. BMCs will have nothing to do with them—in any manner, at any stage or even for that matter, any say, hold or control.[44] The parallel exercise, with the blessings of the very powerful Science and Technology Ministry which, like it did in the Biosafety issue (in which the authorities under this law and the MoEFCC would have little role),[45] will strike at the root of the existence and challenge the utility of BMCs.

The foregoing analysis and critique are indicative and not exhaustive, with intent to buttress the argument that the holders, carriers and practitioners of ETK, concerning biological resources have always had the worst of the bargain and their share in the benefits have remained unjust and iniquitous.

Human intellect, intellectual labour, modern scientific inventions and innovations, have been celebrated and accorded due recognition and protection in the Anglo-Saxon jurisprudential universe, of which India has become a part, right from the time of its colonization by the British and continue to be so, even to this day. The most galling aspect of the situation is that of systematic marginalisation and non-recognition of something associated with Communities and their ways of life and tradition—ETK. Having all the attributes of individual initiatives, as stated above, but some that that evolved over a long period of time, backed by incessant labour among the members of the collective and perhaps, in a most democratic manner, expressed itself into crystallised practice, contributing, in a wholesome way to the human health, well-being, environmental integrity, societal harmony and sustainable development, hardly gets its due, in the legal order. The irony is that it is this, with which the Indians can proudly identify themselves, as their own, but do not feel legally obliged to share the benefits fairly and equitably! In most of the discourses and discussions on ABS, trade in genetic resources and rewarding technological contributions, get all the prominence. But, the communitarian interest and their contributions, that prioritize conservation imperatives do not figure in, at the time of sharing of benefits. The predominant pre-occupation with IPR laws, memorandum of understanding (MoUs) in Public Private Partnership (PPP) mode and contractual arrangements between the state and the private sector, clearly showing a pro-industry bent, cannot feel compelled to appreciate, accommodate or incentivise Communitarian contributions. Till such time, this deeply entrenched colonial mind-set does not make way for that which is truly Indian, in our system, Equity and Fairness in the Sharing of Benefits for the Conservers, Holders and Practitioners of our ETK would remain a pipe dream. These Communities, continue to languish as, Justice V. R. Krishna Iyer had said somewhere, "eternal tenants of an extortionate system"! While imposters can reap

---

[44] GUIDANCE FOR HANDLING AND WORKINGWITH TRADITIONAL KNOWLEDGE: PRIOR INFORMED CONSENT, ACKNOWLEDGEMENT, RECOGNITION AND ACCESS & BENEFIT SHARING—ESSENTIAL ASPECTS—COUNCIL OF SCIENTIFIC AND INDUSTRIAL RESEARCH (CSIR), 2023. The document is currently in public circulation.

[45] See "Rules for the manufacture, use/import/export and storage of hazardous microorganisms/ genetically engineered organisms or cells, 1989".

all the benefits, these would remain, "systematically and systemically, marginalised Civilizational people"![46]

References

Bhat, P. I. (1985). A critical appraisal of the Hindu gains of learning act. *Journal of the Indian Law Institute, 27*(4), 578–593.

De Schutter, O. (2019). *International human rights law*. Cambridge University Press.

Global Commons. https://earthcommission.org/

Guha, R. (1983, October 29). Forestry in British and post-British India: A historical analysis. *Economic and Political Weekly, 18*(44), 1882–1896.

Guha, S. (1999). *Environment and Ethnicity in India, 1200–1991*. Combridge University Press.

http://nbaindia.org/uploaded/pdf/ABS_Factsheets_1.pdf

http://www.nbaindia.org/uploaded/act/BDACT_ENG.pdf

https://www.jstor.org/stable/24482157

https://www.cbd.int/convention/wg8j.shtml

Indian Contract Act. https://www.indiacode.nic.in/bitstream/123456789/2187/2/A187209.pdf

Jodha, N. S. (1986, July 5). Common property resources and rural poor in dry regions of India. *Economic and Political Weekly, 21*(27), 1169–1181.

Langton, M. (Ed.). (2004). *Honour among nations?: Treaties and agreements with Indigenous people*. Academic Monographs.

Lewans, M. (2018). Administrative constitutionalism and the unity of public law. *Osgoode Hall Law Journal, 55*, 515.

Nagoya Protocol. https://www.cbd.int/abs

Preamble to CBD. https://www.cbd.int/doc/legal/cbd-en.pdf

Raghavachariar, N. R., & Venkataraman, S. (1980). *Hindu law: Principles and precedents; case law and statute law brought upto date.*

Secretariat of the United Nations Convention on Biological Diversity. (2001). *Handbook of the convention on biological diversity*. https://www.cbd.int/convention

Sohoni, M. (2021). Equity and the Sovereign. *Notre Dame Law Review, 97*, 2019.

The Protection of Plant Variety and Farmers Right Act 2001. https://plantauthority.gov.in/

Traditional Knowledge in India. https://www.wipo.int/wipo_magazine/en/2011/03/article_0002.html

Weissberg, G. (1975). United Nations movements toward world law. *International & Comparative Law Quarterly, 24*(3), 460–524.

[46] The reference is to those marginalised sections of society, who are unjustly denied the fruits of Economic Development. This is a Paraphrased expression, to the one used by Prof. Upendra Baxi (italics are mine), while teaching a Course on Law and Development, to the students of the Masters' Programme in Law, in Warwick Law School, England, during 1996–1997 Academic Year. I was privileged to work as the Co-Teacher to Prof. Baxi.

IPR in the Context of CBD

Enumeration of IPR-Related Cases of Bio-Prospecting Indian Bio-Resources

C. J. Anson

Indian Bio-Resources and Threats

India boasts a rich tapestry of biodiversity across unique ecosystems, with over 45,000 documented plant species and 91,000 documented animal species (Rahmani, 2011). These bioresources hold immense potential for scientific discoveries and commercial applications, offering significant opportunities for development in sectors like medicine, agriculture, and industry (Gupta & Dasgupta, 2008). However, the journey from raw bioresource to commercial product is intricately woven with complex legal and ethical considerations, particularly surrounding intellectual property rights (IPR) and equitable benefit sharing. This research article delves into the intriguing narrative of bio-prospecting on Indian bio-resources, focusing on landmark IPR-related cases that have shaped the discourse and challenged the existing paradigms. By critically analysing these cases, we aim to illuminate the intricate interplay between scientific exploration, intellectual property claims, and the crucial principle of fair and equitable benefit sharing with local communities and indigenous knowledge holders. Landmark IPR-related cases in India have been crucial turning points in the bio-prospecting landscape.

C. J. Anson (✉)
Inter-University Centre for IPR Studies, Cochin University of Science and Technology, Kochi, India
e-mail: ansoncj@cusat.ac.in

L. Krishna Panicker et al. (eds.), *Biodiversity and Business*,
https://doi.org/10.1007/978-3-031-71674-4_20

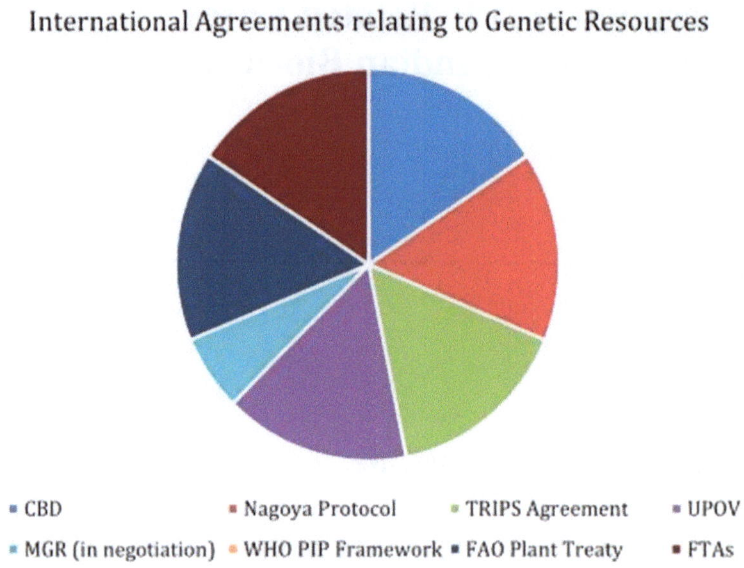

Source www.econstor.eu/bitstream/10419/248629/1/RP-130.pdf

The Neem case (1995), where the US Patent and Trademark Office revoked a patent granted on the fungicidal properties of turmeric extract due to prior art in India, highlighted the importance of recognising traditional knowledge in IPR claims (Shiva, 2001). The Basmati rice case (1997), where an American company was denied a patent on basmati rice lines as they were not distinct inventions, underscored the need to protect indigenous varieties and prevent biopiracy (Gupta, 2006). These cases, along with others like the Honey Bee case (2010) and the Millets case (2019), have established important legal precedents that strive to balance scientific innovation with the rights of local communities and the conservation of biodiversity (Gadgil, 2013; Srinivasan, 2020). This article tries to enumerate the cases in Bioprospecting of Indian bio-resources. It highlights the need for a more stringent Biodiversity Act to protect the interest of traditional knowledge holders.

Bioprospecting in India: A Rich Landscape with Complexities

Due to its vast biological resources, India, a megadiverse nation, is a prime target for bioprospecting. This practice involves exploring flora, fauna, and commercially valuable compounds with applications in medicine, agriculture, and industry (Mittal & Rahman, 2009). India's rich biodiversity offers immense potential for bioprospecting, with researchers focusing on medicinal plants, microbes for antibiotic production, and enzymes for industrial processes (Singh et al., 2011). However, bioprospecting in India is not without its challenges. Concerns regarding biopiracy, the appropriation of genetic resources, and traditional knowledge of local communities without fair compensation have emerged (Shiva, 2001). The Ballia barley case exemplifies this challenge, highlighting the difficulty of protecting germplasm and ensuring equitable benefit sharing (Down To Earth, 2023). To address these concerns, India's Biological Diversity Act (2002) and (2023) establish a regulatory framework for access to biological resources and associated knowledge, promoting bioprospecting while safeguarding the rights of local communities (Mittal & Rahman, 2009). The effectiveness of this BD Act in protecting India's Bio-wealth needs to be analysed based on the legal cases.

The Biological Diversity Act: Safeguarding India's Bio-Wealth

India's Biological Diversity Act (2002) and (2023) are a crucial shield against the misappropriation of its rich biological resources and their traditional knowledge. The Act establishes a regulatory framework for access to these resources, ensuring that benefits arising from their commercialization are equitably shared with local communities. This is achieved through critical provisions:

Prior Informed Consent (PIC): The Act mandates obtaining the explicit consent of local communities before accessing their biological resources and knowledge (Mittal & Rahman, 2009). This ensures communities have a say in how their resources are used and empowers them to negotiate benefit-sharing agreements. PIC is more powerful than individual ownership of intellectual property in all means of securing ownership as it creates cohesion regarding ownership of resources and associated knowledge.

Based on the Biodiversity Act (2002) and (2023)	Prior informed consent	Benefit sharing
Genetic resources/Bio-resources	Required	Required
Traditional Knowledge	Not specified	Not specified
Species	Required	Required
Digital Sequence information	Not Specified in the Act	Not Specified in the Act
Insects	Required	Required
Microbial	Required	Required if the ownership

Source Own Preparation

Benefit Sharing: The Act requires sharing benefits from biological resources, such as royalties or technology transfer, with local communities. This incentivizes their participation in bioprospecting activities and promotes sustainable biodiversity conservation (National Biodiversity Authority, n.d.). These provisions directly address concerns about biopiracy, where companies exploit genetic resources without fair compensation. The Biological Diversity Act, while progressive, faces particular challenges. Critics argue that the bureaucratic process for obtaining permits can be cumbersome, potentially hindering legitimate bioprospecting endeavours (Ghosh, 2015). Additionally, concerns exist regarding the effectiveness of the Act in protecting traditional knowledge, which can be challenging to document and verify (Shiva, 2001).

Prior informed consent (PIC): Prior informed consent (PIC) and benefit-sharing mechanisms are crucial to ethical practices, particularly in contexts involving indigenous communities or vulnerable populations. These mechanisms ensure that individuals or communities are adequately informed about any proposed activities that may affect them and that they receive fair benefits. The following case studies show these two mechanisms and the effectiveness of these mechanisms are analysed in the analysis part.

Case Studies in Action:
The Act has demonstrably protected India's bio-wealth. In the case of neem, the Act's provisions helped challenge a US patent on its pesticidal properties, recognising the long-standing knowledge of Indian communities (Mittal & Rahman, 2009). However, the Ballia barley case highlights ongoing complexities. While the details remain unclear, the Act empowers India to investigate potential biopiracy and pursue fair compensation (Down To Earth, 2023).

1. **The Ballia Barley Case: A Spotlight on Biopiracy Concerns in India**

The Ballia barley situation is a significant example of biopiracy anxieties within the Indian context. Ballia barley, a high-quality barley variety indigenous to the Ballia region of Uttar Pradesh, India, is known for its exceptional malting properties, making it valuable for the brewing industry (Thakur et al., 2018). However, Japanese companies have raised concerns about the potential misappropriation of Ballia barley germplasm (genetic material) (Down To Earth, 2023). The crux of the issue lies in the potential for biopiracy. Biopiracy refers to exploiting genetic resources and related traditional knowledge of local communities without fair compensation or benefit sharing (Shiva, 2001). In the Ballia barley case, anxieties stem from the possibility that Japanese companies may have accessed Ballia barley germplasm, perhaps through research collaborations or seed exchange programs, and subsequently developed improved barley varieties without ensuring benefits to local community.

The lack of clarity surrounding the details of germplasm access in the Ballia barley case makes it challenging to determine the nature of the alleged biopiracy. It is possible that the germplasm was accessed before India became a signatory to the International Treaty on Plant Genetic Resources for Food and Agriculture (ITPGRFA) in 1995. The ITPGRFA aims to facilitate access to plant genetic resources and ensure benefit sharing. However, its provisions might not apply to germplasm accessed before its implementation (International Treaty on Plant Genetic Resources for Food and Agriculture, 2001).

2. **The Neem Case: A Landmark Victory for Bioprospecting and Benefit Sharing in India**

The landmark Neem case is a testament to India's fight against biopiracy and its commitment to fair access and benefit sharing (ABS) in bioprospecting endeavours. In 1995, the United States Patent and Trademark Office (USPTO) granted a patent to the US Department of Agriculture (USDA) on a neem-based fungicide, even though neem's pesticidal properties had been known and used in India for centuries (Shiva, 2003). This sparked outrage, as it represented a clear case of biopiracy—the appropriation of genetic resources and associated knowledge without fair compensation to the source country.

The Indian government and activists like Vandana Shiva vehemently challenged the patent. A key argument centred on the principle of prior informed consent (PIC) (Mittal & Rahman, 2009). The Neem case highlighted the inherent injustice of granting patents on knowledge freely available and utilised by local communities for generations. Following extensive legal battles, the USPTO revoked the patent on Neem in 2000 (Shiva, 2003). This victory safeguarded India's traditional knowledge and set a precedent for fair access to biological resources. The case emphasized the importance of ABS, a cornerstone of the Biological Diversity Act (2002) enacted by India.

The Neem case is a powerful example of how bioprospecting can be conducted ethically and sustainably. By upholding the rights of local communities and ensuring fair benefit sharing, India paves the way for mutually beneficial collaborations that contribute to scientific progress, biodiversity conservation, and economic development for all stakeholders.

3. **The case of Orissa Mining Corporation Ltd. v. Ministry of Environment and Forests (2013) 6 SCC 476**

The case of Orissa Mining Corporation Ltd. v. Ministry of Environment and Forests (2013) 6 SCC 476, while not directly related to bioprospecting, emphasis on access and benefit sharing (ABS) principles in the context of natural resource utilisation by highlighting the importance of recognising the rights of local communities. The case involved the Orissa Mining Corporation's (OMC) petition challenging the Ministry of Environment and Forest's (MoEF) rejection of forest clearance for bauxite mining in Odisha. While not directly related to bioprospecting, the Supreme Court's decision emphasised the need to consider the rights and interests of local communities impacted by such projects (Orissa Mining Corporation Ltd. v. Ministry of Environment and Forests, 2013). The ABS principle within bioprospecting revolves around ensuring fair access to biological resources and equitable sharing of benefits with local communities with knowledge and traditional practices related to those resources. Though not about bioprospecting, the Orissa Mining Corporation case highlights the importance of recognising these same community rights. Mining activities, like bioprospecting ventures, can disrupt local communities' livelihoods and cultural practices relying on natural resources. The court's emphasis on involving the Gram Sabha (village council) in the decision-making process regarding the mining project resonates with the ABS principle of obtaining prior informed consent (PIC) from local communities before accessing their biological resources.

Benefit Sharing: While not explicitly mentioned in the case, ensuring sustainable development and potentially sharing some of the project's benefits with the local communities could be seen as aligning with the spirit of ABS. The Orissa Mining Corporation case, although not directly related to bioprospecting, offers valuable insights into the broader context of ABS. The case highlights the importance of recognising local communities' rights and interests in utilising natural resources, which is a core principle for ethical bioprospecting practices.

4. **The Kani Tribe and Arogyapacha**

The case of the Kani tribe and Arya Vaidyan Patasala (AVP) regarding the herbal medicine "Arogyapacha" is a well-documented example of bioprospecting, access, and benefit sharing (ABS) in India. Here is a breakdown of the case and its connection to ABS principles:

The Kani tribe, an indigenous community in Kerala, possessed traditional knowledge about the medicinal properties of a specific plant. In the 1990s, the Kani tribe shared this knowledge with AVP, a traditional medicine company, to develop the herbal medicine "Arogyapacha" (Shiva, 2006). This case exemplifies bioprospecting, where a company (AVP) accesses biological resources (plant knowledge) from a local community (Kani tribe) for commercial gain. However, concerns regarding ABS principles arose:

Prior Informed Consent (PIC): It is unclear whether the Kani tribe was fully informed about the potential commercial use and benefits of sharing their knowledge (Shiva, 2006).

Benefit Sharing: The initial benefit-sharing agreement between the Kani tribe and AVP was considered inadequate by the Kani community (Shrivastava et al., 2007).

Following intervention from the National Biodiversity Authority (NBA) established under the Biological Diversity Act (2002), a revised benefit-sharing agreement was negotiated (Mittal & Rahman, 2009). This agreement ensured a more equitable distribution of benefits for the Kani tribe.

5. **Access and Benefit Sharing: A Case Study of Seaweed Cultivation in Tamil Nadu**

This research delves into the intricacies of access and benefit sharing (ABS) in India, using the seaweed cultivation industry in Tamil Nadu as a focal point. The study examines how the Convention on Biological Diversity (CBD) principles is applied in this context, mainly through the regulatory framework established by the Biodiversity Act of 2002 and the role of the National Biodiversity Authority (NBA). Through the case of AquAgri Processing, which acquired Pepsico India Holding's seaweed business in 2008, the article illustrates the successes and challenges of ABS implementation. Despite significant contributions to the NBA's National Biodiversity Fund, obstacles such as the lack of biodiversity management committees (BMCs) hinder the effective utilisation of these funds. Furthermore, environmental concerns arise regarding cultivating *Kappaphycus alvarezii*, an alien seaweed species that has invaded coral reefs in marine reserves like the Gulf of Mannar Marine National

Park. This case underscores the importance of rigorous assessment and monitoring mechanisms to mitigate adverse impacts on biodiversity.

In conclusion, the case study of seaweed cultivation in Tamil Nadu sheds light on the complexities surrounding ABS practices in India. While ABS principles aim to ensure equitable benefits and biodiversity conservation sharing, challenges persist in balancing economic interests with environmental sustainability. The article emphasises the need for enhanced regulatory oversight, stakeholder engagement, and scientific research to address these challenges effectively. By analysing the seaweed case, policymakers can gain insights into refining ABS frameworks to promote sustainable resource management and equitable benefit-sharing in India's diverse ecosystems.

6. **Benefit Sharing Agreement for Ayurvedic Anti-Snake Venom Patent: A Case Study**

Dr Geeta Pandurang Pawar, an ayurvedic doctor based in Pune, India, developed a novel anti-snake venom formulation comprising four medicinal plants. Recognising the potential of this invention in addressing the urgent need for effective snakebite treatment, Dr. Pawar applied for a patent under the title "A Process for Preparation of Ayurvedic Anti-Snake Venom Capable of Administering Orally or Intravenously" (Indian Patent Number: 236637). The patent, granted under the jurisdiction of Mumbai, highlights the innovative approach of leveraging traditional Ayurvedic knowledge to create a life-saving medication. Dr. Pawar entered into a benefit-sharing agreement with the NBA through the Biological Diversity Act of 2002 and the Access and Benefit Sharing (ABS) Guidelines. The agreement, available on the official website of the Intellectual Property Office of India, outlines the terms and conditions for the equitable sharing of benefits arising from the commercialisation of the patented product. The NBA fixed the benefit-sharing rate at 2% of the gross sales or gross revenue generated from using biological resources accessed. As part of the agreement, Dr. Pawar committed to paying royalties to the NBA based on the specified benefit-sharing rate. The royalty payments serve as a mechanism for compensating the communities and countries that possess traditional knowledge and biological resources utilised in the invention. Dr. Pawar has adhered to the terms of the agreement by making timely royalty payments to the NBA. By adhering to the ABS Guidelines and fulfilling their obligations under the benefit-sharing agreement, Dr. Pawar demonstrates a commitment to ethical and responsible innovation. Such agreements not only ensure fair compensation for the use of traditional knowledge and biological resources but also contribute to the conservation of biodiversity and the preservation of traditional practices.

7. **Avon Products Has Sparked Attention Due to Its Incorporation of Asian Medicinal Plant**

Avon Products has sparked attention due to its incorporation of Asian medicinal plants into skincare products, evidenced by patents covering 16 plant species, many of which hold traditional significance across Asian cultures (Hammond, 2012). This strategic move aligns with consumer demands for skincare formulations targeting anti-aging concerns, driving Avon's interest in botanical ingredients (Hammond,

2012). However, concerns arise regarding the documentation of plant collection practices and benefit-sharing agreements, highlighting the need for transparency and accountability in Avon's utilisation of biodiversity resources (Hammond, 2012). Avon's utilisation of Asian plants underscores the importance of robust regulatory frameworks and international cooperation to address cross-border biodiversity issues and ensure equitable benefit-sharing arrangements under the Convention on Biological Diversity and the Nagoya Protocol (Hammond, 2012). Efforts should prioritise the documentation of traditional medicinal uses of Asian plants and foster collaboration among countries to uphold biodiversity conservation and ensurc fair access to genetic resources (Hammond, 2012). Avon should be encouraged to adhere to CBD obligations and engage in transparent partnerships with source countries and communities to promote the sustainable use of biodiversity resources (Hammond, 2012).

8. **Divya Pharmacy, a Subsidiary of Patanjali Ayurved vs Uttarakhand State Biodiversity Board (SBB)**

Divya Pharmacy, a subsidiary of Patanjali Ayurved, contested the claim by the Uttarakhand State Biodiversity Board (SBB) that they were required to share profits derived from bioprospecting activities (Ministry of Environment, Forests and Climate Change, Government of India, 2002). Bioprospecting refers to commercially utilising natural resources, such as the medicinal plants used in Divya Pharmacy's Ayurvedic products. The Biological Diversity Act of 2002 (BDA) mandates that companies utilising biological resources share a portion of their profits with local communities who possess knowledge about these resources and contribute to their conservation, a concept known as fair and equitable benefit sharing (FEBS) (GoI, 2002). Divya Pharmacy argued that the Act only applied to foreign companies and that Indian entities were exempt from profit-sharing provisions (*The Hindu*, 2018). However, the Uttarakhand High Court ruled against Divya Pharmacy in 2018 (High Court of Uttarakhand, 2018). The court's decision clarified that all companies utilising biological resources, irrespective of nationality, must comply with the following provisions of the BDA (GoI, 2002).

This landmark case set a significant precedent in India. The Uttarakhand High Court's ruling (2018) established that the BDA applies to all companies engaged in bioprospecting, and Indian companies are not exempt from benefit-sharing obligations. This ensures that local communities contributing to knowledge and conservation of biological resources are fairly compensated (GoI, 2002).

9. **Conviction of Czech Nationals for Illegal Insect Collection under the Biological Diversity Act in Singalila National Park, India**

An Indian court convicted two Czech nationals, Emil Kucera and Petr Svacha, for illegally collecting insects in Singalila National Park, West Bengal. Kucera was sentenced to three years in jail and fined Rs 50,000, while Svacha was fined Rs 20,000. The Wildlife Trust of India (WTI) aided the prosecution, noting the pressure on the court to expedite the trial because the defendants were foreigners. Despite challenges in identifying confiscated insects, the court found them guilty of violating Section 3 of the Biological Diversity Act and other relevant sections under the Wildlife Protection Act, 1972, regarding collecting biological resources without proper approval. While some scientists argued that the Czechs were conducting scientific research, the court emphasised their obligation to adhere to Indian laws, considering Kucera's prior visits to the country.

10. **Protecting Traditional Knowledge: India Thwarts Colgate's Nutmeg Mouthwash Patent**

In a victory for the preservation of traditional knowledge, India's Ministry of Science and Technology successfully blocked an attempt by Colgate-Palmolive to patent a mouthwash containing nutmeg extract (Sentinel Digital Desk, 2015). The crux of the challenge rested on the prior existence of this knowledge in ancient Indian medicinal practices. The Council of Scientific and Industrial Research's Traditional Knowledge Digital Library (CSIR-TKDL) played a pivotal role by providing evidence from historical texts. These texts documented the use of nutmeg and its extracts for oral care in traditional Indian medicine. This documentation served as "prior art," a concept in patent law that establishes that an invention is not novel and, therefore, ineligible for a patent. Colgate-Palmolive's patent application was hinged on the notion that their mouthwash formula was a novel invention. However, by citing historical evidence, CSIR-TKDL effectively countered this claim. The outcome of this case highlights the significance of protecting traditional knowledge. Here, documented practices from ancient India prevented a corporation from potentially monopolising the use of a common ingredient in a mouthwash formula (Sentinel Digital Desk, 2015). This case serves as a reminder of the importance of safeguarding traditional knowledge systems and ensuring they are not wrongly claimed as new inventions.

11. **The Turmeric Case: Protecting Traditional Knowledge from Patenting**

The patenting of traditional knowledge has been a contentious issue, raising concerns about biopiracy and the exploitation of indigenous knowledge systems. The Turmeric Case is a landmark example of how traditional knowledge can be protected from patenting. Researchers Suman K. Das and Harihar P. Cohly applied for a patent in the United States Patent and Trademark Office (USPTO) for a method of wound healing using turmeric. They claimed to have invented a novel method of administering turmeric, either orally or topically, to promote wound healing in patients. India's Council of Scientific and Industrial Research (CSIR) challenged the patent application. They argued that using turmeric for wound healing was well-established in traditional Indian medicine for centuries. However, due to the informal transmission of this knowledge, readily available published information on this specific use of turmeric was challenging to source. USPTO ultimately revoked the turmeric patent based on the concept of prior art. The evidence presented by CSIR demonstrated an apparent lack of novelty, non-obviousness, and inventive steps in the claimed invention. This implies that using turmeric for wound healing would not be considered a ground breaking invention for someone with a basic understanding of traditional Indian medicine. The Turmeric Case is a significant victory for protecting traditional knowledge. It highlights the importance of documenting and defending traditional practices to prevent their appropriation through patents.

12. **Two Japanese Nationals Arrested for Smuggling Wildlife at Cochin International Airport**

In an incident reported on [date of the article], two Japanese nationals, Suta Shibasaki and MurayiYojka, were arrested at Cochin International Airport in India for attempting to smuggle live snakes, insects, and other exotic wildlife. The wildlife was discovered during a routine X-ray security check. While the bag's contents triggered suspicion due to the presence of scissors and adhesive tape, it was upon closer inspection that security officials found glass jars containing approximately 20 live creatures. These creatures included snakes, spiders, lizards, chameleons, tortoises, and earthworms. According to Shibasaki, he was a wildlife researcher in Japan but claimed ignorance of Indian wildlife protection laws. The report further suggests that Shibasaki and Yojka had been in India for two weeks, explicitly visiting Kochi. Their purpose was to collect exotic wildlife specimens, and they had even rented a room in Athirappilly to facilitate their forays into the forest to capture these animals. The article also mentions the seizure of video footage from the duo, which reportedly depicted them catching these animals in the Athirappilly forest.

13. **Bt. Brinjal and the Battle Against Biopiracy: A Case Study from India**

The development of Bt. Brinjal, India's first genetically modified food crop, has been marred by allegations of biopiracy. This case study examines the ongoing legal battle and the role of the Environment Support Group (ESG) in raising concerns about the protection of India's biodiversity. The ESG stands as a critical player in this controversy. In 2010, they filed a complaint with the National Biodiversity Authority (NBA) accusing several entities, including multinational corporations like Monsanto and prominent Indian agricultural institutions, of biopiracy (Environment Support Group [ESG], n.d.). The complaint alleges that these parties accessed over 16 brinjal varieties native to India without complying with the Biological Diversity Act (BDA) while promoting Bt. Brinjal for commercial release (ESG, n.d.). Further criticism from the ESG centres on the initial inaction of regulatory bodies like the NBA and the Ministry of Environment and Forests. These agencies allegedly failed to verify BDA compliance throughout the relevant period (ESG, n.d.). While the NBA eventually voted to pursue criminal charges against the accused parties in 2012, this decision came after internal resistance and contradicted previous assurances from the Environment Minister (ESG, n.d.). These inconsistencies raise concerns about potential political or economic pressures influencing legal action. The situation becomes even more complex when examining the actions of the Karnataka Biodiversity Board. This board initially mirrored the NBA's approach, deciding to prosecute parties allegedly illegally accessing brinjal varieties from Karnataka. However, this investigation was controversially suspended in 2012, prompting a petition from the ESG against this decision (ESG, n.d.). The Bt. Brinjal's case in India serves as a critical example of the challenges surrounding biopiracy and biodiversity protection. The ESG's actions highlight the difficulties of enforcing environmental regulations and the potential influence of corporations and political interests. As the legal battle continues, it remains to be seen whether India will effectively address these concerns and ensure compliance with its biodiversity protection laws.

Conclusion

The Biological Diversity Act of India is a cornerstone in safeguarding the country's rich biodiversity and preventing the misappropriation of its biological and genetic resources. Through various case studies, including the Ballia Barley Case, the Neem Case, and the Kani Tribe and Arogyapacha, the significance of the Act in addressing biopiracy concerns becomes evident. The Act aims to ensure Prior Informed Consent (PIC) and Fair and Equitable Benefit Sharing (FEBS) with local communities, thereby protecting their traditional knowledge and ensuring they receive fair compensation for commercialising derived products. However, challenges persist in effectively implementing benefit-sharing mechanisms, as highlighted in the seaweed cultivation case in Tamil Nadu. Despite significant contributions to the National Biodiversity Fund, obstacles such as the lack of biodiversity management committees (BMCs) hinder the equitable distribution of benefits to local communities. Additionally, the Turmeric Case and the recent conviction of Czech nationals for illegal insect collection underscore the Act's role in enforcing regulations and holding violators accountable for unauthorised access to biological resources. While the Act represents a crucial step towards biodiversity conservation and equitable benefit sharing, efforts are needed to address gaps in implementation and strengthen regulatory oversight to ensure the Act's objectives are fully realised.

In conclusion, examining the diverse cases presented highlights the successes and shortcomings of India's Biological Diversity Act in protecting traditional knowledge associated with biological resources and digital sequence information (DSI). While the Act has made strides in addressing instances of biopiracy and promoting fair benefit sharing, it grapples with inherent limitations that impede its efficacy in safeguarding traditional knowledge and governing DSI. One of the notable limitations lies in the Act's lack of explicit provisions for protecting traditional knowledge, especially in oral traditions or community practices. The absence of robust mechanisms for documenting and preserving such knowledge hinders the ability of indigenous communities to assert their rights and seek recourse in cases of misappropriation or unauthorised use.

Moreover, the Act's inadequacies extend to digital sequence information, where rapid advancements in biotechnology and genomic research outpace regulatory frameworks. With no explicit provisions addressing the protection of DSI, a significant gap exists in safeguarding genetic information in digital form. Challenges in regulating the access, use, and commercialisation of DSI pose formidable obstacles, particularly in global data sharing and collaborations. The lack of clear guidelines on DSI governance further complicates efforts to prevent unauthorised access and transfer of genetic data, exacerbating concerns surrounding biopiracy and exploitation of genetic resources.

Addressing these limitations requires concerted efforts to strengthen legal frameworks, enhance technological capabilities, and promote international cooperation. Amendments to the Biological Diversity Act must include provisions explicitly addressing the protection of traditional knowledge and DSI, incorporating principles of prior informed consent and fair and equitable benefit sharing. Additionally, investment in technological infrastructure and capacity-building initiatives is essential to facilitate the documentation, storage, and management of traditional knowledge and genetic data. Collaborative efforts between indigenous communities, scientific institutions, and policymakers are crucial in developing digital platforms or databases for sharing and preserving traditional knowledge and genetic resources along with the Traditional Knowledge Protection Act. Furthermore, engagement in multilateral forums and partnerships can foster harmonisation of biodiversity regulations and enhance cross-border cooperation in the governance of DSI. By addressing these challenges and leveraging opportunities for innovation and collaboration, India can strengthen its biodiversity governance framework, uphold the rights of indigenous communities, and promote sustainable development and conservation efforts.

References

Access and Benefit Sharing (ABS) Guidelines, India.

Access and Benefit-sharing Initiative. (2014, July 11). *Fair and equitable benefit-sharing*. https://environment.ec.europa.eu/topics/nature-and-biodiversity/sharing-natures-genetic-resources_en

Bennett, A. B., Wernberg, T., & Connell, S. D. (2019). Local options for achieving sustainability in seaweed farming. *Frontiers in Marine Science, 6*, 149. https://www.seattletimes.com/nation-world/the-next-frontier-in-farming-the-ocean/

Biological Diversity Act, 2002, India. https://www.indiacode.nic.in/bitstream/123456789/2046/4/a2003-18.pdf

CBD Secretariat. (2016, December 01). *Prior informed consent*. Convention on Biological Diversity. https://www.sciencedirect.com/science/article/pii/S0301479721018296

Down To Earth. (2023, February 15). *Japanese biopiracy of our Ballia barley*. https://www.downtoearth.org.in/blog/its-still-open-season-for-biopiracy-1765

Drahos, P. (2000). Indigenous knowledge, intellectual property and biopiracy: Is a global bio. *EIPR*. https://www.anu.edu.au/fellows/pdrahos/articles/pdfs/2000ipandbiopiracy.pdf

Gadgil, M. (2013). *Biodiversity, equity, and sustainability: A framework for policy and action*. Orient Blackswan.

Gebru, A. (2017). Intellectual property and bioprospecting: A model legal framework. *NCJL & Tech*.https://heinonline.org/hol-cgi-bin/get_pdf.cgi?handle=hein.journals/ncjl19&section=11

Gebru, A. (2018). *Intellectual property law and bioprospecting: A model legal framework*. SSRN.

Ghosh, S. (2015). India's Biological Diversity Act, 2002: A critical review. *Journal of Intellectual Property Rights, 20*(1), 70–79.

Government of India, Ministry of Environment, Forests and Climate Change. (2002). *The Biological Diversity Act, 2002 [Act No. 18 of 2003]*. https://www.indiacode.nic.in/handle/123456789/2046?sam_handle=123456789/1362

Gupta, A. K., & Dasgupta, P. (2008). Bioprospecting: A boon for conservation? *Trends in Ecology & Evolution, 23*(6), 393–398.

Gupta, S. (2006). Intellectual property rights and the protection of traditional knowledge: The Basmati rice case. *Economic and Political Weekly, 41*(24), 2443–2447.
Hammond, E. (2012). *The Avon Lady comes collecting Asian medicinal plants.* Third World Network Briefing Paper No. 64. Retrieved from TWN website
High Court of Uttarakhand. (2018). [Case name not publicly available].
International Treaty on Plant Genetic Resources for Food and Agriculture. (2001, November 6). *The International Treaty on Plant Genetic Resources for Food and Agriculture.* https://www.fao.org/plant-treaty/en/
Martens, P. (2012). *Can traditional knowledge owners and producers in developing countries use geographical indications for protection and economic development gain?* https://doi.org/10.2139/ssrn.2083680
Mittal, A., & Rahman, A. (2009). Regulation of bioprospecting in India: An analysis of the Biological Diversity Act, 2002. *Journal of Intellectual Property Rights, 14*(4), 370–378.
National Biodiversity Authority (NBA). (n.d.). http://ipindiaservices.gov.in/PatentSearch/PatentSearch//eRegistrationReport
Orissa Mining Corporation Ltd. v. Ministry of Environment and Forests, (2013) 6 SCC 476.
Patenting of Traditional Knowledge in Light of the Turmeric Case. (2022, September 10). *IIPRD.* https://www.npr.org/2023/09/01/1197321273/turmeric-india-biopiracy-patent-tkdl
Pordié, L. (2012). Hijacking Intellectual Property Rights: Identities and Social Power in the Indian Himalayas. *Tibetan Medicine in the Contemporary World.* https://api.taylorfrancis.com/content/chapters/edit/download?identifierName=doi&identifierValue=10.4324/9780203932544-9&type=chapterpdf
Rahmani, A. R. (2011). Biodiversity in India: Status and conservation strategy. *Journal of the Bombay Natural History Society, 108*(1–2), 121–132.
Sentinel Digital Desk. (2015, July 16). *Colgate's bid to patent nutmeg mouthwash foiled* [Press release].
Shiva, V. (2001). *Biopiracy: The plunder of nature and knowledge.* Zed Books.
Shiva, V. (2003). Protecting our biological and intellectual property rights. *Economic and Political Weekly, 38*(17), 1733–1739.
Shiva, V. (2006). Biodiversity and intellectual property rights. In C. Perrings (Ed.), *The economics of biodiversity* (pp. 221–232). Cambridge University Press.
Shrivastava, A., Kulkarni, N., & Raina, A. (2007). Sharing the benefits of biodiversity: Lessons from the Kani-AVP case. *Economic and Political Weekly, 42*(41), 4244–4249.
Singh, S., Pandey, A., & Popli, S. (2011). Bioprospecting of some Indian medicinal plants for antibacterial activity. *International Journal of Applied Biology and Pharmaceutical Technology, 2*(4), 124–128.
Singh, S. P., & Singh, A. M. (2015). Bioprospecting: Enhancing the Value of Biodiversity and Intellectual Property. In R. K. Salgotra & B. B. Gupta (Eds.), *Plant genetic resources and traditional knowledge for food security* (pp. 231–36). Springer Singapore.
Soejarto, D. D., Fong, H. H. S., Tan, G. T., Zhang, H. J., Ma, C. Y., Franzblau, S. G., Gyllenhaal, C., et al. (2005). Ethnobotany/ethnopharmacology and Mass Bioprospecting: Issues on Intellectual Property and Benefit-Sharing. *Journal of Ethnopharmacology, 100*(1–2), 15–22.
Srinivasan, U. (2020). *Millet is revolutionizing food and nutrition security.* Routledge.
Thakur, M., Singh, G., Pandey, M., & Singh, S. P. (2018). Screening of barley germplasm for malting quality traits under varying temperature regimes. *Physiology and Molecular Biology of Plants, 24*(3), 659–667.
The Hindu. (2018, April 11). Uttarakhand High Court asks Patanjali's Divya Pharmacy to share profits with farmers. https://www.thehindu.com/news/national/supreme-court-issues-notice-to-ramdev-in-patanjali-advertising-case/article67967323.ece

Role of Traditional Knowledge Associated with Plants and Its Bioprospecting: Realities and Challenges

Andrews Baskaran

Introduction

Since ancient times, biological resources, mostly, plants have been the primary source of medicine for indigenous people and local communities. Because of its easy availability and affordability, even today medicinal plants offer primary health care support to the local people around the globe. An estimate by the World Health Organization (WHO) endorsed the statement in its report, which says; around 80% of the world's population uses traditional medicine. Samuel and Andrews (2010) has reported the contributions of local communities' knowledge to ethnopharmacological and ethnomedicinal interventions are noteworthy and it's evidenced in a plethora of ethnobotanical reports. Fabricant and Farnsworth (2001) have listed 88 drugs derived from plants with their ethnomedical correlations and sources. The analgesic, codeine, and morphine from *Papaver somniferum*, the cardiotonic, digitalin, and digoxin from *Digitalis purpurea*, the antimalarial, quinine from *Cinchona ledgeriana* and the antihypertensive, reserpine from *Rauvolfia serpentina* are some of the notable century old drugs.

Hence it has proved that plants are the preferable therapeutic constituent(s) in traditional medicine systems, and they continue to play a pivotal role in modern medicine as well. Notably, their natural origin makes plants a primary choice for individuals seeking complementary and alternative medicine (CAM). Historically, traditional medicinal information and therapeutic value of plants have been preserved in ancient texts such as the Ayurvedic texts of India, Traditional Chinese Medicine texts, and ancient Greek and Roman medical writings.

A. Baskaran (✉)
Former Research Associate II, National Innovation Foundation—India, Grambharti, Amrapur, Gandhinagar-Mahudi Road, Gandhinagar, Gujarat 382650, India
e-mail: andrews.baskaran@gmail.com

L. Krishna Panicker et al. (eds.), *Biodiversity and Business*,
https://doi.org/10.1007/978-3-031-71674-4_21

Recently, modern research often corroborates the efficacy of many of these traditional remedies, leading to the development of pharmaceuticals based on phytochemicals. There are a wide range of chemical substances such as polyphenols, flavonoids, alkaloids, terpenes, and antioxidants contributing to healthcare systems. Modern scientific developments have revolutionized the extraction, identification, isolation, and purification of active substances from plants. These advancements have significantly enhanced the efficiency, precision, and speed of isolation of active substances from plants, thereby accelerating the discovery and development of natural products for various therapeutic applications. This chapter attempts to give an overview of challenges and opportunities in bioprospecting studies of plants and associated traditional knowledge from the perspective of the conservation of medicinal plants.

Promoting traditional knowledge through legislation

There is as yet no accepted definition for traditional knowledge (TK) at the international level (World Intellectual Property Organization, 2001). United Nation's Convention on Biological Diversity (CBD) has explained traditional knowledge in Article 8(j) of the Convention which states "respect, preserve and maintain knowledge, innovations, and practices of indigenous and local communities embodying traditional lifestyles relevant for the conservation and sustainable use of biological diversity and promote their wider application with the approval and involvement of the holders of such knowledge, innovations, and practices and encourage the equitable sharing of the benefits arising from the utilization of such knowledge, innovations and practices".

In general, traditional knowledge associated with plants can either be protected in the form of codified/recorded information or preserved/undisclosed by the families of local communities as a non-codified system. Traditional Knowledge Digital Library (TKDL) is a database protecting codified traditional knowledge. Since non-codified oral traditional knowledge is not recorded, documentation of traditional knowledge associated with plants is necessary to avoid misappropriation.

Acceptance of Herbal Medicine & Regulatory Approval

The value of the herbal medicine market is projected to reach USD 437 billion by 2032. It had shown USD 216.40 billion in 2023 and it's expected to touch 233.08 billion in 2024.[1] Though herbal drugs are important alternatives to modern allopathic medicine, their safety and efficacy remain a major worry. For any stream of the health care system, the medicine recommended for a particular disease has to undergo

[1] Herbal Medicine Market Size, Growth | Industry Trends [2032] (fortunebusinessinsights.com) accessed on 15 April 2024.

thorough clinical testing and obtain regulatory approvals. In the herbal medicine sector, very few medicinal herbs have been scientifically tested for their specific pharmacological actions in the body. United States Food and Drug Administration (FDA) estimates 50,000 adverse events caused by botanical and other medicinal dietary supplements annually in the United States. Marcus and Grollman (2016) have mentioned some of the adverse events caused by the use of traditional folk remedies in Africa and Asia. Higher mortality was observed due to renal and liver dysfunction in South Africa. Likewise, Kenya also has seen mortality due to the use of herbal remedies. These events raised concern about herbs exhibiting toxicity. Biomedical researchers, health authorities, traditional healers, and public representatives have to meet and share information on plants with toxic effects. Allopathic drugs are approved through rigorous scientific studies to prove their safety and efficacy. Its pharmacokinetic and pharmacodynamic characteristics could be altered based on the requirement. Similarly, plant-based herbal medicine also needs scientific evaluation, which shall have no hindrance in drug administration.

Traditional Knowledge and Management of COVID-19

The outbreak of COVID-19 disease was one of the most devastating viral infections the world has faced in the recent past. The disease has had a profound and far-reaching impact on every country. As per WHO data, till 15th March 2024, over 770 million confirmed cases of COVID-19 and more than 7 million COVID-19 deaths were reported globally. The battle against COVID-19 was exhausting for many individuals, families, and healthcare professionals worldwide, and even for Government machinery. Obtaining COVID-19 case confirmation and death data from remote pockets of the countries can pose significant challenges due to limited access to medical facilities, lack of access to testing facilities, geographical location, and logistical constraints. Workable strategies like mobile testing units, deployment of trained community health workers, telemedicine and remote consultations, and other innovative techniques like drones, and remote sensing of population movement could offer opportunities to gather information. The strong establishment of collaboration between stakeholders is key for resource allocation and adaptation to local environments (Ola, 2022).

Though traditional knowledge-based use of plants for health-related issues lacks scientific backing or approval, curative ability is known even for viral diseases. When the confirmation of the COVID-19 pandemic is in high numbers, the hospitals or COVID care centers face a heavy rush. Patients requiring mild care, prevention, and management of COVID-19 could be provided healthcare by registered local traditional medical practitioners by adhering to the guidelines set forth by Government health authorities. Countries like China, India, Africa and Latin America have recommended traditional medicine practices to manage COVID-19. A comprehensive global data specifically focusing on the use of traditional medicine for COVID-19 management is required for future reference. To meet the global requirements during pandemic situations, traditional plant-based medicines' safety and efficacy have to be scientifically proven.

In India, the Ministry of Ayush formed an inter-disciplinary AYUSH R&D task force by collaborating with ICMR, DBT, CSIR, AIIMS, and AYUSH institutions and developed an Ayurvedic polyherbal formulation called AYUSH-64. It's composed of multiple plant-based ingredients with known antiviral, anti-inflammatory, and immunomodulatory properties. Four constituents of AYUSH-64 such as *Saptaparna* (*Alstonia scholaris* R. Br.), *Katuki* (*Picrorhiza kurroa* Royle ex. Benth), *Kiratatikta* (*Swertia chirata* Pexbex. Karst) and *Kuberaksha* (*Caesalpinia crista* L.) are reported to possess immunomodulatory, anti-inflammatory, antipyretic, antioxidant and antiviral properties (Aleem et al., 2018; Panda et al., 2022; Qureshi et al., 2015; Upadhyay et al., 2019). Plant ingredients used in the ayurvedic formulation have ethnobotanical and ethnopharmacological references (Dharshini et al., 2024; Nasim et al., 2022).

Poly-Herbal Formulations for COVID-19

Kabasurakudineer is a traditional polyherbal Siddha medicine used in Tamil Nadu, India, believed to have immune-boosting properties. The concoction contains an equal portion of 15 herbal ingredients such as dried rhizome of *Zingiber officinale*, the fruit of *Piper longum*, the flower bud of *Syzygium aromaticum*, the root of *Anacyclus pyrethrum*, the root of *Tragiainvolucrata*, the root of *Hygrophila auriculata*, fruit rind of *Terminalia chebula*, leaves of *Justicia adhatoda*, leaves of *Plectranthus amboinicus*, the root of *Costus speciosus*, the stem of *Tinospora sinensis*, the the root of *Clerodendrum serratum*, whole plant of *Andrographis paniculata*, root tuber of *Cyperus rotundus* and root of *Cissampelos pareira*. Though the medicine has been used to cure fever and respiratory diseases its efficacy data on the management of COVID-19 was not available before the pandemic. Recently, it was reported that the polyherbal Siddha medicine Kabasurakudineer, significantly reduced the viral load of SARS-CoV-2 after a week and it prevented the progression of asymptomatic COVID-19 patients to symptomatic. No adverse events were recorded during the study (Natarajan et al., 2021). A further detailed investigation is required to enhance the therapeutic action of the formulation suitable for symptomatic cases. A new polyherbal Ayurvedic formulation, NOQ19 (a combination of 19 herbal ingredients) has also been investigated for antiviral efficacy with Vero E6 cell lines (Kanchibhotla et al., 2022). It eliminates virus 100% at 0.9 mg/ml concentration.

Poly-herbal formulation/medicine requires more numbers/quantities of medicinal herbs. For example, Kabasurakudineer requires fifteen herbal ingredients of known action. More importantly out of fifteen ingredients nine are underground parts or whole plants. Ayurvedic formulation NOQ19 comprises nineteen herbal ingredients of which, ten are underground parts or whole plants. Ensuring raw material quality and continuous availability from the cultivation or wild is the cornerstone of medicine manufacturing. Traditional harvesting methods followed by traditional people ensure a continuous supply of plant materials. For example, traditional communities in coastal districts of Karnataka follow religious practices while collecting stem/bark of *Alstonias cholaris*. They pray before gathering the bark and use a stone or blunt instrument like a hammer to peel the bark in the morning time. Sharp objects like knives or sickles are not allowed. Traditional communities of coastal Karnataka use stem bark of *A.scholaris* for the treatment of fever, leucorrhoea, asthma, eczema, indigestion/constipation & spider bites (Bhandary, 2020). Likewise, tribal and non-tribal people who dwell in different parts of India harvest medicinal plants sustainably. Globally, the demand for medicinal plant-based products is increasing. When the requirement increases, the users may either face a shortage of plant material supply or unsustainable harvesting may increase which in turn reduces the raw material quality. If the medicinal plants are harvested unsustainably in bulk quantities, those plant species could face extinction or endangerment.

Medicinal plant-based industries are facing major challenges such as over-exploitation, habitat loss, unscientific harvesting practices, and biopiracy. To overcome these challenges collaborative efforts, from different stakeholders i.e., Government & Non-Governmental organizations, private agencies, cultivators, and traditional knowledge holders could be encouraged to develop conservation measures such as increasing suitable areas for medicinal plant cultivation, proper post-harvesting practices and infrastructure for storage of plant material. Ministry of AYUSH, Government of India has developed an operational guideline to promote *in-situ* and *ex-situ* conservation of medicinal plants under the multi-pronged strategy.

National Medicinal Plant Board (NMPB) & Its Conservation Initiatives

According to NMPB, more than 90% of formulations are plant-based in Ayurveda, Siddha, and Unani systems of medicine. About 22% of the medicine production is sourced through cultivation. Both domestic and export demand are expected to increase post-COVID, and medicinal plant cultivation and sustainable utilization are to be promoted. The board has collaborated with growers of medicinal plants through State Medicinal Plant Boards (SMPBs) and helped in storing, transporting, and marketing their produce. The board has support to keep herbal gardens, school herbal gardens, and home herbal gardens both in plains and forest areas. About 1349 (35 Nos. during 2018–2019) JFMCs/Panchayats/Biodiversity Management Committees (BMCs) have been supported for value addition activities like drying, grading, storage, primary processing, etc. NMBP has launched a mobile application 'e-Charak' which helps in the marketing of medicinal plant produce and knowledge sharing among the users. To combat the challenges and identify new opportunities, the board has supported R&D projects covering areas such as multiplication of RET plants through tissue culture, database on medicinal plants, inter-cropping, chemical and molecular profiling, quality planting materials, germplasm conservation, authentication and standardization of crude drugs.[2]

Identification and Isolation of 'Active Principles'

Plants contain myriad chemical constituents occurring in low quantities, making it difficult to isolate the active ingredient(s) responsible for a particular pharmaceutical activity. Conventional extraction methods of phytochemicals are facing challenges i.e., time-consuming steps, requirement of more solvents (because most bioactive substances are water insoluble), boiling might degrade target substances, and loss of thermolabile compounds, and yield is also low. Therefore, the requirement for a large quantity of plant samples and solvents led researchers to find innovative techniques for extraction.

Primarily, extraction of plants as a whole or their parts (dried/fresh) with suitable solvents plays a key role in every investigation leading to the separation of active compounds. The conventional extraction procedures involve boiling, digestion, decoction, percolating, or macerating the plants usually with water, ethanol, methanol, or other solvents to release biologically active constituents from the cell matrices of the plant into the solvents. Soxhlet extraction is also a widely used conventional method that involves continuous boiling of solvent.

Development of various new and innovative extraction techniques have been introduced to aid novel drug development processes (Pirintsos et al., 2022). The semi-bionic extraction is one such technique where biological materials such as cells or proteins, are partially combined with synthetic or bionic components for various applications. This method doesn't require organic solvents and the extraction temperature is lower. The supercritical fluid extraction method uses carbon dioxide which behaves both as a gas and a liquid under specific pressure and temperature. Minimal solvent usage and ability to extract heat-sensitive compounds without leaving behind solvent residues are preferred. The use of microwave energy and ultrasonic waves has the advantage of shorter extraction time, higher yield, and reduced solvent consumption compared to traditional extraction methods. Microwave-assisted extraction increases the solubility of target molecules through the generation of heat energy whereas ultrasonic energy creates localized pressure and temperature that facilitates the release of target compounds into the solvents. Enzyme-assisted breakdown of cell walls to release active compounds from plant materials also has advantages such as higher selectivity, milder extraction conditions, and the ability to target specific compounds. Molecular distillation is employed for extracting delicate compounds under vacuum conditions utilizing the differences in boiling points of substances to separate them based on their molecular weights. Membrane separation technology

[2] Achievements_NMPB_31_March_2019.pdf (accessed on 4 May 2024).

is used to separate and concentrate active principles from complex mixtures based on differences in molecular size, shape, charge, or affinity.

Chromatography is a powerful analytical method suitable for the separation and quantitative determination of a considerable number of compounds. Integration of chromatography with spectroscopy offers several advantages in the analysis and characterization of diverse samples. Sophisticated new methodologies and instrumentation such as HPLC-MS, LC-MS, GC-MS, NMR, and crystallography has led to the parallel development of the biological and pharmaceutical field. These technologies are continuously evolving with advancements in instrumentation, software, and methodologies enabling researchers to tackle increasingly complex analytical challenges and push the boundaries of modern analytical chemistry.

Ethnobotanical Bioprospecting & Other Drug Discovery Approaches

The search for new pharmaceutically active agents from ethnomedicinal plants is increasing. To overcome the challenges of conventional screening methods, a more sophisticated high throughput screening method is used which could reduce the extraction time and also minimize the requirement of plant samples (Najmi et al., 2022).

Cultivation, harvesting, and processing of plant materials procured are time-consuming. Nowadays, phytopharmaceutical drug manufacturers rely mainly on combinatorial chemistry and rational drug-designing options rather than sources from natural habitats. Molecular modeling approaches are employed to target specific binding sites to inhibit the action of biological molecules. This approach involves no natural plant materials. However, as of now small and structurally simple molecules can be synthesized.[3]

Medicinal Plant Conservation and Policy Obligations

Traditional plant-based medicine requires raw material supply. In India, about 90% of medicinal plants are found in forest areas, and only 10% of it found in other ecosystems. The medicinal plants are expected to be well conserved and preserved in a special ecological niche having protectable climatic, demographic, and social factors.

United Convention on Biological Diversity (CBD) is a crucial international treaty aimed at conserving biodiversity, promoting sustainable use of biological resources, and ensuring fair and equitable sharing of benefits arising from genetic resources.

[3] Bioprospecting—Biodiversity and conservation (inflibnet.ac.in) accessed on 4 May 2014.

It's a significant step toward addressing the global challenges of biodiversity loss and ecosystem degradation.

Researchers shall consider the regulatory approvals for access and utilization of biological resources under The Biological Diversity Act (2002) as one of the additional requirements like institutional clearances and obligations. Now the Ministry of Law and Justice has amended the aforesaid Act and published "The Biological Diversity (Amendment) Act (2023)" in Official Gazette.[4] The same comes into force from enacted law is come into force from April 1st, 2024. Section 3 (aa) states "benefit claimers" means the conservers of biological resources, their by-products, creators or holders of traditional knowledge associated thereto (excluding codified traditional knowledge only for Indians), and information relating to the use of such biological resources, innovations and practices associated with such use and application.

Conclusion

There is a renaissance of traditional medicinal knowledge among mankind post-COVID. Traditional medicinal information is time-tested practices followed by traditional community knowledge holders or practitioners. Once the information gathered is scientifically validated and authenticated, it could benefit ailing people. To tap more uses from ethnomedicinal knowledge, screening, and bioprospecting studies are to be performed in large numbers. Sometimes the knowledge/information associated with medicinal plants may fall into the hands of individuals/groups that may utilize it for their benefit without compensating traditional knowledge holders/creators. Leakage of non-codified traditional medicinal information may cause revenue loss to the local knowledge holders/communities, which in turn hamper conservation activities. Hence regulated access to medicinal plants and/or associated traditional knowledge and systematic documentation is necessary. Commercial utilization of ethnomedicinal knowledge associated with medicinal plants may lead to environmental problems, which start from over-exploitation and subsequent species loss. IUCN Red list provides the conservation status of medicinal species which helps in prioritizing conservation activities. *In-situ* conservation measures and restoration of habitat including water bodies could be implemented in a phased manner by categorizing ecological zones/regions. Commercial cultivation and sustainable harvesting of medicinal plants could also be promoted to ensure continuous raw material supply. Coordinated research and development work to help medicinal plant cultivators and end users. Once species loss is restored then its associated traditional medicinal knowledge also gets rebirth. Hence, species and/or their associated traditional knowledge restoration needs a conducive ecosystem.

[4] http://nbaindia.org/uploaded/pdf/BDAct_2023.pdf (accessed on 4 May 2024).

References

Aleem, A., & Kabir, H. (2018). Review on *Swertia chirata* as traditional uses to its Phytochemistry and Pharmacological activity. *Journal of Drug Delivery and Therapeutics, 8*, 73–78. https://doi.org/10.22270/jddt.v8i5-s.1957

Bhandary, M. (2020). Alstoniascholaris in the ethnomedicinal and religious tradition of Coastal Karnataka, India. *Biodiversitas Journal of Biological Diversity, 21*. https://doi.org/10.13057/biodiv/d210438

Bioprospecting—Biodiversity and conservation. inflibnet.ac.in. Accessed on 4 May 2014.

Dharshini, M. D., Sreelakshmi, M. S., Adithya, J., Lakshmi, P. K., & Nath, L. R. (2024). A systematic analysis of the ethnopharmacological relevance of an Indian traditional plant, *Hemidesmus indicus* (L.) R.Br. for the past 10 years. *Journal of Applied Pharmaceutical Science, 14*(1), 037–044.

Fabricant, D. S., & Farnsworth, N. R. (2001). The value of plants used in traditional medicine for drug discovery. *Environmental Health Perspectives, 109*(Suppl. 1), 72.

Herbal Medicine Market Size, Growth I Industry Trends [2032]. fortunebusinessinsights.com. Accessed on 15 April 2024.

http://nbaindia.org/uploaded/pdf/BDAct_2023.pdf. Accessed on 4 May 2024.

https://nmpb.nic.in/sites/default/files/Achievements_NMPB_31_March_2019.pdf. Accessed on 4 May 2024.

Intellectual property needs and expectations of traditional knowledge holders. WIPO Report on Fact-finding Missions on Intellectual Property and Traditional Knowledge (1998–1999), WIPO, Geneva, Switzerland, April 2001.

Kanchibhotla, D., Subramanian, S., Ravi Kumar, R. M., Venkatesh Hari, K. R., & Pathania, M. (2022). An *In-vitro* evaluation of a polyherbal formulation, against SARS-Cov-2. *Journal of Ayurveda and Integrative Medicine, 13*(3), 100581. https://doi.org/10.1016/j.jaim.2022.100581

Karunyal Samuel, J., & Andrews, B. (2010). Traditional medicinal plant wealth of Pachalur and Periyur hamlets, Dindigul district, Tamil Nadu. *Indian Journal of Traditional Knowledge, 9*(2), 264–270.

Marcus, D. M., & Grollman, A. P. (2016). Toxicity of botanical medicines: An overlooked global health problem. *American Journal of Public Health, 106*(1), 16–17. https://doi.org/10.2105/AJPH.2015.302937

Najmi, A., Javed, S. A., Al Bratty, M., & Alhazmi, H. A. (2022). Modern approaches in the discovery and development of plant-based natural products and their analogues as potential therapeutic agents. *Molecules, 27*, 349.

Nasim, N., Sandeep, I. S., & Mohanty, S. (2022). Plant-derived natural products for drug discovery: Current approaches and prospects. *Nucleus, 65*, 399–411. https://doi.org/10.1007/s13237-022-00405-3

Natarajan, S., Anbarasi, C., Sathiyarajeswaran, P., Manickam, P., Geetha, S., Kathiravan, R., Prathiba, P., Pitchiahkumar, M., Parthiban, P., Kanakavalli, K., & Balaji, P. (2021, September 15). KabasuraKudineer (KSK), a poly-herbal Siddha medicine, reduced SARS-CoV-2 viral load in asymptomatic COVID-19 individuals as compared to vitamin C and zinc supplementation: Findings from a prospective, exploratory, open-labeled, comparative, randomized controlled trial, Tamil Nadu, India. *Trials, 22*(1), 623. https://doi.org/10.1186/s13063-021-05583-0.PMID:34526104;PMCID:PMC8441246

Ola, K. (2022, March 6). Role of traditional knowledge in the COVID-19 battle. *The Journal of World Intellectual Property*. https://doi.org/10.1111/jwip.12220

Panda, A. K., Kar, S., Rai, A. K., Rao, B. C. S., & Srikanth, N. (2022). AYUSH- 64: A potential therapeutic agent in COVID-19. *Journal of Ayurveda and Integrative Medicine, 13*(2), 100538. https://doi.org/10.1016/j.jaim.2021.100538

Pirintsos, S., Panagiotopoulos, A., Bariotakis, M., Daskalakis, V., Lionis, C., Sourvinos, G., Karakasiliotis, I., Kampa, M., & Castanas, E. (2022, June 24). From traditional ethnopharmacology to modern natural drug discovery: A methodology discussion and specific examples. *Molecules, 27*(13), 4060. https://doi.org/10.3390/molecules27134060.PMID:35807306; PMCID:PMC9268545

Qureshi, H., Masood, M., Arshad, M., Qureshi, R., Sabir, S., Amjad, M. S., & Tahir, Z. (2015). *Picrorhizakurroa*: An ethnopharmacologically important plant species of the Himalayan region. *Pure and Applied Biology.*

Upadhyay, P., Joshi, B., Sundriyal, A., & Uniyal, S. (2019). *Caesalpinia crista* L.: A review on traditional uses, phytochemistry, and pharmacological properties. *Current Medical and Drug Research, 3.* https://doi.org/10.53517/CMDR.2581-5008.312019191

IPR and Bioprospecting in the Context of Post-CBD Developments

Narendran Thiruthy and Rachel Thomas

Introduction

The relationship between the Convention on Biological Diversity (CBD) and the TRIPS regime is complex and sometimes contentious. The central point of this conflict is the treatment of intellectual property rights (IPRs) related to genetic resources. The CBD recognizes the importance of IPRs. But at the same time seeks to ensure they do not undermine the rights of countries that provide genetic resources. On the other hand, the TRIPS-based IP Regime strongly emphasizes protecting IPRs and has been criticized for not addressing the concerns of developing countries in this area. Many efforts have been made to reconcile the CBD and TRIPS regimes, including establishing the WIPO Intergovernmental Committee on Intellectual Property and Genetic Resources, Traditional Knowledge and Folklore. However, progress has been slow, and the approaches still differ significantly. Digital Sequence Information (DSI), which refers to the dematerialized form of genetic resources, is a new entrant to this discord. The 15th Conference of Parties (COP 15) to the CBD decided that using DSI must result in fair and equitable sharing of benefits. This means that any research or innovation activities involving DSI must comply with the Access and Benefit Sharing (ABS) framework established by the Convention. The ABS framework requires that the benefits of research and innovation activities that use genetic resources be shared fairly and equitably with the countries and communities that provide those resources. The unauthorised utilization of biological resources and knowledge results in misappropriation, infringing the rights of the knowledge holders. The TRIPs and the CBD forums have always witnessed the tussle between developing and developed countries on issues concerning Genetic Resources, Traditional Knowledge, and IPR. The standpoints of both sides have been the subject of academic discourse for a long time. Increased transparency and accountability

N. Thiruthy (✉) · R. Thomas
RGSOIPL, Indian Institute of Technology Kharagpur, Kharagpur, India
e-mail: narendranthiruthy@gmail.com

L. Krishna Panicker et al. (eds.), *Biodiversity and Business*,
https://doi.org/10.1007/978-3-031-71674-4_22

concerning the use of genetic resources is a long-standing demand from developing countries. The impact of the COP 15 decision on regional or national IP regimes is still unclear. However, it is possible that the decision could have implications for how DSI is protected under IP laws in some countries. It may also lead to greater scrutiny of DSI-based patent applications to ensure they do not violate the ABS framework. Some countries have been very proactive in challenging patent applications on the grounds of non-compliance with ABS requirements. The decision on DSI can also result in an escalation of such objections. This paper will examine the steps taken at national and regional levels for implementing the COP 15 decision and the possible intersections it can create with the IP regime.

Intellectual Property Rights and DSI

The developments in biotechnology and the digital revolution have transformed the way genetic information is stored, accessed, and analysed. Digital Sequence Information (DSI) refers to the representation of genetic material in digital form. This includes DNA and RNA sequences stored in digital databases around the world. DSI has now become an essential element for research, especially in commercial R&D.[1] The development of sequencing technologies and increased use of tools like bioinformatics and computational biology have resulted in an explosion of genetic information generated and shared globally. This has led to a scenario where direct access to tangible genetic material is not necessary to engage in industrial R&D. The access and benefit-sharing requirements created under the CBD and national legal frameworks predominantly address issues related to obligations triggered by access to tangible materials. Even these legal frameworks lack effective measures to ensure benefit-sharing from patents obtained on innovations. The lack of consonance between the objectives of two treaty regimes, CBD and TRIPs, has resulted in this remaining unsettled issue even after three decades. One of the numerous obstacles to the successful application of ABS procedures is the traceability of genetic resource utilisation. These issues will multiply significantly with the expanding use of DSI.

Intellectual Property Rights play a crucial role in incentivizing innovations and thereby creating an environment conducive to creativity. Innovations in life sciences especially related to the biotech sector and pharma are mainly protected through patents. Some countries have introduced disclosure requirements, which make it mandatory to disclose the source and geographical origin of genetic resources used in the invention. This requirement strengthens the claims for ABS and serves as the backbone for a transparent IP system. However, implementing such a measure for DSI will be challenging and may be impractical as well. Traceability and tracking of DSI is a significant challenge as identified during COP negotiations. Since genetic data is information by nature, its intrinsic qualities make it nonrival and enable unrestricted faster dissemination of knowledge. The patenting trends in the biotech

[1] Rohden and Scholz (2022).

sector reveal increased use of DSI for research. Even though DSI as such is not patentable, innovations resulting from the use of DSI may be patentable if they satisfy the independent criteria of novelty, inventive step, and industrial application. The mega biodiverse countries in the developing world are concerned about the increasing exploitation of DSI without complying with the ABS principles. The dematerialized form of DSI and its digital nature create challenges for sovereign control and imposition of regulatory norms. This necessitated extensive discussions in the CBD forums to look for ways to effectively regulate the use of DSI.

Cop 15—Negotiating History of DSI

The Convention on Biological Diversity (CBD) was concluded in 1992 as an international framework for facilitating sustainable utilization of biological diversity and ensuring fair and equitable sharing of benefits. Before the CBD, there existed a predominant view that genetic resources were part of the common heritage of mankind and anybody could obtain and utilize the resources as they deemed fit.[2] However, the scenario completely changed when CBD acknowledged the country's sovereign right over genetic resources. The Convention on Biological Diversity (CBD) has three main objectives: the conservation of biological diversity, the sustainable use of the components of biological diversity, and the fair and equitable sharing of the benefits arising out of the utilization of genetic resources. In the new context, the potential advantages of resource consumption provide states with an incentive to preserve resources and encourage sustainable resource usage within their borders. The Nagoya Protocol on Access to Genetic Resources and the Fair and Equitable Sharing of Benefits further streamlined the implementation and provided for a robust regulatory framework to ensure the equitable and impartial distribution of benefits when genetic resources and associated traditional knowledge are utilized. This involves the need to secure the prior informed consent of the local communities and establish mutually agreed terms (PIC and MAT) prior to any use.[3] The post-CBD period saw many national governments enacting domestic legislation to give effect to the principles of CBD. In the initial years, the main challenge for actualizing ABS was the intersection biodiversity governance has with the IP system. The TRIPS agreement resulted in the standardisation of IP norms in both the developed and developing world. The introduction of IP rights to encourage biotechnological innovations resulted in the propertisation of GR and TK. The conflicts between CBD and TRIPS systems created hurdles in effectuating the principles of ABS. However, some governments were able to effectively balance these two interests at the national level through regulatory frameworks. The most significant challenge in the post-CBD period is the

[2] Pisupati and Sathyarajan (2020).

[3] Article 6-Access to genetic resources, Nagoya Protocol 2010. https://www.cbd.int/abs/text/articles/default.shtml?sec=abs06#:~:text=In%20the%20exercise%20of%20sovereign,the%20country%20of%20origin%20of.

technological advancements like synthetic biology and next-generation sequencing (NGS), which have transformed our understanding of the nature of genetic resources. Now, it is possible to develop innovations using 'dematerialized' forms of genetic resources. DNA sequencing technology allows individuals to develop products identical to the original without physically accessing the bioresource, and it is utilized for both materialistic and non-materialistic activities.[4] This has, in effect, removed the need for accessing physical biological resources for research and development of biotechnological innovations. Gene editing technologies like "CRISPR" have further opened the arena for developing new products.[5]

Modern-day genomics research gives greater emphasis to data. The technical developments thus lead to a scenario where research becomes possible without the difficulty of accessing tangible genetic resources. However, accessing a dematerialized form of genetic resource like DSI creates new challenges in regulating access to resources and ensuring biodiversity governance. In the international forum, the discussions on addressing these challenges began as early as 2016 with COP 13. DSI thus became an active subject of discussion in the various forums of the Convention on Biological Diversity (CBD). The COP 13 acknowledged the developing trends in biotechnology and the need to address the issues related to DSI. It was also accepted that a 'coordinated and non-duplicative approach'[6] is required to address the matter, thus leading to the initiation of a fact-finding and scope study to elucidate the concept, terminologies, scope, extent, and condition of DSI usage. The parties also agreed to hold discussions in the Ad Hoc Technical Expert Group on Digital Sequence Information on Genetic Resource, 2018 (AHTEG) and to guarantee the participation of local communities, governments, stakeholders, and corresponding organizations. The discussions were intended to concentrate on the potential implications of using digital sequence information considering the goals of both the CBD and Nagoya Protocol. AHTEG established a group to discuss (a) appropriate terminology, including the technical, legal, and scientific implications of existing words, (b) the potential implications of using DSI on genetic resources for biodiversity conservation and sustainable use of its components and (c) possible effects of DSI on genetic resources for equitable and impartial benefit sharing resulting from the use of genetic resources.

During the initial stage, member countries accepted 'Digital Sequence Information' as a placeholder term for further negotiations. However, there were some disagreements as well over the use of this term. Some alternative words were also suggested that could cover the different information categories related to the use of genetic resources. There was a lack of consensus about the applicability of the CBD framework as well. Some experts took the position that DSI falls under the Nagoya Protocol's definition of a 'genetic resource' because DSI is a part of genetic

---

[4] Prathapan et al. (2018).

[5] Behjati and Tarpey (2013).

[6] DSI on Genetic Resources, thirteenth meeting of COP to CBD, CBD/COP/DEC/XIII/16, dated 16 December 2016, Cancun, Mexico. https://www.cbd.int/doc/decisions/cop-13/cop-13-dec-16-en.pdf.

resources. However, some differed with this argument and took the position that since DSI is intangible and genetic resources are physical substances, they do not fall within the definition.

In the CBD framework, an open-ended intergovernmental scientific advisory body known as the Subsidiary Body on Scientific, Technical, and Technological Advice (SBSTTA) was established under Article 25 of the CBD to provide the Conference of the Parties (COP) timely advice relating to the implementation of the Convention. The SBSTTA has, on various occasions, provided assistance in the technological and scientific assessment of biological diversity and given recommendations for the effective implementation of the CBD. SBSTTA, also provided advice on technology transfer, and answered queries on methodological and scientific inquiries from the COP and other subsidiary bodies.[7] During its 22nd meeting in 2018, the SBSTTA reviewed the recommendations made by AHTEG, and fact-finding and scoping studies on digital sequence information about genetic resources. The role that DSI can play in protecting biodiversity, ensuring public health, and ensuring food security was widely recognised. DSI represents the greater understanding mankind has gained about biodiversity. The SBSTTA acknowledged the necessity of a balanced strategy for free and unhindered access to genetic sequence information as well as equitable and impartial distribution of the advantages that arise from the use of DSI on genetic resources.[8] The SBSTTA recognized that the CBD and Nagoya Protocol principles will ensure that the benefits of the commercial use of digital sequence information are distributed fairly and equitably. The role of domestic laws in governing the use of DSI for non-commercial causes was also highlighted.

The Ad hoc Technical Expert Group met again in Montreal, Canada, in 2020, after COP 14, to discuss the distribution of gains from the use of DSI with reference to genetic resources. The parties agreed on using a 'science and policy-based process' to reach a consensus.[9] The primary concerns while developing a regulatory framework are the challenges related to traceability and the use of DSI in life science research. Finally, it has been recommended that the AHTEG submit its research results and reports to the COP 14 Open-ended Working Group on the post-2020 Global Biodiversity Framework. The working group was given the task of reviewing the Ad Hoc Technical Expert Group's findings and recommending changes to COP 15.

The third meeting of the Open-Ended Working Group on the post-2020 Global Biodiversity Framework, 2021 (Part 1) saw extensive discussions on issues related to DSI. Multiple stakeholders were invited to negotiate on the terminology, scope, and extent of DSI, the need for capacity-building initiatives, and the technology

---

[7] Article 25 of the Convention of Biological Diversity, 1992. https://www.cbd.int/sbstta#:~:text=Article%2025%20of%20the%20Convention,other%20subsidiary%20bodies%2C%20with%20timely.

[8] Subsidiary Body on Scientific, Technical and Technological Advice, CBD/SBSTTA/REC/22/1, Dated 2–7 July 2018 Montreal, Canada. https://www.cbd.int/doc/recommendations/sbstta-22/sbstta-22-rec-01-en.pdf.

[9] Ad Hoc Technical Expert Group on Digital Sequence Information on Genetic Resources, CBD/DSI/AHTEG/2020/1/7, Dated 17–20 March 2020, Montreal, Canada. https://www.cbd.int/doc/c/ba60/7272/3260b5e396821d42bc21035a/dsi-ahteg-2020-01-07-en.pdf.

transfer issues about access, generation, and use of DSI. The parties to the Convention, governmental organizations, aboriginal communities, local people, and other stakeholders became part of the discussions. The findings and recommendations of the OEWG meeting included the need for conservation efforts to be more realistic than the theoretical. OEWG also recommended following an equitable, and impartial methods of access and benefit sharing; the inclusion of additional DSI activities to integrate post-2020 biodiversity conservation activities, and the need to take open access, funding distribution, etc. into account.[10] A few issues about DSI and its governance were not resolved during the OEWG meeting and were agreed to be discussed further in the next meetings.

The third meeting of the Open-Ended Working Group on Post 2020 Global Biodiversity Framework was held in Geneva in March 2022 (Part 2) which also discussed issues related to DSI. The discussions resulted in identifying some indicators of equitable and impartial benefit distribution which include; (a) the creation of more financial and non-financial measures, (b) the absence of obstacles to research and innovation, (c) the preservation of open access to data consistency, (d) harmonization with other international instruments, etc. The OEWG acknowledged that CBD's goals would be met by utilizing incentives that are both financial and non-financial. The meeting also discussed possibilities of establishing a 'Multilateral benefit sharing mechanism'. The funds created in this way will be used to support biodiversity conservation and sustainable use. It was also proposed that the Global Biodiversity Fund will be utilised to provide project-based support for grassroots conservation initiatives undertaken by IPLC.

The fifth meeting of the Open-Ended Working Group on the Post-2020 Global Biodiversity Framework was held in December 2022 in Montreal. The requirement to incorporate measures for creating revenue through equitable and impartial distribution of benefits was discussed.[11] The OEWG welcomed the addition of disclosure requirements on geographic origin in the articles to public databases like INSDC. The consensus in OEWG was that the bilateral benefit sharing will not be a suitable mechanism for DSI because, in the case of a bilateral mechanism, it may not be possible to follow the identifiers for fair and impartial sharing. The OEWG was, therefore, in favour of a multilateral benefit-sharing mechanism. During this discussion, it was decided that no measures should be introduced that would infringe upon the rights of local communities or indigenous people. The OEWG suggested creating a multilateral benefit-sharing mechanism and offered policy solutions for equitable and impartial benefit-sharing from the use of DSI on genetic resources. These recommendations on DSI were placed for further negotiation and decision in the 15th COP of CBD.

---

[10] Open Ended Working Group on the Post 2020 Global Biodiversity Framework, CBD/WG2020/3/4, dated 5 July 2021, Online https://www.cbd.int/doc/c/afd4/4df3/d2d62f5f6a1bfe367c7448f4/wg2020-03-04-en.pdf.

[11] Open Ended Working Group on Post 2020 Global Biodiversity Framework, CBD/WG2020/5/L.1, dated 5 December 2022. https://www.cbd.int/doc/c/948c/0049/b45adb5f28e8b76bb17cda5f/wg2020-05-1-01-en.pdf.

Recognising the Benefit-Sharing Obligation for the Use of DSI

In 2022, the fifteenth meeting of the COP to CBD was held in Montreal, Canada. The COP discussed a wide range of issues about sustainable use and biodiversity conservation. The new and emerging areas like digital sequence information and the need for equitable sharing of benefits were also discussed in the COP. The COP-15 of the CBD thus took a significant step by adopting decision no 15/9 on Digital sequence information on genetic resources. Based on the results of the science and policy-based process, as outlined in COP 14 decision 14/20, the parties at COP 15 reached a final agreement on DSI. The parties had differing opinions during the CBD negotiations about DSI, particularly about its permissible use under the Convention on Biological Diversity. Thus, any solution for the just and equitable distribution of benefits resulting from the use of DSI on genetic resources must complement one another and be flexible enough to work with other tools, as acknowledged by the COP 15 decision. The decision recognises that the creation, availability, and use of DSI on genetic resources, along with the just and equitable distribution of advantages resulting from its application, will encourage innovation and research and help to fulfil the goals of the CBD. The member nations' technical capacities to apply the latest advancements in DSI technology differ significantly from one another. The necessity of technology transfer, capacity-building, and scientific cooperation was thus underlined by the parties to facilitate the creation, availability, and utilisation of DSI on genetic resources. Deposition of DSI-related information in private databases will hinder public access and can have implications on the ability of the developing world to use this information.[12] Thus, there is a need to encourage depositing data in public databases. The tracking and tracing of DSI is not practical and is the biggest hurdle in facilitating ABS. Thus, the COP-15 decision welcomed the efforts by different organisations to encourage the tagging of records with information on geographical origin. During the negotiations, the parties recognised the need to look for innovative solutions for addressing benefit sharing obligations resulting from the utilization of DSI.[13] The benefit sharing from DSI will help in the mobilization of resources to support the conservation and sustainable use of biological diversity. However, there are also several challenges in designing a viable ABS model for DSI. Thus, the parties recognised that a solution on benefit-sharing from the use of DSI may include innovative revenue generation measures. The differences between public and private databases will have to be considered in the development of a solution.

In the COP-15, the parties agreed to continue the use of the term 'digital sequence information' for further discussions. The tracking and tracing of all DSI on genetic resources is not practical. The parties agreed on the need to encourage the depositing of more DSI on genetic resources, with appropriate information on geographical

[12] Rohden and Scholz (2022).

[13] Watanabe (2019).

origin and other relevant metadata, in public databases. As a significant step, the parties agreed that the benefits from the use of DSI on genetic resources should be shared fairly and equitably. However, considering the distribution of DSI on genetic resources and distinctive practices in its use, a distinctive solution is required for benefit-sharing. Thus, the parties agreed to continue the negotiations to develop a solution for the sharing of benefits arising from the use of DSI on genetic resources. The parties identified some criteria that need to be considered while developing the solution for benefit sharing. The identified criteria are (a) Be efficient, feasible and practical; (b) Generate more benefits, including both monetary and non-monetary, than costs; (c) Be effective; (d) Provide certainty and legal clarity for providers and users of DSI on genetic resources; (e) Not hinder research and innovation; (f) Be consistent with open access to data; (g) Not be incompatible with international legal obligations; (h) Be mutually supportive of other access and benefit-sharing instruments; and (i) Take into account the rights of indigenous peoples and local communities, including with respect to the traditional knowledge associated with genetic resources that they hold.

The parties recognised that a multilateral approach to the sharing of the benefits arising from the use of DSI will have the potential to meet these identified criteria. The parties also decided to establish, as part of the Kunming-Montreal Global Biodiversity Framework, a multilateral mechanism for benefit-sharing for DSI with a global fund.[14] It was decided to establish a fair, transparent, inclusive, participatory and time-bound process to further develop and operationalize the mechanism. While designing the ABS model, both the options of monetary and non-monetary sharing shall be considered for supporting conservation and providing benefits to indigenous peoples and local communities. The COP 15 decision also highlights that the approach to fair and equitable benefit-sharing from the use of DSI as set out in the present decision 15/9 shall not affect existing rights and obligations under the CBD and Nagoya Protocol, especially the rights of indigenous peoples and local communities. The decision calls for specific and targeted capacity-building, technology transfer, and scientific cooperation to support developing countries. Even though there was general consensus regarding following a multilateral approach to benefit sharing, parties also agreed to look for scenarios where exceptions could be created. The decision also stresses that the new model shall be without prejudice to national access and benefit-sharing measures.

The COP 15 also saw the member countries adopting the Kunming-Montreal Global Biodiversity Framework (GBF).[15] The Global Biodiversity Framework's primary goals are to stop the loss of biodiversity, restore the ecosystem, and defend the rights of indigenous people. The Global Biodiversity Framework (GBF) includes measures aimed at augmenting the financial ability of developing nations to meet the objectives and recommendations of the framework by 2030. There is a specific

---

[14] Bagley (2022).

[15] COP 15 ends with landmark biodiversity agreement, United Nations Environmental Program, 20 December 2022, Nature Action. https://www.unep.org/news-and-stories/story/cop15-ends-landmark-biodiversity-agreement.

mention of benefit sharing in the context of DSI in Goal C of the GBF. The use of DSI on genetic resources and associated TK should result in both pecuniary and non-pecuniary benefits, which should be shared fairly and equitably as appropriate with IPLC, and which should be substantially increased by 2050, according to Goal C. By giving the providers these advantages, they take part in and support the conservation of biodiversity by adhering to rules governing benefit sharing and access.[16] To improve policy implementation, Goal D of the GBF calls for adequate financial resources, capacity-building programs, technical and scientific cooperation, and the transfer of technologies. It seeks to close the funding gap on biodiversity between developed, developing, and least developed nations. To ensure the fair and equitable sharing of benefits resulting from the use of DSI on GR and related TK, Target 13 of the GBF calls for taking policy, administrative, and legal measures at all levels. It also requires facilitating appropriate access to resources and increasing benefit-sharing by 2030 by international access and benefit-sharing norms. Even though the COP 15 decision stressed the need for a multilateral mechanism for benefit sharing for DSI, the crucial issues like triggering points of benefit sharing, the extent of multilateral benefit-sharing mechanism, non-pecuniary benefits, monitoring, evaluation and review of effectiveness, relationship with Nagoya Protocol and governance of funds needs to be settled through further negotiations.

Multilateral Mechanism for Benefit Sharing from the Use of Digital Sequence Information

In COP 15, the parties decided to follow a multilateral approach to the sharing of the benefits arising from the use of DSI. The establishment of a multilateral mechanism for benefit sharing was also decided as part of the Kunming-Montreal Global Biodiversity Framework. The COP-15 decision proposes to constitute a global fund for this purpose. It was also decided to establish a fair, transparent, inclusive, participatory, and time-bound process to further develop and operationalize the mechanism. However, many aspects of the multilateral system need to be determined through further negotiations. National legislations in many countries provide the option of both monetary and non-monetary benefit sharing. Thus, both these options will have to be considered while developing the multilateral system as well. The formulation of the new system shall be done in such a way that it does not affect the existing rights and obligations under the CBD and Nagoya Protocol, especially the rights of indigenous peoples and local communities.

Article 10 of the Nagoya Protocol provides for the establishment of a multilateral benefit-sharing mechanism to govern genetic resources and traditional knowledge occurring in trans boundary situations or for which it is not possible to grant or obtain prior informed consent. Thus, Article 10 applies to scenarios where prior informed

[16] Kuming-Montreal GBF, COP to CBD, CBD/COP/15/L.25, dated 18 December 2022, Montreal, Canada. https://www.cbd.int/doc/c/e6d3/cd1d/daf663719a03902a9b116c34/cop-15-l-25-en.pdf.

consent can't be granted or obtained.[17] Up until recently, the 'global multilateral benefit-sharing mechanism', as defined by the Nagoya Protocol, was a dormant and non-functional entity. However, in the COP-15, members of the Conference of Parties have resolved to establish a comparable benefit-sharing system for DSI on GR. As a result, the multilateral system is expected to be functional soon. Creating the global multilateral benefits-sharing mechanism under the NP was a major topic of discussion from the COP-MOP 1 in 2014 to the COP-MOP 3 in 2018. A global multilateral benefit-sharing system, as stipulated by Article 10 of the Nagoya Protocol, was considered by the delegates of the most recent COP-MOP meeting in 2022 as well.[18] During the Nagoya Protocol's fourth meeting, the participants welcomed COP 15 decision to establish a multilateral system for benefit-sharing from the use of DSI on GR.[19]

Another international treaty that provides for a multilateral benefits-sharing mechanism is the International Treaty for Plant Genetic Resources for Food and Agriculture (ITPGRFA), 2001 (Seed Treaty). Its goals include protecting farmers' rights and promoting sustainable use, conservation, and exchange of plant genetic resources for food and agriculture, as well as equitable and fair benefit sharing from their use.[20] This treaty is one of the major attempts to forge a global consensus around plant genetic resources that acknowledge farmers as a crucial component for the conservation of biodiversity.[21] Articles 10 to 13 of the Seed Treaty deal with a multilateral benefits-sharing mechanism, which works through a Standard Material Transfer Agreement (SMTA). The contracting parties to the multilateral benefit-sharing system consent to share benefits, facilitate technology transfer, share commercial benefits, contribute to information exchange, and strengthen capacity-building project.[22] The fundamental goals of this Seed Treaty, like those of CBD, are to regulate genetic resources.

A similar mechanism is also there under the World Health Organization, which has developed the Pandemic Influenza Preparedness Framework for the Sharing of Influenza Viruses and Access to Vaccines and Other Benefits (PIP Framework), 2011, a worldwide tool for public health. By establishing a global multilateral benefit-sharing method to manage access and benefit-sharing, this framework enhances the

[17] Article 10 Nagoya Protocol on access to genetic resources and fair and equitable sharing of benefits arising from their utilisation to the Convention of Biological Diversity, 2010. https://www.cbd.int/abs/doc/protocol/nagoya-protocol-en.pdf.

[18] Global multilateral benefits sharing mechanisms, COP to CBD serving as MOP to Nagoya Protocol on access to genetic resources and fair and equitable sharing of benefits arising from their utilisation, CBD/NP/MOP/DEC/4/10, dated 19 December 2022, Montreal, Canada. https://www.cbd.int/doc/decisions/np-mop-04/np-mop-04-dec-10-en.pdf.

[19] Digital Sequence Information on Genetic Resources, COP to CBD serving as MOP to Nagoya Protocol on Access to genetic resources and the fair and equitable sharing of benefits arising from their utilisation, CBD/NP/MOP/DEC/4/6, dated 19 December 2022. https://www.cbd.int/doc/decisions/np-mop-04/np-mop-04-dec-06-en.pdf.

[20] Aubry (2019).

[21] Halewood et al. (2018).

[22] Article 12 of International Treaty on Plant Genetic Resources for Food and Agriculture, 2001. https://faolex.fao.org/treaty/docs/tre000005E.pdf.

GISR.[23] For the exchange of biologics, this framework also follows the Standard Material Transfer Agreement (SMTA).[24] This clause was added to provide benefits to the providers equitably. The revenue gathered will enhance readiness and reactions, particularly in nations with limited resources.[25]

Even though some models, as discussed above, are available, developing a multilateral benefit-sharing mechanism, as suggested by the COP 15 meeting, has several challenges. This needs to be addressed through further negotiations and needs careful consideration of many critical issues.

Issues for Further Consideration

Even though the COP-15 has resulted in the adoption of a decision to create a multilateral benefit-sharing mechanism for DSI, the characteristics of the system need to be formulated through further discussions. The COP decision thus indicates that further negotiations will be required to materialize the multilateral system and make it operational. One of the crucial issues is how far it would be possible to control the use of DSI on GR. Even though sovereign rights over genetic resources were recognised after CBD, the developing world has struggled to effectively regulate their utilization. A part of the problem comes from the fact that tracking and monitoring across several industry sectors is a serious challenge even with regard to tangible resources. The challenge is going to increase by many folds when it comes to regulating DSI. The use of DSI for R&D will ultimately lead to new patent applications being filed which rely on the data downloaded from databases. Patent law in many countries requires full disclosure of source and origin in case a genetic resource is used for developing the invention. The effectiveness of the disclosure requirement in the case of DSI is questionable and the solution for this need to be identified through further deliberations. The character and scope of benefit sharing under the multilateral system also need to be formalized through future negotiations. The COP-15 decision has listed some of the areas that need further consideration in negotiations.

A. **Triggering Point for ABS Obligation**

Under the CBD-Nagoya framework, the nature of a person's use of genetic material determines his obligations. It is the clearly defined triggering points that make a user responsible for taking regulatory approvals and sharing fair and equitable benefits. 'Commercial utilization' is a commonly accepted triggering point for benefit sharing. Some countries like India consider 'research' and 'filing application for

[23] GSIR means, Global Influenza Surveillance and Response System.

[24] PIP framework; for more details. https://apps.who.int/gb/pip/pdf_files/pandemic-influenza-preparedness-en.pdf.

[25] Rourke (2019).

patent' as trigger points for ABS obligations.[26] However, this position is not acceptable for many countries that don't consider patenting as an activity regulatable under the CBD-Nagoya framework. They take the position that the imposition of regulations on patenting will run counter to the objectives of TRIPs. The position that patenting doesn't qualify as a regulatable activity makes it difficult to track the use of genetic resources in inventions. In the context of DSI, it needs to be deliberated as to what types of uses shall constitute triggering points. While the hardline approach is to regulate even research, as in the Indian case, the lenient view is to regulate only commercial use, as done in the EU's approach.[27] The regulation on access to tangible resources for all purposes was justifiable to some extent, as sustainable utilization had to be ensured. However, in the case of DSI, there is no direct conservation threat, and thus, regulating access on the grounds of sustainable utilization becomes unjustifiable. Thus, there exist sound arguments for facilitating open access to DSI for research. However, granting open access to research will have implications for how commercial utilization and patenting are regulated. Regulating the use of DSI in patenting requires strict implementation of disclosure norms. However, the imposition of disclosure norms for tangible resources itself is a contentious issue.[28] In this context, it is doubtful as to how far a disclosure requirement could be introduced concerning DSI. Thus, there is a need to explore the possible challenges and identify solutions through deliberations before the triggering points for ABS obligation with respect to DSI are finalized.

2 Establishment and Governance of Funds

The decision taken at COP 15 provides for establishing a multilateral benefit-sharing fund. This will be an essential requirement for the operationalisation of a multilateral benefit-sharing mechanism. The multilateral system is proposed based on the assumption that obtaining PIC would be difficult as tracking and monitoring is a challenge. But even in the case of the MLS system, the identification of the source is essential for facilitating the sharing of benefits. In the present context, it is undeniable that users are finding it extremely difficult to identify the information source. The MLS system is expected to simplify the procedural hurdles for ABS and thus facilitate easier access to DSI. The multilateral framework needs to be in line with similar systems in other international agreements. A conflict between the provisions will lead to unequal benefit sharing and possibly higher transaction costs.[29]

3 Collection, Management, and Disbursal of Benefit Sharing

Since benefit-sharing collection and management were not covered in the 15th COP meeting, more discussions among member nations are necessary to resolve this issue. In line with the CBD and the Nagoya Protocol, the benefits could be non-monetary

---

[26] Section 3 and Section 6 of the Biological Diversity Act, 2002 (India).

[27] The EU Access and Benefit Sharing (ABS) Regulation (Regulation (EU) No 511/2014).

[28] 2024 WIPO Diplomatic Conference on Genetic Resources and Associated Traditional Knowledge Geneva, May 13–24, 2024 will negotiate on the disclosure requirements.

[29] Sara et al. (2021).

or monetary. The benefits that are obtained, how they are managed, and how they can be dispersed among the provider nations, especially the low and middle-income countries, are all left unclear by the CBD framework and the COP meetings.

4 Options for Non-Monetary Benefit Sharing

The COP 15 decision calls for the sharing of both monetary and non-monetary benefits in the case of DSI. However, the discussions during the negotiations stage highlight the increased scope for non-monetary benefits. Non-monetary measures could include sharing research findings, collaborating and coordinating research projects, and working with the less technologically advanced provider nation on biotechnological research projects. They also include raising awareness and educating people about biodiversity, involvement of the provider country in product development and technology transfer, institutional capacity-building initiatives, and supporting the local economy. Some of the non-monetary options identified have great potential in enhancing the capacity of the developing world to reap the benefits from DSI.

5 Tracking and Monitoring

According to the scientific community, using DSI in research significantly reduces the amount of time needed for the actual sample collection. However, the users of DSI have raised reservations about how the benefits of using DSI in research will be shared. Many consider DSI as a free good and take the position that benefit sharing is not required as there is no physical access to genetic material. The identification of geographical origin for DSI is often impossible and impractical. Thus, it is extremely difficult to identify the actual source of genetic material used for developing DSI. The DSI is often stored in publicly accessible databases that are freely accessible without complex restrictions. The multilateral benefit-sharing mechanism recommended by COP 15 becomes relevant when the source of DSI cannot be identified.

The disclosure requirement in patent law about genetic resources helped a long way in facilitating ABS. Some countries like India require mandatory disclosure of the source and geographical origin of genetic resources. Non-compliance with this provision can lead to invalidation of the patent. However, insisting on a similar requirement for DSI is an impractical and challenging one. The parties need to look for innovative solutions to assist in tracking and monitoring of DSI.

Conclusion

CBD, in its original form, was formulated to protect and conserve genetic resources usually identifiable in tangible form. However, the advancements in the scientific field and biotechnology sector have a significant impact on the regulatory framework for biodiversity. Trends at the global level indicate that researchers, scientists, and multinational corporations favour open access to sequencing data and push for information sharing through public databases. However, the provider group, mainly consisting of indigenous people and the local community, supports the benefit-sharing commitment with regard to DSI. In order to share both monetary and non-monetary benefits with those who provide genetic resources, a systematic method of ABS needs to be implemented. The non-traceability of the country of origin for DSI and the challenges in regulating access and use are the two significant issues that need immediate attention.

The increased use of DSI for patenting is going to cause concerns in the IPR system as well. The countries that have an effective disclosure requirement will have to see whether the same needs to be extended to DSI as well. The tracking and monitoring of DSI will require innovative solutions and improvements in the patent system. There are still unsettled matters like triggering points of obligations, management of the funds, rights of the aboriginal and local people, relation with the Nagoya protocol, etc., that need to be negotiated further and resolved through discussions in CBD forums.

Conflict of Interest The authors have no conflict of interest to declare.

References

Aubry, S. (2019, August 30). The future of digital sequence information for plant genetic resources for food and agriculture, policy and practice review. *Frontiers in Plant Science.*https://doi.org/10.3389/fpls.2019.01046

Bagley, M. A. (2022). "Just" sharing: The virtues of digital sequence information benefit-sharing for the common good. *Harvard International Law Journal, 63*, 1.

Behjati, S., & Tarpey, P. S. (2013, August). What is next generation sequencing? *Archives of Disease in Childhood: Education and Practice Edition, 98*(6), 236–238. https://pubmed.ncbi.nlm.nih.gov/23986538/

Halewood, M., et al. (2018, October). Using genetic sequence information to increase conservation and sustainable use of crop diversity and benefit sharing, biopreservation and biobanking. *National Library of Medicine National Centre for Biotechnology Information.* https://doi.org/10.1089/bio.2018.0043

Pisupati, B., & Sathyarajan, S. (2020). The need for a Nagoya "plus"—Access and benefit sharing in the context of digital sequence information. In T. Ganesh & B. Pisupati (Eds.), *Biodiversity conservation through Access and Benefit Sharing (ABS) Himalayas and Indian Sub-Continent.* Springer.

Prathapan, K. D., Pethiyagoda, R., Bawa, K. S., Raven, P. H., Rajan, P. D., & 172 co-signatories from 35 countries. (2018). *When the cure kills—CBD limits biodiversity research* (pp. 1405–1406). https://www.science.org/doi/full/10.1126/science.aat9844

Rohden, F., & Scholz, A. H. (2022). The international political process around Digital Sequence Information under the Convention on Biological Diversity and the 2018–2020 intersessional period. *Plants, People, Planet, 4*(1), 51–60.

Rourke, M. F. (2019, March). *Access by Design, benefit if convenient: A closer look at the pandemic influenza preparedness framework's standard material transfer agreements national library of medicine.* National Centre for Biotechnology Information. https://doi.org/10.1111/1468-0009.12364. Accessed 28 February 2024.

Sara, R., Hufton, A. L., & Scholz, A. H. (2021, December 22). *Compatible or Incompatible? DSI, Open Access and Benefit sharing*. https://osf.io/preprints/socarxiv/nw8g9/

Watanabe, M. E. (2019). The Nagoya Protocol: The conundrum of defining digital sequence information. *BioScience, 69*(6), 480–480.

Exploring Traditional Knowledge: Bio-Prospecting and Biopiracy in India and Southeast Asian Mega-Diversity Nations

Ghanashyam Sharma and Bharat Kumar Pradhan

Introduction

The indigenous and local communities (ILCs) possess a wealth of undocumented traditional knowledge which can be categorized into traditional medicinal knowledge (TMK), traditional agricultural knowledge (TAK), and traditional ecological knowledge (TEK) based on centuries of living in harmony with nature. Such knowledge is not static but is continuously evolving as they respond to new challenges and needs. It would not be incorrect to opine that the ILCs are the real custodian and holders of traditional knowledge (TK); in other words, traditional knowledge is the identity of the ILCs (Baby & Suriyaprakash, 2022). The scientific application of their plant and animal-based TMK through bioprospecting has led to the discovery of many life-saving and life-changing novel drugs for treating malaria, tuberculosis, cancer, and HIV/AIDS, the treatment of which was once, next to impossible. Bioprospecting, involves the methodical and sustainable exploration of biochemical and genetic resources from nature, followed by their subsequent development into valuable products for trade and commerce. The concept of bioprospecting is rooted in the recognition of the significance of natural product discovery for developing new crops and medicines, often drawing from TK. Nevertheless, bioprospecting has the potential to transform into biopiracy, as it is the act of asserting property rights over biodiversity and its derivatives through intellectual property rights regimes and patents (Shiva, 2001). Bioprospecting can impact any industry reliant on genetic resources for developing commercially viable products for the global market (Kate & Laird, 2020).

G. Sharma (✉)
The Mountain Institute India, Tadong Daragaon, Gangtok 737102, Sikkim, India
e-mail: banstolag@gmail.com

B. K. Pradhan
Sikkim Biodiversity Board, Forest and Environment Department, Government of Sikkim, Deorali, Gangtok 737102, Sikkim, India

L. Krishna Panicker et al. (eds.), *Biodiversity and Business*,
https://doi.org/10.1007/978-3-031-71674-4_23

The discourse on 'Bioprospecting' emerged in the late 1980s or early 1990s, focusing on the search for biological resources that could contribute to conservation and the discovery of beneficial products (Robinson, 2012). This exploration of biodiversity for new resources with social and economic value is conducted by various industries, including pharmaceuticals, agriculture, manufacturing, engineering, and construction (Beattie et al., 2011; Mgbeoji, 2005). Recent years have seen increased attention on bioprospecting which is driven by the growing need for new drugs to address incurable diseases and replace ineffective treatments. As a consequence, multinational corporations from industrialized nations are engaged in the overexploitation of biological resources and associated TK in developing countries. This poses a significant threat to biodiversity and associated TK, notably due to escalating bio-prospecting by entities like pharmaceutical companies in indigenous territories. Today, pharmaceutical bioprospecting is been criticized for 'biopiracy,' as they use and patent the genetic resources and associated TK of the ILCs without their prior informed consent (Beattie et al., 2011; Mgbeoji, 2005). The bioprospecting companies are prospering based on medicinal plants discovered by the ILCs, but those who created this knowledge, the ILCs neither get recognition nor any compensation.

The term biopiracy was coined by Pat Mooney in the 1990s to describe the theft or misappropriation of genetic resources (GRs) and TK through the intellectual property system and the unauthorised and uncompensated collection of genetic material for commercial purposes (Silva, 2020). Robinson (2010) defined biopiracy, as "the unauthorised extraction of biological resources and/or associated TK from developing countries, or the patenting of spurious inventions based on such knowledge or resources without compensation". Jamil (1998) defined biopiracy as "the manipulation of intellectual property rights laws by corporations to gain exclusive control over national genetic resources, without giving adequate – if any – recognition or remuneration to the original possessors of those resources". As per Robinson (2012), there are two distinct categories of biopiracy. The patent-based biopiracy involves securing patents for (often questionable) inventions derived from biological resources or TK, obtained without proper authorization and benefit-sharing from other (typically developing) countries or indigenous communities. Non-patent biopiracy encompasses other forms of intellectual property control related to biological resources or TK, similarly acquired without appropriate authorization and benefit-sharing from other (usually developing) countries or indigenous communities.

Biopiracy is historically associated with colonialism which exploited the valuable commodities from resource-rich countries leading to environmental destruction (Silva, 2020). In their quest for treasure, the European explorers navigated both east and west and discovered medicinal plants as the unexpected treasures. In his travelogue "The Customs of the Kingdoms of India," Marco Polo (1254–1324) writes, *"We shall tell you next of the great kingdom of Malabar [Southwestern India] where there is a great abundance of pepper and ginger, alongside plentiful cinnamon and coconuts"* (Polo & Latham, 1958a, 1958b). The travel route of Marco Polo was instrumental in the development of the Spice Route (Robinson, 2010) which was later utilized in bio-colonialism that began in 1400 AD and peaked in the sixteenth and seventeenth centuries (Whittle, 1970). Following the Spice Route,

Christopher Columbus voyaged across the Atlantic Ocean explored new routes to the east, and obtained plant-based wealth of knowledge. During his voyage, Columbus also collected many valuable plants including Tobacco from Cuba (Charlton, 2004).

Today, the term biopiracy is used to denote the illegal or improper appropriation of TK and biological materials. It has emerged as a 'silent disease', which is hardly detectable. This silent pillaging is depriving countries like Africa, Latin America, and Asia which lack proper advancement in biotechnology (Yoonus et al., 2021). Biopiracy, actually is disrupting global biodiversity conservation efforts (Dunagan, 2009; Ramirez Garcia, 2014; Sharma et al., 2018) and is depriving the indigenous communities by not only excluding them from global markets but also allowing corporations to have legitimate control over scarce biological resources and associated TK (Oli & Sharma, 2010; Singh, 2003). Today biopiracy is flourishing under capitalism and globalization. Addressing biopiracy, safeguarding biodiversity, and promoting sustainable practices are recognized as significant challenges for the twenty-first century (Ahoyo et al., 2021; Banerjee, 2019).

Global Biopiracy

The covert and elusive nature of biopiracy, often referred to as "a silent disease," stems from its lack of frequent traces, thereby erasing local identity. On the other hand, giant enterprises patent indigenous medicinal plants, seeds, GRs, and traditional formulae employing intellectual property rights which have led to instances of biopiracy (Yoonus et al., 2021). Notable examples from the Indian context include the transfer of basmati rice varieties to boost the rice economy of the United States and the export of neem seeds from Indian farms by major corporations, as emphasized during protests in the late 1990s (Shiva, 2020; Singh, 2003).

The issuance of patents for non-patentable knowledge, particularly in the realm of traditional medicines, whether rooted in existing TK from the developing world or representing minor variations thereof, raises significant concerns for developing nations. Table 1 underscores instances of bio-piracy of TK, with many countries having to engage in protracted battles for the revocation of granted patents. However, revocation may not be a viable option for all patents associated with TK due to substantial costs and time constraints (https://www.tkdl.res.in/tkdl/langdefault/common/BioPiracy.asp?GL=).

The extensive TK, particularly in the treatment of various diseases, represents a valuable resource vulnerable to misappropriation due to its significant potential. Ongoing biopiracy of codified TK persists, primarily because this information is available in local languages, creating a language barrier that hinders patent offices from effectively searching it as prior art before granting patents. The formulations derived from TK for treating human ailments have stood the test of time, having been in practice for centuries. The reliability of traditional medicine systems, combined

Table 1 Global biopiracy incidents

Country of origin	Incident	Country (Company) involved in biopiracy
India	Neem (*Azadirachta indica*)	USA (US-American firm W.R. Grace and US Department of Agriculture; (*US Patents Nos. 5420318, 5391779 and 5371254*)
	Turmeric (*Curcuma longa*)	USA (*US Patent 5401504*)
	Basmati rice (*Oryza sativa*)	*US Patent No. 6274183 and 5663484*
	Commiphora mukul	*USA (US Patent No. 6113949)*
China	*Momordica charantia*	USA (*US Patent No. 5484889*)
	Camptotheca lowreyana	USA (*US Patent No. PP11959*)
Sri Lanka	Heen Bavila (*Sida cordifolia*)	Japan
	Komarika (Aloe vera)	USA
	Kothala Himbutu (*Salacia reticulata*)	Japan, USA, Europe
	Magul Karanda (*Pogamia glabra*)	Japan
	Masbadda (*Gymnema sylvestre*)	Japan
	Weniwalgata (*Coscinium fenestratum*)	Japanese, European, and USA pharmaceutical manufacturers
South Asian countries including Sri Lanka	Karawila (*Momordica charantia*)	USA
Amazon Basin	Ayahusca (*Banisteriopsis caapi*)	USA (Plant Medicine Corporation); (*US patent PP 05751*)
Philippines	Banaba (*Lagerstroemia speciosa*)	Japan (Itoen KK) and USA (*US Patent No. 5980904*)
	Ilang-ilang (*Cananga odorata*)	France (Yves St. Laurent);*N/A but used by several perfumeries in France*
Mexico	Enola Bean (*Phaseolus vulgaris*)	USA (POD-NERS LLC); (*US patent no. 5894079*)
South Africa (San indigenous people)	Hoodia plant (*Hoodia gordonii*)	South Africa (South African Council of Scientific and Industrial Research), UK (*WO 9846243*)
Australia	Kakadu Plum (*Terminalia ferdinandiana*)	USA
Pacific, especially in Fiji and Vanuatu	Kava (*Piper methysticum*)	France (L'Oreal); USA (*US Patents Nos. 6405948, 6277396, 6080410, 6025363, etc.*)

(continued)

Table 1 (continued)

Country of origin	Incident	Country (Company) involved in biopiracy
Thailand	Ma-rakheenok (*Momordica* spp.)	USA
	Plao-noi (*Croton sublyratus*)	Japan (Sankyo)
Andes, especially Quechua and Aymara people	Quinoa (*Chenopodium quinoa*)	USA (*US patent no. 5304718*)
Samoa	*Homalanthus nutans*	USA (*US Patent No. 5599839*)
Brazil	Rubber seeds (*Hevea brasiliensis*)	UK

(Afreen & Abraham, 2008; Avantika et al., 2015; Chang et al., 1996; Dwivedy et al., 2019; Grain, 2006; Rattray, 2002

with the lack of this information within patent offices, creates an opportune situation for individuals to obtain patents on therapeutic formulations originating from traditional medicine systems (ENVIS, 2011).

During the examination of patentability for any claimed subject matter, patent examiners in the international patent office utilize available resources to search suitable non-patent literature sources. Patent literature, typically consolidated in distinct databases, is more efficiently searchable and retrievable, while non-patent literature prior art is often scattered across numerous and diverse sources. Consequently, there is a recognized need to establish easily accessible non-patent literature databases specifically focused on TK.

In the realm of intellectual property law, the issuance of patents for naturally occurring GRs and TK from developing countries has given rise to a contemporary form of colonization. Unlike past invasions of land, the current peak of invasion pertains to life forms through genetic engineering, involving the unauthorized utilization of TK from less affluent nations. It is unfortunate that Intellectual Property Rights (IPRs) do not require the holder to share benefits, with the provider of genetic resources and/or TK or to recognize any contribution of indigenous peoples and local communities leading to the misappropriation of TK (Aguilar, 2001). This presents a significant concern for developing countries which have a long tradition of using traditional medicine. The biotechnology revolution has resulted in the colonization of entities that are inherently "independent, self-regenerative, and freely available" (Shiva, 1997). Biotechnology has become a tool to legally exploit the genetic resources and the associated TK and gain exclusive control over it through patents thereby ignoring the contributions of the ILCs as well as threatening the traditional rights of the original possessor of the resource and the associated knowledge.

Haldi, Turmeric (*Curcuma longa*); Photo: Ghanashyam Sharma

Neem (*Azadirachta indica*); Photo: India Biodiversity Portal

HeenBavila (*Sida cordifolia*); Photo: India Biodiversity Portal

Ghritakumari (*Aloe vera*); Photo: Ghanashyam Sharma

KothalaHimbutu (*Salacia reticulata*); Photo: https://www.fortheloveofnature.in/products

Magul Karanda (*Pogamia glabra*); Photo: https://www.flowersofindia.net/catalog/slides/Pongam%20Tree.html

Masbadda (*Gymnema sylvestre*): India Biodiversity Portal

The Enola Bean is an alleged case of biopiracy, where Larry Procter, a Colorado executive in the bean industry cultivated yellow beans he bought in Mexico on vacation for which he received a US patent two years later on all yellow beans of this variety. Larry's company, Pod-Ners, admits that its Enola bean, (named after Larry Proctor's wife), is a descendant of the traditional Mexican bean from the Andes, the Mayacoba, but that it has a better yellow color and a more consistent shape. By obtaining a patent and a U.S. Plant Variety Protection Certificate, he secured what amounted to a legal monopoly over yellow beans sold in the United States. Under the terms of the patent, he can therefore sue anyone in the United States who sells or grows a bean that he considers to be his particular shade of yellow. Procter also profits from yellow beans imported from Mexico by imposing on them a six-cent-per-pound royalty. As a result, both farmers in the United States and particularly in Northern Mexico have suffered great economic hardship. The case has stimulated great debate over whether TK and/or GRs should be patentable in the first place. As the number of patents filed by large corporations for native crops increases, activists become more concerned about the adverse effects of these patents on developing countries and particularly indigenous people.

Weniwalgata (*Coscinium fenestratum*); Photo: https://herbalplantslanka.blogspot.com

Karawila (*Momordica charantia*); Photo: Ghanashyam Sharma

Ayahusca (*Banisteriopsis caapi*); Photo: Courtesy of Rafael G. dos Santos

Banaba (*Lagerstroemia speciosa*)

Ilang-ilang (*Cananga odorata*); Photo: India Biodiversity Portal

Enola Bean (*Phaseolus vulgaris*); Photo: Ghanashyam Sharma

Hoodia plant (*Hoodia gordonii*); Photo: https://pza.sanbi.org/hoodia-gordonii

Kakadu Plum (*Terminalia ferdinandiana*); Photo: https://www.google.com/

Kava (*Piper methysticum*); Photo: https://ntbg.org/database/plants/detail/piper-methysticum

Ma-rakheenok (*Momordica* spp.); Photo: Ghanashyam Sharma

Plao-noi (*Croton sublyratus*); Photo: https://www.google.com/

Rubber (*Havea brasiliensis*)

Quinoa (*Chenopodium quinoa*); Photo: Ghanashyam Sharma

Legal Frameworks for Protection Against Biopiracy

Recent years have seen increasing attention given to the protection of TK of the ILCs (Table 2). There are multilateral systems in place to check biopiracy, but some of those regulations have pitfalls that can increase the risk of biopiracy. For example:

- **The United Nations Convention on Biological Diversity (CBD)**: The CBD provides a greater opportunity for the resource-rich countries to conserve their biological diversity, protection of associated TK, and the necessity for equitable sharing of benefits. However, it leaves a window open to individual parties, concerning interpretation of its provision (Aguilar, 2001).
- **The Bonn Guidelines**: It is the most detailed ABS guidelines at the international level that were drafted before the Nagoya Protocol came into force. Although this seems to be a beneficial approach in terms of the national ABS frameworks' compatibility with the international ABS principles; however, neither the regulators nor the users exactly know what actions some of these benefit-sharing options entail (Sirakaya, 2019).
- **The Nagoya Protocol**: Although the Nagoya Protocol covers important issues like global biodiversity, environmental and health justice, and international IPR Frameworks, they fail to address the issues of representation, equitable compensation, and drug access which are key concerns for indigenous communities (*novotech-cro.com/faq/bioprospecting-and-biopiracy*).
- **Trade-Related Aspects of Intellectual Property Rights (TRIPS)**: At present TRIPS provides limited options for developing countries. While it ostensibly establishes a common global standard on intellectual property rights, in reality, its provisions accommodate the specific levels of industrialization and development that characterize developed countries. Consequently, the IPR regimes and standards created end up protecting the corporate interests of the developed countries (Jamil, 1998).

Globally, numerous countries are enacting legislation to regulate access to their biological and genetic resources for bioprospecting purposes. The primary legal concern revolves around the absence of globally recognized policies regulating the nexus among intellectual property, TK, GRs and biodiversity. Existing environmental, trade, and region-specific agreements present provisions that are either inadequate, vague or in conflict concerning bioprospecting activities. As a result, there is a dearth of explicit regulations governing ownership, access, benefit-sharing, and environmental stewardship in the context of TK bioprospecting. The absence of precise rules and guidelines, coupled with evident loopholes in current legislation, creates an environment conducive to biopiracy, as corporations exploit the fragmented legal frameworks and policies for their commercial gains. Efferth et al. (2016) contended that when pharmaceutical companies patent herbs or natural products, they overlook the indigenous communities' knowledge of these substances and how they function. Despite multiple court rulings in the U.S. and Europe, as well as various international agreements (e.g., from the United Nations, World Health

Table 2 Instruments within the international legal framework that operates to safeguard TK, either directly or indirectly

Institutions and organizations	Primary objective
WIPO—The IGC on Intellectual Property Genetic Resources and Traditional Knowledge, 2014	A global platform for addressing intellectual property matters concerning access to genetic resources, equitable benefit sharing, and safeguarding and expressions of folklore
Convention on Biological Diversity, 1993	Pioneering global effort to acknowledge: a. the preservation of biological diversity, b. the sustainable utilization of its elements, and c. the just and fair sharing of benefits derived from its exploitation
The Nagoya Protocol, 2010	Facilitate fair and equitable sharing of benefits derived from genetic resources, while ensuring appropriate access to genetic resources and technology transfer for the protection of biodiversity
ITPGRFA, 2004	International Treaty on Plant Genetic Resources for Food and Agriculture acknowledges the right of farmers to save, use, exchange, and sell farm-saved seed through the establishment of a multilateral system
Doha Ministerial Declaration, 2001	Analyze the connection between the Trade-Related Aspects of Intellectual Property Rights (TRIPS) Agreement and CBD, as well as the safeguarding of TK and folklore
The International Union for the Protection of New Varieties of Plants (UPOV) Convention	Granting exclusive rights on the production, sale, and trade to safeguard new plant varieties
UN Declaration on Rights of Indigenous Persons, 2007	Highlights the importance of preserving the rights of indigenous individuals to uphold and reinforce their institutions, cultures, and traditions to promote their development

(continued)

Table 2 (continued)

Institutions and organizations	Primary objective
Regional initiatives	
a. Swakudump Protocol	Establishes provisions for the rights and acknowledgment of TK holders, introduces the concept of equitable benefit sharing, and outlines licensing mechanisms
b. Cusco Declaration, 2002	The Cancun Declaration of February 18, 2002, resulted in the establishment of the 'Group of Like-Minded Megadiverse Countries.' Bolivia, Brazil, China, Colombia, Costa Rica, Ecuador, the Philippines, India, Indonesia, Kenya, Malaysia, Mexico, Peru, South Africa, and Venezuela collaboratively formed this mechanism to facilitate consultation and cooperation. The primary goal is to advance shared interests and priorities concerning the conservation and sustainable use of biological diversity

Organization, World Trade Organization, African Unity, and others), a fair distribution of benefits between pharmaceutical producers and indigenous communities remains far from being achieved. Efferth et al. (2016) highlighted that their responsibilities extend beyond the norm, stressing the importance of involving scientists from the countries of origin when medicinal and aromatic plants from developing nations are studied in industrialized countries' laboratories. Their participation in the scientific progress concerning these plants is essential. Many laboratories globally have already embraced this approach.

Policies and Initiatives for Preventing Biopiracy in India

The Indian subcontinent, a megadiverse region covering only 2.4% of the world's land area, is home to 7–8% of all recorded species, including over 148,600 species of plants and animals. The country's varied physical landscapes and climatic conditions have created a wide range of ecosystems, such as forests, wetlands, grasslands, deserts, mangroves, and coastal and marine areas, all of which support rich biodiversity and enhance human well-being. India is also home to four of the 35 globally recognized biodiversity hotspots: the Himalayas, the Western Ghats, the North-East, and the Nicobar Islands (http://nbaindia.org/uploaded/pdf/IDB_ABS.pdf). India is recognized as one of the world's megadiverse countries, containing nearly 7–8% of all recorded species and featuring four of the 34 globally identified biodiversity hotspots (Himalayas, Indo-Burma, Western Ghats and Sri Lanka, Sundaland). The country is also a significant repository of traditional knowledge related to biological resources. To date, over 91,200 animal species and 45,500 plant species have been documented across its ten biogeographic regions, with continuous surveys and exploration leading to new discoveries. Alongside species richness, India boasts high levels of endemism, ranking tenth globally for birds (69 endemic species), fifth for reptiles (156 species), and seventh for amphibians (110 species). Endemism is particularly high in amphibians (61.2%) and reptiles (47%). India is also one of the eight Vavilovian centers of origin and diversity for crop plants, with more than 300 wild ancestors and close relatives of cultivated plants, still evolving naturally (www.cbd.int/countries/profile?country=in). No other country than India has vast experiences of misappropriation of its bioresource-based TK by corporate entities and foreign research institutions. Numerous patents have been granted to inventions and discoveries on TK associated with biological resources from India (Table 3).

Aware of the fact that biopiracy is a complex issue and involves multiple stakeholders, including ILCs, corporations, and government, India has developed a legal framework and taken many initiatives to address the issue of biopiracy (Table 4). These laws provide a legal mechanism to protect TK and biodiversity and ensure that benefits derived from the use of biological resources and associated TK is shared in fair and equitable manner with the ILCs (Sen, 2023).

Table 3 Examples of biopiracy of genetic resources and traditional knowledge from India

Name of biological resource	Botanical name	Patents granted
Neem	*Azadirachta indica*	European Patent
Basmati Rice	*Oryza sativa*	US patent 5663484
Amla	*Phyllanthus emblica*	USPTO has granted five patents; four patents filed by Japanese Patent Office
Kala Jeera	*Cuminum cyminum*	US 5653981
Ghiu Kumari	*Aloe barbadensis*	US 5652265
Amaltas	*Cassia fistula*	US 5411733
Pomegranate	*Punica granatum*	US 5411733
Harad	*Terminalia chebula*	US 5529778
Aswagandha	*Withania somnifera*	US 5466452
Turmeric	*Curcuma longa*	US patent 5401504
Wheat	*Triticum aestivum*	EP 0445929 B1

TKDL Outcome Against Biopiracy

Initially, an estimated 2,000 patents associated with Indian medicinal systems were being granted globally each year, leading to the unjust appropriation of knowledge and restricting Indian producers in foreign markets. It was because India's traditional medicinal knowledge existed in languages such as Sanskrit, Tamil, Telugu, Kannada, Malayalam, Hindi, Assamese, Nepali, Arabic, Urdu, and several others and often in ancient local dialects that are no longer in practice which made it difficult for the patent examiners at patent office worldwide from effectively searching it as prior art before granting patents.

In response, TKDL was conceptualized to overcome the language barrier by systematically and scientifically converting and structuring the available contents of the ancient texts—currently focussed on Indian Systems of Medicines i.e. Ayurveda, Siddha, Unani, and Sowa Rigpa as well as Yoga practices—into five international languages, namely, English, Japanese, French, German, and Spanish. As of date, more than 4.54 lakh formulations/practices have been transcribed into the TKDL database. So far, 324 patent applications have either been set aside/withdrawn/amended, based on the prior art evidence present in the TKDL database with minimal costs and without expending undue time and effort. Today, TKDL serves as a collective resource for managing intellectual property rights (Bala & Verma, 2020; UNEP, 2018). Public access to the TKDL database is limited to prevent the commercial exploitation of Indian TK. It has proven to be an effective deterrent against biopiracy and has set a benchmark in TK protection around the world (www.csir.res.in/documents/tkdl#).

Table 4 Legislation and initiatives to prevent biopiracy in India

Sl. No.	Legislation/initiatives	Objectives
1	Protection of Plant Varieties and Farmer's Rights Act, 2001 (PPVFRA, 2001)	This act recognizes the contribution of farmers and local communities in the conservation and development of plant genetic resources. It provides for the registration of plant varieties and farmers' rights over their TK and resources. The act also includes provisions for compensation to farmers and local communities for the use of their resources. Additionally, it promotes seed development, facilitating commercialization by farmers (FRA 2001; GOI 2016)
2	1. Biological Diversity Act, 2002 2. Biological Diversity (Amendment) Act, 2023	This act aims to conserve India's biological diversity and ensure equitable sharing of benefits arising out of the use of biological resources. The act provides for access and benefit-sharing agreements, requiring individuals or organizations to obtain prior informed consent from local communities before accessing biological resources or TK. The act also establishes the National Biodiversity Authority (NBA), State Biodiversity Boards (SBBs) and Biodiversity Management Committees (BMCs) to regulate access to biological resources and ensure the equitable sharing of benefits. The Act also warrants individuals or entities to obtain consent from NBA before applying for intellectual property, both within and outside India, for any invention based on research or information concerning a biological resource obtained from India or repositories outside India
3	1. Guidelines on Access to Biological Resources and Associated Knowledge and Benefits Sharing Regulations, 2014 2. Revised (Draft) Guidelines on Access to Biological Resources and Associated Knowledge and Benefits Sharing Regulations, 2019	The National Biodiversity Authority has developed guidelines for ABS agreements, providing a framework for the protection of TK and the equitable sharing of benefits. The guidelines provide for the identification of local communities and their TK, negotiation of benefit-sharing agreements, and monitoring and enforcement of agreements
4	Traditional Knowledge Digital Library (TKDL)	The TKDL is a database of TK and resources, developed to prevent the misappropriation of TK by patent offices around the world. The database contains information on TK related to medicinal plants, yoga, and Ayurveda

Kani Model of Access and Benefit Sharing in India

The Kani tribe, residing in the Southern Western Ghat region of Kerala State in India, traditionally uses *Trichopus zeylanicus* var. *travancoricus*, known as '*Arogya-pacha*,' to treat fatigue and stress. The proactive contribution of this tribal community has led to the creation of a scientifically validated drug called "*Jeevani*" by the Tropical

Botanic Garden and Research Institute (TBGRI). In the process of transferring the technology for drug production to a pharmaceutical company Arya Vaidya Pharmacy, TBGRI committed to sharing the license fee and royalty with the Kani tribe on a fifty-fifty basis, marking the world's first benefit-sharing model (Anuradha, 2008). However, the Kani case has faced criticism regarding whether the commercialization obtained informed consent from the tribal community and whether financial benefits were shared equitably. Despite being the first of its kind, this benefit-sharing model has been scrutinized for not achieving the intended outcomes (Chaturvedi, 2008; Shiva, 2007; *The Hindu*, 2012).

Arogyapaccha Trichopus zeylanicus var. *travancoricus; Photo: India Biodiversity Portal*

Policies for Preventing Biopiracy in Malaysia

Malaysia is ranked 12th in the world on the National Biodiversity Index, reflecting its rich biodiversity and high endemism across terrestrial vertebrates and vascular plants. Despite rapid economic development, largely driven by the exploitation of natural resources, about 60% of Malaysia's land area remains forested, including permanent reserves, national parks, and wildlife sanctuaries. This supports Malaysia's pledge to maintain at least 50% forest cover, made at the 1992 Rio Earth Summit. Additionally, 10.6% of the country's land is designated as terrestrial protected areas. Malaysia hosts an estimated 15,000 species of vascular plants, 306 mammals, 742 birds, 242 amphibians, 567 reptiles, over 449 freshwater fish, more than 500 marine fish, and over 150,000 invertebrate species (https://www.cbd.int/countries/profile?country=my). Being one of the megadiverse countries, Malaysia is a party to various biodiversity related multilateral agreements like CBD, CITES, Ramsar Convention, Nagoya Protocol, etc. It signed CBD in 1992 and became a party to it in 1994. Subsequently, a countrywide survey on biodiversity was conducted and produced in the form of the document "Assessment of biological diversity in Malaysia" in 1997 (Baskaran et al., 2003). Following this, it developed and launched the National Policy on Biological Diversity in 1998 and formulated National Policy on Biological Diversity 2016–2025, Goal 4 of which emphasizes working towards institutionalising a national regulatory framework in line with CBD and Nagoya Protocol on access and benefit sharing, taking into account the rights of the ILCs. The enactment of Access to Biological Resources and Benefit Sharing Act 2017 (Act 795) guided by Access to Biological Resources and Benefit Sharing Regulations 2020 demonstrates Malaysia's commitment to safeguarding its rich biodiversity, promoting its sustainable use, ensuring fair and equitable sharing of benefits from the utilization of biological resources (BRs) and protecting the TK of the ILCs, as well as their innovations and practices associated with biological resources by bringing it under the ABS framework.

Act 795 regulates access to BRs and associated TK for research and commercial purpose and provide a mechanism for sharing of benefits arising out of the use of the BRs and associated TK in a fair and equitable manner thereby contributing to the conservation of biological diversity and the sustainable use of its component and also to the welfare of the ILCs.

As per the Act, any local or foreign individual or corporation intending to access BRs or associated TK for commercial or potential commercial or non-commercial purposes must obtain permission from the competent authority (CA). The application should be accompanied by the prior informed consent (PIC) of the ILCs and benefit sharing agreement with them or the resource provider, based on mutually agreed terms (MAT). No such permit/agreement is required for accessing the BRs or associated TK for non-commercial purposes if (1) the research is being done under or the exchange of BRs is between persons within a public higher education institution, research institution, or Government agency within Malaysia, or (2) the access to a BR is by any person outside Malaysia or in a private institution within Malaysia from a permit holder who possesses a valid permit to access to carry out or continue any research for non-commercial purpose.

The Act restricts (1) the permit holders from transferring the BRs or associated TK or the result of research to a third party without the permission of the CA, (2) the permit obtained to access BRs or associated TK for non-commercial purposes to be used for commercial purposes without the approval of the CA and without entering into benefit sharing agreement. It further necessitates the individual to inform the CA regarding the filing of patent applications within and outside Malaysia, on the BRs or associated TK accessed. It is obligatory on the part of the applicant to pay a percentage of monetary benefit derived, from the Government, as determined and the non-monetary benefit are to be used for the welfare of the ILCs.

The ABS Act is implemented by the National Competent Authority (NCA) at the Federal level and Competent Authorities (CA) at the State level under the scientific, technical, and legal guidance of the Advisory Committee. The CA constitutes an Advisory Body comprising representations from the ILCs, to deal with matters relating to ILCs and their TK with respect to the use of BRs.

The NCA acts as the national agency for implementing ABS law, maintaining the national clearing house mechanism, establishing measures for monitoring and tracking of BRs accessed to ensure resource users comply with the regulatory process, determining application fees, supporting customary laws and practices of ILCs, develop community protocols and ABS agreement, and maintain register of permits issued by the CAs. The NCA also functions as a CA for all *ex-situ* collections where the origin of BRs cannot be ascertained or in the case where access to BRs and associated TK doesn't fall under the jurisdiction of any other CA. On the other hand, the CA functions under the NCA and processes applications for access of BRs and associated TK for research and commercial purposes in addition to collecting fees, record-keeping, and monitoring compliance.

Interestingly, the States of Sabah and Sarawak in Malaysia enacted their biodiversity law which incorporated access and benefit-sharing about genetic resources (Singh, 2012) way before the Central Act came into existence. The various laws enacted by the State of Sarawak are:

1. Sarawak Biodiversity Centre Ordinance 1997
2. Sarawak Biodiversity Centre (Amendment) Ordinance 2003
3. Sarawak Biodiversity Centre (Amendment) Ordinance 2014
4. Sarawak Biodiversity (Action, Collection and Research) Regulation 1998
5. Sarawak Biodiversity (Action, Collection and Research) Amendment Regulation 2001
6. Sarawak Biodiversity Regulations 2004
7. Sarawak Biodiversity Regulations 2016
8. Sarawak Biodiversity Regulations (Amendment) 2018.

Pursuant to Ordinance 1997, Sarawak Biodiversity Council was established in 1998 which served as a competent authority to regulate access to genetic resources for commercial and non-commercial purposes and protect biodiversity. Subsequently, the Sarawak Biodiversity Centre was established in 1998 to assist the Council in the implementation of the legislation. Later, Ordinance 1997 was amended in 2003 and subsequently in 2014 which took into account the TK of the ILCs and the need to obtain their PIC. Similarly, Regulation 1998 amended in 2001 was revoked and replaced by the Sarawak Biodiversity Regulations 2004 (Nordin et al., 2012) and subsequently by Sarawak Biodiversity Regulations 2016 and the amended Regulations 2018.

Similarly, the various laws enacted by the State of Sabah are:

1. Sabah Biodiversity Enactment 2000
2. Sabah Biodiversity (Amendment) Enactment 2017
3. Sabah Biodiversity (Amendment) Bill 2023 (Bill passed)
4. Sabah Access to Biological Resources and Benefit Sharing Regulations (Draft 2018).

Sabah Biodiversity Enactment (SBE) 2000 covers access to BRs and associated TK for research and development activity. The Enactment mandates the applicant to obtain PIC from the resource provider or the ILC and undergo a benefit sharing agreement based on MAT. To implement the Enactment, Sabah Biodiversity Council was established which acts as a state's competent authority to regulate access to Sabah's BRs and associated TK. Additionally, it oversees Sabah Biodiversity Centre established in 2008, which deals with the applications for access to/ensures fair and equitable sharing of benefits arising out associatedof the use of the BRs and associated TK, and maintains records of applications, related decisions, and permits issued. It also acts as the secretariat and implementer of all the decisions made by the Council.

The SBE 2000 was amended in 2017 which made it necessary for users of BRs to share the benefits arising from the use of BRs and associated TK for research and development, in a fair and equitable manner with the resource owner and the knowledge holders. Subsequently, the state drafted Sabah Access to Biological Resources and Benefit Sharing Regulations in 2018. At the latest, the Bill 2023 to amend SBE 2000 was passed to empower conservation activities, sustainable use, management, enforcement, and matters related to ABS including TK.

The State of Sabah, to conserve the rich biological diversity and to ensure sustainable use of its components for the continued progress and socioeconomic development of the state and well-being of the its people, prepared the Sabah Biodiversity Strategy 2012–2022 that builds upon the Sabah Conservation Strategy 1992.

Policies for Preventing Biopiracy in the Philippines

The Philippines is one of the world's 18 mega-biodiverse countries, holding two-thirds of Earth's biodiversity and 70–80% of its plant and animal species. It ranks fifth in plant species diversity and maintains 5% of the world's flora. Endemism is notably high, with at least 25 plant genera and 49% of terrestrial wildlife unique to the country, making it fourth globally in bird endemism. The Philippines is also a biodiversity hotspot, home to over 700 threatened species, making it a key area for global conservation. The 2004 national list of threatened faunal species includes 42 land mammals, 127 birds, 24 reptiles, and 14 amphibians. The country also has around 3,214 fish species, of which 121 are endemic and 76 are threatened. In 2007, the Department of Environment and Natural Resources established a list of threatened plants, with 99 species critically endangered, 187 endangered, 176 vulnerable, and 64 others threatened (https://www.cbd.int/countries/profile?country=ph). The Philippines has a long history of regulating bioprospecting and access and benefit sharing. It was one of the earliest signatories to the CBD (ratified in 1993) and the first country in the world to enact national ABS policy in 1995 through Executive Order (EO) 247. It regulates access to bio-genetic resources, their by-products and derivatives, and associated TK for scientific research and commercial purposes by local and foreign individuals or entities. To implement the EO 247, guidelines on regulating research and bioprospecting of biological and genetic resources were issued by the Department of Environment and Natural Resources (DENR) vide Department Administrative Order No. 96-20. The Philippines also signed the Nagoya Protocol in 2010 and acceded to it in 2015.

The EO 247 mandates the resource user to execute bioprospecting undertaking (BU) with the Philippines Government, obtain PIC from the ILCs by their customary law, and undergo benefit sharing agreement. All benefits resulting directly or indirectly from bioprospecting activities are to be shared equitably and fairly between the Philippines Government, the concerned communities, and the researcher. The EO necessitates the foreign researcher or entity undertaking research in the Philippines to collaborate with the local scientists from the government agency or government-recognized local institutions in the Philippines.

Nevertheless, several difficulties were encountered during the implementation of EO 247 which includes EO's scope and coverage, the requirements for obtaining PIC, fair and equitable sharing of benefits, and the processing period and requirements for the research agreement. Thus, a joint DAO No. 01, Series of 2005 was issued setting guidelines for the bioprospecting activities in the Philippines by the DENR and the Department of Agriculture, Palawan Council for Sustainable Development, and National Commission on Indigenous Peoples. The guidelines applied to bioprospecting of any biological resources found in the Philippines including species listed in CITES (Conventions on International Trade in Endangered Species of Wild Fauna and Flora) and the IUCN (International Union for Conservation of Nature) Red List (PCAF, 2019).

Biopiracy in the Philippines arises when pharmaceutical companies from developed nations patent endemic animals and plants, which have long been utilized by indigenous communities for their medicinal properties. This exploitation of TK is facilitated by gaps in our legal framework, including the ambiguity of the Intellectual Property Code, the absence of specific biopiracy laws, inadequate enforcement of international treaties, and the lack of a structured system for documenting indigenous innovations.

To strengthen ABS in the Philippines, the government through the Department of Environment and Natural Resources (DENR)—Biodiversity Management Bureau (BMB) is implementing a 06-year (2021–2027) UNDF supported GEF funded ABS Project to reinforcing the national ABS framework, boosting national and local capacities, and cultivating essential expertise in ABS agreements. The ABS Project further seeks to conserve its rich biological diversity, promote its sustainable use, and increase economic opportunity for ILCs through ABS.

Policies for Preventing Biopiracy in Indonesia

Indonesia is the most ecologically diverse country in the world having a domain on terrestrial, freshwater, and marine ecosystems, and has a wealth of abundant biodiversity and biological resources (Baines & Hendrom, 2022). The Indonesian archipelago comprises two of the world's biodiversity hotspots (von Rintelen et al., 2017). It has the second highest number of indigenous medicinal plants, after the Amazon rain forests (Elfahmi et al., 2014) and the importance of biodiversity in traditional medicine and agriculture is deeply rooted in Indonesian society (von Rintelen et al., 2017). Indonesia has the constitutional basis for the management of its rich GRs and the rights of the indigenous people (Article 33 Paragraph (3) and Article 18B Paragraph (2) of the Constitution of the Republic of Indonesia). Indonesia's archipelago consists of approximately 17,000 islands, with around 990 being permanently inhabited. The country is divided into seven major biogeographic regions, centered on its main islands and surrounding seas. Conservation International recognizes Indonesia as one of the 17 "megadiverse" countries, home to two of the world's 25 biodiversity "hotspots," 18 of the World Wildlife Fund's "Global 200" ecoregions, and 24 of BirdLife International's "Endemic Bird Areas." Indonesia also harbors 10% of the world's flowering species, with an estimated 25,000 flowering plants, 55% of which are endemic. It is considered one of the world's centers for agrobiodiversity in plant cultivars and domesticated livestock. In terms of fauna diversity, Indonesia is home to about 12% of the world's mammals (515 species), ranking second globally after Brazil. It also has 16% of the world's reptiles (781 species) and 35 species of primates, placing it fourth worldwide. Additionally, Indonesia ranks fifth globally for bird species, with 17% of the world's birds (1,592 species), and sixth for amphibians, with 270 species (https://www.cbd.int/countries/profile?country=id).

In recognition of its rich biological heritage, it was the first country to develop a National Biodiversity Action Plan in 1993 (Baines & Hendrom, 2022) which later was updated to Indonesian Biodiversity Strategy and Action Plan (IBSAP) 2003–2020. It ratified the Convention on Biological Diversity in 1994 by enacting Law Number 5/1994, followed by other measures over time. In 2015, Indonesia launched a new IBSAP 2015–2020 to protect and sustainably use its biological resources to improve its economic and development opportunities (UNDP, 2016).

Indonesia was one of the first countries worldwide to sign the Nagoya Protocol (NP) in May 2011, ratified it in September 2013 through Act No. 11/2013 (ABS Clearing House, 2016), prepared an amendment draft for the Patent Act, and has further prepared a Draft Act on GRs (Yulia dan, 2013) basically to ensure national sovereignty over its genetic/biological resources, protection of TK associated with GRs, protection of its economic interest as well as to profit from the potential value of technological advances ensuing from bioprospecting of its GRs (Latifa, 2015). However, the institutional arrangement for ABS in Indonesia is still in the infancy stage (Table 5), and the legal instruments created from it are yet to be developed. Many other legal instruments though exist, but do not specifically address issues in ABS (Table 6). Indonesia also has regulations for the legal protection of TK related to genetic resources; however, the regulations only cover in sectorial scale and are still considered as rules, hence lack protection for TK and anything related to it (Mardiastuti, 2019).

Yulia dan (2013) has listed many constraints in implementing ABS in Indonesia, all linked to the biodiversity, local people, information, and regulations. Nevertheless, there are cases that have proven that ABS can happen with strong political will and mutual understanding. During the 1990s, when ABS was not even formulated, the mutual understanding between Indonesia (provider country) and Japan (user country) resulted in sustainable utilization of neglected wild flowers of *Impatiens* sp. for developing garden plants (Sun Patiens) by Sakata Seed Corp. Japan. Later, Sun Patiens was commercially launched in 2006 and Sakata Seed Corporation is paying a share of the Sun Patiens royalty proceeds to the Indonesian Government

Table 5 Ministries and institutions related to ABS in Indonesia

Name of the ministry/institution	Roles and responsibilities related to ABS
Ministry of Environment and Forestry	Manage and conserve genetic resources in terrestrial ecosystem, especially of wild origin; National Focal Point for Convention on Biological Diversity and Nagoya Protocol
Ministry of Agriculture	Manage cultivated genetic resources, including for plants and animals; under this ministry there are many research centres
Ministry of Fisheries and Marine Affairs	Manage and conserve genetic resources in marine ecosystem, especially of wild origin
Ministry of Law and Human Rights	Granting patents
Ministry of Research, Technology and Higher Education	Granting permit to foreign researchers to conduct research and take samples
Ministry of Foreign Affairs	Responsible for political relationships between/among countries
Ministry of Health	Responsible for national health, including related to viruses and pathogens
Indonesian Institute of Sciences	Responsible for the scientific decisions related to national matters; approval to give a permit for taking samples, Mutually Agreed Terms, database

Source Mardiastuti (2019)

and in addition, performs technology transfer as a non-monetary benefit sharing (Mardiastuti, 2019).

Policies for Preventing Biopiracy in China

China is one of the world's 12 mega-biodiverse countries. Its forests cover 20.36% of the land, with vast forest plantations supporting diverse wildlife. As a center of crop origin, China is home to nearly 10,000 plant species, including 528 genera and 1,339 cultivated plant species, half of which originated there. The country also boasts rich wetlands, with 13.7 million hectares of marshes, 8.35 million hectares of lake wetlands, and 8.2 million hectares of river wetlands. Grasslands make up 41.7% of the land, with 6,704 forage plant species, 320 of which are endemic. China's natural grasslands support over 2,000 wild animal species and numerous plants with economic and medicinal value. In marine environments, over 20,000 species have been recorded, making up more than 10% of the world's total. China has 2.6 million km^2 of deserts, but these areas have limited biodiversity. In total, China is home to over 35,000 plant species (17,300 of which are endemic), 6,445 vertebrates (667 endemic), and 10% of the world's invertebrates. Notably, China

Table 6 Some existing regulations in Indonesia that are relevant to ABS and can be adopted as transitional measures

Law/regulation	Topic
Law Number 5/1990	Conservation of living resources and its ecosystem
Law Number 11/2013	Ratification of the Nagoya Protocol on Access to Genetic Resources and Benefit Sharing
Government Regulation Number 41/2006	Permits for foreign universities, R&D institutions, companies and individuals
Government Regulation Number 7/1999	Preservation of wild plants and animal species
Government Regulation Number 8/1999	Utilization of wild plant and animal species
Decree of Minister of Forestry Number 447/Kpts-II/2003	Administration directive for harvest or capture and the distribution of the specimens of wild plant and animal species
Decree of Minister of Forestry Number 284/Menhut-II/2007	Delegation of authority granting permission or transport of samples for parts of wild plant and animal species and its results for research purposes
Decree of Minister of Forestry Number 447/Kpts-11/2003	Regulates the harvest, capture, and transportation of wildlife specimens
Decree of Minister of Agriculture Number 15/Permentan/OT.140/3/2009	Guidance of the making of Material Transfer Agreement
Decree of Minister of Agriculture Number 37/Permentan/OT.140/7/2011	Conservation and utilization of plant genetic resources
Decree of Minister of Environment and Forestry Number 755/Menlhk/KSDAE/KUM.0/9/2016	Establishment of Working Group on Biodiversity Clearing House (CBD, Cartagena Protocol, and Nagoya Protocol); the members of the working group are representatives from related ministries or institutions that relevant to biodiversity issues
Presidential Decree Number 100/1993	Research permit for foreigners
Governmental Regulation Number 41/2006	Permit to conduct research activity and development for foreign universities, foreign research and development institutions, and foreign business entities and foreigners

Source Mardiastuti (2019)

has the world's largest bird population, with 1,371 species, and 3,862 fish species, or 20.3% of the global total (https://www.cbd.int/countries/profile?country=cn). China is one of the most biologically diverse countries in the world with a wide range of ecosystems, including forests, grasslands, wetlands, marine areas, and deserts. There are 55 officially recognized ethnic minorities in China and many of them have still maintained their traditional lifestyle and have developed a rich pool of TK over centuries concerning the traditional medicinal use of floral and faunal resources. This has made China vulnerable to biopiracy and its concern for growing misappropriation of its rich GRs and associated TK led to the incorporation of ABS in various Chinese laws and regulations. For instance, the 2008 amendment of the Patent Law required disclosure of 'direct and original sources' of GRs in cases where the patent application is related to GRs. It also provided that no patent will be granted for an invention based on GRs that is accessed or utilized in violation of any law or administrative regulation. The issue of ABS was also highlighted in China's 13th Five-Year National Plan for Social and Economic Development (2016–2020). Moreover, the 2017 Law on Traditional Chinese Medicine explicitly required PIC to be obtained from, and benefits to be shared with, the TK holders (Zheng, 2019). Further, it developed the Draft Regulations (Regulation of Access to Genetic Resource and Benefit-Sharing) in 2017 governing the entire use of GRs and associated TK. The Draft Regulations has a total of 7 Articles and 48 Provisions with objectives of contributing to the conservation and sustainable use of biodiversity promoting the ABS principles of the CBD, and fulfilling the obligations of the Nagoya Protocol.

In the Draft Regulation, Article 3 and Article 4 are the main provisions specifically laying out procedures for dealing with access to GRs and associated TK and fair and equitable sharing of benefits. Under the current Draft Regulation, all access to GRs and associated TK would be subject to (i) the PIC of the holders of the resources/ knowledge; (ii) ABS agreements based on MAT; and (iii) registration or approval procedures. There would be simplified arrangements for domestic users. Both access and utilisation of GRs by foreign entities must be conducted within Chinese territory with Chinese collaborators. One exception to the registration and approval procedures would be access and use of GRs by 'farmers, pastoralists, and fisherman according to their traditional way of life'. The subsequent transfer of GRs and TK accessed before or after the Regulation enters into force will require registration and approval under the Draft Regulation (Humphries et al., 2021). The Draft Regulations have specific provisions concerning pandemic outbreaks and emergencies.

The Draft Regulation has established the most comprehensive piece of legislation for ABS in China (Xia, 2023). However, it has failed to adequately fulfill some critical obligations listed under the global norms and is in conflict with China's international obligations required specifically by the Nagoya Protocol. It has also failed to recognizing the rights of the ILCs ignoring the significance of customary laws, community protocols, and ILCs' collective access to a fair and equitable share of benefits arising from their ancestral resources and knowledge. Moreover, China's Draft Regulation barely pays attention to its citizens and corporations using GRs and associated TK from other countries, or to China's role in complying with these countries' ABS regulations. In contrast, the compliance mechanisms established in the Draft Regulation are created purely from the provider's perspective, which is in contradiction with users' compliance measures mandated by the Nagoya Protocol. Also, the breadth and fragmentation of the decision-making arrangements of the bureaucracy are severe structural problems identified in the Draft Regulation, with serious implications for those wishing to access GRs and associated TK (Xia, 2021).

According to Zheng (2019), GRs and associated TK which is the subject of ABS, are not clearly defined and their legal implications remain obscure. The complexity of the ethnic makeup of Chinese populations and their established rights under the current legal system adds to the difficulty in establishing legitimate ABS beneficiaries to receive fair and equitable benefit-sharing. Potential conflicts among overlapping authorities over natural resources, as well as rights-holders of TK, challenge the functionality of the envisaged ABS framework in China and its potential efficiency. Further, its provisions, especially on the mechanism of benefit-sharing, remain problematic for both interpretation and implementation.

Technologies for Preventing Biopiracy

The genetic identification methods outlined in Table 7 are presently employed in plant identification within translational research, presenting both advantages and disadvantages. Anticipated enhancements and innovations in sequencing methodologies are likely to enable cost-effective genotyping-by-sequencing for various non-model plant species. However, acquiring comprehensive genomic sequence data for all individuals in a study would be superfluous in most scenarios, leading to unnecessary cost escalation. The utilization of next-generation sequencing (NGS) for identifying microsatellite markers emerges as a rapid and cost-effective approach, increasingly becoming economically viable.

Megadiverse Countries as Providers of GRs and Digital Sequence Information (DSI)

Currently, digital sequence information (DSI) has become a focal point of numerous discussions regarding its role in storing genetic data. Its significance was underscored by the agreement among the 196 participating countries at the meeting to explore potential instances of digital sequence information leading to new forms of biopiracy (Böll Heinrich Fundación, 2016). DSI enables what is termed "digital biopiracy," wherein genetic sequences of plants, microorganisms, and seeds can be downloaded from the Internet and utilized to recreate physical DNA through synthetic biology techniques. This process often occurs without due consideration for the potential benefits to the countries and communities of origin, where the genetic information originates (Böll Heinrich Fundación, 2016).

DSI encompasses a wide array of data, including sequences of nucleotides forming part of DNA, RNA sequences, protein-forming amino acids, genetic-information-derived chemical compounds (metabolites), environmental data, and information concerning ecological interactions between sequences (epigenetics), among others. Presently, public and private databases contain millions of DSI entries, which can be utilized and altered for commercial purposes and patented without adherence to the fundamental principles outlined by the Nagoya Protocol (CBD, 2020a; Houssen et al., 2020). Consequently, their utilization often does not involve PIC or Mutually Agreed

Table 7 Available genetic barcoding methods for the prevention of biopiracy

Methods	Advantages	Disadvantages	References
Next-generation sequencing (NGS)	i. No prior sequence information or probe generation is required	i. High in cost and not available in all laboratories ii. Time-consuming and high data store required	Shokralla et al. (2014)
Simple sequence repeats (SSR) and intersimple sequence repeats (ISSR)	i. Readily use for taxonomic studies	i. Allele sizes often differ from different laboratories ii. Lack of potential in discriminating species in hypervariables iii. Occurrence of null alleles complicate the interpretation of data because heterozygotes cannot be identified, and reaction failures cannot be detected	Varshney et a.l (2005) and Mosa et al. (2019)
Amplified fragment length polymorphism (AFLP)	i. No prior sequence information or probe generation is needed	i. AFLP generates huge quantities of information technically demanding in the laboratory and, especially, in data analysis ii. Interspecies genetic hybrids in the genus are difficult to identify	Cornell University (2003)
Random amplified polymorphic DNA (RAPD)	i. Producing DNA patterns that allow comparison of many loci simultaneously ii. Low cost	i. It may not be practical to identify the species of origin in products containing mixtures of species ii. It does not seem to be adequate for analysis of severely degraded material, as in autoclaved samples	Palacios and González-Candelas (1997) and Asensio et al. (2002)

Adapted from Yoonus et al. (2021)

Terms (MAT), let alone ensuring equitable sharing of benefits derived from genetic resources' utilization. DSI is inherently linked to "physical" genetic resources, and their separation is practically impossible. Accessing DSI without adhering to the primary regulations of the NP encourages biopiracy and results in unilateral economic gains, undermining the core objective of the CBD: the conservation and sustainable utilization of biological diversity (CBD, 2020b).

Recognizing DSI's significance as equivalent to any "physical" genetic resource is crucial to addressing these challenges effectively. Advances in DNA sequencing and synthesis are leading to disruptions in the intricate ABS agreements between those who utilize GRs ('users') and those who provide them ('providers'). For some observers, this signals the dawn of a new era characterized by the establishment of open-source gene banks aimed at addressing global challenges. However, for others, it raises concerns regarding the emergence of unjust digital biopiracy. The topic of DSI was a focal point during the 2016 Cancun negotiations of the UN Convention on Biological Diversity and its Nagoya Protocol on ABS, underscoring its ongoing relevance today. While previous studies have explored potential remedies for digital sequencing and ABS, limited attention has been devoted to critically examining the underlying issue itself (Adelita & Tello, 2020).

India and South Africa serve as notable examples of countries that have implemented measures to guarantee that the utilization of their domestic GRs and associated TK aligns with their respective ABS laws. In both nations, the processing of a domestic patent application may be halted until compliance with ABS laws is confirmed, even in cases involving subject matter related to DSI. In South Africa, details regarding the origin of GRs are relayed to the domestic National Focal Point (NFP), which then verifies whether the patent applicant has fulfilled the necessary ABS obligations. This mandatory disclosure requirement can be seen as a mechanism to facilitate adherence to the domestic ABS framework. In India, the requirement for benefit-sharing concerning DSI is currently assessed on a case-by-case basis. This entails that benefit-sharing obligations are typically linked to a percentage of the commercial gains derived from DSI utilization. However, instances involving non-commercial use of biological resources or associated knowledge usually do not lead to commercial profits. Nevertheless, there is no legal obstacle preventing the imposition of "non-monetary" benefits from DSI. These non-monetary benefits may include sharing scientific information relevant to the conservation and sustainable management of biological diversity, such as biological inventories and taxonomic studies (Bagley et al., 2020).

The question of whether DSI heralds a new era of digital biopiracy or marks the beginning of a transformative open-source movement in life engineering is a topic of intense debate. The implications at the intersection of the digital and the material, where biology intersects with economics, and where capitalism grapples with ontological complexities, are significant and warrant continual scrutiny (Bond & Scott, 2020).

Conclusion

The prevailing prevalence of biopiracy in the contemporary world can be interpreted as a consequence of the political and economic ideology of neoliberalism. This concept is intricately connected to the notion of intellectual property rights for biological inventions, a highly contentious topic. Patenting biotechnology inventions becomes significantly more intricate due to the involvement of numerous international agreements in the process.

Biopiracy involves the unauthorized patenting of GRs or TK from developing countries, lacking prior consent, recognition, or fair compensation. This act can be seen as a contemporary form of colonialism over the biological resources of Third World nations. Notably, inventions deemed immoral or harmful to plant life, animals, humans, or the environment are ineligible for patents. Ethical biopiracy, characterized by prior consent and fair recognition, can be justifiable. Despite some countries excluding TK from patentability, it is still exploited. Developed nations should seek prior consent, recognize indigenous communities, and share equitable benefits to address this issue. Granting patents is appropriate if these countries meet all three patentability requirements. Essentially, ethical bioprospecting, viewed as responsible biopiracy, can be considered justifiable.

The exploitation of monopoly rights by developed countries in biopiracy has adverse effects on competitors, consumers, and indigenous communities, leading to market blockage, elevated prices for patented products, and the unauthorized utilization of biological resources and TK. Moreover, the environmental impact of patenting biological products is detrimental. To address these concerns, it is crucial to regulate rather than prevent biopiracy. Developed countries should be permitted to patent inventions using biotechnological resources and TK, provided they adhere to ethical practices.

Granting patents to developed countries in the context of biopiracy can be considered legally and ethically justifiable if conducted in an ethical manner. This involves obtaining prior informed consent, recognizing indigenous resources and knowledge, and ensuring fair and equitable benefits are shared with developing countries. Fulfilling obligations outlined in bioprospecting contracts is crucial. Ethical conduct reduces potential issues in securing patents, emphasizing the importance of ethical practices in the biopiracy process.

To proactively address potential problems, developing countries are advised to establish a searchable database of TK and GRs like TKDL employed by India. This measure facilitates global access for patent examiners, minimizing the risk of granting patents for undeserving claims. Ultimately, this initiative aims to safeguard the heritage of developing countries from exploitation by foreign companies.

Conflict of Interest The authors state no conflict of interest.

References

ABSCH. (2016). *The Access and Benefit-Sharing Clearing-House (ABSCH) by the Secretariat of the Convention on Biological Diversity*. https://absch.cbd.int. Accessed on 01 June 2024.

Adelita, A. J. M., & Tello, S. V. (2020). Beyond green gold: Megadiverse countries as providers of genetic resources and digital sequence information. In *SECTION 2 BIOPIRACY: THE PLUNDER OF BIODIVERSITY AND KNOWLEDGE*. https://navdanyainternational.org/wp-content/uploads/2020/10/SECTION-2-BIOPIRACY-BG-REPORT.pdf

Afreen, S., & Abraham, B. P. (2008, September). *Biopiracy and protection of traditional knowledge: Intellectual property rights and beyond* (Working Paper Series No. 629). Indian Institute of Management. https://ir.iimcal.ac.in

Aguilar, G. (2001). Access to genetic resources and protection of traditional knowledge in the territories of indigenous peoples. *Environmental Science & Policy, 4*, 241–256.

Ahoyo, C. C., Houéhanou, T. D., Yaoïtcha, A. S., Prinz, K., Kakai, R. G., Sinsin, B. A., & Houinato, M. R. (2021). Traditional medicinal knowledge of woody species across climatic zones in Benin (West Africa). *Journal of Ethnopharmacology, 265*, 113417. https://doi.org/10.1016/j.jep.2020.113417

Anuradha, R. V. (2008). Sharing the benefits of biodiversity the Kani TBGRI deal in Kerala, India. *Journal of International Wildlife Law and Policy, 3*(2), 125–151. https://doi.org/10.1080/13880290009353952

Asensio, L., González, I., Fernández, A., et al. (2002). Application of random amplified polymorphic DNA (RAPD) analysis for identification of grouper (*Epinephelusguaza*), Wreck fish (*Polyprion americanus*), and Nile Perch (*Lates niloticus*) fillets. *Journal of Food Protection, 65*(2), 432–435. https://doi.org/10.4315/0362-028x-65.2.432

Avantika, G., Vinil, T., & Swati, S. (2015). Bio-piracy in India: A decline in cultural values. *International Research Journal of Environment Sciences, 4*(9), 80–82.

Baby, B. T., & Suriyaprakash, T. N. K. (2022). Intellectual property rights: Bioprospecting, biopiracy and protection of traditional knowledge—An Indian Perspective. In S. L. Prabu & T. N. Kuppusami (Eds.), *Intellectual property*. IntechOpen. https://doi.org/10.5772/intechopen.99596, https://www.intechopen.com/chapters/78249

Bagley, M., Karger, E., Muller, M. R., Welch, F. P., & Thambisetty, S. (2020). *Fact-finding study on how domestic measures address benefit-sharing arising from commercial and non-commercial use of digital sequence information on genetic resources and address the use of digital sequence information on genetic resources for research and development* (CBD/DSI/AHTEG/2020/1/5). www.cbd.int/dsi-ahteg-2020-01-05-en.PDF

Baines, G., & Hendrom, M. (2022). Indonesia. In J. Carew-Reid (Ed.), *Biodiversity planning in Asia* (p. 131–158). IUCN.

Bala, M., & Verma, D. (2020). National intellectual property rights policy of India—A review. In M. Jain & A. Moudgil (Eds.), *Sensitizing and imparting awareness about intellectual property rights among students* (1st ed., pp. 125–132). National Press Associates.

Banerjee, M. (2019). Biopiracy in India: Seed diversity and the scramble for knowledge. *Phytomedicine, 53*, 296–301.

Baskaran, K., Adenan, M. I., Ali, M. R. M., & Nimura, S. (2003). Tropical rainforest: A cradle for biological resources and the Malaysian policy on CBD. *Actinomycetol, 17*(2), 50–53.

Beattie, A. J., Hay, M., Magnusson, B., de Nys, R., Smeathers, J., & Vincent, J. F. V. (2011). Ecology and bioprospecting. *Austral Ecology, 36*(3), 341–356. https://doi.org/10.1111/j.1442-9993.2010.02170.x

Böll Heinrich Fundación. (2016). *Four steps forward and one step back in the global regulation of synthetic biology*. Heinrich Böll Stiftung. https://mx.boell.org/es/2016/12/21/cuatro-pasos-adelante-y-uno-hacia-atras-en-la-regulacionglobal-de-la-biologia-sintetica. Accessed on 4 July 2020.

Bond, M. R., & Scott, D. (2020). Digital biopiracy and the (dis)assembling of the Nagoya Protocol. *Geoforum, 117*, 24–32.

CBD. (2020a). *Digital sequence information on genetic resources: Concept, scope and current use*. Ad-Hoc Technical Expert Group on Digital Sequence Information On Genetic Resources, Montreal, Canada, 17–20 March 2020. CBD/DSI/AHTEG/2020/1/3 29 January 2020.

CBD. (2020b). *Fact-finding study on how domestic measures address benefit-sharing arising from commercial and non-commercial use of digital sequence information (DSI) on genetic resources and address the use of DSI on genetic resources for research and development*. Ad-Hoc Technical Expert Group on Digital Sequence Information On Genetic Resources, Montreal, Canada, 17–20 March 2020, CBD/DSI/AHTEG/2020/1/3 29 January 2020.

Chang, M. K., Conkerton, E. J., Chapital, D. C., Wan, P. J., Vadhwa, O. P., & Spiers, J. M. (1996). Chinese melon (*Momordicacharantia* L.) seed: Composition and potential use. *Journal of the American Oil Chemistsi Society, 73*(2), 263–265.

Charlton, A. (2004). Medicinal uses of tobacco in history. *Journal of the Royal Society of Medicine, 97*(6), 292–296.

Chaturvedi, S. (2008). *ABS and Kani Case in India* (pp. 1–36).

Cornell University. (2003). *Amplified fragment length-crop genebank knowledge base*. Cornell University. https://cropgenebank.sgrp.cgiar.org/images/file/learning_space/molecular_markers/volume1/AFPLs.pdf.

Dunagan, M. (2009). Bioprospection versus biopiracy and the United States versus Brazil: Attempts at creating an intellectual property system applicable worldwide when differing views are worlds apart-and irreconcilable. *Law and Business Review of the Americas, 15*, 620.

Dwivedy, A. K., Singh, V. K., Das, S., Chaudhari, A. K., Upadhyay, N., Singh, A., Singh, A., & Dubey, N. K. (2019). Biodiversity bioprospection with respect to medicinal plants. In J. K. Patra, G. Das, S. Kumar, & H. Thatoi (Eds.), *Ethnopharmacology and biodiversity of medicinal plants* (1st ed., pp.3–30). Apple Academic Press.

Efferth, T., Banerjee, M., Paul, N. W., et al. (2016). Biopiracy of natural products and good bioprospecting practice. *Phytomedicine, 23*(2), 166–173. https://doi.org/10.1016/j.phymed.2015.12.006

Elfahmi, Y., Woerdenbag, H. J., & Kayser, O. (2014). Jamu: Indonesian traditional herbal medicine towards rational phytopharmacological use. *Journal of Herbal Medicine, 4*(2), 51–73. https://doi.org/10.1016/j.hermed.2014.01.002

ENVIS. (2011). Biopiracy. *ENVIS Newsletter, 19*, 3–15. http://deskuenvis.nic.in/pdf/Newslet19.pdf

Grain, K. (2006). *Traditional knowledge of biodiversity in Asia-Pacific: Problems of piracy and protection*. http://www.grain.org. Accessed on 27 February 2024.

Houssen, W., Sara, R., & Jaspars, M. (2020). *Digital sequence information on genetic resources: Concept, scope and current use* (CBD/DSI/AHTEG/2020/1/3). www.cbd.int/dsi-ahteg-2020-01-03-en.PDF

Humphries, F., Benzie, J. A. H., Lawson, C., & Morrison, C. (2021). A review of access and benefit-sharing measures and literature in key aquaculture-producing countries. *Reviews in Aquaculture, 13*(3), 1531–1548. https://doi.org/10.1111/raq.12532

Jamil, U. (1998). *Biopiracy: The patenting of Basmati by RiceTec* (Working Paper Series 37. Commission on Environmental, Economic and Social Policy). South Asia and Sustainable Development Policy Institute.

Kate K. T., & Laird, S. A. (2020). *The Commercial Use of Biodiversity: Access to Genetic Resources and Benefit Sharing*. Routledge Library Editions: Ecology. Routledge Taylor & Francis Group. https://www.routledge.com/Routledge-Library-Editions-Ecology/book-series/RLEEC

Latifa, E. (2015). Access to genetic resources in Indonesia: need further legislation? *Oklahoma Journal of Law and Technology, 11*(1), Article 2. https://digitalcommons.law.oe.edu/okjolt/vol11/iss1/2

Mardiastuti, A. (2019). Implementation of access and benefit sharing in Indonesia: Review and case studies. *Jurnal Manajemen Hutan Tropika, 25*(2), 35–43. https://doi.org/10.7226/jtfm.5.1.35

Mgbeoji, I. (2005). *Global biopiracy: Patents, plants, and indigenous knowledge*. University of British Columbia Press.

Mosa, K., Gairola, S., Jamdade, R., et al. (2019). The promise of molecular and genomic techniques for biodiversity research and DNA barcoding of the Arabian Peninsula flora. *Frontiers in Plant Science, 9*. https://doi.org/10.3389/fpls.2018.01929

Nordin, R., Hassan, Z. A., Zainol, R. S., Shapiee, R., Jalil, F., Witbrodt, M. A., & Hassan, M. S. (2012). Regulating access to genetic resources and the sharing of benefit through the Sarawak access benefit-sharing legislation. *Research Journal of Applied Sciences, 7*(3), 185–193.

Oli, K.P., & Sharma, G., (2010). Issues and challenges of access and benefit sharing mechanism in the Hindu Kush Himalayan countries. *Triggering synergies between intellectual property rights and biodiversity. Deutsche GesellschaftfürTechnischeZusammenarbeit (GTZ) GmbH, Postfach, 5180* (65726), 219–233.

Palacios, C., & González-Candelas, F. (1997). Lack of genetic variability in the rare and endangered *Limonium cavanillesii* (Plumbaginaceae) using RAPD markers. *Molecular Ecology, 6*(7), 671–675. https://doi.org/10.1046/j.1365-294x.1997.00232.x

PCAF. (2019). Access and Benefit Sharing (ABS) for genetic resources in the Philippines. *Philippine Council for Agriculture and Fisheries (PCAF) e-Express, 3*(1).

Polo, M., & Latham, R. E. (1958a). *The travels of Marco Polo*. Penguin Group.

Polo, M., & Latham, R. E. (1958b). *The customs of the kingdoms of India*. Penguin Group.

Ramirez Garcia, H. S. (2014). Biopiracy. In H. Ten Have (Ed.), *Encyclopedia of global bioethics* (pp. 1–7). Springer.

Rattray, G. N. (2002). The Enola bean patent controversy: Biopiracy, novelty, and fish-and-chips. *Duke Law & Technology Review, 1*(1), 1–8.

Robinson, D. F. (2010). *Confronting biopiracy: Challenges, cases and international debates* (208 pp.). Taylor and Francis. https://doi.org/10.4324/9781849774710

Robinson, D. F. (2012). Biopiracy and the innovations of indigenous peoples and local communities. In P. Drahos & S. Frankel (Eds.), *Indigenous people's innovations: Intellectual property pathways to development* (pp. 77–93). Griffin Press, Australian National University E Press Book. https://www.nintione.com.au/?p=3493. Accessed on 14 February 2024.

Sahai, S., Pavithran, P., & Barpujari, I. (2007). Biopiracy—Imitations, Not Innovations. Khanpur, New Delhi, India: Gene Campaign.

Sen, A. (2023, March 16). *Biopiracy in India: Exploring the laws and regulations to protect traditional knowledge and biodiversity*. http://legallyflawless.in/biopiracy-in-india-exploring-the-laws-and-regulations-to-protect-traditional-knowledge-and-biodiversity/#. Accessed on 05 February 2024.

Sharma, B., Maurya, S. S., & Brahmacharimayum, J. (2018). India's fight against agricultural and medicinal plants' biopiracy: Its implications on food security, traditional rights and knowledge degradation. *International Journal of Agriculture Environment and Biotechnology, 11*(6): 881–887, https://doi.org/10.30954/0974-1712.12.2018.9

Shiva, V. (1997). *Biopiracy: The plunder of nature and knowledge*. South End Press. https://archive.org/details/isbn_9780896085565

Shiva, V. (2001). Recovering biodiversity. *Social Change, 31*(1–2), 21–37. https://journals.sagepub.com/doi/10.1177/004908570103100204

Shiva, V. (2007). Bioprospecting as sophisticated biopiracy. *Signs, 32*(2), 307–313. https://doi.org/10.1086/508502

Shiva, V. (2020). *The neem tree—A case history of biopiracy*. Third World Network. https://twn.my/title/pir-ch.htm

Shokralla, S., Gibson, J. F., Nikbakht, H., Janzen, D. H., Hallwachs, W., & Hajibabaei, M. (2014). Next-generation DNA barcoding: Using next-generation sequencing to enhance and accelerate DNA barcode capture from single specimens. *Molecular Ecology Resources, 14*(5), 892–901. https://doi.org/10.1111/1755-0998.12236

Silva, D. (2020, December 15). *Biopiracy: The largely lawless plundering of Earth's genetic wealth*. Landscape News. Accessed on 29 January 2024.

Singh, D. (2003). Corporate hijacking of food is the most important health hazard of our time. *BMJ, 327*(7420), 890. https://doi.org/10.1136/bmj.327.7420.890-d

Singh, G. S. (2012). *Malaysia's implementation of multilateral system of access and benefit sharing*. Biodiversity International and Malaysian Agricultural Research Development Institute.

Sirakaya, A. (2019). Balanced options for access and benefit sharing: stakeholder's insight on provider country legislation. *Frontiers in Plant Science, 10*, Article 1175 (10 pp.). https://doi.org/10.3389/fpls.2019.01175

The Hindu. (2012). A benefit-sharing model that did not yield desired results. https://www.thehindu.com/news/national/A-benefit-sharing-model-that-did-not-yield-desired-results/article12561312.ece. Accessed on 20 January 2024.

UNDP. (2016). *Indonesia launches national blueprint to protect its biodiversity*. UNDP Press Release. http://www.id.undp.org/content/indonesia/en/home/presscenter/pressreleases/2016/01/21/indonesia-launches-national-blueprint-to-protect-its-biodiversity.html

UNEP. (2018). *The Clearing-House Mechanism of the Convention on Biological Diversity: 6th National Report—India* (pp. 1–270). https://chm.cbd.int/database/record?documentID=245938

Varshney, R. K., Graner, A., & Sorrells, M. E. (2005). Genic microsatellite markers in plants: Features and applications. *Trends in Biotechnology, 23*(1), 48–55. https://doi.org/10.1016/j.tibtech.2004.11.005

von Rintelen, K., Arida, E., & Häuser, C. (2017). A review of biodiversity-related issues and challenges in megadiverse Indonesia and other Southeast Asian countries. *Research Ideas and Outcomes, 3*, e20860. https://doi.org/10.3897/rio.3.e20860

Whittle, T. (1970). *The plant hunters*. Chilton Book Co.

Xia, N. (2021). *Implementing the Nagoya Protocol on Access and Benefit-Sharing in China: The emerging legal issues*. www.preprints.org. https://doi.org/10.20944/preprints202109.0404.v1

Xia, N. (2023). Access and benefit-sharing in China: Exploring the extent to which China fulfils the obligations of the Nagoya Protocol. *Queen Marry International Journal of Intellectual Property, 13*(1), 97–120. https://doi.org/10.4337/qmjip.2023.01.05

Yoonus, I., Nalaka, W., Lakmal, G., Yu-Chung, C., & Di Salva, K. R. D. (2021). Biopiracy: Abolish corporate hijacking of indigenous medicinal entities. *The Scientific World Journal, 2021*, Article 8898842 (8 pp.). https://doi.org/10.1155/2021/8898842

Yulia dan, Z. A. Z. (2013). Melindungi keanekaragaman hayati dalam rangka Protokol Nagoya. *Mimbar Hukum, 25*(2), 271–283. https://doi.org/10.22146/jmh.16084

Zheng, X. (2019). Key legal challenges and opportunities in the implementation of the Nagoya Protocol: The case of China. *Review of European Comparative & International Environmental Law, 28*(2), 175–184. https://doi.org/10.1111/reel.1228

Traditional Knowledge and Bio Economy

Pradip Vithal Sarmokadam and Anna Neena George

Introduction

At its core, Traditional Knowledge reflects the cumulative wisdom passed down through generations within indigenous societies. This knowledge is deeply intertwined with the local biodiversity, encompassing the properties and uses of plants, fungi, animals, and other biological resources unique to specific ecosystems. Through centuries of close interaction with their environment, traditional societies have honed their understanding of medicinal plants, herbal formulations, and sustainable agricultural and forest practices.[1] Here, the term 'bioprospecting' has garnered significant negative connotations, often implying an inherent link to biopiracy. Additionally, some traditional communities may perceive bioprospecting negatively, viewing it as an offensive exploitation of sacred biological resources or traditional knowledge essential for their livelihoods. Hence, it is important to understand all its aspects to know its depth and actual worth (Banerjee, 2019).

Traditional wisdom translating into successful products for human well-being results from the contemplation of the profound manifestation of mankind with nature. Since the beginning of time, humanity has been inspired by nature. The main factor in our ancestors' peaceful coexistence and lower conflict rate was their environmentally conscious way of life. Food included all the nutrients because it was natural and unadulterated. The way of living was in perfect harmony with the natural world. Our ancestors have experimented, explored, discovered, invented, reinvented, innovated, improvised, and transmitted their traditional knowledge to subsequent generations,

[1] WIPO. (2000). *Traditional knowledge*. https://www.wipo.int/tk/en/tk/.

P. V. Sarmokadam (✉)
Goa State Biodiversity Board, Saligao, Goa, India
e-mail: pradipgsbb@gmail.com

A. N. George
GVM's Dada Vaida College of Education, Ponda 403401, Goa, India

L. Krishna Panicker et al. (eds.), *Biodiversity and Business*,
https://doi.org/10.1007/978-3-031-71674-4_24

mostly orally and rarely through means of writings or by way of folklore, songs, poems, rituals, taboos, etc. The wisdom pool, which generally lies with communities or peoplescapes, gets assimilated when TS translates into products that are found extremely useful for life sustenance.

Humanity has studied nature's life support systems via experimentation. Humans have investigated the biodiversity around them for anything from food to medicine, timber, materials, and fulfilment of nearly every need. Investigating every potential application for bioresources has improved many cultures over centuries and produced a body of knowledge known as traditional knowledge or wisdom. However, the quality of life in the modern world is more dependent on medications and external air conditioning, which stresses resources because of the growing population, unplanned and unsustainable growth, climate change, extreme heat, stress, rainfall extremities, pollution, etc.

Discovering the Notion of Traditional Knowledge

To grasp the entirety of TK, it is imperative to comprehend the term itself. TK serves as a testament to the deep-rooted connection between local communities and their surrounding environments. As aptly stated by WIPO, "This knowledge is an outcome of their connection with local biodiversity, including plants, fungi, animals, and other endemic biological materials." Spanning centuries of observation, experimentation, and cultural practices, TK encompasses a wealth of insights into the intricate workings of ecosystems and the diverse array of life forms they sustain. In light of modernization and globalization, the preservation and utilization of TK have emerged as crucial endeavours, not only for the well-being of indigenous communities but also for the sustainable handling of innate means and the advancement of the bio-economy.[2]

Some conventional knowledge-based solutions are used everywhere as a result of globalisation. Due to the proliferation of information and worldwide sharing of platforms in the current era, some uses are known outside their original location. In rare instances, the origin itself may be difficult to determine. The majority of foods that are widely recognised as healthy and significant from a medicinal standpoint are the product of numerous trials and experiments. Examples—milk, eggs & several other items known as nutritious are accepted almost worldwide, and lots of research has validated the same. The use of *basil* (tulsi), *vasaca* (adulsa), *gloriosasuperba* (waghchapho) etc., are well known as medicinally important almost everywhere in Asia and India. The economy predominantly based on the sustainable utilisation of biological resources is bioeconomy.

From the perspective of humanity, we can consider this from the following viewpoints in a nutshell, which we are going to discuss in detail in the later part of this chapter.

[2] WIPO. (2000). *Traditional knowledge*. https://www.wipo.int/tk/en/tk/.

Firstly, ensuring the preservation and transmission of traditional knowledge (TK) is crucial. It stands as a vital aspect of cultural identity and community resilience. TK acts as a repository of wisdom. It sustains age-old customs. It fosters a sense of belonging within indigenous societies. However, TK encounters numerous challenges in contemporary times, including the erosion of cultural traditions and the encroachment of external influences. Secondly, efforts aimed at safeguarding TK should be complemented by initiatives to strengthen local capacities and empower young indigenous individuals as custodians of their heritage. Integrating TK into formal education systems and vocational training programs can bridge the divide between traditional and scientific knowledge systems, fostering mutual respect and facilitating cross-cultural dialogue. Thirdly, TK is deeply interwoven with the cultural fabric of indigenous communities, reflected in rituals, ceremonies, and storytelling, which reinforce social unity and connections across generations. It also provides valuable insights into sustainable resource management practices, promoting ecological resilience and reducing the risk of environmental degradation. Fourthly, TK holds significant economic value within the bio-economy sector, with indigenous resources supporting local livelihoods and contributing to global markets. However, challenges such as biopiracy and the commodification of TK pose threats to its integrity and cultural significance, necessitating community-centered governance structures that prioritize local ownership and participation. Lastly, knowledge comes under it, and offers adaptive strategies for climate change resilience, rooted in local ecological wisdom and community resilience, highlighting its relevance in addressing contemporary environmental challenges.

Still, we observed, "Several issues regarding the safeguarding of Indigenous wisdom and its cultural expressions must be talked about" (Ahoyo et al., 2021; Van Overwalle, 2005). "It is concerning to observe the growing attentiveness among research scholars from various universities and huge industries in the contemporary use of indigenous plants, leading to the pursuit of patents for financial gain. This trend puts the livelihoods of hundreds of millions of people in traditional communities worldwide at risk" (Ahoyo et al., 2021; Tedlock, 2006; Van Overwalle, 2005), while utilizing the prevailing patent law arrangement to safeguard these valuable compounds and establish private rights, it is imperative to also value the rights and interests of TK holders, who serve as the foundation for these noteworthy patents. Additionally, the potential for patent-resulting inventions from forefathers' medicines presents a significant incentive for pharmaceutical companies, as it potential long-term benefits that increase over time. Therefore, such information ultimately contributes to the betterment of mankind. Sometimes, "the attempt to patent indigenous plants in certain regions where they have been used for generations has sparked considerable criticism. This is primarily because such patents often rely on pre-established indigenous or TK and, hence, fail to meet the criteria of being 'new' for patent approval. A notable example of this is the controversy surrounding patents allied to the neem tree. The European Patent Office (EPO) revoked a contentious patent concerning the use of antifungal agents obtained from the neem tree, prompting global awareness about the issue of biopiracy" (Brondizio et al., 2019, Hellerer & Jayaraman, 2000).

I recently came across a summary from the 2019 IPBES" Global Assessment on Biodiversity and Ecosystem Services; it emphasizes the importance of acknowledging "the knowledge, innovations, practices, and values of indigenous peoples and local communities. Their inclusion and participation in environmental governance not only enhance their quality of life but also contribute to the conservation, restoration, and sustainable use of nature, which benefits society at large. Governance structures incorporating customary institutions, management systems, and co-management regimes involving indigenous peoples and local communities prove effective in safeguarding nature and its contributions to people by integrating locally attuned management systems and indigenous knowledge. Recognizing land tenure, access, and resource rights, along with applying principles of free, prior, and informed consent, and promoting collaboration and fair benefit-sharing are essential for facilitating the positive contributions of indigenous peoples and local communities to sustainability" (Brondizio et al., 2019).

Threat to Traditional Knowledge

The threat is also noticed usually, using TK for making money can sometimes lead to stealing it and ignoring the rights of the people who own it. To deal with this serious problem, we need to find ways to protect and support traditional knowledge for long-term benefits for those who own it. Preserving, protecting, and promoting innovations and practices rooted in it within local communities' holds significant importance, especially for developing nations. Their abundant reserve of traditional knowledge and biodiversity contributes significantly to various aspects of their well-being, including healthcare, food security, cultural heritage, religious practices, identity, environmental resilience, trade, and overall development.[3] The international focus on safeguarding and conserving this knowledge has become increasingly prominent, especially among developing nations, driven by a range of concerns and viewpoints. Many nations endowed with abundant genetic resources, cultural heritage, and traditional practices argue that local communities have been denied the benefits stemming from their knowledge and innovations. These assets have often been exploited and monopolized by external entities, particularly large corporations, without proper consent or recognition, resulting in a significant disparity in benefits. Developed countries bear an ethical commitment to ensure that indigenous and local populations obtain just and fair shares of the advantages arising from the utilization of their cultural heritage and the commercialization of hereditary resources.

[3] WIPO. *Traditional knowledge and intellectual property—Background brief*. https://www.wipo.int/pressroom/en/briefs/tk_ip.html.

Protection of Traditional Knowledge

We can see that, the World Intellectual Property Organization (WIPO) is actively engaged in drafting a global legal instrument for the safeguarding of traditional knowledge through the Intergovernmental Committee on Genetic Resources, Traditional Knowledge, and Folklore (IGC-GRTF). This initiative aims to enable access to TK for those freestanding the community or country of its authentic holders. WIPO defines it as a dynamic body of understanding that evolves within a community, often forming an integral part of its cultural or spiritual identity and passed down through generations. WIPO's efforts in this area encompass three main domains: traditional knowledge, traditional cultural expressions, and genetic resources, which are interconnected. The pursuit of two types of intellectual property protection for traditional knowledge is a key aspect of WIPO's work"[4] (Javed et al., 2020).

Continuing the discussion on patents, a significant step has already been taken in India; the Traditional Knowledge Digital Library (TKDL) is a combined effort among the Council of Scientific and Industrial Research (CSIR) and the Ministry of Agriculture, Food and Public Health (AYUSH). Its prime goal is to prevent patent offices worldwide from granting patents for inventions based on India's ancient traditional knowledge (TK). "The notion for instituting TKDL emerged following India's efforts to challenge patents issued by the United States Patent and Trademark Office (USPTO) for the wound-healing properties of turmeric (Curcuma longa) and by the European Patent Office (EPO) for the antifungal properties of neem (*Azadirachta indica*). To address this, concrete steps have been taken to develop a program for documenting the knowledge contained in ancient Ayurveda, Siddha, and Unani texts, along with creating a comprehensive database on medicinal plants and their traditional uses. The TKDL initiative has successfully transcribed over 2.90 lakh medical formulations from Ayurveda, Unani, and Siddha into five internationally recognized languages. It serves as a collective resource for managing intellectual property rights"[5] (Gaudillière, 2014).

Biopiracy Relevance to Traditional Knowledge

When discussing bio-piracy, certain aspects are never overlooked. Biopiracy, the unlawful commercial exploitation of biological resources and associated indigenous wisdom, poses significant ethical and legal challenges. Often, indigenous communities hold valuable knowledge about the utilizations and properties of local plants, animals, and other biological resources, accumulated over generations. However, external entities may exploit this knowledge and resources without proper consent,

[4] WIPO. *Traditional knowledge and intellectual property—Background brief*. https://www.wipo.int/pressroom/en/briefs/tk_ip.html.

[5] UNEP. (2018). *The Clearing-House Mechanism of the Convention on Biological Diversity: 6th National Report-India* (pp. 1–270). https://chm.cbd.int/database/record?documentID=245938.

acknowledgment, or fair compensation, leading to the marginalization and manipulation of indigenous peoples. The issue of biopiracy highlights the importance of protecting TK and ensuring equitable benefit-sharing between the holders of such knowledge and those who seek to profit from it. Efforts to address biopiracy involve legal frameworks, international agreements, and community-led initiatives aimed at safeguarding it and promoting respect for indigenous rights and sovereignty.

The rich biodiversity and diverse cultural heritage of nations like India represent significant strengths, offering a wealth of TK. However, the emergence of international regulations, particularly Intellectual Property Rights (IPR), has raised concerns about the misappropriation of this it. Despite these challenges, the vast reservoirs of TK across different cultures hold immense potential for bioprospecting, leading to the development of novel drugs, pharmaceuticals, phytochemicals, and valuable industrial genes. The global market for goods derived from the falsification of genetic resources, coupled with it, is calculated to range between US $500 to 800 billion annually. Notably, sales of herbal medicine derived from traditional knowledge in biodiversity-rich nations alone exceeded US$12.5 billion in 1994 and flowed to US$30 billion in 2000, with annual escalation rates averaging between 5 and 15% based on the region.

This term encompasses breaches of contractual agreements related to the access and utilization of biological resources and traditional knowledge, resulting in harm to the providers. Examples of biopiracy masquerading as bioprospecting in the Asia–Pacific region are outlined in the below table.

Country	Biological resource	Biopirate country	Notes
China	Bitter Melon (*Momordica charantia*)	US	US Patent No. 5484889
China	Xi Shu/Happytrees (*Camptothecalowreyana*)	US	US Patent No. PP 11959
Malaysia	Bintangor tree (*Calophyllum lanigerum*)	Singapore US	US Patents including Nos. 6420571, 6369241, 6160131 and 6277879
Pacific	Kava (*Piper mythesticum*)	US	US Patents including Nos. 6405948, 6277396, 6080410, 6025363, 5977120, 5976550 and 5770207
Pacific	Nonu (Indian Mulberry Morindacitrifolia)	Europe US	In 1995 Nonu Samoa Enterprises began export of nonu, a tree with medicinal properties, to the US with US collaboration
Pakistan	Basmati Rice	US	US Patent Nos. 6274183 and 5663484
PNG	Coral reef sponges	US	US Patent Nos. 6281196, 6153590, 5646138 and 5494893

(continued)

(continued)

Country	Biological resource	Biopirate country	Notes
Philippines	Soil microbes	US	The multinational company Eli Lily has earned billions of dollars from the erythromycin antibiotic, which was developed from a bacterium isolated from a soil sample that Filipino scientist Abelardo Aguilar collected in his home province of Lloilo. Neither Aguilar nor the Philippines received any royalties
Philippines	Llang-Ilang (*Canangaodorata*)	France	The use of the extracts from llang. Llangin the cosmetic industry is perhaps as old as perfume in France. Several perfumeries in France have used and continue to use it in their products
Philippines	Banaba (*Lagerstroemia sp.*)	Japan, US	US Patent No. 5980904
Philippines	Nata de coco	Japan, US	US Patent Nos. 6280767, 6140105, 5962277 and 5795979
Philippines	Snails (Conus)	US	US Patent Nos. 6369193, 6344551, 6197535, 6153738, 6077934, 5633347, 5595972, 5589340 and 5514774
India	Basmati Rice	US	US Patent Nos. 5663484 and 4522838
India	Turmeric (*Curcuma longa*)	US	US Patent Nos. 5401504, 5135796 and 5047100
India	Neem (*Azadirachta indica*)	US	Several US Patent including Nos. 5420318, 5391779 and 5371254; the US multinational company W.R Grace's EPO Patent No. 0426257
India	Guggul (*Commiphora mukul*)	US	US Patent No. 6113949 and US Patent Application 20020018757
Thailand	Jasmine Rice	US	A US plant geneticist has developed a strain of Jasmine Rice to be able to grow it in the US; he received the original seeds of the Thai KhaoDok Mail 105 (KDM 105) jasmine rice variety from the international Rice Research Institute (IRR) in 1995
Thailand	Plao-noi (*Croton sublyratus*)	Japan	In 1975 Sankyo of Japan extracted the active ingredient of the Thai local plant to produce the patented product Kelnac
Samoa	Mamala tree (*Homalanthus nutans*)	US	US Patent No. 5599839

(continued)

(continued)

Country	Biological resource	Biopirate country	Notes
Sri Lanka	Kothalahimbutu (*Salacia reticulata*)	Japan, US	Takama System, Ltd. (Yamaguchi, JP)'s US Patent No. 6376682

Ironically, the valuable traditional knowledge that underpins much of recent scientific research is often overlooked as a scientific contribution, leaving traditional communities with minimal benefits. The UN Convention on Biological Diversity (CBD) recognizes the importance of respecting indigenous knowledge and ensuring fair profit distribution. In response, India has implemented laws like the Protection of Plant Varieties and Farmers Act of 2001 and the Biodiversity Act of 2002. Furthermore, India has pioneered a benefit-sharing model, known as the Pushpangadan model, where TK undergoes scientific study in government laboratories, leading to equitable sharing of benefits between research institutes and indigenous tribes (Pushpangadan et al., 2018).

Traditional Knowledge and Benefit Sharing

According to S. Kannaiyan (2007) in India, a notable instance of benefit-sharing can be observed in the study of Arogyapacha. While an ethno-botanical excursion in the tribal region of the Western Ghats in Kerala, scientists encountered the Kani tribe consuming seeds of the wild plant *Trichopus zeylanicus*, locally known as 'Arogyapacha', for energy. This traditional practice has been employed by the Kani tribe for years to endure physical exertion. Subsequently, the plant's properties were investigated; leading to the development of a standardized drug named "Jeevani" depending on the Kani knowledge. Upon commercial production of the drug in 1995, the Tropical Botanic Garden and Research Institute (TBGRI) agreed to split the license fee equally with a pharmaceutical firm. Additionally, 2% of the royalties from sales were allocated to the tribal community. Following this, the Kani community established a trust, fully owned and managed by them, with around 60% of Kani families in Kerala becoming members. As per the trust's regulations, the license fee and royalties from "Jeevani" sales contribute to the trust's fixed assets, with only the interest generated being used for Kani welfare activities. This model was developed over 12 years from 1987 to 1999 in collaboration with the Kani tribe.

Traditional Knowledge and Agriculture

Over generations, "indigenous communities have cultivated various cash crops, food items, and horticultural produce from wild plants found in forests, known as landraces or indigenous varieties. These serve as the cornerstone of modern plant breeding and contribute significantly to global food security. Additionally, indigenous farming communities have identified and managed a plethora of genes through selective breeding and crossbreeding, which possess traits such as pest and disease resistance, drought tolerance, salt tolerance, cold tolerance, and waterlogging tolerance. International scientists seeking to develop crops resilient to global warming and climate change across agricultural zones often rely on traditional knowledge and local farmers for drought-tolerant or resistant crop varieties. Through the breeding and selection process involving local varieties, they aim to develop high-yielding crop varieties with combinations of traits suitable for drought or salinity tolerance, as well as resistance to pests and diseases"[6] (Erbisch & Maredia, 1998).

Bioprospecting

While discussing the concept of bioprospecting, it's crucial to understand the core essence of this term. The term 'bioprospecting' emerged in the late 1980s or early 1990s, referring to the study of biological resources that contribute to both safeguarding efforts and the recognition of valuable products. Essentially, bioprospecting involves seeking significant genetic and biochemical knowledge present in plants, wild animals, or microbial organisms for specific product development, both scientifically and commercially. It encompasses the exploration of biodiversity to uncover new biological resources with significant social and economic value, a pursuit undertaken by various industries, including pharmaceuticals, agriculture, manufacturing, engineering, and construction. Fundamentally, bioprospecting underscores the importance of discovering natural products for developing new crops and medicines, often drawing from TK. However, the practice, particularly in pharmaceutical bioprospecting, has faced criticism for instances of 'biopiracy,' where major international pharmaceutical corporations exploit indigenous medicinal knowledge without duly recognizing it as indigenous intellectual property (Beattie et al., 2011, Mgbeoji, 2006). Bioprospecting has gained increased attention in recent years due to the growing wakefulness of the urgent need for new drugs. This is driven by the necessity to address presently incurable diseases affecting a progressively expanding global population, as well as the demand for replacements for increasingly ineffective drugs used to tackle health issues. The impact of bioprospecting extends across industries that rely, either entirely or partially, on accessing, processing, sourcing, or utilizing genetic resources to design and develop profitably feasible products for the global consumer and market[7] (Baby & Suriyaprakash, 2022).

---

[6] Sharma, A. (2020). *India: Geographical indication tagging of agricultural products and foodstuffs*. https://www.mondaq.com/india/trademark/898690/geographical-indication-tagging-of-agricultural-products.

[7] Resource Manual for Bioprospecting. http://ipbio.org/pdfs/papers/ResourceManual_Oct14_2004.pdf.

'Bioprospecting is a systematic and coordinated search for beneficial products from bioresources, such as plants, microorganisms, animals, etc., with social and commercial intent'. The Bioprospecting process includes identifying, collecting, and extracting genetic material from biodiversity as the source to create products for sectors such as pharmaceutical and allied, agricultural, industrial, or chemical, including premium applications, such as the processing of food and beverages, paper, pulp, textiles, animal feed, detergents, cosmetics, and chemical synthesis processes, enzymes originating from microorganisms. Scientific and methodical bioprospecting by high-end technological applications could be carried out at genetic, species, and ecosystem levels and could be based on traditional wisdom associated with biodiversity or could be based on an innovative approach not even inspired by traditional wisdom.

Although bioprospecting as the term is relatively new, the practice of exploring bioresources for the presence of useful consumptive values is not new. It has been done ever since humans started evolving. But with a high degree of scientific and technological advancements in investigation at the molecular level, higher is the possibility of returns from innovative products. The exploration of biological resources for novel commercial applications has been integral to global economic and social progress. However, issues arise when bioprospecting ventures veer into biopiracy or environmentally unsustainable practices, like excessive sample collection from specific areas. The term 'bioprospecting' has garnered significant negative connotations, often suggesting an inherent link to biopiracy (Baby & Suriyaprakash, 2022).

Exploring Biodiversity: Opportunities in Bioprospecting

Biodiversity is something where every species hides valuable secrets waiting to be found. As we explore bioprospecting, we uncover nature's riches and unlock countless possibilities for the benefit of humanity. We are also trying for discovery and invention, tapping into nature's riches to create limitless opportunities for the betterment of humankind. The investigation into biodiversity spans several crucial fields, including herbal technology, biotechnology, and nanobiotechnology, all of which are pivotal for advancing knowledge-based industrial ventures, fostering international trade, and safeguarding intellectual property rights (IPR). Biodiversity is renowned for its multifaceted value, encompassing economic, ecological, aesthetic, and intrinsic dimensions, serving as a reservoir of resources essential for human sustenance and well-being, such as food, medicine, and biopesticides. Herbal technology involves the development of enhanced products through accessible technologies, while biotechnology and nanobiotechnology utilize sophisticated interventions at the subcellular and ultramolecular levels, unlocking new possibilities and innovations. Biodiversity not only delivers tangible benefits to humanity but also upholds the delicate ecological balance crucial for the biosphere's integrity and resilience. Within this framework, bioprospecting emerges as a strategic approach, entailing the systematic study of genes, natural compounds, and organisms found in natural habitats. Through chemical, gene, and bionic prospecting, bioprospecting offers promising pathways for product development, harnessing the diverse array of life forms to propel various industries and scientific pursuits forward.

Biodiversity and Bioprospecting

Bioprospecting involves exploring biodiversity to discover commercially valuable biological and genetic resources, aiming to identify new commercial applications for biological materials. This process is essential across various industries, including pharmaceuticals, biotechnology, agriculture, horticulture, cosmetics, and food. Bioprospectors obtain their materials through two main methods: directly from biological resources and through traditional knowledge associated with these resources. For instance, a significant study conducted in 1985 revealed that 122 plant-derived compounds were globally used as drugs, with 80% of them having ties to traditional medicinal practices. Surprisingly, these compounds originated from just 94 plant species, despite the existence of around 250,000 higher plant species worldwide. Many scientists argue that screening plants based on traditional medical knowledge is the most efficient way to discover new medicinal compounds. However, there is an ongoing debate, with some advocating for a random screening approach for drug discovery.

One of the primary methods applied in bioprospecting is the direct extraction of biological resources from their natural habitats. This approach involves sampling various organisms, such as plants, animals, fungi, and microorganisms, to uncover unique compounds or genetic traits that possess commercial value. For instance, pharmaceutical companies often search for bioactive compounds in medicinal plants, which can lead to the discovery of new drugs for treating various diseases. Similarly, agricultural biotechnology companies may seek out genetic traits in crop plants that confer resistance to pests or environmental stresses, enhancing crop productivity and resilience. This knowledge passed down through generations, serves as a valuable source of information for bioprospectors seeking to identify potential commercial applications. For example, traditional herbal remedies used by indigenous peoples may inspire the development of new pharmaceutical drugs, providing insights into the medicinal properties of specific plant species. However, the ethical and legal considerations surrounding the unbiased and equitable distribution of profits resulting from long-established knowledge remain a subject of debate and contention in bioprospecting initiatives.

One such example is the traditional classical method of 'preparing over 20 formulations from Vasaka (*Adhatoda vasica*) by subjecting bolus of crushed leaves using heat treatment and other methods for extracting specific bioactive ingredients. Studies have shown that *Adhatoda vasica* contains pyrroloquinazoline alkaloids, vasicine, vasicol, adhatonine, vasicinone, vasicinol, and vasicinolone. Each of these phytochemicals has a specific role in treating certain ailments. For example, vasicine has been found to have broncho dilatory, respiratory stimulant, and uterine stimulant effects. Other known components include curcumin from turmeric, 'allicin, saponinsetc from garlic. *Mucuna pruriens* beans contain protein, starch, (l-Dopa) 3, 4-dihydroxy-l-phenylalanine, and seeds contain alkaloids, glycosides, saponins, reducing sugars, and tannins' (Soni et al., 2008).

Extracts of specific constituents known to cure or have applications in the market have very high economic opportunities. Bioprospecting finds high applications in pharmaceutical formulations due to proven curative properties and global market opportunities.

Adhatoda vasica (Adulsa) (*Source* Soni et al. [2008])

Gloriosa Superba (Waghchapho) (*Source* Soni et al. [2008])

TK serves various purposes across different commercial sectors, serving as a guiding force in the initial stages of discovery as well as in subsequent research phases. Typically obtained from published academic literature or secondary sources, rather than directly from field studies, it provides valuable insights into various domains. It can aid in the identification of new leads and compounds during the early stages of research, offering a foundation upon which further investigations can be built. Moreover, traditional knowledge often plays a crucial role in guiding subsequent research efforts once a promising compound or discovery has been made (Baby & Suriyaprakash, 2022).

The Great Bioprospecting Race

In the nineteenth century, a notable surge in botanical science occurred, alongside the establishment of institutes committed to the systematic pool of medicinal plants to serve the interests of capitalist economies and national agendas (Global Bioeconomy Summit, 2018; Mgbeoji, 2003). This era, often referred to as the Great Bioprospecting Race, was characterized by intense competition among nations, leading to incidents of smuggling and conflicts over the control of valuable bioresources. One such instance occurred in 1876 when the British clandestinely transported a staggering 70,000 rubber seeds from Brazil to its Asian colonies, thereby significantly undermining Brazil's dominant position in the global rubber industry (Mulligan & Stoett, 2000).

In contemporary times, bioprospecting has purportedly transitioned from its exploitative colonial origins to a more ethical and legally regulated approach to harnessing biodiversity for commercial ventures. However, a closer examination reveals persistent contentious issues surrounding this practice. Since the 1990s, the diversity of germplasm has emerged as a lucrative potential resource, attracting the interest of biotech firms predominantly from the global North. This interest stems from the escalating costs associated with chemical screening and the imminent expiration of patents, motivating companies to explore biodiversity-rich regions (Pordié & Gaudillière, 2014).

Likewise," the term 'bioprospecting' evokes parallels with the historical pursuit of gold during the 1840s Wild West era. It encompasses the accumulation of germplasm and traditional knowledge to develop commercial pharmaceutical, agricultural, industrial, or chemical products" (Christian, 2007). Despite its evolution into a more regulated practice, bioprospecting continues to raise ethical, legal, and environmental concerns, highlighting the need for comprehensive frameworks to govern the sustainable exploitation of biodiversity for commercial purposes.

Bioeconomy

Bioeconomy encompasses protection, sustainable production and utilisation, conservation of bioresources, and leveraging science and technology innovation, besides generating bio-based revenue. So, bioeconomy is a broader term that encompasses economic activities related to sustainable development. It discusses explicitly innovation, traditional wisdom, and conservation, including products and services across all economic sectors. This term has a very broad scope, and the future seems bleak if bioeconomy is not given priority. Enhancing the bioeconomy has to be the new way of addressing economic planning, and if fair and equitable benefit sharing in the circular economy is also made part of our economy, then it will be an important driver in shaping the global future.

Keeping bioeconomy at the centre of sustainable development and assuring that sovereign rights of communities over resources will be unviolated, requires efforts for institutionalization of bioprospecting and traditional knowledge associated with biodiversity.

"Bioeconomy is the production, utilization, and conservation of biological resources, including related knowledge, science, technology, and innovation, to provide information, products, processes, and services across all economic sectors aiming toward a sustainable economy" (*Source* Global Bioeconomy Summit [2018]).

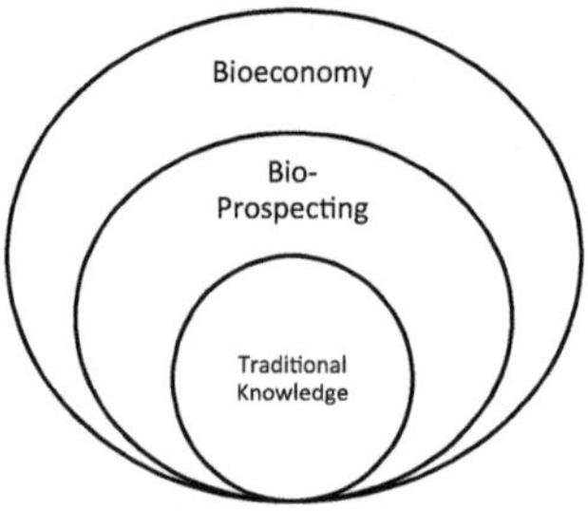

In the preceding discussion, I examined how the interplay of embedding and abstracting processes shape the diverse geographical dimensions of value creation. Central to this dynamic are cooperation and power. Initially, cooperation emerges from the integration of bio-knowledge within unique institutional frameworks, leading to localized scalar effects. This integration allows actors to pursue strategies that complement rather than compete with each other, unconstrained by scarcity concerns. Moreover, the generation of value through market transactions relies on these actors as sources of inherent value, as their knowledge and status are inseparable from their products. Consequently, their practices engender localized or differentiated scalar effects by promoting proximity, facilitating increased interaction, and reducing barriers to knowledge exchange.

However, these scalar effects, stemming from horizontal interactions, are insufficient on their own for value creation. Secondly, power imbalances occur from the abstraction of institution-specific practices in particular discourses, notably through the standardization of bio-knowledge in decontextualized and systematized assessments. These assessments generate globalizing scalar effects by disseminating international standards, conventions, and property rights. The institutionalization of standardization and commodification via intellectual property (IP) is crucial for appraising the value of bio-knowledge in market terms and facilitating their function in market exchanges. This, in turn, enables the realization of worth through the comparison of "standardized" commodities. Such standardization fosters globalizing scalar effects by establishing competitive markets, enticing transnational corporations (TNCs) operating in high-value sectors, such as the bioeconomy, where investment costs are substantial.

Cooperation and power, despite being opposing forces, are both essential for value creation in the bioeconomy, highlighting their dynamic nature in this context. Firstly, without the integration of bioknowledge into specific practices, there would be no differentiation to distinguish between various practices and abstract discourses. This dissimilarity is crucial as it prevents everyone from engaging in the same activities, underscoring the importance of social relations in shaping these processes. Secondly, without the abstraction of institutionally explicit practices, there would be no standardization or establishment of property rights, making bio-knowledge non-commodifiable. Hence, neither of these methods alone is sufficient; the formation of value in the bioeconomy relies on the interplay between difference and similarity, as discussed in the literature on economic anthropology. In essence, the localization of bio-knowledge must be complemented by the standardization of practices through global discourses to enable the systematic evaluation of new bio-knowledge across diverse political economies and scales, thereby facilitating value creation.

The implications of this study for the life sciences are significant. Firstly, the bioeconomy heavily relies on complementary, cooperative, and transparent knowledge relationships, often supported by government funding rather than private sector initiatives. Consequently, the privatization of knowledge through intellectual property rights (IP) negatively impacts the generation of new knowledge, as evidenced by the growing emphasis on "open science" initiatives like the Innovative Medicines Initiative in biomedical research. Secondly, to capitalize on the bioeconomy, private sector firms often resort to coercive tactics, hindering their ability to foster open relationships. While these strategies are understandable—small firms need to safeguard their intellectual assets, while large firms aim to protect their market dominance—reconciling these conflicting demands, often among multiple firms, necessitates innovative incentive structures such as royalties instead of straightforward payments, and alternative forms of competition beyond pricing alone. Thirdly, the inherent flaws between these two necessities of the bioeconomy frequently lead to conflicts between technological advancement and value creation, with the former often sacrificed for the latter. Consequently, the bioeconomy finds itself entangled in competing demands that may delay the emergence of new developments, knowledge, and products until these conflicting imperatives are resolved (Birch, 2012).

Opportunities and Relevance to Goa

Bioeconomy aims to converge all the activities, ideas, concepts, accounting procedures, research and development, and innovation in the biological or biodiversity sector. If this is institutionalized and incorporated at all levels, it will define a new strategy for the world that everybody wants to live in. Originally, Goa's economy was mostly focused on agriculture, but it also benefited greatly from the tourism industry and export of mineral resources. Linking the aforementioned economic endeavours and sources of income to ecological sustainability, however, was more counter-productive than beneficial. Therefore, greater integration and institutionalization of biodiversity and ecological protection for livelihood and economic bolstering through a sustainable bioeconomy-based development paradigm is necessary to maximize the opportunities. The mere presence of rare herbs like *Mappia foetida (narkya or amrute as known locally), Salacia species, Gloriosa superba*, etc. or commonly available plants like *Laurus nobilis (bay leaf), Mucuna pruriens (Kapikacchu), Asian pigeonwings (Aparajita)* flower and their present use provides indicative opportunity. However, it will take time to transform the ideas into workable economic models predicated on fair and equitable sharing; these kinds of projects will create the foundation for the bioeconomy.

Goa has unique applications of traditional knowledge in prospering its economy, such as the heritage drink of Goa known as 'feni'—an alcoholic beverage made from cashew apple juice which is fermented and then distilled using copper vessels. Here, the method of juice extraction involves the crushing of cashew apples by trampling underfoot in the lateritic pit. Broader stick support is also used in this process. The most interesting part is that after extraction of the main portion of juice, a strong natural thread made from the *Careya arborea* plant (locally known as kumyo) is tied to the fibre residue of the cashew apple, which still contains some juice and clean laterite stone is kept on this pallet as weight. After a few minutes, a clear juice obtained from this process is known as 'neera' and is relished as a sweet non-alcoholic fresh beverage and is a delicacy for locals. Similarly, a coconut toddy tapper is known to possess traditional wisdom of collecting fresh toddy for making alcoholic beverages which in its fresh form before fermentation is used for preparing coconut fresh, jaggery, vinegar, and alcoholic beverage known as 'madachifeni'. But in this case, also the fresh sap captured immediately in an ice box was recently developed by the Central Food Technology Research Institute and is sold under the brand name 'kalparasa' by a private enterprise which is a very deliciously sweet, nutritious, and healthy drink kept in a cold chain—this is further research on traditional knowledge and is strength for bioeconomy.

Hemidesmus indicus

Cashew Apple

Process of Cashew Feni Making

Cashew Apple being collected, crushed by trampling under foot and then residual fibre using thread tied (Photo Credit: NV.Eco-Farms)

Typical copper vessel used for distillation of cashew feni from fermented cashew apple juice (Above Photos: Shri. Arnold Dominic Noronha)

Copper vessel in series for distillation of cashew feni (Above Photos: Shri. Arnold Dominic Noronha)

The use of herbs like *Hemidesmus indicus* to impart additional properties and spices for flavouring alcoholic beverages is an important part of innovation in traditional beverage-making styles. Producing over 15 types of wines from various local fruits is also a well-known art possessed by a few communities.

Festival delicacies, condiments, and preparations are unique to the forest and agricultural produce available at that time. The abundance of natural resources and produce is celebrated in Goa like it is done everywhere. The counterpart equivalent of *'idli'* (rice-based savoury cake, a steamed delicacy belonging to south India) in Goa is known as *'sanna'* which is made from specific native rice varieties mainly during festivals and served in different combinations including sweet type made using local jaggery and coconut milk. In addition to Cashew Feni, a few other bioresources like

Myndoli Banana, two varieties of chillies Khola and Harmal, a traditional condiment and sweet dish, namely GoanKhaje (chickpea flour crunchy stick encrusted with jaggery infused with ginger and sprinkled of sesame seeds) has been recognized in the list of Geographical Indications of Goa. They are the basis of the Goan bioeconomy.

Significant Issues

When it comes to the problems surrounding the application of the bioeconomy, there is a dispersed, compartmentalized approach that does not converge to satisfy the equilibrium needs. On one side, authorities like the National Biodiversity Authority are trying to achieve conservation through sustainable utilization and a fair and equitable mechanism of Access and Benefit Sharing (ABS). However, because their raw materials are derived from biodiversity, companies are unable to see or accept that biodiversity protection and sustainability will also guarantee the sustainability of their industries. Although this is a clear fact, people still do not embrace it, and compliance is more in the nature of enforcement. Since compliance will help the companies to maintain their business, it should be voluntary. However, today, we see that it is more of a compliance-based approach; they will try to comply with all enforced laws. Therefore, the government must also change its strategy and stop depending solely on the legal system to punish, prosecute, or imprison people. This may not be the only way; we will have to try other methods, such as incentivizing and rewarding conservationists in a very significant manner, which will make a difference to their business. Sincerity and recognition will propel them into more exclusive markets, and moral behaviour ought to be promoted in a way that demonstrates their dedication to the world's premium markets.

> If the production, processes, products, marketing, and branding of a business are based on sustainable, bio-economic principles that prioritize ethics, fairness, and equitable sharing of benefits with communities, then such initiatives and activities should be recognized, rewarded, and organized in a way that boosts their business goals while being inclusive.

The Opportunity of Bioprospecting for Bioeconomic Growth in Goa

Traditional knowledge (TK) refers to the cumulative body of knowledge, practices, and beliefs passed down from generation to generation within a particular community or culture. It encompasses various disciplines, including agriculture, medicine, ecology, and resource management. Bioprospecting, on the other hand, is the exploration and utilization of biological resources, such as animals, plants, and microorganisms, for commercial reasons, comprising the development of new products, processes, or services.[8]

The intersection of traditional knowledge and bioprospecting holds immense potential for the bioeconomy, which is an emerging economic model that aims to create value from renewable biological resources. By harnessing the wealth of traditional knowledge and combining it with modern scientific techniques, we can unlock the potential of biological resources and develop sustainable solutions for various industries, including pharmaceuticals, agriculture, and energy.[9]

In the context of the bioeconomy, traditional knowledge can provide valuable insights into the properties and uses of biological resources, which can inform the development of new products and processes. For instance, many traditional medicinal practices have been the starting point for discovering new drugs and therapies, highlighting the importance of preserving and valuing this knowledge (Getgen Kestenbaum, 2006).

However, the value of traditional knowledge in bioprospecting raises ethical concerns, such as ensuring fair and even-handed sharing of benefits with the communities that hold this knowledge, as well as respecting their cultural practices and intellectual property rights. Striking the right balance between economic development and the preservation of traditional knowledge and biodiversity is a critical challenge that needs to be addressed.[10]

Goa, with its rich biodiversity and diverse cultural heritage, presents interesting possibilities for leveraging traditional knowledge and bioprospecting for the bioeconomy. Here are some potential areas of exploration:

1. **Ayurvedic and traditional medicine**: Goa has a long-standing tradition of practicing Ayurvedic medicine, which relies on plant-based remedies and therapies. Bioprospecting efforts could focus on identifying and exploring the medicinal properties of local flora, potentially leading to the development of new drug candidates or natural health products, the potential of trees like *Salacia sp, Hemidesmus indicus, Ashoka, Holarrhena antidysentrica, Zanthoxylum rhetsa*, etc. can be studied. It is possible to collaborate in Goa with our certified traditional community healers to understand the potential of bioprospecting (Okagu et al., 2021; Sawant & Rodrigues, 2015).
2. **Sustainable agriculture and agroforestry**: Traditional farming practices in Goa, such as organic farming and agroforestry systems, can provide valuable insights into sustainable agricultural methods. Bioprospecting could involve

[8] https://www.wipo.int/tk/en/tk/#:~:text=Traditional%20knowledge%20%28TK%29%20is%20knowledge%2C%20know-how%2C%20skills%20and,forming%20part%20of%20its%20cultural%20or%20spiritual%20identity.

[9] United Nations Environment Programme. (2000). *Traditional knowledge and biological diversity* (UNEP/CBD/COP/5/INF/7). https://www.cbd.int/doc/meetings/cop/cop-05/information/cop-05-inf-07-en.pdf.

[10] https://www.cbd.int/abs.

identifying and developing new crop varieties or exploring the potential of locally adapted plant species for applications in agriculture, food production, or biofuel production. Focus on Kulagar method of multilayer plantations, Khazan lands, and salt-tolerant species such as "korgut" can also make a market during a climate change scenario (Gomes, 2022; Miguel, 2023; Sonak & Sonak, 2014).

3. **Coastal and marine resources**: Goa's coastal region is rich in marine biodiversity, including various algae, fish, and other marine organisms. Traditional knowledge related to the use of these resources, such as in food, medicine, or other applications, could be explored for bioprospecting opportunities in the marine biotechnology sector (Demunshi & Chugh, 2010).
4. **Handicrafts and natural dyes**: Traditional Goan handicrafts, such as pottery, weaving, and natural dye production, often rely on locally sourced biological materials. Bioprospecting could involve investigating the properties of these materials and exploring their potential applications in various industries, such as textiles, cosmetics, or sustainable packaging. *Indigo feratinctoria* (indigo) or *Lawsonia inermis* (henna) in textile dyeing could be explored for developing eco-friendly and sustainable dye alternatives[11] (ShetVerenkar, 2021).

Proposed Systems and Models

It is now essential to devise a system or an innovative method to deal with these problems. The proposed system is that in the case of traditional wisdom, the knowledge holders may be identified and properly recognized through a system of evaluation. Goa State Biodiversity Board certified the Traditional Community Health Providers (TCHPs) through the Quality Council of India. Likewise, it is necessary to identify and certify TK holders through the verification process by involving local Biodiversity Management Committees (BMCs).

TK documentation is already provisioned in Peoples Biodiversity Registers (PBRs). PBR is the first-hand documentation containing information about local biodiversity and associated traditional knowledge with due credit to the knowledge-holding individuals and communities. Several formats, techniques, methods, and innovative approaches are deployed in preparing PBRs. Goa State Biodiversity Board has added a few more formats in addition to those prescribed by the National Biodiversity Authority, like recording practices of using bioresources from birth to death in various communities, which also includes bioresources used in the infant stage, nutrition of babies, lactating mother, those used in rituals, unique recipes, etc. Another format is prescribed for human-animal conflict and its management, and likewise, the PBRs are dynamic documents that are updated periodically and validated at three levels in Goa i.e. at BMC level, Gram Sabha Level, and State Level by State PBTR Validation Committee comprising of experts from various domains, departments including TK. TK recorded in PBRs could be further validated using scientific methods.

[11] https://goa-tourism.com/culture/handicrafts-of-goa-2/.

The process of certifying traditional knowledge holders is tricky. Validation is essential to ensure that genuine and recognised TK holders get the due credit they deserve. It is necessary to ensure that the assessing mechanism will respect the theological and ethnic aspects associated with traditional wisdom holders. Even if it initially seems like there is deviation from the scientific norms, it may lead to new findings. Some TK holders may come out with, unbelievably effective methods or solutions. Hence, we must respect them until they are verified, validated, and found genuine or otherwise. In this system, once it is established that a particular community possesses traditional wisdom, it can further inspire bioprospecting or is ready-to-use formulation. Respecting TK avoids parallel channels by accessing their knowledge.

There could be two systematic approaches to institutionalize, record, and ensure accreditation.

1. The First approach is to provide one-time monetary compensation to deposit their TK in case their future generations are not interested in taking ahead TK-based practices. In such a case, in addition to one-time benefits, they should be provided with a certain portion of the revenue generated as royalty, which will also ensure that their family will always get recognition and financial support in the future. However it should be case-specific based on the degree of TK utilized in a new product, process, or solution. If it is just inspiration, then a one-time settlement may be decided.
2. The second strategy involves the community of traditional knowledge holders and practitioners who have demonstrated the strength to spread and use TK-based practices, either by institutionalizing or systematizing the handholding process or by being willing to accept it. Such custodians of traditional knowledge should be acknowledged and institutionalized further so that specialists from the community and all subsequent generations can be acknowledged through a thorough scientific process. They should also be disseminating this information and reaping the advantages in a morally responsible manner.

Such TK holders should be acknowledged and institutionalized further so that specialists from the community and all subsequent generations can be acknowledged through a thorough scientific process. They should also be disseminating this information and reaping the advantages in a morally responsible manner. An ethical organization or a State entity may be entrusted with the further research and development of developing fool proof solutions. Then, they may be permitted to keep a major chunk of benefits as mentors to ensure that while the TK holder community derives fair and equitable sharing of benefits, society at large also benefits from TK-inspired research.

A State entity or an ethical organization may be tasked with conducting additional research and development to create fool proof solutions. After that, they may be allowed to retain a sizable portion of the benefits in their role as mentors to ensure that, in addition to the community of TK holders receiving a fair and equitable share of the benefits, society as a whole also benefits from TK-inspired research. There must be precise rules, and in cases when arbitration is required, a State-level body with the necessary authority should be established to handle the cases.

In general, a model should be created to give direct managerial opportunity to future generations, who have already tapped into and conserved such information over time. We will take this chance to carefully preserve everything. For the bioeconomy to be included in state and national budgets, planned or policy-level documentation is needed at the State, union territory, and federal levels. This is because, as stated in the 2018 Global Bio-Economic Summit, bioeconomy leads to the convergence of conservation, innovation, and technology production processes across all economic sectors.

Conclusion

In nutshell, the intersection of traditional knowledge and bioprospecting holds immense potential for shaping the bioeconomy of the future. As we discussed this complex but needed subject, it is imperative to recognize and respect the contributions of indigenous and local communities to scientific advancements. Upholding principles of equity, fair benefit sharing, and sustainable resource management will be crucial in harnessing the full potential of bioprospecting while safeguarding the rights and interests of traditional knowledge holders. By adopting collaborative partnerships and embracing innovative models of benefit sharing, we can create a bioeconomy that not only drives economic growth but also promotes cultural preservation and environmental sustainability for generations to come.

Conflict of Interest The authors state no conflict of interest.

References

Ahoyo, C. C., Houéhanou, T. D., Yaoïtcha, A. S., Prinz, K., GlèlèKakaï, R., Sinsin, B. A., et al. (2021, September). Traditional medicinal knowledge of woody species across climatic zones in Benin (West Africa). *Journal of Ethnopharmacology, 26*, 1134175.

Baby, T. B., & Suriyaprakash, T. N. K (2022). *Intellectual property rights: Bioprospecting, biopiracy and protection of traditional knowledge—An Indian perspective*. IntechOpen. https://doi.org/10.5772/intechopen.99596

Banerjee, M. (2019, July). Biopiracy in India: Seed diversity and the scramble for knowledge. *Phytomedicine, 53*, 296–301. https://doi.org/10.1016/j.phymed.2018.10.017

Beattie, A. J., Hay, M., Magnusson, B., de Nys, R., Smeathers, J., & Vincent, J. F. V. (2011). Ecology and bioprospecting. *Austral Ecology, 36*(3), 341–356.

Birch, K. (2012). Knowledge, place, and power: Geographies of value in the bioeconomy. *New Genetics and Society, 31*(2), 183–201. https://doi.org/10.1080/14636778.2012.662051

Christian, N. D. (2007). *From biopiracy to bioprospecting: An historical sociologyof the search for biological resources*. PhD dissertation submitted to the Department of Sociology, University of Warwick, Coventry.

Definition adopted in Global Bioeconomy Summit 2018 in report of FAO 2019.

Demunshi, Y., & Chugh, A. (2010). Role of traditional knowledge in marinebioprospecting. *Biodiversity and Conservation, 19*, 3015–3033.

Erbisch, F. H. & Maredia, K. M. (1998). *Intellectual property rights in agricultural biotechnology* (pp. 4–6). CAB International.

Gaudillière, J. P. (2014). An Indian path to biocapital? The traditional knowledge digital library, drug patents, and the reformulation regime of contemporary ayurveda. *East Asian Science, Technology and Society: An International Journal, 8*(4), 391–415.

Getgen Kestenbaum, J. (2006). *Bioprospecting and indigenous cultural rights: Creating authorship, agency, empowerment and the next generation of biopirates?* Agency, Empowerment and the Next Generation of Biopirates.

Gomes, V. (2022, May). *Understanding Goa's kulagars*. Gomantak Times. https://www.gomantaktimes.com/news/goa/understanding-goas-kulagars

Hellerer, U., & Jayaraman, K. S. (2000). Greens persuade Europe to revoke patent on neem tree. *Nature, 405*(6784), 266–267.

https://goa-tourism.com/culture/handicrafts-of-goa-2/

https://www.cbd.int/abs

https://www.wipo.int/tk/en/tk/#:~:text=Traditional%20knowledge%20%28TK%29%20is%20knowledge%2C%20know-how%2C%20skills%20and,forming%20part%20of%20its%20cultural%20or%20spiritual%20identity

IPBES. (2019). *Global assessment report on biodiversity and ecosystem services of the Intergovernmental Science-Policy Platform on Biodiversity and Ecosystem Services* (E. S. Brondizio, J. Settele, S. Díaz, & H. T. Ngo, Eds.), IPBES Secretariat.

Javed, G., Priya, R., & Deepa, V. K. (2020). Protection of traditional health knowledge: International negotiations, national priorities and knowledge commons. *Society and Culture in South Asia, 6*(1), 98–120. https://doi.org/10.1177/2393861719883069

Kannaiyan, S. (2007). Paper circulated for discussion at the National Consultation Workshop on Agro biodiversity hotspots and Access and benefit sharing held during July 19–20, 2007 at Annamalai University, Annamalai Nagar.

Mgbeoji, I. (2003). Beyond rhetoric: State sovereignty, common concern, and the inapplicability of the common heritage concept to plant genetic. *Society and Culture in South Asia, 16(*4), 821–837.

Mgbeoji, I. (2006). *Global biopiracy: Patents, plants, and indigenous*. University of British Columbia Press.

Miguel, B. (2023, August). *From North to South Goa, rice cultivation is on the upswing*. https://www.gomantaktimes.com/lifestyle/from-north-to-south-goa-rice-cultivation-is-on-the-upswing#goog_rewarded

Mulligan, S., & Stoett, P. (2000). A global bioprospecting regime: Partnership or piracy? *International Journal, 55*(2), 224–246.

Okagu, I. U., Ndefo, J. C., Aham, E. C., & Udenigwe, C. C. (2021). Zanthoxylum species: A comprehensive review of traditional uses, phytochemistry, pharmacological and nutraceutical applications. *Molecules, 26*(13), 4023.

Pordié, L., & Gaudillière, J.-P. (2014). The Reformulation regime in drug discovery: Revisiting polyherbals and property rights in the ayurvedic industry. *East Asian Science, Technology and Society: An International Journal, 8*(1), 57–79.

Pushpangadan, P., George, V., Ijinu, T. P., & Chithra, M. A. (2018). Biodiversity, bioprospecting, traditional knowledge, sustainable development and value-added products: A review. *Journal of Traditional Medicine and Clinical Naturopathy, 7*, 256. https://doi.org/10.4172/2573-4555.1000256

Resource Manual for Bioprospecting. http://ipbio.org/pdfs/papers/ResourceManual_Oct14_2004.pdf

Sawant, A. S., & Rodrigues, B. F. (2015). Documentation of some medicinal plant species from Goa. In *Advances in plant sciences and biotechnology* (pp. 10–16). Goa University.

Sharma, A. (2020). *India: Geographical indication tagging of agricultural products and food-stuffs*. https://www.mondaq.com/india/trademark/898690/geographical-indication-tagging-of-agricultural-products

ShetVerenkar, N. G. (2021). *Screening, evaluation, characterization of potential biological stains and textile dyes of plant origin* (Doctoral dissertation). Goa University.

Sonak, S. M., & Sonak, S. M. (2014). An introduction to Goa and Khazan ecosystems. In *Khazan ecosystems of Goa: Building on indigenous solutions to cope with global environmental change* (pp. 15–31). Springer.

Soni, S., Anandjiwala, S., Patel, G., & Rajani, M. (2008, January–February). Validation of different methods of preparation of *Adhatoda vasica* leaf juice by quantification of total alkaloids and vasicine. *Indian Journal of Pharmaceutical Sciences, 70*(1), 36–42. National Library of Medicines, India. https://www.ncbi.nlm.nih.gov/pmc/articles/PMC2852058/

Tedlock, B. (2006). Indigenous heritage and biopiracy in the age of intellectual property rights. *Explore: The Journal of Science & Healing, 2*(3), 256–259.

UNEP. (2018). *The Clearing-House Mechanism of the Convention on Biological Diversity: 6th National Report-India* (pp. 1–270). https://chm.cbd.int/database/record?documentID=245938

United Nations Environment Programme. (2000). *Traditional knowledge and biological diversity* (UNEP/CBD/COP/5/INF/7). https://www.cbd.int/doc/meetings/cop/cop-05/information/cop-05-inf-07-en.pdf

Van Overwalle, G. (2005). Protecting and sharing biodiversity and traditional knowledge: Holder and user tools. *Ecological Economics, 53*, 585–607.

WIPO. (2000). *Traditional knowledge*. https://www.wipo.int/tk/en/tk/

WIPO. *Traditional knowledge and intellectual property—Background brief*. https://www.wipo.int/pressroom/en/briefs/tk_ip.html

The Quest for Shangri-La: The Expectations and Challenges Under the New BBNJ Treaty

Arindam Basu and Sharda Mandal

Introduction

The treaty for the conservation of marine biological diversity in the areas beyond national jurisdiction (hereinafter BBNJ Treaty or simply Treaty)[1] is convened to address serious concerns in international law, *i.e.* the continuous declining of marine biodiversity in Areas Beyond National Jurisdiction (ABNJ). To date, these areas have received weak legal protection, mostly through the provisions of the United Nations Convention on the Law of the Sea, 1982 (UNCLOS or hereinafter Convention),[2] the principal international legal instrument for ocean governance. The new Treaty maintains the aspiration of the participating nations to deal with the limitations primarily of the UNCLOS and also other pertinent international or regional legal instruments.

The acknowledgment of the importance of the ocean is almost universal. Other than hosting numerous life forms, oceans are the cradles of life support systems, living and non-living. On top of it, the science of climate change is directly connected to the complex and delicate ocean chemistry.[3] Therefore, what happens to the ocean and its vital resources in the areas beyond 200 nautical miles from the shore[4] is of

[1] The Agreement was adopted in New York on 19 June 2023. It is now open for signature.

[2] Convention on the Law of the Sea, Dec.10, 1982, 1833 U.N.T.S.397.

[3] According to IUCN, in the last five decades, the oceans have absorbed between 24 and 33% of carbon dioxide emissions. This enormous intake has caused 'ocean acidification', changing ocean chemistry drastically; For more information on ocean acidification visit https://www.iucn.org/our-work/topic/oceans-and-climate-change/ocean-acidification#:~:text=The%20massive%20input%20of%20CO2,(i.e.%20increases%20its%20acidity) (Last accessed 25 February 2024).

[4] Part V of UNCLOS deals with Exclusive Economic Zone and Part VI deals with rules regarding continental shelf.

A. Basu (✉) · S. Mandal
Rajiv Gandhi School of Intellectual Property Law, IIT Kharagpur, Kharagpur, India
e-mail: arindam@rgsoipl.iitkgp.ac.in

L. Krishna Panicker et al. (eds.), *Biodiversity and Business*,
https://doi.org/10.1007/978-3-031-71674-4_25

great importance. Because we epitomize the biodiversity loss at the high seas as a problem of global commons, it is important to understand the nature of the problem carefully.

As a global common, ABNJ covers the water column and the seabed of the high seas. In principle, this extension of seabed beyond the continental shelf and after 200 nautical miles Exclusive Economic Zone (EEZ) is also known as the 'Area'.[5] The difference between the Area and the high seas is not straightforward. Two provisions of UNCLOS are illustrative. Article 87 clearly states that 'the high seas are open to all States, whether coastal or land-locked'. Accordingly, states shall use their freedoms of the high seas, as enumerated in Clause (a) to (f) in that Article, under the conditions laid down by UNCLOS along with other rules of international law.[6] In addition, the freedoms available under Article 87 'shall be exercised by all States with due regard for the interests of other States'.[7] Article 140, on the other hand, provides that 'activities in the Area shall….be carried out for the benefit of mankind as a whole, irrespective of the geographical location of States, whether coastal or land-locked'.[8] What we can approximately infer from these broadly worded provisions is that activities on the high seas are to be carried out by the states with the 'due regard for the interests of other States'. But the activities in the Area are to be 'carried out for the benefit of mankind as a whole'. Beyond this understanding, any other suggestion will be non-literal.

However, there remains uncertainty about the regulation of the high seas. In that respect, the idea to govern the high seas by application of the Common Heritage of Mankind (CHM) principle, having roots in historic *res communis* standard,[9] is odd. Moreover, the 'community of interest' ideology remains at the core of CHM[10] in general and is viewed as a contrast to the state sovereignty doctrine. In other words, states, enjoy complete freedom on the high seas,[11] are allowed to make arguments to preserve their self-interests without undertaking any serious obligation. This is a restatement of sovereign exercise of the right. However, CHM modifies this absolute freedom marginally. Yet, the regulation of ocean resources in ABNJ is essentially conditioned by the belief of the user states, suitable to their self-interests. This typical self-interest preserving norm gradually had been understood by scholars as a problem

[5] The Area is defined as 'the seabed and ocean floor and subsoil thereof beyond the limits of national jurisdiction.' See Article 1(1) and Part XI of UNCLOS.

[6] Article 87(1) of UNCLOS.

[7] Article 87(2) of UNCLOS.

[8] Article 140 (1) of UNCLOS.

[9] An area of land or water that lacks sovereign ownership by any state. Examples include the high seas, as outlined in Article 2 of the Geneva Convention on the High Seas and Article 89 of the 1982 Convention on the Law of the Sea, as well as outer space. These areas are governed by the principle of the common heritage of mankind. See, Oxford Reference, *Overview: Res Communis*; Available at https://www.oxfordreference.com/display/10.1093/oi/authority.20110803100408305 (Last accessed 19 February 2024).

[10] For a detail discussion on transformative history, see Tanaka (2023).

[11] Subject to minimal regulation imposed by UNCLOS.

of global commons, giving birth to a remarkable array of literature. The major inspiration came from Garett Hardin who in 1968 published his classic account. *The Tragedy of the Commons* (The Tragedy of the Commons 1968).[12] Over the years, the theory has been refined in the hands of other scholars and influenced different disciplines wherever the management of public goods is involved. Hence, a claim is not far off, reaffirming the 'problem' as the 'tragedy' of global commons where adequate regulation of freely available resources is an issue.

For the matter of regulation, UNCLOS is operational in all parts of the ocean, covering various activities. However, Part XII, spreading across 45 Articles,[13] of the UNCLOS is of particular interest to the marine environment as it cuddles general obligations for all states. These obligations to protect and preserve the marine environment, ideally should cover the biodiversity as well. However the ambiguity flows from the fact that Part XII does not cover ABNJ. In this context, Part XI of the UNCLOS, containing provisions for regulating polluting activities in the Area, can also be referred to.[14] Oddly, both Part XII and Part XI are silent about the conservation of marine biodiversity on high seas, even though, the possibility of damage to any such ecosystem from mining activities carried on in the Area or polluting activities from the ships are well-acknowledged. For sure, the wording of Article 192 of UNCLOS can be considered as a possible reminder for the states,[15] whether parties or not, to take appropriate action to protect the marine environment as a whole. Yet, this argument leaves a key question unanswered—How, in the absence of obligatory norms, should states escape the collective bargaining problem when they are primarily driven by self-motivation?

BBNJ Treaty is aimed at answering that question. Quietly, the farmers of the Treaty have returned to Hardin's basic premise—freedom in commons brings ruin to all. The ensuing 'tragedy' marks an irrational phenomenon of overconsumption of resources (The Prisoner's Dilemma 2015).[16] As the stock of biodiversity on high seas is dwindling, one can condemn the choices made by the material world to exploit more and care less. Stopping the collapse of marine biodiversity in the global commons is a must and after UNCLOS, that task is now left to the BBNJ Treaty. But BBNJ Treaty in its current format has left several issues unanswered. It is one thing to say that the structure of the Treaty will be developed dynamically based on the primary framework agreed by the parties. But it is an entirely different claim that backs Treaty's strength in a hurry. This begs further clarification whether provisions of BBNJ Treaty are conceived with enough flexibility to address deficiencies that exist under UNCLOS regime. Alternatively, this has also to be seen if the same

---

[12] It was originally, conceptualized by British scholar William Foster Lloyd in 19th Century. See Hardin (1968).

[13] Article 192-237 of UNCLOS.

[14] Articles 209 and 215 of UNCLOS.

[15] Article 192 states that States must protect and preserve the marine environment and its content is informed by more specific obligations in the subsequent provisions, including Articles 194, 204, 205, and 206. See, Liao (2021).

[16] Economists call it Prisoners' dilemma. For further details see Peterson (2015).

provisions allow state parties too much freedom to allocate their risks, following their motivation. It appears that the identification, prioritization, and subsequent actions taken by the states under BBNJ Treaty will follow the power configuration similar to UNCLOS. Therefore, the BBNJ Treaty is already standing on a questionable normative foundation where the rigid form of state sovereignty remains as a controlling mechanism. Additionally, the agreement epitomizes extravagant expressions, thinning the commitments of the participating nations. This makes the future opaque as there appear to be very few incentives for the states to deviate from their cherished freedom of the high seas.

In this chapter, we build up on the idea that UNCLOS has proved inadequate to provide a practical solution to the problem of commons and in that backdrop, we examine how far motivation of the states to use and exploit high seas resources has shaped the BBNJ Treaty. We argue that in general, participants have stopped short of advancing comprehensive solution towards the problem of commons, and the promise, whispering around during the negotiation stage, has folded without delivering a sufficient blow to the strong sense of state sovereignty or sovereign interests of the states. We intend to examine this aspect closely with an essential rider that environmental issues like conservation and management of common resources leave a narrow space of elbowing for all nations irrespective of their positions in international politics. Hence, there is a pressing need to revisit the pitch of negotiation to avoid any further 'tragedy' that already has turned up high seas into a battleground for interest negotiation.

This chapter is divided into six parts. Part I introduces the theme. In Part II we delve into the legal schemes that apply to global commons. We explore the theoretical premise to see how UNCLOS has assimilated the concept within its framework. Parts III and IV are built on this understanding. In Part III we examine the provisions of UNCLOS and BBNJ Treaty to see how the open language of both these international agreements supports a 'system in progress'. This is of particular importance as the conservation process of marine environment, in general, should lend support for biodiversity protection, within and beyond the agreed legal structure. This discourse has to be tested at the altar of cooperative model in international law. We argue that alongside the outcome, it is imperative to look into the process as well. Part IV is the logical extension of Part III where we identify common markers between UNCLOS and BBNJ Treaty, only to settle on diverging points in the end that can have a significant impact on BBNJ Treaty's future development. In Part V we revisit the key premise of our arguments, *i.e.* impediments that rigid state sovereignty doctrine poses against the development of a cooperative legal framework. The discussions reveal that BBNJ Treaty has not yet done enough to modify the incentives of the nations' so that willingly they will shy away from displaying their narrow interests when it comes to protecting marine biodiversity on the high seas. Part VI concludes the chapter.

Global Commons and Rules of Engagement—A Debate Within UNCLOS

International law presents before us stimulating questions, foremost of which are related to the creation and functioning of treaties. Scholars often wonder when and why states enter into multilateral treaties and how these treaties work.[17] Theories tell us that one of the important functions of international law is to improve upon state relationships where externalities are involved.[18] In general, the process of treaty-making is a complex affair, involving five phases—pre-negotiation, negotiation, ratification, implementation, and renegotiation. Disputes may arise at or regarding any stage of this sufficiently lengthy process. Any linear assumption about complete compliance with all the provisions of the treaty by a state is misplaced. Even the conventional interpretations acknowledge two contrasting phenomena. The legal strength of a treaty[19] leads to normativity and thereby, enlarges the scope of conveying a sense of obligation to the state parties. Yet, a party to a treaty sometimes moves away from compliance as its self-interests outplay the sense of broader obligation.[20]

Evaluating the 'self-interest' of a state and forming any rational judgment about the reason for non-compliance depends on how international externalities are taken care of by the states and sometimes by a private entity. Posner and Sykes offer an understanding that can be useful in absorbing the process. According to them, when a firm operating within a country causes pollution in another country, the former has no incentive to control such pollution. This is primarily because the effect of pollution is felt in another country and the originating state will be reluctant even though the regulation is justified from a global perspective. Alternatively, any excessive regulation by the state may also have a negative effect when the global costs of such regulation exceed its benefits.[21] In this example, the behaviour of the state and the firm is independent of any existing treaty obligation. It may be possible that the originating state is already regulating the cross-border pollution, the source of which is located within its territory (Stockholm Declaration 1972).[22] Even so, any action taken by the originating state is essentially conditioned by its incentive to take action, the cost of which must not exceed its benefits. Any treaty, reversing this

[17] Goldsmith and Posner (2005, p. 81).

[18] Posner and Sykes (2010, p. 571).

[19] This often is related to the choice of expressions in the treaty text.

[20] Goldsmith and Posner (2005, p. 83).

[21] This may particularly happen when the firm is a foreign company and lowers the price of its products to keep its customer base intact. However this internalization of the costs of regulation may be of no use as the overall costs of regulation surpass the benefits. Posner and Sykes (2010, p. 571).

[22] In the absence of any specific treaty obligation states have a non-binding general obligation to do so under some of the landmark international environmental conventions. See, Principle 21 of the Stockholm Declaration, 1972 [UN Doc. A/RES/2994(XXVII)] and Principle 2 of the Rio Declaration [31 I.L.M. 874 (1992)].

process by including strong compliance provisions, runs the risk of failure in the form of aggregate or collective non-compliance.[23]

Nonetheless, it can also be argued that the existence of externalities should justify the need for global regulation. Moreover, when the overall benefits exceed the cost to one nation, the cooperative model presents a strong case for international regulation. But this argument is a precursor to norm-creation process in international law and coveys very little about the choice mechanism, preferred by a state under a specific situation. The issue gets further complex when common resources are involved. By description, such resources give rise to free-riding problem,[24] leading to market failure and externalities. Many activities carried on at the high seas are about using resources freely available to all. Therefore, there always remains a doubt about the efficient use of such resources. Accordingly, it appears that international cooperation may be the only solution to address these externalities.[25] Recalling Garett Hardin's hypothesis that individuals with their narrow self-interest are likely to over-utilize public resources, it appears that regulation of global commons has to be a consequential response that is supposed to serve two-fold purposes. Firstly, any such effort must offer participants a reason to act collectively and secondly, identification of such reason must be instilled into a model of action that ideally should explain the rules of engagement by keeping sufficient disincentive against non-compliance. Though, Hardin's theory has been worked upon by many, these basic principles are never altered.

Nonetheless, the presence of social dilemma in Hardin's theory also indicates towards the need for control. But what form of control should be appropriate to avoid the tragedy that may ensue from the overuse of high seas resources? Paternalistic intervention by international law to allocate property rights to individual states may be a way out. This is similar to the argument that the externality problem desirably should to be addressed by international law through the rules of cooperation. But still, it does not tell us anything about the prerequisite for cooperative solutions for those nations that do not wish to take part in the treaty negotiation process. Hence, the agreeable solution, if any, will only benefit a close group, having incentives to agree with the bargaining outcome. Elinor Ostrom, has provided a different perspective for the cooperative model. Putting much emphasis on experimental works, she has established the presence of variable outcomes relating to the governance of commons. These differences largely originate from the unpredictable degrees of governmental interference and the inevitable influence of a market.[26] Ostrom particularly cautions us against deceitful self-interested decision-making model and 'rampant opportunistic

[23] Example is Kyoto Protocol to the United Nations Framework Convention on Climate Change, Dec. 10, 1997, 2303 U.N.T.S. 162.

[24] The free riding problem and the logic of collective action can be understood in many different contexts. In international law, free riding means benefiting from a collective good, such as global commons, without incurring the costs of its conservation. In modern times the problem was first systematically analysed by Mancur Olson. See Olson (1971).

[25] Posner and Sykes (2012, pp. 17–20).

[26] See generally, Ostrom (2015).

behaviour' that in certain situations "severely limits what can be done jointly without major investments in monitoring and sanctioning arrangement".[27]

Also, we must not overlook the voluntary cooperative model that private actors can always come up with to avoid externality problems. But voluntary initiatives involve uncertainties and lack permanency.[28] However, governments can distribute the resources appropriately among the interest groups, such as allocating the volume of fish caught in an area or auctioning mineral rights[29] or prohibit polluting activities. In any case, it amounts to maximize the value of identifiable common resources.[30]

However, the contemporary environmental problems are very complex, resolutions to which rely on multiple factors. This requires additional interpretation of the classical theory of commons. Consequently, any rules for collective action must take into consideration two sets of interests—the bargained-for interests of participants and also of those entities who can likely to influence the operation of the rules of participation. This opens up to one of the dominant concepts of public international law—state sovereignty that has deeply influenced most of the international agreements and UNCLOS is no exception. Remarkably, UNCLOS has preferred a scheme of territorial allocation based on state sovereignty in coastal regions. This is similar to acknowledging private property rights on land. However, a different approach in the case of high seas is followed wherein all states are allowed to access the areas subject to certain rules.[31] There are multiple reasons for this divergence. Firstly, the remoteness of the high seas makes monitoring costly and states felt reluctant to get into it. Secondly, the scientific and commercial benefits of resources available in seabed and beyond were not perfect during the negotiation of UNCLOS, leaving the management issues unresolved. Thirdly, the gravity of environmental issues of international dimensions was also not entirely understood during the time of UNCLOS negotiation. Therefore, UNCLOS negotiators were unsure about the externalities involved and in its unreliable remedial cost. Fourthly, shipping activities on high seas were mostly individual and state-regulated, a system that was the extension of an unequivocal recognition of state sovereignty doctrine within UNCLOS system. In the long run, this has not posed any distribution problem as states themselves can decide the rules of engagement after following some general guidelines of UNCLOS.[32] Sure enough, this has been a deficient arrangement as UNCLOS over the years fail to resolve some of the difficult problems related to the usage of common resources.[33]

---

[27] Ostrom (2015, p. 36).

[28] There are instances in international law where non-binding UN General Assembly Resolution is accepted by the states and private players voluntarily. For example, Resolution establishing a moratorium on high seas driftnet fishing [G. A. Res. 46/215, U.N. Doc. A/RES/46/215 (December 20, 1991)].

[29] The functioning of the International Seabed Authority is based on this understanding.

[30] Posner and Sykes (2012, pp. 6–7).

[31] For additional discussion see Ostrom and Hess (2008).

[32] Posner and Sykes (2012, pp. 19–23).

[33] This deficiency is present despite UNCLOS's mandate for a compulsory dispute settlement mechanism. Posner and Sykes (2012).

Nonetheless, the emergence of the UNCLOS text was the outcome of conflict, interaction, and assimilation of countless majority and minority interests. Thus, it's dynamic and to some extent open structure in a few parts (as in the case of high seas) was tolerable. Conceivably, the parties thought that the framework they agreed upon would later eventually accommodate the negative attitude of other states or maybe the will of the majority would be strong enough to influence the decisions of those states.[34] In the end, it would be about recognition of myriad rights for which a change in perspective should be necessary. Philip Allot puts it in accurate perspective by saying that "[t]here are no freedoms from the law. There are no freedoms apart from the law. Freedoms are powers conferred by and under the law."[35] For sure, the notion of state sovereignty in the context of developing a separate treaty for high sea biodiversity is to be worked upon further with this theoretical basis.

At the moment, the consensus is that BBNJ Treaty is an improvement over UNCLOS. This is fairly acceptable with a caveat that UNCLOS itself has left enough opening for the states to develop any additional or even completely new ocean jurisprudence. But then, the weight of state sovereignty as an enduring principle was felt during the negotiation of BBNJ Treaty. We know that BBNJ Treaty is aimed to solve the problem of commons.[36] But, the text of the treaty, confirmed after the fifth session of the Intergovernmental Conference (IGC 5.2),[37] does not contain any clear reference to global commons. Yet, it is clear that the outcomes of IGC 5.2 resolve to enlarge the traditional boundary of the commons management by accommodating the claims of both, the participants and non-participants. The open language in the BBNJ text preferred by the parties may be interpreted as a 'system in progress', holding the keys to put up with unique claims that may come up in the future. Because the cooperative model in international law can be looked into from two different ways, a process and an outcome, it is the latter that ideally should interest us more. But the outcome is very much dependent on the process. Therefore, there is a need to evaluate the process as well.[38] In the next segment, we delve into that matter, firstly with UNCLOS and then specifically in the case of BBNJ Treaty.

[34] Allott (1983, pp. 6–8).

[35] Allott (1983, p. 27).

[36] This was emphasized by Jessica Battle, Senior Global Ocean Governance and Policy Expert, who was leading WWF's team at the negotiations: "Overfishing and illegal fishing, habitat destruction, plastic and noise pollution as well as climate change impacts are all rife in the high seas. Heavily subsidized, industrial fishers seek to exploit and profit from ocean resources that, by law, belong to everyone. It's a tragedy of the commons." See *WWF: High Seas Treaty Critical to Achieving 30% Global Ocean Protection Goal*; Available at https://wwf.panda.org/wwf_news/?7704966/high-seas-treaty-2023. (Last accessed19 February 2024).

[37] Details are available at https://www.un.org/bbnj/ (Last accessed 24 February 2024).

[38] Barrett (1998, p. 318).

UNCLOS, Marine Environment and Global Commons–Legal Discourse:

A series of provisions in UNLCOS provides a legal framework for the states to cooperate among themselves for the protection of the marine environment. This is the result of a long process of codification where various treaties and customary norms are fused. Unsurprisingly, the 1982 version of UNCLOS developed into a complex and multidimensional text.[39] The success story of UNCLOS is hardly disputed, though, in the beginning, there were discordant notes among some participants. For them, self-preservation was the best option in the absence of clear recognition of their rights. What emerged as a result of a prolonged negotiation was a 'package deal', covering jurisdictional entitlements, shipping activities, resource utilization, and marine pollution.[40] The area-wise divisions like territorial waters, contiguous zone, continental shelf, exclusive economic zone, and the high seas set the new standard for international ocean governance.[41] This demarcation underpinned the presence of a strong state sovereignty principle and states agreeing on the delineation arrangements had settled on compromises. Of course, this can also be termed as 'cooperation' that sounds a little more constructive. But the change in expression does not play down the constant tension that exists among the states about the scope of the territorial sovereignty and the spatial aspects of their sovereign rights. In this context, a reference to Article 74 of UNCLOS may be insightful. Clause (1) of Article 74 provides that "[t]he delimitation of the exclusive economic zone between States with opposite or adjacent coasts shall be effected by agreement based on international law, as referred to in Article 38 of the Statute of the International Court of Justice, to achieve an equitable solution."[42] In addition, Article 83 is also noteworthy for delimitation of continental shelf as its wordings are identical to Article 74.[43] Both these provisions give prominence to effect mutual agreements in accordance with international law. But critics often point out the inequitable geographical conditions across the global oceans that may cause serious practical difficulty in developing any harmonious practice (Geographical Change and the Law of the Sea 2019).[44]

At least, it is clear that high seas freedom available under UNCLOS is sanctioned from the relative standpoints of different states. Articles 116–120, dealing with 'Conservation and Management of the Living Resources of the High Seas' validate this fact. Freedoms available under these provisions are drawn on the historic tradition of navigation through the sea routes. In antiquity, notions like *res communis*

[39] The text is available at https://www.un.org/depts/los/convention_agreements/texts/unclos/unclos_e.pdf (Last accessed 19 February 2024).

[40] Strati et al. (2006, p. 19).

[41] Yet, the Agreement has failed to bring everyone on board. A major holdout is the United States.

[42] Article 74(1) of UNCLOS.

[43] Article 83 of UNCLOS.

[44] For additional discussion see, Purcell (2019).

and *res nullius* gradually took root and put negative obligations on the states.[45] These concepts contained flavour of public law that prominently resonated in Article 87 of UNCLOS, providing a mandatory requirement for all states to have due regard towards the interests of other states while exercising freedoms.[46] It is difficult to comprehend how the risk of subjective interpretation can be avoided in such a situation as it is left entirely to the states to determine what is 'due' or 'reasonable'. This normative vagueness has been a structural reality of UNCLOS.

Noticeably, the urge for relying on cooperative model is strong in Part XII of UNCLOS, dealing with the 'Protection and Preservation of the Marine Environment'. A vast array of provisions, spanning eleven sections and 45 Articles (Articles 192–237) are proof of participating nations' sincerity in addressing diverse issues connected to the marine environment. It is logical to hold that the expression 'marine environment' is supposed to be understood in the widest possible sense and the conservation of marine biodiversity has to be a subset of that setting. Remarkably, Part XII of UNCLOS fails to provide any guidance regarding how states should carry on their activities in the high seas without harming biodiversity. This omission is not unintentional though and is attributed to a difficulty faced by the parties in articulating CHM as a guiding principle within the final text of UNCLOS. The swift advancement of technological prowess after the industrial revolution allowed mankind to explore oceanic areas that were untouched and unknown to mankind before. Simultaneously, for the general interest of the global community, there was increasing demand to give more credence to ocean conservation, especially the vast resources of the Antarctic region. In 1926, Argentine jurist José León Suárez forcefully argued against the restrictive nature of the law of the sea which thus far, was more attentive to 'reciprocity and commerce'.[47] Following this, in 1948 draft World Constitution, a tacit reference to CHM was made.[48] In 1967, Maltese ambassador Arvid Pardo delivered his famous speech to the United Nations where he argued for CHM.[49] Since then; the principle has stayed in its course and gradually gained prominence.

Yet, the centrality of the debate has always been not only about the measure of diminishing motivation to protect commons but also about developing a clear

---

[45] *Res communis* means belonging to a community and *res nullius* means belonging to no one. See Park (1976).

Jus gentium means 'the law of nations', referring to the idea that there are universal moral standards that are obtained within any civilized society; see Holst (2022, pp. 90–91).

[46] Article 87 of UNCLOS recognises the freedoms of the high seas, including the right to fish, but stresses that each state must use these freedoms while taking other states' interests into account when using the high seas.

[47] Schrijver (2016, p. 1254).

[48] This draft was prepared in 1948 by the Committee to Frame a World Constitution, involving Robert M. Hutchins, G. A. Borgese, Mortimer J. Adler, Stringfellow Barr, Albert Leon Guérard, Harold A. Innis, Erich Kahler, Wilber G. Katz, Charles H. Mellwain, Robert Redfield, and Rexford Guy Tugwell.

[49] Pardo's speech is available at https://www.un.org/depts/los/convention_agreements/texts/pardo_ga1967.pdf (Last accessed 24 April 2024).

methodology that should underlie a consequential and significant international legal protection. But during the post-Hardinian era, the controversy over global commons was raging mainly because of prolonged negotiations in search of a comprehensive treaty to protect the ocean. The difficulty was to isolate a normative value for which CHM can be given an obligatory veneer. Because CHM has its basis in both the traditional sovereignty doctrine and freedom of seas principle, a politically agreeable legal framework was not easy to come by.[50] As a minimum, in the final text of UNCLOS, the states settled on the necessity of collective control as CHM finds its place in Part XI of UNCLOS which to some extent marks a break with the traditional 'sovereignty versus freedom' narrative. This also reflects the 'institutionalised' common interests' framework.[51] After all, this recognition of 'common' interests in the Area, primarily is conditioned by states' interest in accessing resources therein, instead of a priori 'common' interest to conserve or use them sustainably. Therefore, it is possible to distinguish CHM principle under UNCLOS 'as a jurisdictional principle that provides a basis for the allocation of exploitation rights.'[52] But it cannot be claimed that UNCLOS reflected a linear relationship between international law and global commons. If on the one hand, it has emerged as an instrument to forward equitable solutions to commons, on the other it has left enough openings for the states to argue otherwise, effectively undermining the importance of collective actions.[53]

But to protect biodiversity in high seas using CHM presents a fundamental challenge as therein 'sovereignty' does not collide with 'freedom', rather ratifies it. This eventually is tantamount to two clear interpretations—the absence of rights of any one state or the presence of rights of all states as a prearranged abstraction. This is a matter of semantics and in both cases, the application criteria are hardly required to be changed. We argue that the overall structure of UNCLOS has carefully maintained a balance when it comes to recognizing sovereign rights and the subsequent dilution of such rights on the high seas. A comparison can be drawn with the deep-sea bed mining regime as the process of recognizing common heritage principles was a long arduous process. States more or less were aware of potential economic gain and their ability to exploit resources available.[54] But high sea biodiversity was not a 'heritage' of mankind but merely a 'concern' at that time. Parties did not wish to fill this gap and left it to future development.

Even during the negotiation of BBNJ Treaty G-77 argued for recognising marine genetic resources (MGR) as CHM. UNCLOS negotiators knew very little about MGR and felt no compulsion to give any importance to that. For instance, Article 136 provides that "[t]he Area and its resources are the common heritage of mankind." This textual clarity leaves the matter for interpretation, indicating that the freedom of the seas should continue to be applied without any restriction for MGR.[55] It is

[50] See Holst (2022, p. 99).

[51] Holst (2022).

[52] Holst (2022, p. 100).

[53] Ranganathan (2014, p. 149).

[54] Ranganathan (2014).

[55] Nordquist et al. (2019, pp. 102–104).

easy to believe that the vastness and resource abundance on the high seas can lead to complacent thought in favor of uninhibited exploitation. BBNJ Treaty's primary aim is to curb this tendency. Its Preamble highlights 'the need for a comprehensive global regime' for the conservation and sustainable use of marine biological diversity in ABNJ.[56] It is further encouraging to notice the parties' decision to accept CHM as a guiding principle to achieve the objectives of the Treaty.[57] Moreover, this may also be cited as an effort to erode state sovereignty in environmental law-making. But a closer look at the text of the Treaty tells us that the devil is in the detail.

BBNJ Treaty-UNCLOS-Commonality and Divergence

BBNJ Treaty is for the conservation of marine biodiversity in ABNJ. In one sense, it is a product of conflicts and differences that existed between global North and South for many years. But it also conveys much more than that. The final text of the Treaty has enmeshed the principles of the freedom of high seas and CHM to lay down rules how marine biological diversity will be accessed, utilized, and shared.[58] In addition, the growing concern over climate change has been adequately reflected in its provisions.

There exists a debate about how BBNJ Treaty should be interpreted in conjunction with UNCLOS. Two options are available. Firstly, the Treaty may be seen as an implementing agreement of UNCLOS. This is similar to the Agreement Relating to the Implementation of Part XI of UNCLOS (New York Agreement), adopted through a resolution passed in 1994.[59] Article 2(1) of the New York Agreement makes it clear that for the implementation of Part XI of UNCLOS, its provisions shall be interpreted and applied in unison. In other words, the New York Agreement and PART XI of UNLCOS are to be read as one instrument.[60] The second option supports the broader view. Accordingly, BBNJ Treaty is not only a timely improvement over UNCLOS but also its ambit should grow beyond the limitations of the classical law of the sea regime. Hence, the integration of available principles and concepts from international environmental law is necessary.[61] Both arguments have their own merits. In the end, there is a need to evaluate how the process is supposed to contribute to the overall growth of marine environmental law. Article 197 of the UNCLOS is important in this regard as it obliges the states "to cooperate on a global basis and, as appropriate, on a

---

56 The BBNJ Treaty text is available at https://treaties.un.org/doc/Treaties/2023/06/20230620%2004-28%20PM/Ch_XXI_10.pdf (Last accessed 24 February 2024).

57 Article 7(b) of BBNJ Treaty.

58 Hey (2022, p. 368).

59 Text is available at https://www.un.org/depts/los/convention_agreements/texts/agreement_part_xi/agreement_part_xi.htm (Last accessed 24 February 2024).

60 Article 2(1) of the New York Agreement provides that "[t]he provisions of this Agreement and Part XI shall be interpreted and applied together as a single instrument. In the event of any inconsistency between this Agreement and Part XI, the provisions of this Agreement shall prevail."

61 Payne (2023, p. 6).

regional basis, directly or through competent international organizations, in formulating and elaborating international rules, standards and recommended practices and procedures consistent with this Convention, for the protection and preservation of the marine environment.....".[62]

An alternative way to look into this issue is the state's participation in the negotiation process. It is noticeable that a few important countries, such as the United States, Colombia, Turkey, and Venezuela had taken active role in shaping the BBNJ Treaty. But they are not the parties to UNCLOS. For these countries, any formula to protect high sea biodiversity should preferably be outside UNCLOS' mandate. Therefore, they may be willing to take affirmative action which should likely to be different from what UNCLOS demands.[63] Put differently, these countries were not ready to accept anything that would bind them under customary international law. This bold demonstration of sovereignty triumphs as Article 5(3) of the BBNJ Treaty provides that the legal status of the non-parties to UNCLOS is not affected by the Treaty.[64]

Similar to UNCLOS, BBNJ Treaty has been shaped through a long process of bargaining. More than 20 years of negotiation culminated in June 2023 when the final text of the Treaty was agreed upon. Again, the basic structure of the Treaty manifests a 'package deal' for the parties. The principal focus is on the conservation and sustainable use of marine biological diversity in ABNJ. But the overall process includes four components—(i) establishing area-based management tools, including marine protected areas (MPAs), (ii) assessing activities under Environmental Impact Assessments (EIAs), (iii) endorsing the freedom of marine scientific research with fair and equitable sharing of benefits of MGRs found in areas beyond national jurisdiction, and (iv) focusing on capacity-building and technology transfer. These tasks are to be streamlined further by the Conference of the Parties (COPs). This is a break from the UNCLOS structure which does not operate through COPs. Instead, resemblances may be drawn from the climate change regime which now operates through COP decisions.[65] Moreover, there is another area where UNCLOS and BBNJ Treaty differ. This is already pointed out that CHM principle will be a guiding force for the BBNJ Treaty. The same CHM principle finds its place within UNCLOS too. But it is noteworthy that UNCLOS text has mentioned 'mankind' and BBNJ text has preferred 'humankind' while recognising the principle. It is difficult to tell whether this change in expression will make any significant difference in application except validating the importance of gender equity that may be relevant in terms of access and benefit sharing.[66]

[62] Article 197 of the UNCLOS.

[63] Ibid.

[64] Article 5(3) of BBNJ Treaty; Also refer Payne (2023, p. 6).

[65] Latest COP 28 took place in Dubai, UAE on November 2023.

[66] See Payne (2023, p. 4).

Sure enough, with an elaborate institutional framework,[67] BBNJ Treaty characterises an international cooperative model. Because the conservation of biodiversity and its components is the central theme of the Treaty, its Preamble explicitly recognizes 'the need to address, coherently and cooperatively, biological diversity loss and degradation of ecosystems of the ocean.'[68] This looks promising on paper. But there are at least two areas of concern. Firstly, the Treaty surprisingly has omitted any definition of 'biodiversity' or 'biological diversity'.It defines 'biotechnology' though, which is similar to the definition available in the Convention on Biological Diversity (CBD).[69] At the top of it, biological diversity and biological resources both are to be protected from the impacts of climate change, and other anthropogenic litter, such as plastic pollution and unsustainable use.[70] This widens the scope of BBNJ Treaty enormously. Ideally, if the principal aim of the Treaty is regulation and sustainable use of biological resources, then what the parties are required to do is to optimally set the terms of use of those resources only which historically has been a matter of terrific contest among the nations. Nonetheless, BBNJ Treaty looks beyond that objective and wants to deal with the impacts of climate change on the world's oceans as well.[71] But none of these provisions is specific. Instead, the parties set their sights based on liberal ideology, the magnanimity of which is amply reflected in the words of Article 1(6), describing what are the 'cumulative impacts' of different activities and 'consequences of climate change, ocean acidification, and related impacts' that parties should consider.[72] This is, however, a common template in almost all international legal documents, binding and non-binding. Therefore, it is logical to have some reservations about the future of the BBNJ Treaty, especially when the goal is to improve upon the old system.

For BBNJ Treaty, the second most difficult problem originates from the implementation challenge. UNCLOS negotiations were gravely affected by the controversy about the extension of the continental shelf beyond 200 nautical miles. Once again, the rights and freedoms of states were at the center of the debate. To make everyone happy, Article 82 was drafted with a typical sense of equity. Under this provision, a coastal state is required to "make payments or contributions in kind in respect of the exploitation of the non-living resources of the continental shelf beyond 200 nautical miles."[73] No reference to biodiversity is made. However, how a state should use the international seabed in an area beyond EEZ has immense indirect significance on the conservation of high sea biodiversity. The entire arrangement is 'enabling' in

[67] Other than creating COPs, the treaty also houses a Scientific and Technical Body, a dispute settlement mechanism with a feature to settle disputes of a 'technical nature'.

[68] Preamble of BBNJ Treaty.

[69] Article 1 (3) of BBNJ Treaty.

[70] Preamble of BBNJ Treaty.

[71] For example, see Preamble, Article 1(8), Article 5(g), Article 14(c), clause (f) of indicative criteria for identification of areas mentioned in Annex I and clause (iv) of Annex II identifying, types of capacity-building and transfer of marine technology.

[72] Article 1(6) of the BBNJ Treaty.

[73] Article 82 of UNCLOS.

nature and framed to incentivise the states to proceed with their mineral exploitation plans. This empowerment of the parties over the years has given birth to overlapping sovereignty claims around 200 nautical miles maritime boundaries.[74] An example of this is the dispute between China and Philippines in South China Sea. It is to be noted further that as per Articles 74(3) and 83(3) of UNCLOS, the basic requirement for states concerned, 'in a spirit of understanding and cooperation', is to enter into 'provisional arrangements of a practical nature and, during this transitional period, not to jeopardize or hamper the reaching of the final agreement'.[75] But, these provisions appear to be 'almost empty'.[76] At the same time, this outcome was also obvious as these provisions were agreed upon through a process of significant compromise, lacking precision to encourage more participation.

BBNJ Treaty's response to this complex problem is entirely exclusionary. In Article 5 the Treaty makes it clear that its interpretation should be in 'a manner consistent with the Convention'. Besides, the Treaty shall not "prejudice the rights, jurisdiction, and duties of States under the Convention, including in respect of the exclusive economic zone and the continental shelf within and beyond 200 nautical miles."[77] On top of it, the Treaty further emphasizes a situation when its COPs may request the International Tribunal for the Law of the Sea (ITLOS) to give an advisory opinion on a legal question. But such request shall not be "sought on a matter within the competences of other global, regional, subregional or sectoral bodies, or on a matter that necessarily involves the concurrent consideration of any dispute concerning sovereignty or other rights over continental or insular land territory or a claim thereto, or the legal status of an area as within national jurisdiction."[78] Buttressing this, Clause (9) of Article 47 even more explicitly provides that "[n]othing in this Agreement shall be interpreted as conferring jurisdiction upon a court or tribunal over any dispute that concerns or necessarily involves the concurrent consideration of the legal status of an area as within national jurisdiction, nor over any dispute concerning sovereignty or other rights over continental or insular land territory or a claim thereto of a Party to this Agreement….."[79]

Additionally, a concise reference comes in Article 19 of the Treaty. As a provision in Part III of the BBNJ Treaty, dealing with area-based management tools and marine protected areas, it trades with measures that could 'affect or could reasonably be expected to affect the superjacent water above the seabed and subsoil of submarine areas over which a coastal State exercises sovereign rights in accordance with

[74] For additional discussion on overlapping maritime claims see, Schofield (2021); also see, Dupont (2018).

[75] Article 74(3) and 83(3) of UNCLOS.

[76] *UNCLOS: The Law of the Sea in the 21st Century*, 2nd Report of Session 2021–22, International Relations and Defence Committee, House of Lords, at p. 69; Available at https://committees.parliament.uk/publications/9005/documents/159002/default/ (Last accessed 24 February 2024).

[77] Article 5 of the BBNJ Treaty.

[78] Article 47(7) of the BBNJ Treaty.

[79] Article 47(9) of the BBNJ Treaty.

the Convention'.[80] But how should we construe the phrase 'in accordance with the Convention' in the presence of clear language of Article 5 and Article 47(7) and (9)? Clearly, in this context BBNJ Treaty is incomplete. Instead, looking carefully, we realise that any measures that its COPs take 'shall have due regard to the sovereign rights of such coastal States' and much will depend further on 'consultation' in accordance with the provisions of Part III which is mandatory.[81] These gaps had surfaced in the development stage of the BBNJ Treaty. Scholars often point out the pacifying impact of international environmental treaties on the rigid state sovereignty doctrine. But, BBNJ Treaty, considering the importance of the global environment as a whole, is not entirely convincing in that regard. This thought further gets consolidated when we read its Preamble, emphatically providing that the parties should respect 'the sovereignty, territorial integrity and political independence of all States'.[82]

Finally, there is another drawback in BBNJ Treaty that is worth mentioning. One of the momentous achievements of the BBNJ Treaty is the inclusion of basic rules for the equitable sharing of marine genetic resources. However, serious disagreements took place during the time of negotiations. The main controversy was about whether there should be a mandatory benefit sharing obligation on the developed countries. Hindside, it was the interest of some of the world's leading biotech companies, already enjoying or would probably be enjoying strong intellectual property protection, that were expected to benefit enormously by sequencing genetic information from resources found in the area beyond national jurisdiction. So, participants were uncertain about whether relying on the CHM principle would be wise or not.[83] One difficulty is that the development of CHM principle in international law has been slow, putting developing nations on guard. For them, solicitation for the applicability of CHM to marine genetic resources, found in a part of the seabed not under the jurisdiction of any particular state, was a tricky issue. There was also uncertainty about the clarification of the amount of freedom that a state should enjoy in this context.[84]

Nonetheless, participants finally settled on encouraging outcomes. In Article 5, the Treaty makes it clear that the parties shall be guided by the principle of the common heritage of humankind set out in the Convention.[85] The Treaty also provides that 'activities concerning marine genetic resources of areas beyond national jurisdiction may be carried out by all Parties'[86] and 'no State shall claim or exercise sovereignty or sovereign rights over marine genetic resources of areas beyond national jurisdiction.'[87] But most remarkably it has been agreed by the parties that "activities concerning marine genetic resources of areas beyond national jurisdiction are in the

[80] Article 19 of BBNJ Treaty.

[81] Ibid.

[82] Preamble BBNJ Treaty.

[83] Patrick (2023).

[84] Epps and Chazot (2023, p. 8).

[85] Article 5(b) of BBNJ Treaty.

[86] Ibid at Article 9(1).

[87] Ibid at Article 9(4).

interests of all States and for the benefit of all humanity, particularly for the benefit of advancing the scientific knowledge of humanity and promoting the conservation and sustainable use of marine biological diversity, taking into particular consideration the interests and needs of developing States."[88] Together, these provisions appear to be strong and equitable. At the same time, it can be seen that rigid sovereignty doctrine survives strongly as considerable doubts exist regarding what 'activities' should or shouldn't be counted 'in the interests of all States and for the benefit of all humanity'. This assumes much significance when we recall that BBNJ Treaty cannot 'prejudice the rights, jurisdiction and duties of States' available under UNCLOS. For sure, the benefit sharing aspect, if challenged in the future, will face similar complications that affected UNCLOS in the past. Because the Treaty has made benefit sharing mandatory, it cannot be said that parties have lost their rights enough that they would not act contrary to the interest of the larger community. For that matter, the meaning of 'cooperation' itself has not been given any functional content in the Treaty, leaving a wide-open territory yet to be enclosed.

Final Reflections

The deep-rooted features of UNCLOS are the outcome of a long developmental discourse. UNCLOS survives and so is the possibility of reading more new treaties in conjunction with it. The drafters of BBNJ Treaty by linking it to UNCLOS have only exhibited due respect to that extended historical process. Hence, the means and methods of balancing mechanisms in both these international agreements are by their very nature going to be parallel events. It can be argued that the rigid concept of state sovereignty has generated enough response through which such a possibility has emerged. In other words, there is an 'openness' in-built the sovereignty concept which is not a bad thing. As a constitutive element of UNCLOS, *i.e.* state as an actor in the international legal order, the concept of sovereignty is supposed to be a dynamic one.[89]

Goldsmith and Levinson argue from a different point of view. To date, to resolve the problem of state sovereignty in international law, evaluating the nature of state consent has been a standard norm. The duo argues that securing state consent is not the only tactic that international law should rely on. An alternative approach can be to adjust state sovereignty by adding 'international law constraints into its definition.'[90] To support this argument, they cited the example of intervention made by international law on the grounds of humanitarian aspects at the time of large-scale human rights violations or serious atrocities. In such situation, the responsibility to protect ideology should modify the concept of state sovereignty to the extent that the "rights of the sovereign — understood as a shield from intervention — depend upon

88 Ibid at Article 9(5).

89 See Holst (2022, pp. 102–104).

90 Goldsmith and Levinson (2009, p. 1850).

its capacity to fulfil [certain] responsibilities" in the international order as outlined in the U.N. Charter.[91]

Goldsmith and Levinson's argument allows us to look beyond the meta-question—what state sovereignty mean to the international treaty-making process? Identifying certain markers, such as the 'need for a new regime' or 'requirement for stringent legal norms', is just the beginning of the reformative process. States may differ on what humankind needs. Hence, normative configurations will also likely alter. Philip Allott has aptly described it as a process of 'metamorphosis.'[92] UNLCOS has already gone through substantial changes. Now BBNJ Treaty will certainly experience the same.

Hardin once suggested that to induce cooperation, the government should allocate property rights by actively intervening and altering the behaviour of non-cooperative entities. UNCLOS and BBNJ Treaty, have both followed somewhat the similar pattern but only in the 'rights' context. It is a matter degree how far the states should feel content with this paternalistic solution and decide to remain within the legitimate boundary, laid down by these Treaties. Thus far, it appears that BBNJ Treaty has provided enough flexibility, that states may enjoy to protect their interests. Therefore, the agreed formula almost certainly will go through a transformation. Following Ostrom's cautionary note, an unease settles about the decisional complexities that shared resource users may face while designing rules of access through their COP meetings. Because it is impossible to overlook the influence of the market and the diverse interests of the participants, the outcome may not be always consistent and if that happens, the cooperative model of law-making will always be affected negatively.

Conclusion

The BBNJ Treaty presents an opening to understand what ocean biodiversity means to humankind, not only from an empirical vantage point but also from a normative perspective. It is a great opportunity for the parties to advance solutions to some of the enduring international law problems in general and also to specific issues such as biodiversity loss in the high seas. Its open structure may be seen as a repetition of the typical international treaty-making process. But at the same time, the 'openness' should allow states to alter their behaviour and incentives. Any future improvement the parties make through COPs must constantly emphasize the importance of cooperative model for the conservation of ocean biodiversity. At the same time, there is a need to break free from the legacy of UNCLOS, with which BBNJ Treaty shares commonality. To put it simply, there is a need to build consensus among signatories and non-signatories both, to accept the urgency of forward-looking actions. For that matter, narrow sovereign interests must not be allowed to win over community interests that can spoil decades of hard work. The more comprehensive and

[91] Goldsmith and Levinson (2009, p.1851) [Bali (2012)].

[92] See Allott (1983, p. 8).

neater its structure becomes, the erosion of holding power of any one nation or small group of nations can be ensured in favour of developing a strong international ocean governance.

Conflict of Interest The authors have no conflict of interest to declare.

References

Allott, P. (1983). Power sharing in the law of the sea. *The American Journal of International Law, 77*(1), 1–30.

Bali, A. (2012). Legality and legitimacy in the international order: The changing landscape of nuclear non-proliferation. In R. Falk et al. (Eds.), *Legality and legitimacy in global affairs.* Oxford University Press.

Barrett, S. (1998). On the theory and diplomacy of environmental treaty-making. *Environmental and Resource Economics, 11*, 317–333.

Convention on the Law of the Sea, Dec.10, 1982, 1833 U.N.T.S.397.

Dupont, P. E. (2018). Contested sovereignty over land territory and maritime zones. In S. Minas & H. J. Diamond (Eds.), *Stress testing the law of the sea: Dispute resolution, disasters & emerging challenges* (pp. 31–54). Brill Nijhoff.

Epps, M., & Chazot, C. (Eds.). (2023). *The high seas biodiversity treaty: An introduction to the agreement under the United Nations Convention on the Law of the Sea on the conservation and sustainable use of marine biological diversity of areas beyond national jurisdiction.* IUCN. https://www.iucn.org/sites/default/files/2024-01/iucn-bbnj-treaty-policy-brief.pdf. Last accessed 24 February 2024.

Goldsmith, J. L., & Levinson, D. J. (2009). Law for states: International law, constitutional law, public law. *Harvard Law Review, 122*(7), 1791–1868.

Goldsmith, J. L., & Posner, E. A. (2005). *The limits of international law.* Oxford University Press.

Hardin, G. (1968). The tragedy of the commons. *Science, 162*, 1243–1248.

Hey, E. (2022). Informal law making and the future of the law of the sea: Developing legal infrastructure and regulating human activity. In N. Klein (Ed.), *Unconventional lawmaking in the law of the sea.* Oxford University Press.

Holst, R. J. R. (2022). *Change in the law of the sea: Context, mechanisms and practice.* Martinus Nijhoff.

Liao, X. (2021). The road not taken: Submission of disputes concerning activities in undelimited maritime areas to UNCLOS compulsory procedures. *Ocean Development & International Law, 52*(297), 297–324.

Nordquist, M. H., Moore, J. N., & Long, R. (Eds.). (2019). *Cooperation and engagement in the Asia-Pacific region.* Martinus Nijhoff.

Olson, M. (1971, January 1). *The logic of collective action: Public goods and the theory of groups* (Revised ed.). Harvard University Press.

Ostrom, E. (2015). *Governing the commons: The evolution of institutions for collective action.* Cambridge University Press.

Ostrom, E., & Hess, C. (2008). *Private and common property rights* (Research Paper No. 2008-11-01). School of Public & Environmental Affairs, Indiana University, Bloomington. https://ssrn.com/abstract=1304699. Last accessed 24 February 2024.

Park, Y. O. (1976). Res Communis versus Res Nullius: The International Legal, Political and Strategic Context. *Global Economic Review, 5*(1), 77–97. https://www.tandfonline.com/doi/pdf/10.1080/12265087609432820. Last accessed 20 February 2024.

Patrick, S. (2023). *The high seas treaty is an extraordinary diplomatic achievement*. https://carnegieendowment.org/2023/03/08/high-seas-treaty-is-extraordinary-diplomatic-achievement-pub-89228. Last accessed 20 February 2024.

Payne, C. R. (2023). *The new high seas biodiversity treaty offers conservation, equity, and regulatory certainty*. https://www.asil.org/sites/default/files/ASIL_Insights_2023_V27_I6.pdf. Last accessed 19 February 2024.

Peterson, M. (Ed.). (2015). *The prisoner's dilemma*. Cambridge University Press.

Posner, E. A., & Sykes, A. O. (2010). Economic foundations of the law of the sea. *The American Journal of International Law, 104*(4), 569–596.

Posner, E. A., & Sykes, A. O. (2012). *Economic foundations of international law*. NYU School of Law. http://www.law.nyu.edu/sites/default/files/ECM_PRO_073779.pdf. Last accessed 24 February 2024.

Principle 21 of the Stockholm Declaration, 1972 [UN Doc. A/RES/2994(XXVII)] and Principle 2 of the Rio Declaration [31 I.L.M. 874 (1992)].

Purcell, K. (2019). *Geographical change and the law of the sea*. Oxford University Press.

Ranganathan, S. (2014). *Strategically created treaty conflicts and the politics of international law*. Cambridge University Press.

Schofield, C. (2021). Options for overcoming overlapping maritime claims: Developments in maritime boundary dispute resolution and managing disputed waters. *The Journal of Territorial and Maritime Studies, 8*(2), 21–41. https://www.jstor.org/stable/48617339. Last accessed 17 February 2024.

Schrijver, N. (2016). Managing the global commons: Common good or common sink? *Third World Quarterly, 37*(7), 1252–1267.

Strati, A., Gavouneli, M., Skourtos, N., & Strate, A. (Eds.). (2006). *Unresolved issues and new challenges to the law of the sea: Time before and time after*. Martinus Nijhoff.

Tanaka, Y. (2023). *The international law of the sea*. Cambridge University Press.

Use of New Approach Methodologies in Education, Toxicology and Biomedical Science: Biodiversity Conservation and Business Opportunities

Mohammad Abdulkader Akbarsha, Parameswarapanicker Sreejith, and Krishan Kumar Sharma

Introduction

Animals are used for various purposes, other than they are biologically intended viz., prey-predator, foods and feeds, companion/farm animals to humans including, pets, etc. The made-up purposes include use as a tool in education (dissections to understand anatomy, including correlation with the human; and conduct cytological, genetic, physiological, embryological, ecological, etc., experiments, ultimately extrapolating to the understanding of human/human-related biology), and as the tools of experimentation in toxicology and biomedical science, including pharmacology, cosmetics, vaccines, medical devices, etc. Millions of animals from the wild and those that are bred in the laboratory are used for these purposes. The removal of animals from the wild as a tool in education, for risk assessment of chemicals, and drug testing disrupts the biodiversity, very often leading to population depletion and disruption of the food chain, and threatened with extinction as in the case of many amphibians (Akbarsha, 2015). The laboratory-bred animals are used extensively in toxicology and biomedical science experiments with the ultimate objective of correlating the outcome to the human. Over some time there has been societal objection

M. A. Akbarsha (✉)
National College (Autonomous), Tiruchirappalli, India
e-mail: rc@nct.ac.in

P. Sreejith
University of Kerala, Kariavattom, Thiruvavanthapuram, India

K. K. Sharma
MDS University, Ajmer, Rajasthan, India

M. A. Akbarsha · P. Sreejith
Indian Society for Alternatives to Animal Experiments, Bharathidasan University, Tiruchirappalli, India

L. Krishna Panicker et al. (eds.), *Biodiversity and Business*,
https://doi.org/10.1007/978-3-031-71674-4_26

to these practices from ethical, environmental, and scientific perspectives. The use of alternatives in the place of live animals has been patronized. In as much as the alternatives offer scope for safeguarding the interest of the animals, they also provide opportunities for business for the newer commodities and services. This chapter aims at elaborating on these two-sides of the same coin- dispensing with animal use and encouraging business based on the alternatives.

The Alternatives

As a project of the Universities Federation for Animal Welfare, England, initiated in 1955, W.M.S. Russell, a zoologist, and Rex Burch, a microbiologist, brought up "*The Principles of Humane Experimental Technique*" in 1959. They referred to these principles as the '3Rs,' i.e., Replacement, Reduction, and Refinement. The 3Rs are defined as follows: (1) ***Replacement*:** Animals must be substituted with non-sentient test systems when such a system exists or with poorly sentient simple model organisms. As it emerged later, in vitro cell systems, computational tools and model organisms of lesser sentience are gaining momentum as replacements for in vivo experimentation. (2) ***Reduction*:** If replacement is not possible, the number of animals in respect of the particular experiments may be reduced wherever possible if the same result can be obtained with fewer animals. (3) ***Refinement*:** Where replacement is not practicable, and reduction is a matter of choice, the suffering and distress of the animals may be avoided or minimized by using, for example, analgesics or working under narcosis, and maintaining the animals in good animal house conditions and on nutritious foods (Hubrecht & Carter, 2019).

Replacement was an imagination 65 years ago at which time cell culture and computer programs were in their infancy. Few scientists could imagine that such methods might lead to predictive tools. Over the last few decades, however, industry, science, and politics have demonstrated a commitment to the 3Rs to end animal experiments today rather than wait for tomorrow. The 3Rs became the basis for a credible investment in overcoming reliance on animal testing. Animal experiments decreased until the turn of the century by an estimated two-thirds from their peak in the mid-1970s. Since then, however, animal experiments have been increasing, since the new techniques of gene manipulation of lab animals in general, and in mice in particular, which have become very popular scientific models. Rex Burch once said, "*Humane science is good science, and the best way to achieve that is rigorous application of the 3Rs*".

Despite receiving an overall welcome, Russell and Burch's work remained largely ignored well into the 1970s. In 1978, physiologist D.H. Smyth restored the 3Rs to the light of day and encompassed them under the concept of ***Alternatives*,** which he defined as "*all procedures which can completely replace the need for animal experiments, reduce the numbers of animals required, or diminish the amount of pain or distress suffered by animals in meeting the essential needs of man and other animals*". Since then, there have been many technological developments such as *in vitro* and in silico methods built on the back of the developments in technology, computing, and molecular biology, which have found connection with the concept and practice of alternatives. Animal taxa such as zebrafish, Drosophila, *Caenorhabditis elegans*, etc., have come to contribute greatly as alternative model organisms lesser in sentience than the mammalian models.

These associations have been consolidated into ***New Approach Methodologies*** (NAMs). NAMs are defined as "*any technology, methodology, approach, or combination that can provide information on chemical hazard and risk assessment without the use of animals, including in vitro, in silico, in chemico, and ex vivo approaches*" (Miller-Holt et al., 2022). With NAMs, the idea is to promote the best science possible by using all of the tools in our technological toolbox. The goal is the same whether we are talking about toxicological research, such as chemical safety, and drug efficacy testing, or other types of biomedical research that examine biological mechanisms of disease, for example. Traditionally, scientists have relied heavily on animal models, which have provided insights into biological pathways and toxicological mechanisms. But in recent years, there have been many advances in in vitro [cell-based] research, bioengineering technologies, chemical screening methods, computational approaches including Artificial Intelligence [AI], etc.[1]

Thus, the future of alternatives is NAMs.

Digital and Other Alternatives in Education

In education and training, wild-collected as well as laboratory-bred animals in millions are used in the name of knowledge acquisition and/or skill development in basic life sciences, as well as in professional education such as medical, veterinary, pharmacology, etc. Dissections of and experiments using animals have been practiced as laboratory exercises in Zoology and allied Life Sciences, medical sciences, and veterinary science, since the 1920s, at a time when the subjects of animal anatomy, taxonomy, and evolution were emphasized in theory courses (Balcombe, 2001). These classical subjects have now been relegated to the background in light of emerging subjects such as biochemistry, molecular biology, and biotechnology, warranting lectures and laboratory exercises in these subjects (Akbarsha, 2007). Further, the demand for animals for the purpose of dissections has increased to such an extent that animals are no longer available in large enough numbers to meet the demand. The large-scale removal of animals from their natural habitats for an unnatural purpose can potentially lead to biodiversity loss and an ecological crisis (Akbarsha, 2015). Laws have been enacted protecting selected animal species, but these laws are often not heeded, and animals continue to be used in education. Social groups and scientists have questioned both the logic and the relevance of animal dissections in education in the contemporary scenario. In most contexts animal dissection is unnecessary—and even counter-productive—to achieve valid

[1] https://factor.niehs.nih.gov/2023/3/feature/4-new-approach-methodologies.

educational goals, especially higher-order goals. It has been strongly recommended that the new alternative learning modalities arising from information and computer technology (ICT) and other areas can effectively replace dissections and experiments using animals (Akbarsha et al., 2013). A variety of digital tools and simulation programs have been developed to enhance the learning methodology and to provide students with a better alternative in terms of learning outcomes (Balcombe, 2001). This includes (i) multimedia presentations viz., Neotek's Frog Dissection Lab; Operation Frog Deluxe by Tom Snyder Productions (Scholastic); VisiFrog by Ventura Educational Systems; Boreal Labs' Digital Dissection Series; DryLab by Tangent Scientific; BioLab by Carolina; ProDissector FROG, a Schneider & Morse Group Production, etc. These products offer very similar functionality, presenting a student with text, artwork, photographs, and in some cases, video clips. While they present the student with an integrated and self-paced learning environment, they only offer linear and/or restricted learning paths, and every student ultimately observes the same thing, usually in the same sequence. (ii) The enhanced multimedia presentations such as Digital Frog by Digital Frog International; CyberEd Dissection Series by Plato; DissectionWorks by ScienceWorks; and Frogguts by Frogguts, Inc. These products allow a student to select a tool, such as a scalpel, and float the tool over a 2D representation of a frog. V-Frog™, a more recent addition to this list, is an innovative virtual reality-based dissection simulator developed using virtual surgery technology, and it is highly efficient in achieving quality learning (Figs. 1, 2, 3, 4, 5, 6, 7 and 8).

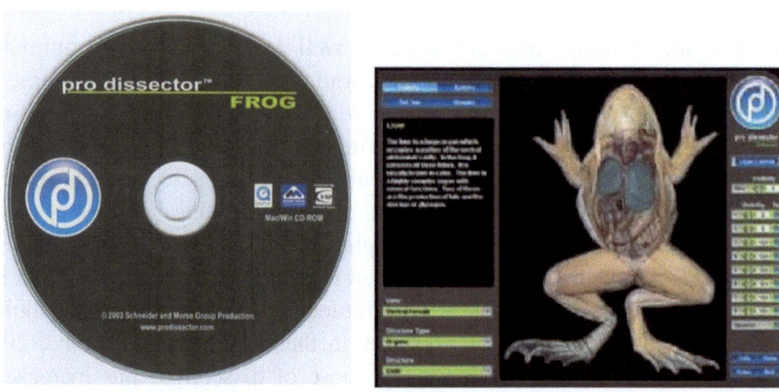

Price: Single User: US$99.95; Lab Pack of 5: US$199.95; Home School, Single User: US$34.95

Author: Roy E. Schneider & Dennis E. Morse

Supplier:
Schneider & Morse Group (SMG)
artimiss@bex.net

Fig. 1 Digital alternatives for dissection and experiments. Pro-Dissector Frog

Fig. 2 Digital alternatives for dissection and experiments. Dissection Works

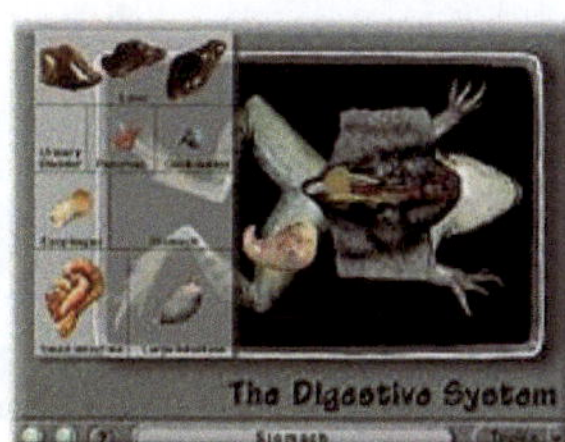

Fig. 3 Digital alternatives for dissection and experiments. BioLab Frog- Dissection

Of course, there are other simpler alternatives such as:

- Models, mannequins, and mechanical simulators
- Films, interactive videos
- Non-invasive self-experimentation & human study
- Plant experiments
- Observational & field studies
- Materials from slaughter houses & fisheries
- Permanently preserved specimens

Froguts Inc.
601 Union St., Ste. 4200-202
Seattle, WA 98101
www.froguts.com

Fig. 4 Digital alternatives for dissection and experiments. Froguts

Fig. 5 Digital alternatives for dissection and experiments. The Digital Frog 2

- *In vitro* studies on cell lines
- Animal cadavers from an animal-friendly and ethical source such as from animals euthanized for medical reasons
- Supervised clinical experience

Given that many countries have dropped the dissection of purpose-killed animals from the curriculum, the pace of change in India has been initially rather slow,

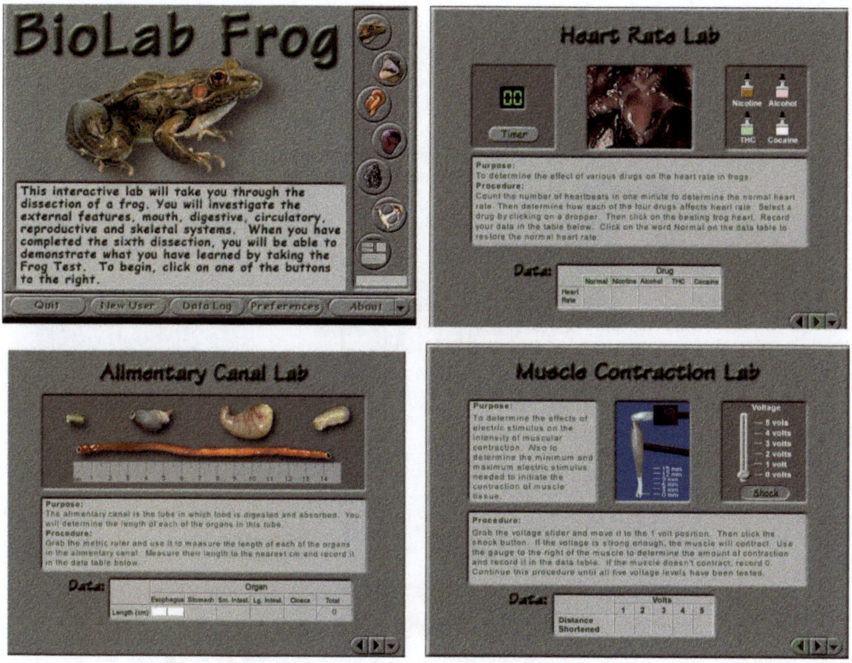

Fig. 6 Digital alternatives for dissection and experiments. BioLab Frog- Physiology Experiments

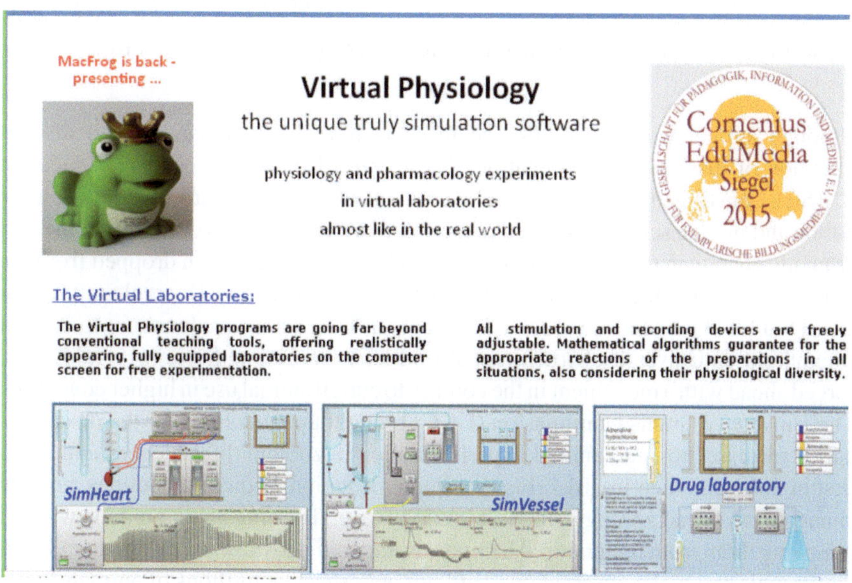

Fig. 7 Digital alternatives for dissection and experiments. Virtual Physiology

Fig. 8 Digital alternatives for dissection and experiments. Virtual Animal Lab for Experiments

although animal welfare groups and teachers have been demanding the discontinuation of dissections in education. The Central Board of Secondary Education (CBSE), constituted by the Union Government of India, a Board of Education for public and private schools, stopped prescribing dissections in its curriculum, the first dramatic change in this connection in India. Dissections have been dropped from the school curriculum of almost all State Boards. Mahatma Gandhi-Doerenkamp Center (MGDC) for Alternatives to Use of Animals in Life Science Education, located at Bharathidasan University, Tiruchirappalli, which draws strength from several NGOs, forged ahead with a movement in the country to curtail animal use in higher education. The University Grants Commission (UGC) brought up guidelines for the discontinuation of dissection and animal experimentation in zoology/life sciences education first in 2011 and then in 2014 (Fig. 9). Implementation of these guidelines, and the change from "dead zoology" to "live zoology", i.e., learning about animals without removing them from their natural habitats, is imperative. Allowing the learner to visit the animals and make observations, will not only save the millions of animals being killed/sacrificed for the purpose of higher education in India but also will facilitate the introduction of ICT tools, the "digital alternatives", in the pedagogy of laboratory

University Grants Commission
New Delhi

No.F.14-6/2014 (CPP-II) July, 2014

Subject: Dissection and Animal Experimentation in Zoology /Life Sciences and allied disciplines in undergraduate, postgraduate and research programmes.

In exercise of powers conferred by Section 12 (j) of the UGC Act, 1956, the UGC hereby issues the following instructions in regard to animal experimentation in Zoology/Life Sciences and allied disciplines.

7.1.1 For undergraduate and postgraduate programs, both at major and allied levels, no animal from any species shall be dissected, either by teachers or students for any purposes. The teachers shall demonstrate one or more aspects of anatomy to students with the help of digital alternatives, models and charts etc.

7.1.2 The curriculum, both theoretical & practical for Invertebrates and Chordates, as the case may be, shall be oriented towards levels and patterns of organization, biosystematics, biodiversity, adaptations, developmental stages, population dynamics, ecological implications, etc. The laboratory exercises should make use of museum specimens and microscopic preparations, photographs, video clippings, models, charts, plastinated specimens, field observations combined with photography and/or videography, and so on. Intermittent field visits should be arranged.

7.1.3 Digital alternatives such as ProDissector Frog, BioLab Frog, DigiFrog, Dissection Works etc. are available in plenty on various platforms and market. They should be suitably procured from commercial sources or internet. Many of these digital learning devices have modules for testing, which can be used to evaluate the students at the examination. Separate budgetary provision should be made by the concerned institution for the procurement of digital material and technology.

Fig. 9 UGC notification on dissection ban in India

exercises. The Medical- and Pharmacy Councils of India notified the introduction of computer-assisted modules in the place of animal experiments (Fig. 10).

NAMs for Experimental Toxicology and Biomedical Science

Toxicology is a field of science that helps us understand the harmful effects that chemicals, substances, or situations, can have on people, animals, and the environment. Chemical substances need to be safe for human health (for example consumers and workers) and the environment. Chemical substances are used in industrial processes.

2 THE GAZETTE OF INDIA : EXTRAORDINARY [PART III—SEC. 4]

PHARMACY COUNCIL OF INDIA

NOTIFICATION

New Delhi, the 25th August, 2014

"Wherever animal experimentations are prescribed in the curriculum, the required knowledge and skill should be imparted by using computer assisted modules. Animal hold area shall be as per the Committee for the Purpose of Control and Supervision of Experiments on Animals (CPCSEA) guidelines."

MEDICAL COUNCIL OF INDIA

NOTIFICATION

New Delhi, the 18th March, 2014

"For teaching Physiology and Pharmacology in UG curriculum, the required knowledge and skills should be imparted by using Computer Assisted modules. Only an animal hold area, as per CPCSEA Guidelines, is required."

Fig. 10 Pharmacy and Medical Council notifications prescribing computer-assisted modules in the pharmacy and medical education programs, respectively.

They can also be found in food as food additives (dyes, preservatives) and as contaminants (pesticide residues, chemical substances released from food contact materials). Chemical substances can also be used for the production of consumer products such as cleaning products, paint, cosmetics, and many types of articles (clothes, furniture, and electronics). Toxicology is the "Science of Safety". It has evolved from a science focused on studying poisons and adverse effects of chemical exposures to a science devoted to studying safety. Toxic chemicals leave alone those chemicals that could potentially get into utilization in multiple ways, are those that are potentially present in our foods, consumer goods, preservatives, colorants, agricultural chemicals, and even drugs. Toxicology uses the power of science to predict what and how chemicals may cause harm and then shares that information to protect public health.[2]

Biomedical Science is a set of sciences applying portions of natural science and/or physical science to develop knowledge, interventions, or technology that are of use in healthcare or public health. It has many branches. **Pharmacology** is the study of the origin, chemistry, and uses of drugs and their effects on the body. Pharmacology deals with the origin of drugs, their characteristics, and their effects, such as biological, chemical, or therapeutic, on a living system. Two major aspects of pharmacology are pharmacokinetics and pharmacodynamics. **Pharmacokinetics** describes the **a**bsorption, **d**istribution, **m**etabolism, and **e**xcretion of drugs, abbreviated as ADME. **Pharmacodynamics** refers to the effects of drugs in the body and the mechanism of their action, including interaction with the targets and the receptor molecules to understand how the drug acts to manifest its therapeutic properties. **Disease models** are ones that are induced in animals and used for understanding the disease and to test drugs for therapeutic inferences. The induced disease models are artificial in the sense the etiology of the diseases follows a course very different from

[2] https://www.niehs.nih.gov/health/topics/science/toxicology.

that in human. Take, for example, type 1 diabetes mellitus. It is juvenile diabetes caused by to autoimmune pathology of beta cells of islets of Langerhans. In the animal models, it is a chemical-induced pathology produced due to treatment of, e.g., alloxan. Alloxan diabetes has nothing to do with autoimmune disorders. Alloxan directly hits the beta cells of islets of Langerhans. Type 2 diabetes is produced due to defects in insulin receptors and insulin transporters. Streptozotocin-induced diabetes mellitus in rodent models is a chemical pathology affecting islet cells and has nothing to do with receptor and transporter. Therefore, a scientifically relevant approach to disease models should be the ones produced in human-derived systems and not in rodent models. **Cosmetic** is any of several preparations (excluding soap) that are applied to the human body for beautifying, preserving, or altering the appearance or for cleansing, coloring, conditioning, or protecting the skin, hair, nails, lips, eyes, or teeth. The animal testing of cosmetics, and the ingredients in them, is painful and affects millions of animals that may ultimately die. Vaccine is a preparation that is used to stimulate the body's immune response against diseases. Vaccines are usually administered through needle injections, but some can be administered through oral route or sprayed into the nose. Testing novel vaccine candidates in animal models is a critical step in the development of modern vaccines. Animal models are being used to assess the quality and quantity of the immune response, to identify the optimal route of delivery and formulation, to determine protection from infection and disease transmission, and to evaluate the safety and toxicity of the vaccine formulation. A **medical device** can be any instrument, apparatus, implement, machine, appliance, implant, software material, or other similar or related article, intended by the manufacturer to be used, alone or in combination for a medical purpose. The medical devices are animal tested to validate the feasibility, safety, and efficacy of medical devices. Thus, the various chemicals man is exposed to, the pharmaceuticals, the cosmetics, vaccines, medical devices, etc., intended for human use, need to be tested for the purpose of safety. Pharmaceuticals, cosmetics, vaccines, and medical devices need to be tested to find if they fulfil the intended application and to exonerate the toxicity; this is called pre-clinical testing. These toxicological and pharmacological tests are technically due to be conducted on humans because they are intended for use by humans but it is not permitted given ethical issues.

Therefore, toxicity and drug testing are traditionally conducted in animals such as mice, rats, hamsters, rabbits, fishes (examples—zebra fish, trout), birds (mainly chicken), guinea pigs, amphibians, primates, dogs, cats, etc., are being used in research with the hope of extrapolating the data/inference to the human. Essentially, the animals are surrogates for the human.

Drug testing and toxicological screenings which are useful in the development of new treatments for infectious and non-infectious diseases are the main purpose of such studies. Animals also serve as a tool to understand the effects of medical procedures and surgical experiments. Moreover, they are used to obtain products like vaccines, antibiotics, etc., which are used in diagnostics as well as treatments.

Every year, millions of experimental animals are used all over the world but there is no absolute data. For example, in the UK, 3.71 million animals were used for research in the year 2011.[3] The total number of animals used in the USA in the year 2009 was estimated to be 1,131,076 (Doke & Dahwale, 2015) while that in Germany reached up to 2.13 million in 2001 (Rusche, 2003). Taylor and Alvarez (2019) provided an update for the year 2015. Data from 37 countries that publish national statistics were standardised against the definitions of 'animals' and 'procedures' used in the European Union (EU) Directive 2010/63/EU. They, then, applied a prediction model, based on publication rates, to estimate animal use in a further 142 countries. This yielded an overall estimate of global animal use in scientific procedures of 79.9 million animals. They further extrapolated this estimate to obtain a more comprehensive final global figure for the number of animals used for scientific purposes in 2015, of 192.1 million. This figure included animals killed for their tissues, normal and genetically modified (GM) animals without a harmful genetic mutation that are used to maintain GM strains, and animals bred for laboratory use but not used. This huge population of experimental animals usually comes from breeding centers located in various universities and national breeding centers. All of these are known as class-A dealers, while the brokers who acquire the animals from miscellaneous sources (like auctions and animal shelters) are identified as class-B dealers. In a few instances, use of the wild animals such as monkeys and birds is also practiced. In clinical testing laboratories, animals are isolated from their groups and used as a tool irrespective of their natural instincts. For the experimental procedures, either a whole animal or its organs and tissues are used. For this purpose, animals are euthanized (killed) by established methods. Many times, the animals surviving the clinical testing are euthanized at the end of an experiment to avoid later pain and distress. In some cases (for example in LD 50 analysis) animals die as a result of the experiment. Thus, the animals are surrogates for the human.

The pain, distress, and death suffered by the animals during scientific experiments have been a debated issue for a long time. A section of people are passionate about animals because being alive, animals have the right against pain and distress and hence, their use for experimentation is unethical and must be stopped. Various acts and laws have been passed to bring the control over unethical use of animals and minimize the pain to animals during experimentation. For example, in 1824, the organization for animal rights was formed by the Royal Society for the Prevention of Cruelty to Animals, UK. In 1876, an act for prevention of cruelty to animals was formed in the UK. It came into existence in India, France, and the USA in the year 1960, 1963, and 1966, respectively. At present, many rules and acts are followed at the international level, to protect animals against cruelty and misuse. The organizations like ICH (International Conference on Harmonization of Technical Requirements for Registration of Pharmaceuticals for Human Use), CCSEA (Committee for Control and Supervision on Experiments on Animal), NIH (National Institute of Health), and OECD (Organization for Economic Co-operation and Development) provide the guidelines for animal housekeeping, breeding, feeding, transportation, and mainly for their use in scientific experiments. Besides the major concern of ethics, a few more disadvantages of animal experimentation are a requirement of skilled/trained

[3] www.rspca.org.uk.

manpower and time-consuming protocols. Moreover, the very high cost involved in breeding, housing, and lengthy protocols of animal experiments is another drawback.

Subjecting animals to confinement, pain and cruelty through lab experiments naturally raises ethical concerns. Additionally, the significant differences between animal and human anatomy and physiology expound that animal studies may not reliably predict the outcome of substances being tested on humans. There is poor predictive value of animal experimentation in crucial fields such as toxicology and pharmacology. For example, many potential drugs fail to show any results in humans, although they work on mice. Research suggests that the outcomes of animal studies can sometimes even be misleading, and may be false positives or false negatives. This is what is known as species difference. In the present context, it could be explained in the phase I metabolizing enzymes. Cytochrome P450 monooxygenases (CYP450) catalyze the oxidation and metabolism of a large number of xenobiotics and endogenous compounds. CYP450 enzymes evolved as the primary defense against xenobiotics and in this process are also responsible for the bioactivation of drugs and toxicants to more reactive intermediates. Cytochrome P450 enzymes are primarily found in liver cells but are also located in cells throughout the body. A large body of knowledge exists on the role of the complex CYP450 isoenzymes in xenobiotic metabolism of activation or inactivation. It has been shown that the isoenzymes differ between species, so an enzyme that activates a chemical, would act in a manner opposite to another. E.g., most isoenzymes are expressed in hamster, mouse, monkey and human, whereas rats, pigs, rabbits, and dogs lack several isozymes. It explains why 90–95% of potential drugs tested in animals are found to be safe and therapeutically positive, do not pass through the clinical trial which means these drugs (and the chemicals which they are made of) perform differently between animals and men, and even between different species of animals, say rat and mouse. This and other data raise concern about the relevance of animal experimentation in toxicology and biomedical science and strongly support the concept of alternatives/NAMs that would generate human-relevant data.

Regulatory Change in Requirement of Animal Experiments

The anti-animal testing movement has gained huge support, and countries have brought about rules prohibiting/not requiring animal testing in toxicology and drug testing. A recent US legislation (December 2022) no longer requires new medicines to be tested on animals, and FDA is allowed to promote a drug to human trials after either animal or non-animal tests (Wadman, 2023). The EU legislation in Directive 2010/63/EU contains "*on the protection of animals used for scientific purposes*". The final aim of the Directive is to replace all animal research with non-animal methods. EU Directive 76/768/EEC ("Cosmetics Directive") provides a regulatory framework for phasing out animal testing for cosmetics purposes across the EU. It prohibits testing of a finished cosmetic product and its ingredients on animals and also prohibits marketing of finished cosmetic products or ingredients if they are tested on animals (Marinou & Dontas, 2023).

In India Rule 148C of the Drugs and Cosmetics rule prohibits the testing of cosmetics on animals; and Rule 135B prohibits the import of cosmetics that have been tested on animals (Koduri & Akbarsha, 2014). The New Drugs and Clinical Trial Rules, 2019 (CT Rules), have been amended in 2023 to replace the use of animal tests in drug development. The amendment allows researchers to use non-animal and human-relevant methods, including technologies like 3D organoids, organs-on-chip, and advanced computational methods, to test the safety and efficacy of new drugs (Akbarsha & Mascarenhas, 2019). The changes in regulatory frameworks around the world are positive and reassuring, but a complete abolition of animal testing is yet to be achieved. In India, for example, there is still no legislative ban on animal testing for drugs, which means that the current drug approval process may not change noticeably at least shortly. A long road lies ahead, but chipping away at regulation and continuing research on alternatives may eventually lead to the use of animals in experiments being significantly reduced, if not eliminated (Obhan et al., 2023).

Alternative Methods/NAMs

NAMs are broadly defined as any technology, methodology, approach, or combination thereof that can be used to replace, reduce, or refine animal toxicity testing and allow for more rapid or effective prioritization and/or assessment of chemicals. These methods may include the use of computer-based (computational) models, modernized whole organism assays or assays with biological molecules, cells, tissues, or organs, as well as exposure prediction approaches. Assays with biological molecules can include "omics" studies that determine the chemical dose at which molecular changes begin to occur. While some NAMs may still make use of animals, the methods are refined to provide new mechanistic knowledge and associated dose–response data. This is an important step towards reducing the total number of animals used in research, product development, and chemical risk assessment until the necessary NAMs become available to replace animal toxicity testing. Two of the more recognized descriptions of NAMs are that they include in silico approaches, *in chemico* and in vitro assays, as well as the inclusion of information from the exposure of chemicals in the context of hazard assessment. They also include a variety of new testing tools, such as "high throughput screening" and "high-content methods" such as genomics, proteomics, and metabolomics, and the term NAM is a broadly descriptive reference to any technology, methodology, approach, or combination thereof that can be used to provide information on chemical hazard and risk assessment that avoids the use of intact animals.

NAMs can be used to improve understanding of the mechanisms underpinning adverse effects and to identify doses below which effects are not expected to occur from a human or ecological toxicity perspective. Data generated by NAMs are increasingly used to provide hazard, exposure, and risk information for prioritizing chemicals for further action and can contribute to the weight-of-evidence in chemical risk assessments. Additionally, NAM data can highlight where further testing may be needed in an integrated or tiered testing strategy.

There is an ongoing international effort to replace, reduce, or refine the use of traditional whole animal-based tests. NAM data can be used to augment the datasets considered in both ecological and human health risk assessments. The mechanistic nature of NAM data can improve our understanding of how a chemical or mixture causes toxicity. Additionally, NAMs can provide data with greater human or ecological relevance compared to standard toxicity tests conducted using animals. For example, in contrast with animal studies, cell- and organ-based approaches allow for the use of human tissues and, therefore, toxicity data obtained from these NAMs may be more relevant to inform human health risk assessments. NAMs are especially useful in data-poor situations, as the generation of data may be done using computational models, with cell-based tests, or using alternative approaches that are often less resource intensive and have higher-throughput (meaning more data can be generated in the same amount of time) than traditional animal toxicity methods. The NAM-based approaches used in chemical risk assessment have been demonstrated to be as protective of human health and the environment as traditional animal methods. The incorporation of NAMs into the risk assessment framework is important to reduce reliance on the use of traditional animal tests while maintaining scientific rigor to ensure that regulatory decisions will continue to keep people and the environment safe from harmful substances.

Advantages associated with the use of NAMs are time- efficiency, requires less manpower, and cost-effectiveness. The following are the commonly practiced methods.

Cells and Tissue Cultures

In vitro cell and tissue cultures, which involve the growth of cells outside the body in a laboratory environment, are an important alternative for animal experiments. The cells and tissues from the liver, kidney, brain, skin, etc. are removed from an animal and can be kept outside the body, in a suitable growth medium, for a few days to several months or even for a few years. In vitro culture of animal/human cells includes their isolation from each other and growing as a monolayer over the surface of culture plates/flasks. Cellular components like membrane fragments, cellular enzymes, etc. can also be used. Various types of cultures like cell culture, callus culture, tissue culture, and organ culture are used for various purposes. The benefits associated with these techniques are easy to follow, less time-consuming, and less expensive. These methodologies are routinely used for preliminary screening of potential drug molecules/chemicals to check their toxicity and efficacy.

Monolayer/2D Culture Systems

The most fundamental and simple approach to in vitro test systems is to use the monolayer or 2D culture system. This being essentially a human cell, the outcome of the test is immediately human-relevant, and there is no scope for species difference. Nevertheless, 2D cell cultures do have some inherent limitations such as growing on a flat surface is not a good way to understand how cells grow and function in a human body, where they are surrounded by other cells in three dimensions. Further, 2D cell testing is not always predictive, which increases the cost and failure rate of new drug discovery clinical trials, and development every year. Despite these disadvantages, 2D cell cultures are still practiced for the majority of in vitro tests, particularly for large-scale screening to eliminate the toxic chemicals and to find tangible leads. The reasons are that it is less expensive, it is well established, there is already enormous data and literature to compare, this is one which researchers and scientists understand, and it is easier to culture the cells and for analysis (Doke & Dahwale, 2015).

3D Culture Systems

Culturing cells in 3D format allows the cells to adopt morphology and a migration pattern similar to that found in vivo. Recapitulating the normal morphology of cells is important as their shape can directly affect their biological activity. Thus, 3D cell culture is a better way of representing human tissue outside the body. 2D cell cultures only exist in two dimensions. That's not an accurate representation of how cells grow or how they are affected by disease and injury. 3D cultures now dominate the test systems, especially in cancer treatment and stem cell research. Reasons for the increasing acceptance and use of 3D cell cultures include more relevant cell models, interaction between different types of cells, integration of flow, establishment of cell–cell and cell-ECM interactions, better simulation of conditions in a living organism, reduced use of animal models, and more realistic way to grow and treat cells including the cell environment can be manipulated to mimic that of a cell in vivo and provide more accurate data about cell-to-cell interactions, tumor characteristics, drug discovery, metabolic profiling, stem cell research, and other types of diseases.[4]

Almost all cosmetics, drugs, and chemicals are tested for their toxicity and efficacy, using these tests. For example, to check the irritancy of chemicals, previously Draize test was used which requires animals (mainly rabbits). It is very painful and every time a new animal is used. An alternative that uses reconstructed human epidermis (RHE) replaces animal to test irritation and/or corrosion. Bovine corneal organ culture is another model used to evaluate the toxicological effect of chemical irritancy of the eye in vitro. Many stem cell-based preparations/in vitro

[4] https://www.mimetas.com/en/blogs/315/2d-versus-3d-cell-cultures-advantages-and-disadvantages.

models of the brain, liver, kidney, lung, intestine, etc., have now come into practice. These are human *in vitro* systems, providing for cell–cell/tissue interactions that obviate species differences. There is a great evolution of 3D cell culture systems, such as scaffold cultures, scaffold-free cultures, spheroids, organoids, organs-on-chips, 3D bioprinting, and microfluidic/micro physiological systems, in increasing complexities but closer to in vivo human, each with its advantages and utilities.

Scaffold Culture

The most conventional approach for 3D cell culture is the growth of cells on a structural scaffold that mimics the physiological extracellular matrix (ECM) through biopolymers. These scaffolds can be inserted into conventional cell culture workflows or can be the starting point of more complex tissue engineering projects, such as developing artificial organs. The scaffold element provides physical support to cells, allowing them to aggregate, proliferate, and migrate. The available scaffold materials are made up of collagen, cellulose, alginate, agarose, chitosan, gelatine, matrigel, or biocompatible polymers. This model is useful in several diseases including cancer.[5]

Scaffold-free Culture, Spheroids, Organoids

The advances in culturing mini organs outside of the body have made it possible to use 3D culture techniques as an alternative to in vivo models. These mini organ cultures, termed 'spheroids/organoids', originate from a variety of sources including neonatal tissues, pluripotent/induced pluripotent stem cells, tissue biopsies, and adult stem cells. Organoids are tiny, self-organized three-dimensional tissue cultures that are derived from stem cells. Such cultures can be crafted to replicate much of the complexity of an organ or to express selected aspects of it like producing only certain types of cells. Organoids grow from stem cells—cells that can divide indefinitely and produce different types of cells as part of their progeny. Scientists have learned how to create the right environment for the stem cells so they can follow their genetic instructions to self-organize, forming tiny structures that resemble miniature organs composed of many cell types. Spheroids are 3D cell cultures that self-aggregate into sphere-like formations during cell proliferation.[6] Organoids can range in size from less than the width of a hair to five millimetres.[7] The resulting spheroids/organoids self-organize and recreate the physiology of organs, as well as accurately represent clinical diseases in remarkable detail. The culturing of scaffold-free cultures is a simple and cheap process, even possible in Petri dishes, enabling the mass production of spheroids of a uniform and controllable size for high-throughput applications. Organoids initially were used to model tissue development and stem cell fate.

---

[5] https://www.cellgs.com/blog/what-is-scaffold-based-3d-culture.html.

[6] https://www.cellink.com/blog/what-are-spheroids-and-why-are-they-important/.

[7] https://hsci.harvard.edu/organoids.

Genes of interest were marked or removed, and the resulting organoids followed in real time to identify lineage specifications and cell fates. Today, organoids can be used in many experimental approaches developed for cell lines (Temple et al., 2022).

Organs on Chip

These are advanced cell cultures combined with microfluidics. The chips are simple and miniature devices that contain parallel upper and lower microfluidic channels, each seeded with organ-specific cells. The channels are separated by thin porous membranes, which create an interphase for cell–cell communication. The membrane is coated with a specific extracellular matrix. This technology is nowadays used by academia and the pharmaceutical industry. Within academia, it enables scientists to develop new models closer to humans to better understand the human physiology and disease mechanisms.[8]

Traditionally, spheroids, organoids, and organs on chips have been extensively used for disease modeling in an academic setting. They are now also being translated towards more pharmacological applications, such as drug screening, toxicity and pharmacokinetics testing (Fang & Eglen, 2017). There are several formats and materials available that help the culturing of cells in 3D. For example, there are some hydrogel substrates, e.g., beads, injectable gels, moldable gels, and macroporous structures. Other techniques are also available that help prepare 3D cultures, e.g., hanging drop, low-binding plastic, and pyramid plates. There are also macroporous scaffolds, e.g., meshes, foams, or fibrous patches that allow spatial organization of cells and seeding cells throughout the thickness of the matrix culture materials (Andersen et al., 2015).

3D Bioprinting

3D bioprinting is an additive manufacturing process that uses bioinks to print living cells to develop structures layer-by-layer that imitate the behavior and structures of natural tissues. Bioinks that are used as materials in bioprinting are made of natural or synthetic biomaterials that can be mixed with living cells. The technology and bioprinted structures enable researchers to study the functions of the human body in vitro. Mostly, 3D bioprinting can be used for several biological applications in the fields of tissue engineering, bioengineering, and materials science. The technology is also increasingly used for pharmaceutical development and drug validation. Clinical settings such as 3D-printed skin and bone grafts, implants, and even full 3D-printed organs are currently at the center of bioprinting research.[9] At Microtia Congenital Ear Deformity Institute, San Antonio, Texas, USA, Dr. Arturo Bonilla in 2022 carefully implanted an outer bioprinted ear on a 20-year-old woman born without one.

[8] https://emulatebio.com/an-introduction-to-organ-on-A-chip-technology/.

[9] https://www.upmbiomedicals.com/solutions/life-science/what-is-3d-bioprinting/.

The ear on the woman's right side had been constructed in the size and shape of her left. At Wake Forest Institute, Winston-Salem, North Carolina, USA, scientists have been lab-growing organs and tissues for years. They've used 3D printing to create in the laboratory essentially a mini-kidney and a mini-liver.[10] Even in the short term, progress is well-marked. Researchers at Harvard, Cambridge, Massachusetts, USA, generated cardiac cells from human pluripotent stem cells, then seeded them on a bioengineered chip with integrated sensors that can track the beating tissue. This 3D-printed-heart-on-a-chip can be used to test various cardiac drugs for potentially toxic side effects and may alleviate the need for animal testing. Researchers in Poland bioprinted a functional prototype of a pancreas in which a stable blood flow was achieved in pigs during an observed two-week period.[11] Dr. P. R. Anilkumar, at Sree Chitrathirunal Institute for Medical Sciences and Technology (SCTIMST), Thiruvananthapuram, is currently working, with great hopes to bioprint a 3D liver lobe that can potentially be useful for regenerative medicine and laboratory testing.

Microphysiological Systems (MPS)

Microphysiological system (MPS) broadly refers to any advanced in vitro system that allows cells to be cultured in a more natural, physiologically relevant environment. MPS exposes cells to structural, material, or biophysical elements that replicate their native tissue physiology. Two of the most common MPS include (i) three-dimensional (3D) spherical clusters of stem-cell-derived and organ-specific cell types, known as organoids, that are typically grown in an extracellular matrix (ECM), where they replicate certain structural and functional features of the organs they represent; and (ii) Micropatterned tissue constructs that are static systems typically designed to grow heterogeneous cell populations in a 3D scaffold. Microengineering technology allows for cells and their tissue-specific ECM proteins to be precisely positioned with a scaffold structure that closely resembles in vivo tissue.

By culturing cells in a more complex microenvironment that mimics their natural setting, MPS encourages cells to behave as they would in vivo, both in terms of transcriptomic profiles and phenotypic responses to various stimuli. Organ chips in particular have been demonstrated to recapitulate organ-level function for the liver, producing characteristic metabolites, metabolizing xenobiotics, and displaying signs of tissue damage in response to hepatotoxic drugs.[12] As a result, the use of MPS in industries like pharma and academia has greatly increased in recent years, with applications ranging from the general study of human physiology to drug safety testing.

---

[10] https://fortune.com/well/2023/02/15/3d-printed-organs-may-soon-be-A-reality/.

[11] https://fortune.com/well/2023/02/15/3d-printed-organs-may-soon-be-A-reality/.

[12] https://emulatebio.com/microphysiological-systems-the-future-of-research-and-drug-development/.

It is envisioned that a future in which, if sufficient effort is made, tests based on human cell systems will be able to replace, completely or almost totally, testing in animals.

Computer Models or in Silico Tools

A popular tool is the Structure Activity Relationship (SARs) computer program. It predicts the biological activity of a drug candidate based on the presence of chemical moieties attached to the parent compound. Computers can help to understand the various basic principles of biology. Specialized computer models and software programs help to design new medicines. Computer-generated simulations are used to predict the various possible biological and toxic effects of a chemical or potential drug candidate without animal dissection. Only the most promising molecules obtained from primary screening are used for in vivo experimentation. For example, to know the receptor binding site of a drug, in vivo experimentation is necessary. Software known as Computer Aided Drug Design (CADD) is used to predict the receptor binding site for a potential drug molecule. CADD works to identify probable binding sites and hence avoids testing of unwanted chemicals having no biological activity. Also, with the help of such software programs we can tailor make a new drug for the specific binding site, and then in the final stage animal testing is done to obtain confirmatory results. Hence, the total number of experimental animals is lowered and the objectives of 3R concept are nearly achieved. Quantitative Structure Activity Relationship (QSAR) is the mathematical description of the relationship between physicochemical properties of a drug molecule and its biological activity. The activities like carcinogenicity and mutagenicity of a potential drug candidate are well predicted by the computer database. The QSAR software shows more appropriate results while predicting the carcinogenicity of any molecule. The advantages of computer models over conventional animal models are the speed and relatively inexpensive procedures. In silico is today a huge area of prediction of chemical toxicity and drug design. The in silico approach has become indispensable for the drug discovery process. The in silico methods have been proven beneficial in the estimation of biological activities of chemical compounds against a target. Moreover, it has been used to examine the binding affinities towards the target and also used in the prediction of physicochemical properties depending upon their molecular and structural features of a broad range of chemical compounds. Besides, computational screening seems an important method to limit the cost, time, and manpower in comparison to the traditional drug discovery process (Karthikeyan et al., 2015; Lin et al., 2020).

Notwithstanding, in silico models have been used for a multitude of applications, from predicting the toxicity of pollutants or agrochemicals to environmental species to optimising drug candidates. Thousands of models and hundreds of software packages are now available. The following few tools are listed:

- Sequence analysis tools: BLAST, ClustalW, T-Coffee, MEME, MEGA, PHYLIP, etc.
- Structure analysis tools: CN3D, PyMOL, RasMol, MODELLER, ETC.
- Function analysis tools: GEO, InterProScan, COBRA Toolbox, Pathway tools, etc.

Many databases, allow for the organization of vast amounts of biological data in a structured manner; which are important resources for researchers that can aid in their research; which can be used to develop new bioinformatics tools and methods to drive further research; and they enable collaboration between researchers and facilitate data sharing and resources, are in existence reference to which can obviate new animal-based experiments. Following are some of the most popular biological databases: GenBank, EMBL, DDBJ, PDB, PIR, PROSITE, Pfam, KEGG, OMIM, etc.[13]

In toxicogenomics (or cellular toxicology), based on the analysis of the reactions of genes contained in human cells or cell lines tested to the chemicals, a revolution is taking place in biology. At its center is the progress being made in the elucidation of cellular-response networks. Those networks are interconnected pathways composed of complex biochemical interactions of genes, proteins, and small molecules that maintain normal cellular function, control communication between cells, and allow cells to adapt to changes in their environment.

A series of experiments with the new CompTox (Computational Toxicology), based on the use of "robots" that allow an assessment of the toxicity of chemicals at high speed. "*The program, initially started at EPA as ToxCast to assess 1,000 chemicals (and known as* ***Tox21*** *in its expanded form), employs a robot to speed chemical screening in which "wells" replace the old standby of toxicology-animal testing. In addition to being slow and controversial, animal tests do not reveal how a chemical might impact humans, nor do they deliver any insight into the mechanisms by which a given chemical produces toxic outcomes. Simply by running the robotic tests, the EPA and its partner agencies will generate more information*". One of the most ambitious studies in the field is the Human Toxome Project which will comprehensively map the pathways of endocrine disruption (ED), as a first step toward mapping the whole Human toxome (Bouhifd et al., 2015).

Artificial Intelligence Application in Animal Testing

Artificial intelligence and machine learning are known for generating reliable outcomes efficiently and effectively throughout the year if the training data is accurate. AI models can save the lives of millions of animals with computer vision and accurate datasets. This is one of the true alternatives to animal models that hold huge potential to generate reliable and safe outcomes for drug discoveries. The emergence

[13] https://microbenotes.com/bioinformatics-databases-software-tools/.

of quantum computing is creating a massive way with breakthroughs and experiments. Thus, there is no need for utilizing different animals for not-so-reliable animal testing. Thomas Hartung at, CAAT, John's Hopkins, led some researchers to successfully develop an artificial intelligence algorithm that can determine substance toxicity after comparing it to similar databases and predictions from previously conducted animal testing (Luechtefeld et al., 2018). This software project showed this group of researchers that testing on animals showed inconsistencies and different animals can show different results in the same experiment. There is a concern that laboratories cannot use animals for these experiments but all kinds of testing are not possible to be completed by computers. Start-ups like Verisim Life have started utilizing the power of artificial intelligence and machine learning in biosimulation to replace animal testing in the near future. It is a San Francisco biotechnology start-up focused on building digital animal simulations to reduce animal testing for drug discoveries. When animal testing is slow and unreliable, this AI model can eradicate the cruelty as well as boost the process of drug discovery to supply at a faster rate.[14]

Alternative Model Organisms

The ethical issues have posed many restrictions over the experimental use of higher model vertebrates like guinea pigs, rats, dogs, monkeys, etc. Therefore, the use of alternative model organisms has been proposed (Zeeshan et al., 2014). Lower vertebrates and invertebrates are an attractive option because:

1. The genetic relatedness to the higher vertebrates including mammals.
2. There are fewer ethical problems involved in the experimental use of these models since the animals concerned are poorly sentient; and
3. These are organisms cultured on a large-scale and so there is no loss whatsoever to biodiversity.

Example—Danio rerio

Danio rerio, commonly called zebra fish, is a small freshwater fish with an approximate length of 2–4 cm. It has a nearly transparent body during early development, which helps easy visual access to the internal anatomy. The optical clarity allows direct observation of developmental stages, identification of phenotypic traits during mutagenesis, easy screening, assessment of endpoint of toxicity testing, and direct observation of gene expression through light microscopy. Small size, short life cycle and high fecundity favor its laboratory use. The working space, cost of laboratory solutions, test chemicals and the manpower involved are reduced by opting

[14] https://www.analyticsinsight.net/potential-of-artificial-intelligence-replacing-animal-testing-in-the-future/.

D. rerio as an alternative to animals. Its embryos and larvae can be developed and used for testing in cell culture plates and Petri dishes. Whole genome sequence availability makes Zebra fish an attractive option for molecular and genetic research. From infancy to the adult stage it is used in a variety of applications, mainly for the detection of various toxicological studies of chemicals and pharmaceuticals. It also has wide applications in the investigation of cancer, heart diseases, neurological malfunctions, behavioral diseases and to observe the mutations and problems in organ development due to exposure to test molecules. Modeling certain human diseases in zebra fish could be used to ameliorate the disease phenotype and malfunctions in organ development (Akbarsha et al., 2013).

Example—Drosophila melanogaster

Drosophila melanogaster, also known as fruit fly, is one of the most widely studied invertebrates in research. It has a well-studied genome which enables the study of molecular mechanisms underlying human diseases. Its complete genome, sequenced and annotated, has more than 13,600 genes on four chromosomes. Nearly 75% of the genes involved in human diseases are believed to have a functional homolog in the fly. *D. melanogaster* requires an extremely low cost of maintenance, propagation and screening as compared to the other mammal-based models. It also produces the results very rapidly due to a short life cycle. Fruit fly possesses four stages in the life cycle—the embryo, the larva, the pupa, and the adult. Each stage of fly has its advantage, hence they are considered as a multiple model organism to study the various concepts. The embryo is frequently used to study the cell fate determination, neuronal development, axon path finding, organogenesis, fundamental developments and to examine pattern formation. The larva is used to study the physiological and developmental processes and behaviors like foraging. The adult fly is a very complex organism. The functions of various structures like the heart, lungs, gut, kidney, and reproductive tract are equivalent to that of mammals.

The response of flies to many drugs that are acting on CNS is similar to that observed in mammals. The brain of the adult fly is quite extraordinary because more than 100,000 neurons form the discreet circuits, which mediate various complex behaviors like circadian rhythms, learning and memory, feeding, sleep, courtship, aggression, grooming and flight navigation Number of molecular and genetic tools has been made available to study *Drosophila.* Due to many similarities in development and behavioral activities, fruit fly served as a unique and sensitive model for the study of human genetics and diseases. It is also used to express the protein products found in human diseases and to compare the resulting pathologic conditions. Fruit fly serves as an important tool to investigate neurodegenerative diseases like Alzheimer's disease, Parkinson's disease, and Huntington's disease. It is used in primary small molecule discovery validation as well as in the target discovery processes by taking advantage of the sophisticated genetics available in it. The 1995 Nobel Prize for Physiology and Medicine was awarded to Ed Lewis for his pioneering research defining gene structure in flies, as well as to Eric Weischaus and Christiane Nusslein-Volhard for their studies investigating embryogenesis (Akbarsha et al., 2013).

Example—Caenorhabditis elegans

Caenorhabditis elegans is a eukaryotic nematode. This multicellular organism is approximately 1 mm in length and has a very short generation time. The complete life cycle of this hermaphrodite is about 2–3 weeks. Embryogenesis occurs in 12 h and an adult form is developed in 2.5 days. It is transparent, genetically amenable, and has simple cellular complexity. Hence, was selected as a model organism by Nobel laureate Brenner. The life cycle of *C. elegans* proceeds through various complex developmental stages like embryogenesis, morphogenesis, and growth into an adult. This is one of the most commonly used model organisms for research purposes. Information obtained can apply to more complex organisms like humans. As a model, *C. elegans* has been used to study various neurological disorders like Huntington's disease, Parkinson's disease, and Alzheimer's diseases; various immune disorders as well as cancer, and diabetes. It has served the development and testing of therapeutic agents for the treatment of these diseases (Akbarsha et al., 2013).

Example—Hydra, More Recent Entry as a Model Organism into Toxicity Testing

Given the varied manifestations of emerging pollutants, risk assessment of the chemical entities before releasing them into the environment is necessary. Hydra as a model organism is easy to culture, multiplies rapidly by asexual reproduction, cost-effective, and sensitive to inorganic substances, morphological changes can be easily monitored and classified, and genetically identical twins can be made use of for reproducible results at transcriptional levels. Moreover, the whole genome sequence of hydra revealed that most of the genes are conserved which means it offers an advantage over even the most routinely used other invertebrate model organisms. Several assay methods have been standardises which formed the nucleus of several publications. Its robust nature of whole-body regeneration from the dissociated cells or gastric region or body column makes Hydra a suitable organism for teratogenicity testing. A comprehensive study including various end-point analyses and expression of biomarker genes can explain the underlying mechanism of toxicity induced by various chemical substances. Thus, Hydra is an organism highly suited for aquatic eco-toxicity testing (Murugadas et al., 2019).

NAMs/Alternatives- Business Opportunities

Both animal testing and its alternatives are industries in their own right, so offer remarkable business opportunities for biotech and IT companies as well as contract research organizations (Meigs et al., 2018). Business opportunities mean economic pursuits, intentionally or otherwise. Economic aspects constitute an important driver for change (Hartung, 2017). *Markets and Markets* states that, "*Some key factors driving the market growth are the increasing pharmaceutical R&D activities, mandatory government regulations for animal testing, innovations in animal models, and the development of exclusive* in vivo *toxicology tests. Factors such as increasing research in oncology and personalized medicine and the rising demand for humanized animal models are expected to offer lucrative opportunities for players in the global market*". The performance of the new animal-free approaches is often so high that they have made animal experimentation obsolete or reduced it by 80% or more. In addition to the substitution of in vivo testing, the in vitro and in silico approaches are used as enabling technologies in many fields, sometimes complementing data obtained from animals, sometimes providing entirely new information.

The impact of NAM on business success/development of new predictive methods (NAM) leads to: (i) expansion of the business into new areas (e.g., production of cosmetic ingredients or production of medical devices), (ii) capture of new markets (e.g., establishing subsidiaries/spin-offs to produce organoids), (iii) more effective and rapid discovery of new chemical entities (NTE), (iv) increased satisfaction of the end users (also with regard to ethical concerns), and (v) manufacturing of better and safer products. All these together ultimately lead to increased productivity and turnover. Using NAMs for toxicity testing improves the capacity of companies to correctly predict the toxicity of chemicals in their products. The reduction of the number of false negatives (FN, products that are toxic, but are predicted to be safe) directly increases consumer safety. Decreasing the rate of false positives (FP, i.e., products that are in reality safe, but that are predicted to be toxic) has a direct effect on productivity and allows the marketing of innovative products that would otherwise have been filtered out. The combined effect of allowing for safer products (low FN rate) and more marketable products from the discovery process (low FP rate) leads to increased business profit (Meigs et al., 2018).

Digital/Virtual Alternatives in Education

The ban on animal killing for the purpose of education saves several million animals annually. In India, the ban saves not less than 20 million animals each year. Thus, it is a great fillip to the conservation of biodiversity. The various digital alternatives are products of IT companies and are aggressively marketed (Figs. 1–8). Though a comprehensive estimate is not available, the prescription of the digital and virtual alternatives supports market value, and the companies rely on employees who are paid, meaning employment opportunities.

Cell-Based Assays/In Vitro Alternatives

The rising prevalence of chronic disorders acts as a significant driver for the cell-based assay market due to the need for understanding, treatment, and management of these disorders. Chronic disorders such as cardiovascular diseases, diabetes, cancer, and neurodegenerative diseases require a deeper understanding of their underlying mechanisms. Cell-based assays provide a precise tool for studying disease pathology, identifying key cellular targets, and investigating disease progression. Cell-based assays allow researchers to mimic disease conditions in vitro and assess cellular responses to various stimuli or treatments. The rising prevalence of chronic disorders fuels the demand for cell-based assays as researchers seek to gain insights into disease mechanisms.

3D cell culture models provide a more relevant environment. The opportunities for the market are available in leveraging 3D cell culture models to enhance the predictive power and reliability of cell-based assays. 3D cell culture models offer the potential to create disease-specific microenvironments, allowing researchers to study disease mechanisms and develop better models for drug discovery and development. These models can be tailored to replicate specific disease conditions, such as cancer, neurodegenerative disorders, or organ-specific diseases. By integrating 3D cell culture with cell-based assays, researchers can gain insights into disease pathogenesis, identify therapeutic targets, and evaluate personalized treatment strategies.

Here is another perspective: "Based on product, the 3D cell culture market is segmented into scaffold-based 3D cell cultures, scaffold-free 3D cell cultures, microfluidics-based 3D cell cultures, and magnetic and bioprinted 3D cell cultures. The scaffold-based 3D cell cultures segment accounted for the largest share in 2022. The advantages of scaffolds in 3D cell cultures, such as structural rigidity, the availability of attachment points, and support, have greatly driven the preference for the segment and ensured its large share in the market."[15]

The commercial products and services are concerned with the following:

- Reagents
- Assay Kits
  - Cell Viability Assays
  - Cytotoxicity Assays
  - Cell Proliferation Assays
  - Reporter Gene Assays
  - Second Messenger Assays
- Microplates
- Probes & Labels
- Instruments & Software
- Cell Lines

[15] https://www.marketsandmarkets.com/requestsampleNew.asp?id=19107284.

- Primary Cell Lines
- Stem Cell Lines
- Immortalized Cell Lines

According to Precedence Research (2023), Cell-based assays are crucial in evaluating the toxicity of chemicals, drugs, or environmental agents. Cell viability assays, apoptosis assays, or cell cycle analysis can help assess the adverse effects of compounds on cellular health and function. The cell-based assays market plays around the production and services related to cell-based assays. The global cell-based assays market carries a wide range of reagents, consumables, instruments, and services used in conducting cell-based assays for research, drug discovery, and toxicity testing applications. The market includes both the products and services provided by manufacturers, as well as the contract research organizations (CROs) and academic research institutions that utilize cell-based assays for their research.

The global cell-based assays market size was estimated at USD 15.57 billion in 2022 and it is projected to be worth around USD 35.69 billion by 2032, poised to grow at a compound annual growth rate (CAGR) of 8.7% over the forecast period 2023 to 2032. U.S. cell-based assays market was valued at USD 6.7 billion in 2022.[16]

According to a new report from MarketsandMarkets, the market is projected to grow from $18.9 billion in 2023 to $28.2 billion by 2028, expanding at a compound annual growth rate of 8.4%. Leading players operating in the global cell-based assays market include Thermo Fisher Scientific, Danaher Corporation, Merck KGaA, Lonza Group, Charles River Laboratories, PerkinElmer Inc., Eurofins Scientific, Bio-Rad Laboratories, Promega Corporation, Cell Signaling Technology among others. Acquisitions, partnerships, and new product development are major strategies being pursued by these top companies. Some of the major developers of alternative testing technologies in the USA include Emulate Inc, Organovo Holdings, MatTek Corporation & EMD Millipore. The major Japanese developers include J-TEC, Reprocell Inc. and Organ Technologies.[17]

As per the recently published report by Markets and Markets™, "3D Cell Culture Market by Product (Hydrogels, ECM, Hanging Drop, Bioreactor, Microfluidics, Magnetic Levitation), Application (Cancer, Stem Cell, Tissue Engineering), End User (Pharma, Biotech, Research), Region - Global Forecast to 2028", size is projected to reach 2.5 billion by 2028 from 1.3 in 2023, at a CAGR of 14.6% during the forecast period. Major factors driving the market growth at significant CAGR are increasing adoption of 3D cell culture over, restrictions over animal testing and increased use of 3D cell culture in personalized medicine.

Another more recent study by Westmoreland et al. (2022) shows that based on product, the 3D cell culture market is segmented into scaffold-based 3D cell cultures, scaffold-free 3D cell cultures, microfluidics-based 3D cell cultures, and magnetic & bioprinted 3D cell cultures. The scaffold-based 3D cell cultures segment accounted for the largest share in 2022. The advantages of scaffolds in 3D cell cultures, such

[16] https://www.precedenceresearch.com/cell-based-assays-market.

[17] https://www.marketsandmarkets.com/pdfdownloadNew.asp?id=119917269.

as structural rigidity, the availability of attachment points, and support, have greatly driven the preference for the segment and ensured its large share in the market.

Based on region, the 3D cell culture market is segmented into North America, Europe, the Asia Pacific, Latin America, and the Middle East & Africa. In 2022, North America dominated the market, followed by Europe. The increasing incidence of cancer and a well-established pharmaceutical & biotechnology industry are key factors supporting the growth of the North American 3D cell culture market. The Asia Pacific market is projected to grow at the highest CAGR of 16.0% during the forecast period. Most companies in the market focus on organic and inorganic growth strategies, such as product launches, expansions, acquisitions, partnerships, agreements, and collaborations, to increase their product offerings, cater to the unmet needs of customers, increase their profitability, and expand their presence in the global market.[17]

According to Microfluidic Reviews,[18] **3D in vitro modeling for drug discovery** aims at creating 3D cellular models to mimic living tissues so that the bulk of research and drug discovery can be done in vitro instead of on animals (or humans). 3D in vitro models are meant to be used at different stages of drug discovery (Fang & Eglen, 2017). The global cell-based assays market is poised for strong growth over the next five years, driven by rising demand for drug discovery and toxicity screening along with increasing adoption of cell-based models over animal testing.[19]

Conclusion

Animal testing has been the gold standard for the hazard assessment of a range of chemicals and products and drug testing. With advancing science, more efforts are made to develop non-animal-based test systems that mimic or resemble toxicity and disease pathways relevant to humans. Governments as well as different industries around the world have been investing significant amounts of time, money, and effort in NAMs, which has led to significant progress in development and validation efforts in the field of NAMs in the last decade.

Mahony (2019) discussed current efforts and collaborations between international regulators, researchers, and the cosmetic industry to build confidence in using non-animal methods. The publication also outlined the discussion on the "application of NAMs in NGRA of cosmetic products", held at Eurotox 2018 session. Further, the publication has discussed the current practice, progress, and future work plans while applying the threshold of toxicological concern (TTC), read across and integrated safety assessment approaches in cosmetic safety assessment.

[18] https://www.elveflow.com/microfluidic-reviews/organs-on-chip-3d-cell-culture/3d-cell-culture-market-industrial-needs/.

[19] https://www.marketsandmarkets.com/pdfdownloadNew.asp?id=119917269.

Recently there have been efforts to update TTC approach by adding more chemical data into three Cramer class datasets, especially in class II, where there are very limited chemicals. In addition, despite significant challenges, there have been developments to derive route-specific TTC values, i.e., inhalation and systemic TTC values. TTC can be a useful tool for exposure-based risk assessment and metabolism-based read across assessment. While discussing the use of read across as NAM in risk assessment, it is also emphasized that apart from chemical similarity, mechanistic similarity is an equally important parameter among chemicals. Toxicokinetic data can be valuable support to prove the mechanistic similarity of reading across chemicals. Furthermore, the development of exposure and mode of action (MoA) led integrated safety assessment workflow as well as its regulatory acceptance will be a key future work area in NAMs application in NGRA. Overall, it can be concluded that there will be very interesting and promising developments in NAMs application in NGRA. As regulators, industries as well as researchers are working together and in collaboration to help make the toxicity testing paradigm shift in twenty-first century, despite many challenges, it appears that in the coming days more NAMs will be accepted by regulatory authorities around the world.[20]

In summary, there is reason to hope that, in a not-too-distant future, animal testing can be completely overcome or reduced to very rare limited cases. To achieve such goals technological innovations are never enough. It will be necessary to spread among scientists and people a critical awareness of all the ethical and medical implications of animal experimentation, and a strong social and cultural commitment will be needed by those who are already sensitive to these issues. Cell-based assays represent an essential enabling technology for biopharmaceutical innovation and toxicity testing. With drug pipelines expanding and animal testing facing mounting scrutiny, this market is poised for robust growth. Companies that can integrate automation, improve assay reproducibility, and offer turnkey services will be best positioned to capitalize on the opportunities and disrupt this space.

Conflict of Interest The authors have no conflict of interest to declare.

References

Direct Citations

Akbarsha, M. A. (2007). Movement to curtail animal dissections in Zoology curriculum: Review of the Indian experience. *ALTEX, 24*, 163–166.

Akbarsha, M. A. (2015). The state of frogs in India: A case for discontinuation of use of frogs in dissections and experiments. In O. V. Oommen & P. R. Sudhakaran (Eds.), *Laladhas KP*

[20] https://toxminds.com/adopting-new-approach-methodologies-nam-in-the-next-generation-risk-assessment-ngra/.

(pp. 41–52). Biodiversity conservation—Challenges for the future. https://www.eurekaselect.com/chapter/7322

Akbarsha, M. A., & Mascarenhas, B. (2019) Cosmetic regulation and alternatives to animal experimentation in India. In H. Kojima, T. Seidle, H. Speilmann (Eds.), *Alternatives to animal testing. Proceedings of Asian Congress* 2016 (pp. 57–62). Springer Open.

Akbarsha, M. A., Zeeshan, Md., Meenakumari, K. J. (2013). Alternatives to animals in education, research and risk assessment: An overview with special reference to Indian context. *ALTEX: Alternativen zu Tierexperimenten, 2*, 1/13, 5–19.

Andersen, T., Auk-Emblem, P., & Dornish, M. (2015). 3D cell culture in alginate hydrogels. *Microarrays, 4*(2), 133–161.

Balcombe, J. (2001). Dissection: The scientific case for alternatives. *Journal of Applied Animal Welfare Science, 4*(2), 117–126.

Russell, W. M. S., & Burch, R. L. (1959). *The principles of humane experimental technique*. Methuen.

Bouhifd, M., Andersen, M. E., Baghdikian, C., Boekelheide, K., Crofton, K. M., Fornace, A. J., Kleensang, A., Li, H., Livi, C., Maertens, A., McMullen, P. D. (2015). The Human Toxome Project", *ALTEX—Alternatives to Animal Experimentation, 32*(2), 112–124.

Doke, S. K., & Dahwale, S. C. (2015). Alternatives to animal testing: A review. *Saudi Pharmaceutical Journal, 23*, 223–229.

Fang, Y., & Eglen, R. M. (2017). Three-dimensional cell cultures in drug discovery and development. *SLAS Discovery: Advancing the Science of Drug Discovery, 22*(5), 456–472.

Hartung, T. (2016a). Making big sense from big data in toxicology by read-across. *ALTEX, 33*, 83–93.

Hartung, T. (2016b). Making big sense from big data in toxicology by read-across. *ALTEX, 33*, 83–93.

Hartung, T. (2017). Evolution of toxicological science: The need for change. *International Journal of Risk Assessment and Management, 20*, 21–45.

Hubrecht, R. C., & Carter, E. (2019). The 3Rs and humane experimental technique: Implementing change. *Animals, 9*, 754.

Karthikeyan, B. S., Parthasarathy, S., & Akbarsha, M. A., et al. (2015). Technologies to transform toxicity testing in the 21st century: In silico toxicology and its battery of methods. In H. Chandana (Ed.), *Animals and alternatives in life science research* (pp. 91–105). Luminous Books.

Koduri, C., & Akbarsha, M. A. (2014). India: End to animal testing of cosmetics and house-hold products. *ALTEX- Alternatives to Animal Experimentation, 31*(2), 236–237.

Lin, X., Li, X., & Lin, X. (2020). A review on applications of computational methods in drug screening and design. *Molecules, 25*, 1375.

Luechtefeld, T., Marsh, D., Rowlands, C., & Hartung, T. (2018). Machine learning of toxicological big data enables read-across structure activity relationships (RASAR) outperforming animal test reproducibility. *Toxicological Sciences, 165*(1), 198–212.

Mahony, C. (2019). Building confidence in non-animal methods: Practical examples of collaboration between regulators, researchers and industry. *Computational Toxicology, 10*, 78–80.

Marinou, K. A., & Dontas, I. A. (2023). European Union legislation for the welfare of animals used for scientific purposes: Areas identified for further discussion. *Animals (Basel), 13*(14), 2367.

Meigs, L., Smirnova, L., Rovida, C., Leist, M., & Hartung, T. (2018). Animal testing and its alternatives—The most important omics is economics. *ALTEX—Alternatives to Animal Experimentation, 35*(3), 275–305.

Miller-Holt, J., Behrsing, H. P., Clippinger, A. J., Hirn, C., Stedeford, T. J., & Stucki, A. O. (2022) Use of new approach methodologies (NAMs) to meet regulatory requirements for the assessment of tobacco and other nicotine-containing products. *Frontiers in Toxicology, 4*, 943358

Murugadas, A., Zeeshan, M. D., Akbarsha, M. A. (2019). Futuristic approach to alternative model organisms: Hydra stakes its claim. In H. Kojima, T. Seidle, & H. Speilmann (Eds.), *Alternatives to Animal Testing. Proceedings of Asian Congress* 2016 (pp .110–123). Springer Open.

Obhan, E., Yadav, C., & Sharma, S. (2023). India: Experiments with animals: The long road to preventing animal testing in R&D in India. Obhan & Associates: Mondaq

Blog. https://www.mondaq.com/india/life-sciences-biotechnology--nanotechnology/1345374/experiments-with-animals-the-long-road-to-preventing-animal-testing-in-rd-in-india
Rusche, B. (2003). The 3Rs and animal welfare - conflict or the way forward? *ALTEX, 20*(Suppl 1), 63–76.
Smyth, D. H. (1978). *Alternatives to Animal Experiments*. Scholar Press Ltd.
Taylor, K., & Alvarez, L. R. (2019). An estimate of the number of animals used for scientific purposes worldwide in 2015. *Alternatives to Laboratory Animals, 47*(5–6), 196–213.
Temple, J., Velliou, E., Shehata, M., Lévy, R., & Gupta, P. (2022) Current strategies with implementation of three-dimensional cell culture: the challenge of quantification. *Interface Focus, 12*, 20220019.
Wadman, M. (2023). FDA no longer has to require animal testing for new drugs. *Science, 379*(6628), 127–128.
Westmoreland, C., Bender, H. J., Doe, J. E., Jacobs, M. N., Kass, G. E., Madia, F., Mahony, C., Manou, I., Maxwell, G., Prieto, P., & Roggeband, R. (2022) Use of New Approach Methodologies (NAMs) in regulatory decisions for chemical safety: Report from an EPAA Deep Dive Workshop. *Regulatory Toxicology and Pharmacology, 35*, 105261.
Zeeshan, M. D., Murugadas, A., & Akbarsha, M. A. (2014). Alternative model organisms for toxicity testing and risk assessment. In P. P. Mathur (Ed.), *Contemporary topics in life sciences* (pp. 259–275). Narendra Publishing House.

Literature

Artal-Sanz, M., de Jong, L., & Tavernarakis, N. (2006). *Caenorhabditis elegans*: A versatile platform for drug discovery. *Biotechnology Journal, 1*(12), 1405–1418.
Cheng, L., Baonza, A., & Grifoni, D. (2018). Drosophila models of human disease. *BioMed Research International, 2018*, 7214974.
Deveaua, A. P., Bentley, V., & Bermana, J. N. (2017). Using zebrafish models of leukemia to streamline drug screening and discovery. *Experimental Hematology, 45*, 1–9.
Domazet-Losˇo, T., Klimovich, A., & Anokhin, B. (2014). Naturally occurring tumours in the basal metazoan Hydra. *Nature Communications, 5*, 4222.
DuRaine, G. D., Brown, W. E., Hu, J. C., & Athanasiou, K. A. (2015). Emergence of scaffold-free approaches for tissue engineering musculoskeletal cartilages. *Annals of Biomedical Engineering, 43*(3), 543–554.
Franco, N. H. (2013). (2013) Animal experiments in biomedical research: A historical perspective. *Animals, 3*(1), 238–273.
Franzen, N., van Harten, W. H., Retèl, V. P., Loskill, P., van den Eijnden-van Raaij, J., & IJzerman, M. (2019). Impact of organ-on-a-chip technology on pharmaceutical R&D costs. *Drug Discovery Today, 24*(9),1720–1724.
Greenhalgh, T. (2001). Computer assisted learning in undergraduate medical education. *BMJ, 322*(7277), 40–44.
Hartung, T., Pantoja, I. E. M., & Smirnova, L. (2024). Brain organoids and organoid intelligence from ethical, legal, and social points of view. *Frontiers in Artificial Intelligence, 6*, 1307613.
Herrmann, K., & Jayne, K. (Eds.). (2019). *Animal experimentation: Working towards a paradigm change*, Vol 22. Brill.
Hill, A. J., Teraoka, H., Heideman, W., & Peterson, R. E. (2005). Zebra fish as a model vertebrate for investigating chemical toxicity. *Toxicological Sciences, 86*, 6–19.
Höckel, M., & Dornhöfer, N. (2005). The hydra phenomenon of cancer: Why tumors recur locally after microscopically complete resection? *Cancer Research, 65*(8), 2997–3002.
Jabalia, N., Kumar, A., Kumar, V., Rani, R. (2021). In silico approach in drug design and drug discovery: An update. In S. K. Singh (Eds.), *Innovations and implementations of computer*

aided drug discovery strategies in rational drug design (pp. 245–271). Springer. https://doi.org/10.1007/978-981-15-8936-2_10

Kamila, M., Soňa, K., Zuzana, B., Norbert, B., & Alžbeta, K. A. (2023). Applications of a powerful model organism *Caenorhabditis elegans* to study the neurotoxicity induced by heavy metals and pesticides. *Physiological Research, 72*, 149–166.

Loewa, A., Feng, J. A., & Hedtrich, S. (2023). Human disease models in drug development. *Nature Reviews Bioengineering, 1*, 545–559.

Madorran, E., Stožer, A., Bevc, S., & Maver, U. (2020). In vitro toxicity model: Upgrades to bridge the gap between preclinical and clinical research. *Bosnian Journal of Basic Medical Sciences, 20*(2), 157–168.

Marrs, J. A., & Sarmah, S. (2021). The genius of the zebrafish model: Insights on development and disease. *Biomedicines, 9*, 577.

Nass, R., Merchant, K. M., & Ryan, T. (2008). *Caenorhabditis elegans* in Parkinson's disease drug discovery: Addressing an unmet medical need. *Mol. Intervention, 8*, 284–293.

NTP. (2024). *Submissions of test methods for evaluation.* https://doi.org/10.22427/NICEATM-2

Okafor, O. (2022). Hydra vulgaris: A possible disease model for stroke. *Ottawa-Carleton Journal of Neuroethology, 1*(1), 1–3.

Peterson, R. T., Nass, R., Boyd, W. A., Freedman, J. H., Dong, K., & Narahashi, N. T. (2008). Use of non-mammalian alternative models for neurotoxicological study. *Neurotoxicology, 29*, 546–555.

Ryder, A., & Onuma, M. (2018). Viable cell culture banking for biodiversity characterization and conservation. *Annual Review of Animal Biosciences, 6*, 83–98.

Schmeisser, S., Miccoli, A., & von Bergen, M. (2023). New approach methodologies in human regulatory toxicology—Not if, but how and when! *Environment International, 178*, 108082.

Smirnova, L., Scaffo, B., Gracios, D. H., Morales Pantoja, I. E., Tang, B., Zack, D. J., Berlinicke, C. A., Boyd, J. L., Harris, T. D., & Johnson, E. C. (2023). Organoid intelligence (OI): The new frontier in Biocomputing and intelligence-in-a-dish. *Frontiers in Science, 1*, 1017235.

Stucki, A. O., Barton-Maclaren, T. S., Bhuller, Y., Henriquez, J. E., Henry, T. R., Hirn, C., Miller-Holt, J., Nagy, E. G., Perron, M. M., Ratzlaff, D. E., & Stedeford, T. J. (2022). Use of new approach methodologies (NAMs) to meet regulatory requirements for the assessment of industrial chemicals and pesticides for effects on human health. *Frontiers in Toxicology, 10*, 3380.

Temple, J., Velliou, E., Shehata, M., Lévy, R., & Gupta, P. (2022). Current strategies with implementation of three-dimensional cell culture: The challenge of quantification. *Interface Focus, 12*, 20220019.

Fostering Healthy Longevity Through Regenerative and Precision Medicine: Biodiversity's Vital Role and Equitable Benefit Sharing

Pravi Prasad, Abiya Mathew, Shalu Jose, Mohammad Abdulkader Akbarsh, and Sreejith Parameswara Panicker

Introduction

The term "healthy aging" describes the state of aging in which an individual can maintain physical, social, mental, and spiritual well-being throughout the aging process (Castruita et al., 2022). Age is the main factor that increases the chance of death from almost every major cause in developed countries. Nevertheless, the majority of biomedical research focuses on isolated illness pathways, frequently disregarding the complex interrelationships between aging and disease. Geroscience, a subject focused on understanding the mechanics of biological aging, has made significant progress in revealing the molecular processes that contribute to aging. Significantly, these findings provide possible measures to postpone aging and promote good health and long life (Kaeberlein et al., 2015).

In pursuing enhancing human health and longevity, a multifaceted approach encompassing various medical innovations has emerged as a cornerstone of modern healthcare paradigms. Among these innovations, regenerative medicine, precision medicine, nutrigenomics, microbiome research, personalized medicine, and the ancient wisdom of Ayurveda stand out as promising avenues for fostering healthy aging and improving overall well-being (Aruoma et al., 2019; Hood & Friend, 2011; López-Otín et al., 2013; Lynch & Pedersen, 2016; Patwardhan et al., 2005).

P. Prasad · A. Mathew · S. Jose · S. P. Panicker (✉)
Department of Zoology, University of Kerala, Thiruvananthapuram 695581, Kerala, India
e-mail: psreejith@keralauniversity.ac.in

Advanced Centre for Regenerative Medicine and Stem Cell Research in Cutaneous Biology (AcREM-STEM), University of Kerala, Thiruvananthapuram 695581, Kerala, India

M. A. Akbarsh
National College (Autonomous), Tiruchirappalli, India

L. Krishna Panicker et al. (eds.), *Biodiversity and Business*,
https://doi.org/10.1007/978-3-031-71674-4_27

Regenerative medicine is centered around developing innovative therapies to replenish or revitalize lost, impaired, or aging cells within the human body to restore their normal function. Achieving this objective requires collaborative endeavors spanning across non-mammalian and human development, stem cell biology, genetics, materials science, bioengineering, and tissue engineering. Currently, the primary focus of this field lies in comprehending the intrinsic reparative mechanisms present in humans and harnessing the latent capacity for tissue regeneration (Gurtner et al., 2007).

Precision medicine aims to optimize healthcare quality by tailoring the healthcare journey to each patient's distinctive and evolving health condition. This endeavor encompasses various scientific domains such as drug discovery, genetics/genomics, health communication, and causal inference, all geared towards supporting evidence-based, data-driven decision-making processes. Formalized as a treatment regimen, precision medicine consists of a series of decision rules, each corresponding to a specific decision point, which utilizes current patient information to recommend appropriate actions (Kosorok & Laber, 2019).

Intricate epigenetic pathways significantly influence the aging process and longevity. The field of nutrition significantly impacts these epigenetic mechanisms, providing potential strategies for controlling lifespan and aging. Understanding how nutrition affects the epigenome is crucial for creating preventive and interventional approaches to improve well-being and health, promoting sustained longevity (Alam et al., 2019).

Throughout the human body, a diverse array of microbial communities and their associated genes, collectively known as the microbiome, play fundamental roles in shaping human health and influencing disease processes (2012). The gut microbiome plays a crucial role in human health, impacting homeostasis throughout life, especially in old age. With age, there's a gradual decline in organ function related to growth, metabolism, and immunity, making the role of the gut microbiome even more significant. Studies have identified changes in the gut microbiome across different age groups, including maladaptation associated with age-related conditions like frailty, neurodegenerative diseases, and cardiometabolic diseases. Additionally, the gut microbiome interacts bidirectionally with therapies for age-related diseases, affecting and being affected by their efficacy. Considering this, developing integrated microbiome-based intervention strategies to promote a healthy gut microbiome could be an innovative approach to fostering healthy aging and achieving longevity (Barone et al., 2022).

In addition to these modern scientific approaches, the ancient holistic healing system Ayurveda, renowned as one of the world's most comprehensive mind-body-spirit medicinal systems, presents diverse perspectives on the aging process. Embedded within Ayurvedic principles are therapies that foster healthy aging, promoting optimal health and extending an individual's health span by fostering harmony with nature (Rao, 2018).

These advanced fields of study provide new and unique understandings and treatment approaches that are specifically designed for individual requirements, fundamentally changing our understanding and approach to age-related illnesses. Furthermore, incorporating biodiversity and bioactive substances into medical research and practice enhances the capacity to promote good aging and longevity (Carmona & Michan, 2016). We can discover fresh possibilities for preventative and curative treatments using abundant natural resources and ancient knowledge systems. Nevertheless, prioritizing the fair distribution of benefits derived from these unique resources is of utmost importance to promote ideals of social justice and sustainable development. This chapter introduces the idea of combining medical advancements, traditional knowledge, and biodiversity conservation to improve the health and well-being of people as they age.

Healthy Longevity

Aging is a steady physiological deterioration in all living beings as they age. This process exhibits both diversity and variability, characterized by a mixture of different elements and variations in timing. In terms of heterogeneity, the aging process can occur at different rates among different organisms, and even individuals of the same species may experience aging at different speeds. Furthermore, the fact that various cells and tissues within a single organism age at different rates highlights the heterochronic nature of the aging process. From a biological standpoint, aging is characterized by the accumulation of molecular and cellular damage, leading to structural and functional irregularities in cells and tissues. The anomalies encompass disturbances in the balance of mitochondria, impaired communication between cells, cellular arrest (halting development and division), and a reduced ability to regenerate (Holliday, 2004).

There are at least two perspectives on the concept of 'the aging process' (Hayflick, 2004). According to one viewpoint, aging represents a biochemical process that does not directly cause disease but increases the likelihood and severity of age-related ailments over time. This progression reduces the body's capacity to withstand, endure, and recuperate from physiological disruptions. On the other hand, an alternative perspective views the aging process as a conceptualization encompassing all age-related diseases (Masoro, 2005). In various species, including humans, these diseases, whether immediately life-threatening or initially less severe, follow a somewhat predictable pattern within populations. This pattern becomes evident through observable signs such as greying hair, wrinkles, increased body fat, and changes in posture, as well as through the prevalence of specific diseases like atherosclerosis, cancer, and neurodegenerative disorders later in life. Notably, a viewpoint suggests that aging's molecular mechanism involves accumulating unrepaired damage (Hekimi et al., 2001).

In the first perspective, the impact of accumulated damage is widespread: it weakens the body's ability to maintain internal stability (homeostasis), thereby elevating the likelihood of disease development, although it doesn't trigger disease onset directly. Conversely, the second viewpoint asserts that accumulated damage directly triggers disease onset, with various types of damage leading to different diseases (Hekimi, 2006). Enhanced longevity has been linked to the concept of healthy aging. This assertion supports the observation that interventions—whether genetic, dietary, or pharmacological—that foster cellular homeostasis, encourage stress resistance, and provide defence against age-related diseases often result in an extension of lifespan. Conversely, actions that compromise these factors tend to have the opposite effect on longevity. Substantial scientific evidence affirms the assertion that aging does not have a singular cause. Significant strides in the field of aging biology, particularly in recent decades, have led to the recognition of numerous mechanisms influencing the aging process. Despite these advancements, pursuing interventions promoting healthy aging in humans remains challenging. The conserved molecular and cellular mechanisms that form the basis of aging, particularly the intricate interactions among these pathways and how the complex lifestyles and environments impacting humans modify them, remain incompletely understood (Carmona & Michan, 2016).

In genetic studies focusing on human aging, researchers typically utilize various phenotypes. These include lifespan, which refers to the age at death; longevity, denoting long life, often defined as reaching a particular advanced age or older at the time of the study; exceptional longevity, defined as achieving or surpassing a specific remarkable age milestone; and healthy aging, which encompasses a blend of old age and good health, frequently defined as being free from specific disorders or exhibiting desirable levels of performance on functional tests. Research on longevity primarily centers on long-lived individuals (LLI), frequently focusing on centenarians who have reached the age of 100 or beyond. One notable benefit of these studies is the straightforward definition of the phenotype. On the other hand, healthy aging can be characterized in various ways, often emphasizing the attainment of a reasonably old age without specific diseases or disabilities or, alternatively, the presence of favourable attributes like preserved cognitive function or mobility. It is important to distinguish both types of studies from investigations into the fundamental biological mechanisms of aging, such as the study of cellular senescence (Brooks-Wilson, 2013).

The Blue Zones project has pinpointed nine factors linked with longevity (Buettner & Skemp, 2016) including engaging in natural movement, possessing a sense of purpose, regularly practicing downshifting, eating until 80% satiety, adopting a primarily vegetarian diet, moderate alcohol consumption, feeling a sense of belonging, prioritizing loved ones, and being part of social circles that promote good health. Further research has emphasized the importance of maintaining a balanced diet, practicing moderation in food consumption, and engaging in regular physical and social activities to enhance healthy longevity. While exceptional longevity is commonly attributed to a combination of genetic predisposition, environmental influences, resilience, and chance occurrences, it's crucial to recognize that these factors are heavily shaped by cultural and geographical contexts (Pillay et al., 2024).

Regenerative Medicine

The concept of healing wounds and regenerating organs has been ingrained in human history, echoing through the ancient Greek tale of Prometheus. This myth recounts the punishment of Prometheus, a Greek titan, by Zeus for disobedience in bestowing fire and knowledge upon humanity. In the narrative, Prometheus is bound to a rock, and each day, an eagle consumes a portion of his liver, only to witness its regeneration (Kolios & Moodley, 2012). The field of medicine has advanced significantly since the Age of Enlightenment, marked by the widespread use of targeted therapies and genetically engineered mouse models replicating human diseases. Genome sequences for numerous species are now available. However, despite these strides, numerous clinical needs persist, particularly in addressing degenerative diseases. In the case of advanced stages of such conditions, the emphasis extends beyond merely slowing disease progression to reversing its course, a fundamental principle of regenerative medicine. Yet, achieving these goals is challenging without a comprehensive understanding of the underlying disease mechanisms (Chien, 2008).

For instance, while genome-wide association studies rapidly accumulate insights into the molecular pathways associated with cardiovascular diseases, the broader understanding of how these pathways influence the pathogenesis of most diseases remains elusive. Consequently, comprehending the mechanistic logic behind many degenerative diseases in humans and devising effective strategies to disrupt their progression, let alone reverse them, is still relatively nascent (Chien, 2008).

Advancing ideas within the field of aging biology are now elucidating a mechanistic comprehension of crucial signalling targets involved in the aging process. This progress lays the groundwork for substantial advancements in therapeutic interventions (Golpanian et al., 2017; Schulman et al., 2018). Over the past decade, a significant shift has been toward stem cells as a focal point in translational medicine. These cell types possess two distinct characteristics that render them particularly appealing for applications in regenerative medicine. A stem cell is characterized by its ability to self-renew while retaining the capacity to differentiate into various cell types. This sets them apart from progenitor cells, which can differentiate into mature cell types but lack self-renewal capabilities, and somatic cells, which can proliferate but cannot differentiate. Combining self-renewal and differentiation properties in stem cells can generate an extensive array of cell types with clinical significance. Stem cells are also gaining prominence in basic research due to their ability to be expanded *in vitro* while preserving their inherent properties. This enables unfeasible fundamental studies with primary tissue cultures or biopsy materials (Teo & Vallier, 2010).

The realm of regenerative medicine involves diverse approaches, incorporating materials and newly formed cells, individually or in various combinations. These methods aim to substitute absent tissue, restore it in structure and function, or facilitate tissue healing (Bajaj et al., 2014). The body's natural healing mechanisms can also be harnessed to encourage regeneration. However, it's worth noting that adult humans have restricted regenerative capabilities compared to lower vertebrates (Kami & Gojo, 2014).

The primary focus of regenerative medicine revolves around human cells, encompassing somatic, adult stem, or embryo-derived cells. Notably, reprogrammed versions of embryo-derived cells are now from adult cells, allowing both to be conveniently categorized as 'pluripotent cells' (Takahashi et al., 2007; Yu et al., 2007). The evolving interest in this sequence is propelled by challenges related to the limited availability of many specialized somatic cells and the constrained expansion capacity of adult stem cells, coupled with their inherent heterogeneity, particularly from sources like bone marrow. Human embryos, although pluripotent, present technical challenges aside from ethical and moral concerns, making alternative sources for pluripotent cells appealing. This progression involves the introduction of genes into human cells, a development that holds the potential to bring regenerative medicine and gene therapy into closer alignment (Wernig et al., 2007).

Regenerative medicine is revolutionizing the concept of healthy longevity by focusing on repairing and rejuvenating damaged tissues and organs. Unlike traditional medicine, which often manages symptoms, regenerative medicine aims to restore the body's natural functions. This is achieved through innovative approaches such as stem cell therapy, tissue engineering, and gene therapy, all of which have the potential to address the root causes of aging and extend health span. By promoting the repair and regeneration of tissues, regenerative medicine helps maintain functional abilities essential for well-being in old age, ultimately enhancing the quality of life (Mason & Dunnill, 2008). One of the most promising aspects of regenerative medicine is stem cell therapy. Stem cells can uniquely differentiate into various cell types, making them invaluable for repairing damaged tissues and organs. For instance, stem cell therapy can be used to regenerate cardiac tissue after a heart attack, restore neural function in neurodegenerative diseases, and repair cartilage in joints affected by osteoarthritis (Trounson & McDonald, 2015). These applications alleviate the symptoms of age-related diseases and restore vital organs' structural and functional integrity, contributing to prolonged health and vitality.

Tissue engineering and biomaterials are another critical component of regenerative medicine. By creating artificial tissues and organs, these technologies provide solutions for conditions where natural repair processes are insufficient. Tissue-engineered products, such as skin grafts for wound healing and bioengineered organs for transplantation, offer new hope for patients suffering from severe tissue damage or organ failure (Atala, 2012). Additionally, biomaterials that mimic the natural extracellular matrix can support tissue regeneration and repair, enhancing the body's ability to heal itself. These advancements are pivotal in extending the health span and improving the quality of life for aging populations.

Gene therapy, a cutting-edge approach within regenerative medicine, targets the genetic basis of diseases. By correcting genetic mutations that cause age-related conditions, gene therapy can prevent or delay the onset of these diseases. For example, gene-editing technologies like CRISPR have the potential to correct mutations in genes associated with Alzheimer's disease, significantly impacting the prevalence and progression of this debilitating condition (Doudna & Charpentier, 2014). Moreover, gene therapy can enhance the body's natural repair mechanisms, promoting cellular regeneration and longevity.

Integrating personalized medicine with regenerative therapies further enhances the potential for healthy longevity. Personalized medicine involves tailoring medical treatments to the individual characteristics of each patient, including their genetic profile, lifestyle, and health status. This approach ensures that regenerative treatments are effective and precisely targeted, maximizing their benefits. For instance, personalized stem cell therapies can be developed based on a patient's specific genetic makeup, improving the success rates of these treatments (Ashley, 2016). Additionally, personalized medicine can predict and prevent diseases before they occur, ensuring that individuals maintain their functional abilities well into old age.

Regenerative medicine is also crucial in managing chronic conditions and enhancing older adults' overall quality of life. By addressing the underlying causes of chronic diseases, regenerative treatments can reduce dependency on long-term medication and improve mobility and independence. For example, regenerative therapies for osteoarthritis can repair joint cartilage, reduce pain, and increase mobility (Caplan & Correa, 2011). Similarly, treatments for neurodegenerative diseases can preserve cognitive functions, allowing individuals to maintain their mental acuity and engage in meaningful activities throughout their lives. Overall, regenerative medicine holds immense promise for promoting healthy longevity. By focusing on the repair, rejuvenation, and regeneration of tissues and organs, it addresses the fundamental causes of aging and age-related diseases. Integrating personalized approaches ensures that treatments are tailored to individual needs, maximizing their effectiveness. As regenerative medicine advances, it offers the potential to transform aging, allowing individuals to enjoy a longer, healthier, and more fulfilling life.

Precision Medicine

Precision medicine is a well-established approach that leverages patient diversity and data-driven methodologies to improve treatment options, ensuring that each patient receives the most suitable medication at the best possible time. The Precision Medicine Initiative, which President Obama announced during the 2015 State of the Union Address, emphasized the significance of precision medicine at a national level (House, 2015). Precision medicine is philosophically aligned with stratified medicine (Lonergan et al., 2017) and customized medicine (Kosorok & Moodie, 2015).

The idea of precision medicine, which tailor's prevention and treatment approaches to individual differences, is not novel (2011). For over a century, blood typing has guided blood transfusions. However, recent advancements in large-scale biologic databases (e.g., the human genome sequence), advanced patient characterization methods (proteomics, metabolomics, genomics, diverse cellular assays, and mobile health technology), and computational tools have significantly enhanced the potential application of precision medicine. To fully leverage this concept, there is a critical need for a comprehensive research program that fosters innovative precision medicine approaches, subjects them to rigorous testing, and ultimately establishes an evidence base for guiding clinical practices (Collins & Varmus, 2015).

Precision medicine aims to enhance healthcare quality by tailoring the healthcare process to each patient's unique and evolving health status. This pursuit encompasses many scientific domains, including drug discovery, genetics/genomics, health communication, and causal inference, supporting evidence-based, data-driven decision-making. Formalized as a treatment regimen, precision medicine consists of a series of decision rules, each applicable to a specific decision point, mapping current patient information to a recommended action. These actions may involve selecting the appropriate drug, determining the dosage, timing of administration, suggesting a specific diet or exercise, or other aspects of treatment and care (Kosorok & Laber, 2019).

The initiative on Precision Medicine by Collins and Varmus in 2015, comprises two key elements: an immediate emphasis on cancers and a more extended goal to accumulate knowledge relevant to the entire spectrum of health and disease. Advances in fundamental research, encompassing molecular biology, genomics, and bioinformatics, have brought both components within our grasp. Additionally, the initiative aligns with current trends, capitalizing on heightened connectivity facilitated by social media and mobile devices, along with the increasing inclination of Americans to actively engage as partners in medical research. Precision medicine places a significant emphasis on statistical research as it leverages data to guide decision-making in healthcare. Consequently, the scope of precision medicine encompasses various tasks such as drug discovery, biomarker identification, estimating and inferring causal treatment effects, modeling health communication and shared decision-making, and designing studies (Kosorok & Laber, 2019).

Focusing on oncology is the optimal choice for amplifying the immediate impact of precision medicine. Cancers, being prevalent and among the leading causes of death globally, particularly with the aging population, present a compelling target. Their formidable reputation, attributed to lethality, distressing symptoms, and the often harsh treatments, adds urgency to the need for precise interventions. Scientific research has uncovered numerous molecular abnormalities driving cancers, revealing each cancer's unique genomic signature. While cancers result from accumulated genomic damage, inherited genetic variations significantly contribute to cancer risk. This enhanced understanding reshapes risk assessments, diagnostic classifications, and therapeutic approaches, including the growing use of targeted therapies and immunologic methods designed to counter specific molecular drivers, showing promising results (de BONO & Ashworth, 2010; Snyder et al., 2014).

Technological advancements are continually expanding the horizons of healthcare reform, with a noticeable shift from emphasizing precision medicine to precision health. The concept of risk factors underscores the stealthy onset and progression of chronic diseases, emphasizing the need for prognosis well before the manifestation of clinically detectable symptoms. This phenomenon is extensively documented in studies highlighting associations between risk factors in early to midlife and later-life impacts across various chronic disease states. A longitudinal study on successful aging identified factors such as non-smoking, a balanced diet, and regular exercise as linked to favorable outcomes, devoid of major diseases and indicative of high functionality (Britton et al., 2008). Further supporting these findings, a study by Yaffe et al. revealed that cumulative exposure to cardiovascular risk factors in early to midlife, including elevated blood pressure and fasting blood glucose, is associated with adverse outcomes on neuropsychological tests (Yaffe et al., 2014).

For instance, a 45-year-old woman received a diagnosis of HER2-positive breast cancer. Precision medicine allowed her oncologist to use genetic testing to pinpoint the presence of the HER2 gene, resulting in targeted therapy utilizing trastuzumab (Herceptin). The individualized therapy demonstrated a substantial decrease in tumor size and enhanced rates of survival when compared to conventional chemotherapy. Additionally, it resulted in fewer adverse effects and improved the overall quality of life throughout treatment (Slamon et al., 2001). Another example involves a middle-aged person who experienced the advantages of precision medicine by undergoing genetic testing that accurately diagnosed his familial hypercholesterolemia (FH) problem. Administering PCSK9 inhibitors, a new category of medications that lower cholesterol, successfully decreased his cholesterol levels, considerably reducing his susceptibility to heart attacks and strokes (Robinson et al., 2015). The individualized strategy demonstrated more efficacy than conventional statins, promoting an extended and improved lifespan. Likewise, a juvenile male with cystic fibrosis with the mutation observed notable enhancements through the application of precision medicine. We specifically developed medications, ivacaftor, and lumacaftor/ivacaftor, to rectify his impaired protein function, improve pulmonary function, reduce hospitalizations, and enhance the overall quality of life (Wainwright et al., 2015). In addition, a juvenile female diagnosed with acute lymphoblastic leukemia and possessing the Philadelphia chromosome mutation underwent treatment with imatinib (Gleevec), a precision medicine. Implementing this precision medicine strategy significantly decreased cancer cells and successful remission, reducing adverse effects and significantly enhancing her overall health and lifespan (Druker et al., 2001). These stories demonstrate how precision medicine tailors' treatments based on individual genetic profiles, resulting in better results and improved quality of life for people with various illnesses.

Regenerative Medicine in Healthy Longevity

Emerging theories in the biology of ageing are starting to clarify the mechanical knowledge of key signalling targets involved in the ageing process, opening the door to significant therapeutic intervention. Importantly, these ideas align with the field of regenerative medicine, which may provide useful treatments. With several recent developments in ageing research holding out the promise of revolutionising healthcare and extending healthy, productive lifespans, biogerontology, and regenerative medicine fields are expanding quickly. From altered gene expression and DNA methylation patterns to shorter telomeres and failing genome maintenance systems, ageing body cells differ as a huge impact on young organism cells in many ways. Therefore, the antiaging strategies aim to significantly slow down the ageing process, prolonging people's productive, youthful lives rather than rejuvenating them. These strategies also aim to delay or prevent the onset of age-related diseases.

A long-standing human desire is to live as long and healthily as possible (Vellai et al., 2003). There is a huge loss in innate homeostatic and regenerative capacity along with the phenomena of ageing (West et al., 2019). The objectives of illness management are changing from "care to cure" due to the evolving knowledge of regenerative science, which provides instruments to stop or reverse the progression of resistant diseases (Blau & Daley, 2019). Developments in regenerative medicine have started to outline a fresh outlook for healthy longevity in the future. Regenerative medicines are becoming more clinically ready for use in age-related diseases. Regenerative immunotherapies, in particular, have improved cancer treatment choices. Chimeric antigen receptors have been modified for autologous administration in haematological malignancies. These receptors are expressed in immune cells to fight antigen-expressing cancer cells (Liu et al., 2020).

Researchers have studied tissue repair following life-saving resection surgery to improve the physical and emotional health of cancer survivors. Regenerative medicines for ageing populations have not been fully explored to maximize their efficacy, despite their safety and effectiveness in treating age-related illnesses. Regenerative rehabilitation, in this sense, refers to the combination of physical therapy, biomaterial science, and regenerative medicine. It serves as a verified paradigm for designing comprehensive therapeutic approaches. The regenerative paradigm of normative restoration seeks to protect the individual's identity and aid in health recovery, serving as a liberator from weakness and illness (Veres et al., 2019).

Achieving the required level of function and quality of life requires that regeneration feasibility be matched with each person's values, aspirations, and worldview (Rose et al., 2018). Wider adoption would necessitate the implementation of regenerative solutions that have been approved and are scalable, accessible, and reasonably priced. The financial burden that ageing has placed on society has spurred significant investments in the high-tech sector to take advantage of antisenescent and regenerative treatments. Driven by actualized and projected financial gains (Yamada et al., 2021). The continuous conversion from bench to bedside, the value proposition of novel treatments must be carefully considered in light of increased longevity. Notably, research on anti-aging is moving away from traditional symptom mitigation in late, incapacitating diseases and towards early, proactive health management. This shift is being made toward developing preventative and curative therapies (Mason & Dunnill, 2008). Radical curative measures could offset the initial high costs that impede wider adoption and impair utilisation by providing long-term cost-saving and cost-effective benefits (Mkrtchyan et al., 2020). As an example, preliminary real-world evidence indicates that Medicare patients' treatment costs may decrease when regenerative immunotherapies are used to treat blood malignancies such as lymphoma (Driscoll et al., 2017). Additionally, a health economic simulation based on the first clinical experience with cell therapy for Parkinson's disease suggests that treating early-onset degenerative illnesses could result in an attainable ten-year cost savings (Roth et al.,

2018). Similarly, it has been estimated that within an 8-year timeframe, stem cell-derived ß cell therapy for diabetes will have a better cost-benefit ratio than lifetime insulin therapy (Hagell et al., 2002). However, the health economic assessment of cutting-edge treatments is still few and unresolved (Archibald & Williams, 2015).

The basis of all organs and tissues in our body, stem cells, lose their ability to regenerate as we age, which causes tissue deterioration and age-related illnesses. Ill-treatment and environmental factors might also adversely affect the quality of life and health span (de Windt et al., 2017). In addition to providing treatments for various degenerative (Lopez-Otin et al., 2023), autoimmune, haematological, and neurodegenerative diseases, organ failures, and other chronic problems, stem cells can regenerate or repair damaged tissues and organs. Recently, a clinical experiment involving patient-derived induced pluripotent stem cells (iPSCs) and Parkinson's disease patients was approved by the US Food and Drug Administration (FDA). Chemical reprogramming provides a rare chance to direct the destiny of somatic cells and produce desired cell types—including pluripotent stem cells with potential medical uses (Ye et al., 2020). Activating a program similar to regeneration in human somatic cells can result in chemically reprogrammed cells that offer a different approach to generating stem cells for clinical application (Wang et al., 2022). Comprehending the epigenetic factors that determine stemness is essential to actualizing the complete potential of in vitro reprogrammed cells. Notable accomplishments include creating whole human embryo models from stem cells devoid of sperm or eggs and creating human-animal hybrid embryos (Wang et al., 2023). The self-renewal, reprogramming, regeneration, rejuvenation, repair, reversal, rehabilitation, and remodelling of animal use are the eight functional "R"s of stem cell medicine. Stem cell research programmes addressing key clinical concerns and longevity are crucial for achieving the optimal human health span. A unique gut microbiota ecology has been linked to longevity and may help prevent aging-associated diseases (Oldak et al., 2023). The microbiota and their metabolites influence the survival and fate of intestinal stem cells. Young mice's faecal microbiota transplants restore old hematopoietic stem cells by reducing inflammation (Johansen et al., 2023). Studies on intestinal and hematopoietic stem cell ageing and the gut microbiota may suggest potential treatment approaches for aging-related illnesses (Zeng et al., 2023). Throughout life, persistent inflammation is the cause of many chronic diseases and reduces adaptive immunity.

Another category of regenerative medicine, including tissue engineering, involves study of artificial organs or tissues created in a laboratory. This can replace damaged organs, potentially addressing the lack of available organs and enhancing the quality of life for those suffering from organ failure. The ideal state of tissue engineering still needs to be attained, including producing profitable off-the-shelf tissues and organs, developing scaffolds and three-dimensional tissue architectures that can sustain blood flow, choosing the best biomaterials, and figuring out which cell source is most effective for use in tissue engineering and cell therapy. The ultimate goal of tissue engineering is to produce organs and tissues that are readily available, opening up a wide range of possibilities for treating various diseases and lowering the demand for organ donors. In addition to treating millions of patients, this would extend human life and improve quality of life (Ahmad Sophien et al., 2023). The study and therapeutic use of these artificial tissues and organs are hampered by new challenges that arise as tissue engineering advances. The foundations of this multidisciplinary field necessitate addressing dependable cell sources, finding biomaterials, and creating scaffolds for *in vivo* cell proliferation. Therefore, slow progress in the practical application of tissue engineering addresses challenges in various scientific fields, including material science, bioengineering, and cell biology (Vacanti & Vacanti, 2014). A number of the stem cells' functional characteristics are impacted by ageing. For example, the cellular polarity of old stem cells is lost, particularly in brain stem cells, germline stem cells, and muscle satellite cells. They thereby lose their capacity to divide asymmetrically, a crucial characteristic of stem cells that aid in preserving the stem cell repertoires (Langer, 2009).

Cell extrinsic factors such as ageing niche cells, abnormal adhesion molecules, altered morphogen, growth factors signalling, and inflammation are primarily responsible for this loss of polarity (Schultz & Sinclair, 2016). The ultimate aim of figuring out the ageing mechanisms is rejuvenation, or making a senescent cell young again, or geroprotection, which prevents ageing. Three alternatives are available for this purpose: modifying the intrinsic variables, adjusting the extrinsic factors, or adjusting the genetic targets of those modifications. We have three options for working with stem cells: genetically, pharmacologically, or by changing their surroundings. Possibly the most realistic approach is to use pharmaceuticals to slow down or even reverse the effects of ageing. Pharmacological approaches have been used to target several pathways. For example, antioxidants such as vitamin C and N-acetyl cysteine have lowered ROS *in vivo* and *in vitro* (Pan et al., 2007).

Precision Medicine in Healthy Longevity

Precision medicine, sometimes known as personalised medicine, has gained popularity since the Human Genome Project was completed. Precision medicine in the context of healthy longevity involves tailoring medical interventions, treatments, and preventive measures to individuals based on their unique genetic makeup, lifestyle factors, environmental exposures, and other personal characteristics. The goal is to optimize health span and prevent age-related diseases, ultimately promoting healthy aging and longevity. This method allows for individual variability in clinical diagnosis and treatment of disease, and it is patient-centered. According to the theory that only genetics can provide a conclusive explanation for shared aspects of our physiology and pathophysiology, our genome may be able to precisely predict each person's own risk of experiencing a potentially fatal illness over their lifetime. More targeted and efficient treatment modalities have been made possible by our growing understanding of the underlying mechanisms and variables that promote carcinogenesis in a particular person. Longevity medicine is a rapidly developing area that combines advanced tailored preventive medicine with profound indicators of ageing and longevity. The rapidly developing fields of precision, preventative, and functional medicine, as well as geroscience and biogerontology, are all included in this field. Customised training on the newest developments in longevity medicine and how to apply this information in clinical practice is necessary for healthcare professionals. Patients lack access to medical professionals skilled in longevity medicine and can handle their care from this perspective. Promoting longevity education alongside practical application would ultimately improve global healthcare systems and decrease illness prevalence by equipping medical practitioners to tackle the primary and most significant cause of disease: unhealthy aging. The field of longevity medicine employs cutting-edge technologies and best practices from other disciplines, including deep learning and artificial intelligence, to evaluate the patient's biological age throughout the course of life. People who age are more susceptible to most diseases because ageing is a complicated, multifaceted process that results in function loss (Ermolaeva et al., 2018). In model organisms, ageing is linked to numerous genetic and epigenetic modifications that are implicated in lifespan and ageing (Mitchell-Fearon et al., 2015). Many of these alterations are also linked to non-communicable, primarily chronic diseases in humans, even though their processes in human aging are not entirely known (Moskalev et al., 2014). One definition of precision medicine is "the right drugs for the right patients at the right time". The current "precision medicine" paradigm is intended to properly classify patients into subgroups sharing a common biological foundation of diseases to attain this goal, leading to more effective therapy and better care outcomes Kennedy et al., 2014). Illness subtyping divides an illness into separate patient subgroups based on unique patient characteristics. This information can then be used to guide treatment decisions based on which subgroup a patient belongs to. It is a crucial goal of precision medicine (Wu et al., 2017). Personalized medicine tailors healthcare interventions to each individual's unique genetic makeup, lifestyle, and environmental factors. It offers the potential for predicting, preventing, and treating diseases in a highly customized manner, thereby supporting successful aging. Precision medicine encompasses early disease detection, targeted therapies, lifestyle guidance, nutritional support, vaccination strategies, and environmental considerations, all tailored to individual needs.

The attainment of a lengthy existence is referred to as longevity (Saria et al., 2015). Many individuals desire a lengthy lifespan in order to have ample opportunities to see the world and enjoy meaningful moments with their cherished ones. While reaching an advanced age, it is important to note that having a long and fulfilling life is not guaranteed if one is burdened by illness or disability. Therefore, maintaining good health and living long is crucial for a fulfilling existence (Garg, 2024). Various variables influence a healthy lifespan, including a well-balanced diet, frequent physical activity, maintaining a normal weight, abstaining from smoking, and consuming alcohol in moderation. In addition to having a sense of purpose, strong social relationships, and mental stimulation, enhancing sleep quality and practicing intermittent fasting also promote a long and healthy life.

Nutrigenomics and Healthy Longevity

Many multidisciplinary gene-based approaches are available to study health and disease. Nutrigenomics is one such project that combines genomic science and nutrition. Genes are important for defining function, but diet affects how different genes are expressed, affecting whether people reach the potential predetermined by their genetic background (Lozada-Martinez & Anaya, 2024). Thus, the term "nutrigenomics" originally described the study of how nutrients affect a person's genetic expression (Chen et al., 2024).

The nutrigenomics idea implies bioactive food ingredients can directly or indirectly affect the human genome, affecting how genes and gene products are expressed (Rahman & Muhammad, 2023). Hence, dietary patterns and/or particular dietary components may alter a number of cellular processes, including aging. The evident symptoms of ageing might be the deregulation of nutrient sensing and mutated signalling pathways (Klotz & Carlberg, 2023). Nutrigenomics exhibits strong impacts on healthy longevity (Sathyamurthy et al., 2022). Relevant studies have proven a correlation between nutrition, aging and gene-nutrient interactions that continuously shape our phenotype (Fenech, 2010; Ordovas & Mooser, 2004; Riscuta, 2016). Also, the distortions in nutrition pave the way for several diseases, such as cancer, obesity, and diabetes (Garg, 2024). An adverse factor in the advancement and creation of chronic illnesses that can shorten life expectancy is malnutrition. Nine essential nutrients that may have distinct effects on genomic integrity. Six of these nutrients—folate, vitamin B12, niacin, vitamin E, retinol, and calcium—are linked to a decrease in DNA damage when taken in large amounts through food, while three others—riboflavin, pantothenic acid, and biotin—are linked to an increase in DNA damage that is comparable to that seen with occupational exposure to genotoxic and carcinogenic chemicals (Moskalev et al., 2014). In the case study of obese people, the excessive intake of methyl donor micronutrients causes alterations in gene-specific methylation that were linked with weight change, thereby leading to cardiovascular diseases, metabolic syndrome, and diabetes (Garg, 2024). Researches look forward to identifying how prenatal nutrition malfunction may contribute to diseases and reduced life span. Sirtuin 1, Insulin-like receptor (InR), the insulin-like receptor substrate, and sub-group O (foxo) gene are prominent examples of nutrition sensing/signalling genes that are linked to Drosophila aging (Ermolaeva et al., 2018). Anti-Aging studies in Drosophila focus on gene expression, mitochondrial dysfunction, and telomerase activity (Lozada-Martinez & Anaya, 2024). The effect of diet on life span is also evaluated, and it could be concluded that a high-fat diet increases the mortality rate.

In contrast, a glucose diet increases the healthy life span of drosophila (Chen et al., 2024; Rahman & Muhammad, 2023). The effect of the amino acid diet on the drosophila diet is still under study, though some research reported that metabolic restrictions may improve lifespan (Klotz & Carlberg, 2023; Sathyamurthy et al., 2022). While validating the effect of bioactive compounds and their mechanism of healthy aging in drosophila, the polyphenols in berries and fruits, such as apples, increase Drosophila's mean life span by 10% (Mitchell-Fearon et al., 2015, Riscuta, 2016). This is due to the downregulation of mth and Rpn11 and the overexpression of catalase (CAT) and superoxide dismutase (SOD) (Fenech, 2010; Mitchell-Fearon et al., 2015). The maximal longevity of Rosmary was increased by 12% at 3 mg/mL. Additionally, it inhibited lipid peroxidation, which reduced oxidative stress and malondialdehyde (MAD). Moreover, it enhanced SOD and CAT activity (Maegawa et al., 2010; Ordovas & Mooser, 2004). The research conducted by Mitchell-Fearon et al. (2015) revealed that ginger extract markedly improved the lifespan of Drosophila melanogaster. At a dosage of 1 mg/ml, ginger extract enhanced lifespan by 11.97%, correlated with the overexpression of critical antioxidant enzymes, superoxide dismutase (SOD) and catalase (CAT). At a dosage of 3 mg/ml, ginger extract increased lifespan by 4.66%, potentially via the downregulation of the methuselah (mth) gene, which is associated with stress resistance and aging. Likewise, the aronia extract demonstrated a more pronounced enhancement in lifespan, augmenting longevity by 18%. This impact was ascribed to several reasons, including increased locomotor activity, diminished formation of reactive oxygen species (ROS), and reduced accumulation of malondialdehyde (MDA), a biomarker of lipid peroxidation. Moreover, the aronia extract enhanced the expression of various antioxidant and stress-response genes, including SOD1, catalase, glutathione peroxidase, and Hsp68 (heat shock protein 68). The participation of these genes underscores the significance of antioxidant mechanisms and stress responses in mitigating aging and enhancing longevity. Additionally, the study observed the regulation of genes such as Efl (elongation factor-like) and Jafrac2 in relation to neurodegenerative illnesses, including Hereditary Spastic Paraplegia (HSP2). Jafrac2, a constituent of the thioredoxin peroxidase family, is instrumental in the neutralization of reactive oxygen species (ROS), whereas Efl participates in protein translation. The results indicate a possible connection between antioxidant modulation and the reduction of age-related neurodegenerative processes. This study highlights the importance of natural extracts like ginger and aronia in regulating critical aging-related processes, specifically by affecting oxidative stress, locomotor activity, and stress response genes (Mitchell-Fearon et al., 2015). Resveratrol depends on Sir 2 mechanism did not show any significant changes (Maegawa et al., 2010). Ursolic acid extended the lifespan of fruit flies by enhancing SRL Sarcalumenin and climbing activity (Mitchell-Fearon et al., 2015; Riscuta, 2016). And finally, Epigallocatechin gallate extended the lifespan of Drosophila by preventing the accumulation of iron, inhibiting α-amylase and α-glucosidase and suppressing interleukin 5, Phosphoenolpyruvate carboxykinase and upd2 (Mitchell-Fearon et al., 2015).

Microbiome and Healthy Longevity

The term "microbiome" describes microorganisms living in a specific habitat, such as bacteria, viruses, fungus, and other microbes (Alam et al., 2019). A balanced and composed microbiome has a direct connection to healthy longevity. Alterations in the human microbiome induce several kinds of cutaneous, cognitive, digestive, and immunological disorders. Various factors influence microbiome alterations, such as genetics, environmental factors, lifestyle, stress, and many more. Relevant studies documented that a healthy microbiome contributes to healthy ageing (Nisha et al., 2020; Piper et al., 2005). The balanced composition of the microbiome helps maintain overall homeostasis.

In humans, the microbiome in the gastrointestinal tract is one of the most well studied and significant areas, constituting trillions of microorganisms. Researchers discovered that GI microbe profiles in older persons with more distinctive patterns generally resulted in better health and longer lifespans than counterparts with less divergent microbiomes (Grandison et al., 2009; Liao et al., 2021). Gut microbiome is widely used as a modulator of healthy ageing. As we age, alterations may occur in the gut microbiome due to physiological deterioration, diet, physical activity, medication, reduced social interactions, and many more. Likewise, stress, cellular senescence, disordered lifestyle, obesity, and genetics also trigger alterations in the gut microbiome, thereby leading to unhealthy aging and associated diseases among people. So, there is a bidirectional link between the gut microbiome and a person's overall health (Bongiorni et al., 2022; Lee et al., 2014; Peng et al., 2012). The Gut-Brain axis represents the central and enteric nervous system (Wang et al., 2017). The gut microbiome not only exerts properties on intestinal functions but also significantly affects cognitive, skin, hair, and social aspects (Staats et al., 2018). An imbalance in the composition of the gut microbiome leads to neurodegenerative diseases like Alzheimer's, Parkinson's Disease, Multiple Sclerosis, Schizophrenia, and Depression. Dysbiosis in the gut microbial flora triggers the secretion of harmful metabolites and toxic neuroactive molecules. It will increase gut barrier permeability and eventually transport these compounds to the brain via the gut-brain axis. These neurotoxins accumulate in the CNS system, and ultimately, degeneration of neuronal tissues is induced. According to studies, the gut microbiota resulting from AD stimulated C/EBPβ/AEP signalling, which in turn triggered the pro-inflammatory pathway for the brain's metabolism of polyunsaturated fatty acids (PUFAs) (Berg et al., 2020). This, in turn, accelerated the maturation of microglia in response to AD pathologies and cognitive deficits. In Parkinson's Disease, the neurons in the substantia nigra is lost due to the aggregation of abnormally folded alpha-synuclein (α-syn) proteins which form the intracellular inclusions within the neurons and cause tremor in hands, stiffness, and loss of balance (Ragonnaud & Biragyn, 2021). Evaluation in Schizophrenia depicts the alterations in the *lactobacilli* bacteria, which induced the production of inflammatory compounds like interferons, tumour necrosis factor and antibodies in response. Besides, the bacterial translocation marker SCD14 was also enhanced in the system (Santoro et al., 2018; Wilmanski et al., 2021). In Multiple Sclerosis, a significantly higher amount of pro-inflammatory microbial species Methanol *Brevibacter* (phylum *Euryarchaeota*) and *Akkermansia* (phylum *Verrucomicrobia)* was reported. Besides, subsequent reduction of anti-inflammatory microbe *Butyricimonas* was also validated along with this. The distortions induce the transport of immune cells, $CD4^+$, $CD8^+$, T lymphocytes to CNS and production of interleukins causes ultimate damage and loss of tissues (Gomaa, 2020). With regard to depression (psychiatric illness), a sizable drop in microorganisms that produce butyrate has been observed (Voreades et al., 2014).

Besides cognitive disorders, it also causes skin inflammation and hair diseases. The common inflammatory disease Atopic dermatitis patient's profile shows enhanced production of *Faecalibacterium prausnitzii*, which damages the integrity of the epithelial layer of skin. Patients with psoriasis reported a significant increase in *Streptococcus* and *Staphylococcus* species compared to their healthy counterparts (Aziz et al., 2013; Salazar et al., 2023). Data from Acne Vulgaris also documented the higher amount of *Bacteroides* and reduced number of *Firmicutes* (Carabotti et al., 2015). Clinical examination and experimental findings proved that gut microbiota dysbiosis resulted in hair-related disturbances. Consuming the *Lactobacillus paracasei* strain has reduced the severity of seborrheic dermatitis (Hiippala et al., 2018). Clinical cases confirm the strong correlation between disruption of hair follicles in alopecia and GM dysbiosis. The system produces interferons when alopecia manifests, which causes changes in the gut. This results in the aberrant growth of hair follicles, which eventually die.

The oral microbiome is the second largest microbiome, comprising bacteria, fungi, viruses and protozoa (Drew, 2018; Obergasteiger et al., 2023). It resides mainly on the hard surface of teeth and soft tissues of oral mucosa. The microbial community significantly impacts digestion and oral and systemic health. Not only oral diseases a wide range of systemic disorders, including heart disease, pneumonia, rheumatoid arthritis, pancreatic cancer, colorectal cancer, oesophageal cancer, stroke, and unfavorable pregnancy outcomes, are also linked to dysbiosis in oral microbiome. Dental caries, one of the most common oral diseases, is caused by the acidification of *Streptococcus* and *Actinomyces* species (Harrison et al., 2017). Alterations in several microbial strains are associated with cancer. Enhanced production of *Porphyromonas gingivalis* is associated with oral cancer, while increased production of *Fusobacteria nucleatum* is linked with colorectal cancer (Scheurink et al., 2023). Similarly, *Porphyromonas gingivalis* and *Aggregatibacter actinomycetemcomitans* increased production leads to pancreatic cancer (Yamamura, 2023). Case studies reported that *Streptococcus mutans* has a strong impact on the development of atherosclerosis. *S. mutans* can potentially induce atherosclerosis by impairing endothelial cells' function, among the first symptoms of cardiovascular disorders (Nikolova et al., 2021). Likewise, *Porphyromonas gingivalis* could significantly impact the formation and development of atherosclerosis (Park et al., 2020). Oral bacteria could be used as a marker for lung diseases. In the case of ventilator-associated pneumonia, the pathogens methicillin-resistant *S. aureus* and *Pseudomonas aeruginosa* can localize in the oral cavity due to poor oral hygiene of the patient. Following respiration, the bacteria's deposits and toxic products may reach the lungs and make people susceptible to infection (Deo & Deshmukh, 2019; Wade, 2013). Hence, it could be concluded that the human system's microbiome significantly impacts systemic health and well-being.

Personalized Medicine and Ayurveda

Personalized Medicine, also known as precision medicine or individualized medicine, is an innovative approach that uses information about an individual's genomic, environmental, and lifestyle information to guide decisions related to their medical management (Ginsburg & Willard, 2009). This tailor-made medical care methodology emphasizes one's unique molecular and genetic profiling. Unlike conventional medicine, which uses 'One–size-fits-all- approach' in clinical trials, personalized medicine targets genetic makeup. Also, the chances of trial and error in old-fashioned medicinal care are not evident in precision medicine (Asbury & Plomin, 2013). Ultimately, this can prevent adverse effects, predict dosing, improve compliance, and a large group of patients benefit from this medical aspect.

A major objective of personalized medicine is determining the best care for each patient. Identifying mutations in patients with undiagnosed conditions, avoiding unnecessary scanning/imaging/tests and unwanted hospitalization, and detecting genetic risk factors to guide lifestyle/environment recommendations that can improve each patient's health are some of the foremost goals of personalized medicine. The significance of personalized medicine in the modern era is inevitable. First of all, through advanced diagnostics and data analysis, precision healthcare aims to predict disease risks and optimize preventive strategies. Precision healthcare aims to deliver the most effective and least invasive interventions by analyzing patients' unique profiles. Then, it provides customer care through medical interventions based on individual characteristics, such as genetics, lifestyle, and medical history. Subsequently, precision healthcare can contribute to sustainable healthcare systems by optimizing resources and promoting preventive strategies. The cost-efficacy of individualized medicine is also effective by minimizing unnecessary treatments, hospitalizations, and resource allocation. In conclusion, precision healthcare can enhance patient outcomes by personalizing treatments, reducing adverse reactions, and improving response rates (Perlmutter, 2022; Tutton, 2016; Weyers, 2019).

Ayurveda, the ancient medical practice, is closely associated with personalized medicine. Ayurveda has its roots in India and dates back more than 4,000 years. It is grounded on the idea that mental, physical, and spiritual well-being are intricately balanced for overall health and wellbeing. Ayurvedic medicine prognoses and diagnoses patients holistically, emphasizing health promotion above disease prevention. The most common ayurvedic technologies include dietary changes, herbal medicine, acupuncture, massage, meditation, breathing exercises, panchakarma, sound therapy, and yoga (Chopra & Doiphode, 2002; Niemi & Ståhle, 2016).

The fundamental principle of Ayurveda is that every individual is made up of the five elements of nature: air, fire, water, earth, and space. These components combine to generate life forces (dosha) like Vata, Pitta, and Kapha. Everyone possesses a prominent dosha that exhibits mental, emotional, and physical traits. Ayurvedic philosophy relies on balancing the dosha via various means and restoring the balance between body, mind, and spirit for a healthy and harmonious life (Frawley & Ranade, 2004).

Ayurveda and Personalized Medicine principles are closely associated with each other and lead to a community's ultimate well-being. While Personalized Medicine is all about DNA makeup, Ayurveda focuses on ahara (diet), vihara (lifestyle), and aushadhi (medication), which are the three pillars of prakriti-based medicine, making it a holistic science. Reliable sources reported that humans are 99% identical, but the variation in the 1% due to single nucleotide polymorphism determines a person's individuality. In accompanying with personalized medicine, Ayurveda provides treatment in a highly individualized manner since it strongly believes every entity should be different and treated accordingly. Ayurveda is completely independent of racial, ethnic, and geographical aspects, and it strictly emphasizes the Prakriti of an individual. So, it could be concluded that Ayurveda follows a personalized medicinal care approach and is thereby recognized as an experimental science. In a nutshell, Ayurveda and personalized medicine contribute to healthy longevity (Chatterjee & Pancholi, 2011).

Biodiversity and Bioactive Compounds

Biodiversity refers to the variety of life forms found on Earth, encompassing anything from genes to entire ecosystems. The existence of a diverse range of living organisms is critical for our planet's long-term survival. Biodiversity fulfills all living organisms' fundamental requirements, such as sustenance, hydration, and habitation. Biodiversity might be likened to a complex network, where every individual component plays a vital role in sustaining life (Singh et al., 2017). The disruption in the natural environment profoundly impacts the efficient operation and has adverse consequences for living organisms. Bioactive chemicals are mostly present in plants, microbes, and related constituents. They can exert significant biological effects on cells, tissues, and organs. Biologically active substances enhance the state of being healthy and living for a long time. Herbal plants and microbes provide a plentiful reservoir of bioactive chemicals. The prominent instances of bioactive compounds are terpenoids, polyphenols, alkaloids, carotenoids, flavonoids, carnitine, choline, coenzyme Q, dithiolthiones, phytosterols, phytoestrogens, glucosinolates, polyphenols, and taurine. Most of these components exhibit diverse biological features, such as antioxidant, cardioprotective, neuroprotective, anti-inflammatory, anticancer, and antibacterial action (Godlewska-Żyłkiewicz et al., 2020). The importance of these chemicals lies in their ability to inhibit aging and age-related ailments. A comprehensive investigation has been conducted on various fruits and vegetables to ascertain their potential for mitigating the harmful effects of free radicals. These free radicals can cause damage to DNA and cell membranes by inducing oxidative stress. Oxidative stress is the primary factor responsible for most age-related dysfunctions and illnesses, impeding the individual's ability to maintain a healthy lifespan (Tan et al., 2018).

Biodiversity, containing many natural substances to treat various ailments, is essential for finding and creating new drugs. The vast array of different species and habitats provides an exceptional reservoir of bioactive chemicals that can serve as models for creating novel pharmaceuticals. Natural sources are the origin of several contemporary medications, such as penicillin (derived from fungi), aspirin (originally from willow bark), and artemisinin (from sweet wormwood). Natural chemicals frequently exhibit distinctive structures and modes of action that synthetic compounds struggle to imitate, making them valuable in the quest for novel therapeutic medicines (Newman & Cragg, 2016).

Furthermore, investigating biodiversity hotspots, such as tropical rainforests and coral reefs, has resulted in identifying new chemicals with noteworthy therapeutic effects. The genetic variability in these ecosystems increases the probability of discovering distinct compounds with medicinal properties. Nevertheless, the escalating decline in biodiversity due to deforestation, climate change, and habitat destruction presents a substantial menace to this priceless asset. Conserving biodiversity maintains ecological equilibrium and raises the possibility of future pharmacological breakthroughs that could tackle unmet medical requirements and emerging health issues (Harvey, 2000). Therefore, pharmaceutical research and innovation's continuous advancement and prosperity are intrinsically linked to preserving biodiversity.

Plants have historically been a highly productive source of therapeutic chemicals that are fundamental to modern medicine. For example, we obtain aspirin from the bark of the willow tree, specifically from species within the Salix genus. People have long used salicin, the active ingredient, as an analgesic and to reduce inflammation. Afterward, scientists chemically produced this substance to yield acetylsalicylic acid, also known as aspirin, widely used for its anti-inflammatory, analgesic, and antipyretic properties (Vane & Botting, 2003). Quinine, derived from the bark of the cinchona tree, is another noteworthy plant-based medication (Achan et al., 2011). Quinine was the initial efficacious remedy for malaria and is a crucial medication in managing the condition. The opium poppy (Papaver somniferum) yields morphine, a significant plant-derived drug. People use it to control severe pain due to its strong pain-relieving qualities (Gutstein & Akil, 2006).

Animal-derived compounds have made a significant contribution to the field of pharmaceuticals. The salivary secretions of the Gila monster (*Heloderma suspectum*) generate exenatide, a pharmaceutical agent for managing type 2 diabetes. This molecule imitates the function of incretin hormones, which boost insulin release, offering a new method for treating diabetes. The discovery of peptides in the venom of the Brazilian pit viper (*Bothrops jararaca*) led to the development of captopril, a medication for high blood pressure. Ondetti et al. (1977) introduced the first ACE inhibitor in clinical practice, which significantly transformed the management of excessive blood pressure and heart failure. The southeastern pygmy rattlesnake (*Sistrurus miliarius barbouri*) produces a protein that produces eptifibatide, a medication that inhibits platelet aggregation to prevent the formation of blood clots. This demonstrates the therapeutic possibilities of venom constituents.

Microorganisms have proven to be a valuable reservoir for significant medical advancements. Alexander Fleming discovered penicillin, the first widely used antibiotic, from the mold *Penicillium notatum.* This marked the commencement of the antibiotic era and revolutionized the treatment of bacterial infections. The fungus *Tolypocladium inflatum* produces cyclosporine, an immunosuppressant that significantly enhances organ transplant efficacy by inhibiting rejection. Fungi, specifically *Aspergillus terreus*, are the source of statins like lovastatin and simvastatin. Microorganisms significantly influence modern medicine, as evidenced by their vital role in controlling cholesterol levels and reducing the risk of cardiovascular disease (Penders et al., 2013; von Wartburg & Traber, 1988).

The fundamental principle of "natural conservation," which prioritizes the preservation of the region's abundant species, guides biodiversity hotspots. A biodiversity hotspot is a harmonious biological system that supports a diverse array of endangered or threatened flora and fauna. Intransigent conservation efforts are required in biodiversity hotspots. Biodiversity hotspots meet two criteria: they must be under threat, contain unique endemic vascular plants, and maintain 30% of their original natural vegetation. The Himalayan Hotspot, Indo-Burma Hotspot, Sundaland, and Western Ghats, Sri Lanka, are the primary biodiversity hotspots in India. The Indo-Burma hotspot provides habitat to a distinctive freshwater turtle species, while the Himalayan hotspot is home to a variety of vultures, tigers, elephants, rhinoceros, and wild water buffalo. The Sundaland region is the only place where the orangutan lives. Lastly, a variety of distinctive species found nowhere else on Earth call the Western Ghats in Sri Lanka home. These species include the endangered lion-tailed macaque, Asian elephants, and Indian tigers. It is also at risk due to the high prevalence of freshwater fish endemism (Banerjee & Bandopadhyay, 2016; Myers et al., 2000).

The presence of biodiversity hotspots is advantageous to ongoing medicinal research, as it offers a promising environment for the development and investigation of potent medications that are effective against a variety of diseases. The most significant source of vincristine and vinblastine, widely used in chemotherapy, is *Catharanthus roseus* (Dhyani et al., 2022). In addition, the bark of Cinchona contains Quina, which is an effective treatment for malaria (Achan et al., 2011). People commonly employ codeine, derived from the opium poppy, as an analgesic (Gutstein & Akil, 2006). Researchers have demonstrated the effectiveness of turmeric from *Curcuma longa*, piperine from black pepper, cinnamon from Cinnamomum, and ginger from the Zingiberaceae family in treating menstrual-related issues (Utami et al., 2020). The expansive domain of medicine was transformed by the extraction of pharmacologically active substances from fungi (antibiotics such as penicillin and cephalosporins), marine sponges (antiviral agents like nucleoside Ara-A (vidarabine) from Tethya crypta and Hamigeran from Hamigera tarangaensis), and insects (anticancer compounds including sericin, solenopsin, melittin, and cecropin), among others (Sagar et al., 2010). The development and production of innovative compounds are facilitated by sophisticated technologies and instruments. Due to the abundance of species in the biodiversity hotspot, it can be utilized to investigate the biochemical pathways that lead to the production of novel compounds, thereby making a significant contribution to the global health sector (Gjerde et al., 2007). Additionally, the comprehensive examination of biodiversity regions contributes to a novel perspective on disease ecology. The occurrence and infection of outbreaks are always present in these biological centers as a result of climate change (Ostfeld & Keesing, 2000). Therefore, it may be feasible to establish a correlation between the natural life scenario, vectors, and pathogens. In the end, it may be beneficial the development of prevention, or at least for the treatment of the specific diseases.

Benefit Sharing

The application of the legal notion of benefit sharing concerning the Convention on Biological Diversity (CBD) and its support of local and indigenous populations' livelihoods has substantially evolved. Specifically, based on the language of the convention and its Conference of the Parties (COP) rulings, Benefit sharing is a notion that has been developing not only regarding the exploitation of genetic resources but also, with quite distinct legal implications, concerning biodiversity protection and sustainable usage. However, another important CBD clause envisions a quite different interpretation of benefit sharing as a state-to-community contribution to equity and sustainable development. In this instance, benefit sharing refers to forming a relationship based on national law between the community and the State in which it resides.

With the promise of biodiversity providing endless possibilities for regenerative and precision medicine, we are at the cusp of a new age, and with it comes the crucial necessity of fair benefit sharing. The intrinsic connection between the genetic resources found in nature and the advancements in modern healthcare demands a paradigm change in how we think about ownership, access, and sharing the advantages of these valuable resources. The idea of just and equal access to genetic resources and the related traditional knowledge that Indigenous and local groups possess is central to this discussion. This idea was made possible by the 1992 Convention on Biological Diversity (CBD), which recognized national sovereignty over biological resources and stipulated that Access and Benefit Sharing (ABS) required both prior informed permission and mutually agreed terms. However, as the divisive topic of biopiracy demonstrates, there have been many obstacles in the way of achieving this goal. The unapproved and unpaid use of biological resources and traditional knowledge, or "biopiracy," has long clouded relations between industrialised and developing countries. Outrage has been generated by cases like the attempt to patent the *Ayahuasca* plant, which indigenous communities in the Amazon basin have been using for centuries, and the patenting of an anti-cancer compound found in Indian neem trees. These cases have highlighted the urgent need for strong mechanisms to protect these communities' rights and interests (Cooney, 2004). A legal framework for the application of the CBD's ABS provisions was established in 2010 with the adoption of the Nagoya Protocol on Access to Genetic Resources and the Fair and Equitable Sharing of Benefits Arising from their Utilisation in response to these difficulties (Secretariat of the Convention on Biological Diversity, 2011). This protocol aims to guarantee the just and equal distribution of advantages that result from the use of genetic resources, including financial and non-financial advantages like technology transfer, chances for joint research, and capacity building. A potential way forward is provided by the idea of "bioprosperity," which arose in the wake of these international agreements. It imagines a time when environmental conservation and economic development are driven by the equitable sharing of benefits and the sustainable use of biodiversity. Bioprosperity programs promote biodiversity preservation by establishing partnerships among researchers, companies, and local people. This approach creates economic opportunities and enhances the welfare of indigenous and local communities. A prominent illustration of a prosperous bioprosperity endeavor is the Samoan business Seales Nursery, which grows and ships a wide range of decorative plants, among them the well-known "fire plant" (*Nephthytis spp.*). A portion of the proceeds from selling these plants are reinvested in community development projects, including infrastructure upgrades, healthcare initiatives, and education programs, thanks to a benefit-sharing arrangement with the nearby villages (Fecteau, 2001).

Another promising avenue for equitable benefit sharing lies in Biocultural Community Protocols (BCPs). These protocols, developed by indigenous and local communities, outline their customary laws, values, and decision-making processes concerning using their traditional knowledge and genetic resources. By providing a framework for prior informed consent and mutually agreed terms, BCPs empower communities to negotiate with researchers and companies on their terms, fostering transparency and ensuring that benefits are shared equitably. The promise of this strategy is demonstrated by initiatives like the Potato Park in Peru, where Quechua communities have created a BCP to control the usage and exchange of their ancestral potato varieties. These communities have gained access to resources for building capacity, scientific competence, and chances for sustainable revenue production while guaranteeing the preservation of their cultural legacy and traditional knowledge, thanks to cooperative agreements with researchers and seed banks (Krishnan et al., 2020).

It is critical to acknowledge the fundamental importance of traditional knowledge systems and the critical role played by indigenous and local people in protecting and fostering biodiversity as we traverse the intricacies of benefit sharing in the context of regenerative and precision medicine. We can pave the way for bio prosperity—where the benefits of scientific advancement are distributed fairly, and the stewards of biodiversity are enabled to pursue long, healthy lives actively—by adopting a paradigm of cooperation, respect, and understanding (Mgbeoji, 2006).

Conclusion

The relentless march of human innovation has brought us to a pivotal crossroads in our pursuit of healthy longevity. As we stand at the vanguard of regenerative and precision medicine, the boundless potential of biodiversity looms large, beckoning us to unlock its myriad secrets and harness its restorative powers. This convergence of cutting-edge scientific disciplines and the vast tapestry of nature's bounty presents us with a transformative opportunity to redefine the essence of aging and pave the way for a future where vitality and well-being are the hallmarks of our twilight years.

At the core of this endeavour lies the burgeoning field of regenerative medicine, a revolutionary approach that transcends the mere treatment of symptoms and dares to dream of true rejuvenation. Through the artful manipulation of stem cells, the intricate weaving of engineered tissues, and the precise modulation of genetic expression, we can coax the body's innate healing mechanisms into action, reversing the ravages of time and restoring function to damaged or aging tissues. This paradigm shift in medical intervention holds the promise of not only alleviating the physical ailments that plague the elderly but also of addressing the very root causes of age-related decline, ushering in a new era of proactive, restorative healthcare.

Yet, as we stand poised to unlock the full potential of regenerative medicine, we must acknowledge the vital role that biodiversity plays in this quest. Nature's boundless repository of genetic and molecular diversity offers a treasure trove of untapped potential, waiting to be harnessed and leveraged for the betterment of human health. From the depths of the oceans to the lush canopies of ancient forests, every species and organism harbours the secrets to novel therapeutics, nutraceuticals, and regenerative compounds within its genetic code. By embracing the principles of sustainable exploration and equitable benefit-sharing, we can forge a harmonious alliance between modern science and the natural world, ensuring that the fruits of our discoveries are shared justly with the guardians of this precious biodiversity.

Moreover, the advent of precision medicine, with its emphasis on tailoring treatments to an individual's unique genetic and molecular profile, further amplifies the synergy between regenerative therapies and the bounties of biodiversity. By harnessing the power of genomics, proteomics, and metabolomics, we can unlock personalized approaches to disease prevention and treatment, leveraging the vast array of natural compounds to craft bespoke interventions that resonate with each individual's biochemical makeup. This marriage of cutting-edge technology and nature's wisdom promises to revolutionize healthcare, ushering in an era where preventive measures and targeted therapies become the norm rather than the exception.

As we navigate this uncharted territory, we must remain steadfast in our commitment to ethical principles and equitable access. The benefits of regenerative and precision medicine must not be confined to the privileged few. Still, they must instead be made available to all, regardless of socioeconomic status, geographic location, or cultural background. By fostering robust collaborations between researchers, policymakers, and local communities, we can ensure that the fruits of our labour are distributed equitably, empowering underserved populations and bridging the chasm of healthcare disparities.

Furthermore, we must remain vigilant in our pursuit of data privacy and security, safeguarding the sensitive genetic and medical information that underpins these cutting-edge therapies. As we delve deeper into the realms of personalized medicine, we must establish robust frameworks that strike a careful balance between scientific progress and the protection of individual autonomy, fostering trust and enabling the responsible sharing of data for the greater good.

Lastly, we must acknowledge the pivotal role of health literacy in this endeavour. By empowering individuals with knowledge and understanding, we can foster active participation in healthcare decisions, enabling informed choices that align with personal values and preferences. Through robust educational initiatives and open dialogues, we can demystify the complexities of regenerative and precision medicine, fostering a society that embraces these innovations with open arms and a deep appreciation for their transformative potential.

In the grand tapestry of human ingenuity, the convergence of regenerative medicine, precision medicine, and biodiversity represents a stunning masterpiece in the making. As we weave together the threads of scientific brilliance, ethical responsibility, and environmental stewardship, we can create a world where healthy longevity is not merely a fleeting dream but a tangible reality within reach. By embracing this holistic vision, we can usher in a new era of healthcare, one that transcends the limitations of the past and ushers in a future where vitality and well-being are the hallmarks of our twilight years, a future where the wisdom of nature and the marvels of human innovation coalesce in a symphonic harmony, allowing us to savor the richness of life in all its splendor truly.

Conflict of Interest The authors state no conflict of interest.

References

Achan, J., Talisuna, A. O., Erhart, A., Yeka, A., Tibenderana, J. K., Baliraine, F. N., Rosenthal, P. J., & D'Alessandro, U. (2011). Quinine, an old anti-malarial drug in a modern world: Role in the treatment of malaria. *Malaria Journal, 10*, 144.

Ahmad Sophien, A. N., Jusop, A. S., Tye, G. J., Tan, Y. F., Zaman, W. K., & W. S., Nordin, F. (2023). Intestinal stem cells and gut microbiota therapeutics: Hype or hope? *Frontiers in Medicine, 10*, 1195374.

Alam, I., Ali, F., Zeb, F., Almajwal, A., Fatima, S., & Wu, X. (2019). Relationship of nutrigenomics and aging: Involvement of DNA methylation. *Journal of Nutrition & Intermediary Metabolism, 16*, 100098.

Archibald, P. R., & Williams, D. J. (2015). Using the cost-effectiveness of allogeneic islet transplantation to inform induced pluripotent stem cell-derived β-cell therapy reimbursement. *Regenerative Medicine, 10*, 959–973.

Aruoma, O. I., Hausman-Cohen, S., Pizano, J., Schmidt, M. A., Minich, D. M., Joffe, Y., Brandhorst, S., Evans, S. J., & Brady, D. M. (2019). Personalized nutrition: Translating the science of nutriGenomics Into practice: Proceedings from the 2018 American College of Nutrition Meeting. *Journal of the American College of Nutrition, 38*(4), 287–301. https://doi.org/10.1080/07315724.2019.1582980

Asbury, K., & Plomin, R. (2013). *G is for genes: The impact of genetics on education and achievement*. Wiley.

Ashley, E. A. (2016). The precision medicine initiative: A new national effort. *JAMA, 315*(8), 797–798.

Atala, A. (2012). Tissue engineering of human bladder. *British Medical Bulletin, 101*(1), 67–78.

Aziz, Q., Doré, J., Emmanuel, A., Guarner, F., & Quigley, E. M. M. (2013). Gut microbiota and gastrointestinal health: Current concepts and future directions. *Neurogastroenterology & Motility, 25*(1), 4–15.

Bajaj, P., Schweller, R. M., Khademhosseini, A., West, J. L., & Bashir, R. (2014). 3D biofabrication strategies for tissue engineering and regenerative medicine. *Annual Review of Biomedical Engineering, 16*, 247–276.

Banerjee, A., & Bandopadhyay, R. (2016). Biodiversity hotspot of Bhutan and its sustainability. *Current Science, 110*(4), 521–527.

Barone, M., D'Amico, F., Rampelli, S., Brigidi, P., & Turroni, S. (2022). Age-related diseases, therapies and gut microbiome: A new frontier for healthy aging. *Mechanisms of Ageing and Development, 206*, 111711.

Berg, G., Rybakova, D., Fischer, D., Cernava, T., Vergès, M. C. C., Charles, T., Chen, X., Cocolin, L., Eversole, K., Corral, G. H., Kazou, M., Kinkel, L., Lange, L., Lima, N., Loy, A., Macklin, J. A., Maguin, E., Mauchline, T., McClure, R., & Schloter, M. (2020). Microbiome definition re-visited: Old concepts and new challenges. *Microbiome, 8*, 1–22.

Blau, H. M., & Daley, G. Q. (2019). Stem cells in the treatment of disease. *New England Journal of Medicine, 380*, 1748–1760.

Bongiorni, S., Arisi, I., Ceccantoni, B., Rossi, C., Cresta, C., Castellani, S., Forgione, I., Rinalducci, S., Muleo, R., & Prantera, G. (2022). Apple polyphenol diet extends lifespan, slows down mitotic rate and reduces morphometric parameters in drosophila melanogaster: A comparison between three different apple cultivars. *Antioxidants, 11*(11), 2086.

Britton, A., Shipley, M., Singh-Manoux, A., & Marmot, M. G. (2008). Successful aging: The contribution of early-life and midlife risk factors. *Journal of the American Geriatrics Society, 56*(6), 1098–1105.

Brooks-Wilson, A. R. (2013). Genetics of healthy aging and longevity. *Human Genetics, 132*, 1323–1338.

Buettner, D., & Skemp, S. (2016). Blue zones: Lessons from the world's longest lived. *American Journal of Lifestyle Medicine, 10*(5), 318–321.

Caplan, A. I., & Correa, D. (2011). The MSC: An injury drugstore. *Cell Stem Cell, 9*(1), 11–15.

Carabotti, M., Scirocco, A., Maselli, M. A., & Severi, C. (2015). The gut-brain axis: Interactions between enteric microbiota, central and enteric nervous systems. *Annals of Gastroenterology: Quarterly Publication of the Hellenic Society of Gastroenterology, 28*(2), 203.

Carmona, J. J., & Michan, S. (2016). Biology of healthy aging and longevity. *Revista De Investigacion Clinica, 68*(1), 7–16.

Castruita, P. A., Piña-Escudero, S. D., Rentería, M. E., & Yokoyama, J. S. (2022). Genetic, social, and lifestyle drivers of healthy aging and longevity. *Current Genetic Medicine Reports, 10*(3), 25–34.

Chatterjee, B., & Pancholi, J. (2011). Prakriti-based medicine: A step towards personalized medicine. *AYU (An International Quarterly Journal of Research in Ayurveda), 32*(2), 141–146.

Chen, Z., Gong, Y., Miao, R., Chen, S., Zhang, Y., He, X., Huang, X., Chen, N., & Li, S. (2024). Effects of dietary binder sources on growth, digestive enzymes activity, and gut barrier function and microbial composition in larval largemouth bass (Micropterus salmoides). *Aquaculture, 586*,

Chien, K. R. (2008). Regenerative medicine and human models of human disease. *Nature, 453*(7193), 302–305.

Chopra, A., & Doiphode, V. V. (2002). Ayurvedic medicine: Core concept, therapeutic principles, and current relevance. *Medical Clinics, 86*(1), 75–89.

Collins, F. S., & Varmus, H. (2015). A new initiative on precision medicine. *New England Journal of Medicine, 372*(9), 793–795.

Convention on Biological Diversity. (2011). *Nagoya protocol on access to genetic resources and the fair and equitable sharing of benefits arising from their utilization.* Secretariat of the Convention on Biological Diversity.

Cooney, R. (2004). *The precautionary principle in biodiversity conservation and natural resource management: An issues paper for policy-makers, researchers and practitioners*. IUCN.
de BONO, J. S., & Ashworth, A. (2010). Translating cancer research into targeted therapeutics. *Nature, 467*(7315), 543–549.
de Windt, T. S., et al. (2017). Early health economic modelling of single-stage cartilage repair. Guiding implementation of technologies in regenerative medicine. *Journal of Tissue Engineering and Regenerative Medicine, 11*, 2950–2959.
Deo, P. N., & Deshmukh, R. (2019). Oral microbiome: Unveiling the fundamentals. *Journal of Oral and Maxillofacial Pathology, 23*(1), 122–128.
Dhyani, P., Quispe, C., Sharma, E., Bahukhandi, A., Sati, P., Attri, D. C., Szopa, A., Sharifi-Rad, J., Docea, A. O., Mardare, I., Calina, D., & Cho, W. C. (2022). Anticancer potential of alkaloids: A key emphasis to colchicine, vinblastine, vincristine, vindesine, vinorelbine and vincamine. *Cancer Cell International, 22*(1), 206.
Doudna, J. A., & Charpentier, E. (2014). The new frontier of genome engineering with CRISPR-Cas9. *Science, 346*(6213), 1258096.
Drew, L. (2018). An age-old story of dementia. *Nature, 559*(7715), S2–S3.
Driscoll, D., Farnia, S., Kefalas, P., & Maziarz, R. (2017). The high cost of high tech medicine: Planning ahead for market access. *Stem Cells Translational Medicine, 6*, 1723–1729.
Druker, B. J., Talpaz, M., Resta, D. J., et al. (2001). Efficacy and safety of a specific inhibitor of the BCR-ABL tyrosine kinase in chronic myeloid leukemia. *New England Journal of Medicine, 344*(14), 1031–1037.
Ermolaeva, M., Neri, F., Ori, A., & Rudolph, K. L. (2018). Cellular and epigenetic drivers of stem cell ageing. *Nature Reviews Molecular Cell Biology, 19*, 594–610. https://doi.org/10.1038/s41580-018-0020-3
Fecteau, L. M. (2001). The Ayahuasca patent revocation: Raising questions about current US patent policy. *BC Third World LJ, 21*, 69.
Fenech, M. F. (2010). Dietary reference values of individual micronutrients and nutriomes for genome damage prevention: Current status and a road map to the future. *The American Journal of Clinical Nutrition, 91*(5), 1438S–1454S.
Frawley, D., & Ranade, S. (2004). *Ayurveda, nature's medicine*. Motilal Banarsidass Publication.
Garg, R. (2024). *K-LionER: Meta-heuristic approach for energy efficient cluster based routing for WSN-assisted IoT networks* (pp. 1–15). Cluster Computing.
Ginsburg, G. S., & Willard, H. F. (2009). Genomic and personalized medicine: Foundations and applications. *Translational Research, 154*(6), 277–287.
Gjerde, I., Sætersdal, M., & Blom, H. H. (2007). Complementary hotspot inventory–a method for identification of important areas for biodiversity at the forest stand level. *Biological Conservation, 137*(4), 549–557.
Godlewska-Żyłkiewicz, B., Świsłocka, R., Kalinowska, M., Golonko, A., Świderski, G., Arciszewska, Ż, Nalewajko-Sieliwoniuk, E., Naumowicz, M., & Lewandowski, W. (2020). Biologically active compounds of plants: Structure-related antioxidant, microbiological and cytotoxic activity of selected carboxylic acids. *Materials, 13*(19), 4454.
Golpanian, S., DiFede, D. L., Khan, A., Schulman, I. H., Landin, A. M., Tompkins, B. A., Heldman, A. W., Miki, R., Goldstein, B. J., Mushtaq, M., Levis-Dusseau, S., Byrnes, J. J., Lowery, M., Natsumeda, M., Delgado, C., Saltzman, R., Vidro-Casiano, M., Pujol, M. V., Da Fonseca, M., & Hare, J. M. (2017). Allogeneic human mesenchymal stem cell infusions for aging frailty. *Journals of Gerontology Series A: Biomedical Sciences and Medical Sciences, 72*(11), 1505–1512.
Gomaa, E. Z. (2020). Human gut microbiota/microbiome in health and diseases: A review. *Antonie Van Leeuwenhoek, 113*(12), 2019–2040.
Grandison, R. C., Piper, M. D., & Partridge, L. (2009). Amino-acid imbalance explains extension of lifespan by dietary restriction in Drosophila. *Nature, 462*(7276), 1061–1064.
Gurtner, G. C., Callaghan, M. J., & Longaker, M. T. (2007). Progress and potential for regenerative medicine. *Annual Review of Medicine, 58*, 299–312.

Gutstein, H. B., & Akil, H. (2006). Opioid analgesics. *Goodman & Gilman's the Pharmacological Basis of Therapeutics, 11*, 547–590.

Hagell, P., et al. (2002). Dyskinesia following neural transplantation in Parkinson's disease. *Nature Neuroscience, 5*, 627–628.

Harrison, P., Cowen, P., Burns, T., & Fazel, M. (2017). *Shorter Oxford textbook of psychiatry*. Oxford University Press.

Harvey, A. L. (2000). Strategies for discovering drugs from previously unexplored natural products. *Drug Discovery Today, 5*(7), 294–300.

Hayflick, L. (2004). Debates: The not-so-close relationship between biological aging and age-associated pathologies in humans. *The Journals of Gerontology Series A: Biological Sciences and Medical Sciences, 59*(6), B547–B550.

Hekimi, S. (2006). How genetic analysis tests theories of animal aging. *Nature Genetics, 38*(9), 985–991.

Hekimi, S., Burgess, J., Bussière, F., Meng, Y., & Bénard, C. (2001). Genetics of lifespan in C. elegans: Molecular diversity, physiological complexity, mechanistic simplicity. *TRENDS in Genetics, 17*(12), 712–718.

Hiippala, K., Jouhten, H., Ronkainen, A., Hartikainen, A., Kainulainen, V., Jalanka, J., & Satokari, R. (2018). The potential of gut commensals in reinforcing intestinal barrier function and alleviating inflammation. *Nutrients, 10*(8), 988.

Holliday, R. (2004). Debates: The close relationship between biological aging and age-associated pathologies in humans. *The Journals of Gerontology Series A: Biological Sciences and Medical Sciences, 59*(6), B543–B546.

Hood, L., & Friend, S. H. (2011). Predictive, personalized, preventive, participatory (P4) cancer medicine. *Nature Reviews Clinical Oncology, 8*(3), 184–187.

House, W. (2015). FACT SHEET: President Obama's precision medicine initiative. *Press release, 30*.

Johansen, J., Atarashi, K., Arai, Y., Hirose, N., Sorensen, S. J., Vatanen, T., et al. (2023). Centenarians have a diverse gut virome with the potential to modulate metabolism and promote healthy lifespan. *Nature Microbiology, 8*, 1064–1078.

Kaeberlein, M., Rabinovitch, P. S., & Martin, G. M. (2015). Healthy aging: The ultimate preventative medicine. *Science, 350*(6265), 1191–1193.

Kami, D., & Gojo, S. (2014). Tuning cell fate: From insights to vertebrate regeneration. *Organogenesis, 10*(2), 231–240.

Kennedy, B. K., Berger, S. L., Brunet, A., et al. (2014). Geroscience: Linking aging to chronic disease. *Cell, 159*, 709–713.

Klotz, L. O., & Carlberg, C. (2023). Nutrigenomics and redox regulation: Concepts relating to the Special Issue on nutrigenomics. *Redox Biology*, 102920.

Kolios, G., & Moodley, Y. (2012). Introduction to stem cells and regenerative medicine. *Respiration, 85*(1), 3–10.

Kosorok, M. R., & Moodie, E. E. (Eds.). (2015). *Adaptive treatment strategies in practice: Planning trials and analyzing data for personalized medicine*. Society for Industrial and Applied Mathematics.

Kosorok, M. R., & Laber, E. B. (2019). Precision medicine. *Annual Review of Statistics and Its Application, 6*, 263–286.

Krishnan, P., Majumder, R., Mavalankar, D., & Deen, J. (2020). Bioprosperity and health: Key issues, current consequences, and future prospects. *Health Affairs, 39*(2), 251–259.

Langer, R. (2009). Perspectives and challenges in tissue engineering and regenerative medicine. *Advanced Materials, 21*, 3235–3236. https://doi.org/10.1002/adma.200902589

Lee, B. C., Kaya, A., Ma, S., Kim, G., Gerashchenko, M. V., Yim, S. H., Hu, Z., Harshman, L. G., & Gladyshev, V. N. (2014). Methionine restriction extends lifespan of Drosophila melanogaster under conditions of low amino-acid status. *Nature Communications, 5*(1), 3592.

Liao, S., Amcoff, M., & Nässel, D. R. (2021). Impact of high-fat diet on lifespan, metabolism, fecundity and behavioral senescence in Drosophila. *Insect Biochemistry and Molecular Biology, 133*, 103495.

Liu, E., et al. (2020). Use of CAR-transduced natural killer cells in CD19-positive lymphoid tumors. *New England Journal of Medicine, 382*, 545–553.

Lonergan, M., Senn, S. J., McNamee, C., Daly, A. K., Sutton, R., Hattersley, A., Pearson, E., & Pirmohamed, M. (2017). Defining drug response for stratified medicine. *Drug Discovery Today, 22*(1), 173–179.

López-Otín, C., Blasco, M. A., Partridge, L., Serrano, M., & Kroemer, G. (2013). The hallmarks of aging. *Cell, 153*(6), 1194–1217.

Lopez-Otin, C., Blasco, M. A., Partridge, L., Serrano, M., & Kroemer, G. (2023). Hallmarks of aging: An expanding universe. *Cell, 186*, 243–278.

Lozada-Martinez, I., & Anaya, J. M. (2024). Missing centenarians are an international concern. *Nature Aging, 4*(3), 277–278.

Lynch, S. V., & Pedersen, O. (2016). The human intestinal microbiome in health and disease. *New England Journal of Medicine, 375*(24), 2369–2379.

Maegawa, S., Hinkal, G., Kim, H. S., Shen, L., Zhang, L., Zhang, J., Zhang, N., Liang, S., Donehower, L. A., & Issa, J. P. J. (2010). Widespread and tissue specific age-related DNA methylation changes in mice. *Genome Research, 20*(3), 332–340.

Mason, C., & Dunnill, P. (2008). A brief definition of regenerative medicine. *Regenerative Medicine, 3*(1), 1–5.

Masoro, E. J. (2005). Overview of caloric restriction and ageing. *Mechanisms of Ageing and Development, 126*(9), 913–922.

Mgbeoji, I. (2006). *Global biopiracy: Patents, plants, and indigenous knowledge*. UBC Press.

Mitchell-Fearon, K., Waldron, N., Laws, H., et al. (2015). Non-communicable diseases in an older, aging population: A developing country perspective (Jamaica). *Journal of Health Care for the Poor and Underserved, 26*, 475–487.

Mkrtchyan, G. V., et al. (2020). ARDO 2020: From aging mechanisms to interventins. *Aging, 12*, 24484–24503.

Moskalev, A. A., Aliper, A. M., Smit-McBride, Z., Buzdin, A., & Zhavoronkov, A. (2014). Genetics and epigenetics of aging and longevity. *Cell Cycle, 13*, 1063–1077.

Myers, N., Mittermeier, R. A., Mittermeier, C. G., Da Fonseca, G. A., & Kent, J. (2000). Biodiversity hotspots for conservation priorities. *Nature, 403*(6772), 853–858.

Newman, D. J., & Cragg, G. M. (2016). Natural products as sources of new drugs from 1981 to 2014. *Journal of Natural Products, 79*(3), 629–661.

Niemi, M., & Ståhle, G. (2016). The use of ayurvedic medicine in the context of health promotion–a mixed methods case study of an ayurvedic centre in Sweden. *BMC Complementary and Alternative Medicine, 16*, 1–14.

Nikolova, V. L., Hall, M. R., Hall, L. J., Cleare, A. J., Stone, J. M., & Young, A. H. (2021). Perturbations in gut microbiota composition in psychiatric disorders: A review and meta-analysis. *JAMA Psychiatry, 78*(12), 1343–1354.

Nisha, R. K., Pragati, T. S., Chanu, S. I., & Sarkar, S. (2020). Aging: Reading, reasoning, and resolving using drosophila as a model system. *Models, molecules and mechanisms in biogerontology: Cellular processes, metabolism and diseases* (pp. 259–302).

Obergasteiger, J., Castonguay, A. M., Pizzi, S., Magnabosco, S., Frapporti, G., Lobbestael, E., Baekelandt, V., Hicks, A. A., Pramstaller, P. P., Gravel, C., Corti, C., Lévesque, M., & Volta, M. (2023). The small GTPase Rit2 modulates LRRK2 kinase activity, is required for lysosomal function and protects against alpha-synuclein neuropathology. *NPJ Parkinson's Disease, 9*(1), 44.

Oldak, B., Wildschutz, E., Bondarenko, V., Comar, M. Y., Zhao, C., Aguilera-Castrejon, A., Tarazi, S., Viukov, S., Pham, T. X. A., Ashouokhi, S., Lokshtanov, D., Roncato, F., Ariel, E., Rose, M., Livnat, N., Shani, T., Joubran, C., Cohen, R., Addadi, Y., … Hanna, J. H. (2023). Complete human day 14 post-implantation embryo models from naive ES cells. *Nature*.

Ondetti, M. A., Rubin, B., & Cushman, D. W. (1977). Design of specific inhibitors of angiotensin-converting enzyme: New class of orally active antihypertensive agents. *Science, 196*(4288), 441–444.

Ordovas, J. M., & Mooser, V. (2004). Nutrigenomics and nutrigenetics. *Current Opinion in Lipidology, 15*(2), 101–108.

Ostfeld, R. S., & Keesing, F. (2000). Biodiversity series: The function of biodiversity in the ecology of vector-borne zoonotic diseases. *Canadian Journal of Zoology, 78*(12), 2061–2078.

Pan, L., Chen, S., Weng, C., Call, G., Zhu, D., Tang, H., Zhang, N., & Xie, T. (2007). Stem cell aging is controlled both intrinsically and extrinsically in the Drosophila ovary. *Cell Stem Cell, 1*, 458–469. https://doi.org/10.1016/j.stem.2007.09.010

Park, Y. M., Lee, S. Y., Kang, M. J., Kim, B. S., Lee, M. J., Jung, S. S., Yoon, J. S., Cho, H. J., Lee, E., Yang, S. I., Seo, J. H., Kim, H. B., Suh, D. I., Shin, Y. H., Kim, K. W., Ahn, K., & Hong, S. J. (2020). Imbalance of gut Streptococcus, Clostridium, and Akkermansia determines the natural course of atopic dermatitis in infant. *Allergy, Asthma & Immunology Research, 12*(2), 322.

Patwardhan, B., Warude, D., Pushpangadan, P., & Bhatt, N. (2005). Ayurveda and traditional Chinese medicine: A comparative overview. *Evidence-Based Complementary and Alternative Medicine, 2*, 465–473.

Penders, J., Stobberingh, E. E., Savelkoul, P. H., & Wolffs, P. F. (2013). Cyclosporins, fungal metabolites with immunosuppressive activities. *Progress in Medicinal Chemistry, 25*, 1–33.

Peng, C., Zuo, Y., Kwan, K. M., Liang, Y., Ma, K. Y., Chan, H. Y. E., Huang, Y., Yu, H., & Chen, Z. Y. (2012). Blueberry extract prolongs lifespan of Drosophila melanogaster. *Experimental Gerontology, 47*(2), 170–178.

Perlmutter, R. (2022). *Pharmaceutical contributions to personalized medicine. Can precision medicine be personal; can personalized medicine be precise?* (p. 253).

Pillay, S. S., Candela, P., Croghan, I. T., Hurt, R. T., Bonnes, S. L., Ganesh, R., & Bauer, B. A. (2024). Leveraging the metaverse for enhanced longevity as a component of health 4.0. *Mayo Clinic Proceedings: Digital Health, 2*(1), 139–151.

Piper, M. D., Skorupa, D., & Partridge, L. (2005). Diet, metabolism and lifespan in Drosophila. *Experimental Gerontology, 40*(11), 857–862.

Ragonnaud, E., & Biragyn, A. (2021). Gut microbiota as the key controllers of "healthy" aging of elderly people. *Immunity & Ageing, 18*, 1–11.

Rahman, M. N. A., & Muhammad, N. H. (2023). Precision nutrition: Using nutrigenetic and nutrigenomic concepts in personalized nutrition. *Nutrition, 6*, 7.

Rao, R. V. (2018). Ayurveda and the science of aging. *Journal of Ayurveda and Integrative Medicine, 9*(3), 225–232.

Riscuta, G. (2016). Nutrigenomics at the interface of aging, lifespan, and cancer prevention. *The Journal of Nutrition, 146*(10), 1931–1939.

Robinson, J. G., Farnier, M., Krempf, M., et al. (2015). Efficacy and safety of alirocumab in reducing lipids and cardiovascular events. *New England Journal of Medicine, 372*(16), 1489–1499.

Rose, L. F., et al. (2018). The convergence of regenerative medicine and rehabilitation: Federal perspectives. *NPJ Regenerative Medicine, 3*, 19.

Roth, J., et al. (2018). Cost-effectiveness of axicabtagene axicabtagene ciloleucel for adult patients with relapsed or refractory large B-cell lymphoma in the United States. *Journal of Medical Economics, 21*, 1238–1245.

Sagar, S., Kaur, M., & Minneman, K. P. (2010). Antiviral lead compounds from marine sponges. *Marine Drugs, 8*(10), 2619–2638.

Salazar, J., Durán, P., Díaz, M. P., Chacín, M., Santeliz, R., Mengual, E., Gutiérrez, E., León, X., Diaz, A., Bernal, M., Escalona, D., Hernández, L. A. P., & Bermúdez, V. (2023). Exploring the relationship between the gut microbiota and ageing: A possible age modulator. *International Journal of Environmental Research and Public Health, 20*(10), 5845.

Santoro, A., Ostan, R., Candela, M., Biagi, E., Brigidi, P., Capri, M., & Franceschi, C. (2018). Gut microbiota changes in the extreme decades of human life: A focus on centenarians. *Cellular and Molecular Life Sciences, 75*, 129–148.

Saria, S., & Goldenberg, A. (2015). Subtyping: What it is and its role in precision medicine. *IEEE Intelligent Systems, 30*, 70–75.

Sathyamurthy, V. H., Jayaraj, I., Velraja, S., & Parvathi, V. D. (2022). Nutrigenomics studies to explore antiaging: Drosophila approach: Nutrigenomics. *Journal of Microbiology, Biotechnology and Food Sciences, 11*(4), e4814.

Scheurink, T. A. W., Borkent, J., Gangadin, S. S., El Aidy, S., Mandl, R., & Sommer, I. E. (2023). Association between gut permeability, brain volume, and cognition in healthy participants and patients with schizophrenia spectrum disorder. *Brain and Behavior,* e3011.

Schulman, I. H., Balkan, W., & Hare, J. M. (2018). Mesenchymal stem cell therapy for aging frailty. *Frontiers in Nutrition, 5,* 108.

Schultz, M. B., & Sinclair, D. A. (2016). When stem cells grow old: Phenotypes and mechanisms of stem cell aging. *Development, 143,* 3–14. https://doi.org/10.1242/dev.130633

Singh, P. K., Chauhan, R. S., & Singh, P. (2017). Biodiversity and its conservation. *Principles and Applications of Environmental Biotechnology for a Sustainable Future,* 315–340.

Slamon, D. J., Leyland-Jones, B., Shak, S., et al. (2001). Use of chemotherapy plus a monoclonal antibody against HER2 for metastatic breast cancer that overexpresses HER2. *New England Journal of Medicine, 344*(11), 783–792.

Snyder, A., Makarov, V., Merghoub, T., Yuan, J., Zaretsky, J. M., Desrichard, A., Walsh, L. A., Postow, M. A., Wong, P., Ho, T. S., Hollman, T. J., Bruggeman, C., Kannan, K., Li, Y., Elipenahli, C., Liu, C., Harbison, C. T., Wang, L., Ribas, A., … Chan, T. A. (2014). Genetic basis for clinical response to CTLA-4 blockade in melanoma. *New England Journal of Medicine, 371*(23), 2189–2199.

Staats, S., Wagner, A. E., Kowalewski, B., Rieck, F. T., Soukup, S. T., Kulling, S. E., & Rimbach, G. (2018). Dietary resveratrol does not affect life span, body composition, stress response, and longevity-related gene expression in Drosophila melanogaster. *International Journal of Molecular Sciences, 19*(1), 223.

Takahashi, K., Tanabe, K., Ohnuki, M., Narita, M., Ichisaka, T., Tomoda, K., & Yamanaka, S. (2007). Induction of pluripotent stem cells from adult human fibroblasts by defined factors. *Cell, 131*(5), 861–872.

Tan, B. L., Norhaizan, M. E., Liew, W. P. P., & Sulaiman Rahman, H. (2018). Antioxidant and oxidative stress: A mutual interplay in age-related diseases. *Frontiers in Pharmacology, 9,* 402374.

Teo, A. K., & Vallier, L. (2010). Emerging use of stem cells in regenerative medicine. *Biochemical Journal, 428*(1), 11–23.

The Human Microbiome Project Consortium. (2012). A framework for human microbiome research. *Nature, 486,* 215–221. https://doi.org/10.1038/nature11209

Trounson, A., & McDonald, C. (2015). Stem cell therapies in clinical trials: Progress and challenges. *Cell Stem Cell, 17*(1), 11–22.

Tutton, R. (2016). *Genomics and the reimagining of personalized medicine.* Routledge.

Utami, R. B., Damayanti, D. F., & Rodiah, D. (2020). The effectiveness of curcuma longa drink in decreasing the intensity of dysmenorrhea. *Biomedical and Pharmacology Journal, 13*(4), 2055–2060.

Vacanti, J. P., & Vacanti, C. A. (2014). The history and scope of tissue engineering. In *Principles of tissue engineering* (4th ed., pp. 3–8). Elsevier. https://doi.org/10.1016/B978-0-12-398358-9.00001-X

Vane, J. R., & Botting, R. M. (2003). The mechanism of action of aspirin. *Thrombosis research, 110*(5–6), 255–258.

Vellai, T., Takacs-Vellai, K., Zhang, Y., Kovacs, A. L., Orosz, L., & Muller, F. (2003). Genetics: influence of TOR kinase on lifespan in C. elegans. *Nature, 426,*620.

Veres, A., et al. (2019). Charting cellular identity during human in vitro β-cell differentiation. *Nature, 569,* 368–373.

von Wartburg, A., & Traber, R. (1988). The human microbiome as a reservoir of antimicrobial resistance. *Frontiers in Microbiology, 4,* 87.

Voreades, N., Kozil, A., & Weir, T. L. (2014). Diet and the development of the human intestinal microbiome. *Frontiers in Microbiology, 5,* 108740.

Wade, W. G. (2013). The oral microbiome in health and disease. *Pharmacological Research, 69*(1), 137–143.

Wainwright, C. E., Elborn, J. S., Ramsey, B. W., et al. (2015). Lumacaftor-Ivacaftor in patients with cystic fibrosis homozygous for Phe508del CFTR. *New England Journal of Medicine, 373*(3), 220–231.

Wang, H. L., Sun, Z. O., Rehman, R. U., Wang, H., Wang, Y. F., & Wang, H. (2017). Rosemary extract-mediated lifespan extension and attenuated oxidative damage in drosophila melanogaster fed on high-fat diet. *Journal of Food Science, 82*(4), 1006–1011.

Wang, J., Sun, S., & Deng, H. (2023). Chemical reprogramming for cell fate manipulation: Methods, applications, and perspectives. *Cell Stem Cell, 30*, 1130–1147.

Wang, Y., Fang, J., Liu, B., Shao, C., & Shi, Y. (2022). Reciprocal regulation of mesenchymal stem cells and immune responses. *Cell Stem Cell, 29*, 1515–1530.

Wernig, M., Meissner, A., Foreman, R., Brambrink, T., Ku, M., Hochedlinger, K., Bernstein, B. E., & Jaenisch, R. (2007). In vitro reprogramming of fibroblasts into a pluripotent ES-cell-like state. *Nature, 448*(7151), 318–324.

West, M. D., et al. (2019). Toward a unified theory of aging and regeneration. *Regenerative Medicine, 14*, 867–886.

Weyers, W. (2019). "Personalized excision" of malignant melanoma—Need for a paradigm shift in the beginning era of personalized medicine. *The American Journal of Dermatopathology, 41*(12), 884–896.

Wilmanski, T., Diener, C., Rappaport, N., Patwardhan, S., Wiedrick, J., Lapidus, J., Earls, J. C., Zimmer, A., Glusman, G., Robinson, M., Yurkovich, J. T., Kado, D. M., Cauley, J. A., Zmuda, J., Lane, N. E., Magis, A. T., Lovejoy, J. C., Hood, L., Gibbons, S. M., … Price, N. D. (2021). Gut microbiome pattern reflects healthy ageing and predicts survival in humans. *Nature Metabolism, 3*(2), 274–286.

Wu, P. Y., Cheng, C. W., Kaddi, C. D., Venugopalan, J., Hoffman, R., & Wang, M. D. (2017). Omicand electronic health record bigdata analytics for precision medicine. *IEEE Transactions on Biomedical Engineering, 64*, 263–273.

Yaffe, K., Vittinghoff, E., Pletcher, M. J., Hoang, T. D., Launer, L. J., Whitmer, R. A., Cooker, L. H., & Sidney, S. (2014). Early adult to midlife cardiovascular risk factors and cognitive function. *Circulation, 129*(15), 1560–1567.

Yamada, S., Behfar, A., & Terzic, A. (2021). Regenerative medicine clinical readiness. *Regenerative Medicine, 16*, 309–322.

Yamamura, T. (2023). Time to reconsider the classification of multiple sclerosis. *The Lancet Neurology, 22*(1), 6–8.

Ye, Z., Zhang, Z., Li, Z., Huang, C., & Zhang, Y. (2020). Toward wiping out osteoarthritis in China: Research highlights. *Chinese Medical Journal, 133*, 883–885.

Yu, J., Vodyanik, M. A., Smuga-Otto, K., Antosiewicz-Bourget, J., Frane, J. L., Tian, S., Nie, J., Jonsdottir, G. A., Ruotti, V., Stewart, R., Slukvin, I. I., & Thomson, J. A. (2007). Induced pluripotent stem cell lines derived from human somatic cells. *Science, 318*(5858), 1917–1920.

Zeng, X., Li, X., Li, X., Wei, C., Shi, C., Hu, K., et al. (2023). Fecal microbiota transplantation from young mice rejuvenates aged hematopoietic stem cells by suppressing inflammation. *Blood, 141*, 1691–1707.

MIX
Papier aus verantwortungsvollen Quellen
Paper from responsible sources
FSC® C105338

If you have any concerns about our products,
you can contact us on
ProductSafety@springernature.com

In case Publisher is established outside the EU,
the EU authorized representative is:
Springer Nature Customer Service Center GmbH
Europaplatz 3, 69115 Heidelberg, Germany

Printed by Libri Plureos GmbH
in Hamburg, Germany